Handbook of Research on Ambient Intelligence and Smart Environments:

Trends and Perspectives

Nak-Young Chong
Japan Advanced Institute of Science and Technology, Japan

Fulvio Mastrogiovanni
University of Genova, Italy

A volume in the Advances in
Computational Intelligence and Robotics
(ACIR) Book Series

Senior Editorial Director:	Kristin Klinger
Director of Book Publications:	Julia Mosemann
Editorial Director:	Lindsay Johnston
Acquisitions Editor:	Erika Carter
Development Editor:	Christine Bufton
Production Editor:	Sean Woznicki
Typesetters:	Keith Glazewski, Natalie Pronio, Jennifer Romanchak, Milan Vracarich, Jr.
Print Coordinator:	Jamie Snavely
Cover Design:	Nick Newcomer

Published in the United States of America by
Information Science Reference (an imprint of IGI Global)
701 E. Chocolate Avenue
Hershey PA 17033
Tel: 717-533-8845
Fax: 717-533-8661
E-mail: cust@igi-global.com
Web site: http://www.igi-global.com

Library of Congress Cataloging-in-Publication Data

Handbook of research on ambient intelligence and smart environments: trends and perspective / Nak-Young Chong and Fulvio Mastrogiovanni, editors.
    p. cm.
  Includes bibliographical references and index.
  Summary: "This book covers the cutting-edge aspects of AMI applications, specifically those involving the effective design, realization, and implementation of a comprehensive ambient intelligence in smart environments"-- Provided by publisher.
  ISBN 978-1-61692-857-5 (hardcover) -- ISBN 978-1-61692-858-2 (ebook) 1. Ambient intelligence. 2. Telematics. I. Chong, Nak-Young, 1965- II. Mastrogiovanni, Fulvio, 1977-
  QA76.9.A48H36 2011
  004.01'9--dc22
                                    2010041637

This book is published in the IGI Global book series Advances in Computational Intelligence and Robotics (ACIR) Book Series (ISSN: 2327-0411; eISSN: 2327-042X)

British Cataloguing in Publication Data
A Cataloguing in Publication record for this book is available from the British Library.

Advances in Computational Intelligence and Robotics (ACIR) Book Series

ISSN: 2327-0411
EISSN: 2327-042X

MISSION

While intelligence is traditionally a term applied to humans and human cognition, technology has progressed in such a way to allow for the development of intelligent systems able to simulate many human traits. With this new era of simulated and artificial intelligence, much research is needed in order to continue to advance the field and also to evaluate the ethical and societal concerns of the existence of artificial life and machine learning.

The **Advances in Computational Intelligence and Robotics (ACIR) Book Series** encourages scholarly discourse on all topics pertaining to evolutionary computing, artificial life, computational intelligence, machine learning, and robotics. ACIR presents the latest research being conducted on diverse topics in intelligence technologies with the goal of advancing knowledge and applications in this rapidly evolving field.

COVERAGE

- Adaptive & Complex Systems
- Agent Technologies
- Artificial Intelligence
- Cognitive Informatics
- Computational Intelligence
- Natural Language Processing
- Neural Networks
- Pattern Recognition
- Robotics
- Synthetic Emotions

IGI Global is currently accepting manuscripts for publication within this series. To submit a proposal for a volume in this series, please contact our Acquisition Editors at Acquisitions@igi-global.com or visit: http://www.igi-global.com/publish/.

Titles in this Series

For a list of additional titles in this series, please visit: www.igi-global.com

Intelligent Technologies and Techniques for Pervasive Computing
Kostas Kolomvatsos (University of Athens, Greece) Christos Anagnostopoulos (Ionian University, Greece) and Stathes Hadjiefthymiades (University of Athens, Greece)
Information Science Reference • copyright 2013 • 349pp • H/C (ISBN: 9781466640382) • US $195.00 (our price)

Mobile Ad Hoc Robots and Wireless Robotic Systems Design and Implementation
Raul Aquino Santos (University of Colima, Mexico) Omar Lengerke (Universidad Autónoma de Bucaramanga, Colombia) and Arthur Edwards-Block (University of Colima, Mexico)
Information Science Reference • copyright 2013 • 347pp • H/C (ISBN: 9781466626584) • US $190.00 (our price)

Intelligent Planning for Mobile Robotics Algorithmic Approaches
Ritu Tiwari (ABV – Indian Institute of Information, India) Anupam Shukla (ABV – Indian Institute of Information, India) and Rahul Kala (School of Systems Engineering, University of Reading, UK)
Information Science Reference • copyright 2013 • 320pp • H/C (ISBN: 9781466620742) • US $195.00 (our price)

Simultaneous Localization and Mapping for Mobile Robots Introduction and Methods
Juan-Antonio Fernández-Madrigal (Universidad de Málaga, Spain) and José Luis Blanco Claraco (Universidad de Málaga, Spain)
Information Science Reference • copyright 2013 • 497pp • H/C (ISBN: 9781466621046) • US $195.00 (our price)

Prototyping of Robotic Systems Applications of Design and Implementation
Tarek Sobh (University of Bridgeport, USA) and Xingguo Xiong (University of Bridgeport, USA)
Information Science Reference • copyright 2012 • 321pp • H/C (ISBN: 9781466601765) • US $195.00 (our price)

Cross-Disciplinary Applications of Artificial Intelligence and Pattern Recognition Advancing Technologies
Vijay Kumar Mago (Simon Fraser University, Canada) and Nitin Bhatia (DAV College, India)
Information Science Reference • copyright 2012 • 784pp • H/C (ISBN: 9781613504291) • US $195.00 (our price)

Handbook of Research on Ambient Intelligence and Smart Environments Trends and Perspectives
Nak-Young Chong (Japan Advanced Institute of Science and Technology, Japan) and Fulvio Mastrogiovanni (University of Genova, Italy)
Information Science Reference • copyright 2011 • 770pp • H/C (ISBN: 9781616928575) • US $265.00 (our price)

Particle Swarm Optimization and Intelligence Advances and Applications
Konstantinos E. Parsopoulos (University of Ioannina, Greece) and Michael N. Vrahatis (University of Patras, Greece)
Information Science Reference • copyright 2010 • 328pp • H/C (ISBN: 9781615206667) • US $180.00 (our price)

www.igi-global.com

701 E. Chocolate Ave., Hershey, PA 17033
Order online at www.igi-global.com or call 717-533-8845 x100
To place a standing order for titles released in this series, contact: cust@igi-global.com
Mon-Fri 8:00 am - 5:00 pm (est) or fax 24 hours a day 717-533-8661

Editorial Advisory Board

List of Contributors

Table of Contents

Detailed Table of Contents

Chapter 1

Lorenzo Magnani, University of Pavia, Italy
Emanuele Bardone, University of Pavia, Italy

We will introduce some cognitive and epistemological foundational aspects related to Ambient Intelligence (AmI). We will show how three concepts which derive from the tradition of cognitive science may be of help in deepening some of the main theoretical aspects concerning Ambient Intelligence: the notion of distributed cognition, the concept of cognitive niche, and the concept of affordance. We contend that this theoretical perspective will shed new light on some of the most promising and interesting developments recently brought about by Ambient Intelligence.

Chapter 2

Christine Leignel, Université Libre de Bruxelles, Belgium
Jean-Michel Jolion, Université de Lyon, France

This chapter presents a survey of methods used for tracking in video sequence. We mainly focus this survey on tracking persons. We introduce three main approaches. First, we present the graph based tracking approach where the sequence of tracked objects are embodied in a graph structure. Then we introduce the features (extracted from the images) based tracking and matching with a model. We survey the main primitives and emphasize the approaches based on 2D and 3D body model. We present the particular case of tracking in a network of cameras with the particle filtering method. Finally, As a generalization, we focus on the single vs. stereo approaches.

Chapter 3

Axel Steinhage, Future-Shape GmbH, Germany
Christl Lauterbach, Future-Shape GmbH, Germany

The following chapter describes two systems, both are perfect examples for ambient intelligence. The first system is, sensor electronics, which is invisibly integrated into the floor. This system is able to detect people walking across the floor and can be used to recognize peoples' location and movement behavior. The main application domains are Ambient Assisted Living (AAL), health care, security systems and home automation. The second system serves for localizing moving objects such as robots, wheelchairs or hospital beds by means of RFID tags in the floor. In the following, we describe the technical details of the two systems and possible applications.

In this chapter, a case study on augmenting a daily object, mirror, for a contextual ambient display is presented. The mirror presents information relevant to a person who is standing and utilizing unshareable objects, e.g. a toothbrush, in front of it on the periphery of his/her field of vision. We investigated methods of interaction with the mirror by analyzing user preferences against contrastive functionalities. Experiments were conducted by a Wizard-of-Oz method and an in-situ experiment. The results showed that a short absence of the mirror function was not a big issue for the majority of participants once they were interested in presented information. The analysis also allowed us to specify requirements and further research questions in order to make an augmented mirror acceptable.

No Ambient Intelligence can survive without human-computer interactions. Over ninety percent of information in our communication is verbal and visual. The mapping between one-dimensional words and two-dimensional images is a challenge for visual information classification and reconstruction. In this Chapter, we present a model for the image-word two-way mapping process. The model applies specifically to facial identification and facial reconstruction. It accommodates through semantic differential descriptions, analogical and graph-based visual abstraction that allows humans and computers to categorize objects and to provide verbal annotations to the shapes that comprise faces. An image-word mapping interface is designed for efficient facial recognition in massive visual datasets. We demonstrate how a two-way mapping of words and facial shapes is feasible in facial information retrieval and reconstruction.

This Chapter is a survey dealing with the use of emotions and mood to characterize an Ambient Intelligence system. In particular, a key aspect of the described research is the assumption that each level of an Ambient Intelligence infrastructure (e.g., sensing, reasoning, action) can benefit from the introduction of emotion and mood modelling. The Chapter surveys well-known models (e.g., OCC, Big Five – Five Factor Model, PAD, just to name a few) discussing for each one Pros and Cons. Next, architectures for emotional agents are discussed, e.g., Cathexis (assuming the somatic marker hypothesis proposed by Damasio), Flame, Tabasco and many others. Finally, specific implementation examples of emotional agents in Ambient Intelligence scenarios are described.

Chapter 7

    Carole Adam, RMIT University, Australia
    Benoit Gaudou, UMI 209 UMMISCO, IRD, IFI, Vietnam
    Dominique Login, University of Toulouse, CNRS, IRIT, France
    Emiliano Lorini, University of Toulouse, CNRS, IRIT, France

Ambient Intelligence (AmI) is the art of designing intelligent and user-focused environments. It is thus of great importance to take human factors into account. In this chapter we especially focus on emotions, that have been proved to be essential in human reasoning and interaction. To this end, we assume that we can take advantage of the results obtained in Artificial Intelligence about the formal modeling of emotions. This chapter specifically aims at showing the interest of logic as a tool to design agents endowed with emotional abilities useful for Ambient Intelligence applications. In particular, we show that modal logics allow the representation of the mental attitudes involved in emotions such as beliefs, goals or ideals. Moreover, we illustrate how modal logics can be used to represent complex emotions (also called self-conscious emotions) involving elaborated forms of reasoning, such as self-attribution of responsibility and counterfactual reasoning. Examples of complex emotions are regret and guilt. We illustrate our logical approach by formalizing some case studies concerning an intelligent house taking care of its inhabitants.

Chapter 8

    Tibor Bosse, Vrije Universiteit Amsterdam, The Netherlands
    Mark Hoogendoorn, Vrije Universiteit Amsterdam, The Netherlands
    Michel Klein, Vrije Universiteit Amsterdam, The Netherlands
    Rianne van Lambalgen, Vrije Universiteit Amsterdam, The Netherlands
    Peter-Paul van Maanen, Vrije Universiteit Amsterdam, The Netherlands
    Jan Treur, Vrije Universiteit Amsterdam, The Netherlands

In this chapter, we propose to outline the scientific area that addresses Ambient Intelligence applications in which not only sensor data, but also knowledge from the human-directed sciences such as biomedical science, neuroscience, and psychological and social sciences is incorporated. This knowledge enables the environment to perform more in-depth, human-like analyses of the functioning of the observed humans, and to come up with better informed actions. A structured approach to embed human knowledge in Ambient Intelligence applications is presented an illustrated using two examples, one on automated visual attention manipulation, and another on the assessment of the behaviour of a car driver.

This book chapter provides a review of the assistive technologies deployed in smart spaces with a variety of smart home or house examples. In the first place, home networking technologies and sensing technologies are surveyed as fundamental technologies to support smart environment. After reviewing representative smart home projects from across the world, concrete assistive services related with the fundamental technologies in smart environment are deployed not only for the elderly and handicapped but for people in ordinary families as well. Adaptability is one of the key essences in the assistive technologies in smart environment and, for this purpose, human-ware studies including man-machine interfaces, ergonomics and gerontology are needed to be linked with the hardware specific fundamental technologies.

There are some important requirements to build effective smart spaces, like human aspects, sensing, activity recognition, context awareness, etc. However, all of them require adequate system support to build systems that work in practice. In this chapter, we discuss system level support services that are necessary to build working smart spaces. We also include a full discussion of system abstractions for pervasive computing taking in account naming, protection, modularity, communication, and programmability issues.

Chapter 11

Activity-oriented computing (AOC) is a paradigm promoting the run-time realization of applications by composing ubiquitous services in the user's surroundings according to abstract specifications of user activities. The paradigm is particularly well-suited for enacting ubiquitous applications. However, there is still a need for end-users to create and control the ubiquitous applications because they are better aware of their own needs and activities than any existing context-aware system could ever be. In this chapter, we give an overview of state of the art ubiquitous application composition, present the architecture of the MEDUSA middleware and demonstrate its realization, which is based on existing open-source solutions. On the basis of our discussion on state of the art ubiquitous application composition, we argue that current implementations of the AOC paradigm are lacking in end-user support. Our solution, the MEDUSA middleware, allows end-users to explicitly compose applications from networked services, while building on an activity-oriented computing infrastructure to dynamically realize the composition.

Chapter 12

Jochen Meis, Fraunhofer Institute for Software and System Engineering, Germany

Manfred Wojciechowski, Fraunhofer Institute for Software and System Engineering, Germany

This Chapter deals with the important process related to smart environments engineering, with a specific emphasis on the software infrastructure. In particular, the Chapter focuses on the whole process, from the initial definition of functional requirements to the identification of possible implementation strategies. On the basis of this analysis, a context model as well as the possible choice of relevant sensor types is carried out.

Chapter 13

Rebecca L. Willard, Waynesburg University, USA

Baoying Wang, Waynesburg University, USA

Ambient Intelligence is the concept that technology will become a part of everyday living and assist users in multiple tasks. It is a combination and further development of ubiquitous computation, pervasive computation, and multimedia. The technology is sensitive to the actions of humans and it can interact with the human or adjust the surroundings to suit the needs of the users dynamically. All of this is made possible by embedding sensors and computing components inconspicuously into human surroundings. This paper discusses the middleware needed for dynamic ambient intelligence networks and the ambient intelligence network architecture. The bottom-up middleware approach for ambient intelligence is important so the lower layers of all ambient intelligence networks are interchangeable and compatible with other ambient intelligence components. This approach also allows components to be programmed to be compatible with multiple ambient intelligence networks. The network architecture discussed in this paper allows for dynamic networking capabilities for minimal interruptions with changes in computer components.

Chapter 14

Weishan Zhang, China University of Petroleum, P.R. China

Klaus Marius Hansen, University of Copenhagen, Denmark

Abstract Context-awareness is an important feature in Ambient Intelligence environments including in pervasive middleware. In addition, there is a growing trend and demand on self-management capabilities for a pervasive middleware in order to provide high-level dependability for services. In this chapter, we propose to make use of context-awareness features to facilitate self-management. To achieve self-management, dynamic contexts for example device and service statuses, are critical to take self-management actions. Therefore, we consider dynamic contexts in context modeling, specifically as a set of OWL/SWRL ontologies, called the Self-Management for Pervasive Services (SeMaPS) ontologies. Self-management rules can be developed based on the SeMaPS ontologies to achieve self-management goals. Our approach is demonstrated within the Hydra pervasive middleware. Finally, our experiments with performance, extensibility, and scalability in the context of Hydra show that the SeMaPS-based self-management approach is effective.

A general infrastructure that can facilitate the development of context-aware applications in smart homes is proposed. Unlike previous systems, our system builds on semantic web technologies, and it particularly concerns the contexts from human-artifact interaction. A multi-levels' design of our ontology (called SS-ONT) makes it possible to realize context sharing and end-user-oriented customization. Using this infrastructure as a basis, we address some of the principles involved in performing context querying and context reasoning. The performance of our system is evaluated through a series of experiments.

The work is motivated by the expanding demand and limited supply of long-term personal care for People with Dementia (PwD), and assistive technology as an alternative. Telecare allows PwD to live in the comfort of their homes for a longer time. It is challenging to have remote care in smart homes with ambient intelligence, using devices, networks, and activity and plan recognition. Our scope is limited to mostly related work on existing execution environments in smart homes, and activity and plan recognition algorithms which can be applied to PwD living in smart homes. PwD and caregiver needs are addressed in a more holistic healthcare approach, domain challenges include doctor validation and erroneous behaviour, and technical challenges include high maintenance and low accuracy. State-of-the-art devices, networks, activity and plan recognition for physical health are presented; ideas for developing mental training for mental health and social networking for social health are explored. There are two implications of this work: more needs to be done for assistive technology to improve PwD's mental and social health, and assistive software is not highly accurate and persuasive yet. Our work applies not only to PwD, but also the elderly without dementia and people with intellectual disabilities.

In order to provide adequate assistance to cognitively impaired people when they carry out their activities of daily living (ADLs) at home, new technologies based on the emerging concept of Ambient Intelligence (AmI) must be developed. The main application of the AmI concept is the development of Smart Homes, which can provide advanced assistance services to its occupant when he performs his ADLs. The main difficulty inherent to this kind of assistance services is to be able to identify the ongoing inhabitant ADL from the observed basic actions and from the sensors events produced by these actions. This chapter will investigate in details the challenging issues that emerge from activity recognition in order to provide cognitive assistance in Smart Homes, by identifying gaps in the capabilities of current approaches. This will allow to raise numerous research issues and challenges that need to be addressed for understanding this research field and enabling ambient recognition systems for cognitive assistance to operate effectively.

In this chapter we discuss intention recognition in general, and the use of logic-based formalisms, and deduction and abduction in particular. We consider the relationship between causal theories used for planning and the knowledge representation and reasoning used for intention recognition. We look at the challenges and the issues, and we explore eight case studies.

Chapter 19

Most of context models have limited capability in involving human intention for system evolvability and self-adaptability. Human intention in context aware systems can evolve at any time, however, context aware systems based on these context models can provide only standard services that are often insufficient for specific user needs. Consequently, evolving human intentions result in changes in system requirements. Moreover, an intention must be analyzed from tangled relations with different types of contexts. In the past, this complexity has prevented researchers from using computational methods for analyzing or specifying human intention in context aware system design. The authors investigated the possibility for inferring human intentions from contexts and situations, and deploying appropriate services that users require during system run-time. This chapter first focus on describing an inference ontology to represent stepwise inference tasks to detect an intention change and then discuss how context aware systems can accommodate requirements for the intention change.

     Jit Biswas, Institute for Infocomm Research, Singapore
     Andrei Tolstikov, Institute for Infocomm Research, Singapore
     Aung-Phyo-Wai Aung, Institute for Infocomm Research, Singapore
     Victor Siang-Fook Foo, Institute for Infocomm Research, Singapore
     Weimin Huang, Institute for Infocomm Research, Singapore

This chapter provides examples of sensor data acquisition, processing and activity recognition systems that are necessary for ambient intelligence specifically applied to home care for the elderly. We envision a future where software and algorithms will be tailored and personalized towards the recognition and assistance of Activities of Daily Living (ADLs) of the elderly. In order to meet the needs of the elderly living alone, researchers all around the world are looking to the field of Ambient Intelligence or AmI (see http://www.ambientintelligence.org).

     Björn Gottfried, University of Bremen, Germany

This Chapter provides an introduction to the emerging field of Behaviour Monitoring and Interpretation (BMI in short). The study of behaviour encompasses both social and engineering implications: on one hand the scientific goal is to design and represent believable models of human behaviours in different contexts; on the other hand, the engineering goal is to acquire relevant sensory information in real-time, as well as to process all the relevant data. The Chapter provides a number of examples of BMI systems, as well as discussions about possible implications in Smart Environments and Ambient Intelligence in a broad sense.

     Hans W. Guesgen, Massey University, New Zealand
     Stephen Marsland, Massey University, New Zealand

Identifying human behaviours in smart homes from sensor observations is an important research problem. The addition of contextual information about environmental circumstances and prior activities, as well as spatial and temporal data, can assist in both recognising particular behaviours and detecting abnormalities in these behaviours. In this chapter, we describe a novel method of representing this data and discuss a wide variety of possible implementation strategies.

     Claus Moebus, University of Oldenburg, Germany
     Mark Eilers, University of Oldenburg, Germany

The Human or Cognitive Centered Design (HCD) of intelligent transport systems requires digital Models of Human Behavior and Cognition (MHBC) enabling Ambient Intelligence e.g. in a smart car. Currently MBHC are developed and used as driver models in traffic scenario simulations, in proving safety assertions and in supporting risk-based design. Furthermore, it is tempting to prototype assistance systems (AS) on the basis of a human driver model cloning an expert driver. To that end we propose the Bayesian estimation of MHBCs from human behavior traces generated in new kind of learning experiments: Bayesian model learning under driver control. The models learnt are called Bayesian Autonomous Driver (BAD) models. For the purpose of smart assistance in simulated or real world scenarios the obtained BAD models can be used as Bayesian Assistance Systems (BAS). The critical question is, whether the driving competence of the BAD model is the same as the driving competence of the human driver when generating the training data for the BAD model. We believe that our approach is superior to the proposal to model the strategic and tactical skills of an AS with a Markov Decision Process (MDP). The usage of the BAD model or BAS as a prototype for a smart Partial Autonomous Driving Assistant System (PADAS) is demonstrated within a racing game simulation.

Chapter 24

In this chapter, we first briefly introduce the setting: mobility assistants (the wheelchair Rolland and iWalker) and smart environment control in the Bremen Ambient Assisted Living Lab. In several example scenarios, we then outline our contributions to the state of the art, focussing on spatial knowledge representation, reasoning and spatial interaction (multi-modal, but with special emphasis on natural language dialogue) between three partners: the user, a mobility assistant, and the smart environment.

Chapter 25

The main goal of this Chapter is to introduce SAM, an integrated architecture for concurrent activity recognition, planning and execution. SAM provides a general framework to define how an intelligent environment can assess contextual information from sensory data. The architecture builds upon a temporal reasoning framework operating in closed-loop between physical sensing and actuation components in a smart environments. The capabilities of the system as well as possible examples of its use are discussed in the context of the PEIS-Home, a smart environment integrated with robotic components.

Chapter 26

Geunho Lee, Japan Advanced Institute of Science and Technology (JAIST), Japan
Nak Young Chong, Japan Advanced Institute of Science and Technology (JAIST), Japan

Ambient Intelligence (AmI) is a multidisciplinary approach aimed at enriching physical environments with a network of distributed devices, such as sensors, actuators, and computational resources, in order to support humans in achieving their everyday task. Within the framework of AmI, this chapter presents decentralized coordination for a swarm of autonomous robotic sensors building intelligent environments adopting AmI. The large-scale robotic sensors are regarded as a swarm of wireless sensors mounted on spatially distributed autonomous mobile robots. Therefore, motivated by the experience gained during the development and usage for decentralized coordination of mobile robots in geographically constrained environments, our work introduces the following two detailed functions: self-configuration and flocking. In particular, this chapter addresses the study of a unified framework which governs the adaptively self-organizing processes for a swarm of autonomous robots in the presence of an environmental uncertainty. Based on the hypothesis that the motion planning for robot swarms must be controlled within the framework, the two functions are integrated in a distributed way, and each robot can form an equilateral triangle mesh with its two neighbors in a geometric sense. Extensive simulations are performed in two-dimensional unknown environments to verify that the proposed method yields a computationally efficient, yet robust deployment.

Chapter 27

Mária Bieliková, Slovak University of Technology in Bratislava, Slovakia
Marián Hönsch, Slovak University of Technology in Bratislava, Slovakia
Michal Kompan, Slovak University of Technology in Bratislava, Slovakia
Jakub Šimko, Slovak University of Technology in Bratislava, Slovakia
Dušan Zeleník, Slovak University of Technology in Bratislava, Slovakia

Increasing energy consumption requires our attention. Resources are exhaustible, so building new power plants is not the only solution. Since residential expenditure is of major parts of overall consumption, concept of intelligent household has potential to participate on energy usage optimization. In this chapter, we concentrate on software methods, which based on inputs gained from an environment monitor, analyze and consequently reduce non-effective energy consumption. We gave a shape to this concept by description of real prototype system called ECM (Energy Consumption Manager). Besides active energy reduction, the ECM system also has an educative function. User-system interaction is designed to teach the user how to use (electric, in case of our prototype) energy effectively. Methods for the analysis are based on artificial intelligence and information systems fields (neural networks, clustering algorithms, rule-based systems, personalization and adaptation of user interface). The system goes further and gains more effectiveness by exchange of data, related to consumption and appliance behaviour, between households.

Chapter 28

Raffaele De Amicis, Fondazione Graphitech, Italy
Giuseppe Conti, Fondazione Graphitech, Italy

The unprecedented success of 3D geobrowsers, mostly due to the user-friendliness typical of their interfaces and to the extremely wide set of information available, has undoubtedly marked a turning point within the geospatial domain, clearly departing from previous IT solutions in the field of Geographical Information Systems (GIS). This technological great leap forward has paved the way to a new generation of GeoVisual Analytics (GVA) applications capable to ensure access, filtering and processing of large repositories of geospatial information. Within this context we refer to GeoVisual Analytics as the set of tools, technologies and methodologies which can be deployed to increase situational awareness by helping operators identify specific data patterns within a vast information flow made of multidimensional geographical data coming from static databases as well as from sensor networks.

    Peter Mikulecký, University of Hradec Králové, Czech Republic
    Kamila Olševičová, University of Hradec Králové, Czech Republic
    Vladimír Bureš, University of Hradec Králové, Czech Republic
    Karel Mls, University of Hradec Králové, Czech Republic

The objective of the chapter is to identify and analyze key aspects and possibilities of Ambient Intelligence (AmI) applications in educational processes and institutions (universities), as well as to present a couple of possible visions for these applications. A number of related problems are discussed as well, namely agent-based AmI application architectures. Results of a brief survey among optional users of these applications are presented as well.

    Shishir K. Shandilya, Devi Ahilya University, India
    Suresh Jain, KCB Technical Academy, India

E-commerce is an increasingly pervasive element of ambient intelligence. Ambient intelligence promotes the user-centered system where as per the feedback of user, the system changes itself to facilitate the transmission and marketing of goods and information to the appropriate e-commerce market. Ambient Intelligence ensures that the e-commerce activities generate good confidence level among the customers. The confidence occurs when the customers feel that the product can be relied upon to act in their best interest and knowledge. It affects the decision that whether a customer decides to buy the product or not.

Foreword

We are living with various types of intelligent system such as personal computer, cell phone, IC tag, web cameras and so on. Once we accept these intelligent systems, they evolve with the human society. We, humans, adapt to the systems and change our life style. The internet and cell phone have dramatically changed our life style. At the same time, the systems have been improved in order to adapt to the human society more. That is, the intelligent systems evolve human itself.

With the intelligent system, we, humans, are able to overcome constrains of the physical bodies and extend the ability beyond time and space. Human cannot be defined based on the fresh body anymore.

What we are looking for with the intelligent systems is human abilities that we do not know yet and deeper definition of human. By accepting the idea of human evolution with the intelligent systems, we can excite with possibilities of human.

The intelligent systems themselves also evolve in the human society. Various companies develop various intelligent systems for various purposes in the huge market. In the beginning, they have appealed the novelty. However, once the human society accepts the systems, they are going to be embedded in our life as the sustainable systems.

The systems are going to be invisible and tightly coupled with human and the human society. Then, they will be indispensable ones for our daily life. For such systems, higher level intelligence is required such as context-awareness and reasoning in addition to sensing and actuation. The higher level intelligence of the system and the human intelligence harmonize each other.

This book just includes the important and futuristic topics for realizing the intelligent systems embedded in the human society, namely Ambient Intelligence. The concept of Ambient Intelligence leads our society to the next stage where we, human, enhance our ability beyond time and space.

Hiroshi Ishiguro
Professor of Department of Systems Innovation, Osaka University Fellow of Advanced Telecommunications Research Institute International

Hiroshi Ishiguro *received a D.Eng. in systems engineering from the Osaka University, Japan in 1991. He is currently Professor of Department of Systems Innovation in the Graduate School of Engineering Science at Osaka University (2009–). He is also Visiting Group Leader (2002–) of the Intelligent Robotics and Communication Laboratories at the Advanced Telecommunications Research Institute, where he previously worked as Visiting Researcher (1999–2002). He was previously Research Associate (1992–1994) in the Graduate School of Engineering Science at Osaka University and Associate Professor (1998–2000) in the Department of Social Informatics at Kyoto University. He was also Visiting Scholar (1998–1999) at the University of California, San Diego, USA. He was Associate Professor (2000–2001) and Professor (2001–2002) in the Department of Computer and Communication Sciences at Wakayama University. He then moved to Department of Adaptive Machine Systems in the Graduate School of Engineering Science at Osaka University as a Professor (2002-2009). His research interests include distributed sensor systems, interactive robotics, and android science.*

Preface

WHY IS AMBIENT INTELLIGENCE CHALLENGING?

Ambient Intelligence (AmI in short) is a multidisciplinary approach aimed at enriching physical environments with a network of distributed devices, such as sensors, actuators, and computational resources, in order to support humans in achieving their everyday goals. These can range from helping businessmen in increasing their productivity to supporting activities of daily living for elderly and people with special needs, such as people suffering from cognitive impairments or physical injuries.

From a technological perspective, AmI represents the convergence of recent achievements in communications technologies (e.g., sensor networks and distributed computing), industrial electronics (e.g., the miniaturization process affecting computational devices, such as wearable sensors), Pervasive Computing, intelligent user interfaces (e.g., alternative communication media such as eyes' gaze and speech) and Artificial Intelligence (e.g., knowledge representation and context or situation awareness), just to name but few. AmI is not only the superimposition of these disciplines, but it is a major effort to integrate and make them really useful for the everyday human life.

The Handbook is aimed at covering cutting-edge aspects involved in the effective design, realization and implementation of comprehensive AmI applications. In particular, it can be considered as a conceptual path originating from the very basic consideration of human aspects that are involved in AmI systems design, heading to the effective reasoning and acting capabilities that AmI systems must be characterized with to operate in the real-world.

In particular, eight key aspects have been identified as being fundamental in the design of AmI systems (see Figure 1).

The first one is *human aspects*: AmI systems are expected to work in symbiosis with humans, and therefore they must understand not only human actions, but also moods, desires and feelings. In other words, AmI systems should encode models of human behaviour (in the mid term possibly in well-defined scenarios, such as people with dementia or typical home activities) and cognition in order to make predictions and to take purposive actions. These models can be either encoded by designers or learned by the system itself through observation.

Models of behaviour and cognition must be grounded with respect to actual data provided by sensors monitoring the given physical space: *sensing* is therefore a major topic to address when classifying actual human behaviour with respect to encoded models. Many sensing techniques have been investigated in literature. However, what is important is the relationship between sensory information and the algorithms adopted to process it: among the different issues involved in this process, we can identify data

Figure 1. Information flow in ambient intelligence

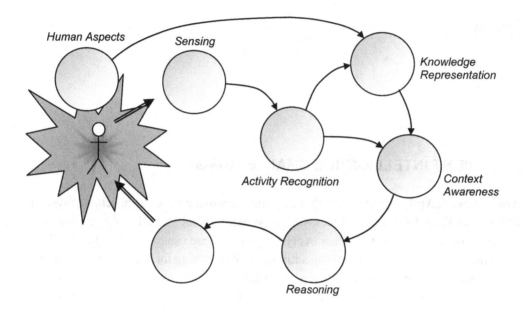

fusion, reliability, real-time responsiveness, symbol grounding and anchoring with respect to *knowledge representation* structures.

Knowledge representation (KR in short) is a central topic in AmI applications and intelligent systems in general. Borrowing techniques developed by research in Artificial Intelligence, it is now considered a core part of any AmI system. In particular, it is at the basis of many other important aspects to consider: models of human behaviours and cognition can naturally be represented using KR techniques, topological and physical aspects of the environment can be easily encoded, reasoning schemes are structures whose templates are stored in *knowledge bases* (otherwise called *ontologies*), just to name a few. However, the main feature that makes KR techniques suitable for use in AmI applications is the fact that they can associate *semantic* information to represented knowledge, i.e., a sort of *meaning* characterizing symbols and their relationships.

Many techniques for detecting human activity have been developed and appeared in literature in the past few years. *Activity recognition* is strictly related to sensing and knowledge representation: on the basis of sensory data and structured knowledge, a major dependence exists with respect to sensory modalities and knowledge representation techniques used to manage information. This issue aroused – and still foments – many philosophical and technological debates about the very notion of "activity", from which completely different approaches originated.

The recognition of human activities, coupled with semantic knowledge represented within the system, forms the basis of *context-awareness*. A context-aware system is "something" that is aware of what sensory data mean. In other words, a context-aware system is able to associate meaning (useful for humans as well as for automated systems) to observations, and to make the best use of sensory data once the meaning has been assessed. During the past decade, context-aware applications have been one of the hottest research topics worldwide. Several techniques have been proposed to design aware AmI systems: from this experience, it emerged a tight relationship with respect to the capabilities of activity recognition and knowledge representation.

Once a situation has been assessed (possibly involving many humans, their performed activities and related events), the AmI system is expected to purposively act in order to reach a given objective. Depending on the techniques used to provide the AmI system itself with context-aware capabilities (and, ultimately, on knowledge representation strategies), many methods to support *reasoning* have been proposed in literature. However, research in AmI has been geared more towards representation than action and task planning: as a consequence, the use of reasoning techniques in AmI applications is still in its infancy.

Achieving a *sensible interaction* level with humans is probably the main goal in AmI. This objective involves a tight – yet well-defined – coupling between *reasoning* (and the involved context awareness techniques) and the considerations of relevant *human aspects*. In a sense, it can be argued that sensible interaction is what is actually perceived by humans when dealing with AmI systems, therefore encompassing not only what kind of information is presented or what action is performed by the system, but above all *how* this is achieved. Without any doubt, this is considered the most challenging part of an AmI system, and therefore it is far from being clearly addressed in literature.

It is not possible to implement real-world AmI systems (involving all or part of the preceding topics) without taking *architectural aspects* into account. These refer to the underlying sensory machinery, to the middleware, to real-time and operating systems requirements, to guarantee up-to-date information when needed, and to enforce scalability and robustness, just to name a few. One of the goal of this Handbook is to provide the reader with a comprehensive overview of an AmI system as a whole (although made-up of many interacting parts) thus showing how the various modules can effectively interact.

THE AMBIENT INTELLIGENCE SLOGANS

Ambient Intelligence is a novel and groundbreaking paradigm emerged during the past few years as a consequence of important achievements in various research and more technical fields. As a matter of fact, AmI has tremendous implications in such fields as design, healthcare, tele-medicine and Robotics, just to name a few.

Defining AmI is a rather difficult task, for it encompasses a broad range in the spectrum of interaction modalities between humans and their environment. In spite of this fuzzyness, it is possible to roughly define AmI as a computational paradigm aimed at designing, developing and realizing "augmented" environments (i.e., environments provided with electronic devices, such as sensors and actuators) that prove to be sensitive, responsive and aware of both humans and their activities. In a world where the AmI vision is realized in its entirety, distributed devices seamlessly cooperate with each other and with humans, in order to support them in better carrying out their everyday activities.

The AmI vision is characterized by a number of key ideas which distinguish AmI itself from similar paradigms aroused in the past few years, all of them closely related to the relationships among humans, their activities and their environment.

No more users, but humans! Traditionally, the generic notion of "computation" is meant at providing support for activities people must carry out either for work or leisure. However, it requires purposive and well-defined interaction modalities on the user side and, specifically, users should be skilled enough to benefit from the interaction. On the contrary, AmI envisions a future where purposive interaction on the user side will be kept to a minimum, whereas the "system" will provide information and assistance when needed. In other words, people will no longer need to be users: in order to interact with intelligent systems they will simply *act*, and the system will be able to autonomously understand their needs and goals.

Non only smart homes, but smart environments! Early research in AmI concentrated on specific types of augmented environments, and in particular on Smart Homes (intelligent apartments devoted to enforce specific aspects of domestic life). As a matter of fact, this approach helped in assessing first results and in conceiving new ideas to pursue. However, smart homes are only one side of the whole AmI vision: outside homes, AmI systems should be silent companions to human activities when travelling, sitting in a cafeteria, at work or in the pool. In other words, each accessed service in everyday life should be personalized and tailored especially for the single individual.

No more visible computational devices!. The fact that AmI shares with other paradigms, such as Ubiquitous and Pervasive Computing, the idea that computational devices (including sensors and actuators) should disappear from sight is an important concern. In particular, service provisioning and specific information should be conveyed through a hidden network connecting all these devices: as they will grow smaller, more connected and more integrated into our environments, the technology will disappear into human surroundings until only the (possibly implicit) interface will be directly perceivable. However, differently from Ubiquitous and Pervasive Computing, the network is the means by which data are gathered by a system that intelligently operate on them in order to provide directed, specific and targeted assistance.

Ambient Intelligence can have a tremendous impact on the future of society: from disaster assessment to elderly care, from surveillance activities to tele-medicine, from the tracking of patients in hospitals to package tracking in factories, just to name a few. AmI techniques can be exploited for the benefit of humans.

FEW KEYWORDS

Human aspects. Humans are the central components of AmI. In a sense, an AmI system is meant to wrap humans within an augmented physical space, where implicit and sensible interaction can occur and made effective. Modelling human capabilities is therefore a central topic that AmI must address in depth. Human models can be considered under many perspectives. Among them, we can think of cognitive models, interaction models, or preference models. This is a topic becoming increasingly central in AmI and in Artificial Intelligence in particular, thereby deserving great attention in future treatises.

Sensing. In order to realize the aforementioned implicit, purposive and sensible interaction, an AmI system must be provided with the necessary sensory information. Different sensors have been traditionally used to detect human activity: from the rich, highly expressive but prone to errors information provided by cameras, to the simple, poorly expressive but extremely reliable information provided by binary sensors, a plethora of techniques have been presented in literature. It seems reasonable to assume that different approaches will be required in different scenarios, thereby enforcing modularity, scalability and integration.

Activity recognition. Sensory information must be directed and manipulated in order to make sense of it. Specifically, recognizing human activity within a given physical environment is the first step to support *purposive interaction*. Many techniques have been proposed in literature during the past few years, all of them characterized by benefits and drawbacks: probabilistic, logic-based or ontology-based techniques to achieve activity recognition are currently hot topics in this area. Cross-fertilization among different approaches is expected to greatly improve the overall system capabilities.

Knowledge representation. It is widely recognized that AmI needs to efficiently and reliably represent the information gathered by distributed sources. During the past few years, in the Artificial Intelligence community, many techniques have been proposed to this aim: ontologies (conceptualization of a given domain) clearly emerged as a powerful modelling tool to encode possibly the overall behaviour of an AmI system, but other approaches, possibly taking relational as well as temporal information into account, are suitable as well. To date, there is no agreement about standard techniques and models, therefore requiring a deep analysis in AmI research.

Context-awareness. A system that is aware of what is going on within a given physical environment is closely part of the AmI vision. Assessing a situation means to characterize the relevant properties of an environment, the activities currently carried out within the environment and the possible consequences. In particular, the notions of "context" and "situation", when related to humans, aroused many harsh debates about philosophical and practical aspects of AmI since, from them, originate completely different approaches to *user modelling*, *activity recognition* and *knowledge representation*. In particular, context-awareness identifies modalities under which AmI system components cooperate with each other: it requires, from the designer's perspective, a description of what the system can do, and how to do it. This is probably the hottest topic of research in AmI in the latest few years.

Reasoning. Once the current situation has been assessed, the AmI system is expected to purposively interact with humans. In AmI systems, this interaction ultimately depends on many factors, such as the kind of requested interaction (e.g., physical for people suffering from physical disabilities or injuries or at the information level for businessmen with busy schedule), the kind of information needed (and the related reliability, obtained through merging different data sources), and associated priorities (e.g., understanding which, among many information request, is more relevant for the user given his/her current situation). Reasoning in AmI systems poses unique challenges to techniques developed in Artificial Intelligence so far: real-time requirements, human constraints, plausibility and compliance with common sense requirements. This will be for sure the next hot topic to deal with in AmI systems.

Sensible interaction. This topic describes how to couple *context-awareness* with *reasoning* in order to unleash the full potential of AmI applications. Sensible interaction policies must be either encoded within the system or learned by the system itself over time. In both cases, these policies should represent when and why the AmI system should operate, what to do and where to do it. Since this research direction is aimed at encoding how the AmI system interacts with humans, further research in this direction is expected to push the field to its very limits.

Architectural aspects and infrastructure. During the past few years it clearly emerged that AmI applications are complex systems composed of many different interacting parts. This difference is related to both conceptual and technological aspects: information flow should be clearly modelled in order to enforce reliability and efficiency of the whole system, whereas many technologies must be put together, thereby requiring complex interfaces at the software level. For instance, think about providing differentiated services to a businessman travelling from home to the airport: different information must be made available at different times, whereas how information itself is conveyed changes according to businessmen location. To date, these aspects have been mostly disregarded (due to the fact that AmI systems have been rather small), but nonetheless it is for sure one of the main challenges to face.

WHAT IS THIS BOOK FOR?

This Handbook is thought of as the book about AmI "the Editors would have read if it had been available when Ambient Intelligence began to be an interesting research field in its own right". In spite of the many publications labelled as "AmI-related", the need arises for an exhaustive treatise of AmI research "as a whole", and not as being originated by many small (and expensive) gadgets or *ad-hoc* systems considering mutual interactions as a *plus*.

In the Editors' perspective, the Handbook is meant at providing the skilled reader with a reasoned view of what is currently pursued by researchers and practitioners worldwide. In principle, it should be possible to read a chapter, to think about it, and to contact its authors in order to discuss about timely topics. More specifically, it is possible to identify the two following main objectives:

- *Comprehensive view of AmI systems.* The Handbook is intended to provide a clear view of an AmI system in its entirety, from human modelling to architectural aspects related to system engineering.
- *Oriented towards current research and future perspectives.* In the spirit of the "Handbook of Research" series, each Chapter provides the reader with a close look at what researchers are doing "here and now".

In the Editors' opinion, a book on AmI based on the previous objectives is both timely and needed. It is widely recognized that the next step in AmI will pass through a comprehensive description of *human aspects*, *knowledge representation*, *context-awareness* and *reasoning*. In a sense, this Handbook could be the first attempt in systematizing the field with a look at future developments in the meanwhile. From human perspective, this is a dramatic paradigm shift with respect to how humans themselves relate to technology and information technology in general. From old Science Fiction books and movies in the Seventies down to European Union strategies for the new century, AmI-related technologies are considered fundamental.

Information Science is perhaps the field that can benefit more from the AmI vision. Information will have to be managed in completely different ways: it will have to be stored for quick retrieval, to be lightweight in order to allow quick delivery across many heterogeneous networks, and to be reliable in that humans will depend on it on many activities. To achieve all these goals, research in AmI is expected to enforce the development of dependable infrastructures, to understand and reason upon human needs, to act proactively in the real world.

Technology development will surely be driven by the AmI vision. From the miniaturization of microelectronic components down to innovations in user interfaces, AmI is expected to provide designers with general guidelines as well as practical directions to follow in the mid and long-term period. Furthermore, AmI is not limited to specific environments and scenarios; on the contrary, its application is only limited by human acceptance and imagination. Automotive industry, building and factory automation, smart homes, intelligent building, energy saving policies, hospitals, tele-medicine, home assistance, gerontechnology, remote assistance, and disaster management are the first fields that push up after a brief brainstorming session. Understanding the potential offered by AmI that can be unleashed on these fields is fundamental for the well-being of our society.

From a management perspective, understanding what technology can offer can lead to better decisions in the mid and long-term period. When dealing with the AmI vision, this need is even more important, since AmI is expected to dramatically change our society in unexpected ways and with unprecedented benefits. As soon as AmI applications will be part of our society, new business models (taking AmI principles into account) will for sure be established.

The present Handbook is targeted to researchers and practitioners in Ambient Intelligence. However, since AmI *per se* is the conjunction of many research and technological fields, it is possible to assume that the Handbook can be useful for researchers and practitioners in Ubiquitous Computing, Pervasive Computing, Artificial Intelligence, Sensor Network and Sensor Technologies, Knowledge Representation, Automated Reasoning, Automated Learning, System Engineering, Software Engineering, Man-Machine Interfaces. Furthermore, for its practical implications, it is possible to argue that interested readers should also be acquainted with eldercare, tele-medicine, gerontechnology, assisted living technologies for people with disabilities.

Many other books have been published on AmI topics. As previously pointed out, these are most of the time edited collections of chapters devoted to specific aspects, such as privacy, human aspects, architecture, distributed computing, and so on. Although these contributions represent cutting-edge achievements in the field, they are not grounded with respect to a more abstract and looking-forward view of Ambient Intelligence as a whole.

We hope you will enjoy reading this Book as much as we enjoyed collecting all the material presented herein. It has been a daunting task: a special thanks to all the Authors, the Reviewers and the IGI-Global personnel, especially Elizabeth Ardner and Christine Buffon for their endless support and patience.

Fulvio Mastrogiovanni
University of Genova, Italy

Nak Young Chong
Japan Advanced Institute of Science and Technology, Japan

Chapter 1
Ambient Intelligence as Cognitive Niche Enrichment:
Foundational Issues

Lorenzo Magnani
University of Pavia, Italy

Emanuele Bardone
University of Pavia, Italy

ABSTRACT

We will introduce some cognitive and epistemological foundational aspects related to Ambient Intelligence (AmI). We will show how three concepts which derive from the tradition of cognitive science may be of help in deepening some of the main theoretical aspects concerning Ambient Intelligence: the notion of distributed cognition, the concept of cognitive niche, and the concept of affordance. We contend that this theoretical perspective will shed new light on some of the most promising and interesting developments recently brought about by Ambient Intelligence.

INTRODUCTION

We will introduce some cognitive and epistemological foundational aspects related to Ambient Intelligence (*AmI*). We will show how three concepts which derive from the tradition of cognitive science may be of help in deepening some of the main theoretical aspects concerning Ambient Intelligence: the notion of distributed cognition, the concept of cognitive niche, and the concept of affordance. We contend that this theoretical perspective will shed new light on some of the most promising and interesting developments recently brought about by Ambient Intelligence.

The main idea we are going to argue on in this paper is that Ambient Intelligence strongly favors the development of some crucial and structural aspects of human cognition, in so far as it creates – given the tremendous technological advance-

DOI: 10.4018/978-1-61692-857-5.ch001

ments in the recent decades – a new way in which humans can cognitively live their environments.

An important related aspect resorts to the emphasis on the distributed nature of human cognition. Technologies can be considered as resulting from those activities in which humans delegate cognitive functions to the environment by transforming and manipulating it. In doing so, cognition is defined as the complex network of interactions coming from the continuous interplay between humans and their environment, appropriately designed and constructed so as to meet human needs. Ambient Intelligence surely represents a novelty with respect to how humans can re-distribute and manage the cognitive resources delegated to the environment. Although massive cognitive delegations have already been successfully completed, Ambient Intelligence drastically favors the establishment of new kinds of environment which present novel theoretical and cognitive features, worthy of analysis.

We contend that, in analyzing the cognitive significance of Ambient Intelligence, the notion of *cognitive niche* is crucial. Such a notion would allow us to clear up possible misunderstandings about the role played by technological innovation in evolution. Besides, the notion of the cognitive niche will be of interest, as we will claim that Ambient Intelligence can be considered a new stage in the history of cognitive niche construction, given its "eco-cognitive" impact. Accordingly, we will claim *AmI* can be considered a form of *cognitive niche enrichment*.

The idea of Ambient Intelligence as a form of cognitive niche enrichment will be developed also taking advantage of the concept of *affordance*. By defining a cognitive niche as a set of affordances, we will point out that Ambient Intelligence provides humans with a special kind of affordances we will call *adapting affordances*. That is, Ambient Intelligence considered as a sophisticated cognitive niche activity structures our environments so as to populate them with devices able to *adaptively* meet our needs. Basically, the

affordances furnished by intelligent environments display a higher degree of adaptability which cannot be exhibited by other "things" that are already present in the human cognitive niches.

AMBIENT INTELLIGENCE AND DISTRIBUTING COGNITION

The Distribution of Human Cognition

The main novelty introduced by Ambient Intelligence is that it brings about a different way of distributing cognitive functions not only to some particular objects, but to the entire environment we live in (Remagnino, Foresti, & Ellis, 2005). That is, it is not only about how to design objects and tools that make us smarter, it is also about a new way of looking at our environment as a structure that deeply mediates our cognitive performances.

The distributed cognition approach is relevant to looking into this point (Hutchins, 1995, Clark & Chalmers, 1998, Magnani, 2007, Clark, 2008, Magnani, 2009): it explicitly acknowledges the mutuality between humans and their environment. According to this approach, it is the interaction between the two that provides humans with additional cognitive capabilities. Cognitive activities like, for instance, problem solving or decision-making, cannot only be regarded as internal processes that occur within the isolated brain.

Humans extend their minds into the material world, exploiting various external resources. For "external resources" we mean everything that is not inside the human brain, and that could be of some help in the process of deciding, thinking about, or using something. External resources can be artifacts, tools, objects, and so on. Basically, the exploitation of external resources is the process which allows the human cognitive system to be shaped by environmental (or contingency) elements. According to this statement, we may argue that detecting suitable cognitive chances play a pivotal role in almost any cognitive process.

The various external objects and tools we use are not merely memory aids: they can give people access to knowledge and skills that are unavailable to internal representations, help researchers to easily identify aspects and to make further inferences, they constrain the range of possible cognitive outcomes in a way that some actions are allowed and others forbidden. The mind is limited because of the restricted range of information processing, the limited power of working memory and attention, the limited speed of some learning and reasoning operations; on the other hand the environment is intricate, because of the huge amount of data, real time requirement, uncertainty factors. Consequently, we have to consider the whole system, consisting of both internal and external representations, and their role in optimizing the whole cognitive performance of the distribution of the various subtasks. In this case humans are not "just bodily and sensorily but also cognitively permeable agents" (Clark, 2008, p. 40). From the point of view of everyday situations external objects function like cognitive mediators that elaborate a simplification of the reasoning task and a redistribution of effort across time (Hutchins, 1995), when we need to manipulate concrete things in order to understand structures which are otherwise too abstract (Piaget, 1974), or when we are in presence. From this perspective the expansion of the minds is in the meantime a continuous process of disembodiment of the minds themselves into the material world around them. In this regard the evolution of the mind is inextricably linked with the evolution of large, integrated, material cognitive systems.

What is quite unique about human beings is their ability to modify the environment in order to exploit it as a mediating structure (Hutchins, 1995). As already mentioned, humans do not bear in their mind a complete representation of what surrounds them (Zhang, 1997). Instead, they use the environment itself as a mediating structure, which embeds cognitive chances – some of them purposefully manufactured that they can pick up

upon occasion. The ability of turning the environment – inert from a cognitive point of view – into a part of our cognitive system is basically *mimetic* (Magnani, 2009). It is mimetic in the sense that humans are constantly engaged in a process of *mimetically* externalizing fleeting thoughts, private ideas, etc., into external supports

A basic mimetic dynamics describing the distributed character of cognition involves the continuous delegation of cognitive roles to external objects so as to the continuous recapitulation of what occurring outside, over there, after having manipulated the external invented structured model. Humans constantly organize their brains through an eco-cognitive activity that reifies them in the external environment and then re-projected and reinterpreted through new configurations of neural networks and chemical processes. External representations are formed by external materials that express (through reification) concepts and problems already stored in the brain or that do not have a natural home in it; internalized representations are internal re-projections, a kind of recapitulations (learning), of external representations in terms of neural patterns of activation in the brain (Magnani, 2009). They can sometimes be "internally" manipulated like external objects and can originate new internal reconstructed representations through the neural activity of transformation and integration. In this sense, the external representations as main products of mimetic processes do not only mirror concepts and problems that are already represented in the brain, but they sometimes can creatively give rise to new concepts and meanings.

As regards the computational devices – which are indeed a product of highly sophisticated mimetic activities – humans have reached a point never reached before. They have been able to create what Magnani called "mimetic minds" (Magnani, 2009). As just mentioned, the process of externalizing cognitive functions is mimetic, as we represent something internal into external supports. In the case of sophisticated computational devices, the mimetic activities allowed us to build something that seems to occur "completely" outside, although it is still operating at the level of an on-line environment that also includes a human agent (Magnani, 2009). The various computing devices like for instance the ones from Artificial Intelligence are mimetic minds because they have representations that mimic general cognitive performances that we usually attribute to our minds: calculating, planning, solving problems, reasoning, just to quote some of the more sophisticated cognitive activities humans used to be involved in.

Ambient Intelligence as Context-Aware Computing

The advantages of having such devices that mimic a human mind are not difficult to acknowledge. Basically, we are talking about the possibilities of having external artifacts able to *proactively* and sensibly assist people in a number of tasks. Ambient Intelligence adds up a new layer to the traditional ways of disembodying the mind: Ambient Intelligence basically puts those sophisticated and smart devices – mimicking our mind – into our environments. In doing so even our familiar objects may embed high-level computing power. More generally, we argue that Ambient intelligence deals not only with reproducing some kinds of sophisticated human cognitive performances, but also with paying attention on an eco-cognitive dimension of computing – what is called context-aware computing (Cook & Das, 2007).

Ambient Intelligence enriches the experience of our environment in many ways. The most striking aspect related to Ambient Intelligence as a form of distributed intelligence is the level of autonomy that smart environments can reach (Hildebrandt, 2008a). A smart environment has an amazing power of monitoring and subsequently keeping track not only of our actions – what we do – but also of our preferences – what we desire (Remagnino et al., 2005, Cook, Augusto, & Vikramaditya, 2009). Collecting such an amount of data

– and aggregating it – allows smart environments to provide us with feedback that exhibit a degree of adaptability that cannot be compared with any other traditional environment (or cognitive niche, as we will show in the following sections). First of all, a smart environment adapts itself to infer one's preferences. It can act on the basis of past interactions that have been appropriately stored and then exploited by various tools mimicking some sophisticated forms of reasoning. This means that in smart environments high-level customization is possible, relying on the collection and aggregation of data about our behavior. These environments can also be creative in as far as they can anticipate user preferences, even before they become aware of them themselves (Hildebrandt, 2008a). In this case, smart environments exhibit what Verbeek calls "composite intentionality" (Verbeek, 2008). Basically, composite intentionality refers to situations in which the intentionality resulting from an action we take is made up of our own in coordination with that emerging from the interaction with an artifact. The intentionality resulting from interaction in smart environments is indeed highly composite, as *AmI* is designed specifically for augmenting – and thus making accessible – some experience of the world with respect to various modalities, namely, sensing, acting, and making decisions. We will come back to this issue in the section devoted to "adapting affordance".

Thus, to summarize, the idea behind distributed cognition is that human cognitive capabilities are fundamentally shaped by environmental chances that are ecologically rooted. Cognitive processes do not happen in a vacuum, so the context and the resources one has at one's disposal are crucial for describing and also explaining human cognition. Ambient Intelligence can certainly be considered one of the most sophisticated ways humans have invented to distribute cognitive functions to external objects. In this case, the massive cognitive delegation contributes to a radical re-distribution of the cognitive load humans are subjected to.

Basically, Ambient Intelligence improves people's experience in their environments (Cook & Das, 2007). That is, it increases the number and range of tasks one can accomplish. As the result of a massive cognitive delegation, humans are provided with environments that bring sophisticated interactions into existence, in which the cognitive load is partly carried out by intelligent devices displaying an unprecedented level of autonomy and transparency.

AMBIENT INTELLIGENCE AS COGNITIVE NICHE ENRICHMENT

Evolution and the Cognitive Niche

In order to better assess the *eco-cognitive discontinuity* brought about by *AmI* we introduce the concept of cognitive niche construction. Our main take is that *AmI* enhances the experience of our local environment by means of what we call *cognitive niche enrichment*.

The activity of cognitive niche construction places attention upon the fact that humans extend their cognition in the environment by transforming and manipulating it. In this sense, technological innovations are meant to be very powerful and successful activities of cognitive niche construction, as they bring into existence new cognitive chances and new ways of behaving. The theory of niche construction has been recently advocated by a number of biologists, who revisited some basic aspects concerning Darwinism in order to acknowledge the agency of living organisms and its impact on evolution. It is crucial to this theory the revision of the notion of *adaptation*.

Basically, evolution is a response to changes in selection pressures all living organisms are subjected to (Laland & Brown, 2006). Traditionally, the environment has been always meant to be the only source of selection pressures so that adaptation is considered a sort of top-down process that

goes from the environment to the living creature (Godfrey-Smith, 1998).

In contrast to that, a bunch of biologists (Laland, Odling-Smee, & Feldman, 2000, Day, Laland, & Odling-Smee, 2003, Odling-Smee, Laland, & Feldman, 2003) has recently tried to provide an alternative theoretical framework by emphasizing the active role played by the living organisms. The environment is meant to be a sort of "global market" that provides creatures with unlimited possibilities. Indeed, not all the possibilities that the environment offers can be exploited by tan organism. For instance, the environment provides water to swim in, air to fly in, flat surfaces to walk on, and so on. However, no creatures are fully able to take advantage of all of them. Therefore, all organisms try to modify their surroundings in order to better exploit those elements that suit them and eliminate or mitigate the effect of the negative ones.

According to this view, adaptation is meant to be *symmetrical*. Organisms (humans, in our case) adapt to their environment *and viceversa* (Laland & Brown, 2006). That is, environments cause some features of living creatures to change. But also creatures cause some features of the environments to change. In altering their local environments, living creatures contribute to construct and modify the so-called ecological niches.

In any ecological niche, the selective pressure of the local environment is drastically modified by organisms in order to lessen the negative impacts of all those elements which they are not suited to. Indeed, this does not mean that natural selection is somehow halted. Rather, this means adaptation cannot be considered only by referring to the agency of the environment, but also to that of the organism acting on it. In this sense, animals and other living creatures are ecological engineers, because they do not simply live their environment, but they actively shape and change it (Day et al., 2003).

From an evolutionary perspective, the ubiquitous presence of the cognitive niches contribute to introducing a second and non-genetic inheritance system insofar as the modifications brought about on the environment persist, and so are passed on from generation to generation (Odling-Smee et al., 2003). As for humans, which had become extremely successful ecological engineers, the main advantage of having a second inheritance system is that it enabled them to access a great variety of information and resources never personally experienced, resulting from the activity of previous generations (Odling-Smee et al., 2003). That is, the information and knowledge humans can draw on are not simply transmitted, but they can also be accumulated in the so-called cognitive niches. Indeed, the knowledge we are talking about embraces a great variety of resources including knowledge about nature, social organization, technology, the human body, and so on. In any cognitive niche the relevant aspects of the local environment are appropriately selected so as to turn the surroundings – inert from a cognitive point of view – into a mediating structure delivering suitable chances for behavior control (Odling-Smee et al., 2003).

Ecological inheritance system is different from the genetic one in the following way (Odling-Smee et al., 2003): 1) genetic materials can be inherited only from parents or relatives. Conversely, modifications on the environment can affect everyone, no matter who he/she is. It may regard unrelated organisms also belonging to other species. There are several global phenomena such as climate change that regard human beings, but also the entire ecosystem; 2) genes transmission is a one way transmission flow, from parents to offspring, whereas environmental information can travel backward affecting several generations. Pollution, for instance, affects young as well as old people; 3) genetic inheritance can happen once during one's life, at the time of reproductive phase. In contrast, ecological information can be transferred during the entire duration of life. Indeed, it depends on the eco-engineering capacities at play; 4) genetic inheritance system leans on the presence of repli-

cators, whereas the ecological inheritance system leans on the persistence of whatsoever changes made upon the environment.

Niche construction is therefore a form of feedback in evolution that has been rarely considered in the traditional evolutionary studies (Day et al., 2003). On this basis a co-evolution between niche construction and brain development and its cognitive capabilities can be clearly hypothesized, a perspective further supported by some speculative hypotheses given by cognitive scientists and paleoanthropologists (Donald, 1998, Mithen, 1999, Richerson & Boyd, 2005).

New Eco-Cognitive Smart Environments

As already mentioned, the outstanding characteristics of human beings are that they have progressively become *eco-cognitively* dominant, as their abilities as eco-cognitive engineers out-competed the ones of other species in occupying and then modifying according to their needs the shared environments (Flinn, Geary, & Ward, 2002). The notion of cognitive niche acknowledges the artificial nature of the environment humans live in, and their active part in shaping it with relation to their needs and goals. It is within this context that we locate the innovative character of Ambient Intelligence. That is, the creation of smart environments can certainly be viewed as a new way of constructing the cognitive niche. More precisely, Ambient Intelligence constitutes an eco-cognitive activity in which our pre-existing cognitive niches are dramatically enriched with objects and tools that re-configure our capacities for extending human cognition and its boundaries.

It is worth noting – even though we will not be dealing with this issue – that niche construction activities do not halt natural selection. This certainly has some consequences that should be acknowledged. Niche construction activities may help humans reduce the negative impacts of pre-existing niches but that does not mean that they might not produce even worse consequences. This is because, even when a niche has been modified, selection pressures continue to act upon us and other organisms as well. Certain cognitive niches resulting from intensive eco-cognitive activity may turn out to be "maladaptive" or – at least – they might actually endanger us and other species as well. Here, the term "maladaptive" is not intended in a Darwinian sense.

In Ambient Intelligence we have many examples of potential maladaptive outcomes resulting from cognitive niche enrichment activity. For example, the problem of the possible negative consequences of Ambient Intelligence in the case of agency and (criminal) liability is discussed in (Hildebrandt, 2008a). Hildebrandt brilliantly points out that the emergence of ambient technologies that are able to monitor and anticipate human behavior can become a threat to a number of values that are crucial in Western democracies, like for instance the values that are at play in the case of criminal liability. She argues that, as far as we cannot say whether a certain action has been carried out by us or by an artificial device, then we will also have severe problems in attributing criminal liability. This may also cause a person to exploit the ambiguity resulting from the *AmI* establishment of a hybrid agency, in their favor.[1] Another possible negative consequence is related to the notion of identity. As argued by Gudwirth (2009), *AmI* technologies empower humans in the process of profiling. Profiling basically deals with the possibility of assigning an identity to a user relying on data and information gathered from the behaviors of the same user as well as of others. The pervasiveness of *AmI* drastically changes our ability and the effectiveness of such a task, given the continuous and detailed monitoring of the user's behavior made possible by the smart devices available. *Being profiled*, as Gudwirth argued, could, however, easily become a threat to the development of our own identity because it can be assigned automatically, and even without our consent. In this sense, *AmI* potentially and

dangerously induces us, first of all, to adopt an identity we did not have the opportunity to choose. This is clearly a limitation of our freedom insofar as we would be obliged – more or less tacitly – to match in with some arbitrary categories generated by the profiling algorithms. Secondly, it enforces us to adapt "to a context moulded by other actors" (Gutwirth, 2009) favoring various dynamics related to standardization and, sometimes, even group tyranny.[2]

In order to further and better assess the eco-cognitive contribution of Ambient Intelligence the notion of *affordance* is fundamental. The idea we are going to present in the following section is that the novelty brought about by *AmI* lies on the fact that Ambient Intelligence enriches human cognitive niches by providing new affordances. We will show how affordances are cognitive chances embedded in the interaction between a (human) organism and its environment, and how they are organized in cognitive niches, which make them easily accessible. Ambient Intelligence populates our cognitive niches with objects and devices that are to some extent intelligent objects. More precisely, through the activity of cognitive niche enrichment it delivers new kinds of affordances, which preexisting technologies could not furnish. This is the a creation of a kind of affordance that we call *adapting affordances*, as they exhibit *adaptability*.

AMBIENT INTELLIGENCE AND ADAPTING AFFORDANCES

Affordances as Eco-Cognitive Interactional Structures

During the last decades the concept of affordance has attracted a lot of scholars and researchers coming from different disciplines; its richness is demonstrated by the breadth of ideas and applications it brought about or, at least, inspired. From visual perception to Human Computer

Interaction (*HCI*) and distributed cognition, the concept of affordance provided valuable insights about how to manage the intricate relationship between humans and their environment (Magnani & Bardone, 2008). The way humans interact with their environment is the central feature of what a theory of affordance has to deal with. From a general perspective the main thesis held by its proponents is that the representation we have of the world includes also information about those offerings provided by the environment (Gibson, 1979, Scarantino, 2003, Vicente, 2003). That is, an object is directly perceived in *action terms* (Natsoulas, 2004). And those possible actions are specified by what Gibson called "affordances" (Gibson, 1979).

The theory of affordance potentially re-conceptualizes the traditional view of the relationship between action and cognition according to which we extract from the environment those information which build up the mental representation that in turn guides action (Marr, 1982). From an eco-cognitive perspective, the distinction between action and cognition is questioned. The notion of affordance contributes to shed light on that issue fairly expanding it.

According to the distributed cognition theory mentioned above, humans do not retain in their memory rich representations of the environment and its variables, but they actively manipulate it by picking up information and resources upon occasion (Thomas, 1999), already available, or extracted/created and made available: information and resources are not only given, but they are actively sought and even manufactured. In this sense, humans are constantly involved in a process of seeking chances (Magnani & Bardone, 2008). In our terminology, chances are not simply information, but they are "affordances" arranged in cognitive niches, which are environmental anchors that allow us to better exploit external resources (Magnani & Bardone, 2008, Magnani, 2009, Abe, 2009).

Gibson defines "affordance" as what the environment offers, provides, or furnishes. For instance, a chair affords an opportunity for sitting, air breathing, water swimming, stairs climbing, and so on. By cutting across the subjective/objective frontier, affordances refer to the idea of agent-environment mutuality. Gibson did not only provide clear examples, but also a list of definitions (Wells, 2002):

- affordances are opportunities for action;
- affordances are the values and meanings of things which can be directly perceived;
- affordances are ecological facts;
- affordances imply the mutuality of perceiver and environment.

The Gibsonian ecological perspective originally achieves two important results. First of all, human and animal agencies are somehow hybrid, in the sense that they strongly rely on the environment and on what it offers (Zhang & Patel, 2006). Secondly, Gibson provides a general framework about how organisms directly perceive objects and their affordances. His hypothesis is highly stimulating: "[…] the perceiving of an affordance is not a process of perceiving a value-free physical object […] it is a process of perceiving a value-rich ecological object", and then, "physics may be value free, but ecology is not" (Gibson, 1979, p. 140).

In the examples given by Gibson – a chair affords sitting, water affords swimming, stairs climbing, etc. – it seems that affordances only deal with pragmatic actions. However, the notion of affordance can be fairly extended beyond its mere pragmatic sense embracing a variety of situations that involve higher and more sophisticated cognitive dimensions. According to our view, the term "action" refers to the fact that one can act on the environment in such a way to exploit what the surroundings tacitly offer in terms of cognitive chances. Action is therefore interpreted in a cognitive sense. We are referring here to the distinction between epistemic action and pragmatic action introduced by (Kirsh & Maglio, 1994). Pragmatic actions are the actions that an agent performs in the environment in order to bring itself physically closer to a goal. In this case the action modifies the environment so that the latter acquires a configuration that helps the agent to reach a goal which is understood as physical, that is, as a desired state of affairs. Epistemic actions are the actions that an agent performs in a semiotic environment in order to discharge the mind of a cognitive load or to extract information that is hidden or that would be very hard to obtain only by internal computation. Consider for instance a graph representing the so called supply and demand curve. In this case, the graph conveys information that facilitate to uncover some economic consequences, which otherwise would be not available. Basically, the graph affords *understanding*: it provides visual clues facilitating some kind of interpretations rather than others. It is clear here that the term action cannot be considered in a mere pragmatic sense, but it opens up to a more sophisticated cognitive dimension.

The notion of affordance may be of help in understanding how *AmI* enriches our cognitive niches. The idea we are going to describe is that smart environments provide us with *adapting affordances*, which – to some extent – *mimic* those usually furnished by other human beings.[3] In order to introduce this topic, let us spend a few words on the issue related to affordance detection.

In Magnani and Bardone (2008), we argued that affordance detection can be seen as a result of an inferential activity – mainly abductive – in which a person infers from the presence of signs or clues the way she can cope with an object or even a situation (Magnani & Bardone, 2008).[4] It follows that an affordance is detected when one can make use of those signs and clues that specify it. In the case of a chair, we detect the affordance of sitting because it is rigid, flat, and robust. Rigidity, robustness, and flatness are all signs that

make us able to infer that an action possibility is to sit on it.

Some affordances are wired by evolution: that is, the detection of those signs specifying the simplest affordances of an object are granted by dispositions wired by evolution.[5] Other affordances require a person to learn how to detect them. By means of learning and rehearsing, a person acquires the proper knowledge about how to detect additional signs and clues. In doing so, one becomes attuned to more complex affordances. This is the case of affordances that require the exploitation of more plastic endowments. The list of examples could be almost endless. Just think of the manipulations expert physicians perform on a patient's body. Physicians are indeed extremely good at detecting and then making use of symptoms, which an untrained eye could not even perceive or be aware of.

Part of the affordances available are delivered by other human beings. Other human beings are so important because they deliver a special kind of affordance, which Ambient Intelligence tries to mimic. These affordances can be called *adapting affordances*. As just stated, affordance detection relies on abductive skills that allow us to infer possible ways to cope with an object or situation. In the case of affordances furnished by other people, the process of detection may be further mediated by social interaction. That is, other people may provide us with additional clues to help us better exploit environmental chances, namely, hidden or latent affordances (Gibson & Pick, 2000).

Care-giving is an example of what we are talking about. As a matter of fact, babies are heavily dependent on other human beings from the very beginning of their existence. They would not survive without constant care administrated by their parents. More precisely, caregivers assist infants in turning the environment into something they can handle by means of learning (Gibson & Pick, 2000). In doing so they constantly adapt their behavior to create suitable action possibilities that infants can be easily afforded by (Gibson &

Pick, 2000). Caregivers, for instance, contribute to expanding the basic repertoire a newborn is equipped with by manipulating her/his attention. In doing so they prevent the baby from expensive and exhaustive trial and error activities. For example, caregivers:

- act for the infant *embodying motions* that she/he should perform;
- they *show* what the infant is supposed to do;
- they *offer demonstrations* providing gestures so that
- they *direct the infant's attention* to relevant aspects of the environment (Zukow-Goldring & Arbib, 2007).

The influence people have is not limited to care-giving activities. There are a number of situations in which people exhibit the capacity to adaptively alter the representation other fellow humans may have of something. Another example is the so-called *intentional gaze* (Frischen, Bayliss, & Tipper, 2007, Frischen, Loach, & Tipper, 2009). Recent findings have shown that intentional gaze confers additional properties to a given object, which would not be on display if not looked at (Becchio, Pierno, Mari, Lusher, & Castello, 2007). Gaze-following behavior affects the way an object is processed by our motor-cognitive system. More precisely, intentional gaze changes the representation of an object, not merely by shifting attention, but by enriching it with "motor emotive and status components" (Becchio et al., 2007, p. 257).

From a theoretical perspective we may argue that human beings function as a kind of *adapting task-transforming representation*. The term "task-transforming representation" was introduced by Hutchins (Hutchins, 1995) to refer to the fact that external artifacts shape the structure of a task – its representation – helping people solve the problem they are facing. A tool may transform the structure of a task:

1. redistributing the cognitive load;
2. rearranging constraints and action possibilities;
3. unearthing additional computational abilities;
4. increasing the number of operations while reducing mental costs.

In the case of *adapting affordances*, the cognitive load is reduced by means of a *transformation* (Leon, 2002), which adapts the structure/representation of the task to allow a person to detect latent environmental chances. Caregivers and the intentional gaze are fair examples, as they show how people adaptively manipulate the representations their fellows have of the environment to favor or facilitate the exploitation of latent affordances.

Adapting Affordances in Ambient Intelligence Environment

As already argued, Ambient Intelligence can be considered as "cognitive niche enrichment" because of the way it enriches our cognitive niches, that is, by populating them with devices able to keep track of what we do, and then adjust their response adaptively. These various devices are thought to deliver affordances that are somehow *adapting*.

Adapting affordances are those affordances that help the agent exploit latent environmental possibilities providing additional clues. As we have already pointed out, an affordance may be hidden for a number of reasons. For instance, one may lack the proper knowledge required to detect an affordance at the moment of acting. On the other hand, it might remain hidden because of the ambiguity of certain sign configuration so that the agent's function is not immediately intuitive. Finally, affordances may prove unavailable just because a certain person suffers from some impairment – temporary or otherwise – that prevents her from exploiting some particular environmental offerings.

AmI may enrich one's experience of her environment in a number of ways. In the following we are going to present three main cases in which our experience is enriched with respect to three "modalities". *AmI* may eco-cognitively enrich our ability to 1) sense, 2) act, and 3) reason or make decisions. This general classification is indeed arbitrary and it is not meant to cover all the possible ways in which an agent can interact with her environment. However, it may be useful when showing some examples in which "adapting affordances" are provided, and looking at how they enrich the experience of the agent's environment.

The first modality concerns our senses. Sensing something is the basic modality through which we can interact with our environment, and it can basically be related to haptic perception, visual perception, auditory perception, and tactile perception. At times some of our senses may be impaired or less effective for a number of reasons, consequently rendering a person less able to cope with her environment the way she would like to. This happens for example to elderly people, who often suffer from pathologies that impoverish their senses, and thus their wellbeing. Assisted living technologies are meant to give support to those people that have various problems related to aging. An interesting example of Ambient intelligence environment supporting sensing is provided by ALADIN. ALADIN is lighting system designed to assist elderly people in the process of adjusting indoor lighting (Maier & Kempter, 2009).

Generally speaking, light has an important impact on the wellbeing of a person. It affects a number of activities that are indeed fundamental for us. Cognitive performances like reading or concentrating may be drastically affected by light. Light also affects sleep quality, the metabolic system and changes of mood. The impact of light may acquire even greater importance for elderly people, who often have impaired vision or suffer from limited mobility so that they remain at home for most of their time (Maier & Kempter, 2009).

ALADIN is meant to assist them in designing and maintaining indoor lighting.

Basically, ALADIN is capable of acquiring and analyzing information about "the individual and situational differences of the psycho-physiological effects of lighting" (Maier & Kempter, 2009, p. 1201) so as to provide adaptations specific to the user's needs in her domestic environment. More precisely, the system constantly monitors the user's state through biosensors which measure various parameters – elecrodermal, electrocardiographic, respiratory, and those related to body movement. The data are transferred via bluetooth to the computer system and analyzed. Then, according to the adaptive algorithms, the system tries to find the best light modification. In fact, the best light modification is reached by simple feedback provided by the user and from the biofeedbacks acquired through biosensors attached to the user's body.

As already mentioned, lighting may affect a person in a variety of ways, for instance, in the preservation of an active and independent lifestyle or physical and mental fitness (Heschong, 2002). For instance, elderly people are often unable to correctly assess some internal states with great precision and to then act accordingly. Reduced cognitive performances due to aging may also have a negative impact on their sensing capacity. This, in turn, limits their ability to detect those affordances enabling them to design and maintain suitable indoor lighting to enhancing some activities like reading or relaxing. Biosensors monitor some physiological parameters – as we have just mentioned – and provide some information. This information may help the user modify their domestic light setting in a way that may eventually make them aware of some physiological needs, for instance, the need to rest or have more light in the room. In this sense, the various adjustments the system carries out during a session clearly function like a "mediator" which unearths new environmental possibilities. That is, the system alters the representation of the user's body in a way that allows them to be better afforded by the lights.

The experience of our environment could also be improved, when we act on it. In order to better grasp this point, we refer to the case of the guide sign system. Navigating effectively through certain spaces such as university or hospital buildings could become a problem, sometimes even involving negative emotions like anxiety and tension. Generally, losing the way is annoying and it wastes so much time. Therefore, having guide sign systems that can effectively assist a person to go where they are supposed to go is extremely important.

Moving through a space can be considered as a cognitive task (Conlin, 2009), in which one tries to make sense of the various detectable clues. This is especially true for those spaces that can be labeled as *meaningful*, as they are intentionally designed to bring people to certain places just like in the case of a hospital building. Meaningful spaces furnish people with a variety of clues, which should help them find the way. For instance, in hospitals the use of the uniform design or the use of same materials and colors. Generally speaking, these clues and guide signs furnish affordances that users can exploit in order to reach the place they are supposed to.

However, the information and clues that guide sign designers can scatter are limited given the various physical as well as cognitive constraints. Physical constraints are given, for instance, by the fact that if space is limited so is the possibility of locating signs. Cognitive constraints are due to the fact that sometimes having more information may cause cognitive overload, and thus waste time. Therefore, affordance detection may become a problem. Ambient intelligence can assist people, for instance, in navigating space providing adapting affordances. Hye described and tested a prototype of an interactive guide sign system meant to help people navigate a hospital, constructing smart environments by means of various devices like *RFID* (Radio Frequency

Identification), smart card, and wireless communication (Hye, 2007). In this case, the use of such devices provide additional affordances that have the main role of helping users exploit latent environmental possibilities, which in this case are basically guide signs.

Adapting affordances can assist people in contexts involving more complex cognitive activities, for instance, reasoning and decision-making. An interesting example about how Ambient Intelligence may help people is illustrated by Eng, Douglas, & Verschure (2005). Using a learning model called "distributed adaptive control" (DAC), Eng and colleagues have designed intelligent environments supporting an interactive entertainment deployed at the Swiss national exhibition Expo.02. DAC was designed to influence the behavior of visitors by learning how to scatter cues so as to make some areas of the exhibition – often avoided or disregarded by the visitors – more visible and easier to reach. Here again, the contribution of the so-called smart environment is to adaptively adjust the system response with respect to what the visitors and users know by furnishing the proper signs.

Another example is provided by the Glass Bottom Float project (*GBF*), which aims to inform beach visitors about water quality so as to increase the pleasure of swimming (Bohlen & Frei, 2009). Basically, *GBF* is a beach robot monitoring a number of parameters: water and air temperature, relative humidity, pH, conductivity, salinity, dissolved oxygen, etc., but also the speed of boats passing by, people playing on the float, and, more generally, underwater marine life. In addition, these parameters are validated by opinions gathered from interviews with swimmers at the beach.

It is worth noting that *GBF*, coupled with the use of *SPM*, is completely different from the other projects aiming to deliver information about water and air quality. We know that both government agencies and official news releases provide the user with similar data. However, what is interesting in the case of *GBF* is the frequency and quality of the updates it delivers to the public. The way this is done is completely different, and enriches one's experience. For instance, the use of beach visitor opinions potentially increases the reliability of the recommendations. This enables a completely new way of distributing knowledge among beach visitors.

The data gathered are then processed using the so-called "swimming pleasure measure" (*SPM*). Basically, the *SPM* makes predictions about the expected pleasure people may have swimming on a certain beach. The recommendations generated by the system are then delivered in real time to the public via mobile phones. In this way, people can obtain important information before driving to the beach. Here we have an example showing how our ability to formulate judgments and make decisions can be supported and improved by information that would not otherwise be available. More precisely, *SPM* provides us with additional clues that permit us to be *afforded* by the local environment in order to make decisions and carry out reasoning.

CONCLUSION

In this paper we have illustrated some foundational aspects related to Ambient Intelligence. We analyzed three main issues. First of all, Ambient Intelligence represents a very sophisticated way of *distributing cognitive functions* to the environment. We have argued that the novelty brought about by Ambient Intelligence is related to its level of autonomy and transparency. In this sense, Ambient Intelligence can be considered as a general re-configuration of our cognitive delegations that mainly pays attention to the eco-cognitive dimension of computing – a context-aware computing, we contended.

These eco-cognitive aspects have been developed through the notion of cognitive niche construction. More precisely, we have argued that

Ambient Intelligence plays a fundamental role in enriching our pre-existing cognitive niches in a completely new way. Given its impact on our environments, Ambient Intelligence is a form of *cognitive niche enrichment*. This means that Ambient Intelligence populates our ecologies with devices that re-define some crucial human cognitive activities.

In order to better assess the impact and relevance Ambient Intelligence has on our cognitive niches, we have introduced the concept of affordance, seen as kinds of "environmental anchors". Our main idea is that Ambient Intelligence enriches our cognitive niches insofar as it provides us with what we called *adapting affordances*. Adapting affordances are not only chances that are ecologically delivered, they are also resources that furnish support in the process of affordance detection. Relying on a couple of examples, we have illustrated how adapting affordances can adaptively assist people by proving them with additional clues to exploit further latent or hidden environmental chances.

REFERENCES

Abe, A. (2009). Cognitive chance discovery. In S. C. (Ed.), *Universal Access in HCI, Part I, HCII2009 (LNCS5614)* (pp. 315–323). Berlin/New York: Springer.

Becchio, C., Pierno, A., Mari, M., Lusher, D., & Castello, U. (2007). Motor contagion from eye gaze. the case of autism. *Brain, 130*, 2401–2411. doi:10.1093/brain/awm171

Bohlen, M., & Frei, H. (2009). Ambient Intelligence in the city. overview and new perspectives. In Nakashima, H., Aghajan, H., & Augusto, J. (Eds.), *Handbook of Ambient Intelligence and Smart Environments* (pp. 911–938). Heidelberg: Springer.

Clark, A. (2008). *Supersizing the Mind: Embodiment, Action and Cognitive Extension*. Oxford: Oxford University Press.

Clark, A., & Chalmers, D. J. (1998). The extended mind. *Analysis*, *58*, 10–23. doi:10.1111/1467-8284.00096

Conlin, J. (2009). Getting around: making fast and frugal navigation decisions. In Markus Raab, M., Johnson, J., & Heekeren, H. (Eds.), *Mind and Motion: The Bidirectional Link between Thought and Action* (*Vol. 174*, pp. 109–117). Elsevier. doi:10.1016/S0079-6123(09)01310-7

Cook, D., Augusto, J., & Vikramaditya, R. (2009). Ambient Intelligence: Technologies, applications, and opportunities. *Pervasive and Mobile Computing*, *5*, 277–298. doi:10.1016/j.pmcj.2009.04.001

Cook, J., & Das, S. (2007). How smart are our environments? an updated look at the state of the art. *Pervasive and Mobile Computing*, *3*(2), 53–73. doi:10.1016/j.pmcj.2006.12.001

Day, R. L., Laland, K., & Odling-Smee, J. (2003). Rethinking adaptation. The niche-construction perspective. *Perspectives in Biology and Medicine*, *46*(1), 80–95. doi:10.1353/pbm.2003.0003

de Leon, D. (2002). Cognitive task transformations. *Cognitive Systems Research*, *3*, 349–359. doi:10.1016/S1389-0417(02)00047-5

Donald, M. (1998). Hominid enculturation and cognitive evolution. In Renfrew, C., Mellars, P., & Scarre, C. (Eds.), *Cognition and Material Culture: the Archeology of External Symbolic Storage* (pp. 7–17). Cambridge: The McDonald Institute for Archaeological Research.

Eng, K., Douglas, R., & Verschure, P. (2005). An interactive space that learns to influence human behavior. *Systems, Man and Cybernetics, Part A, IEEE Transactions on, 35*(1), 66–77.

Flinn, M., Geary, D., & Ward, C. (2002). Ecological dominance, social competition, and coalitionary arms raceswhy humans evolved extraordinary intelligence. *Evolution and Human Behavior*, *26*(1), 10–46. doi:10.1016/j.evolhumbehav.2004.08.005

Frischen, A., Bayliss, A., & Tipper, S. (2007). Gaze-cueing of attention: Visual attention, social cognition and individual differences. *Psychological Bulletin*, *133*(4), 694–724. doi:10.1037/0033-2909.133.4.694

Frischen, A., Loach, D., & Tipper, S. P. (2009). Seeing the world through another person's eyes: Simulating selective attention via action observation. *Cognition*, *111*(2), 212–218. doi:10.1016/j.cognition.2009.02.003

Gibson, E., & Pick, A. (2000). *An ecological approach to perceptual learning and development*. Oxford: Oxford University Press.

Gibson, J. J. (1979). *The Ecological Approach to Visual Perception*. Boston, MA: Houghton Mifflin.

Godfrey-Smith, P. (1998). *Complexity and the Function of Mind in Nature*. Cambridge: Cambridge University Press.

Gutwirth, S. (2009). Beyond identity. *IDIS*, *1*, 123–133. doi:10.1007/s12394-009-0009-3

Heschong, L. (2002). Daylighting and human performance. *ASHRAE Journal*, *44*(6), 65–67.

Hildebrandt, M. (2008a). Ambient intelligence, criminal liability and democracy. *Criminal Law and Philosophy*, *2*(2), 163–180. doi:10.1007/s11572-007-9042-1

Hildebrandt, M. (2008b). A vision of ambient law. In Brownsword, R., & Yeung, K. (Eds.), *Regulating Technologies: Legal Futures, Regulatory Frames and Technological Fixes* (pp. 175–191). Oxford: Hart Publishing.

Hutchins, E. (1995). *Cognition in the Wild*. Cambridge, MA: The MIT Press.

Hye, P. (2007). *A design study of pedestrian space as an interactive space*. (presented at IASDR 2007)

Kirsh, D., & Maglio, P. (1994). On distinguishing epistemic from pragmatic action. *Cognitive Science*, *18*, 513–549. doi:10.1207/s15516709cog1804_1

Laland, K., & Brown, G. (2006). Niche construction, human behavior, and the adaptive-lag hypothesis. *Evolutionary Anthropology*, *15*, 95–104. doi:10.1002/evan.20093

Laland, K., Odling-Smee, J., & Feldman, M. (2000). Niche construction, biological evolution and cultural change. *The Behavioral and Brain Sciences*, *23*(1), 131–175. doi:10.1017/S0140525X00002417

Magnani, L. (2007). *Morality in a Technological World. Knowledge as Duty*. Cambridge: Cambridge University Press. doi:10.1017/CBO9780511498657

Magnani, L. (2009). *Abductive Cognition. The Epistemological and Eco-Cognitive Dimensions of Hypothetical Reasoning*. Berlin, Heidelberg: Springer-Verlag.

Magnani, L., & Bardone, E. (2008). Sharing representations and creating chances through cognitive niche construction. The role of affordances and abduction. In Iwata, S., Oshawa, Y., Tsumoto, S., Zhong, N., Shi, Y., & Magnani, L. (Eds.), *Communications and Discoveries from Multidisciplinary Data* (pp. 3–40). Berlin: Springer. doi:10.1007/978-3-540-78733-4_1

Maier, E., & Kempter, G. (2009). Aladin - a magic lamp for the elderly? In Hideyuki Nakashima, H., Aghajan, H., & Augusto, J. (Eds.), *Handbook of Ambient Intelligence and Smart Environments* (pp. 1201–1227). Heidelberg: Springer.

Marcus, G. (2004). *The Birth of the Mind: How a Tiny Number of Genes Creates the Complexities of Human Thought*. New York: Basic Books.

Marr, D. (1982). *Vision*. San Francisco, CA: Freeman.

Mithen, S. (1999). Handaxes and ice age carvings: hard evidence for the evolution of consciousness. In Hameroff, A. R., Kaszniak, A. W., & Chalmers, D. J. (Eds.), *Toward a Science of Consciousness III. The Third Tucson Discussions and Debates* (pp. 281–296). Cambridge: MIT Press.

Natsoulas, T. (2004). To see is to perceive what they afford: James J. Gibson's concept of affordance. *Mind and Behaviour*, *2*(4), 323–348.

Odling-Smee, F., Laland, K., & Feldman, M. (2003). *Niche Construction. A Neglected Process in Evolution*. New York, NJ: Princeton University Press.

Peirce, C. S. (1967). *The Charles S. Peirce Papers: Manuscript Collection in the Houghton Library*. Worcester, MA: The University of Massachusetts Press. (Annotated Catalogue of the Papers of Charles S. Peirce. Numbered according to Richard S. Robin. Available in the Peirce Microfilm edition. Pagination: CSP = Peirce / ISP = Institute for Studies in Pragmaticism.)

Piaget, J. (1974). *Adaption and Intelligence*. Chicago: University of Chicago Press.

Remagnino, P., Foresti, G., & Ellis, T. (2005). *Ambient Intelligence*. New York: Springer. doi:10.1007/b100343

Richerson, P. J., & Boyd, R. (2005). *Not by Genes Alone. How Culture Transformed Human Evolution*. Chicago, London: The University of Chicago Press.

Scarantino, A. (2003). Affordances explained. *Philosophy of Science*, *70*, 949–961. doi:10.1086/377380

Thomas, N. J. T. (1999). Are theories of imagery theories of imagination? An active perception approach to conscious mental content. *Cognitive Science*, *23*, 207–245. doi:10.1207/s15516709cog2302_3

Verbeek, P.-P. (2008). Cyborg intentionality: Rethinking the phenomenology of human–technology relations. *Phenomenology and the Cognitive Sciences*, *7*, 387–395. doi:10.1007/s11097-008-9099-x

Verbeek, P.-P. (2009). The moral relevance of technological artifacts. In Sollie, P., & Duwell, M. (Eds.), *Evaluating New Technologies* (pp. 63–77). Berlin: Springer. doi:10.1007/978-90-481-2229-5_6

Vicente, K. J. (2003). Beyond the lens model and direct perception: toward a broader ecological psychology. *Ecological Psychology*, *15*(3), 241–267. doi:10.1207/S15326969ECO1503_4

Wells, A. J. (2002). Gibson's affordances and Turing's theory of computation. *Ecological Psychology*, *14*(3), 141–180. doi:10.1207/S15326969ECO1403_3

Zhang, J. (1997). The nature of external representations in problem solving. *Cognitive Science*, *21*(2), 179–217. doi:10.1207/s15516709cog2102_3

Zhang, J., & Patel, V. L. (2006). Distributed cognition, representation, and affordance. *Cognition & Pragmatics*, *2*, 333–341. doi:10.1075/pc.14.2.12zha

Zukow-Goldring, P., & Arbib, M. (2007). Affordances, effectivities, and assisted imitation: Caregivers and the directing of attention. *Neurocomputing*, *70*(13-15), 2181–2193. doi:10.1016/j.neucom.2006.02.029

ENDNOTES

[1] As regards the relationship between modern law and technology, more generally, Hildebrandt (2008b) has recently argued that they are indeed coupled together, as technologies always exhibit what she called "technological normativity", which should not be confused, however, with so-called "legal normativity". Basically, technological normativity refers to the fact that every technology has a certain "normative" impact on our behavior insofar as it permits or facilitates us to do something (*regulative normativity*) and, at the same time, prohibits us from doing something else, thereby constraining our behavioral chances (*constitutive normativity*). We should thus acknowledge that modern law is embedded in certain practices which are shaped by pre-existing devices and tools. That is, modern law did not emerge in a technological vacuum, but it is a response to specific societal needs. However, as technological innovation brings new tools and devices into existence, we should acknowledge that modification of the legal framework is of urgent need to preserve those democratic values that our societies are – and continue to be – imbued with.

[2] A more exhaustive treatment of the relationship between ethics and technology is provided in Magnani (2007). Magnani discussed at length that morality is extended, and that recent technological advancements are clearly re-configuring some of the crucial aspects of our moral life. On this topic, Verbeek too points out that it is basically "a mistake to locate ethics exclusively in the 'social' realm of the human, and technology exclusively in the 'material' realm of the non-human" (Verbeek, 2009, p. 65). Technologies are indeed part of our moral endowments

3 in the way that our moral response can be shaped by technological devices.

3 The term "adapting" does not refer to a strict Darwinian/evolutionary meaning; we are simply referring to the fact that a certain device or mechanism may exhibit some flexible behavior in interacting with the environment. This means that it can plastically and flexibly adjust its response to a given situation: that is, it can adapt. It indirectly enters the Darwinian framework at the level of co-evolution between organisms and the environment, as we have mentioned above.

4 According to Peirce (1967) and his semiotic approach, signs can be of different kinds: images, diagrams, and icons, spoken as well as written words, but also feelings, conceptions, bodily schemata, and so on. More details on this issue can be found in Magnani (2009).

5 The term "wired" can be easily misunderstood. Generally speaking, we accept the distinction between cognitive aspects that are "hardwired" and those which are simply "pre-wired". By the former term we refer to those aspects of cognition which are fixed in advance and not modifiable. Conversely, the latter term refers to those abilities that are built-in prior the experience, but that are modifiable in later individual development and through the process of attunement to relevant environmental cues: the importance of development, and its relation with plasticity, is clearly captured thanks to the above distinction. Not all aspects of cognition are pre-determined by genes and hardwired components. For more detailed account on this problem, see Marcus (2004).

Chapter 2
Tracking Persons:
A Survey

Christine Leignel
Université Libre de Bruxelles, Belgium

Jean-Michel Jolion
Université de Lyon, France

ABSTRACT

This chapter presents a survey of methods used for tracking in video sequence. We mainly focus this survey on tracking persons. We introduce three main approaches. First, we present the graph based tracking approach where the sequence of tracked objects are embodied in a graph structure. Then we introduce the features (extracted from the images) based tracking and matching with a model. We survey the main primitives and emphasize the approaches based on 2D and 3D body model. We present the particular case of tracking in a network of cameras with the particle filtering method. Finally, As a generalization, we focus on the single vs. stereo approaches.

INTRODUCTION

Computer vision reflects a growing interest because of lower cost of new technology whose skills are growing. The video flow, traditionally processed by a human operator, is gradually being replaced by an automatic processing either to detect abnormal events, or to track a person into a scene for teleconferencing applications.

The pioneers in the field of tracking people are (Siebel, 2003), (O'Rourke & Badler, 1980) and (Hogg, 1983). In the area of surveillance, there are many tracking algorithms. The first step in any images sequence processing system for tracking people is to detect the movement of mobile regions in the image. Such regions are classified as individuals, groups and other classes of objects, and are grouped in a graph tracking to facilitate the tracking of individuals over a long period (case for persons who join or leave a group). We can classify the tracking methods in six categories:

DOI: 10.4018/978-1-61692-857-5.ch002

- **Category 1:** methods, sometimes without a model, based region or « blobs » (set of pixels connected and grouped according to a criterion) tracker, based on color, texture, ponctual primitives and contours ((Bremond, 1997), (Cai, Mitiche, & Aggarwal, 1995), (Khan, Javed, Rasheed, & Shah, 2001), (Lipton, Fujiyoshi, & Patil, 1998), (Wren, Azarbayejani, Darrell, & Pentland, 1997));
- **Category 2:** methods using a human body 2D appearance model ((Baumberg, 1995), (Haritaoglu, Harwood, & Davis, 2000), (Johnson, 1998)), with or without explicit model of the shape ;
- **Category 3:** methods with a 3D articulated model ((Gavrila & Davis, 1996), (Sidenbladh, Black, & Fleet, 2000));
- **Category 4:** methods by background removal ((Haritaoglu, Harwood, & Davis, 1998), (Wren, Azarbayejani, Darrell, & Pentland, 1997)). The system can be more robust in textured environments by combining color, texture and movement to segment the foreground;
- **Category 5:** The temporal difference (two or three images) (Anderson, Burt, & Van Der Wal, 1985) yielding a binary map of motion (such as category 4) where motion pixels are grouped into « blobs » ((Haritaoglu, Harwood, & Davis, 2000), (Jabri, Duric, Wechsler, & Rosenfeld, 2000), (Zhao, Nevatia, & Lv, 2001)). The movements and interactions between individuals are obtained by the tracking of the « blobs »;
- **Category 6:** Another complementary approach to that of category 5 is the differential approach based on estimation of the velocity field at all points of the image, also known by motion detection. It calculates the velocity vector in the scene, making the invariance assumption between t and t+d. It defines an error function called

DFD « Difference Deplaced Frames ». It seeks to minimize the DFD for all points of the image at time t. This family includes the method by « optical flow » (Barron, Fleet, & Beauchemin, 1994). The motion estimation by « optical flow » in terms of spatial and temporal variation of the function of intensity is a way of understanding the movement in a scene. Motion detection highlights mobile regions in the current image.

Background subtraction can extract moving objects but the background must be well modeled (by Gaussian or mixtures of Gaussian). This method, faster than other methods, is well suited for indoor environments where brightness is stable and the movements of the background low. The temporal difference is well suited to dynamic environments but suffers from the « problem of aperture» due to the uniform colors of objects in motion and does a poor extraction of primitives. The optical flow is a very robust technique to textured environments with movements in the background or camera movement, but is very expensive in computational cost and thus little used for real-time applications.

In the outdoor scenes, usually a single camera is enough to track an object for a long time. Objects can be in occlusion by the external elements: trees and buildings. A promising solution is to use a network of cameras for tracking objects in a cooperative and coordinated manner from a camera to another. (Matsuyama, 1998) presented such an approach in indoor environment where four cameras were tracking a moving object on the ground. (Chleq & Thonnat, 1996) have made a decision support system for the monitoring of human activity, triggering an alarm in a risky situation. For the detection of dangerous behavior in subways, (Cupillard, Avanzi, Bremond, & Thonnat, 2004) proposed an approach with multiple cameras to recognize individuals, groups or the crowd. Other applications such as teleconferencing, video indexing, virtual reality require a real-time robust tracking in real environments. The main difficulties in analyzing the movement of the human body come from the nature of non-rigid 3D motion, brightness changes, occlusions, changes of background and loose clothing.

We present in this chapter an overview of video surveillance systems through three major subjects: graph based tracking, features based models matching, and tracking in a network of cameras with particle filters.

GRAPH BASED TRACKING

The understanding of image sequences has given rise to numerous studies ((Choi, Seo, Kim, & Hong, 1997), (Nagel, 1998), (Pentland, 1995)). The work of (Rota, 1998) aims to detect, identify and track multiple people in a subway station with a single camera. Each person crosses the scene over time, and the monitoring system should follow its track. A track is a set of points corresponding to the positions of objects over time. The difficulties concerning the tracks are the initialization, ending, mixing and fragmented tracks (see figure 1).

Initializing a track is the first feature characteristic of an object in the scene. The ending of a track is the last element in the following positions of the object trajectory, corresponding to the disappearance of the object in the scene. But in the case of an occlusion, the disappearance of the

Figure 1. Initialization, ending, mixing and fragmented tracks. (Adapted from (Rota, 1998))

object does not correspond to a real loss. The fusion of tracks is for example the case of a group of people who come together, in this case their tracks merge. In contrast, the fragmentation of tracks occurs when a group of people are separating.

The project ESPRIT PASSWORDS[1] is the starting point for this work carried out in 1996, whose main contribution is to provide external information on people and the contribution of external knowledge on 3D scene. The extraction of moving regions by combining images with a reference image, and recognition of people via a model of three parameters (speed, height, width) matching is done via a temporal graph of the objects of the scene and filiations between detections over time. A criterion of spatial overlapping between two regions in successive images allows determining if the two regions belong to the same subject. Filiations between successive detections highlights the correspondence of an object from one image to the next. The temporal graph matching is compromised in heavily populated scenes, highlighting the problems of mixing and fragmented tracks.

To make the system more robust (see figure 2), contextual information (a 3D human model) and the « static context » (the context of the scene or on the persons to follow) were introduced. Indeed, without contextual information, the occlusions are not managed and the temporal graph is untraceable, while with contextual information, the temporal graph solves the problems of occlusions.

The nodes of the graph represent the results of the detection of objects. The edges of the graph are based on the similarity between two nodes corresponding to two identical objects detected. Tracking multiple objects gives a global decision on the objects trajectories by selecting the most likely hypothesis which has accumulated over time. So, abnormal behavior for video surveillance are detected.

To track a person, (Han, Xu, & Gong, 2007) propose a multi hypotheses tracking integrating

Figure 2. Tracking results with and without context. (Adapted from (Rota, 1998))

a detection process in the tracking process, the overall trajectory being searched in the numerous hypothesis detected. The tracking of multi hypothesis trajectories is based on an Hidden Markov model that maximizes the joined probability between the sequence of states and the sequence of observations. Each object tracked is represented by its index and its state at time t is represented by a vector including its location, speed, appearance and scale. The joined probability of a sequence of states given X and a sequence of observations Z is assumed under the Markovian assumption. The space of possible trajectories is very large and the problem is solved by the multi hypotheses tracking (MHT) algorithm for small targets ((Cox & Hingorani, 1996), (Reid, 1979)), by finding all possible combinations of the current observations and existing paths within groups of points. The trajectories tracking can handle temporal difficulties caused by textured background, multi-objects interaction and occlusions.

The monitoring module accumulates the detection results in a graph structure and maintains multiple hypotheses of objects trajectories. The monitoring module consists of three steps:

- **The hypotheses generation:** a graph structure is maintained in the multiple objects tracking algorithm for each trajectory. The nodes are associated to detections. Each node is labelled by the objects detection probability, its size and scale, its location and appearance. The appreance of the object is modeled by an histogram inside a surrounded box. The strength of each edge of the graph is based on proximity, similarity in size and appearance between two nodes. The graph is continuously extended over time during tracking. For each image, given the objects detection results, the hypotheses generation calculates the connections between maintained nodes and nodes in the current image. The hypotheses generation avoids occlusions by separation and grouping of nodes, because if an object reappears after an occlusion, the previous node splits into two objects tracks. Conversely, if an object is in occlusion, the corresponding node is combined with the occluded node (see figure 3). This module also handles missing data, and false detections;

- **The likelihood calculation:** the likelihood of each of the hypotheses is calculated according to the probability of detection, and the trajectory analysis. The graph structure makes it possible to include the most recent detected objects and generates multiple hypotheses about the trajectories. An image of likelihood is calculated to provide a probability to each hypothesis. The probabilities calculated at each time on the entire sequence correspond to the hypotheses likelihood, and provide a global description of the detection results. Hypotheses with the highest likelihood are the best object detections. A likelihood function composed of a likelihood term based object for the original image, a likelihood term based image for the foreground mask and a map detection, are proposed. The advantage of a likelihood based image is if the tracking is in the wrong location, such as a textured background, the likelihood is low because the real target cannot be explained with other objects. By combining the three terms of likelihood, the algorithm for tracking multi path objects chooses the tracks which are made of connections of objects detected with big scores of detection, similar appearances over time, and explains the foreground regions. The strong visual clues make the sequence configuration robust to missing or false detections thanks to the global view of the image sequence, to occlusions and backgrounds textured (see figure 4);

- **The hypotheses management:** The hypotheses management module ranks the hypotheses based on their likelihood. To avoid a combinatorial explosion in the number of hypotheses, the structure of graphs manage multiple hypotheses and make a pruning to get reasonable performance. Successive detections are checked by predicting the location of objects in successive images. This verification gives a better probability score to detected objects that have satisfied the prediction. A limited number of hypotheses is maintained in the structure graph. The tracking module provides a prediction to the objects detection module to enhance performance of local detection.

FEATURES BASED TRACKING AND MATCHING WITH A MODEL

The approaches based primitives and based 2D or 3D body model, a lot of papers have been visited, among them two thesis which give a global view ((Leignel, 2006), (Noriega, 2007a)).

Methods Based Primitives

The main characteristics of the image used to detect low-level indices, for tracking the body, are color, contour, texture, movement and depth.

The goal can also be to identify descriptors related to specific points, and describing the object by a set of geometric attributes (points, segments, parametric curves, edges, contours), or regions of the image. These methods have the advantage of a good robustness to **occlusions** thanks to the use of multiple clues.

(Du, Sullivan, & Baker, 1993) and (Koller, Daniilidis, & Nagel, 1993) detect vehicles by extracting the angles of the roof and the hood. Some primitives, obtained by contour extraction ((Deriche, 1987), (Shen & Castan, 1992)) and analysis of the gradient norm in the image, can also be the size, position, speed, the ratio of two axes of the main surrounded shape. This approach has the advantage of having a semantic content (precise points on the objects, as the roof of a car). Tracking punctual primitives proceeds by mapping an image to the next, and the « blobs » extracted in the entire sequence constitute the trajectory, as in (Megret, 2003). This approach identified only a few points on the object tracked and not the whole, what contour and region approaches do.

In the tracking approaches by the appearance based contour, the object of interest is tracked over time thanks to its contour, or by mapping the contour of the object, or by tracking the contour. By mapping the contour, this is a top-down approach, either by minimizing the distance between

Figure 3. Graph structure of a multi objects trajectory. (Han, Xu, & Gong, 2007) © 2007 SPRINGER. Used with permission

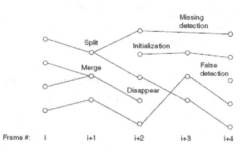

Figure 4. Tracking results for persones trajectories with missing and false detection. (Han, Xu, & Gong, 2007) © 2007 SPRINGER. Used with permission

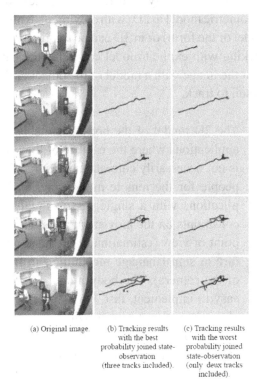

(a) Original image.　(b) Tracking results with the best probabilty joined state-observation (three tracks included).　(c) Tracking results with the worst probabilty joined state-observation (only deux tracks included).

the positions of the contour between two successive moments, or by mapping between image and template (Mori & Malik, 2002). A mixture of tree describe the body to manage occlusions (Ioffe & Forsyth, 2001), or spatio-temporal patterns to detect a walking person (Dimitrijevic, Lepetit, & Fua, 2005) (see figure 5).

Among the methods of contour tracking, active contours called « snakes » consider the boundary of the object at any moment but they are sensitive to the initialization of the contour and are limited in terms of tracking accuracy. The contours tracking can also be achieved by a Hidden Markov Model or HMM, in which the contour is represented by an ellipse, and each point represents a state of the HMM (Chen, Rui, & Huang, 2001).

Tracking by appearance based regions is characterized by the extraction of « blobs » in the current image, and their tracking during the sequence. This method is based on the change of motion in the regions of the image. It does not take into account occlusions. It is assumed that within a region, the appearance is invariant and the motion is homogeneous, such as in « blobs » tracking by Kalman filtering (Crowley & Demazeau, 1993). (Baumberg & Hogg, 1995) and (Chleq & Thonnat, 1996) use the absolute difference between the current image I_t and a reference image I_0: $I_{result} = |\, I_0 - I_t \,|$. The disadvantage of this method is the required updating of the reference image I_0. An other well known method is the difference of successive images (Jain, Martin, & Aggarwal, 1979): $I_{result} = Max\,(|I_t - I_{t-1}|, |I_t - I_{t-1}|)$, where there is no reference image, but it cannot take into account the movements of uniformly colored areas. Some authors model the textural or color distribution by an average color (Fieguth & Terzopoulos, 1997) or a mixture of Gaussians, in a box surrounding the tracked object. (Perez, Hue, Vermaak, & Gangnet, 2002) calculate the

Figure 5. Template matching on contours. (Dimitrijevic, Lepetit, & Fua, 2005) © 2005 IEEE. Used with permission

 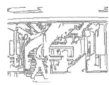

(a) Original image.　　(b) Contours according to Canny algorithm.　(c) Pattern used to detect a walking person.

Bhattacharya coefficient to assess similarities between the histograms in HSV space, and tracking is achieved with the algorithm of « condensation » (Isard & Blake, 1998), evaluating the state vector density of the object tracked.

In the context of the video surveillance, the system W⁴ (Haritaoglu, Harwood, & Davis, 2000), with one grayscale or infrared camera, for detection, tracking and real-time monitoring, analyses what people do (what), where (where), when (when) and which (who) does it. The system tracks the head, torso, arms and legs of a person in real-time, and an appearance model is learned at the same time that people are being tracked. The detection of moving people is done by subtracting the current image with a background bimodal Gaussian model. The appearance model consists of a shape prototype representing the probability that a pixel belongs to the person, and a texture prototype containing the intensity and texture information. Along the sequence, the appearance model is dynamically trained integrating the temporal aspect of tracking. The advantage is its genericity as it can detect and track various postures in real-time. However, as tracking is based on blobs, they must be precisely detected. Tracking is lost in case of shadows, noise and brightness changes. In the latter case the background model is recalculated.

Lee & Cohen (2004) combine a face detector, active contours to detect shoulders, blobs of flesh hue, and the median axis for the legs localization (see figure 6).

The Approaches Based 2D or 3D Body Model

To detect and identify different limbs of the body, a geometric model in 2D (with or without explicit model of the form) or in 3D can be required. The tracking with explicit model compares the data from the image with a model of the object or the person to track.

- The 2D model of the body is suitable for applications where the capture of the pose is not necessarily correct, such as tracking people for the remote monitoring, or applications with a single person involving constraints on the movement and a simple point of view (estimation of the hand posture in sign language recognition in front of the camera). The body model in 2D is easy to implement, fast, but subject to occlusions, and not precise with changes in posture, angle, and appearance;
- 3D approaches correspond to applications for tracking complex movements without constraints. The pose of the human body represented by 3D angles is independent of the point of view. 3D approaches are more accurate and resolve occlusions and collisions, but they are not suited to real-time applications.

An *a priori* knowledge of a shape model makes these methods more robust to occlusions compared

Figure 6. Association of detectors. (Lee & Cohen, 2004) © 2004 IEEE. Used with permission

(a) Face detection   (b) Shoulders detection after active contours   (c) Flesh hue « blobs »   (d) Median axis for the legs

to methods without model. The structural information of the model is mapped onto the image data, either by a bottom-up approach involving hypotheses images or by a top-down approach where the model with the maximum correlation with the image data is desired.

Matching Image/Geometric Model

In a top-down approach, the geometric model is directly used. A model and *a priori* information at the highest level of hierarchy explain the low level observations by mapping between a model and body image. In a bottom-up approach, body limbs are searched from low level features extracted in each image, without *a priori* model. The model parameters are modified to better match the image characteristics, and identify candidate limbs. (Ren, Berg, & Malik, 2005) (see figure 7), in a bottom-up strategy similar to that of (Mori, Ren, Efros, & Malik, 2004) (see figure 8), detect contours, which are broken into segments and a Delaunay triangulation based on these segments is established. Sets of parallel ridges produce limbs hypotheses. Only those that satisfy the constraints of the model, allowing to label limbs, are retained.

Matching Image/Model by Visual Tracking

Appearance is another clue to match the objects over time in a sequence of images. The appearance may be color, shape or texture characteristic. These tracking approaches by texture are named « Visual Tracking », among which we can distinguish the region and the contour approaches. The system of Leeds University, « People Tracker » by Adam Baumberg (see figure 9), under the supervision of David Hogg (Baumberg & Hogg, 1995), is based on an appearance 2D model of external contours. The tracking algorithm, with a single camera, is a model of the active form which hold on the contours of a pedestrian, generated through a training stage. The contours extracted from the model are analyzed by principal component analysis. It works well until the person is visible and with little occlusions. This 2D method is fast enough to be used in real-time.

2D Approach with Explicit Shape Model

The 2D approaches with explicit shape model have an *a priori* knowledge of the human body in 2D. The model can be a « wire » stick figure (Karaulova, Hall, & Marshall, 2000), surrounded by ribbons or « blobs ». The silhouette of the body is detected by background substraction, assuming a stationary background and the camera fixed. The homogeneous regions are identified by color or texture. The 2D model contains the constraints of the joints between the regions corresponding to the limbs of the human body. The movement is detected and the background is separated from the moving objects in each image. The flesh hue allows detection of the face and hands. The

Figure 7. Detection of candidate limbs and recognition of the limbs by a bottom-up approach. (Ren, Berg, & Malik, 2005) © 2005 IEEE. Used with permission

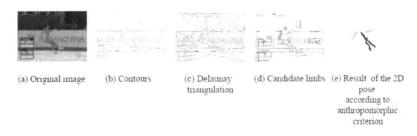

(a) Original image   (b) Contours   (c) Delaunay triangulation   (d) Candidate limbs   (e) Result of the 2D pose according to anthropomorphic criterion

Figure 8. Segmentation of the limbs. (Mori, Ren, Efros, & Malik, 2004) © 2004 IEEE. Used with permission

(a) Original image   (b) Contours   (c) Segmentation by
                                                       normalized cuts

Figure 9. The appearance model of the non-rigid object. (Baumberg & Hogg, 1995)

Figure 10. « Cardboard » model. (Ju, Black, & Yacoob, 96) © 1996 IEEE. Used with permission

Kalman filter (Crowley & Demazeau, 1993) can estimate the parameters of the model throughout the sequence. Markov models (Rigoll, Eickeler, & Muller, 2000) and particle filtering (Chen & Rui, 2004) are techniques of statistical modeling of model parameters.

The contours of 2D models such as proposed by (Ju, Black, & Yacoob, 96) modelise the human body by a « cardboard » model (see figure 10).

The limbs of the body are modeled by connected planar « patches » (rectangular « cardboard » for each limb) or « blobs ». Their projections in the image yield estimates of the 2D likelihood of the hypotheses. (Felzenszwalb & Huttenlocher, 2000) and (Forsyth & Fleck, 1997) have proposed a 2D articulated model in a top-down approach to identify candidate limbs with a similarity measure to detect limbs based appearance, and a strong assumption that limbs wear flesh hue clothes. (Ronfard, Schmid, & Triggs, 2002) propose to replace this assumption by a learning based on a Support Vector Machine (SVM).

In (Leignel & Viallet, 2004) (see figure 11), limbs are detected as segments, leading to 2D skeleton, but the approach is bottom up.

2D Approach without Explicit Shape Model

2D approaches without explicit shape model describe the human movement by low-level 2D features from the regions of interest. Models of the body from these primitives are low-level statistics. The characteristics extracted from the image are matched with the pose of the tracked person. The « structure from motion » is used to find the 3D coordinates of a tracked person over time from 2D moving points in a series of images taken from different angles. The coding of the contours of a silhouette extracted from the image according to a shape descriptor of type « shape context » (see figure 12) allows to compare the image with a base learned.

Figure 11. Detection of the complete superior limbs candidates. (Leignel & Viallet, 2004) © 2004 SPRINGER. Used with permission

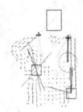

(a) Original image and the complete superior limbs.

(b) Image of the « robust » gradients – Hand detection by flesh hue (square), robust gradients in tight dashes giving the principal direction (segments) of the arms and forearms, from the shoulders (crosses) and the hands, which intersect at the elbow (circle).

Methods with a 3D Articulated Model

The 3D models represent the articulated structure in three dimensions, removing the ambiguities of the 2D models depending on the pose. Limbs are modeled by cylinders (Hogg, 1983) or cones (Sminchisescu & Triggs, 2001), while (Sminchisescu & Telea, 2002) make the choice of super quadrics (see figure 13).

Sometimes 3D Gaussian or « metaspheres » modelise each muscle of the body (Plänkers & Fua, 2003) (see figure 14).

Among the methods using a 3D articulated model of the human body to track people, the model of (Gavrila & Davis, 1996) at the University of Maryland propose 22 degrees of freedom including cylinders and ellipsoids, described by the angles of joints. The measurements are taken with two calibrated cameras; in each orthogonal view, the segmentation of a 2D shape is the result of contour detection. This model gives good results but it is not suitable for monitoring applications because it takes two orthogonal stereo cameras and occlusions are not taken into account. In addition, the contour detector required that people wear colorful clothing to differentiate the different body parts. Finally, the real-time is not reached.

The Sidenbladh model « 3D people tracker » (Sidenbladh, Black, & Fleet, 2000), edited at Brown University, is also a complex 3D articulated, with a single camera. The body model is represented by a set of articulated cylinders containing 25 degrees of freedom. A probabilistic model is estimated via a set of 3D motion data. Repetitive movements such as walking, are decomposed into a temporal model sequence, the « cycles of movement ». This probabilistic model is introduced as « prior » of a Bayesian distribution in a particle filtering.

Figure 12. « Shape context » - localisation of the articulations. The sampled points along the exemplar silhouette (left) and the test (center) are matched. (Mori & Malik, 2002) © 2002 SPRINGER. Used with permission

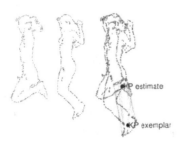

Figure 13. Limbs modeling by ellipsoidal primitives. (Sminchisescu & Telea, 2002)

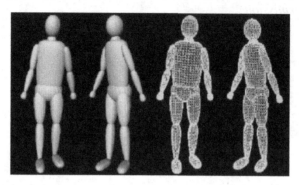

TRACKING IN A NETWORK OF CAMERAS WITH PARTICULE FILTERING

Visual surveillance has become an area of active research in recent years ((Hu, Tan, Wang, & Maybank, 2005), (Valera & Velastin, 2005)). (Remagnino, Shihab & Jones, 2004) have introduced the concept of « intelligent agents », independent modules combining information from several cameras and incrementally constructing the model of the scene. Multiple cameras with covering fields of view offer a larger coverage of the scene, providing a richer 3D information and authorizing occlusions. On the other hand, cameras with views that do not overlap each other may provide a coverage of a large area without losing resolution. One of the problems of these applications is to re-identify the persons that are out of the scope of a camera and come back again, either into the field of the camera, either into the field of another camera. People tracking and re-identification are often more complex than object tracking because people are articulated and move freely. (Javed, Rasheed, Shafique & Shah, 2003) use multiple primitives and the appearance in a probabilistic network to identify the best set matches. (Chang & Gong, 2001) develop a bayesian network to combine information from different cameras. (Wu & Matsuyama, 2003) get a reconstruction of the shape based on the voxels in real-time with multiple cameras.

The Graph Model

When the size of the space increases, it is recommended to divide the problem into a structured graphical model (Jordan, Sejnowski, & Poggio, 2001), whose basic components are nodes of a graph. When nodes are elements of an image, the neighbors may be spatially close in the image, adjacent levels in a multi-scale representation, or in time in a sequence. The exact inference on models of graphs is possible in specific circumstances, in other cases two recent methods are used to approach the inference:

- « Loopy belief propagation » « LBP » (Yedidia, Freeman, & Weiss, 2001) applicable to graphs with cycles. The « belief Propagation » or « BP » is fashionable to calculate inferences in bayesian networks and was applied to graphs with cycles under the name « Loopy belief propagation ». The method consists in sending messages between nodes. When the graph is a chain, a particle filtering may be used. It represents the marginal probabilities in a nonparametric form which is the set of all the particles. The particle filter, widely used in computer vision, works well with models of likelihood images. (Sudderth, Ihler, Freeman, & Willsky, 2003) have shown that the Gibbs sampler is very efficient in

Figure 14. Modeling of muscle tissue with primitive Gaussian. (Plänkers & Fua, 2003) © 2003 IEEE. Used with permission

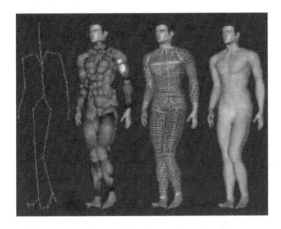

textured scenes to handle the exponential explosion of produced messages;

• « Particle filtering » (Doucet, De Freitas, & Gordon, 2001) authorizing the use of more general distributions of random variables with continuous values, but applied only on graphs with a linear chain structure. For the problems of temporal inference, particle filtering (Isard & Blake, 1996) forms the basis of many tracking algorithms (Sidenbladh & Black, 2003), and are appropriate for temporal problems whose corresponding in graphs are Markov chains (see figure 15).

(Sudderth, Ihler, Freeman, & Willsky, 2003) use LBP to infer relationships between the coefficients in a PCA model of the face based components. Local appearance models based on limbs have common clues with the articulated models used for tracking.

With a « Weak Limbs » Model

Most approaches for the detection and tracking are based on articulated models of the body, in which the body is modelised as a kinematic tree in two dimensions such as the « cardboard » model

(Ju, Black, & Yacoob, 96) or in three dimensions ((Bregler & Malik, 1998), (Deutscher, Blake, & Reid, 2000), (Sidenbladh, Black, & Fleet, 2000), (Sminchisescu & Triggs, 2001)), leading to a high parametric dimension. The search for the solution in such an area is impossible, and when such stochastic search algorithms lose track, the size of the search space makes it difficult to cover tracking. Moreover, this method cannot incorporate low-level treatment or automatic initialization.

One way to avoid the problem of high dimension is to use a hierarchical representation of the person such as the tree structure model (MacCormick & Isard, 2000), or a body model with « weak limbs » (Sigal, Isard, Sigelman, & Black, 2004), e.g. not connected rigidly but rather one attraction to the other (see figure 16), where the likelihood of each limb is evaluated independently. In this way, a particle filtering (propagated through a Bayesian network) can be combined with each limb reducing the size of the search space to the number of degrees of freedom of the limb (Bernier & Cheung-Mon-Chang, 2006). The influence between the limbs is taken into account by the belief propagation on the limbs through a factor graph (Kschischang, Frey, & Loeliger, 2001). Using the training data of the known limbs in the image, a new model of likelihood that captures the statistics of the joints can be learned.

(Sigal, Bhatia, Roth, Black, & Isard, 2004) proposed a probabilistic method to detect and track a person's body in 3D automatically by a

Figure 15. Markov chain and graphical models. (Sudderth, Ihler, Freeman, & Willsky, 2003) © 2003 IEEE. Used with permission

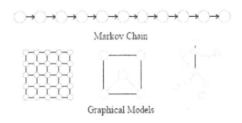

Markov Chain

Graphical Models

Figure 16. Model of tronc of cones with independent limbs called « weak limbs ». (Sigal, Isard, Sigelman, & Black, 2004) © 2004 MIT PRESS. Used with permission

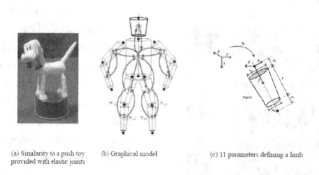

(a) Similarity to a push toy provided with elastic joints   (b) Graphical model   (c) 11 parameters defining a limb

model of « weak limbs » (see figure 17). The inference on this model is driven by the belief propagation on a set of particles. The body is represented by a graphical model in which each of the nodes in the graph corresponds to a limb of the body (torso, arms, etc.). Each limb is parameterized by a vector defining its position and orientation in 3D, and each limb is treated independently. The whole body is assembled by global inference on the entire graphical model. The spatial constraints between limbs of the body (spatial and angular relationships between adjacent limbs) are treated in the edges of the graphs. Each limb is modeled by a cylinder with some fixed parameters. Each edge has a conditional probability distribution that modelises the probabilistic dependencies between adjacent limbs, and is estimated as a Gaussian mixture. Each of the nodes has a likelihood function that modelises the probability of observing the images conditioned on the position and the orientation of limbs. A probability is associated to each limb in the downward direction in the graph (from the thigh to the calf for instance) or bottomward (from the thigh towards the torso for instance). The probabilistic inference combines a body model with a probabilistic likelihood model. This approach could be extended to images with single camera or cameras in motion. We assume that the variables of a node are conditionally independent of the

nodes not immediately neighbors, which can be a disadvantage. In fact, the assumption of independence of limbs of the same nature between right and left conditionally to the position of the torso, omits postures when a limb is hidden by another. The problem would be handled more easily with a kinematic model of the body such as a tree shape. In this case the model « weak limbs » would be an intermediate step between the low level detectors and the complete kinematic model.

This work, compared to the previous (Sigal, Isard, Sigelman, & Black, 2004), does not only deal with the estimation of the poses but also with their tracking. The potential functions are learned from training data and generally approximated by a Gaussian mixture. Each edge between two limbs has an associated potential function that encodes the compatibility between the configurations of limbs pairs and it can be seen as the probability of the configuration of a limb conditionally to the configuration of another limb.

The messages sent in the standard belief propagation are approximated by a set of particles, and the conditional distribution used in the standard particle algorithm is replaced by a product of an incoming message, required for the belief propagation. The Gibbs sampler allows to evaluate the product of messages.

Figure 17. Graphical model for a person. Nodes represent limbs and arrows represent conditional dependencies between limbs. The temporal dependencies are shown between two images in this figure. Actually, each limb is connected by a bow arrow in the same limb in the previous and next images. (Sigal, Bhatia, Roth, Black, & Isard, 2004) © 2004 IEEE. Used with permission

Figure 18. Messages product. (Sigal, Bhatia, Roth, Black, & Isard, 2004) © 2004 IEEE. Used with permission

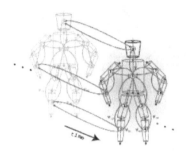

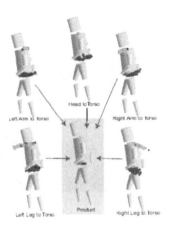

The messages of the head, two arms and two legs are sent to the torso (see figure 18). These messages are distributions represented by a set of samples with weights as in the particule filtering.

Figure 19 shows a tracking example with a model of « weak limbs ».

With a Single Camera

(Noriega, 2007b) describes a method with an articulated graphical model, representing the articulated structure of the human body, for the tracking of the upper body in an unconstrained environment (clothing and light) in monocular color scenes (see figure 20).

The upper body is modeled by a graph including the limbs represented by nodes, edges corresponding to the joints, and constraints of non-collision between limbs (see figure 21). The belief propagation on factor graphs allows the calculation of limbs marginal probabilities.

The model of the body is composed of « weak limbs » (Sigal, Isard, Sigelman, & Black, 2004) including articulatory constraints easily integrable into attraction factors. To resolve the ambiguities

related to monocular tracking, clues are: strong contours, colors, and a map of motion energy (see figure 20). The use of a « weak limbs » and articulated model and the belief propagation provide a good way to incorporate information from various detectors of limbs. The large number of clues selected by (Noriega, 2007b) increases the robustness of tracking. The head and hands are tracked thanks to color and grayscale clues, background subtraction, motion energy, map of contours orientation. Limb interaction factors are computed with a Gaussian of the distance between two articulated points on the limbs.

The complete factor graph includes the previous states to take into account the temporal coherence (see figure 21). The temporal coherence are simple Gaussian, independent for each parameter, centered on the value of the previous image. For hands that can move very quickly, the temporal coherence are a mixture of two similar Gaussian, one centered on the previous parameters and the other centered on the prediction of the current parameters using the previous speed of the hand. The marginal likelihood of the limb states is obtained by using the belief propagation on a factor graph (Bernier & Cheung-Mon-Chang, 2006).

Figure 19. Tracking with model of « weak limbs » on a sequence. (Sigal, Bhatia, Roth, Black, & Isard, 2004) © 2004 IEEE. Used with permission

Frame 7   Frame 12   Frame 17   Frame 22

Figure 20. Monitoring of the upper body. Top line: the original image, in front, side and top, the positions obtained from limbs, with one camera. Bottom line: the background subtraction, the contours, the color map of the face, and the distance map of the motion energy. (Noriega, 2007b)

Figure 21. Factor graph. The circles correspond to nodes that are variables i.e the states of the limbs, and the dark squares to factor nodes (temporal consistency and interaction or non-collision factorial). Two consecutive frames are represented. (Noriega, 2007b)

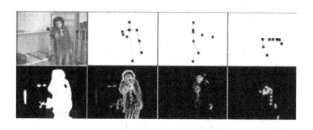

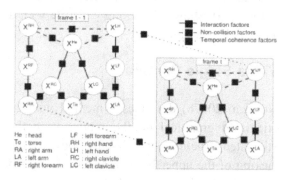

Messages are represented by sets of weighted samples. From one image to the next, they are calculated via a particle filtering algorithm consisting of a re-sampling step followed by a prediction step based on temporal coherence factors (Blake, & Isard, 1998). The loopy belief propagation algorithm is reduced for the current image to loopy belief propagation algorithm in the space of discrete states, the state space of each limb limited to its samples. The algorithm is equivalent to a set of interacting particles filters, where the weighted samples are reassessed at each image

through a belief propagation taking into account the interactions between limbs.

Some examples of tracking in monocular (Noriega, 2007b) are presented below in figure 22.

With Stereo Cameras

(Bernier, 2006) presents a statistical model for fast 3D tracking of articulated upper body with a stereo camera in real-time, similar to the model of

Figure 22. Monocular tracking 3D with difficult poses including occlusions, textured backgrounds and unconstrained environments (brightness and clothing). (Noriega, 2007b)

Figure 23. Graphical model: the lines represent the links between limbs, the dashes represent the constraints of non-intersection. (Bernier, 2006) © 2006 IEEE. Used with permission

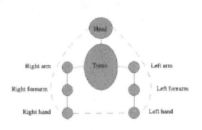

« weak limbs » (Sigal, Isard, Sigelman, & Black, 2004) but where the inter images coherence is taken into account, via the marginal probability of each limb in the previous image as a « prior » information. Belief propagation is used to estimate the current marginal probability of each limb. The resulting algorithm corresponds to a set of particles, one for each limb, where the weight of each sample was recalculated taking into account the interactions between limbs.

Similarly to (Sigal, Bhatia, Roth, Black, & Isard, 2004), a graphical model represents the upper body, consisting of M limbs, each in a given state X (see figure 23). Each limb generates an observation, an image Y, and the model is composed of links between limbs representing the joints but also the constraints of non-intersection. Each limb state is dependent on its state at the previous instant. The model parameters are the conditional probabilities $P(Y/X)$, the prediction of an *a priori* probability of the limbs states $P(X_t/X_{t-1})$, and the interaction potential between limbs for each of the links.

The previous marginal probability is represented by a set of weighted samples. The local belief of each limb is estimated by the method of the particle filtering. The belief is then represented by a weighted sum of samples.

Initially, a face detector based NN detects the face. The depth information is taken into ac-

count. The predicted probabilities are Gaussian, independent for each parameter, centered on the value of the previous image. The observations are estimated 3D points with a confidence factor and a colored probability for the face. The observations are assumed independent for each limb and each pixel. For each limb, the likelihood is proportional to a score S:

- For the head, the score S is a Gaussian distance to a sphere, multiplied by the colored probability of the head;
- For the torso, the score S is a Gaussian distance whose shape is composed of two flat cylinders;
- For the arm and the forearm, the score S is the distance to a rectangular « patch », parallel to the image plane in the direction of the smallest contour;
- For the potential of interaction of the links, a Gaussian of the distance between two points connected is used.

The system correctly tracks the limbs (see figure 24) even in the presence of auto-occlusions and this method can be generalized to other tracking systems, monocular or the whole body. The limitations of the method are related to the inability of the prediction stage of samples to generate samples in regions with high likelihood.

Figure 24. Results of tracking on part of the sequence. (Bernier, 2006) © 2006 IEEE. Used with permission

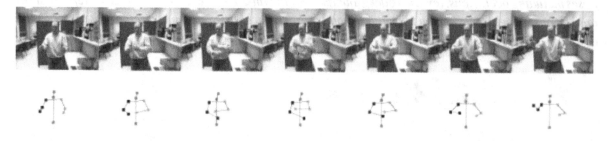

To solve this problem, the prediction step should be conducted with « proposal maps » (Lee & Cohen, 2004) for each limb, generating samples for the regions of high likelihood, especially for the hands and the forearms.

CONCLUSION

Referring to all the works outlined in this literature, the Bayesian network seems to be promising. Each camera is represented by a node in the network. Messages are sent from one camera to another by belief propagation, symbolizing the belief that a person seen in a camera can be a moment later in the field of the other camera, depending on configuration cameras, and the scene analysis.

A « low level » module should detect the movement, based on background subtraction and motion detection, by optical flow for instance. Another module should proceed with the re-identification of people during their transition from one camera to another. It should be able to identify each person and each track individually, with the help of a model of appearance, shape and color of clothing and flesh hue. Once the re-identification established, the trajectories of each person through the network of cameras should be tracked, thanks to a model of the scene. It contains the positioning of cameras in the scene, the positioning of the shelves and aisles, « blind » areas, etc. It is also useful to have a « high-level management », the supervisor, containing a semantic description of the action of each individual tracked at each time. We must in fact be able to combine a global approach by tracking from one camera to another (between areas of the store) with a local analysis in each camera. The overall approach would be managed by the supervisor, taking into account the total trajectory of the person tracked and obtained from low-level information from a local analysis in each camera. Thus, a score (probability) would be assigned to each action, determining the degree of « dangerousness » for instance. A system of « active perception », where the cameras would be activated (zoom, translation, rotation) individually, and in turn, depending on the semantic description of the characters in the scene is not an option. Indeed, it seems difficult to mobilize a camera without risking, by a zoom, to lose a part of the observation of the scene.

REFERENCES

Anderson, C., Burt, P., & Van Der Wal, G. (1985). Change detection and tracking using pyramid transformation techniques. Proceedings from SPIE CIRC '85: *The Conference on Intelligent Robots and Computer Vision*, 579, 72-78.

Barron, J. L., Fleet, D. J., & Beauchemin, S. S. (1994). Performance of Optical Flow techniques. Proceedings from IJCV '94. *International Journal of Computer Vision, 12*(1), 43–77. doi:10.1007/BF01420984

Baumberg, A., & Hogg, D. (1995). An adaptive eigenshape model. Proceedings from BMVC '95: *British Machine Vision Conference*, 87-96, Birmingham, UK.

Baumberg, A. M. (1995). *Learning Deformable Models for Tracking Human Motion*. Doctoral dissertation, University of Leeds, Leeds, UK.

Bernier, O. (2006). Real-Time 3D Articulated Pose Tracking using Particle Filters Interacting through Belief Propagation. Proceedings from ICPR ' 06: *The 18th International Conference on Pattern Recognition*, 1, 90-93.

Bernier, O., & Cheung-Mon-Chang, P. (2006). Real-time 3D articulated pose tracking using particle filtering and belief propagation on factor graphs. Proceedings from BMVC '06: *British Machine Vision Conference*, 01, 5-8.

Blake, A., & Isard, M. (1998). Active Contours. Springer-Verlag. Bogaert, M., Chleq, N., Cornez, P., Regazzoni, C., Teschioni, A., & Thonnat, M. The passwords project. Proceedings from ICIP '96: *The International Conference on Image Processing*, Vol 3, 675-678.

Bregler, C., & Malik, J. (1998). Tracking people with twists and exponential maps. Proceedings from ICCVPR '98: *The International Conference on Computer Vision and Pattern Recognition*, 8-15.

Bremond, F. (1997). *Environnement de résolution de problèmes pour l'interprétation de séquences d'images*. Doctoral dissertation, University of Nice Sophia-Antipolis, France.

Cai, Q., Mitiche, A., & Aggarwal, J. K. (1995). Tracking human motion in an indoor environment. Proceedings from ICIP '95: *The Second International Conference on Image Processing*, 215-218.

Chang, T. H., & Gong, S. (2001). Tracking multiple people with a multicamera system. Proceedings from IEEE ICCV '01: *The International Conference on Computer Vision Workshop on Multi-Object Tracking*, 19-26, Vancouver.

Chen, Y., & Rui, Y. (2004). Real-time Speaker Tracking Using Particle Filter Sensor Fusion. *Proceeding of the IEEE '04*, 92(3), 485-494.

Chen, Y., Rui, Y., & Huang, T. S. (2001). JPDAF based HMM for real-time contour tracking. Proceedings from ICCVPR '01: *The International Conference on Computer Vision and Pattern Recognition*, Vol. 1, 543-550.

Chleq, N., & Thonnat, M. (1996). Realtime image sequence interpretation for videosurveillance. Proceedings from ICIP '96: *The International Conference on Image Processing*, 801-804.

Choi, S., Seo, Y., Kim, H., & Hong, K. (1997). Where are the ball and players? soccer game analysis with color-based tracking and image mosaik. Proceedings from ICIAP '97: *The Ninth International Conference on Image Analysis and Processing*, 196-203.

Cox, I. J., & Hingorani, S. L. (1996). An Efficient Implementation of Reid's Multiple Hypothesis Traking Algorithm and Its Evaluation for the Propose of Visual Traking. *IEEE Transactions on Pattern Analysis and Machine Intelligence, 18*(2), 138–150. doi:10.1109/34.481539

Crowley, J. L., & Demazeau, Y. (1993). Principles and Techniques for Sensor Data Fusion. *Journal of Signal Processing Systems, 32*, 5–27.

Cupillard, F., Avanzi, A., Bremond, F., & Thonnat, M. (2004). Video understanding for metro surveillance. *The IEEE International Conference on Networking, Sensing and Control*, 1, 186-191.

Deriche, R. (1987). Using Canny's criteria to derive a recursively implemented optimal edge detector. *International Journal of Computer Vision*, 2, 167–187. doi:10.1007/BF00123164

Deutscher, J., Blake, A., & Reid, I. (2000). Articulated body motion capture by annealed particle filtering. Proceedings from ICCVPR '00: *The International Conference on Computer Vision and Pattern Recognition*, 2, 126-133.

Dimitrijevic, M., Lepetit, V., & Fua, P. (2005). Human body pose recognition using spatiotemporal templates. Proceedings from ICCV: *The International Conference on Computer Vision workshop on Modeling People and Human Interaction*, Beijing, China.

Doucet, A., De Freitas, J. F. G., & Gordon, N. J. (2001). *Sequential Monte Carlo methods in practice*. Springer.

Du, L., Sullivan, G., & Baker, K. (1993). Quantitative analysis of the view point consistency constraint in mode-based vision. Proceedings from ICCV '93: *The International Conference on Computer Vision*, 632-639.

Felzenszwalb, P. F., & Huttenlocher, D. P. (2000). Efficient Matching of Pictorial Structures. Proceedings from ICCVPR '00: *The International Conference on Computer Vision and Pattern Recognition*, 66-75.

Fieguth, P., & Terzopoulos, D. (1997). Color-based tracking of heads and other mobile objects at video frame rates. Proceedings from ICCVPR '97: *The International Conference on Computer Vision and Pattern Recognition*, 21.

Forsyth, D. A., & Fleck, M. M. (1997). Body plans. Proceedings from ICCVPR '97: *The International Conference on Computer Vision and Pattern Recognition*, 678-683.

Gavrila, D. M., & Davis, L. S. (1996). 3-D Model-Based Tracking of Humans in Actions: A Multi-View Approach. Proceedings from ICCVPR '96: *The International Conference on Computer Vision and Pattern Recognition*, 73-80.

Han, M., Xu, W., & Gong, Y. (2007). Multi-object trajectory tracking. *Journal of Machine Vision and Applications*, 18, 221–232. doi:10.1007/s00138-007-0071-5

Haritaoglu, I., Harwood, D., & Davis, L. S. (1998). Ghost: A human body part labeling system using silhouettes. Proceedings from ICPR '98: *The Fourteenth International Conference on Pattern Recognition*, 1, 77-82.

Haritaoglu, I., Harwood, D., & Davis, L. S. (2000). W^4: Real-Time Surveillance of People and Their Activities. *IEEE Transactions on Pattern Analysis and Machine Intelligence*, 22(8), 809–830. doi:10.1109/34.868683

Hogg, D. (1983). Model-based vision: A program to see a walking person. *Image and Vision Computing*, 1, 5–20. doi:10.1016/0262-8856(83)90003-3

Hu, W., Tan, T., Wang, L., & Maybank, S. (2005). A survey on visual surveillance of object motion and behaviors. *IEEE Transactions on Systems, Man, and Cybernetics. Part C*, 34(3), 334–352.

Ioffe, S., & Forsyth, D. A. (2001). Human tracking with mixtures of trees. Proceedings from ICCV '01: *The International Conference on Computer Vision*, 690-695.

Isard, M., & Blake, A. (1996). Contour tracking for stochastic propagation of conditional density. Proceedings from ECCV '96: *The European Conference on Computer Vision*, 343-356.

Isard, M., & Blake, A. (1998). Condensation-Conditional Density Propagation for Visual Tracking. *International Journal of Computer Vision, 29*(1), 5–28. doi:10.1023/A:1008078328650

Jabri, S., Duric, Z., Wechsler, H., & Rosenfeld, A. (2000). Detection and location of people in video images using adaptive fusion of color and edge information. Proceedings from ICPR '00: *The International Conference on Pattern Recognition, 4*, 627-630.

Jain, R., Martin, W., & Aggarwal, J. (1979). Segmentation throught the detection of changes due to motion. *Computer Graphics and Image Processing, 2*, 13–34. doi:10.1016/0146-664X(79)90074-1

Javed, O., Rasheed, Z., Shafique, K., & Shah, M. (2003). Tracking across multiple cameras with disjoint views. Proceedings from ICCV '03: *The International Conference on Computer Vision*, 1-6.

Johnson, N. (1998). *Learning Object Behaviour Models*. Doctoral dissertation, University of Leeds, Leeds, UK. Retrieved from http://www.scs.leeds.ac.uk/ neilj/ ps/ thesis.ps.gz

Jordan, M. I., Sejnowski, T. J., & Poggio, T. (2001). *Graphical Models: Foundations of Neural Computation*. MIT Press.

Ju, S., Black, M., & Yacoob, Y. (1996). Cardboard people: A parameterized model of articulated image motion. Proceedings from ICAFGR '96: *The International Conference on Automatic Face and Gesture Recognition*, 38-44.

Karaulova, I. A., Hall, P. M., & Marshall, A. D. (2000). A hierarchical model of dynamics for tracking people with a single video camera. Proceedings from BMVC '00: *The British Machine Vision Conference*, 352-361.

Khan, S., Javed, O., Rasheed, Z., & Shah, M. (2001). Human tracking in multiple cameras. Proceedings from ICCV '01: *The International Conference on Computer Vision*, 331-336.

Koller, D., Daniilidis, K., & Nagel, H.-H. (1993). Model-based object tracking in monocular image sequence of road trafic scenes. *International Journal of Computer Vision, 3*(10), 257–281.

Kschischang, F. R., Frey, B. J., & Loeliger, H.-A. (2001). Factor graphs and the sum-product algorithm. *IEEE Transactions on Information Theory, 47*(2), 498–519. doi:10.1109/18.910572

Lee, M. W., & Cohen, I. (2004). Proposal maps driven MCMC for estimating human body pose in static images. Proceedings from ICCVPR '04: *The International Conference on Computer Vision and Pattern Recognition, 2*, 334-341.

Leignel, C. (2006). *Modèle 2D du corps pour l'analyse des gestes par l'image via une architecture de type tableau noir: Application aux interfaces homme-machine évoluées*. Doctoral dissertation, University of Rennes 1, Rennes, France.

Leignel, C., & Viallet, J.E. (2004). A blackboard architecture for the detection and tracking of a person. Proceedings from RFIA '04: *Reconnaissance de Formes et Intelligence Artificielle*, 334-341.

Lipton, A. J., Fujiyoshi, H., & Patil, R. S. (1998). Moving target classification and tracking from real-time video. Proceedings from WACV '98: *The Fourth IEEE Workshop on Applications of Computer Vision*, 129-136.

MacCormick, J., & Isard, M. (2000). Partitioned sampling, articulated objects, and interface-quality hand tracking. Proceedings from ECCV '00: *The European Conference on Computer Vision, 2*, 3-19.

Matsuyama, T. (1998). Cooperative distributed vision. Proceedings from DIU '98: *The Darpa Image Understanding Workshop, 1*, 365-384.

Megret, R. (2003). *Structuration spatio-temporelle de séquences vidéo*. Doctoral dissertation, INSA Lyon, Lyon, France.

Mori, G., & Malik, J. (2002). Estimating human body configurations using shape context matching. Proceedings from ECCV '02: *The European Conference on Computer Vision*, 666-680.

Mori, G., Ren, X., Efros, A. A., & Malik, J. (2004). Recovering human body configurations: Combining segmentation and recognition. Proceedings from ICCVPR '04: *The International Conference on Computer Vision and Pattern Recognition*, 2, 326-333.

Nagel, H.-H. (1998). The representation of situations and their recognition from image sequences. Proceedings from RFIA '98: *Reconnaissance de Formes et Intelligence Artificielle*, 1221-1229.

Noriega, P. (2007a). *Modèle du corps pour le suivi du haut du corps en monoculaire*. Doctoral dissertation, University of Nancy 1, Nancy, France.

Noriega, P. (2007b). Multiclues 3D Monocular Upper Body Tracking Using Constrained Belief Propagation. Proceedings from BMVC '07: *The British Machine Vision Conference*, 10-13.

O'Rourke, J., & Badler, N. (1980). Model-based image analysis of human motion using constraint propagation. *IEEE Transactions on Pattern Analysis and Machine Intelligence*, 2(6), 522–536.

Pentland, A. P. (1995). Machine understanding of human action. Proceedings from IFFTT '95: *Seventh International Forum on Frontier of Telecommunication Technology*, 757-764, Tokyo: ARPA Press.

Perez, P., Hue, C., Vermaak, J., & Gangnet, M. (2002). Color-based probabilistic tracking. Proceedings from ECCV '02: *The European Conference on Computer Vision*, 661-675.

Plänkers, R., & Fua, P. (2003). Articulated soft objects for multiview shape and motion capture. *IEEE Transactions on Pattern Analysis and Machine Intelligence*, 25(9), 1182–1187. doi:10.1109/TPAMI.2003.1227995

Reid, D. B. (1979). An algorithm for tracking multiple targets. *IEEE Transactions on Automatic Control*, 24(6), 843–854. doi:10.1109/TAC.1979.1102177

Remagnino, P., Shihab, A., & Jones, G. (2004). Distributed intelligence for multi-camera visual surveillance. *Pattern Recognition: Special Issue on Agent-Based Computer Vision*, 37(4), 675–689.

Ren, X., Berg, A. C., & Malik, J. (2005). Recovering human body configurations using pairwise constraints between parts. Proceedings from ICCV '05: *The International Conference on Computer Vision*, 1, 824-831.

Rigoll, G., Eickeler, S., & Muller. (2000). Person Tracking in Real-World Scenarios Using Statistical Methods. Proceedings from ICAFGR '00: *The Fourth International Conference on Automatic Face and Gesture Recognition*, 342-347.

Ronfard, R., Schmid, C., & Triggs, B. (2002). Learning to Parse Pictures of People. Proceedings from ECCV '02: *The European Conference on Computer Vision*, 700-714.

Rota, N. A. (1998). *Système adaptatif pour le traitement de séquences d'images pour le suivi de personnes*. Master's thesis, University of Paris IV, DEA IARFA Laboratoire d'informatique, INRIA Sophia-Antipolis.

Shen, J., & Castan, S. (1992). An optimal linear operator for step edge detection. Proceedings from CVGIP '92. *Computer Vision Graphics and Image Processing*, 54(2), 13–17.

Sidenbladh, H., & Black, M. (2003). Learning the statistics of people in images and video. *International Journal of Computer Vision, 54*(13), 183–209.

Sidenbladh, H., Black, M. J., & Fleet, D. J. (2000). Stochastic tracking of 3D human figures using 2D image motion. Proceedings from ECCV '00: *The European Conference on Computer Vision*, 702-718.

Siebel, N. T. (2003). *Design and Implementation of People Tracking Algorithms for Visual Surveillance Applications*. Doctoral dissertation, University of Reading, Reading, UK.

Sigal, L., Bhatia, S., Roth, S., Black, M. J., & Isard, M. (2004). Tracking loose-limbed people. Proceedings from ICCVPR '04: *The International Conference on Computer Vision and Pattern Recognition*, 1, 421-428.

Sigal, L., Isard, M., Sigelman, B. H., & Black, M. (2004). Attractive people: Assembling loose-limbed models using non-parametric belief propagation. *Journal of Advances in Neural Information Processing Systems, 16*, 1539–1546.

Sminchisescu, C., & Telea, A. (2002). Human pose estimation from silhouettes - A consistent approach using distance levels sets. Proceedings from WSCG '02: *International Conference on Computer Graphics, Visualization and Computer Vision*, 413-420.

Sminchisescu, C., & Triggs, B. (2001). Covariance scaled sampling for monocular 3D body tracking. Proceedings from IEEE CCVPR '01: *The IEEE Conference on Computer Vision and Pattern Recognition*, 1, 447-454.

Sudderth, E. B., Ihler, A. T., Freeman, W. T., & Willsky, A. S. Nonparametric belief propagation. Proceedings from ICCVPR '03: *The International Conference on Computer Vision and Pattern Recognition*, 1, 605-612.

Valera, M., & Velastin, S. (2005). Intelligent distributed surveillance systems: a review. Proceedings from IEE-VISP '05: *Vision. Image and Signal Processing, 152*(2), 192–204. doi:10.1049/ip-vis:20041147

Wren, C. R., Azarbayejani, A., Darrell, T., & Pentland, A. P. (1997). PFINDER: Real-time tracking of the human body. *IEEE Transactions on Pattern Analysis and Machine Intelligence, 19*(7), 780–785. doi:10.1109/34.598236

Wu, T., & Matsuyama, T. (2003). Real-time active 3D shape reconstruction for 3D video. Proceedings from ISPA '03: *The third International Symposium on Image and Signal Processing and Analysis*, 1, 186-191.

Yedidia, J. S., Freeman, W. T., & Weiss, Y. (2001). *Understanding belief propagation and its generalizations*. Technical Report TR2001-22, Mitsubishi Electric Research Laboratory.

Zhao, T., Nevatia, R., & Lv, F. (2001). Segmentation and Tracking of Multiple Humans in Complex Situations. Proceedings from ICCVPR '01: *The International Conference on Computer Vision and Pattern Recognition*, 2, 194-201.

KEY TERMS AND DEFINITIONS

3D/2D Articulated Model: The 3D models represent the articulated structure in three dimensions, removing the ambiguities of the 2D models depending on the pose.

Weak Limbs: A body model with weak limbs is composed of limbs not connected rigidly but rather one attraction to the other.

Graphical Model: A graph is composed of nodes connected by links. In a probabilistic graphical model, each node represents a random variable, and the links represent probabilistic relationships between these variables.

Bayesian Network: A bayesian network is a directed graphical model, which is useful when one want to express causal relationships between variables.

Belief Propagation: The Belief propagation algorithm express exact inference on directed graphs without loops.

Particle Filtering: The particle filtering is a sampling method whose aim is to find a tractable inference algorithm.

Re-Identification: This network has to deal with the problem of re-identification of people during their transition from one camera to another. People tracking and re-identification are often more complex than object tracking because people are articulated and move freely.

Scene Analysis: The scene analysis is a model of the scene useful for the re-identification of people from one camera to another.

Occlusion: In the outdoor scenes, objects can be in occlusion by the external elements: trees and buildings.

Intelligent Agents: Intelligent agents are independent modules combining information from several cameras (or several level of information) and incrementally constructing the model of the scene.

ENDNOTE

[1] This is a European project (1994-1997) whose objective is to ensure the transition from engineering experts ((Chleq & Thonnat, 1996), (Bogaert, Chleq, Cornez, Regazzoni, Teschioni, & Thonnat, 1996)).

Chapter 3
SensFloor® and NaviFloor®:
Large–Area Sensor Systems beneath Your Feet

Axel Steinhage
Future-Shape GmbH, Germany

Christl Lauterbach
Future-Shape GmbH, Germany

ABSTRACT

The following chapter describes two systems, both are perfect examples for ambient intelligence. The first system is, sensor electronics, which is invisibly integrated into the floor. This system is able to detect people walking across the floor and can be used to recognize peoples' location and movement behavior. The main application domains are Ambient Assisted Living (AAL), health care, security systems and home automation.

The second system serves for localizing moving objects such as robots, wheelchairs or hospital beds by means of RFID tags in the floor. In the following, we describe the technical details of the two systems and possible applications.

SENSFLOOR®: A LARGE-AREA CAPACITIVE SENSOR SYSTEM

Conventional functions of room floors range from mechanical support, convenience, heating and noise reduction to the expression of individual style and design. However, considering the fact that during the day we are mostly in direct contact with the floor one may ask whether it is possible to exploit this close relationship for even more advanced funtions.

In the following we will introduce a sensor system called SensFloor®, which transforms a room's floor into a sensor plane that detects and monitors peoples' behaviour and allows for a col-

DOI: 10.4018/978-1-61692-857-5.ch003

lection of novel supportive functions. Although the result of these functions is obvious to the user, the sensor system itself remains totally invisible and does not interfere with the material or design of the floor covering in any way. In this respect, the technology we present here is an example for a new class of systems summarized under the expression *Ambient Assisted Living* (AAL).

The basis for the functions offered by the sensor system is the detection and tracking of people moving around within the room. Whereas the detection of general movement can be achieved with conventional infrared or ultrasonic motion sensors, for instance, the acquisition of peoples' exact location requires more advanced systems usually based on camera image processing or wireless identification tags. In addition to the technical problems caused by varying lighting conditions, blind spots caused by furniture in the room and the still unsolved computational task of robustly detecting arbitrarily dressed persons in a video image, cameras installed in every room may interfere with the inhabitant's desire for privacy. The latter does also hold for wireless identification tags as they allow for a labeled behavioural protocol of individuals. In addition, systems like these are not ambient as they require either a visible installation which interferes with the room's design or the user is forced to carry around specific sensor- or identification tags.

In particular in the case of elderly or handicapped persons knowing their current location or behavioural status is crucial. Many wearable sensor systems have been developed for this purpose already. There exist alarm buttons, for instance, which can be pressed in emergency situations. This requires, however, that the person is conscious and still able to press the button. Automatic sensors, such as accelerometers, can detect specific situations such as a fall, for instance. All these wearable sensor require, however, that the user carries them all the time, even in the bathroom. This may become cumbersome for the user and

in addition, other people might directly associate these devices with disabilities of their carriers.

SensFloor® relies on a much more direct way of detection: a grid of sensors underneath the flooring detects local capacitive changes in the environment brought about by humans walking on the floor. By design, this method does not allow for an identification of individuals. However, the persons' locations can be acquired very accurately based on the spatial resolution of the sensor grid. By collecting and processing the sensor patterns over time it is possible to assign movement trajectories to the persons based on which several applications can be realized.

The capacitive measurement principle allows for a unique advantage of the system compared to conventional pressure sensors, for instance, which can be found in fall detection mats or several other smart floor projects ((Richardson, Paradiso, Leydon, Fernstrom, 2004),(UKARI)): as the sensors within the SensFloor® underlay react from a certain distance without direct touch, there is nearly no restriction on the material that covers the sensors. SensFloor® works under carpet, linoleum, laminate, wood and even tiles or stone floors except for conductive material. However, the system is not restricted to the floor alone: it dissolves seamlessly into any large surfaces such as floors, glass panels and walls equipping these surfaces with an interactive functionality. In this chapter, however, we will describe the implementation within the floor only.

SYSTEM DESCRIPTION

The SensFloor® system integrates a grid of capacitive sensor plates and microelectronic modules into a textile composite underlay by using a roll-to-roll manufacturing process that can be combined with the textile manufacturing process itself.

Whenever a person walks across the floor, the integrated sensors are activated and a sequence of location- and time-specific sensor events is generated. This information can be received by a central control unit or directly by actuators such as lamps or motors.

Based on data processing and pattern recognition, the position and movement direction of people can be reconstructed. This allows for countless applications in the domains of security, healthcare, comfort, energy saving, entertainment and market research.

A schematic of the SensFloor® system is shown in Figure 1. The textile underlay contains a dense grid of triangular shaped sensor plates made of conductive textile. Eight triangles together with an electronic module in the center build a sensor panel of 50cm x 50cm. This leads to a spatial sensor resolution of 32 plates per square meter such that each plate has about the size of a small foot. When a foot comes into a range of about 10cm above a sensor plate, the electrical capacity of that plate is slightly enlarged. This change in capacity can be detected by the electronics module by means of a specific Future-Shape software that runs on the microcontroller of the module.

By means of a wireless transceiver on each microelectronics module, the sensor events are broadcasted together with a unique address that identifies the sensor plate where the event has been detected. The broadcasted messages can be received by one or more radio receivers in the room. One example is a central receiver, called *Smart Adapter* (SA). The function of the SA is to process the sensor events coming from the sensor modules. Its task is to analyse the time series and to reconstruct the movement trajectories of people walking on the floor. Based on this information, the SA is able to control wireless switches which can operate automatic doors, alarm devices, lights, heating, traffic counters etc. It is also possible to send commands from the SA to an already existing building control network.

For simple applications like the switching of light or the automatic opening of doors, signals from the microelectronic modules can operate actuators directly, without further computing by means of an SA. In this case, the receivers are relays that are directly connected to the corresponding appliances. Compared with earlier ver-

Figure 1. Schematic of the system: Footsteps on the floor trigger sensor events which are broadcasted wirelessly and can be received by a processing unit, which wirelessly controls home-automation devices. Direct control of actuators is also possible

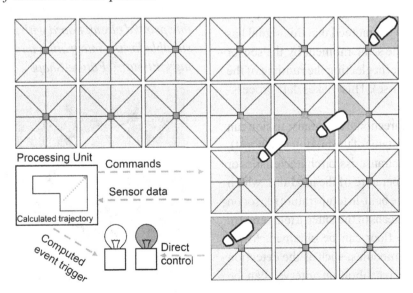

sions of a capacitive sensor floor where the sensor data was transmitted over wires woven into the textile underlay (Li, Hayashi, 1998), the wireless transmission makes the installation of additional applications very easy, once the SensFloor® underlay is installed. In principle, the SensFloor® messages can be received and interpreted by any wireless device that works on the same frequency. However, a proprietary protocol is used and the messages can be encrypted.

In the standard configuration, wireless messages are only generated when there is activity within the room. However, due to the high spatial density of the sensor plates, a large number of messages is generated when many people walk across the floor. Although every single message is very short and the wireless bandwidth is high, it is, in principle, possible that some messages are lost due to collisions on the wireless channel. However, all applications of the SensFloor® are based on evaluating whole populations of sensor events rather than single signals. Therefore, missing some events does not impair the overall function of the system and we explicitly allow for messages being lost. As the sender does not wait for an acknowledgement or whatsoever from the receiver, there is no protocoll overhead and no delay between the detection and the reception of a sensor signal. SensFloor® is a real time system: when a sensor message is received by the SmartAdapter it is guaranteed that this signal has been detected at the same moment by the sensor module in the floor (neglecting the runtime of the radio waves of course). This makes data processing a lot easier.

If required, bidirectional data transmission can be used to read the status of the modules (self test), to reconfigure their transmission characteristics or to request the temperature of the modules, which is determined by the built-in temperature sensor on each module. By means of the latter, a temperature map of the entire floor can be obtained. Many parameters of the modules, such as the detection sensitivity or the recalibration interval can be changed wirelessly even after being completely encapsulated under the flooring.

Figure 2 shows a radio module. We use a flexible substrate for the printed circuit board to let the mechanical characteristics of the electronics be as close as possible to the textile base material. All electronic components have been carefully selected in order to minimize the mechanical strain that is applied to the module after installation.

The transmission frequency is 868 MHz and can easily be changed to 915 MHz. The relatively large contact pads are specially adapted to the integration in the textile underlay. The interconnect between the microelectronic modules and the conductive textiles is done by using conductive adhesive. The electrical properties and the mechanical robustness of this interconnect are crucial parts in design and development of textiles with electronically enhanced functionality. Therefore, extensive degradation tests and analysis are performed to ensure the long-term stability of the system.

For the textile underlay (see Figure 3) we use a conductive copper/nickel-plated fleece with a sheet resistance below 10 mΩ/□, which is laminated on a non-conductive Polyester fleece. This textile composite is structured by removing the

Figure 2. The radio module (868 MHz) on a flexible printed circuit board with printed antenna is suitable for the integration into textile surroundings. Photo: ©Future-Shape

conducting fleece within all areas intended to be isolating. The textile represents a large-area, single-layer circuit board with power supply lines and sensor areas. The width is 100 cm and there are four radio modules integrated per square meter. For redundancy, the modules are connected to two supply and ground lines at opposite sides. A single connection to a 9 Volts switched power adapter at one corner of the underlay is sufficient for powering all radio modules in the reel. As a next step towards mass-producible composite textiles with electronically enhanced functionality, printed conductive patterns are under investigation.

Figure 4 shows an installation of SensFloor® beneath laminate, the graphical user interface at the PC and the Smart Adapter. The person walking on the floor activates sensors which are depicted as orange triangles on the screen.

For several application scenarios the capacitive proximity sensors have distinct advantages compared to the already existing pressure sensors (see e.g. (Richardson, Paradiso, Leydon, Fernstrom, 2004) or the UKARI project (UKARI)). For achieving a high spatial resolution, a pressure sensor requires a soft floor covering. In contrast,

Figure 3. Textile SensFloor® underlay with microelectronic modules. The triangular sensor plates and the power lines are structured into conductive fleece (pattern grid: 50 cm). Photo: ©Future-Shape

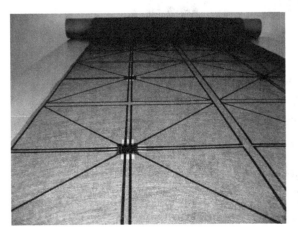

the SensFloor® underlay can be installed even up to approximately 5cm beneath stone or wood; it has no mechanical parts and features self-test abilities. Just by adapting the software on the SmartAdapter, the system is able to run applications within domains as different as comfort, safety, healthcare, emergency, marketing or energy saving.

However, one of the most appropriate fields of application is the assistance of elderly or disabled persons living alone in their appartment. SensFloor® can recognize hazards like falls or immobility and, by long term analysis of the movement patterns of the inhabitants, the system can even predict unusual changes in their behavior. As described in the introduction already one of the main advantages over conventional in-house emergency call systems such as cameras or wearable emergency buttons is its invisibility: the user does not have to carry a device and the design of the appartment is not impaired. In the following, we will show a couple of application scenarios in detail.

APPLICATIONS OF SENSFLOOR®

As mentioned already, one of the intrinsic characteristics of SensFloor® is the fact that the technology remains completely invisible to the user. It is only the function that becomes apparent. From a user's perspective, SensFloor®'s functions can be classified into several categories which we will describe in the following.

Probably the most obvious application for SensFloor® is the control of heating, air condition and illumination. Unlike conventional infrared or ultrasonic sensors which just detect general movement, SensFloor® can count the number of persons in a room and is able to track this number even when people remain static. It is even possible to dim the illumination at empty parts of a large room or of an open-plan office.

Figure 4. Graphical user interface of the SensFloor® system showing the activated sensor areas (high-lighted triangles) beneath laminate flooring. Photo: ©Future-Shape

Using a piece of SensFloor® to control an automatic door, can also contribute to energy saving: opening the door only for a person whose movement trajectory indicates the intention to pass the door prevents its unintended opening whenever someone enters the range of the movement detector without intending to pass the doorway (see Figure 5).

In addition to opening automatic doors, the SensFloor® can support more comfort functions within public buildings.

Initializing the movement tracking to a person at the check in, it is possible, for instance, to give individualized direction information in hotels, hospitals or other large public buildings.

In private homes, comfort facilities such as the hot water pump can be activated already when the floor detects the inhabitants' first activity in the morning. Likewise, home appliances can be switched off when the last person has left the house or gone to bed.

Figure 5. Equipped with SensFloor®, an automatic door opens only when people intend to enter

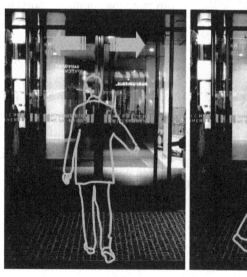

SensFloor® installed in front of the bed can decide whether a person leaves or enters the bed and control the room lights accordingly (see figure 7 right panel).

By analyzing trajectories, SensFloor® can act as an intelligent burglar alarm: in contrast to trajectories starting at the room's door, footsteps starting at a window are associated with an intrusion (see Figure 6).

Implemented in a sally port (security entrance), the floor can detect people entering without using the identification terminal. Likewise, the floor can detect when two persons enter with only one carrying an identification tag.

At night, security personnel can remotely monitor the activity of rooms in which the sensor floor is installed.

In the case of an emergency, SensFloor® can support first responders by providing statistics about how many people had been in which part of a building, at the time when the fire had started.

In areas where people live or work together with robots or other mobile machines, the floor can detect the danger of potential collisions between people and machines. When people enter the workspace of robots, for instance, the machines can slow down their movements. Likewise, transport carts can be warned and slowed down at intersections when there is the danger of a collision with people even when the vision is impaired by walls or furniture (see (Lauterbach, Glaser, Savio, Schnell, Weber, Kornely, Stöhr, 2005) for similar ideas based on pressure sensors).

As the demographic structure of the society changes towards a larger percentage of elderly people, health care becomes a very important field of application for SensFloor®. In particular

Figure 6. Footsteps starting at the window trigger an intrusion alarm

Figure 7. By analyzing the movement trajectories and static sensor patterns, the floor can detect a fall and trigger an emergency call (left and middle panel). When getting up in the dark, the floor can switch on light to prevent accidents (right panel)

  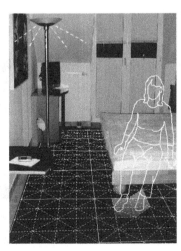

for people who want to avoid moving to a nursery home but want to live in their own apartment as long as possible, SensFloor® can offer various assistive functions. A simple example is a sensor patch besides the bed which switches the room lights off when the user steps into bed and which switches on an orientation light when the person steps out of bed. Already with such a simple system, many accidents of confused dementia patients who cannot find the light switch at night can be prevented (see Figure 7 right panel).

Analyzing the inhabitant's movement trajectories, the floor can detect when a person falls and cannot get up anymore (Figure 7, left panel). Likewise, the floor can detect when the inhabitant has not left the house but remains inactive for a long time. In both cases, a call for help can be triggered automatically over the telephone.

In this field of application, the floor has many advantages compared to conventional systems. A wearable alarm device, for instance, does not work when the user is unable to activate it after a fall. Further, many people put the devices away when showering such that accidents in the bathroom remain undetected. Observing the condition of elderly people by means of surveillance cameras usually requires a human to analyze and interpret the video images which impairs peoples' privacy.

In particular for these AAL-applications another version of the SensFloor® is available, the SensFloor® mat (see Figure 8). In contrast to the large area system, the mat is put on top of the floor covering such that installation, removal and replacement is simple. The mat has the same spatial sensor resolution and the same wireless data transmission as the SensFloor® underlay such that it can be easily integrated in a large area installation if required. However, primarily the mats are meant to be placed at critical locations such as the area besides the bed or the front door (see above). In combination with small wireless embedded receivers that can be plugged into a wall socket, a simple and flexible alert system can be installed. Custom-made designs and coverings of the mats allow for a seamless and unobtrusive integration into the user's interior design.

An interesting application of the sensor floor in the medical field is the long term activity monitoring of patients with mental diseases. From rapid changes of the activity level above or below the average, phase transitions of certain diseases such as depression or schizophrenia can be predicted and treated in advance.

It is also possible to derive information about the general state of health from analysing a person's gait pattern.

Figure 8. SensFloor® mat with customized carpet covering for fall detection and night light control

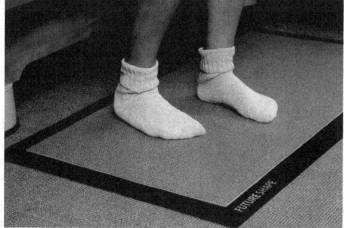

A fascinating application is the coupling of peoples' location with acoustic signals. For Sens-Floor® a MIDI interface exists which can generate different sounds depending on the coordinate of the sensor signals. This way it is possible to provide a blind person with an *acoustic map* of the environment. The access of interesting or dangerous locations can be indicated by specific sounds. Even a potential collision with dynamic obstacles, such as other people, for instance, can be indicated to the blind person by means of special warning tones.

By analyzing peoples' movement behavior in public buildings, SensFloor® can provide important information for marketing purposes building planners and facility management. Visitor streams can be analyzed without interfering with peoples privacy as the floor, by design, cannot detect the identity of people.

Some of the first customers of the system come from the domain of trade fairs and entertainment. Here, SensFloor® can equip an installation with interactivity: illumination and multimedia is controlled by the visitors as they move through the site.

SENSOR DATA PROCESSING TECHNIQUES

Most of the described applications are based on higher level information such as the number, location, velocity and movement direction of people. To derive this information from the low-level sensor information, the receiver must employ certain data processing algorithms which are very similar to image processing techniques. However, as the data arrives sequentially from the sensor modules the first step is to obtain and to update a representation of the current sensor situation on the floor. The sensor modules scan their associated sensor plates ten times per second and as they transmit changes directly after detection the maximum delay when updating the representation is about 1/10 sec.

To derive the higher level parameters several mathematical concepts are used: clustering methods for separating the sensor signals of different persons, probabilistic and dynamic modelling of walking patterns, neural fields, pattern recognition and even neural networks.

Which methods are used depends not only on the application but also on the available hardware platform: for most applications we aim at embedded receivers that can be plugged into a wall socket. On these platforms, cheap embedded processors with limited memory space are common. Therefore, integer-based low complexity versions of the processing algorithms are preferred.

One of the very first public installations of the SensFloor® system is shown in the InHaus2, Duisburg, Germany in the Future Hotel scenario (Figure 9). There, SensFloor® controls the LED illumination of the room according to the behaviour of the guest.

With SensFloor® the presence of people can be detected. However, it is not easily possible to identify individuals. In the following a system is presented which can localize *and* identify objects moving across the floor.

NAVIFLOOR®: A LARGE AREA NAVIGATION SYSTEM

Many applications require information about the exact position of mobile objects. Knowing the location of expensive appliances or the inpatients' beds in a hospital, for instance, can substantially enhance the quality of service and logistics. Likewise, keeping track of the trolleys in a supermarket can support market research, advertising and even energy saving efforts. In particular autonomous robots require exact position information for efficient navigation. Whereas outdoor a modern GPS system can fulfill this task with reasonable accuracy (Li, Hayashi, 1998), keeping track of robots or movable objects in enclosed spaces is no easy task. Mostly, onboard sensors such as

Figure 9. Future-Hotel setup at the InHaus2 in Duisburg, Germany. Photo: © Fraunhofer Gesellschaft

laser scanners (see e.g. (Victorino, Rives, 2004)), ultrasonic sensors (see e.g. (Stormont, Abdallah, Byrne, Heileman, 1998)) or visual systems (see e.g. (Steinhage, Schöner, 1997)) have been used. Recently, localization techniques based on active external beacons such as WLAN transponders (see e.g. (Röhrig, Kunemund, 2007)) are under investigation. Both technologies have in common that substantial structural changes within the environment, such as the rearrangement of furniture, for instance, require a complete remapping of the external cues onto the internal landmark representation of the robot.

One possibility to overcome this problem is to embed stationary and unchangeable beacons ambiently into the environment which are not affected by any rearrangement. Examples are lines painted on the floor or the famous induction loops which are both used for autonomous transport robots. However, compared to the navigation solutions mentioned before, those fixed pathways, by design, lack flexibility with respect to the trajectories that can be generated by the robot.

A new and simple method which combines the advantages of both methods is to place RFID tags (Radio Frequency IDentification) as invisible landmarks in a regular grid beneath the flooring.

The unique ID together with the position of each RFID tag allows the generation of a fixed digital map. Using these maps vehicles equipped with an RFID tag reader are able to calculate their actual position in a robust, reliable and computationally efficient way. Whereas the grid of RFID tags remains stationary, arbitrary pathways can be planned and acted out as in contrast to fixed induction loops, the vehicle does not have to follow a fixed predefined trajectory. The only requirement is that the grid of RFID tags is dense enough so that frequently a recalibration of the robot's position estimation can be triggered.

This mechanism is illustrated in Figure 10 for the example of a cleaning robot: without position estimation, the robot uses fixed movement patterns such as spirals to cover the entire workspace (left side). In the presence of obstacles, these patterns are destroyed and the robot generates a pathway to a random position where a new spiral is driven. As the robot is not aware of its real position within the workspace, the cleaning patterns may overlap and there may occur regions which are never cleaned at all. In particular for autonomous robots which have limited energy resources, this is an inefficient strategy.

With a precise position estimate by means of RFID tags, however, the robot can drive perfectly straight trajectories even when being perturbed by obstacles. The navigation algorithm can keep track of all the areas that have been accessed before such that the cleaning process becomes very efficient.

Already, several manufactures of autonomous vehicles have commercialized the RFID localization method. The RFID tags are inserted into a grid of boreholes in the floor and sealed afterwards. This installation method, including the creation of a coordinate system, drilling, sealing and mapping is quite laborious.

An easy way to generate a grid of RFID tags in the floor is given by NaviFloor®. It is an integrated RFID textile for flooring based on glass fibre reinforcement fabric (see Figure 11) and is fabricated in rolls of 100 cm width and 50 m length. The RFID tags could be integrated for example in grid sizes of 50 cm or 25 cm. Customized patterns are also possible. Before integration, recesses are made in the textile by laser cutting in the region where the RFID tags are placed.

A unique encapsulation technique has been developed for the passivation of the RFID tags and their mechanical protection during fabrication and installation. During assembly of the Navi-Floor® textile, RFID readers are used for testing to ensure the functionality of the integrated tags. As a novelty an electronic map with the IDs of all integrated tags is generated simultaneously. This map is delivered with each roll of NaviFloor®, which simplifies the mapping of the RFID floor at the installation site. As a high position accuracy is required, low-range HF tags operating on a frequency of 13.56 Mhz which have a wireless range of about 5-10cm are used by default. However, our encapsulation method is compatible with other types as well (e.g. UHF).

Based on a newly developed integration technology using a sequence of different synthetic resins, the RFID tags in the NaviFloor® underlay on the floor become pressure-resistant up to 45N/mm² even with a coating of solely 2 mm thickness. The tested value corresponds to a roll weight of industrial trucks of 324kN (32,4 tonnes), which is several times higher than the pressure applied by typically used fork lift trucks.

Figure 10. Without position estimation, fixed movement patterns (e.g. spirals) are perturbed by obstacles (left panel). With NaviFloor®, the robot knows its position at all times and efficient trajectories are generated

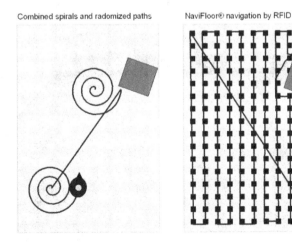

Figure 11. Glass fibre reinforcement fabric with integrated RFID tag (tag size: 5x5 cm², frequency 13.53 MHz). Photo: ©Future-Shape

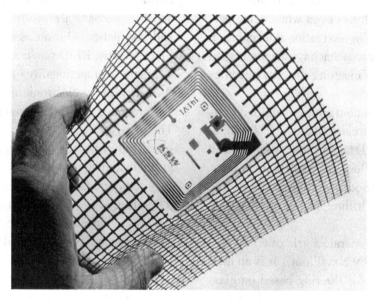

Figure 12 shows the first covering step of the NaviFloor® underlay with synthetic resin (bonding agent) during one of the first installations at the "Arcone Technologie Center", Höhenkirchen-Siegertsbrunn, Germany.

After this step it is possible to apply further layers of synthetic resin to produce a high quality joint less and non-porous flooring which is an exquisite choice for the pharmaceutical industry and clean room technologies. However, it is also

Figure 12. First covering step during the installation of NaviFloor® in artificial resin (grid pattern: 50cm). Photo: ©Future-Shape

feasible to continue with standard processes for tiles, stone flooring, parquet, elastic flooring or carpet. Therefore, NaviFloor® is the first RFID underlay for localization purposes, suitable for all kind of flooring except metallic or other highly conductive materials.

A first application is an autonomous cleaning robot (see Figure 13), which navigates efficiently using NaviFloor®. This robot has a large RFID antenna underneath its base by means of which it can read the HF tags within the floor. Frequent recalibration of the robot's internal navigation engine ensures that the machine remains on the selected trajectory. The latter can be generated free-hand in a graphical user interface on a PC which is connected wirelessly with the robot. The robot's current position estimate as well as its real position in the room can be observed live on the computer screen.

Of course, the navigation of autonomous robots is only one application case for NaviFloor®. Another example is the equipment of non-autonomous mobile objects with a positioning feature. A small embedded device has been developed which contains an RFID reader, a battery and a UHF radio transceiver module similar to the ones used together with SensFloor®. When being moved across the floor, this device broadcasts the exact position information from the RFIDs beneath the floor covering together with the unique ID of the reader. This way, a receiver can display the identity and location of all readers in its environment. This system can be directly used for locating mobile objects to which the readers are attached, such as wheelchairs, hospital beds or even shopping trolleys.

The RFID tags within the NaviFloor® underlay allow even the permanent storage of information imprinted by the RFID reader. This feature implements a memory which is both, distributed and localized. It is possible, for instance, to store information about the cleaning interval or the cleaning chemicals that have been applied exactly to this part of the floor. This information can be read out by portable RFID readers on site.

Figure 13. The wet cleaning robot "Robo 40" possesses an integrated RFID reader and is able to locate itself using a NaviFloor® map. Photo: ©Future-Shape

SUMMARY

SensFloor® and NaviFloor® are examples for intelligent systems which are integrated invisibly into our environment. Unlike conventional sensor systems, both use the floor as functional surface. As people are in contact with the floor almost all the time, a close connection between the user and the sensor systems is guaranteed. However, no active interaction is required by the user making the systems completely ambient. It is our ordinary behaviour only which triggers the systems' functions so that only this function becomes visible. This unique feature is supported even more by the fact that the systems do not impose any specific characteristics on the workspace: any conventional floor covering can be used, the receivers can be hidden anywhere in the room and no cables have to be installed.

Although SensFloor® and NaviFloor® are unobtrusive after installation, they support an enormous variety of applications some of which have been presented within this chapter. As demonstrated, some applications offer enhanced functionality compared to already existing systems such as the intelligent burglar alarm or the efficient control of automatic doors, for instance. In particular the AAL-applications, however, allow for completely new functions such as the long-term activity monitoring by means of gait analysis using SensFloor® or the local storage of cleaning information using NaviFloor®.

REFERENCES

Haritaoglu, I., Harwood, D., & Davis, L. (2000). W4: Real-Time Surveillance of People and Their Activities. *IEEE Trans. on Pattern Recognition and Machine Intelligence, 22*(8).

Lauterbach, C., Glaser, R., Savio, D., Schnell, M., Weber, W., Kornely, S., & Stöhr, A. (2005). *A Self-Organizing and Fault-Tolerant Wired Peer-to-Peer Sensor Network for Textile Applications, Engineering self-organizing applications* (pp. 256–266). Berlin Heidelberg, Germany: Springer-Verlag.

Li, S., & Hayashi, A. (1998, October). *Robot navigation in outdoor environments by using GPS information and panoramic views,* International Conference on Intelligent Robots and Systems, (Volume 1, Issue, 13-17 Oct 1998 Page(s): 570 - 575 vol.1)

Richardson, R., Paradiso, J., Leydon, K., & Fernstrom, M. (2004). Z-tiles: Building blocks for modular pressure-sensing. In *Proc. of Conf. on Human Factors in Computing Systems Chi04*, (pp. 1529-1532).

Röhrig, C., & Kunemund, F. (2007). Mobile Robot Localization using WLAN Signal Strengths. In *Proceedings of the 4th IEEE Workshop on Intelligent Data Acquisition and Advanced Computing Systems: Technology and Applications*, IDAACS 2007, (pp 704-709).

Steinhage, A., & Lauterbach, C. (2008). *(n.d.).* *Monitoring Movement Behaviour by Means of a Large-Area Proximity Sensor Array in the Floor* (pp. 15–27). BMI.

Steinhage, A., & Schöner, G. (1997). Self-calibration based on invariant view recognition: Dynamic approach to navigation. *Robotics and Autonomous Systems, 20,* 133–156. doi:10.1016/S0921-8890(96)00072-3

Stormont, D. P., Abdallah, C. T., Byrne, R. H., & Heileman, G. L. (1998). *A Survey of Mobile Robot Navigation Methods Using Ultrasonic Sensors.* In *Proceedings of the Third ASCE Specialty Conference on Robotics for Challenging Environments.* Albuquerque, New Mexico, April 26-30, 1998, (pp. 244-250).

UKARI. (n.d.). *Website of the UKARI project.* Retrieved from http://open-ukari.nict.go.jp/Ukari-Project-e.html

Victorino, A. C., & Rives, P. (2004). *Bayesian segmentation of laser range scan for indoor navigation,* Intelligent Robots and Systems, 2004. (IROS 2004). In *Proceedings. 2004 IEEE/RSJ* (pp. 2731- 2736 vol.3).

KEY TERMS AND DEFINITIONS

SensFloor: Textile underlay with built-in wireless capacitive sensors for locating people indoor

Textile-Electronics-Integration: Mechanically embedding microelectronics into textile material and contacting it electrically

Ambient Assisted Living (AAL): Research field for evaluating technology and services to allow elderly or people in need of care to stay longer in their private homes instead of moving to nursery homes

Capacitive Proximity Sensors: Detectors which remotely measure the change of the electric field induced by objects or body parts

Embedded Receiver: Small devices that plug into a wall outlet and receive and process the sensor data from the SensFloor directly without the use of an external PC

Trajectory: Movement path of walking people reconstructed from the sensor signals

Activity Monitoring: Continuously measuring behaviourally relevant parameters of people with the aim to detect changes of their state of health

Position Estimation: Determining the position of an object based on either internal or external information

Chapter 4

Embedding Context–Awareness into a Daily Object for Improved Information Awareness:
A Case Study Using a Mirror

Kaori Fujinami
Tokyo University of Agriculture and Technology, Japan

Fahim Kawsar
Lancaster University, UK

ABSTRACT

In this chapter, a case study on augmenting a daily object, mirror, for a contextual ambient display is presented. The mirror presents information relevant to a person who is standing and utilizing unshareable objects, e.g. a toothbrush, in front of it on the periphery of his/her field of vision. We investigated methods of interaction with the mirror by analyzing user preferences against contrastive functionalities. Experiments were conducted by a Wizard-of-Oz method and an in-situ experiment. The results showed that a short absence of the mirror function was not a big issue for the majority of participants once they were interested in presented information. The analysis also allowed us to specify requirements and further research questions in order to make an augmented mirror acceptable.

INTRODUCTION

One of the consequences of the convergence of pervasive technologies, including the miniaturization of computer-related technologies, the proliferation of wireless Internet, and advances in short-range radio connectivity, is the integration of processors and miniature sensors into everyday objects. Computing devices now pervade everyday life to the extent that users no longer think of them as separate entities. This development has revolutionized our

DOI: 10.4018/978-1-61692-857-5.ch004

perception of computing to the extent that we can now communicate directly with our belongings, including watches (Raghunath, 2000), umbrellas (Ambient Devices, 2004), clothes (Baurley, Brock, & Geelhoed, 2007), furniture (Kohtake et al., 2005) and shoes (Paradiso, Hsiao, & Benbasat, 2000). This allows us to acquire personalized information (meeting schedules, exercise logs) and use proactive information services (weather forecasting, stock prices, transportation news) in a timely fashion.

Major challenges in such technology-rich settings include perceiving information, facilitating sustained attention to information of interest, and maintaining awareness of changes in information in an unobtrusive way. Massive amounts of information have the potential to degrade daily living convenience or even create unsafe living conditions. Because of this, it is important to provide information in an appropriate way. This includes taking into consideration the proper media, timing, ease of understanding, location, identity, lifestyle, privacy concerns, and other factors where context-awareness plays a key role. A system that is aware of the context of a user's activities senses information related to a user (contextual information), after which it may (1) present information and services to a user, (2) automatically execute a service, or (3) tag context to information for later retrieval (Dey, 2000). Numerous studies have addressed the methods and effects of context-aware computing (Addlesee et al., 2001; Beigl, Gellersen, & Schmidt, 2001; Brumitt, Meyers, Krumm, Kern, & Shafer, 2000; Dey, 2000; Want, Hopper, Falcao, & Gibbons, 1992). We have been examining the information overload issues, and investigating sensing technologies, middlewares, and service models for many years (Fujinami & Nakajima, 2005; Kawsar, Fujinami, & Nakajima 2005; Fujinami & Nakajima, 2006; Kawsar, Nakajima, & Fujinami, 2008; Fujinami & Inagawa, 2009), where a daily object is used to obtain user contextual information and/or to present information or provide a service.

People normally use objects to perform specific tasks. For example, they use doors to enter or exit rooms. Sensing the usage state of an object by equipping it with a dedicated sensor (i.e., "door opened with inside doorknob"), allows a system to infer a user's current or upcoming action, such as "exiting the room". Possessing an understanding of a user's intention allows a system to take appropriate action that supports the task itself, or related tasks, without additional instructions from the user. The input to such a system is a user's normal activity. A human-computer interaction style that does not require an explicit user command is called implicit interaction (Schmidt, 2000) or background interaction (Hinckley et al., 2005). Furthermore, the sensing method is self-contained in an augmented object, and is thus lightweight, and no external infrastructure is required to determine a user's context. In contrast, other technologies that enable the same service, e.g., location tracking and video-based activity recognition (Harter, Hopper, Steggles, Ward, & Webster, 1999; Brumitt et al., 2000), require the addition of dedicated sensing devices to the environment.

We believe an augmented daily object not only has the potential to recognize the context of a user's activities, it can also assist by providing related information efficiently and effectively. In the case of the above-mentioned door, a display installed near the inside doorknob could be programmed to show the user a reminder to take mail to be posted. In this situation, a display would be situated at the periphery of the user's eyesight, thus allowing him or her to focus on the primary task at hand, which is exiting the room. The display only attracts primary attention when it is appropriate and necessary. Displays of this type are often referred to as peripheral (Matthews et al., 2004) or ambient displays (Wisneski et al., 1998). The challenge is to design the information display function in a way that avoids distracting or annoying the user. Accordingly, the usual manner of execution of a primary task should be respected while the context-enhanced value is being provided.

In this chapter, we present a case study on embedding context-aware features into a daily object, in this case a mirror, which has been augmented to support decision making at home, especially, in the morning. The design process, and evaluations on basic functions and interaction methods are then presented.

A MIRROR AS A PERIPHERAL INFORMATION DISPLAY

Humans have utilized mirrors since prehistoric times (Gregory, 1997). While normally used to examine the physical appearance of the person or object placed in front of it, a mirror can also be used to recognize and monitor objects and activities taking place in the background, such as people passing behind the user, or a boiling kettle. Additionally, people often spend a few moments in front of a mirror daily performing routine tasks such as drying hair, brushing teeth, etc., while thinking about topics relevant to their immediate concerns. These often include their schedule for the day, weather conditions at a travel destination, etc. This suggests that information could be conveyed by the surface of a mirror that incorporates a peripheral information display.

We propose a personalized peripheral information display called AwareMirror. AwareMirror is an augmented mirror that presents information relevant to the person in front of it, in anticipation of the fact that users may decide to modify their behavior once an unwanted/unknown situation is understood. This could be explained by a psychological notion called "objective self-awareness" (Duval, 1972), which is caused by the self-focusing stimulus induced by a mirror. Objective self-awareness is a state in which a person's consciousness is directed inward and the person begins evaluating his or her current state of being against an internal standard of correctness.

In this state, the user tries to either avoid self-focusing stimuli, or to reduce a sensed negative discrepancy through changes in attitude or action. In addition to this natural and prominent effect of a mirror, we believe that information presented on the mirror surface could enhance objective self-awareness. This conclusion came from an assertion made by Pinhanez et al. in the context of a frameless display, which states, "positioning information near an object causes immediate, strong associations between the two" (Pinhanez & Podlaseck, 2005). Although their frameless display projected digital information on or near a physical object, we believe we could extend this idea to include information superimposed on the surface of a mirror because both the information and the user's reflected image would coexist on the same surface. Furthermore, we believe such superimposed information could affect a user more effectively than presenting information using a computer display installed separately near a mirror.

A similar mirror-based information system for supporting decision-making was proposed by Lashina (2004), using gestures and touch contact to manipulate presented information. Miragraphy, a mirror system developed by Hitachi Human-Interaction Laboratory, (Hitachi Human-Interaction Laboratory, 2006), provides involvement that is more active. Information and coordination options are displayed when the radio-frequency identification (RFID) tag of an individual piece of clothing is detected, after which clothing combinations that were previously captured by a video camera, are presented. To the best of our knowledge, a user must explicitly use the functions provided by these systems, which could be visualized as a gap between the user and the system. However, in our interaction method, the gap would be significantly smaller because AwareMirror is used as an extension of the normal activities that take place in front of a mirror. Philips Electronics N.V. released a commercial product, Mirror/TV, which consists of a versatile LCD display integrated into a mirror and which provides the core of an intelligent personal-care environment (Philips, 2007). Various ideas have been proposed for addition to the hardware platform, such as oral care coaching and mirror-integrated TV, "make-up" support. Although AwareMirror focuses on useful information for morning preparations, presenting information that could effectively increase health awareness or encourage improvements to bad habits might also be suitable for a mirror-based peripheral information display (Andrés del Valle & Opalach, 2005; Fujinami & Riekki, 2008).

In terms of information representation, peripheral displays are also used in workgroups during group awareness applications (Ferscha, Emsenhuber, & Schmitzberger, 2006), in physical space settings (Streitz, Rocker, Prante, Alphen, Stenzel, & Magerkurth, 2005), in enterprise applications for knowledge dissemination (McCarthy, Costa, Huang, & Tullio, 2001) or social applications like instant messaging (De Guzman, 2004), and disseminating information regarding family members (Mynatt, Rowan, Jacobs, & Craighill, 2001). The system developed by Rodenstein (1999) can also be categorized as an ambient display. These displays utilize transparency, which is the opposite feature of a mirror and the most significant characteristic of a window, to display information (such as short-term weather forecasts) concurrently with the image of an outdoor setting that would actually be visible if a window was installed at that location. Information visualization by artistic expressions such as Informative Art (Holmquist & Skog, 2003) and InfoCanvas (Stasko, Miller, Pousman, Plaue, & Ullah, 2004) can be considered "augmented paintings" that can be easily fitted into daily life. However, information presentation is the primary function of these systems, while AwareMirror has the specific restriction of maintaining a mirror's original function while offering additional value, i.e., personalized information.

A unique characteristic of AwareMirror is that it utilizes low-level sensing technologies and augments the function of a traditional mirror in a natural way. This is expected to eliminate feelings of obtrusiveness while minimizing the burden of information retrieval and comprehension. AwareMirror illustrates how computing capabilities can be incorporated into daily objects. In next section, the AwareMirror system design is explained.

DESIGNING AWAREMIRROR SYSTEM

Requirements

It is important to consider the usage context of a target object when proposing its augmentation. As the name implies, a daily object is one that has already been integrated into our daily lives, and large deviations involving such objects could cause users to reject the system. The system requirements we have specified, which are specific to a mirror and its installation location, are provided below (R1-R4):

R1: Automatic start/finish presentation when certain conditions are satisfied

R2: User identification that is implicit and natural, yet sensitive to feelings of privacy violation

R3: Easy installation into daily living space

R4: Information presentation without interfering with primary tasks and habitual activities

Information should appear and disappear automatically, so as to maintain the mirror's primary function and avoid turning it into a simple "display device" (R1). Accordingly, the system should identify the user without requiring any explicit action on his or her part. This calls for implicit interaction with the system. Additionally, this higher level of contextual information should be extracted without using privacy-intrusive sensing technologies (R2). While users naturally prefer to be presented with relevant contents at appropriate times, most would avoid using a vision-based service if it incorporated a visual imagery sensing technology because the video cameras used in such technologies cause feelings of privacy violation regardless of the actual data usage. This is because the area around a mirror is used for private activities, such as grooming and personal hygiene. Therefore, privacy requirements must be taken into consideration.

As for the third requirement (R3), easy installation is important because the system is intended for everyday use in locations where a large-scale installation would not be suitable. Finally, but also important, it is necessary to consider the characteristics of tasks or habitual activities conducted in front of a mirror when specifying the required information and display methods (R4). AwareMirror is a mirror, first of all, which means that users often perform their primary tasks, such as face washing, engaging in introspection while viewing themselves, or examining their faces, at times when they are not wearing their eyeglasses. In those situations, they might become aggravated if irrelevant information is displayed during the primary task. Furthermore, information displays should not disrupt the task at hand by taking excessive time to present information or by obscuring the user's mirror image with the superimposed information.

Design Principles and Approaches

The design principles necessary to fulfill the above requirements are (1) leveraging a usage state of an "unshared" object, (2) superimposing information with physical images, and (3) two-level information presentation. These are explained in subsequent subsections.

Utilization of an "Unshared" Object

Controlling the initiation and termination of presentation through the usage state of an "unshared" daily object satisfies the first two requirements (R1 and R2). Popular methods for identifying a person, or the timing at which to provide a ser-

vice are as follows: (1) face recognition or gaze/head orientation detection by video imagery, (2) presenting an RFID tag that represents the user (or a target service), and (3) using biometric information such as finger/voiceprint. Biometric-based identification was not deemed appropriate, in this case, because a user would have to input a fingertip or speak into a sensor, which would violate the first requirement, "implicit and natural identification". The tag-based approach was rejected for the same reason. Video image captures were also deemed unsuitable due to the privacy violation concerned mentioned above. Since the popular methods were deemed unsuitable, we decided to introduce a fourth method. That is, monitoring the usage state of a daily object that is rarely shared with others, and which is normally used in conjunction with a mirror. Such daily objects include safety razors or electric shavers, toothbrushes, combs, etc. With these objects, the user is identified implicitly through its intended purpose once it is determined to be "in use" by an analyzing sensor, such as an accelerometer. In our study, we decided to use toothbrushes to detect users because they are gender neutral and are utilized by almost everyone. In later section, we will discuss a user survey regarding alternative identification methods such as image-based, tag-based, biometric-based, and sensor-augmented detection.

To present information effectively, a system of this nature also needs to ensure the toothbrush user is in front of a mirror. In our system, three types of information are integrated to detect a particular user's presence in front of the mirror. They are, (1) verifying a user's presence in front of the mirror with an infrared range finder (or a distance sensor), (2) proximity of the personal object (toothbrush) with the mirror, and (3) use of the personal object. When a person is detected in front of a mirror while a toothbrush is in use in the near vicinity of the same mirror, we assume that the person utilizing the toothbrush is identical to the person in front of a mirror. External information about the owner of the toothbrush enables identification of the person. Here, we also assume that a person in front of a mirror is looking into the mirror. Tracking a user's eyes might provide more precise detection points for timing purposes, but due to difficulties in installation and related costs, this feature is not feasible for most living environments. Furthermore, although significant effort has been applied to investigate practical tracking (Ohno & Mukawa, 2004), such sensing systems also require careful installation and calibration.

Figure 1 shows the four possible situations in which the AwareMirror system detects users. As shown in the figure, information is only provided to the user in Case 2, which satisfies the three above-mentioned conditions. Case 1 is not appropriate for presenting information because the user in front of the mirror is not using a toothbrush, while in Cases 3 and 4, the subject is using the toothbrush but is not in front of the mirror. Note that in Case 3, information for the person using the toothbrush near the mirror would be visible to any other person coming in front of the mirror later because the final condition (existence of a person in front of the mirror) would be satisfied.

However, we consider this is a "rare case" since people tend to avoid close proximity with others during hygiene-related activities.

Superimposing Information with Physical Image Using a Two-Way Mirror

We considered three setup methods for displaying information on a mirror, i.e., superimposing information using reflected imagery. They were: (1) simulating a mirror using a video camera and a computer screen, (2) reflecting information on the opposite side of a mirror (such as the wall behind the user), and (3) using a two-way mirror and projecting information onto it from behind. Figure 2 shows these setup methods, each of which has advantages and disadvantages.

The simulation-based approach reproduces the function of a mirror by using image-processing techniques (Figure 2-(a)). This allows flexible mixing of information with a person's image and processing of the captured image. Applications for such mirrors include a simulation that makes the user look younger or older to encourage healthy-living (Andrés del Valle & Opalach, 2005)

Figure 1. User identification using low-level contextual information

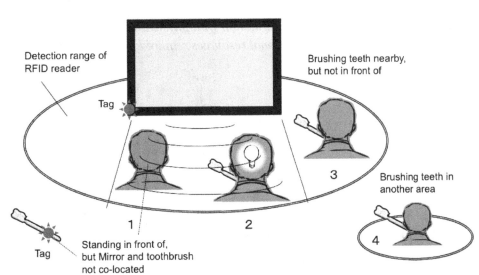

and a makeup support system with automatic zoom/pan, lighting simulation, and other features (Iwabuchi, Nakagawa, & Siio, 2009). However, when considering screen resolution and real-time image rendering, along with the need to make images of the observed world consistent with the viewer's position, it is difficult to make a "virtual" mirror that is realistic enough for actual use (François & Kang, 2003).

The reflection-based method does not require a special mirror (Figure 2-(b)). We developed this alternative version first using an array of LEDs that acts as indicators of two types of information attached to the front wall of an ordinary mirror. However, appropriate positioning of the display was required to prevent the LED display from being hidden by the user and persons crossing behind him or her. Furthermore, forcing the user to assume an unnatural position to see the content could induce the feeling of "being used by the system". Flexible projection systems like those proposed with steerable interfaces (Pingali et al., 2003) could allow the location of the presented information to be changed dynamically, but require a camera-based positioning system and extensive mechanical supports that are not suitable for daily living spaces.

The projection-based method requires a two-way mirror, with which the image and/or text displayed behind the mirror can be seen through the panel while physical objects in front of it are reflected as with an ordinary mirror (Figure 2-(c)). A two-way mirror is a translucent panel that reflects light from the brighter side when the light intensity difference between the two sides of the panel is significant. In other words, bright colors projected onto the panel from behind can be seen from the front, while a reflection of an object placed in front of the panel is seen in the areas that are dark behind the panel. Unlike the reflection-based method, we need not consider a person's position, even though a computer display needs to be installed behind the mirror panel. We adopted this approach to fulfill the third requirement (R3) and after considering recent advancements in thin, inexpensive computer display technologies.

Two-Level Information Presentation

If excessive or hard to understand information is presented, it could disturb a user and interfere with his or her primary task. Furthermore, it is important to ensure private information is safeguarded. Because of these factors, we decided to adopt a two-level information presentation

Figure 2. Three methods of superimposing information with physical images. Note that the size of the "PC" boxes indicate the amount of computational resources required for overall processing

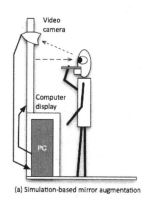
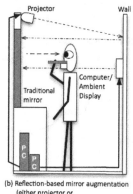
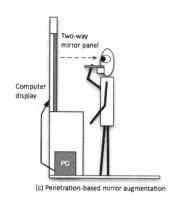

(a) Simulation-based mirror augmentation

(b) Reflection-based mirror augmentation (either projector or computer/ambient display is utilized.)

(c) Penetration-based mirror augmentation

system. At the first level, the AwareMirror system presents information in an abstract manner that the user can perceive in a glance. A second level, which provides information that is more detailed, is presented as text messages. In addition to the two presentation levels, AwareMirror has a default mode that does not display any information.

The transition between these three modes is illustrated in Figure 3, where the solid and dotted lines indicate automatic and manual transitions, respectively. When a user stands in front of an AwareMirror and the system detects a toothbrush in use, the system initiates the abstract mode. In this mode, AwareMirror assumes the role of a peripheral display and projects abstract information on the periphery of a user's field of view. Thus, in the abstract mode, AwareMirror can also be used as an ordinary mirror. Since some users do not wear their eyeglasses while brushing their teeth or washing their faces, and thus may not be able to read detailed information, abstract information is represented using familiar figures, colors or movements that can be readily understood, and which takes into account the characteristics of the location, i.e., in front of a mirror. However, we believe that it is useful for users to, at least, notice the availability of important information.

In the detailed mode, AwareMirror is explicitly used as a display rather than as a mirror. The system displays detailed information using text. However, presenting detailed information auto-matically runs the risk of violating our design philosophy, which is, keeping the original usage purpose of the mirror intact. Furthermore, if text messages are displayed suddenly, the detailed mode could also interrupt a user's primary task. To avoid this, AwareMirror requires the user to change modes manually. Only when the detailed mode is active does an AwareMirror become an information display instead of a mirror. We contend that this does not interrupt the user since the mode shift choice is voluntary.

Although the detailed mode has more information than the abstract mode, the AwareMirror system is not intended to provide a complete solution for our information acquisition requirements. In our daily lives, various types of media have already been developed to supply detailed information, such as TV/radio programs and the World Wide Web. We contend that presenting an excess amount of information would make the system excessively complex and thus unusable.

Once a user is identified and the information is presented, the information display remains visible even after the user finishes brushing his or her teeth. This is because while the toothbrush is used as a trigger to identify the user and start information presentation, the infrared sensor continues to detect the presence of the user even when he or she has stopped using the personal object. Both modes automatically return to the default mode when the infrared sensor no longer detects the

Figure 3. Transitions between default, abstract and detailed presentation modes: The solid and dashed lines indicate automatic and manual transitions, respectively

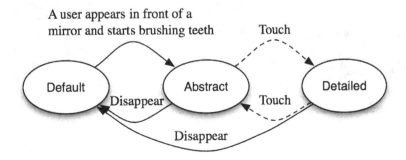

user. When describing the design heuristic of an ambient display, Mankoff et al. (2003) stated that a multi-level presentation approach should make it quick and easy for users to obtain information that is more detailed. The transition from abstract to detailed mode is achieved by an action with which a user can associate a separate action, in this case, "retrieving detailed information". Because of these features, the display allows easy access to information while minimizing interruptions to a user's primary task and while fitting naturally into daily living spaces.

PROTOTYPE IMPLEMENTATION

After presenting a usage scenario, the details of implementation, i.e., internal software components, augmentation of a mirror and a toothbrush, and presented contents, are explained.

Using the AwareMirror System

Figure 4 demonstrates an AwareMirror usage scenario, where information related to the person in front of the mirror is displayed. In Figure 4-(a), a user is engaged in teeth brushing while

perusing information displayed on the mirror at the periphery of his or her line of sight. This abstract representation of the implicit mode is initially displayed. In this case, which is based on the input schedule, it shows that (1) the weather forecast at the user's next destination is rainy (by displaying an image of an umbrella), (2) the maximum temperature at the destination is below 15 degrees Celsius (by displaying an image of a girl wearing gloves and a snow cap), (3) there are problems with rail transportation on the way to the appointment (by displaying a red train), and (4) the time remaining before the next scheduled appointment. Input scheduled appointment information is utilized as contextual information about future as well as present activities.

If the user wants to know more about the information presented in an abstract manner, he or she will suspend the current task (brushing teeth), and activate a touch sensor located the bottom of the mirror to further details (Figure 4-(b)). Then, the following text information scrolls from right to left repeatedly:

"The next appointment is [a meeting] with [Mr. Yama] to [discuss a mirror] at [12:00] at Tokyo station]. The weather (around Tokyo) is [rainy],

Figure 4. AwareMirror's (a) abstract and (b) detailed modes of presentation. Three types of information that support decision-making in front of a mirror are presented: (1) weather forecasting at a user's destination (state and maximum temperature), (2) state of the public transportation system to be taken, and (3) information about a user's next appointment (time to the event)

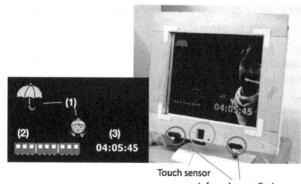

Touch sensor
Infrared range finder
(a) Image-based abstract presentation

Detailed textual messages
(scrolling from right to left)
(b) Text-based detailed presentation

the maximum temperature is [10] degree's Celsius, and the probability of rainfall is [90]%. On your way to the next appointment, [the Central Line] is currently [not operating due to an accident]."

Note that the messages can be presented either in Japanese or in English, and the words in brackets are retrieved and set into template sentences from dedicated information sources described later. The information supports a person's ability to make decisions that are often made in front of a mirror in the morning. With it, users can be reminded to take an umbrella, locate a document necessary for a meeting, decide on an alternate route to a destination, or hurry to catch a train. Thus, AwareMirror provides a person with contextual information about present to future activities, first seamlessly (implicitly) and then explicitly, while preserving its traditional look and interaction mode of the mirror to the greatest extent possible.

System Components

As described in previous section, superimposing rendered information on the reflected physical image is achieved by attaching a two-way

mirror to a computer display connected to a Windows-based PC that operates the functional components explained below. The relationship among the functional components (rectangle) and their implementation (clouds) is illustrated in Figure 5. The system consists of the sensing system, the AwareMirror Controller, personalized content retrieval as well as content rendering and presentation mode controlling components. The core function is the AwareMirror Controller, which accepts three source inputs from the sensing system to identify a person in front of the target mirror.

First, the existence of a possible user is detected by a change in the distance between the person and the mirror, which is determined by two infrared rangefinders. Such rangefinders can measure the distance to an object up to 80 cm from and within 5 degrees of the sensor. Accordingly, we situated two rangefinders side by side to extend the sensing area in a horizontal direction. The sensors are connected to the controlling PC with USB cables via Phidgets Interface Kit. As a second source input, a change in usage state of an accelerometer-augmented toothbrush is observed. The toothbrush is augmented with a Phidgets' 2-axes accelerometer connected to the

Figure 5. Functional components and AwareMirror implementation

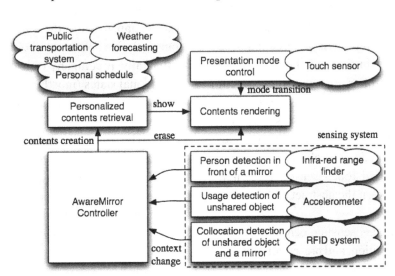

same PC. Here, we defined the movement of a toothbrush with a certain variance in acceleration for a certain period of time as "brushing teeth". To make the toothbrush a more reliable object for personal identification, the fact that the head of a toothbrush is actually inside the user's mouth should be confirmed. However, after considering the difficulties involved in researching an ultra-small and safe chemical sensor to detect specific material in the mouth, we adopted a more approximate method for identifying teeth brushing. Since this component, the usage detector of an unshared object, hides the detail of sensing (such as chemical sensor and accelerometer), any unshared object that is augmented with a sensor could be easily incorporated into the system. Finally, when collocation of the mirror and the toothbrush is detected, a report is made to the controller component. In this situation, RFID tags are attached to both the mirror and a toothbrush, and an RFID tag reader is installed near the mirror. In our experimental environments, we utilized an active Spider III RFID system from RF Code Inc., which has a detection range about 3 m.

The AwareMirror Controller determines if the conditions for identifying a user are satisfied each time an event is updated. If they are satisfied, the component instructs the personalized content acquisition component to retrieve information about the owner of the toothbrush, after which, content is rendered in an abstract manner. If the person detection component reports that the person is no longer in front of the mirror, the AwareMirror Controller instructs the content rendering component to clear the presentation, which means AwareMirror reverts back to service as an ordinary mirror, which is the default mode. The presentation mode-controlling component uses a Phidgets-based touch sensor, to switch back and forth between the abstract and detailed information modes.

The abstraction, i.e. sensing, central controlling, contents acquisition, and presentation mode controlling components, make the AwareMirror system extensible with minimal alterations when a new sensor, information source, and/or presentation mode controller is needed. In actuality, different policies for the detailed mode rendering, as well as the controlling method of the presentation, were implemented and recorded for a later comparative study.

Contents

As shown in the usage scenario, we selected three types of information that can affect our behavior and support decision-making for display: (1) weather forecasts (forecast and temperature) at an intended destination, (2) the state of the user's public transportation route, and (3) information on the user's next appointment. All information types depend on the user's personal schedule.

We then combined these content types to provide actual experiences during the in-situ experiments described in next section. In the prototype implementation phase, a user registers her schedule using an original Web-based scheduler in a structured manner. Once a user is identified, his or her next scheduled destination and the time of the appointment are extracted from the scheduler data. This data are then utilized as keys to obtain the weather forecasts. The system parses information from the weather portal service of Yahoo! Japan while the state of related public transportation systems is obtained by parsing actual railway company websites that report their operational states (mostly accidents).

The contents rendering component delivers the information using Adobe Flash Movie, which was selected because of its aesthetic features and ease of animation building. During the actual rendering, it was necessary to consider content colors because, due to the characteristics of a two-way mirror, dark colors are harder to see when projected through a translucent panel.

USER PREFERENCES FOR USER IDENTIFICATION METHOD

We then carried out a survey to determine the most preferable identification methods.

Profiles of Participants

The survey was conducted using questionnaires distributed among 50 people, 34 of which were members of the author's laboratory including one professor, four PhD students and 29 Masters Course and undergraduate students. Six of the 34 were female and approximately 10 were engaged in work on ubiquitous computing projects. The remaining 16 participants included five undergraduate students, five housewives, four company employees, and two artists ranging in age from 20 to 69.

Procedure

The four previously mentioned identification types were listed: (a) carrying an RFID tag that has a detection range of approximately 3 m (tag-based), (b) face recognition by video imagery (image-based), (c) usage state of a daily object that is rarely shared with others (augmented object-based), and (d) fingerprint or voice recognition (biometric-based). Since some participants were not familiar with these methods, we provided brief description of their general features, including principles, benefits and risks. Then, we asked them to compare the identification methods and assign a rank from 1 to 4, where 1 was the least preferred and 4 was the most preferred. Persons who choose (c) were also asked to identify which object would be most suitable for augmentation.

Results and Implications

Table 1 shows the median user preference score for each identification method. Note that the two central values are averaged if the number of the data is even. Larger values indicate that more participants preferred that method. The results indicate that a biometric-based method is the most preferred, while the imagery-based method is the least preferred. The biometric-based approach was preferred because the participants believe it would perform accurate identification and would not require them to manipulate anything. However, the method also requires explicit interaction with the system, such as putting a finger on a reader to identify a user. In contrast, as anticipated, the image-based method had the lowest preference total, because it engenders an obtrusive feeling of being observed, even though the respondents know that it requires no user input and that it can identify human activities, such as looking at a mirror. The natural characteristics of a washroom are considered likely to aggravate that sensitivity. As for the tag-based approach, participants liked the ability to control the timing of the information display, but they pointed out the possibilities to forgetting or misplacing the tag as a drawback of that approach.

They identified the merits of the augmented object-based approach as its implicit identification of the user through a daily object (such as a toothbrush) that is rarely shared with others. However, participants who specified a lower score emphasized the possibilities of intentional or unintentional use by other persons. We believe that utilizing the system in a reliable closed group, such as a family, would suppress the "misidentification" issue. Further, we believe that intentional misuse

Table 1. Participants' preferences for person identification method

Person identification method	Median of the preference score
Tag-based	2.5
Image-based	2
Augmented object-based	3
Biometric-based	4

rarely occurs and that potential feelings of privacy violation would be low if the object was used unintentionally.

Regarding daily objects that are seldom shared with others, toothbrushes were listed by most of the participants. Other objects include safety razors or electric shavers, towels, hairbrushes, and cosmetics. As described above, AwareMirror Controller separates the detection of an object from the underlying sensing technology. However, as some participants pointed out, a number of these personal objects, such as hairbrushes, might be habitually shared among family members. This indicates that the system needs to have a mechanism to gain a consensus on potential risks of sharing the object, which could be done when the object is associated with a person identifier for the first time.

COMPARATIVE STUDY ON INTERACTION METHODS WITH AWAREMIRROR

We conducted a comparative study with contrastive functions to confirm our augmentation rationale for a mirror, i.e. maintaining a mirror's primary function to the greatest extent possible while providing additional information.

Points of Comparison and Contrastive Functionalities

Contrastive functions were implemented, and tested along with the original ones. The points of the comparison are as follows:

C1: Automatic vs. manual activation of the first (abstract) scene

C2: Automatic vs. manual transition from abstract to detailed presentation

C3: Presence vs. absence of a mirror's primary function during the detailed presentation

C4: One-phase (textual message only) vs. two-phase information provision

Regarding C1, to activate the first scenario, manual activation is realized by touching the same area as that used for mode transition, i.e. the touch sensor. This contrasts with the automatic activation that is invoked when the three conditions for user identification are satisfied. In the presentation mode transition (C2), the automatic transition presents the textual information 10 seconds later. In C3, a version that severely degrades the mirror function is achieved by displaying all messages simultaneously with black characters on a white background (Figure 6). This reduces the surface reflective characteristics of a two-way mirror. Therefore, the user's view is similar to that of a word processing or drawing application on a traditional computer screen, rather than that of a mirror. Finally, in C4, a two-phase information provision is compared to a textual message only function. Users indicate their preferences through voluntary utilization of these functions.

Methodology

The experiments were conducted using two approaches: a simulation-based experiment and an in-situ experiment. The former is often called the Wizard of Oz (WOz) approach (Kelley, 1984). An experimenter simulates the user identification process to obtain a comprehensive view from a relatively large number of participants at an early stage of prototype development. The simulation allows the elimination of the effect of the incomplete user identification component. In contrast, the in-situ experiment is intended to capture more detailed information through actual utilization of the system in a user's home, over a relatively long period of time. We adopted this approach to gain an understanding of the true utility and other qualitative values of our augmented system.

Fourteen participants were invited to the WOz-based experiment: 12 students who were

Figure 6. Contrastive rendering of the detailed presentation. Here, static text messages are displayed in black on a white background. (1) and (2) corresponds to the message in Figure 4 (b), while (0) indicates the details of a user's next appointment

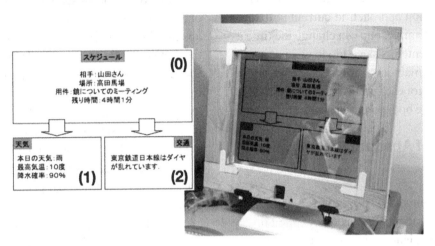

not members of the author's laboratory (6 males and 6 females, in ages ranging from 18 to 29), one non-IT engineer (male, age 23), and one IT saleswoman (age 30). The simulation was conducted as follows. First, an experimenter provided a brief introduction of the system's functions. Next, participants were asked to imagine themselves standing in front of their washroom mirror in the morning. Then, each participant used the system by following instructions with the experimenter controlling situations where the participant experienced contrastive functions that were not fully operational. At the end of the experiment, the participants were asked to complete survey forms and were then interviewed regarding their answers.

Four people participated in the in-situ experiment. They included a family consisting of a 30-year-old female pianist, her 70-year-old artist father and her 63-year-old housewife mother, as well as an IT engineer (male, age 29) who was living alone. A working AwareMirror prototype was installed at both the family home and the single participant's home for 17 and 22 days, respectively. The prototypes were installed in the living room because of limited washroom space and because normal mirrors were already installed

in both washrooms. During the experiment, the participants were asked to brush their teeth in front of AwareMirror, rather than in front of their washroom mirrors. They were further instructed to use all the system functions throughout the experimental period. The participants were also encouraged to change the properties for the different experimental conditions by themselves by means of a simple GUI and a user manual, in order to avoid intervention by the experimenter. Additionally, they were told to change the comparative functions immediately after the last utilization. This was done so that the subjects could focus on performing their primary task or using the function that was subject to evaluation. Thus, preparation for the experiment had to be completed beforehand. Finally, after the testing period, the participants' attitudes were evaluated using a questionnaire and through memos that they had written at various times during the experimental period.

Results and Implications

The results of both experiments are summarized in Figure 7. These results suggest that there was not a one-sided preference in the functionalities. For practical use, a preference management function needs to be integrated into the system. This is a very common approach in current software systems. For example, one can change the time a laptop PC will enter its sleep mode when it is not in use. In the remainder of this section, we show the preferences for each point of comparison (C1-C4), and explore both potential needs and issues.

Activation of Abstract Presentation Mode (C1)

Seven Participants (59.2%) preferred automatic activation that does not require any explicit invocation. The participants who supported automatic activation said that even simply touching the sensor was messy during a busy morning, and was easy to forget sometimes. Interestingly, all four participants from the in-situ experiments preferred this version. In contrast, the participants who preferred manual activation pointed out that they did not always need decision-making information. This normally arose from situations where they

had no specific plans that required the decision-making information provided by AwareMirror. In such situations, the presented information became obtrusive even if it was restricted to image-based abstract information. Therefore, we believe that the system should filter out irrelevant information for users based on their preference, or, more intelligently, by analyzing the contents of information in terms of relevance and usefulness. For instance, it should be possible to set the system so that only a serious accident on a planned railway would be shown. These facts suggest that preserving a mirror's original function is of primary importance, at least initially.

The other interesting reason for not supporting automatic activation is that unreliable behavior of the system caused participants to reject proactive information provision. This situation arose due to a sensing subsystem failure. One participant complained that he had been surprised by the sudden activation of the abstract presentation mode at midnight, since he had not been brushing teeth, but simply looking at his face in the mirror at that time. Because he knew that the utilization of a

Figure 7. Summary of experimental results. Each bar indicates the number of the respondents who preferred a particular contrastive function for both WOz-style and in-situ experiments. The white bar shows the results of the in-situ experiment only

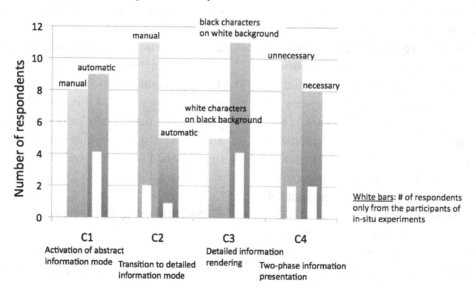

toothbrush was a condition of activation, he did not expect AwareMirror to display information at the time. This made him feel as if he was being monitored. Implicit interaction, an interaction paradigm that AwareMirror is employed, depends on various sensors and context recognition systems, which cannot always prevent misdetection of a user's context. We believe it would be more problematic if the system were activated improperly than would be the case if it did not activate when proper conditions were satisfied because users might experience anxiety about privacy violations. Also, it is more likely to be perceived by a user as a failure of a system than the case of "no-activation" since "no-activation" provides a user with a chance to initiate manually and thus might appease her frustration.

Transition to Detailed Presentation Mode (C2)

Eleven participants (68.8%) supported manual transition. The major reason for this preference was the feeling of being in control of the system. This requires active involvement with an (augmented) mirror that does not exist in relationships with normal mirrors. These participants indicated that they did not want to wait for the automatic transition (10 seconds) when there was something they wanted to know quickly, or that they did not want to see detailed information. Such user preferences were the reason for our two-level design philosophy.

In contrast, the participants who liked the automatic transition gave several reasons for not preferring the manual transition, i.e. not wanting to touch anything with wet hands or the feeling it was bothersome to change the mode by hand. We believe these were not affirmative preferences for automatic transition. Instead, they indicate the need for a new manual control method that takes into consideration the constraints of a washroom. Note that the system should be controlled in a robust manner with less burdensome controls than

the present touch control, since some participant considered even such simple operation to be messy.

Detailed Information Rendering (C3)

The version displaying static textual messages in black characters on a white background (Figure 6) was preferred by 11 participants (68.8%) because they thought the information could be accessed easily. While some participants pointed out the loss of a mirror's reflective function during the full screen presentation, they also indicated that it did not matter to them since the time needed to check the detailed information was short due to its "at a glance" nature.

In contrast, the scrolling text version (Figure 4-(b)) requires a user to follow the message to the end, which means the user must pay attention for a longer period of time. Such users had to follow the scrolling text until the desired information appeared. This could take up to approximately 40 seconds. This suggests that the full screen version contributes to reducing the interruption into the primary task. This is important because, in the case of a manual transition from the abstract presentation mode, a user would become impatient if the system does not immediately supply the requested detailed information.

On the other hand, the participants who had negative impressions of the full screen version noted the superior clarity of the rendered information of the scrolling text version, i.e. black background version, compared to the white background version. This was because of the low contrast of the characters against the white background, which resulted from the characteristics of the two-way mirror used in our experiment. Generally, a two-way mirror does not transmit light completely, which made the text hard to read for some people. However, this situation would be improved if a high quality two-way mirror such as the one used in Philips Mirror TV (which "transfers close to 100 percent of the light through the reflective surface") was adopted.

Another participant preferred the scrolling text version because she wanted to see her face during the entire time she brushed her teeth. The scrolling text version presents information on the bottom of the display using white characters on a black background to maximize the reflective area. It seems the highest priority for this participant was to maintain mirror's original function. This emphasizes the need for a preference management facility that respects the user's relationship with the mirror.

Two-Phase Information Presentation (C4)

Ten participants (55.6%) considered the two-phase information presentation unnecessary. The major reasons were that either (1) they thought the information could sufficiently be represented by images, or (2), they felt that the presented images were of limited value. Four participants thought that only the first scene was needed. However, we believe it would be very difficult to render detailed information, such as the time, location and estimated recovery time for a train accident, with images alone. Accordingly, even the participants who considered the two-phase presentation to be useless might eventually be convinced of its necessity.

The people who participated to the in-situ experiment turned to the textual presentation mode when they actually wanted to know further details. Two of them changed their behavior twice based on information contained in the text displayed (once because of low temperatures and once because of an accident on public transportation). Such decision making support during an ordinary task is what AwareMirror was designed to achieve. However, the small number of supported events (twice) does not mean that the system did not support the users' decision-making process. We believe the small number came from the fact that we utilized real information sources rather than simulated ones. Additionally, the other two participants (the older parents) seldom rode trains or had appointments.

Other Implications

One participant looked forward to seeing the novel changes in the images. She was pleased with the fact that could find something different, rather than with the information itself. This aspect, pleasurability, has been ignored in favor of traditional "efficiency and productivity-focused" computing paradigms for work environments. However, pleasurability is gradually gathering attention as an important aspect for objects designed for daily use (Jordan 2002; Norman, 2003; Pousman et al., 2008). We believe that this additional value makes augmented objects more acceptable and may allow them to be more commercially successful. Therefore, other elements of acceptable "change" that increase a user's engagement with the system need to be investigated. The design space for such elements includes (1) level of expectation in the change itself, i.e. regular vs. irregular change, (2) level of anticipation in the content of changes, (3) level of ambiguity in the representation of changes, and (4) the relationship between the system and a user. A study on a "relational agent" (Bickmore & Picard, 2005) could contribute to fourth element investigations. Bickmore and Picard designed a relational agent as a computational entity that aims at establishing and maintaining long-term social-emotional relationships with users. A relational agent could be equipped with strategies that are often utilized in face-to-face communication, such as social dialogue, empathy dialogue, and humor. In our context, we believe a relational agent would actually be able to attract users to the augmented mirrors in anticipation of pleasurable or amusing experiences. This might also to make him or her more forgiving of unreliable behavior and help to maintain/regain the credibility of the system.

CONCLUSION

In this chapter, an augmented mirror, AwareMirror, was presented. AwareMirror is a mirror that not only reflects the appearance of a person who is standing in front of it, it also supports morning decision-making by providing information in context with the user's present or future actions. Three types of information were selected for inclusion based on the analysis of the cognitive tasks users perform when using a mirror, especially in the morning: (1) weather forecasts at an intended destination, (2) state of the user's public transportation system, and (3) information on the user's next appointment. The information reflects an individual schedule, which requires the system's ability to identify the user of the mirror without disturbing his or her primary task and without causing him or her to feel monitored. Based on these requirements, a combination of (1) a user's presence in front of the mirror, (2) coexistence of an unshared object (e.g., toothbrush) with the mirror, and (3) utilization of an unshared object in front of a mirror using noninvasive sensors was adopted. Hence, AwareMirror system embodies context-awareness both in sensing and in presenting contextual information. We designed user interaction with AwareMirror with the aim of balancing passive utilization of the normal mirror function with active involvement with its augmented functions. Two-level information presentation, the core of the balancing technique, was realized by an implicit activation with abstract representation (triggered by the implicit user identification) and explicit transition to detailed textual presentation (triggered by user input).

To confirm our augmentation rationale for a mirror, we fabricated a prototype system and carried out a survey for investigating preferable identification methods, and a comparative study with contrastive functions. The lessons learned as a result of the experiments are as follows:

L1: Minimum information should be presented in a meaningful fashion.

L2: The environmental constraints of the location where the system is installed should be taken into account when determining sensing and presenting modalities.

L3: A mirror's reflective nature can become secondary once a user becomes interested in the presented information, and the system should be easy to use and useful in order for its eye-disruption characteristics to be accepted.

L4: Improper detection of an activating event can make a user feel monitored.

L5: Appealing element offers pleasurable experiences.

We need to evaluate the impact of context-aware objects on our daily lives over long periods of time. Various forms of informative media, such as TV and radio programs, are used on a daily basis and the relationship between these media and their users can be considered "steady." During the in-situ experiment, two of the four participants already received daily weather forecasting information prior to using AwareMirror. Thus, it would be interesting to investigate whether long-term use of peripheral displays would have an effect on their habits of utilizing those existing media. Further studies on peripheral display systems are required to investigate changes in (1) peripheral comprehension, (2) their life styles after utilizing the system for a long period of time, and (3) impression of the users on the systems when appealing/affective elements are properly adopted under imperfect context recognition conditions.

REFERENCES

Addlesee, M., R., C., Hodges, S., Newman, J., Steggles, P., Ward, A., et al. (2001). Implementing a Sentient Computing System. *IEEE Computer, 34* (8), 50-56.

Ambient devices: Information at a Glance. (2004). Retrieved 08 16, 2008, from http://www.ambientdevices.com.

Andrés del Valle, A. C., & Opalach, A. (2005). The Persuasive Mirror: Computerized Persuasion for Healthy Living. In *Proceedings of the Eleventh International Conference on Human-Computer Interaction.*

Baurley, S., Brock, P., Geelhoed, E., & Moore, A. (2007). Communication-Wear: User Feedback as Part of a Co-Design Process. In I. Oakley & S. Brewster (Eds.), *the Second International Workshop on Haptic and Audio Interaction Design* (pp. 55-68). Springer-Verlag Berlin Heidelberg.

Beigl, M., Gellersen, H. W., & Schmidt, A. (2001). MediaCups: Experience with Design and Use of Computer-Augmented Everyday Objects. *Computer Networks. Special Issue on Pervasive Computing, 35*(4), 401–409.

Bickmore, T. W., & Picard, R. W. (2005). Establishing and Maintaining Long-Term Human-Computer Relationships. *ACM Transactions on Computer-Human Interaction, 12*(2), 293–327. doi:10.1145/1067860.1067867

Brumitt, B., Meyers, B., Krumm, J., Kern, A., & Shafer, S. (2000). EasyLiving: Technologies for Intelligent Environments. In P. Thomas & H.-W. Gellersen (Eds.), *the Second International Symposium on Handheld and Ubiquitous Computing* (pp. 12-29). Springer-Verlag Berlin Heidelberg.

De Guzman, E. S. (2004). *Presence Displays: Tangible Peripheral Displays for Promoting Awareness and Connectedness.* Unpublished master thesis, University of California at Berkeley.

Dey, A. (2000). *Providing Architectural Support for Building Context-Aware Applications.* Doctoral dissertation. Georgia Institute of Technology.

Duval, T. S. (1972). *A theory of objective self-awareness.* Academic Press.

Ferscha, A., Emsenhuber, B., & Schmitzberger, H. (2006). Aesthetic Awareness Displays. In T. Pfifer, A. Schmidt, W. Woo, G. Doherty, F. Vernier, K. Delaney, B. Yerazunis, M. Chalmers, & J. Kiniry (Eds.), *Adjunct Proceedings of the 4th International Conference on Pervasive Computing* (pp. 161-165). Osterreichische Computer Gesellschaft.

François, A. R. J., & Kang, E.-Y. E. (2003). A handheld mirror simulation. In *Proceedings of 2003 International Conference on Multimedia and Expo* Vol.1. (pp. 745-748), IEEE Computer Society, Washington DC, USA.

Fujinami, K., & Inagawa, N. (2009). Page-Flipping Detection and Information Presentation for Implicit Interaction with a Book. *International Journal of Ubiquitous Multimedia Engineering, 4*(4), 93–112.

Fujinami, K., & Nakajima, T. (2005). Sentient Artefact: Acquiring User's Context Through Daily Objects. In T. Enokido, L. Yan, B. Xiao, D. Kim, Y. Dai, & L. T. Yang (Eds.), *the Second International Symposium on Ubiquitous Intelligence and Smart Worlds* (pp. 335-344). International Federation for Information Processing.

Fujinami, K., & Nakajima, T. (2006). Bazaar: A Middleware for Physical World Abstraction. *Journal of Mobile Multimedia, 2*(2), 124–145.

Fujinami, K., & Riekki, J. (2008). A Case Study on an Ambient Display as a Persuasive Medium for Exercise Awareness. In H.O.-Kukkonen, P. Hasle, M. Harjumaa, K. Segerståhl, & P. Øhrstrøm (Eds.), *the Third International Conference on Persuasive Technology* (pp. 266-269). Springer-Verlag Berlin Heidelberg.

Gregory, R. (1997). *Mirrors in Mind.* W. H. Freeman and Company.

Harter, A., Hopper, A., Steggles, P., Ward, A., & Webster, P. (1999). The Anatomy of a Context-Aware Application. In *Proceedings of the Fifth Annual ACM/IEEE International Conference on Mobile Computing and Networking* (pp. 59-68). ACM, New York, USA.

Hinckley, K., Pierce, J., Horvitz, E., & Sinclair, M. (2005). Foreground and Background Interaction with Sensor-Enhanced Mobile Devices. *ACM Transactions on Computer-Human Interaction*, *12*(1), 1–22. doi:10.1145/1057237.1057240

Hitachi Human-Interaction Laboratory. (2006). *Miragraphy*. Retrieved 08 17, 2008, from http://hhil.hitachi.co.jp/products/miragraphy.html (in Japanese)

Holmquist, L., & Skog, T. (2003). Informative art: Information visualization in everyday environments. In *Proceedings of the First International conference on Computer graphics and interactive techniques in Australasia and South East Asia* (pp. 229-235). ACM.

Iwabuchi, E., Nakagawa, M., & Siio, I. (2009). Smart Makeup Mirror: Computer-Augmented Mirror to Aid Makeup Application. In J. A. Jacko (Ed.), *the Thirteenth International Conference on Human-Computer Interaction. Part IV: Interacting in Various Application Domains.* (pp. 495-503). Springer-Verlag Berlin Heidelberg. Jordan, P. W. (2002). *Designing Pleasurable Products: An Introduction to the New Human Factors.* Taylor & Francis.

Kawar, F., Fujinami, K., & Nakajima, T. (2005). Augmenting Everyday Life with Sentient Artefacts. In *Proceedings of the 2005 Joint Conference on Smart Objects and Ambient Intelligence: Innovative Context-Aware Services: Usages and Technologies* (pp. 141-146). ACM, New York, USA.

Kawsar, F., Nakajima, T., & Fujinami, K. (2008). Deploy Spontaneously: Supporting End-Users in Building and Enhancing a Smart Home. In *Proceedings of the Tenth International Conference on Ubiquitous Computing* (pp. 282-291). ACM, New York, USA.

Kelley, J. F. (1984). An iterative design methodology for user-friendly natural language office information applications. *ACM Transactions on Office Information Systems*, *2*(1), 26–41. doi:10.1145/357417.357420

Kohtake, N., Ohsawa, R., Yonezawa, T., Matsukura, Y., Masayuki, I., Thakashio, K., et al. (2005). u-Texture: Self-Organizable Universal Panels for Creating Smart Surroundings. In M. Beigl, S. Intille, J. Rekimoto, & H. Tokuda (Eds.), *the Seventh International Conference on Ubiquitous Computing* (pp. 19-36). Springer-Verlag Berlin Heidelberg.

Lashina, T. I. (2004). Intelligent Bathroom. In *European Symposium on Ambient Intelligence, Workshop "Ambient Intelligence Technologies for Wellbeing at Home".* Eindhoven University of Technology, The Netherlands.

Mankoff, J., Dey, A. K., Hsieh, G., Kientz, J., Lederer, S., & Ames, M. (2003). Heuristic evaluation of ambient displays. In *Proceedings of the SIGCHI Conference on Human Factors in Computing Systems (pp.* 169—176). ACM, New York, USA.

Matthews, T., Dey, A. K., Mankoff, J., Carter, S., & Rattenbury, T. (2004). A Toolkit for Managing User Attention in Peripheral Displays. In *Proceedings of the Seventeenth annual ACM symposium on User Interface Software and Technology* (pp. 247-256). ACM, New York, USA.

McCarthy, J., Costa, T., Huang, E., & Tullio, J. (2001). Defragmenting the organization: Disseminating community knowledge through peripheral displays. *Workshop on Community Knowledge at the 7th European Conference on Computer Supported Cooperative Work* (pp. 117-126).

Mynatt, E., Rowan, J., Jacobs, A., & Craighill, S. (2001). Digital Family Portraits: Supporting Peace of Mind for Extended Family Members. In *Proceedings of the SIGCHI conference on Human factors in computing systems* (pp. 333-340). ACM, New York, USA.

Narayanaswami, C., & Raghunath, M. T. (2000). Application design for a smart watch with a high resolution display. In *Proceedings of the Fourth IEEE International Symposium on Wearable Computers* (pp. 7 - 14), IEEE Computer Society, Washington DC, USA.

Norman, D. A. (2003). *Emotional Design: Why We Love (or Hate) Everyday Things*. Basic Books.

Ohno, T., & Mukawa, N. (2004). A free-head, simple calibration, gaze tracking system that enables gaze-based interaction. In *Proceedings of the 2004 Symposium on Eye Tracking Research & Applications* (pp. 115-122). ACM, New York, USA.

Paradiso, J., Hsiao, K. Y., & Benbasat, A. (2000). Interfacing to the foot: Apparatus and applications. In *Extended Abstracts on Human Factors in Computing Systems* (pp. 175–176). New York, USA: ACM. doi:10.1145/633292.633389

Philips Electronics, N. V. (2007). Mirror TV Multi-media Display with latest high definition LCD display. Retrieved November 19, 2009, from http://www.p4c.philips.com/files/3/30hm9202_12/30hm9202_12_pss_eng.pdf

Pingali, G., Pinhanez, C., Levas, A., Kjeldsen, R., Podlaseck, M., Chan, H., & Sukaviriya, N. (2003). Steerable Interfaces for Pervasive Computing Spaces. In *Proceedings of the First IEEE International Conference on Pervasive Computing and Communications*, (pp. 315-322). IEEE Computer Society, Washington DC, USA.

Pinhanez, C., & Podlaseck, M. (2005). To frame or not to frame: The role and design of frameless display in ubiquitous applications. In M. Beigl, S. Intille, J. Rekimoto, & H. Tokuda (Eds.), *the Seventh International Conference on Ubiquitous Computing* (pp. 340-357). Springer-Verlag Berlin Heidelberg.

Pousman, Z., Romero, M., Smith, A., & Mateas, M. (2008). Living with Tableau Machine: a longitudinal investigation of a curious domestic intelligence. In *Proceedings of the Tenth International Conference on Ubiquitos Computing* (pp. 370-379). ACM, New York, USA.

Rodenstein, R. (1999). Employing the Periphery: The Window as Interface. In *Extended abstracts on SIGCHI Conference on Human Factors in Computing Systems (*pp. 204-205). ACM, New York, USA.

Schmidt, A. (2000). Implicit human computer interaction though context. *Personal Technologies*, *4*(2-3), 191–199. doi:10.1007/BF01324126

Stasko, J., Miller, T., Pousman, Z., Plaue, C., & Ullah, O. (2004). Personalized Peripheral Information Awareness through Informative Art. In N. Davis, E. Mynatt, & I. Siio (Eds.), *the Sixth International Conference on Ubiquitous Computing* (pp. 18-35). Springer-Verlag Berlin Heidelberg.

Streitz, N. A., Rocker, C., Prante, T., Alphen, D. V., Stenzel, R., & Magerkurth, C. (2005). Designing Smart Artifacts for Smart Environments. *IEEE Computer*, *38*(3), 41–49.

Want, R., Hopper, A., Falcao, V., & Gibbons, J. (1992). The Active Badge Location System. *ACM Transactions on Information Systems*, *10*(1), 91–102. doi:10.1145/128756.128759

Wisneski, C., Ishii, H., Dahley, A., & Gorbet, M. B., S., U., et al. (1998). Ambient Displays: Turning Architectural Space into an Interface between People and Digital Information. In N.A.Streitz, S.Konomi, & H.-J. Burkhardt (Eds.), *the First International Workshop on Cooperative Buildings* (pp. 22-32). Springer-Verlag Berlin Heidelberg.

KEY TERMS AND DEFINITIONS

Context Awareness: A system property that defines the capability of a system to handle context information of a user, a device, and/or a surrounding environment. Examples of context are the location and current activity of a person; remaining battery time of a mobile phone; the ambient light level of a room, etc. A context-aware system should support a user's decision making by extracting relevant and necessary information from complex and huge amount of data, should perform automated tasks on behalf of a user, and should make a user feel inconvenient and uncomfortable without it.

Implicit Interaction: A style of human-computer interaction that does not require explicit commands from a user, i.e., a system can take appropriate actions to support a user's primary task or related tasks autonomously. The input to such a system is usually the activity context of a user, and thus it is very natural to a user. In the implicit interaction paradigm, it is essential to consider the unobtrusiveness of the context acquisition as well as service provision mechanisms.

Augmented Daily Objects: A daily object that is augmented with a computational capability to obtain user context and/or present contextual information to a user. People normally use objects to perform specific tasks. For example, they use

doors to enter or exit rooms. Sensing the usage state(s) of an object by equipping it with one or multiple dedicated sensors allows a system to infer a user's current or upcoming action. An augmented daily object is considered as the key element in an implicit interaction.

Peripheral/Ambient Display: A display situated at the periphery of human attention with abstract and aesthetic expressions. The display only attracts attention of a person when it is appropriate and necessary, which allows him or her to focus on the primary task at hand. The primary challenge is to design the display in a way that conveys important and just-in-time information without distracting and annoying the user.

Two-Way Mirror: A translucent panel that reflects light from the brighter side when the light intensity difference between the two sides of the panel is significant. In other words, bright colors projected onto the panel from behind can be seen from the front, while a reflection of an object placed in front of the panel is seen in the areas that are dark behind the panel.

Wizard of Oz Experiment: A simulation-based user experiment. A subject interacts with a system that he or she believes to be autonomous, while an unseen experimenter actually operates or partially operates the system. The functionality that an experimenter performs might be implemented later, yet the complete implementation is considered irrelevant to the goal of the study.

Personalization: In Human Computer Interaction, personalization is used to denote the system feature that enables a user to customize the behavior of a system in his/her own personalized way. It involves using technology to accommodate difference between individuals and to represent their uniqueness.

Chapter 5
Image–Word Mapping

Yang Cai
Carnegie Mellon University, USA

David Kaufer
Carnegie Mellon University, USA

ABSTRACT

No Ambient Intelligence can survive without human-computer interactions. Over ninety percent of information in our communication is verbal and visual. The mapping between one-dimensional words and two-dimensional images is a challenge for visual information classification and reconstruction. In this Chapter, we present a model for the image-word two-way mapping process. The model applies specifically to facial identification and facial reconstruction. It accommodates through semantic differential descriptions, analogical and graph-based visual abstraction that allows humans and computers to categorize objects and to provide verbal annotations to the shapes that comprise faces. An image-word mapping interface is designed for efficient facial recognition in massive visual datasets. We demonstrate how a two-way mapping of words and facial shapes is feasible in facial information retrieval and reconstruction.

INTRODUCTION

Although the original goal of Ambient Intelligence is to make computers fade from our persistent awareness, we still need to communicate with computers *explicitly*, for example, to search videos, enter a location, or simply communicate with others. Over ninety percent of information transfer in our communication is verbal and visual.

For many years, cognitive scientists have investigated visual abstraction using psychological ex-

periments. For example, visual search using foveal vision (Wolfe, 1998; Theeuwes, 1992; Treisman and Gelade, 1980; Verghese, 2001; Yarbus, 1967; Larkin and Simon, 1987; Duchowski, *et al.*, 2004; Kortum and Geisler, 1996; Geisler and Perry, 1998; Majaranta and Raiha, 2002) and mental rotation (Wikipedia, "Mental Rotation"). Visual abstraction models have been developed, for example, Marr's cylinder model of human structure (Marr, 1982) and the spring-mass graph model of facial structures (Ballard, 1982). Recently, scientists began to model the relationship between words

DOI: 10.4018/978-1-61692-857-5.ch005

and images. CaMeRa (Tabachneck-Schijf *et al.*, 1997), for example, is a computational model of multiple representations, including imagery, numbers and words. However, the mapping between the words and images in this system is linear and singular, lacking flexibility. An Artificial Neural Network model has been proposed to understand oil paintings (Solso, 1993), where Solso remarks that the hidden layers of the neural network enable us to map the words and visual features more effectively. With this method, Solso has argued, we need fewer neurons to represent more images. However, the content of the hidden layers of the neural network remains a mystery (See Figure 1).

Because of the two- or three-dimensional structure of images and the one-dimensional structure of language, the mapping between words and images is a challenging and still undertheorized task. Arnheim observed that, through abstraction, language categorizes objects. Yet language, through its richness, further permits humans to create categorizations and associations that extend beyond shape alone (Arnheim, 1969). As a rich abstractive layer, language permits categorizations of textures, two- and three-dimensions, and sub-shapes. As an abstractive layer, language seems to be the only method we have to satisfactorily describe a human subject. To explore this insight further, Roy developed a computerized system known as Describer that learns to generate contextualized spoken descriptions of objects in visual scenes (Roy, 1999). Describer illustrates how a description database could be useful when paired with images in constructing a composite image.

DESCRIPTIONS FOR HUMANS

Our framework has focused on the mapping between words and images for human facial features. Why do we focus on human faces? Humans in general and human faces in particular provide among the richest vocabularies of visual imagery in any modern language. Imaginative literature is a well-known source of such descriptions, where human features are often described in great detail. In addition, reference collections in the English language focused on visual imagery, such as description and pictorial dictionaries, never fail to have major sections devoted to descriptions of the human face. These sections are typically devoted to anatomical rather than social and cultural descriptions of faces based on cultural stereotypes and analogies. The mappings between images and faces we have been exploring are built upon such stereotypical and analogical associations.

In the following sections, we briefly overview a variety of semantic visual description methods, including multiple resolution, semantic differentiation, symbol-number, and analogy. Then, we introduce the computational implementation of these human descriptions in form of the visual, verbal and the interaction between them.

MULTIPLE RESOLUTION DESCRIPTIONS

Human descriptions are classifiers for shape, color, texture, proportion, size and dynamics in *multiple resolutions*. For example, one may start to describe a person's torso shape, then her hairstyle, face, eyes, nose, and mouth. Human feature descriptions have a common hierarchic

Figure 1. The two-way mapping neural network model

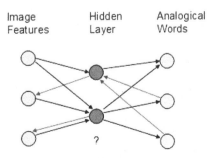

structure (Cai, 2003). For example, figure, head, face, eye, et al. Like a painter, verbal descriptions can be built in multiple resolutions. The words may start with a coarse global description and then 'zoom' into sub components and details. See Figure 2 for a breakdown of a global description of a human head.

In our research to date, we have collected from imaginative literature over 100 entries of multi-resolution descriptions, from global descriptions to components and details. Due to limitations of space, we will only list a few samples, where the underlined sections represent the global levels of description, the **bolded** show the component-based descriptions, and the *italicized* sections are the details:

- "For a **lean face**, pitted and scarred, very **thick black eyebrows** and **carbon-black eyes** with *deep grainy circles of black* under them. A *heavy five o'clock shadow*. But the skin under all was pale and unhealthy-looking. (Doctorow, 1980)"
- "Otto has a **face** like very ripe peach. **His hair** is fair and thick, growing low on his **forehead**. He has small sparkling **eyes**, full of naughtiness, and a wide, disarming **grin** which is too innocent to be true. When

he grins, *two large dimples* appear in his peach **blossom cheeks**.(Isherwood, 1952)"

- "Webb is the oldest man of their regular foursome, fifty and then some- a lean thoughtful gentleman in roofing and siding contracting and supply with a calming gravel voice, his **long face** broken into *longitudinal strips by creases* and **his hazel eyes** almost lost under an amber **tangle of eyebrows**. (Updike, 1996)"

SEMANTIC DIFFERENTIAL REPRESENTATION

The Semantic Differential method measures perceptual and cognitive states in numbers or words aligned on a scale. For example, the feeling of pain can be expressed with adjectives, ranging from weakest to strongest. Figure 3 shows a chart of visual, numerical and verbal expressions of pain in hospitals: No Hurt (0), Hurts Little Bit (2), Hurts Little More (4), Hurts Even More (6), Hurts Whole Lot (8) and Hurts Worst (10). This pictorial representations are very useful in patient communication where descriptions of pain type (e.g., pounding, burning) and intensity (e.g., little, a lot) lack a robust differentiated vocabulary.

Figure 2. Multi-resolution representation of a face

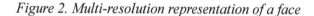
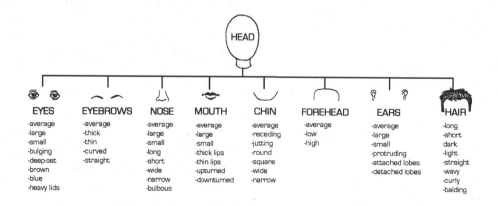

The physical feeling can be quantified with mathematical models. When the change of stimulus (*I*) is very small, we won't detect the change. The minimal difference (*ΔI*) that is just noticeable is called perceptual threshold and it depends on the initial stimulus strength I. At a broad range, the normalized perceptual threshold as a constant, $\Delta I/I = K$. This is Weber's Law (Li *et al.*, 2003).

Given the perceptual strength *E*, as the stimulus *I* changes by *ΔI*, the change of *E* is *ΔE*. We have the relationship $\Delta E = K * \Delta I/I$. Let *ΔI* be *dI* and *ΔE* be *dE*, thus we have the Weber-Fechner's Law:

$$E = K * \ln(I) + C$$

where, *C* is constant and *K* is the Weber Ratio, *I* is stimulus strength and *E* is the perceptual strength. Weber-Fechner's Law states that the relationship between our perceptual strength and stimulus strength is a logarithmic function.

Weber-Fechner's Law enables us to describe responses of signal strength in a focused point perceptual basis, such as feeling of pain, color, pixel intensity, and so on. However, it becomes significantly more complicated when we describe facial features. First of all, facial features contain one-dimensional (e.g. nose length), two-dimensional (e.g. mouth), or three dimensional measurements (e.g. chin). Second, facial features are organized in geometric structures. We not only perceive individual components but also the relationship among features. For example, a long

nose may be associated with long face. Therefore, all measurements are relative. Third, we have different sensitivity thresholds toward different facial features. For example, our experiment results show that people are more sensitive to the changes in mouth size than others features. Figure 4 and Figure 5 are examples of our lab experiments. We use a computer to generate a synthetic 3-D face and we change the facial features.

We found that the resolution for our semantic descriptions of a shape is very limited. *Every time, our brain is searching for the differences with respect to the norm.* We call the phenomena 'spatial clustering'.

We have conducted an experiment to illustrate the effect of the similarity of shapes on human perception. Assume we have seven circular objects with different sizes. The diameters are 100, 200, 300, 400, 500, 600, and 700. They are illustrated in Figure 6.

A computer program is developed to show one of the seven shapes at a random order. Before experiments, subjects are shown all seven shapes. Each type is reviewed three times so that subjects get familiar with those shapes. When subject determines the shape size, the subject pushes the stop button and computer records the reaction time. During the experiment, subjects are shown 14 stimuli. The results show that typical mistaken objects are shape D and E, 7% to 14% of total trials. The response time for recognizing object A or G is significantly less than that for others. As we know, A is the smallest object and

Figure 3. Expressions of pain in pictures, numbers and words

0	2	4	6	8	10
No Hurt	Hurts Little Bit	Hurts Little More	Hurts Even More	Hurts a Whole Lot	Hurts Worse

Figure 4. Variable nose lengths on a synthetic three-dimensional face

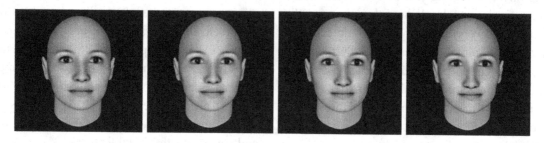

Figure 5. Variable mouth sizes on a synthetic three-dimensional face

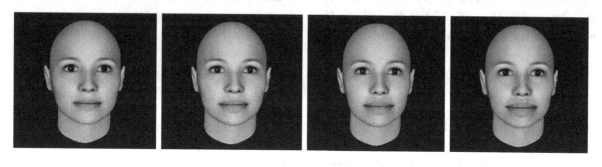

Figure 6. Seven sizes of a shape for perceptual experiment

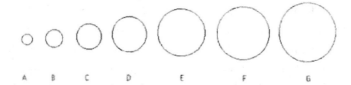

G is the largest object, which is most significant feature. If we arrange the geometric features (e.g. diameter of a circle) in a linear fashion, we would find that the reaction time for the middle size objects is less but more errors occur in that area. We found that there is a 'clustering gravity' at the middle position. Figure 7 shows the general trend of the phenomena.

SYMBOL-NUMBER DESCRIPTIONS

In many cases, numbers can be added to provide even greater granularity. For example, the FBI's Facial Identification Handbook (1988) comes with a class name such as bulging eyes and then a number to give specific levels and types. The FBI has already created a manual for witnesses, victims, or other suspect observers to use in identifying possible suspect features. The catalog presents several images per page under a category such as "bulging eyes"; each image in such a category has bulging eyes as a feature, and the respondent is asked to identify which image has bulging eyes most closely resembling the suspect. See Figure 8 for an example. This book is an extremely efficient and effective tool for both forensic sketch artists and police detectives. It is most commonly used

Figure 7. Spatial clustering curve

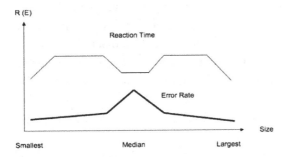

Figure 8. Bulging Eyes from FBI Facial Identification Catalog

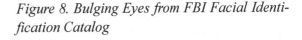

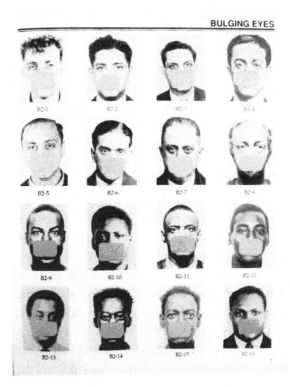

as a tool in helping a witness or victim convey the features of the suspect to the sketch artist in order to render an accurate composite sketch.

ANALOGICAL DESCRIPTIONS

From the multi-resolution point of view, an analogy describes in a coarse way in contrast to symbolic-number descriptions. Instead of describing features directly, people often find it more intuitive to refer a feature to a stereotype, for example, a movie star's face. The analogical mapping includes structural mapping (e.g. face to face), or component mapping (e.g. Lincoln's ear and Washington's nose). Children often use familiar things to describe a person, for example using 'cookie' to describe a round face.

Analogies are culture-based. In the Western world, several nose stereotypes are named according to historical figures. Many analogies are from animal noses or plants. Figure 9 illustrates examples of the nose profiles as described above.

We use a simple line drawing to render the visual presentation.

Analogies are a trigger of experience, which involves not only images, but also dynamics. The far right one in Figure 9 shows a 'volcano nose', which triggers a reader's physical experience such as pain, eruption, and explosion. In this case, readers not only experience it but also predict the consequence. Therefore, it is an analogy of a novel physical process that remains under the visible surface.

Figure 9. Analogical description of noses

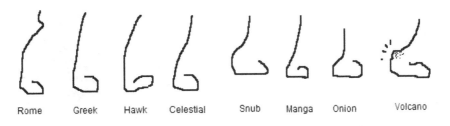

Given a verbal description of the nose, how do we visually reconstruct the nose profile with minimal elements? In this study, we use a set of 5 to 9 'control points' to draw a profile. By adjusting the relative positions of the control points, we can reconstruct many stereotypes of the profiles and many others in between. To smooth the profile contour, we apply the Spline (WIKIPEDIA, "Splines") curve fitting model. See Figure 10.

SHAPE RECOGNITION IN WORDS

On the other hand, mapping shapes to words is a pattern recognition process. In this study, we use the point distribution model (PDM) to describe a new shape from familiar shapes. Point distribution model is a visual abstraction of shape that converts pixels into points on the contour of a shape (Cootes *et al.*, 1992; Sonka, 1998).

We assume the existence of a set of M examples from which to derive a statistical description of the shape and its variation. Some number N of control points $(x_1, y_1, x_2, y_2, ..., x_N, y_N)$ is selected on each boundary; these points are chosen to correspond to feature of the underlying object, for example, the corner points of lips, eyes and face.

Figure 10. Reconstructing a nose profile with points (black) and Spline curve (red)

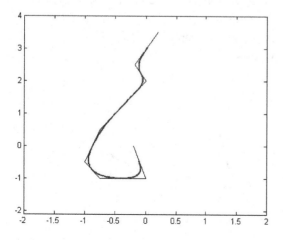

For simple shapes such as face profiles, we use heuristic rules to classify the shapes. For example, in Figure 11, the colored squares represent points that the user can move. The black squares with colored outlines (*a* and *g*) are points that the user cannot move. They are anchored to the appropriate place on the picture of the face. When we use a term such as "line ab", I mean the line from point a to point b.

We use heuristics to classify the noses, for example:

- **Very Straight Nose:** If the slope of line *a-b* is the same as the slope of line *b-c* is the same as line *c-d*, it is a straight nose
- **Convex Nose:** If the slope of line *b-c* is less than the slope of line *c-d* and the slope of line *c-d* is less than the slope of line de, it is a convex nose
- **Concave Nose:** If the slope of line *b-c* is greater than the slope of line *c-d* and the slope of line *c-d* is greater than the slope of line *d-e*, it is a convex nose
- **Snub Nose:** If point *f* is above and to the left of point *g*, and the angle between line *f-g* and the horizontal is greater than 15 degrees, it is a snub nose
- **Hawk Nose:** If point *f* is below and left of point *g*, it is a hawk nose.

We also check the constraints for the element. For example, if a nose does not meet these criteria, it is certainly unrealistic.

Figure 11. Nose classification example

- **Orientation:** points *b-f* must be to the left of points *a* and *g*; points *b-e* must be to the left of the point which comes before them alphabetically. Point *f* may be to the left of right of point e.
- **Ordering:** points *b-f* must be below the point which comes before them alphabetically. Note that it is valid for point g to be above point f; this represents a hawk nose.
- **Nose Tip:** The angle between line fg and the horizontal must be less than or equal to 30 degrees, whether it has a positive or negative slope.

For more complex shapes, we use Point Distribution Model (PDM) to compress control points, we compute the principal components from a training set of shapes. Assume we take the first *k* eingen-values (Sonka, 1998) account for virtually all the variation detected. Thus we can represent a shape as a vector *b*,

$$x = \bar{x} + P\text{b}$$

where x is the 2*N dimensional vector defining the spline, $\bar{x}$ is the mean shape, P is an 2*k*N matrix, and b is a k-dimensional vector parameterizing the shape x. We then can use a Bayesian classifier (Sonka, 1998) to recognition the shapes in words.

INTERACTIVE FACIAL RECONSTRUCTION

We developed a prototype of the interactive system for facial reconstruction on a computer. In the system, a user selects the feature keywords in a hierarchical structure. The computer responds to the selected keyword with a pool of candidate features that are coded with labels and numbers. Once a candidate is selected, the computer will

Figure 12. Interactive facial profile reconstruction based on line-drawing. The code is written in Java so that it is possible to run on the Internet. These descriptions can then be rendered and distributed across the network. http://www.cmu.edu/vis/project9.html

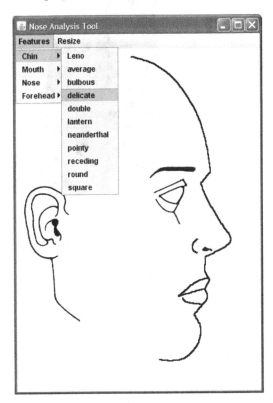

superimpose the components together and reconstruct the face. See Figure 12 and Figure 13.

As we know, a composite sketch of a suspect has often been done by professionals. Our system enables inexperienced users to reconstruct a face using only a simple menu driven interaction. In addition, this reconstruction process is reversible. So it can also be used for facial description studies, robotic vision and professional training.

CONCLUSION

In this study, we assume a hidden layer between the human perception of facial features and refer-

Figure 13. Interactive front facial reconstruction based on image components

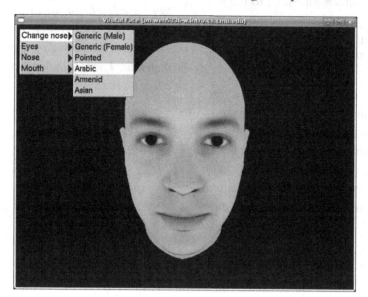

ral expressive words that contain 'control points' that can be articulated mathematically, visually or verbally. This is still a fundamental study of semantic network. Nevertheless, we see its potential for meaningful two-way mapping. At this moment, we only have profile and frontal facial reconstruction models. In the future, we will develop both whole head and body models with far more control points and referral expressions.

Today, we have an overabundance of data but not nearly enough people or bandwidth. Image and video collections grow at an explosive rate that exceeds the capacity of network and human attention. In real-time surveillance systems, over a terabyte per hour are transmitted for only a small number of platforms and sensors. We believe that the visual search network is one of the feasible solutions that can be more thoroughly developed.

ACKNOWLEDGMENT

This research was supported by CyLab at Carnegie Mellon under grants DAAD19-02-1-0389 from the Army Research Office. We are also in debt to Brian Zeleznik, Emily Hart, and Hannah Post for their comments and editing.

REFERENCES

Ballard, D. H., & Brown, C. M. (1982). *Computer Vision*. Prentice-Hall Inc New Jersey.

Cai, Y. How Many Pixels Do We Need to See Things? Lecture Notes in Computer Science (LNCS), ICCS Proceedings, 2003. Arnheim, R. Visual Thinking. University of California Press, 1969 Allport, A. Visual Attention. MIT Press, 1993.

Cootes, T. F., Taylor, C. J., & Graham, J. Training models of shape from sets of examples. In D.C. Hogg and R.D. Boyle, editors, Processings of the Bristish Machine Vision Conference, Leeds, UK, pages 9-18, Springer Verlag, London, 1992

Doctorow, E. L. (1980). *Loon Lake*. New York: Random House.

Duchowski, A. T. (2004). Gaze-Contingent Displays: A Review. *Cyberpsychology & Behavior*, *7*(6). doi:10.1089/cpb.2004.7.621

el al Sonka, M. (1998). *Image Processing, Analysis, and Machine Vision* (2nd ed.). PWS Publishing.

Facial Identification CatalogFBI, Nov. 1988

Geisler, W. S., & Perry, J. S. (1998). Real-time foveated multiresolution system for low-bandwidth video communication. In *Proceedings of Human Vision and Electronic Imaging*. Bellingham, WA: SPIE.

Geisler, W. S., & Perry, J. S. (1998). Real-time foveated multiresolution system for low-bandwidth video communication. In *Proceedings of Human Vision and Electronic Imaging*. Bellingham, WA: SPIE.

http://en.wikipedia.org/ wiki/ Mental_rotation

http://en.wikipedia.org/ wiki/ Visual_search

Isherwood, C. (1952). *Goodbye to Berlin*. Signet.

Kortum, P. and Wilson Geisler, "Implementation of a foveated image coding system for image bandwidth reduction", in SPIE Proceedings, 2657, Page 350-360, 1996.

Larkin, J. H., & Simon, H. A. (1987). Why a diagram is (sometimes) worth 10,000 words. *Cognitive Science*, *11*, 65–100. doi:10.1111/j.1551-6708.1987. tb00863.x

Larkin, J. H., & Simon, H. A. (1987). Why a diagram is (sometimes) worth 10,000 words. *Cognitive Science*, *11*, 65–100. doi:10.1111/j.1551-6708.1987. tb00863.x

Li, Q. (2003). *Rosa, Michael De, Rus, Daniela: Distributed Algorithms for Guiding Navigation across a Sensor Network*. Dartmouth Department of Computer Science.

Majaranta, P., & Raiha, K. J. (2002). Twenty years of eye typing: systems and design issues. In: Eye Tracking Research and Applications (ETRA) Symposium. New Orleans: ACM.

Marr, D. (1982). *Vision*. W.H. Freeman.

Roy, D. (1999). Learning from Sights and Sounds: A Computational Model. Ph.D. In *Media Arts and Sciences*. MIT.

Shell, J. S., Selker, T., & Vertegaal, R. (2003). Interacting with groups of computers. *Communications of the ACM, 46*, 40–46. doi:10.1145/636772.636796

Solso, R. L. (1993). *Cognition and the Visual Arts*. The MIT Press.

Tabachneck-Schijf, H. J. M., Leonardo, A. M., & Simon, H. A. (1997). CaMeRa: A computational model of multiple representations. *Cognitive Science*, *21*, 305–350. doi:10.1207/ s15516709cog2103_3

Theeuwes, J. (1992). Perceptual selectivity for color and form. [Fulltext]. *Perception & Psychophysics*, *51*, 599–606. doi:10.3758/BF03211656

Treisman, A., & Gelade, G. (1980). A feature integration theory of attention. *Cognitive Psychology*, *12*, 97–136. doi:10.1016/0010-0285(80)90005-5

Updike, J. (1996). *The Rabbit is Rich*. Ballantine Books.

Verghese, P. (2001). Visual search and attention: A signal detection theory approach. *Neuron, 31*, 523-535(13).

Web site. 2007: http://en.wikipedia.org/ wiki/ Spline_(mathematics)

Wolfe, J. M. (1998). Visual Search. In Pashler, H. (Ed.), *Attention*. East Sussex, UK: Psychology Press. Fulltext.

Yarbus, A. L. (1967). *Yarbus. Eye Movements during Perception of Complex Objects*. New York, New York: Plenum Press.

Yarbus, (1967) A. L. Yarbus. Eye Movements during Perception of Complex Objects. Plenum Press, New York, New York, 1967

Chapter 6
A Survey on the Use of Emotions, Mood, and Personality in Ambient Intelligence and Smart Environments

Carlos Ramos
Polytechnic Institute of Porto, Portugal

Goreti Marreiros
Polytechnic Institute of Porto, Portugal

Ricardo Santos
Polytechnic Institute of Porto, Portugal

ABSTRACT

This Chapter is a survey dealing with the use of emotions and mood to characterize an Ambient Intelligence system. In particular, a key aspect of the described research is the assumption that each level of an Ambient Intelligence infrastructure (e.g., sensing, reasoning, action) can benefit from the introduction of emotion and mood modelling. The Chapter surveys well-known models (e.g., OCC, Big Five – Five Factor Model, PAD, just to name a few) discussing for each one Pros and Cons. Next, architectures for emotional agents are discussed, e.g., Cathexis (assuming the somatic marker hypothesis proposed by Damasio), Flame, Tabasco and many others. Finally, specific implementation examples of emotional agents in Ambient Intelligence scenarios are described.

DOI: 10.4018/978-1-61692-857-5.ch006

INTRODUCTION

Traditionally, emotions and affects have been separated from cognitive and rational thinking; they have a bad connotation on what is related to the individuals' behaviour, in particular on what is related to the decision-making process. However, in the last years, researchers from several distinct areas (psychology, neuroscience, philosophy, etc) have begun to explore the role of the emotion as a positive influence on human decision making process. Current research in Artificial Intelligence demonstrates also a growing interest in emotional agents. From human-computer interaction, to development of believable agents to the entertainment industry, and to modelling and simulating the human behaviour, there is a wide variety of application areas of emotional agents.

Rosalind Picard enumerates four major reasons to give emotional characteristics to machines (Picard, 1995):

- Emotions may be useful in the creation of robots and believable characters with the capacity to emulate humans and animals. The use of emotion gives agents more credibility;
- The capacity to express and understand emotions could be very useful to a better association between humans and machines, making this relationship less frustrating;
- The possibility to build intelligent machines, although this concept is a little bit vague;
- The possibility to understand human emotions by modelling them.

To achieve the goals identified by Rosalind Picard researchers have been working on the topic, covering several different areas like for example, sociology, psychology, human-machine interaction, computer vision and sensing, virtual environments, and, obviously, Artificial Intelligence. Furthermore, in this area we have a large set of different applications like for example: virtual reality, decision support, computer games, and ambient intelligence.

In this chapter we will focus on the role of emotion, mood, and personality in the development of Ambient Intelligence systems and Smart Environments.

The first part of the chapter will describe methods to deal with emotion, mood, and personality in computer-based systems. Examples are OCC model (Ortony, 2003), Big Five model (McRae and Costa, 1987) and PAD model (Mehrabian, 1996).

Then emotion, mood, and personality can be considered in all the roles covered by Ambient Intelligence systems, and Smart Environments, as proposed in (Ramos et al., 2008), namely in the following roles:

- helping in the interpretation of the environment situation;
- representing the information and knowledge associated with the environment;
- modelling, simulation, and representation of the entities involved in the environment;
- planning about the decisions or actions to take;
- learning about the environment and associated aspects;
- interaction with the human being;
- action on the environment.

Some Ambient Intelligent systems and Smart Environments dealing with emotion, mood, and personality will be presented at the end of the chapter.

EMOTIONAL, MOOD AND PERSONALITY MODELLING

In this section we will describe the main models and methods to deal with emotion, mood, and personality that are adopted by Computer Science community.

Ortony, Clore and Colins Model of Emotions

In 1988, Andrew Ortony and his colleagues published the book "The Cognitive Structure of Emotions" which proposes a computationally tractable model of the cognitive basis of the emotion elicitation. This model is known as the OCC model, named by the first letter of each one of the three authors, Andrew Ortony, Gerald Clore and Allan Colins. The OCC model is the most popular model in the implementation of an environment for simulations of agents with emotions. The authors spent a lot of effort regarding issues related to the artificial intelligence community like for instance the development of cooperative problem-solving systems that they believe will benefit with emotions reasoning. The area of virtual characters is where the OCC model is more widely used to synthesize emotions and express them. Many developers of such characters believe that the OCC model will be all they ever need to equip their characters with emotions.

The OCC model describes a hierarchy that classifies 22 emotions. In Figure 1 it is possible to visualize the original OCC model. In OCC theory, all emotions can be divided into terms according to the emotion-eliciting situation. The OCC model (Ortony et al, 1988) proposes that emotions are the results of three types of subjective appraisals:

- The appraisal of the pleasantness of events with respect to the agent's goals;
- The appraisal of the approval of the actions of the agent or another agent with respect to a set of standards for behaviour;
- The appraisal of the liking of objects with respect to the attitudes of the agent.

Generically in the OCC model emotions are seen as valenced reactions to three different types of stimulus (Ortony et al, 1988): objects; consequence of events and action of agents. These are the three major branches of emotion types. The first branch relates to emotions which are arising from aspects of objects such as love and hate. This constitutes the single class in this branch called

Figure 1. The original OCC model adapted from (Ortony et al, 1988)

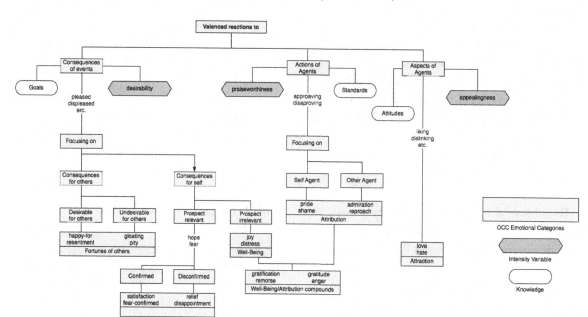

attraction. The second branch relates to emotions which are consequences of events we have the following emotions: happy-for, gloating, pity, resentment, satisfaction, hope, fear, fears-confirmed, relief, disappointment, joy and distress. Four classes constitute this branch, fortunes-of-others, prospect-based, well-being and confirmation. The third branch is related to the actions of agents where we have the emotions: pride, shame, admiration and reproach. This last branch has also only one class namely the attribution. The model considers yet 4 compound emotions, because they are consequence of events and agents actions, which are: gratification, remorse, gratitude and anger.

Ortony later found that 22 distinct emotion types to be too complex for the development of believable agents. A simplified version of this theory was presented later in 2003 by Ortony, where he considered only two different categories of emotional reactions: positive and negative. The emotion types decreased to 12, six positive (joy, hope, relief, pride, gratitude, and love) and six negative (distress, fear, disappointment, remorse, anger, and hate). As in the original model, emotions are the results of three types of subjective appraisals (goal-based, standard-based and taste-based). In Table 1 it is possible to visualize the OCC model reviewed in 2003, after the collapse of some of the original categories.

The OCC model was several times used to model the implementation of emotional agents (Adamatti and Bazan, 2003; Bates, 1994; Elliot, 1992; El-Nasr, 2000). Elliot (Elliot, 1994) describes the construction of the Affective Reasoner that displays/produces behaviours related to a specific emotion generated in that moment. Bates (Bates, 1994) developed credible agents to the OZ project based on the OCC model which consists in emotional agent based micro worlds.

The Five Factor Model of Personality

The Big Five represents a taxonomy of traits that capture the essence of individual personality differences. A trait is a temporally stable, cross-situational individual difference. The Big Five are an empirically based phenomenon, not a theory of personality. These traits were discovered through factor analysis studies. Factor analysis is a technique generally done with the use of computers to determine meaningful relationships and patterns in behavioural data. The original derivations relied heavily on American and Western European samples.

Five-Factor Model denomination is frequently used instead of Big Five. The term "Big Five" was coined by Lew Goldberg and was originally associated with studies of personality traits used in natural language. Allport, Norman and Cattell

Table 1. Five specializations of generalized good and bad feelings (collapsed from (Ortony, 2003))

	Positive Reactions	Negative Reactions
Undifferentiated	…because something good happened (**joy**)	…because something bad happened (**distress**)
Goal-based	…about the possibility of something good happening (**hope**)	…about the possibility of something bad happening (**fear**)
	… because a feared bad thing didn't happen (**relief**)	… because a hoped-for good thing didn't happen (**disappointment**)
Standard-based	… about a self-initiated praiseworthy act (**pride**)	… about a self-initiated blameworthy act (**remorse**)
	… about an other-initiated praiseworthy act (**gratitude**)	…about an other-initiated blameworthy act (**anger**)
Taste-based	… because one finds someone/thing appealing or attractive (**like**)	… because one finds someone/thing unappealing or unattractive (**dislike**)

were influential in formulating this taxonomy which was later refined. Allport compiled a list of 4500 traits. Cattell reduced this list to 35 traits. Others continued to analyze these factors and found congruence with self- ratings, ratings by peers and ratings by psychological staff, that eventually became the Big Five factors. The term "Five-Factor Model" has been more commonly associated with studies of traits using personality questionnaires. The Five-Factor Model and the Big Five in spite of being developed by different researchers and techniques reached to the same results. This conclusion is one of the major strengths of the Big Five/Five-Factor Model.

Although not being universally accepted the five-factor model of personality is the model most widely used to characterize a unique individual (Hampson, 1999). The five-factor model is comprised of five personality dimensions (OCEAN):

- Openness to experience vs not open to experience;
- Conscientiousness vs undirectedness;
- Extraversion vs introversion;
- Agreeableness vs antagonism;
- Neuroticism vs emotional stability.

Openness (O). Sometimes called Intellect or Intellect/Imagination open people are imaginative, intelligent and creative that like to experience new things. "Openness to experience is tendency to be intellectual, interested in the arts, emotionally aware, and liberal" (Ghasem-Aghaee and Oren, 2004). "Openness refers the number of interests to which one is attracted and the depth to which those interests are pursued" (Howard and Howard, 2004).

Conscientiousness (C). Conscientious people have the tendency to show self-discipline and aim for an achievement. They plan very well their actions and take responsibility for them. "Conscientiousness is tendency to set high goals, to accomplish work successfully, and to behave dutifully and morally" (Costa and McCrae, 1992).

Furthermore, "conscience is the awareness of a moral or ethical aspect to one's conduct together with the urge to prefer right over wrong" (Howard and Howard, 2004). Conscientiousness refers to the number of goals on which one is focused (Ghasem-Aghaee and Oren, 2004).

Extraversion (E). Extraverts tend to seek agreement, and appear upbeat and outgoing. They are enthusiastics, action oriented, explore opportunities and energetic in achieving their goals. "Extraversion is a trait associated with sociability and positive affect" (Costa and McCrae, 1992). "It refers to the number of relationships with which one is comfortable" (Ghasem-Aghaee and Oren, 2004). High on extraversion: "Interest in or behavior directed toward others or one's environment rather than oneself" (Howard and Howard, 2004).

Agreeableness (A). Agreeable people have the tendency to be trustworthy and cooperative towards others. They are considered friendly, generous, helpful, and willing to compromise their interests with others. "Agreeableness is tendency to be a nice person" (Costa and McCrae, 1992). "Agreeableness refers to the number of sources from which one takes one's norms for right behaviour" (Ghasem-Aghaee and Oren, 2004).

Neuroticism (N). Neurotic people have the tendency to experience negative emotions, such as anger, anxiety, or depression. Negative Emotionality, neuroticism, or need for stability is the trait associated with emotional instability and negative affect (Costa and McCrae, 1992). "Negative Emotionality refers to the number and strength of stimuli required to elicit negative emotions in a person"(Ghasem-Aghaee and Oren, 2004).

An individual is a combination of these traits, sometimes with an emphasis on one of them. To the purpose of determine the personality of an individual there are available several tools being the most commonly used the Big Five Inventory. The Big Five Inventory (BFI) is a self-report inventory designed to measure the Big Five dimensions. It is quite brief for a multidimensional personality inventory (44 items total), and consists of short phrases with relatively accessible vocabulary.

The Pleasure-Arousal-Dominance Temperament Mood Model

The Pleasure-Arousal-Dominance (PAD) Temperament Model provides a general framework for the study of temperament (emotional state, mood) and personality. The model consists of three nearly independent dimensions: Trait Pleasure-Displeasure (P), Trait Arousal-Nonarousal (A) and Trait Dominance-Submissiveness (D) (Mehrabian, 1995). The pleasure-displeasure trait relates to the emotional state's positivity or negativity. Positively to extraversion, affiliation, nurturance, empathy, and achievement, and negatively to neuroticism, hostility, and depression. The arousal-nonarousal trait shows the level of physical activity and mental alertness. It relates to emotionality, neuroticism, sensitivity, introversion, schizophrenia, heart disease, eating disorders, etc. The last trait is the dominance-submissiveness that indicates the feeling/lack of control. It relates positively to extraversion, assertiveness, competitiveness, affiliation, social skills, and nurturance, and negatively to neuroticism, tension, anxiety, introversion, conformity, and depression.

Experimental findings in this framework have shown, for instance, that, "angry" is a highly unpleasant, highly aroused, and moderately dominant emotional state. "Sleepy" consists of a moderately pleasant, extremely unaroused, and moderately submissive state, whereas "bored" is composed of highly unpleasant, highly unaroused, and moderately submissive components (Mehrabian, 1995).

The PAD temperament traits are independent of each other and form a three-dimensional temperament space. The temperament state of and individual is the intersection between the three dimensions. A description of the various temperament states is facilitated by dichotomizing each of the three temperament-space axes, as follows:

+P and -P for pleasant and unpleasant, +A and -A for arousable and unarousable, and +D and -D for dominant and submissive, temperament, respectively.

Mehrabian defines eight mood types based on the combination of negative and positive values for each trait (Mehrabian, 1996):

- Exuberant (+P+A+D) vs Bored (-P-A-D)
- Dependent (+P+A-D) vs Disdainful (-P-A+D)
- Relaxed (+P-A+D) vs Anxious (-P+A-D)
- Docile (+P-A-D) vs Hostile (-P+A+D)

Mehrabian also defines the relationship between the OCEAN personality traits and the PAD space. This mapping is very helpful for example to define an initial PAD space mood point.

Other Approaches

Psychoticism, Extraversion, and Neuroticism (PEN) Model

The most famous alternative to the five-factor model is the Psychoticism, Extraversion, and Neuroticism (PEN) Model (Eysenck, 1991). The PEN Model was proposed by Eysenk who was a strong oppositor of the five-factor model felt that due to overlaps in the five factor identified a three-factor model was more accurate. The descriptive aspect of the model is a hierarchical taxonomy based on factor analysis. At the top of the hierarchy are the superfactors of Psychoticism, Extraversion, and Neuroticism (PEN). Eysenck found a new trait named Psychoticism that is associated with psychotic episodes and aggressive behaviour.

NEO Personality Inventory-Revised (NEO PI-R)

The Revised NEO Personality Inventory, or NEO PI-R, is a standard questionnaire measure of the Five Factor Model developed by Paul Costa and

Jeff McCrae (Costa and McCrae, 1992). This personality inventory measures 240 items of the FFM personality factors: Extraversion, Agreeableness, Conscientiousness, Neuroticism, and Openness. The test measures also six subordinate dimensions known as facets of each of the FFM personality factors.

Ira Roseman's Model of Emotion

Roseman's model has five appraisal components that can produce 14 discrete emotions (Roseman et al., 1990).

According to this model the emotions are generated by taking based on an association of events. Events are divided into events motive-consistent and with motive-inconsistent events. The first events are defined as being consistent with the objectives of the individual, while the second events, the motive-inconsistent, are events that threaten one of the objectives that and individual proposes to achieve. The events were classified according to cause of the event and can be caused by third parties, by itself or be circumstances. Another category that was used to differentiate emotions was that event have been motivated by the subject you want to claim a reward or avoid

a punishment. An event is definite (certain), only possible (uncertain), or of an unknown probability.

ARCHITECTURES FOR EMOTIONAL AGENTS

Partially due to some of the facts presented above, in recent years there has been an increased interest in developing architectures for emotional agents. Bellow there is a short description of some of the existent architectures.

Cathexis

Velasquez has proposed a model called Cathexis (Velasquez and Maes, 1997) to simulate emotions, moods and temperaments in a multi-agent system. In this architecture only the basic emotions (anger, fear, distress / sadness, enjoyment / happiness, disgust, and surprise) are included. In Cathexis it is presupposed that: for the simulation of emotional mechanisms it is necessary to interpret the neurological structures that support emotions. Cathexis follows the somatic marker hypothesis proposed by Damásio (Damásio, 1994).

In Figure 2 it is showed Cathexis architecture.

Figure 2. Cathexis Architecture adapted from (Velasquez and Maes, 1997)

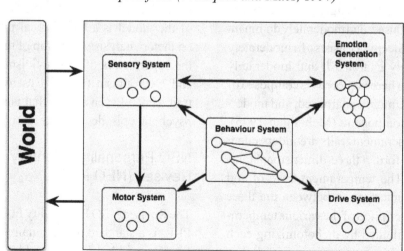

Cathexis was used to implement several synthetic characters like, for instance: Simón the Toddler (synthetic agent representing a young child) and Virtual Yuppy (a simulated emotional pet robot).

Flame

The Flame (Fuzzy logic adaptive model of emotions) emotional model was proposed by El-Nasr (El-Nasr et al. 2000), and is based on fuzzy logic concepts. Flame is composed by three models: emotional, decision making and learning. The emotional model is mainly based on the OCC model. Flame architecture is designed for a single agent, and does not incorporate functionalities related to group behaviour.

Figure 3 illustrate Flame architecture.

Flame was tested in an interactive simulation of a pet, and some of the model details were refined based on user's feedback.

Tabasco

Tabasco (tractable Appraisal-Based Architecture for Situated Cognizers) was proposed by Staller and Petta in 1998 and is based on psychological theories of emotions. TABASCO was designed to support agents situated in a virtual environment.

TABASCO architecture has been used to implement an interactive exhibit, the Invisible Person (Petta 1999; Petta et al. 1999; Psik et al. 2003), as part of a dramatic environment testbed, ActAffAct - Acting Affectively affecting Acting (Rank, 2004) and BehBehBeh (Behaviour Behooving Behaviour (Rank, 2007;Rank and Petta, 2007).

MAMID

MAMID is a cognitive-affective architecture proposed by Eva Hudlicka that implements sequential see-planning-action processing sequence (See

Figure 3. Flame architecture adapted from (El-Nasr et al., 2000)

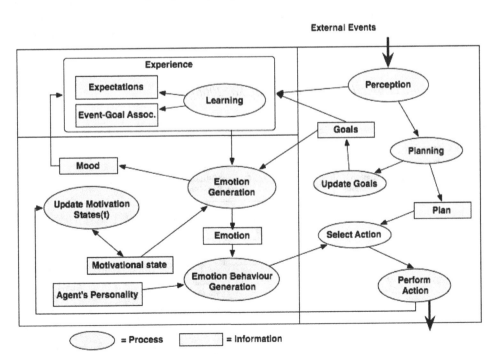

Figure 4). MAMID architecture domain independent and is composed by the following modules:

- sensory pre-processing – translate incoming data in internal representation (cues);
- attention – filter the internal representation, selecting a subset for processing;
- situation assessment – integrate individual cues into a global situation;
- expectation generator – project current situation onto possible future configurations;
- affect appraiser – infer the affective state from a diversity of internal and external elicitors;
- goal selection – select critical goals;
- action selection – select the best action to achieve agent goals.

These modules support the process of transforming incoming data (cues) onto outgoing behaviour (actions), via a set of internal intermediate structures designed by *mental constructs*. This mapping is supported by long term memories associated with each module and implemented as belief nets or rules. Mental constructs in MAMID architecture are characterized by attributes like familiarity, novelty, salience, threat level, and valence.

The Affect appraiser module is a central module of MAMID architecture perform external and internal representations (e.g. situations, expectations), priorities and individual characteristics (e.g. personality traits and existing emotional states) and generates one of the four basic emotions (fear/anxiety, anger/frustration, sadness, joy).

To model the effects of personality traits and emotional states in the cognitive process, MAMID architecture realize a mapping between personality traits and emotional states in architecture parameter values that controls processing inside each of the architecture modules. For instance, in this architecture, a personality trait may for instance affect the emotional decay rate or maximum intensity.

PECS

Urban and Shmidt propose the PECS (Physics, Emotion, Cognition, Social Status) reference model (Urban, 2000; Shmidt, 2002). PECS is an architecture for multi-agent systems, which aims modelling and simulating the human behaviour. PECS agents contain information that's falling into

Figure 4. MAMID architecture adapted from (Hudlicka, 2006)

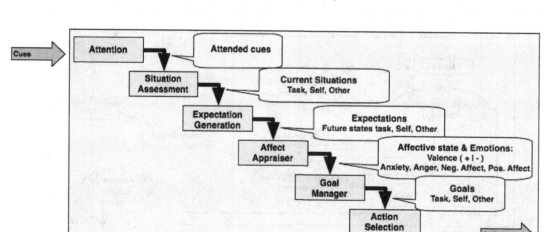

the four categories: physical (the agent's physical condition), emotional (agents feelings), cognitive (agents plans, model of the self and model of the environment) and social status (relations in the community of agents).

A simulation model named Adam was developed to test and demonstrate the PECS architectures capabilities.

Salt&Peper

Salt&Peper architecture was proposed by Luís Botelho and Hélder Coelho (Botelho and Coelho, 2001). Salt&Peper, is an architecture for autonomous agents that aims to implement mechanisms to allow artificial agents being as successful as natural agents. The roots of this architecture are in neuroscience and cognitive science; the authors boost the adaptive role of emotions. Generically we may say that the architecture aims to develop control mechanisms to artificial agents that are emotional based.

The project Safira uses the Salt&Peper architecture for the implementation of its agents (Paiva et al., 2001)

EMA

EMA (EMotion and Adaption) architecture was proposed by Marsella and Gratch (Marsella and Gratch, 2009) (See Figure 5). This architecture is inspired on the appraisal theory of Smith and Lazarus (Smith and Lazarus, 1990). EMA is composed by a set of processes that infer the representation of the relationship between person(s) and the environment in terms of a set of appraisal variables, and a set of coping strategies. Those last ones, the coping strategies, manipulate the representation of the person environment relationship in response to the appraised interpretation.

EMA is based on the SOAR architecture, which is a cognitive architecture that allows to model human though in terms of a set o cognitive operators (e.g. update belief, update intention, update plan, understand speech, wait, monitor goal, monitor expect affect, monitor expect act, Listen to speaker, expect speech, monitor expected event, initiate action and Terminate action).

EMA support a two-level notion of emotional state, namely appraisal and mood. The appraisal level determine the coping strategy that will be

Figure 5. EMA architecture adapted from (Marsella and Gratch, 2006)

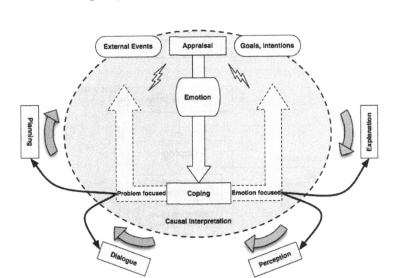

followed by the agent, however the mood state will influence this coping strategy.

Fatima [Paiva]

FAtiMA (FearNot! Affective Mind Architecture) was proposed by Dias and Paiva (Dias and Paiva, 2005). This architecture is inspired on the OCC cognitive structures of emotions (Ortony et al, 1988), where emotions are valence reactions to events. An agent receives a perception, appraises its significance and triggers the appropriate emotion.

For each one of the 22 emotions of the OCC model, in FAtiMA agents may have complete different activation thresholds and decay functions, which implicitly allows the construction of agents with different personalities.

FAtiMA agents emotional state, motivations, priorities and relationships its drive through and OCC-based appraisal model and a coping model.

In Figure 6 it is possible to visualize FAtiMA architecture.

FAtiMA architecture is composed by two layers, either in appraisal as in coping mechanisms, a reactive and a deliberative one. The reactive layer is responsible for a fast mechanism to appraise and react to a given event. The deliberative layer is slower to react, but allow a more complex goal-driven reasoning and behaviour.

Along this section several architectures to support emotional modelling were presented. From the diversity and complexity of the described approaches it is possible to conclude that's model emotions and to build architectures for an emotional agent is not an easy task. In this section we analyse several architectures for emotional agents, there are no perfect approaches, and about this subject Aristotle (Aristotle, The Nichomachean Ethics) states that:

"Anyone can become angry - that is easy. But to be angry with the right person, to the right degree,

Figure 6. Fatima Architecture adapted from (Brisson, Dias and Paiva, 2007)

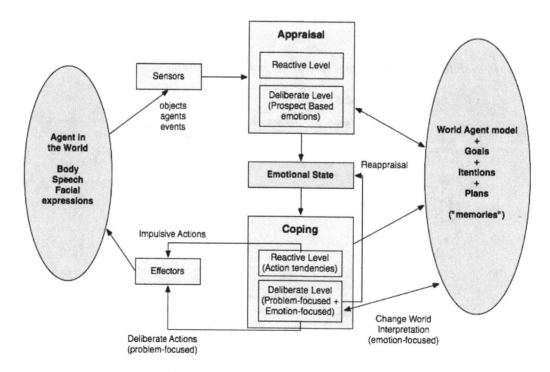

at the right time, for the right purpose, and in the right way - that is not easy."

Under the theme emotional architectures we can adapt the previous sentence and say that:

"Anyone can model emotions. That is easy. But to model emotions in the right context, to the right degree, for the right purpose, and in the right way - that is not easy."

IMPLEMENTATIONS OF EMOTIONAL AGENTS IN AMI ENVIRONMENTS

Some Ambient Intelligence environments are using emotion, mood, and personality models and methods. In this section we will refer some of these systems.

Project eMuu

The project eMuu developed as a cooperation between Philips Research Ambient Intelligence laboratories, the Eidhoven University of Technology and the Advanced Telecommunication Research Institute International (ATR) (Bartneck, 2002). The eMuu is an embodied emotional character for the ambient intelligent home designed as an interface between the AmI environment and its inhabitants.

The character is able to negotiate with the user about stamps. The task is to negotiate international stamps and the goal is to maximize the profit. Stamps were chosen for the object of negotiation because they do not carry individual values for the participants of the experiment. The character used exactly the same actions, such as making a partial offer or refusing a deal, as the user. The character evaluated the negotiation situation from a rational and emotional point of view. Depending on the outcome of these evaluations it made its next game move and expressed its emotional state through its face.

eMuu is capable of speech recognition and distinctive facial expressions. The speech recognition is one of the key components of eMuu because speech is the most natural human interaction. Expressing emotions is also a natural act for humans and eMuu is also able to do it. The emotion model was based on the complete OCC model that is capable of distinguish 22 emotional states. As it is not possible to represent all this emotional states with facial expressions, thus the authors selected a subsection of six emotional states (joy, distress, gratification, remorse, gratitude and anger) that they pretend to represent and made a mapping into three basic states (happiness, sadness and anger) that relate directly to the emotional expression of the character.

eMuu is based on the Muu2 robot developed by Michio Okada from ATR but with the ability to express emotions (emotional Muu).

The main goal of eMuu is to be used in and AmI environment where the interaction between the environment and its inhabitants its made through speech. An embodied home character, such as eMuu, could be the social entity necessary to provide natural dialogues. This character would have to be able to utilize the full range of communication channels, including emotional expressions, to give intuitive feedback to the user.

Project Steve and Jack

The project Steve and Jack is a joint collaboration between the University of Southern California Behavioral Technology laboratories and the Information Sciences Institute. The project main goal is to integrate Steve and Jack independent projects into one common project.

Steve stands for Soar Training Expert for Virtual Environments and is an autonomous animated agent for training in 3D virtual environments that supports the learning process (Marsella et al., 2003). Steve is able to demonstrate skills to students, answer student questions, watch the students as they perform the tasks, and give advice if the students run into difficulties, everything we should expect from an intelligent tutoring system. The system has also the possibility to train students on team tasks supporting dynamic virtual worlds where multiple Steve agents can cohabit with multiple students, always maintaining a plan for completing his current task, and revising the plan to handle unexpected events. The combination of capabilities that Steve is able to perform make him a good substitute for human instructors when they are unavailable.

The architecture of Steve is composed by three different modules to support all the capabilities: the perception, cognition and motor control. The perception module monitors the state of the virtual world maintains a coherent representation of it and provides this information to the cognition and motor control modules. The cognition module interprets its perceptual input and chooses appropriate goals constructs and executes plans to achieve those goals resulting in motor commands. The motor control module implements these motor commands controlling Steve's voice, locomotion, gaze, and gestures, and allowing Steve to manipulate objects in the virtual world.

Jack is an animated human figure that is part of the HeSPI project, a human simulation tool for planning and simulating maintenance tasks in nuclear power plants (Badler et al., 1993). In the HeSPI an experienced operator specifies via a user interface the necessary steps in a tasks that will be varied out by Jack. Inexperienced operators could watch the tasks being accomplished by Jack and use it as a training tool. The problem behind the use of this system as a training tool is that the trainee is limited to watch the task or procedure being impossible to ask questions, get explanations or practice the task.

The integration of Steve and Jack solves the main issues of each system, the lack of a human figure for the demonstrations in Steve and the lack of monitory students, asking questions and giving feedback on theirs actions in Jack (Mendez et al., 2003).

Project FearNot!

The FearNot! (Fun with Empathic Agents to Reach Novel Outcomes in Teaching) is an application that consists in a Virtual Learning Environment inhabited by autonomous agents that create an emergent unscripted narrative (Aylett et al., 2005). The aim of this application is to help reducing bullying problems in school. Bullying is "repeated oppression, psychological or physical, of a less powerful person by a more powerful person" (David Farrington, 1993).

This application was thought and developed for children and allows them to explore what happens in bullying in an unthreatening environment. Children are encouraged to take responsibility of the victims without feeling victimized. The developers spend some effort in the synthetic characters that could, through their appearances, behaviours and features allow the child to create a bond with the virtual character, so that the child user would really care about what happened to the victimized character. The child is asked to act as an invisible friend of the virtual character giving advices that would not influence the autonomy and behaviour of the victim.

In FeatNot! a session is composed by several dramatic episodes divided by periods where the character asks for advices to the child user. The session starts with an introduction to the school and characters following to a dramatic episode in which a bullying incident occurs. The child victim asks for an advice to the user for dealing with this. This sequence can be repeated several times and then is presented a simple educational message followed by an online questionnaire that tries to understand the lessons learned by the child. The goal is to encourage the children to act in a proper way as a spectator or victim.

The virtual characters generate emotions and update their emotional state whenever a goal leads to an intention. The OCC model of emotions is the concept used to model the emotions in the characters. A character perceives an event it the environment, relates it to its goals, standards, and attitudes, and as a result, experiences an emotion.

Project Virtual Humans: Rapport Agent

The Virtual Human Project leaded by Jonathan Gratch integrates research in intelligent tutoring, natural language recognition and generation, interactive narrative, emotional modeling and immersive graphics and audio (Swartout et al., 2004). The main goal of this project is to develop a highly realistic embodied conversational agent that utilizes EMA to model the cognitive and behavioural influences of emotions. This project can be applied in a diversity of areas like training, diagnosis & assessment, health interventions, basic science and business.

The Rapport Agent is an application example of the virtual humans architecture that uses machine vision and prosody analysis to create virtual humans that can detect and respond in real-time to human gestures, facial expressions and emotional cues and create a sense of rapport (Gratch et al., 2006). Rapport is defined as a "harmonious relationship" and can be conceptualized as a phenomenon occurring on three levels: the emotional, the behavioral and the cognitive (Gratch et al., 2007).

The Rapport Agent has an open modular architecture that facilitates the incorporation of different feature detectors, animation systems and behaviour mappings. Uses a stereo camera for the vision tracking features and signal processing of the speech signal to detect features of the human speaker through a microphone. To detect features from the speaker movements the agent focus on the speakers head movements, position and orientation and incorporates motion detection to detect head nods and shakes from head velocities.

Recognized speaker features are mapped into listening behaviours through a set of authorable mapping rules. The rules can be specified and triggered depending on different variables like for example the state of the speaker, the state of the agent, speakers age or temporal constraints. The author can also specify behavioral responses through different animated responses.

The animation system is designed to seamlessly blend animations and procedural behaviors, particularly conversational behavior. The animations are rendered in the popular game engine Unreal Tournament to be displayed to the Speaker.

Project I-Blocks

The project Intelligent Blocks (I-BLOCKS) is based on the concept of building blocks this way users are allowed to build physical constructions able to perform actions (Marti & Lund, 2004). The main objective of this new kind of devices (I-BLOCKS) is to build edutainment solutions to be used in ambient intelligence environments. The I-BLOCKS technology is integrated in the physical blocks embedded with invisible microprocessors for the different purposes. Each block takes the form of a lego duplo brick that contains processing and communication capabilities. Blocks can be attached together in order to build an artefact that can interact with the surrounding environment by sensors able to perceive input process or produce an output. Input sensors include LDR sensors, IR sensors, microphones, switches, potentiometer, and output sensors include servo motor, DC motor, IR emitter, LEDs, sound generator, etc. Each basic block has its own functionality but the overall functionality is obtained by the combination of several blocks that form a physical structure that process and communicates with each other. The overall behaviour of the artefact depends on the physical shape, the processing and the interaction with the surrounding environment.

The flexibility of the I-BLOCKS allows the exploration of several scenarios and different forms of interaction. A very interesting scenario that is presented by the authors is the living tree (Marti & Lund, 2005). The objectives of the living tree is to support the creative processes developed by children creating narratives and externalizing emotions associated to living entities at the level of the construction of characters considering its physical, behaviour and emotional attitudes.

The considered emotions in the living tree scenario are anger, fear, happiness and sadness. There is no emotion generation but directly manipulated by children through the construction of physical shapes. The representation of the emotions is obtained by different types of movement and lights that children use to mean emotional states.

The living tree is able to interact with the environment in which it is situated and its behaviour is defined by the following variables: wind (the tree moves according to the wind), temperature (the tree changes its color according to the temperature) and light (the tree moves its light sensors towards the light source). The physical shape of the tree can be constructed/changed by the user in order to change its behaviour or interaction form.

Project PETEEI

The project PETEEI (A PET with Evolving Emotional Intelligence is a simulator of a pet that makes use of the FLAME model (El-Nasr et al., 1999). The pet learns from its interactions with its user and is capable of acquire emotions such as shame and pride depending on the situation. The pet was chosen instead of a human because pets are simpler than humans; the planning of their goals is much simpler. PETEEI is based on a fuzzy logic model for simulating emotions in agents that uses inference rules and OCC to derive an emotional state based on the character's goals and expectations. The model also incorporates several learning mechanisms so that an agent can adapt its emotions according to its own experience. The emotion generation is associated with certain types of events with positive or negative feelings in order to react with according emotions that will affect the pet emotional state. The emotion triggered is chosen from the values of its appraisal variables such as the desirability of

an event and its probability of occurrence. To represent its emotional state PETEEI filters and expresses emotions according to its own moods and previous experience.

PETEEI has a Role Playing Game (RPG) interface. The interface is composed by five major scenes: a garden, a bedroom, a kitchen, a wardrobe and a living room. The interaction with the pet is made through a graphical user interface environment composed by different controls that represent actions that can be made by the user: introducing objects to the environment, taking objects from the environment, hitting objects, including the pet and talking aloud. Visual feedback from the pet consists of barking, growling, sniffing, etc. There is also a text window describing the pet actions and a graphical display of the pet internal emotional levels.

Project LAID

LAID (Laboratory of Ambient Intelligence for Decision Making) is an Intelligent Environment to support decision meetings (Marreiros et al., 2007; Marreiros et al., 2008; Ramos et al., 2009).

The meeting room supports distributed and asynchronous meetings, so participants can participate in the meeting where ever they are. The software included is part of an ambient intelligence environment for decision making where networks of computers, information and services are shared (Marreiros et al., 2009). In these applications emotions handling are also included. The middleware used is in line with the live participation/supporting on the meeting. It is also able to support a past review in an intuitive way, however the audio and video remember features are still in arrangement. The LAID is composed by the following hardware:

- Interactive 61" plasma screen
- Interactive holographic screen
- Mimio® Note grabber

- Six interactive 26" LCD screens, each one for 1 to 3 persons (red)
- 3 cameras, Microphones and activating terminals controlled by a CAN network.

With this hardware we are able to gather all kind of data produced on the meeting, and facilitate their presentation to the participants, minimizing the middleware issues to the software solutions that intend to catalog, organize and distribute the meeting information.

At the software level LAID is equipped with a system that supports the decision making process. Particularly this system supports persons in group decision making processes considering the emotional factors of the intervenient participants, as well as the argumentation process. This kind of features is not covered by the systems described in the previous sections. For this reason this paper will give a special focus on these characteristics (emotional aspects and argumentation support).

LAID is composed by the following modules: WebMeeting Plus, ABS4GD, WebABS4GD, and the pervasive hardware already referred.

WebMeeting Plus is a tool to help geographically distributed people and organisations in solving multi-criteria decision problems, namely supporting the selection of alternatives, argumentation, voting techniques and meeting setup. WebMeeting Plus is an evolution of the WebMeeting project with extended features for audio and video streaming. In its initial version, based on WebMeeting, it was designed as a Group Decision Support System supporting distributed and asynchronous meetings through the Internet.

ABS4GD (Agent Based Simulation for Group Decision) is a multi-agent simulator system whose aim is to simulate group decision making processes, considering the emotional and argumentative factors of the participants.

WebABS4GD is a ubiquitous version of the ABS4GD tool to be used by users with limited computational power (e.g. PDA) or users accessing the system through the Internet.

CONCLUSIONS

Emotions, mood, and personality had a bad connotation in the past, when Intelligent Systems were oriented for rationality and optimization aspects. However, if the human being is so affected by emotions, mood, and personality, and we want to conceive Ambient Intelligence systems to interact with human beings, then we should consider this reality from the very beginning. In this chapter we have presented some of the main approaches, architectures and systems to deal with emotions, mood, and personality. We do expect that this contribution will be useful for those interested in developing new Ambient Intelligence systems.

REFERENCES

Adamatti, D & Bazzan, A. (2003). AFRODITE – *A Framework for Simulation of Agents with Emotions*. ABS 2003 – Agent Based Simulation 2003. Montpelier, France, 28-30 April.

Aylett, R., Louchart, S., Dias, J., Paiva, A., & Vala, M. (2005). FearNot! - An Experiment in Emergent Narrative. [Springer.]. *IVA, 2005*, 305–316.

Badler, N. I., Phillips, C. B., & Webber, B. L. (1993). *Simulating Humans*. New York: Oxford University Press.

Bartneck, C. (2002). emuu - an embodied emotional character for the ambient intelligent home.

Bates, J. (1994). *The Role of Emotion in Believable Agents*. Communications of the ACM, Special Issue on Agents, July.

Botelho, L. M., & Coelho, H. (2001). Machinery for artificial emotions. *Cybernetics and Systems, 32*(5), 465–506.

Brisson, A., Dias, J., & Paiva, A. (2007). *From Chinese Shadows to Interactive Shadows: Building a storytelling application with autonomous shadows*. Workshop on Agent Based Systems for Human Learning and Entertainment (ABSHLE) AAMAS ACM 2007, 11-02-2007

Costa, P. T. Jr, & McCrae, P. R. (1992). *NEO PI-R Professional Manual*. Odessa, Fla: Psychological Assessment Resources.

Costa, P. T. Jr, & McCrae, R. R. (1992). *NEO PI-R professional manual*. Odessa, FL: Psychological Assessment Resources, Inc.

Damásio, A. (1994). O *erro de Descartes: emoção, razão e cérebro humano*. Publicações Europa-América.

Dias, J., & Paiva, A. (2005). Feeling and Reasoning: a Computational Model for Emotional Agents, in EPIA Affective Computing Workshop, Covilhã, 2005 Springer

El-Nasr, M., Yen, J., & Ioerger, T. R. (2000). FLAME -Fuzzy Logic Adaptive Model of Emotions. *Autonomous Agents and Multi-Agent Systems, 3*, 217–257. doi:10.1023/A:1010030809960

El-Nasr, M. S., Ioerger, T. R., & Yen, J. (1999). *PETEEI: a PET with evolving emotional intelligence*. In Proceedings of the Third Annual Conference on Autonomous Agents (Seattle, Washington, United States). O. Etzioni, J. P. Müller, and J. M. Bradshaw, Eds. AGENTS '99. ACM, New York, NY, 9-15.

Elliot, C. (1992). *The Affective Reasoner A process model of emotions in a multi-agent systems*. PhD dissertation. Northwestern University, USA, (1992)

Eysenck, H. J. (1991). Dimensions of personality: 16, 5, or 3?--Criteria for a taxonomic paradigm. *Personality and Individual Differences, 12*, 773–790. doi:10.1016/0191-8869(91)90144-Z

Farrington, D. P. (1993). Understanding and preventing bullying. In Tonny, M., & Morris, N. (Eds.), *Crime and Justice (Vol. 17)*. Chicago: University of Chicago Press.

Ghasem-Aghaee, N., & Oren, T. I. (2004). Effects of cognitive complexity in agent simulation: Basics. *SCS, 04*, 15–19.

Gratch, J., & Marsella, S. (2006). Evaluating a computational model of emotion. *Journal of Autonomous Agents and Multiagent Systems, 11*(1), 23–43. doi:10.1007/s10458-005-1081-1

Gratch, J., Okhmatovskaia, A., Lamothe, F., Marsella, S., Morales, M., van der Werf, R. J., & Morency, L. (2006). Virtual Rapport. 6th International Conference on Intelligent Virtual Agents, Marina del Rey, CA.

Gratch, J., Wang, N., Gerten, J., Fast, E., & Duffy, R. (2007). Creating Rapport with Virtual Agents. 7th International Conference on Intelligent Virtual Agents, Paris, France.

Hampson, S. (1999). State of the Art: Personality. *The Psychologist, 12*(6), 284–290.

Howard, P. J., & Howard, J. M. (2004). *The BIG FIVE Quickstart: An introduction to the five-factor model of personality for human resource professionals*. Charlotte, NC: Center for Applied Cognitive Studies.

Hudlicka, E. (2006). Depth of Feelings: Alternatives for Modeling Affect in User Models. *TSD, 2006*, 13–18.

John, O. P., & Srivastava, S. (1999). The Big Five Trait Taxonomy: History, Measurement, and Theoretical Perspectives. In Pervin, L. A., & John, O. P. (Eds.), *Handbook of personality: Theory and research (Vol. 2*, pp. 102–138). New York: Guilford Press.

Marreiros, G., Machado, J., Ramos, C., Neves, J.(2009) *Argumentation-based Decision Making in Ambient Intelligence Environments*. International Journal of Reasoning-based Intelligent Systems, 2009.

Marreiros, G., Santos, R., Freitas, C., Ramos, C., Neves, J., Bulas-Cruz, J. (2008). *LAID - a Smart Decision Room with Ambient Intelligence for Group Decision Making and Argumentation Support considering Emotional Aspects*. International Journal of Smart Home Vol. 2, No. 2, Published by Science &Engineering Research Support Center, 2008.

Marreiros, G., Santos, R., Ramos, C., Neves, J., Novais, P., Machado, J., & Bulas-Cruz, J. (2007). *Ambient Intelligence in Emotion Based Ubiquitous Decision Making*, Proc. Artificial Intelligence Techniques for Ambient Intelligence, IJCAI'07 – Twentieth International Joint Conference on Artificial Intelligence. Hyderabad, India.

Marsella, S., & Gratch, J. (2006). EMA: A Computational Model of Appraisal Dynamics. In Trappl, R. (Ed.), *Cybernetics and Systems 2006* (pp. 601–606). Vienna: Austrian Society for Cybernetic Studies.

Marsella, S., Gratch, J., & Rickel, J. (2003). *Expressive Behaviors for Virtual Worlds. Life-like Characters Tools, Affective Functions and Applications*. Springer Cognitive Technologies Series.

Marsella, S. C., & Gratch, J. (2009). EMA: A process model of appraisal dynamics. *Cognitive Systems Research, 10*(1), 70–90. doi:10.1016/j.cogsys.2008.03.005

Marti, P., Giusti, L., Pollini, A., & Rullo, A. (2005). Experiencing the flow: design issues in human-robot interaction. In *Proceedings of the 2005 Joint Conference on Smart Objects and Ambient intelligence: innovative Context-Aware Services: Usages and Technologies* (Grenoble, France, October 12 - 14, 2005). sOc-EUSAI '05, vol. 121. ACM, New York, NY, 69-74.

Marti, P., & Lund, H. H. (2004). Emotional Constructions in Ambient Intelligence Environments. *Intelligenza Artificiale, 1*(1), 22–27.

Mateas, M. (2002). *Interactive Drama, Art and Artificial Intelligence*. Carnegie Mellon University.

McRae, R. R., & Costa, P. T. (1987). Validation of the five-factor model of personality across instruments and observers. *Journal of Personality and Social Psychology, 52*, 81–90. doi:10.1037/0022-3514.52.1.81

Mehrabian, A. (1995). Framework for a comprehensive description and measurement of emotional states. *Genetic, Social, and General Psychology Monographs, 121*, 339–361.

Mehrabian, A. (1996). Analysis of the Big-Five Personality Factors in Terms of the PAD Temperament Model. *Australian Journal of Psychology, 48*(2), 86–92. doi:10.1080/00049539608259510

Mendez, G., Rickel, J., & Antonio, A. (2003). *Steve meets Jack: the integration of an intelligent tutor and a virtual environment with planning capabilities. Computer science school*. Technical University Madrid.

Ortony, A. (2003). *On making believable emotional agents believable*. In R. P. Trapple, P. (Ed.), Emotions in humans and artefacts. Cambridge: MIT Press.

Ortony, A., Clore, C., & Collins, A. (1998). *The Cognitive Structure of Emotions*. Cambridge University Press.

Paiva et. al. (2001). *SAFIRA- Supporting Affective Interactions in Real-time Applications*. CAST - Living in mixed realities, Special Issue of netzpannung.org/journal.

Petta, P. (1999) Principled Generation of Expressive Behavior in an Interactive Exhibit. In Juan D Velásquez (ed.): *Workshop: Emotion-Based Agent Architectures (EBAA'99) at the 3rd International* Conference *on Autonomous Agents, Seattle WA USA, May 1 1999*, pp. 94–98, 1999. 34, 36, 110, 14

Petta, P., Staller, A., Trappl, R., Mantler, S., Szalavári, S., Psik, T., & Gervautz, M. (1999) Towards Engaging Full-Body Interaction. In H.-J. Bullinger and P.H. Vossen (eds.): *Adjunct Conference Proceedings, HCI International '99, 8th International Conference on Human*-Computer *Interaction, jointly with 15th Symposium on Human Interface, Munich Germany, August 22-27 1999*, pp. 280–281. Fraunhofer IRB Verlag, 1999. 36, Picard, R. (1995). *Affective Computing*. M.I.T Media Laboratory Perceptual Computing Section Technical Report N° 321, Nov. 26, (1995). [on-line] Available: http://vismod.media.mit.edu/tech-reports/TR-321.pdf

Psik, T., Matkovic, K., Sainitzer, R., Petta, P., & Szalavári, Z. (2003). The Invisible Person Advanced Interaction Using an Embedded Interface. In Deisinger J. and Kunz A. (eds.): Proceedings of the 7. International Immersive Projection Technologies Workshop and the 9.Eurographics Workshop on Virtual Environments (IPT/EGVE 2003), May 22-23 2003, Zurich Switzerland, pp. 29–37. ACM Press New York USA, 2003. 36, 144

Ramos, C, Augusto, J.C., Shapiro, D. (2008). *Ambient Intelligence: the next step for AI*, IEEE Intelligent Systems magazine, vol 23, n. 2, pp.15-18.

Ramos, C., Marreiros, G., Santos, R., & Freitas, C. (2009). *Smart Offices and Intelligent Decision Rooms. Handbook of Ambient Intelligence and Smart Environments* (Nakashima, H., Augusto, J., & Aghajan, H., Eds.). Springer.

Rank, S. (2004). Affective Acting: An Appraisal-Based Architecture for Agents As Actors. Master's thesis, Institute for Medical Cybernetics and Artificial Intelligence, Medical University of Vienna & Vienna University of Technology, Diplomarbeit, 2004. URL http://www.ofai.at/~stefan.rank/StefanRank-AAAThesis.pdf. 13, 36,145

Rank, S. (2007). Building a computational model of emotion based on parallel processes and resource management. In Proc. of the Doctoral Consortium at ACII2007, Lisbon Portugal EU, Sept. 12-14, 2007.

Rank, S., & Petta, P. (2007). *Basing artificial emotion on process and resource management*. In Paiva A. et al. (eds.): 2nd Int'l. Conf. on Affective Computing and Intelligent Interaction (ACII2007), Lisbon Portugal, Sept.12–14, 2007, LNCS 4738, Springer, 350 361, 2007.

Roseman, I., Spindel, M., & Jose, P. (1990). Appraisals of emotion-eliciting events: Testing a theory of discrete emotions. *Journal of Personality and Social Psychology*, 59, 899–915. doi:10.1037/0022-3514.59.5.899

Schmidt, B. (2002), *How to give Agents a Personality*, In: Proc. 3rd Workshop on Agent- Based Simulation, Passau, Germany, April 7-9, 2002. SCS-Europe, Ghent, Belgium, pp. 13-17.

Smith, C. A., & Lazarus, R. S. (1990). Emotion and adaptation. In Pervin, L. A. (Ed.), *Handbook of personality: Theory and research* (pp. 609–637). New York: Guilford.

Staller, A., & Petta, P. (1998). *Towards a Tractable Appraisal Based Architecture for Situated Cognizers*. In Lola Cañamero, Chisato Numaoka, and Paolo Petta (eds.): Workshop Notes, 5th International Conference of the Society for Adaptive Behaviour (SAB98), Zürich Switzerland, August 21, pp. 56–61, 1998. 36, 149

Staller, A., & Petta, P. (2001). Introducing *Emotions into the Computational Study of Social Norms: A First Evaluation. Journal of Artificial Societies and Social Simulation*, 4(1).

Swartout, W., Gratch, J., Hill, R. W., Hovy, E., Marsella, S., Rickel, J., & Traum, D. R. (2004). Toward virtual humans. In *Working notes of the AAAI Fall symposium on Achieving Human-Level Intelligence through Integrated Systems and Research*, Crystal City, VA, USA.

Urban, C. (2000). PECS – A Reference Model for the Simulation of Multi-Agent Systems. In Suleiman, R., Troitzsch, K. G., & Gilbert, G. N. (Eds.), *Tools and Techniques for Social Science Simulation*. Heidelberg: Physica Verlag.

Velasquez, J., & Maes, P. (1997) *Cathexis: A Computational Model of Emotions*. Proceedings of the First International Conference on Autonomous Agents, pp. 518-519.

KEY TERMS AND DEFINITIONS

Ambient Intelligence: Ambient Intelligence (AmI) deals with a new world where computing devices are spread everywhere, allowing the human being to interact in physical world environments in an intelligent and unobtrusive way. These environments should be aware of the needs of people, customizing requirements and forecasting behaviours.

Context Awareness: Context Awareness means that the system is aware of the current situation we are dealing with.

Intelligent Decision Room: A decision-making space, eg a meeting room or a control center, equipped with intelligent devices and/or systems to support decision-making processes.

Pervasive Computing: Pervasive Computing is related with all the physical parts of our lives, the user may have not notion of the computing devices and details related with these physical parts.

Ubiquitous Computing: Ubiquitous Computing means that we have access to computing devices anywhere in an integrated and coherent way.

Emotion: Emotion is a state of mind accompanied by a certain bodily changes

Personality: Personality is a pattern of behavioral, temperamental, emotional, and mental traits that characterize a unique individual

Chapter 7
Logical Modeling of Emotions for Ambient Intelligence

Carole Adam
RMIT University, Australia

Benoit Gaudou
UMI 209 UMMISCO, IRD, IFI, Vietnam

Dominique Login
University of Toulouse, CNRS, IRIT, France

Emiliano Lorini
University of Toulouse, CNRS, IRIT, France

ABSTRACT

Ambient Intelligence (AmI) is the art of designing intelligent and user-focused environments. It is thus of great importance to take human factors into account. In this chapter we especially focus on emotions, that have been proved to be essential in human reasoning and interaction. To this end, we assume that we can take advantage of the results obtained in Artificial Intelligence about the formal modeling of emotions. This chapter specifically aims at showing the interest of logic as a tool to design agents endowed with emotional abilities useful for Ambient Intelligence applications. In particular, we show that modal logics allow the representation of the mental attitudes involved in emotions such as beliefs, goals or ideals. Moreover, we illustrate how modal logics can be used to represent complex emotions (also called self-conscious emotions) involving elaborated forms of reasoning, such as self-attribution of responsibility and counterfactual reasoning. Examples of complex emotions are regret and guilt. We illustrate our logical approach by formalizing some case studies concerning an intelligent house taking care of its inhabitants.

DOI: 10.4018/978-1-61692-857-5.ch007

INTRODUCTION

Ambient Intelligence (AmI) is the art of designing intelligent and user-focused environments, *i.e.* environments that can adapt their behavior to users and to their specific goals or needs at every moment, in order to insure their well-being in a non-intrusive and nearly invisible way. AmI systems should be embedded, context aware, personalized and adaptive (Jiao, Xu and Du, 2007). AmI is a highly multidisciplinary field of research, at the convergence between as various disciplines as for instance electronics, networks, ergonomics, robotics or computer science. Some examples of applications are ISTAG's research scenarios (Ducatel *et al.*, 2001), industrial ones such as Philips intelligent house HomeLab (Aarts and Eggen, 2002), or academic projects such as MIT's Oxygen project and their MediaLab.

Since the focus of AmI is on the human users, it is of great importance in order to better understand them to have a model of human factors and to take these factors into account (Treuil, Drogoul and Zucker, 2008). In this chapter we focus on a particular human factor, namely emotions, that have been proven to be essential in the human decision making process (Damasio, 1994), but also in the interaction between users and machines (Picard, 1997). Therefore we think that AmI can take advantage of the results obtained in Affective Computing (Picard, 1997) or Artificial Intelligence (AI) about emotions.

In particular, the aim of this chapter is to show how logical tools and methods, traditionally used in the field of AI for the formal design of intelligent systems and agents, can be exploited in order to specify *emotional agents* for AmI applications. The term *emotional agents* refers to intelligent agents endowed with some emotional abilities, *e.g.* identifying and reacting to the users' emotions. These agents may also be able to "feel" emotions[1], to express them, and to behave accordingly. We think that there are two main reasons justifying

the use of logical methods for the design of such emotional agents.

First, although physiological sensors can detect some simple users' emotions such as happiness (Prendinger and Ishizuka, 2001), they are neither able to determine the object of these emotions (*e.g.* happiness about the sun shining or about being in holidays), nor able to differentiate between some close emotions such as sadness, regret and guilt (Oehme *et al.*, 2007). Indeed, to do this, it is necessary to take into account the context and the users' mental attitudes such as beliefs, desires, ideals which are the cognitive constituents of emotions. Now, so-called BDI (Belief, Desire, Intention) logics developed in the field of AI in the last fifteen years (see *e.g.* Cohen and Levesque, 1990; Rao and Georgeff, 1991; Lorini and Herzig, 2008) offer expressive frameworks to represent agents' mental attitudes and to reconstruct on their basis the cognitive layer of emotions (Adam, 2007; Castelfranchi and Lorini, 2003; Steunebrink, Dastani and Meyer, 2007).

Second, in order to adapt to the users, an AmI system must be able to reason about their emotions and to analyze them. In particular, an AmI system must have a model of users' emotions and of the ways in which they affect their behavior, like *action tendencies* (Frijda, 1986) (*e.g.* feeling fear entails escape) and *coping strategies* (Lazarus, 1991) (*e.g.* denying an information causing too much sadness). This model would allow the AmI system to reason about (*e.g.* anticipate) the effects of its own actions on the users' emotions and actions. Now, logical approaches were specifically designed to endow agents with this kind of reasoning capabilities.

The previous two arguments are even more obvious when looking at the cognitive structure of *complex emotions*[2] such as regret, shame, guilt, reproach because they involve complex forms of reasoning. Following (Ekman, 1992; Johnson-Laird and Oatley, 1992), we consider that complex emotions are those that are based on a larger set of appraisal variables[3] than the set involved in *basic emotions* (*e.g.* fear, happiness, sadness). Besides, while basic emotions are related to an individual's basic biological needs and can be elicited after a merely automatic and unconscious analysis of the situation, complex emotions are based on a conscious cognitive activity and involve very sophisticated forms of reasoning such as *self-attribution of responsibility* and *counterfactual reasoning*.

Thus finally, logical approaches seem to be a very well adapted solution for the design of AmI systems which are capable of identifying the users' complex emotions (*e.g.* guilt, shame), determining their object (*e.g.* why is the user feeling guilty? What is (s)he regretting for?), and reasoning about them (*e.g.* will (s)he feel ashamed if I tell him/her that (s)he failed his/her exam?).

The present chapter is organized as follows. In Section 2, we present a very general overview of the role of emotions in some present and future AmI applications; moreover, we discuss in more details the contribution that logical methods could bring to AmI applications. In Section 3, we present a logical framework allowing a formal specification of agents' mental attitudes and of important conceptual components of complex emotions such as the notion of responsibility; we show how this logical framework can be used to define some complex emotions like regret and guilt. Section 4 is devoted to the application of our logical framework for complex emotions to the formalization of some case studies for AmI.

STATE OF THE ART

Applications of Emotions in AmI

Nowadays, more and more AmI systems are endowed with emotional abilities, in order to perform better in various application fields. For instance an emotional agent for AmI can:

- express emotions to reduce users' frustration when interacting with a machine, as proven by Picard (1997), *e.g.* Greta (Dastani *et al.*, 2005), Breazeal's Kismet (Breazeal and Scassellati, 1999), or Philips' i-Cat (van Breemen, 2005);
- favor health and well-being by developing social relationships (proven by Berkman *et al.*, 2000 to have a positive influence on health) with sick or elderly people (*e.g.* Wada and Shibata's Paro robot, 2006);
- favor performance by decreasing stress (Burton, 1988), or by interacting realistically with the user in a virtual world to train him to make decisions under stress (*e.g.* Gratch and Marsella's Mission Rehearsal Exercise, 1991, for military officers, or El Jed *et al.* virtual world for firemen, 2004);
- assist decision making in business (Marreiros *et al.*, 2007) or in health care (Augusto *et al.*, 2007);
- express believable emotions to be more attractive and engaging when employed as a virtual tutor (in an educational application) (Rickel and Lewis Johnson, 1997).

These applications involve different kinds of emotional abilities, such as expressing emotions or being sensitive to the user's ones. In the next section we detail what we assume to be the complete list of emotional abilities that may be useful for AmI applications.

Useful Emotional Abilities for AmI Systems

In this chapter, we consider AmI systems according to the multi-agents paradigm. We thus call *agents* the components of AmI systems. We consider an agent from a very general point of view, *i.e.* as an autonomous entity with some capabilities of perception of its environment, of action on it (including interaction with other agents) and of reasoning. In the following we focus on a particular type of agents, called *cognitive agents* (as opposed to *reactive agents*), whose behavior is guided by motivations and goals. One can consider an AmI system as constituted by both cognitive and reactive agents; for example a cognitive agent can lead a network of reactive sensing agents to get information about the world in order to make adapted decisions. We assume that this representation can fit every AmI system.

Identifying the Users' Emotion

In order to interact with the users, adapt to them or improve their well-being, an agent has to identify their emotional state. According to Searle (1983), we consider emotions as *intentional* mental states, like belief or intention, *i.e.* mental states "directed at or about or of objects or states of affairs in the world" (Searle, 1983, p.1). Thus emotions always have an object; similar affective states but directed at no particular object will be called *moods*. Therefore the identification of the users' emotions comprises two steps: identifying the type of emotion (joy, sadness, anger, *etc.*) and identifying its object (*e.g.* joy about the sunny weather).

On the one hand a great amount of work already allows systems to identify the user's type of emotion from physiological data (Picard, Vyzas and Healey, 2001; Prendinger and Ishizuka, 2005), gestures (Wu and Aghajan, 2008), facial expressions (Susskinda *et al.*, 2007), language and voice intonation (Sebe *et al.*, 2004), *etc*. The challenge for AmI is to perceive this physiological information in a non intrusive way, which will be made possible thanks to technological progress (*e.g.* seamless sensors integrated in clothes).

On the other hand identifying the object of the user's emotion is much more difficult, since it is not physiologically measurable (joy of winning the basket-ball championship final and joy of having an accepted paper in a great conference have the same physiological manifestations); it is thus necessary to reason about the context and the users' mental attitudes to identify this object.

Moreover this additional information allows the agent to identify more precisely the users' emotion (Forest *et al.*, 2006), for example it could allow to distinguish the sadness of having lost the final from the regret for being responsible for this defeat.

Computing the Agent's Emotion

As shown above, various AmI applications use agents able to express emotions, and sometimes able to behave according to their emotions. An agent's emotion should be computed from perceived external stimuli, from the agent's previous emotions, and from its representation of the world as it is or as it should be (*e.g.* in terms of mental attitudes). Moreover, the computation of this emotion should be informed by psychological models of human emotions (Lazarus, 1991; Ortony, Clore and Collins, 1988) to ensure its correctness; formalizations of such models exist to this aim (Adam, 2007; Steunebrink, Dastani and Meyer, 2007). The computation of an agent's emotion is also deeply linked with the agent's hardware capacities.

For example, the embodied conversational agent Greta (Dastani, van Riemsdijk and Meyer, 2005) uses Ortony's model (2003) representing emotions with two dimensions: valence (positive or negative) and time (past, present, future, indicating when the triggering stimulus occurs). A belief network is used to match a situation with an emotion; then the corresponding dimensions are computed.

In the area of pedagogical tutoring, Steve (El Fallh-Seghrouchni and Suna, 2005) is an animated agent who inhabits a 3D environment and teaches students to operate a high pressure air compressor. It is aware of the students' actions and of their effect on the environment. Its emotions, based on Ortony *et al.*'s theory (1988), are a function of its goals and principles, but also of the students' presumed appraisal of the situation and of Steve's relationship with them.

Once the agent is aware of its own emotions or of the users' ones, it must be able to reason about them and to behave accordingly. This is the point of the next paragraphs.

Taking the User's Emotion into Account in Reasoning and Planning

An agent for AmI should be able to take the users' emotion into account in its reasoning process in order to influence its beliefs, goals and future actions. For example an AmI agent aiming at the users' well-being should behave differently depending on the users' emotion, or at least on its valence (positive or negative). Moreover the strategies that it can use to get the users' back to a positive emotion depend on the particular negative emotion that they feel (sadness, anger, *etc.*). We provide examples of various strategies in Section 4.

However only few convincing works are able to accurately identify the users' emotions in a non-intrusive way, and are thus usable in an actual interaction situation. The scenarios of the e-sense project (Oehme *et al.*, 2007) give an insight into how the users' emotions could be used by AmI systems in future applications: e.g. identify and reproduce faithfully the users' emotion on their avatar to improve their immersion in a virtual world or a video game; or monitoring psychotic or paranoiac patients to detect negative emotions (in particular intense fear) and call a doctor in case of need. In this last scenario, the users' emotion drives the behavior of the AmI system by giving it new goals to achieve or actions to perform.

The agent can also use its knowledge about the users' emotion to try to help them to *cope* with it (in the sense of Lazarus' *coping strategies*, 1991), *i.e.* to restore their well-being in stressful situations. For example in (Adam *et al.*, 2006) we have described some case studies involving an AmI agent endowed with a logical formalization of eight basic emotions.

Taking the Agent's Emotion into Account in Reasoning and Planning

Similarly, the agent's own emotion should also impact its reasoning and actions. First the agent should express its emotion to the users, since it is beneficial for the interaction with them (Dastani, Riemsdijk and Meyer, 2005; Picard, 1997). This emotion can be expressed in various ways: facial expressions, color code, gestures, language, *etc.* Second its emotion can influence its reasoning process, in particular by changing its beliefs or goals (for instance it can adopt the goal to interact with users when it "feels" happy). This influence can follow from the simulation of two human psychological processes: *action tendencies* (Frijda, 1986) are associated to emotions to allow individuals to make quick decisions, using their emotions as a kind of heuristic (for instance fear of a spider makes one flee what is interpreted as a danger); *coping strategies* (Lazarus, 1991) can be planfully used to manage stressful situations (for instance one can persuade himself that the frightening spider is finally not big enough to be really dangerous, in order not to be afraid anymore).

Lots of works have been conducted in the particular area of the expression of emotions and of the development of expressive heads and robots, in particular in the Embodied Conversational Agents (ECA) community. For example, Greta (Dastani, van Riemsdijk and Meyer, 2005) was used to dialogue with users in medical applications where empathy with them is crucial. Breazeal's Kismet (Breazeal and Scassellati, 1999) is assumed to be able to make friend with humans and influence them. Similarly, Philips has developed a robot called iCat with an articulated face able in particular to express emotions (van Breemen, 2005).

Besides, some works exist in Artificial Intelligence in order to endow agents with the ability to use coping strategies to manage their own emotions (*e.g.* Gratch and Marsella's EMA agent, 2004, or the logical formalization of coping strategies proposed in Adam and Longin, 2007). However, as far as we know, these works have not been applied to AmI yet.

Contribution of Logic to the Modelling of these Emotional Abilities

While several emotional abilities are subject to a great amount of work (*e.g.* users' emotion recognition, agent's emotional expression), some other ones have nearly not been explored yet. We assume that the tools that are currently used in AmI are not sufficient to explore some points like reasoning about emotions and identifying the user's emotions. We would thus like to show how formal logic can be an adequate tool to manage such unsolved problems.

In particular, determining the object of the emotion is an essential but difficult problem, as pointed by (Zhou *et al.*, 2007). We argue that endowing an AmI agent with mental attitudes, with a logical model of emotions like the one presented in the sequel of this chapter, and with the subsequent ability of logical reasoning and deduction, could be an interesting approach in addition to classical physiological sensing. Indeed when a sensor detects the physiological clues of sadness in the users, it cannot determine alone the object of this sadness. Now if the AmI agent knows the users' beliefs and preferences (*e.g.* users hate rain), is aware of the context (*e.g.* it is raining), and has a model of emotions (*e.g.* sadness comes from infringed preferences), it may be able to determine what has triggered the users' sadness (*e.g.* rain).

Besides the user's emotion may not be exactly sadness but a more complex emotion resembling sadness, like regret or remorse. Thus endowing an AmI agent with a fine-grained model of emotions, along with a knowledge base containing extensive information about norms, responsibilities, actions, context, users' profiles, *etc.* will allow it to finely determine the users' emotions, and thus to finely adapt to the users' needs. Thus we assume that

to precisely and finely recognize both the type and object of the users' emotion, an AmI agent should jointly use physiological sensors (or any other emotional recognition system) and a model of emotions coupled with a representation of their mental attitudes and reasoning.

We can also notice that in practice, very few emotions are used in AmI. This appears as a consequence of the previous point: complex emotions like regret, remorse, shame or pride involve high level reasoning that can hardly be managed without using specific tools (like logic, as it will be shown in the next section). One can object that a few basic emotions are sufficient for the purpose of AmI. But we answer that these complex emotions are felt everyday by the human users in whose life AmI systems aim at being integrated, so they cannot just be ignored by these systems.

Finally it appears that reasoning about emotions or behavioral adaptation from emotions is subject to very little and often preliminary works.

Conclusion

In this section, we have highlighted that emotions begin to be successfully used in AmI systems. Nevertheless much more can be done to integrate this essential human factor in these systems. We have thus argued that using BDI agents in AmI systems, and endowing them with a fine-grained model of emotions, expressed in a non-ambiguous logical language, can have a lot of benefits. In particular, it would allow these agents to identify more precisely the users' emotional state. In the following section we present our logical framework dedicated to the representation of emotions, and specifically of complex ones.

LOGICAL MODELING OF EMOTIONS

Recently, some researchers working in the field of multi-agent systems (MAS) have been interested in developing logical frameworks for the formal specification of emotions (see Adam, Herzig and Longin, 2009, Lorini and Castelfranchi, 2007, Steunebrink, Dastani and Meyer, 2007, for instance). Their main concern is to exploit logical methods in order to provide a rigorous specification of how emotions should be implemented in an artificial agent. The design of agent-based systems where agents are capable to reason about and to display some kind of emotions can indeed benefit from the accuracy of logical methods. These logical approaches also aim at disambiguating the different dimensions of emotions identified in the existing psychological models of emotions (*e.g.* Lazarus, 1991; Orthony, Clore and Collins, 1998; Scherer, 2001).

Most of proposed logical frameworks for the specification of emotions are based on the so-called BDI (Belief, Desire, Intention) logics which have been developed in the last fifteen years in the MAS domain (see Cohen and Levesque, 1990, Lorini and Herzig, 2008, Rao and Georgeff, 1991, for instance). BDI logics are multimodal logics in which agents' attitudes such as beliefs, goals, desires, intentions, *etc.* are formalized. Such concepts are generally expressed by corresponding modal operators and their interaction properties are specified.

In the following Section 3.1, we present a simple BDI language in which three kinds of agents' attitudes can be represented by corresponding modal operators: beliefs, chosen goals (or choices), and moral attitudes called ideals (or imperatives). We show how these modal operators can be combined to provide a formal characterization of some basic emotions such as sadness and joy. Then, in Section 3.2, we show how this basic BDI language can be extended with the STIT operator (operator of the logic of *Seeing To It That*) (Belnap, Perloff and Xu, 2001; Horty, 2001) in order to shift from basic emotions to complex emotions, and in particular to those based on self-attribution of responsibility and on counterfactual reasoning.

A Logic of Agents' Mental Attitudes

The basic BDI language we consider includes two kinds of operators. Operators Bel_i are used to express what a certain agent i believes. Operators $Choice_i$ are used to denote an agent's chosen goals, that is, the goals that the agent has decided and is committed to pursue. Given an arbitrary formula φ of the logic:

- $Bel_i \varphi$ stands for: agent i believes that φ;
- $Choice_i \varphi$ stands for: agent i has the chosen goal that φ (which can be shortened to: agent i wants φ to be true).

These two modal operators have been extensively studied in the field of formal philosophy and AI.

Operators for belief of the form Bel_i are doxastic operators in Hintikka's style[4] (1962) with a standard KD45 axiomatization (Chellas, 1980). This means that every operator Bel_i is supposed to be normal[5], and that an agent cannot have contradictory beliefs:

$\neg (Bel_i \varphi \wedge Bel_i \neg\varphi)$ **DBel**

Moreover, beliefs satisfy the properties of positive and negative introspection:

$Bel_i \varphi \rightarrow Bel_i Bel_i \varphi$ **4Bel**

$\neg Bel_i \varphi \rightarrow Bel_i \neg Bel_i \varphi$ **5Bel**

Operators for chosen goal of the form $Choice_i$ are similar to the operators introduced in (Cohen and Levesque, 1990). These operators are commonly defined as KD operators (Chellas, 1980), that is, every operator $Choice_i$ is normal[6], and an agent cannot have contradictory chosen goals:

$\neg (Choice_i \varphi \wedge Choice_i \neg\varphi)$ **DChoice**

A general hypothesis about the relationship between chosen goals and beliefs is the so-called assumption of *weak realism* (McCann, 1991). According to this hypothesis, an agent cannot choose φ if it believes that φ is an impossible state of affairs. This hypothesis is expressed by the following logical axiom:

Choice$_i$ φ $\rightarrow \neg$ Bel$_i$ $\neg$φ **WeakRealism**

Principles **DChoice** and **WeakRealism** are standard principles of rationality for chosen goals. As we said above, chosen goals (represented by operators Choice$_i$) refer for us to states of affairs that the agent has decided and is committed to pursue: chosen goals are the product of an agent's rational deliberation. This is the reason why they have to be consistent (*i.e.* a rational agent cannot decide to pursue inconsistent states of affairs) and they have to be compatible with the agent's beliefs (*i.e.* a rational agent cannot decide to pursue something that it believes to be impossible).

We add another type of operator to our BDI language to talk about an agent's moral attitudes, after supposing that agents are capable to discern what (from their point of view) is morally right from what is morally wrong. This is a necessary step towards an analysis of complex emotions such as guilt and shame which involve a moral dimension. Our operators for agents' moral attitudes are of the form Ideal$_i$ where Ideal$_i$ φ means: φ is an ideal state of affairs for agent *i*. More generally, Ideal$_i$ φ expresses that agent *i* thinks that it ought to promote the realization of φ, that is, agent *i* conceives a demanding connection between itself and the state of affairs φ. When agent *i* endorses the ideal that φ (*i.e.* Ideal$_i$ φ is true), it means that *i* addresses a command to itself, or a request or an imperative to achieve φ (when φ is actually false) or to maintain φ (when φ is actually true) (Castaneda, 1975). In this sense, *i* feels morally responsible for the realization of φ.

There are different ways to explain how a state of affairs φ becomes an ideal state of affairs

of an agent. A plausible explanation is based on the hypothesis that ideals are just social norms internalized (or adopted) by an agent (Conte and Castelfranchi, 1995). Suppose that an agent believes that in a certain group (or institution) there exists a certain norm (*e.g.* an obligation) prescribing that a state of affairs φ should be true. Moreover, assume that the agent identifies itself as a member of this group. In this case, the agent adopts the norm, that is, the external norm becomes an ideal of the agent. For example, since I believe that in Italy it is obligatory to pay taxes and I identify myself as an Italian citizen, I adopt this obligation by imposing the imperative to pay taxes to myself.

In some particular cases, if an agent *i* endorses the ideal that φ, then agent *i* is committed to promote the realization of φ by acting either in order to achieve φ (when φ is currently false) or in order to maintain φ (when φ is actually true). Note that this property does not reflect the general case, that is, it is not valid for all agents but only for a particular type of agents. This is the reason why we do not suppose it as a logical axiom.

However, not all chosen goals of an agent have a normative origin. A part of an agent's chosen goals are not ideals of the agent, but originate from the agent's desires. In agreement with the Humean conception (Hume, 1978), we conceive a desire as an agent's attitude which consists in an anticipatory mental representation of a pleasant state of affairs φ (*representational dimension* of desires) that motivates the agent to achieve φ (*motivational dimension* of desires). In this perspective, the motivational dimension of an agent's desire is realized through its representational dimension. For example when I desire to be at the Japanese restaurant eating sushi, I imagine myself eating sushi at the Japanese restaurant and this representation gives me pleasure. This pleasant representation motivates me to go to the Japanese restaurant.

It is out of the scope of this chapter to introduce modal operators for characterizing agents' desires and to build a model which explains how an agent's chosen goals originate from its ideals and desires. However, note that some filter of deliberation is needed to obtain a final consistent set of chosen goals which are compatible with the agent's beliefs.

The three operators for beliefs, chosen goals and ideals introduced above can be used to formalize some basic emotions such as joy and sadness.

The fact that an agent i feels joy about a fact φ (or rejoices about φ) can be expressed as follows:

$$\text{Joy}_i\, \varphi =_{def} \text{Bel}_i\, \varphi \wedge \text{Choice}_i\, \varphi$$

According to this definition, agent i feels joy about φ if and only if i believes that φ is true and wants φ to be true. For example, agent i feels joy for having passed the exam because i believes that he has passed the exam and wants to pass the exam. In this sense, i is pleased by the fact that it believes to have achieved what it was committed to achieve. In this sense, joy has a positive valence, that is, it is associated with goal achievement.[7]

The fact that an agent i feels sadness about a fact φ can be expressed as follows:

$$\text{Sadness}_i\, \varphi =_{def} \text{Bel}_i\, \varphi \wedge \text{Choice}_i\, \neg\varphi$$

According to this definition, agent i feels sadness about φ if and only if i believes that φ is true and wants $\neg\varphi$ to be true. Agent i feels sad for not having passed the exam because i believes that he has not passed the exam and wants to pass the exam. In this sense, i is displeased by the fact that it believes not to have achieved what it was committed to achieve. Thus, sadness has a negative valence, that is, it is associated with goal frustration.

In the following section we extend our basic language of beliefs, choices and ideals[8] in order to provide a logical characterization of some complex emotions, namely regret and guilt.

A Logic of Complex Emotions

Although the application of logical methods to the formal specification of emotions has been quite successful (Adam and Longin, 2007; Lorini and Castelfranchi, 2007; Steunebrink, Dastani and Meyer, 2007), we think that there is still much work to be done in the field of computational and logical modeling of complex emotions such as regret, shame, guilt or reproach. By 'complex emotions' we mean those emotions that are based on complex forms of reasoning and on a larger set of appraisal variables than the set involved in 'basic emotions' (*e.g.* fear, happiness, sadness). According to some psychological theories of emotions (see *e.g.* Ekman, 1999, Johnson-Laird and Oatley, 1992), while basic emotions are related to an individual's basic biological needs and can be elicited after merely an automatic and unconscious analysis of a situation, complex emotions are based on conscious cognitive activity and involve very sophisticated forms of reasoning, in particular *self-attribution of responsibility* and *counterfactual reasoning*. For example, an agent regrets something it did, if it is aware to have taken a certain decision which turned out to be not as good as expected, and it believes that if it had decided to do something different, it would have achieved better (hence it feels *responsible*). In this sense, regret is based on the capacity to *imagine* alternative scenarios that could have occurred if it had behaved differently (Kahneman, 1995; Zeelenberg and van Dijk, 2005). Several psychologists working on guilt (see *e.g.* Lazarus, 1991; Tangney, 1999) have stressed that this emotion involves the conviction of being *responsible* for having injured someone or violated some norm or imperative.

Therefore, in order to build a logical framework which allows us to characterize complex emotions, it is necessary to go beyond the standard BDI approach presented in Section 3.1. Indeed, a logic of complex emotions should be expressive enough to characterize not only different types of agents' mental attitudes (beliefs, desires, goals, ideals), but also the concept of responsibility. We think that a promising solution for developing a logic of complex emotions is to combine a BDI logic, as the one presented in Section 3.1, with a logic of cooperation and multi-agent interaction (Alur and Henzinger, 2002; Belnap, Perloff and Xu, 2001; Horty, 2001; Lorini and Schwarzentruber, 2009; Pauly, 2002). Indeed, this kind of logics is well suited to support both counterfactual reasoning and reasoning about responsibility of single agents and of groups of agents. The following section is devoted to present a specific logic of cooperation and multi-agent interaction: STIT logic (the logic of *Seeing To It That*), that we think to be a natural candidate for a logical analysis of complex emotions.

A General Overview of STIT Logic

The modal logic of *Seeing To It That* (STIT) has been proposed in the domain of formal philosophy in the 90ies (Belnap, Perloff and Xu, 2001; Horty and Belnap, 1995; Horty, 2001). More recently, it has been imported into the field of theoretical computer science where its formal relationships with other logics for multi-agent systems have been studied (Broersen, Herzig and Troquard, 2006). STIT is a logic which supports reasoning about actions of agents and joint actions of groups. Moreover, it supports reasoning about individual powers of agents and collective powers of groups, and responsibilities of agents and groups.

The semantics of STIT is based on a branching time structure with ordered *moments*. We here prefer to consider the syntactical aspects of STIT rather than presenting its semantics (that is presented and extensively discussed in Belnap, Perloff and Xu, 2001, Horty, 2001 and Lorini and Schwarzentruber, 2009).

Below AGT denotes the set of all agents. $C \subseteq AGT$ denotes an arbitrary set of agents. $AGT \setminus C$ denotes the complement of the set of agents C with respect to AGT, and $AGT \setminus \{i\}$ denotes the

complement of $\{i\}$ with respect to AGT. In the sequel, we refer to sets of agents as groups.

In STIT logic operators of the form $\text{Stit}_{\{i\}}$ for any individual agent i and Stit_C for any group of agents C are introduced. The modal formulas $\text{Stit}_{\{i\}} \varphi$ and $\text{Stit}_C \varphi$ respectively express that "a certain agent i brings it about that the state of affairs φ is true no matter what the other agents do", and "a group of agents C brings it about that the state of affairs φ is true no matter what the agents outside C do". Indeed, the formulas $\text{Stit}_{\{i\}} \varphi$ and $\text{Stit}_C \varphi$ can be respectively rephrased as follows:

- "agent i has just performed some action δ_i and, for all joint actions $\delta'_{AGT \setminus \{i\}}$ of the group of agents $AGT \setminus \{i\}$, if i does δ_i while the group of agents $AGT \setminus \{i\}$ does $\delta'_{AGT \setminus \{i\}}$, then φ will be true";
- "the group of agents C has just performed some joint action δ_C and, for all joint actions $\delta'_{AGT \setminus \{i\}}$ of the agents in $AGT \setminus C$, if the group C does the joint action δ_C while the group $AGT \setminus C$ does the joint action $\delta'_{AGT \setminus \{i\}}$, then φ will be true";

The STIT expression $\neg \text{Stit}_{AGT \setminus \{i\}} \varphi$ just says that:

- the complement of $\{i\}$ with respect to AGT (i.e. $AGT \setminus \{i\}$) does not see to it that φ (given what the agents in $AGT \setminus \{i\}$ have chosen to do)

which is the same thing as saying that:

- given what the agents in $AGT \setminus \{i\}$ have decided to do, there exists an alternative action of agent i such that, if agent i had chosen this action, the state of affairs φ would not be true now.

If φ is true, the latter sentence just means that *agent i could have prevented the state of affairs φ to be true now*. This expresses a general concept of responsibility. Intuitively, an agent i is said to be *responsible for* a certain state of affairs φ if and only if:

- the state of affairs φ is true now and
- agent *i* could have prevented φ from being true now.

Formally:

$$\text{Resp}_i\varphi =_{def} \varphi \wedge \neg\text{Stit}_{AGT\setminus\{i\}}\varphi$$

where $\text{Resp}_i\varphi$ means that "agent *i* is responsible for φ". For example, imagine that there are two agents $AGT = \{Bill, Bob\}$ who have to take care of a plant. Both Bill and Bob have only two actions available: water the plant or do nothing. If either both Bill and Bob water the plant or both Bill and Bob do nothing, the plant will die. In the former case the plant will die because it does not tolerate too much water. In the latter case it will die because it lacks water. If Bill (resp. Bob) waters the plant and Bob (resp. Bill) does nothing, the plant will survive. Suppose both Bill and Bob decide to water the plant and, as a consequence of their choices, the plant dies. In this situation both Bill and Bob are responsible for the death of the plant: $\text{Resp}_{Bill}deadPlant \wedge \text{Resp}_{Bob}deadPlant$. This means that: the plant is dead and, given what Bill had decided to do (*i.e.* watering the plant), there exists an alternative action of Bob (*i.e.* doing nothing) such that, if Bob did perform this action, the plant would have survived; and given what Bob had decided to do (*i.e.* watering the plant), there exists an alternative action of Bill (*i.e.* doing nothing) such that, if Bill did perform this action, the plant would have survived. In this sense, Bill (resp. Bob) could have avoided the death of the plant. This is the reason why both Bill and Bob are responsible for it.

A Formalization of Regret and Guilt

As said before, an integration of STIT logic with a logic of agents' mental attitudes is particularly suitable to provide a logical characterization of complex emotions based on the concept of responsibility. Here we just consider the emotions of *regret* and *guilt* as case studies.

We add modal operators for beliefs, goals and ideals as the ones presented in Section 3.1 to STIT logic. Then we can come up with the following formal characterization of the concept of *regret* and *guilt*.

We say that *i* regrets for φ if and only if ¬φ is a chosen goal of *i* and *i* believes that it is responsible for φ. Formally:

$$\text{Regret}_i\varphi =_{def} \text{Bel}_i \, \text{Resp}_i \, \varphi \wedge \text{Choice}_i \, \neg\varphi$$

Imagine a situation in which there are only two agents *i* and *j*, that is, $AGT = \{i,j\}$. Agent *i* decides to park its car in a no parking area. Agent *j* (the policeman) fines agent *i* 100 €. Agent *i* regrets for having been fined 100 € (noted $\text{Regret}_i fine$). This means that, *i* wants not to be fined (noted $\text{Choice}_i \neg fine$) and believes that it is responsible for having been fined (noted $\text{Resp}_{i,j} fine$). That is, agent *i* believes that it has been fined 100 € and believes that it could have avoided to be fined (by parking elsewhere).

It is worth noting that, according to the definition of regret given here and the definition of sadness given in Section 3.1, the former emotion entails the latter. That is, if agent *i* regrets for φ then, it feels sad about φ. Formally:

$$\text{Regret}_i\varphi \rightarrow \text{Sadness}_i \, \varphi$$

The concept of responsibility is also relevant for the characterization of the *guilt* emotion. Indeed, guilt involves the belief of being responsible for the violation of an ideal or imperative. Formally:

$$\text{Guilt}_i \, \varphi =_{def} \text{Bel}_i \, \text{Resp}_i \, \varphi \wedge \text{Ideal}_i \, \neg\varphi$$

According to this definition, agent *i* feels guilty for φ (noted $\text{Guilt}_i \, \varphi$) if and only if ¬φ is an ideal state of affairs for i (noted Ideal$_i$ ¬φ) and i believes to be responsible for φ.

For example, imagine a situation in which there are only two agents i and j (that is $AGT = \{i, j\}$). Agent i decides to shoot with gun and accidentally kills agent j. Agent i feels guilty for having killed someone (noted Guilt$_i$ *killedSomeone*). This means that, i addresses an imperative to itself not to kill other people (noted Ideal$_i$ ¬*killedSomeone*) and agent i believes that it is responsible for having killed someone (noted Resp$_i$ *killedSomeone*).

As said above, in some situations it is reasonable to suppose that an agent decides to pursue all its ideals, that is, if φ is an ideal state of affairs for agent i then, i chooses φ. Under this hypothesis, we can infer formally that guilt implies regret, that is:

Guilt$_i$ φ → Regret$_i$ φ

This is not the place to defend the plausibility of such a formal consequence. We think that the only way to validate it is to perform psychological experiments on human subjects.

Related Works on Logical Modeling of Emotions

As emphasized in the previous sections, other researchers have exploited logical methods in order to build formal models of emotions and affective agents. One of the most prominent logical account of emotions is the one proposed in (Steunebrink, Dastani and Meyer, 2007). Meyer *et al.* provide a logical analysis of several types of emotions grounded on existing psychological theories of emotions (*e.g.* Lazarus, 1991; Ortony, Clore and Collins, 1988). To this aim, a logical framework which support reasoning about actions, time and agents' attitudes is used.

It is worth noting that there is a fundamental difference between our approach and Meyer *et al.*'s approach. In Meyer *et al.*'s approach each instance of emotion is represented with a special predicate, or fluent, in the jargon of reasoning

about action and change, to indicate that these predicates change over time. For every fluent a set of effects of the corresponding emotions on the agent's planning strategies are specified, as well as the preconditions for triggering the emotion. On the contrary, in our logic of emotions there are no specific formal constructs which are used to denote that a certain emotion arises at a certain time. We just define emotions from more basic concepts of knowledge and desire. For instance, according to our definition of regret, an agent regrets for φ if and only if it has the *chosen goal* that ¬φ and *believes* that it is responsible for φ. We prefer this kind of approach since it allows to characterize explicitly the cognitive constituents of emotions.

Nevertheless, we are aware that our approach is simplistic. In particular, it misses important psychological aspects such as the fact that the cognitive constituents of emotions (*e.g.* knowledge, desires) are usually joined with bodily activation and components, and these components shape the whole subjective state and determine the nature of the affective reaction (Frijda, 1986). We deal here with the limits of any disembodied mind (and model).

CASE STUDY

In order to illustrate how an agent able to reason about users' complex emotions can be useful in AmI, we present here several scenarios involving such an agent in the context of an intelligent house. There already exist some industrial or academic projects aiming at designing such intelligent houses, for example SGI Japan RoomRender, Philips HomeLab (Aarts and Eggen, 2002) or MIT MediaLab. We suppose that this intelligent house is able to monitor its inhabitants (a family) in order to measure some physiological data that will be useful to identify their emotion. But if this physiological data allows to identify some emotions (Forest et al., 2006; Picard, 1997; Prendinger and Ishizuka, 2005), information about the context and users' mental attitudes allows to be more precise (Forest *et al*., 2006). The house software is thus also composed of several agents integrated in various objects (the fridge, a personal assistant, a car...). Each agent has access to a profile of the inhabitants, including their desires, beliefs, and any useful information; it is also endowed with a logical model of emotions like the one proposed in this chapter. It is then able to reason about all this information in order to precisely identify the users' emotion and its object, and to decide of its subsequent behavior.

To be more precise, an agent first detects an emotion and then tries to refine and explain it *a posteriori* from its knowledge base and its model of emotions; it does not predict emotions, which could lead to unadapted reactions in case of wrong prediction. As a very simple example, an agent can detect a negatively valenced emotion in the mother; then it observes that it is raining and that it is contrary to the mother's preferences; thus it infers from its model of emotions that she may be sad about the weather; and finally it performs some actions to divert her from the rain outside (for instance, display some photos of sunny landscapes on the wall, play music, *etc.*).

The two scenarios detailed below illustrate the two complex emotions that we have formalized in the previous section: regret and guilt.

Scenario 1: Regret

A morning, the father f is in a hurry because of an important meeting with his boss and a big client. He thus decides to use his car instead of the subway he usually takes, in order to gain some time. But he is caught into traffic jam because of an accident that just occurred five minutes ago. He finally arrives later than with the subway, and he is five minutes late for the meeting. He then reasons about his alternative plan (take the subway) and its supposed outcome (arrive on time) that would have been better than the outcome of

his actual plan (arrive late, by car). He thus feels regret about being late because he is responsible for the wrong choice that led to this undesirable situation.

We now assume that the father is equipped with a kind of personal assistant disposing of all useful information about him and able to reason about his mental attitudes and emotions. This agent detects a negative emotion from f's physiological data, and then tries to refine and explain it from its knowledge base and its logical model of emotions. In particular in this context this agent knows[9] that:

- the father f aimed to arrive on time (before 9 a.m.) ($Choice_f \neg isLate$);
- f failed to achieve his goal (he arrived late, at 9.05 a.m.) ($isLate$) and is aware of his failure ($Bel_f isLate$);
- f believes that he could have seen to it not to be late ($Bel_f \neg Stit_{AGT\setminus\{f\}} isLate$): he could have taken the subway which would have made him arrive on time.

The agent possibly also has (in the profile of f) information about the degree to which f is likely to make counterfactual reasoning after a failure (not all people are equally likely to do so).

It is thus able to identify that the father feels regrets about being late ($Regret_f isLate$), because he believes to be responsible for being late ($Bel_f Resp_f isLate$) and he aimed at being on time ($Choice_f isLate$). This precise belief about f's emotion allows the agent to behave in the best manner, for example by adopting adapted strategies to help f to cope with his regret. Our aim is not to formalize these strategies, but rather to present them as illustrations showing the importance for the agent to have as much information as possible about the nature and object of the emotion.

The agent can for example try to free f from his responsibility for being late: indeed he was not able to foresee the accident, and without this accident he would have arrived on time. It can thus switch the responsibility from f to the bad driver who provoked the accident, or to people who did not clean the road quickly enough after the accident; by this way f's emotion switches from regret to anger toward the new supposed accountable people. Moreover the agent can exhibit statistics showing that taking car is usually quicker than taking the subway. Finally it can check if the subway has really arrived on time, since if it is not the case it can inform f that he is not responsible for being late because his other possible decision would have led to the same result.

The agent can also play down the importance of the delay (a delay of only five minutes is quite few) and compare it with the advantages of taking car over taking subway (less stress, calm travel, possibility to listen to music...). The agent cannot imagine what would have gone wrong in subway, but it can refer to an history of past travels in subway that have gone wrong for various reasons: someone spilled coffee on f's suit, subway was late because of strikes, accidents, suicide on the railway... We can even suppose that f's personal assistant is able to communicate with other users' personal assistants to establish statistics about all that can go wrong in the subway.

Scenario 2: Guilt

Tonight it is the son's birthday dinner. He has requested his favorite meal: a roasted chicken. The intelligent fridge has thus proposed a menu taking into account both the dietetic aspects and the family's tastes. Since the father will get back home late tonight due to an important meeting, the mother m was assigned to buy what is needed and to prepare the dinner. After an exhausting day's work, the mother has forgotten to buy the roasted chicken, and she has no more time now to go back to the supermarket. She feels guilty because she was responsible for the shopping and by her fault the dinner will be different from what her son has requested.

We consider that the agent embodied in the intelligent fridge is able to communicate with the cupboards in the whole kitchen in order to know their content. It is also able to prepare the menus for the meals of the week, and to establish the corresponding shopping list. It is thus able to deduce that a roasted chicken is missing to prepare the intended dinner.

Actually, the lack of roasted chicken may have resulted from the mother's decision to change the menu, or from an unavailability of this product at the supermarket, what the agent cannot tell at this point. If the agent tries to predict m's emotion from the limited information it has, it may thus fail and adopt unadapted behavior. However, if it first detect a negative emotion in m, and then uses it as additional information, his behavior will be much more relevant to the situation.

The agent thus detects a negative valence in m informing it of a problem. It then tries to explain it from its knowledge and its model of emotions. In particular this agent knows the following information:

- a roasted chicken is missing (but it does not know why yet), preventing the mother to cook the requested dinner ($\text{Bel}_m \neg dinnerAsRequested$);
- m was assigned to buy the chicken, thus it is her fault that the dinner will not be as expected ($\text{Bel}_m \neg \text{Stit}_{AGT\backslash\{m\}} \neg dinnerAsRequested$);
- there is a rule in the family specifying that the person who celebrates his birthday chooses the menu of the dinner; thus it is ideal for m and for the other inhabitants that the dinner matches the son's request ($\text{Ideal}_m \, dinnerAsRequested$).

From this information, the agent deduces that the dinner will not be the planned one ($\text{Bel}_m \neg dinnerAsRequested$). Moreover, since m is responsible for this situation ($\text{Resp}_m \neg dinnerAsRequested$) that violates an ideal of the family ($\text{Ideal}_m \, dinnerAs$-

Requested), the agent can deduce that m may feel guilty about this (Guilt$_m$ $\neg$ *dinnerAsRequested*). This emotion explains the negative valence measured by the agent. Having identified precisely the mother's negative emotion, the agent can now adopt adequate strategies to improve the mother's well-being. In particular, it can help her to cope with her guilt by proposing either some material strategies (*e.g.* compensating for her forgetting) or emotional ones (*e.g.* looking on the bright side of the situation).

Being aware of everything available in the kitchen, the intelligent agent can propose another ingredient to replace the missing one, for example there is some ham in the fridge. It can also propose another recipe that is possible with what is available at home and that still matches the son's preferences in order to ease his disappointment.

The intelligent fridge could also help the mother to look on the bright side of the situation. For example, she hates roasted chicken and her involuntary omission will lead her to prepare another recipe that the whole family could enjoy. This could also be the occasion to go to the restaurant instead and to discover new meals, or to eat roasted chicken without having to prepare it herself.

Please note again that it is important for the agent to have maximal information about *m*'s mental attitudes and emotion, since the strategy would not be the same if roasted chicken was unavailable at the supermarket, or if she had decided by herself to change the menu.

CONCLUSION

In this chapter, we have shown the importance of emotions for Ambient Intelligence. Moreover, we have argued that existing approaches and tools in AmI are not adapted for reasoning about emotions or identifying the object of a user's emotions. We have shown that these limitations are even more obvious when looking at *complex* emotions

based on sophisticated forms of reasoning such as reasoning about responsibility and counterfactual reasoning.

We have proposed a logical framework for the analysis and formal specification of emotions. In fact, we think that logic is a natural tool for reasoning about the cognitive constituents of complex emotions and for dealing with the identification of the object of emotions. Our logical framework allows to specify the mental attitudes of agents (beliefs, choices, ideals) and to characterize the concept of responsibility which is a fundamental building block of complex emotions. We have also provided a logical characterization of two specific complex emotions based on responsibility: regret and guilt.

Finally, we have presented and formalized two AmI scenarios involving regret and guilt. We have taken into consideration various aspects of emotions relevant to AmI, such as the identification of a user's emotion or the influence of the user's emotion on reasoning and planning.

The present work is a first step towards a general logical model of complex emotions to be applied to Ambient Intelligence. We would like to highlight that what we have proposed in this chapter is a model, in the sense of "an abstract construction that allows to understand a reference system by answering some questions about it." (Treuil, Drogoul and Zucker, 2008). Note that the use of logic as a modeling language allows us to provide an unambiguous model of (complex) emotions and their links with mental attitudes; such a model is also independent of the implementation language. The implementation of a conceptual model is a huge and hard task and this step remains out of the scope of this chapter. Logic has to be considered as a useful specification language to represent Ambient Intelligence concepts. Some dedicated agent programming languages could be used to improve this step such as 3APL (Dastani *et al.*, 1994) or CLAIM (El Fallah-Seghrouchni and Suna, 2005). This could be an interesting continuation of the work presented in this chapter.

Another envisaged continuation will be to extend our analysis to other complex emotions such as shame, reproach, relief or embarrassment.

ACKNOWLEDGMENT

Carole Adam would like to acknowledge the support of the Australian Research Council, and RealThing Pty Ltd. under Linkage Grant LP0882013. Benoit Gaudou would like to acknowledge the support of the French IRD SPIRALES research program and the French 3 Worlds ANR project. Dominique Longin and Emiliano Lorini are supported by the French ANR project CECIL about "Complex Emotions in Communication, Interaction and Language" (www.irit.fr/CECIL), contract No. ANR-08-CORD-005, and by University of Toulouse through the project AmIE "Ambient Intelligence Entities" (scientific call of the Paul Sabatier University – Toulouse III).

REFERENCES

Aarts, E., & Eggen, B. (Eds.). *Ambient Intelligence in Home-Lab*. Neroc, Eindhoven, The Netherlands, 2002.

Adam, C. (2007). *Emotions: from psychological theories to logical formalization and implementation in a BDI agent*. PhD thesis, INP Toulouse, France.

Adam, C., Gaudou, B., Herzig, A., & Longin, D. (2006) A logical framework for an emotionally aware intelligent environment. In *AITAmI'06*. IOS Press.

Adam, C., & Longin, D. Endowing emotional agents with coping strategies: from emotions to emotional behaviour. In *IVA'2007*, (*LNCS* 4722, pp 348–349). Springer, 2007.

Adam, C., Herzig, A., & Longin, D. *A logical formalization of the OCC theory of emotions*. In: *Synthese*, Springer, Vol. 168:2, p. 201-248, 2009.

Alur, R., & Henzinger, T. (2002). Alternating-time temporal logic. *Journal of the ACM, 49*, 672–713. doi:10.1145/585265.585270

Amal El Fallah-Seghrouchni and Alexandru Suna. (2005). CLAIM and SyMPA: A Programming Environment for Intelligent and Mobile Agents. In *Multi-Agent Programming* (pp. 95–122). Languages, Platforms and Applications.

Anand, S. Rao and Michael P. Georgeff. Modeling rational agents within a BDI-architecture. In *Proceedings of KR'91*, pages 473–484. Morgan Kaufmann Publishers, 1991.

Andrew Ortony, G. L. (1988). *Clore, and A. Collins. The cognitive structure of emotions*. Cambridge, MA: Cambridge University Press.

Antonio R. Damasio. *Descartes' error: emotion, reason, and the human brain*. Putnam pub. group, 1994.

Augusto, J.-C., McCullagh, P., McClelland, V., & Walkden, J.-A. Enhanced healthcare provision through assisted decision making in a smart home environment. In *AITAmI07*, 2007.

Belnap, N., Perloff, M., & Xu, M. (2001). *Facing the future: agents and choices in our indeterminist world*. New York: Oxford University Press.

Berkman, L. F., Glass, T., Brissette, I., & Seeman, T. E. (2000). From social integration to health: Durkheim in the new millennium. *Social Science & Medicine, 51*, 843–857. doi:10.1016/S0277-9536(00)00065-4

Broersen, J., Herzig, A., & Troquard, N. (2006). Embedding alternating-time temporal logic in strategic STIT logic of agency. *Journal of Logic and Computation, 16*(5), 559–578. doi:10.1093/logcom/exl025

Burton, D. (1988). Do anxious swimmers swim slower? Reexamining the elusive anxiety-performance relationship. *Journal of Sport & Exercise Psychology, 10,* 45–61.

Castaneda, H. N. (1975). *Thinking and Doing.* Dordrecht: D. Reidel.

Castelfranchi, C., & Lorini, E. Cognitive anatomy and functions of expectations. In *IJCAI03 Workshop on Cognitive Modeling of Agents and Multi-Agent Interactions.* Morgan Kaufmann Publisher, 2003.

Chellas, B. F. (1980). *Modal logic: an introduction.* Cambridge: Cambridge University Press.

Chen Wu and Hamid Aghajan. Context-aware gesture analysis for speaker hci. In *AITAmI'08,* 2008.

Conte, R., & Castelfranchi, C. (1995). *Cognitive and social action.* London: London University College of London Press.

Cynthia Breazeal and B. Scassellati. (1999). *How to build robots that make friends and influence people.* IROS.

de Rosis, F., Pelachaud, C., Poggi, I., Carofiglio, V., & Carolis, B. D. (2003). From Greta's mind to her face: modelling the dynamics of affective states in a conversational embodied agent. *International Journal of Human-Computer Studies, 59*(1-2), 81–118. doi:10.1016/S1071-5819(03)00020-X

K. Ducatel, M. Bogdanowicz, F. Scapolo, J. Leijten, and J.-C. Burgelman. Scenarios for ambient intelligence in 2010. ISTAG report at IPTS Seville, 2001.

Ekman, P. (1992). An argument for basic emotions. *Cognition and Emotion, 6,* 169–200. doi:10.1080/02699939208411068

Ekman, P. (1999). Basic emotions. In Dalgleish, T., & Power, M. (Eds.), *Handbook of cognition and emotion.* John Wiley & Sons.

El Jed, M., Pallamin, N., Dugdale, J., & Pavard, B. Modelling character emotion in an interactive virtual environment. In *AISB 2004 Symposium: Motion, Emotion and Cognition,* 2004.

Elliott, C., Rickel, J., & Lester, J. (1999). Lifelike pedagogical agents and affective computing: An exploratory synthesis. *LNCS, 1600,* 195–211.

Forest, F. (2006). *Astrid Oehme, Karim Yaici, and Celine Verchere-Morice. Psycho-social aspects of context awareness in ambient intelligent mobile systems. In 15th.* IST Summit.

Frijda, N. H. (1986). *The Emotions.* Cambridge University Press.

Helmut Prendinger and Mitsuru Ishizuka. (2001). Social role awareness in animated agents. In. *Proceedings of AGENTS, 2001,* 270–277.

Helmut Prendinger and Mitsuru Ishizuka. (2005). Human physiology as a basis for designing and evaluating affective communication with life-like characters. *IEICE Transactions on Information and Systems. E (Norwalk, Conn.), 88-D*(11), 2453–2460.

Hintikka, J. (1962). *Knowledge and Belief.* New York: Cornell University Press.

Horty, J. F., & Belnap, N. (1995). The deliberative STIT: A study of action, omission, and obligation. *Journal of Philosophical Logic, 24*(6), 583–644. doi:10.1007/BF01306968

Hume, D. *A Treatise of Human Nature.* Clarendon Press, Oxford, 1978. Selby-Bigge, L. A. & Nidditch, P. H. (Eds.).

Jiehan Zhou, J. Riekki, Y. Changrong, and E. Kärkkäinen. AmE framework: a model for emotion-aware ambient intelligence. In *ACII2007: Doctoral Consortium,* 2007.

John, F. (2001). *Horty. Agency and Deontic Logic.* Oxford: Oxford University Press.

John, R. (1983). *Searle. Intentionality: An essay in the philosophy of mind*. Cambridge University Press.

Johnson-Laird, P. N., & Oatley, K. (1992). Basic emotions: a cognitive science approach to function, folk theory and empirical study. *Cognition and Emotion, 6*, 201–223. doi:10.1080/02699939208411069

Jonathan Gratch and Stacy Marsella. (2004). A Domain-independent Framework for modelling Emotions. *Journal of Cognitive Systems Research, 5*(4), 269–306. doi:10.1016/j.cogsys.2004.02.002

Kahneman, D. (1995). Varieties of counterfactual thinking. In Roese, N. J., & Olson, J. M. (Eds.), *What might have been: the social psychology of counterfactual thinking* (pp. 375–396). Mahwah, NJ: Erlbaum.

Kazuyoshi Wada and Takanori Shibata. Robot therapy in a care house -its sociopsychological and physiological effects on the residents. In *ICRA06*, 2006.

Lorini, E., & Castelfranchi, C. (2007). The cognitive structure of surprise: looking for basic principles. *Topoi: an International Review of Philosophy, 26*(1), 133–149.

Lorini, E., & Herzig, A. (2008). A logic of intention and attempt. *Synthese, 163*(1), 45–77. doi:10.1007/s11229-008-9309-7

Lorini, E., & Schwarzentruber, F. A Logic for Reasoning about Counterfactual Emotions. In C. Boutilier (Eds.), Proceedings of the Twenty-first International Joint Conference on Artificial Intelligence (IJCAI'09), AAAI Press, pages 867-872, 2009.

Marreiros, G., Santos, R., Ramos, C., Neves, J., Novais, P., Machado, J., & Bulas-Cruz, J. Ambient intelligence in emotion based ubiquitous decision making. In *AITAmI'07*, 2007.

McCann, H. (1991). Settled objectives and rational constraints. *American Philosophical Quarterly, 28*, 25–36.

Mehdi Dastani, M. (2005). Birna van Riemsdijk, and John-Jules Ch. Meyer. Programming Multi-Agent Systems in 3APL. In *Multi-Agent Programming* (pp. 39–67). Languages, Platforms and Applications.

Oehme, A., Herbon, A., Kupschick, S., & Zentsch, E. Physiological correlates of emotions. In *AISBO07*, 2007.

Ortony, A. (2003). On making believable emotional agents believable. In Trappl, R., Petta, P., & Payr, S. (Eds.), *Emotions in Humans and Artifacts* (pp. 189–212). MIT Press.

Pauly, M. (2002). A modal logic for coalitional power in games. *Journal of Logic and Computation, 12*(1), 149–166. doi:10.1093/logcom/12.1.149

Philip R. Cohen and Hector J. Levesque. Intention is choice with commitment. *Artificial Intelligence Journal, 42*(2– 3):213–261, 1990.

Richard, S. (1991). *Lazarus. Emotion and Adaptation*. Oxford University Press.

Rickel, J., & Lewis Johnson, W. Steve: an animated pedagogical agent for procedural training in virtual environments. *Animated Interface Agents: making them intelligent*, pages 71–76, 1997.

Roger Jianxin Jiao. (2007). *Qianli Xu, and Jun Du. Affective human factors design with ambient intelligence*. HAI.

Rosalind, W. (1997). *Picard. Affective Computing*. MIT Press.

Rosalind, W. (2001). Picard, E. Vyzas, and J. Healey. Toward machine emotional intelligence: analysis of affective physiological state. *IEEE Transactions on Pattern Analysis and Machine Intelligence, 23*(10).

Scherer, K. R. (2001). *Appraisal Processes in Emotion: Theory, Methods, Research, chapter Appraisal Considered as a Process of Multilevel Sequential Checking* (pp. 92–120). New York: Oxford University Press.

Sebe, N., Cohen, I., Gevers, T., & Huang, T. S. Multimodal approaches for emotion recognition: a survey. In *SPIE: Internet Imaging*, pages 56–67, 2004.

Stacy Marsella and Jonathan Gratch. Modeling coping behavior in virtual humans: don't worry, be happy. In *AAMAS-2003*, pages 313–320. ACM, 2003.

Steunebrink, B., Dastani, M., & Meyer, J.-J. A logic of emotions for intelligent agents. In R.C. Holte and A.E. Howe, editors, *Proc. AAAI-07*, pages 142–147, 2007. AAAI Press.

Susskinda, J. M., Littlewortb, G., Bartlettb, M. S., Movellanb, J., & Anderson, A. K. (2007). Human and computer recognition of facial expressions of emotion. *Neuropsychologia, 45*, 152–162. doi:10.1016/j.neuropsychologia.2006.05.001

Tangney, J. P. (1999). The self-conscious emotions: shame, guilt, embarrassment and pride. In Dalgleish, T., & Power, M. (Eds.), *Handbook of cognition and emotion*. John Wiley & Sons.

Treuil, J.-P., Drogoul, A., & Zucker, J.-D. (2008). *Modélisation et simulation à base d'agents: Approches particulaires, modèles à base d'agents, de la mise en pratique aux questions théoriques.* Dunod.

Treur, J. On human aspects in ambient intelligence. In *Proceedings of First International Workshop on Human Aspects in Ambient Intelligence (HAI)*, pages 5–10, 2007.

van Breemen, A. iCat: Experimenting with animabotics. In *AISB*, pages 27–32, University of Hertfordshire, Hatfield, UK, April 2005.

Zeelenberg, M., & van Dijk, E. (2005). On the psychology of "if only": regret and the comparison between factual and counterfactual outcomes. *Organizational Behavior and Human Decision Processes, 97*(2), 152–160. doi:10.1016/j.obhdp.2005.04.001

KEY TERMS AND DEFINITIONS

Agent: Autonomous entity that is able to perceive its environment, reason about it, maintain personal goals, plan appropriate ways to achieve them, and act accordingly to modify in turn its environment. Some agents are also able to communicate with other artificial agents or with humans, or to express and understand emotions.

Emotion: Intentional affective mental state (in the sense of Searle, *i.e.* having an object) that arises automatically as a result of an individual's interpretation of his environment and its relation to his attitudes (*e.g.* goals and interests). Emotions can then influence behavior in various ways, and were thus long considered irrational before research showed that they were actually essential in rational decision making.

Mood: Non-intentional affective mental state (in the sense of Searle, *i.e.* having no object). For example, one can be in a happy mood (about nothing in particular) which is different from feeling the emotion of happiness (about something, for example that the weather is sunny). Moods are also characterized by a longer duration than emotions.

Cognitive Appraisal Theories of Emotions: In these theories, emotions are triggered by a process (called appraisal) determining the significance of the situation for the individual by assessing it *w.r.t.* various criterions called appraisal variables (*e.g.* desirability).

Appraisal Variables: Criteria used to evaluate the significance of a situation *w.r.t.* an individual's goals, ideals or desires, determining the triggered emotion. For example a pleasant situation (the pleasantness variable is measured by comparing the situation with the individual's goals) leads to a positive emotion (*e.g.* joy).

Complex Emotions: Emotions based on sophisticated forms of reasoning such as reasoning about responsibility and counterfactual reasoning. They are thus based on a larger set of appraisal variables than the set involved in basic emotions.

Counterfactual Reasoning: Reasoning based on what the current state of the world would be or would not be if I had acted differently, *i.e.* if I had performed or had not performed a certain action. For example, if I had slept this morning, I would be less tired; if I had not run this morning, I would feel sad; if I had taken the train instead of my car, I would not be caught in a traffic jam; if I had bought petrol earlier, I would not have run out of petrol on the freeway.

Coping Strategies: Strategies (or actions sets) used by an agent to restore its well-being in stressful situations. For example one can deny a piece of information that makes him too sad; one often starts by denying the death of a loved one.

Actions Tendencies: Processes associated to emotions that allow individuals to make quick decisions, using their emotions as a kind of heuristic. For example, fear induces a tendency to escape while anger induces a tendency to attack.

BDI Logic: Formal language used to represent agents' reasoning in terms of their mental attitudes (beliefs, desires, intentions) as described by philosophers of mind (*e.g.* Bratman).

Responsibility: An agent i is responsible for causing p if and only if i has performed a certain action and if i had not performed this action then p would not be true now. For example a driver is responsible for causing an accident by running a red light because if he had not, then the accident would not have occurred.

ENDNOTES

[1] According to Picard: "The computer's emotions are labels for states that may not exactly match the analogous human feelings, but that initiate behavior we would expect someone in that state to display." (Picard, 1997, p. 298)

[2] Some authors (*e.g.* Tangney, 1999) prefer the term *self-conscious emotion* to the term *complex emotion*.

[3] According to cognitive appraisal theories (Frijda, 1986; Lazarus, 1991; Ortony, Clore and Collins, 1988), emotions are triggered by a process (called appraisal) determining the significance of the situation for the individual by assessing it *w.r.t.* various criterions called appraisal variables (*e.g.* desirability).

[4] We highlight the fact that in this formalism an agent belief is something that this agent is sure about: beliefs are the subjective representation of the world for the agent. This sense is quite different from the common use of this concept: to believe a proposition p often means that p is uncertain.

[5] That is, every operator Bel_i satisfies the so-called Axiom K ($Bel_i \varphi \land Bel_i (\varphi \rightarrow \psi) \rightarrow Bel_i \psi$) and the so-called rule of necessitation: if φ is a theorem then $Bel_i \varphi$ is a theorem.

[6] Every operator $Choice_i$ satisfies Axiom K ($Choice_i \varphi \land Choice_i (\varphi \rightarrow \psi) \rightarrow Choice_i \psi$) and the rule of necessitation: if φ is a theorem then $Choice_i \varphi$ is a theorem.

[7] The terms *positive valence* and *negative valence* are used by Ortony *et al.* (1988), whereas Lazarus (1991) uses the terms *goal congruent* vs. *goal incongruent* emotions.

[8] Note that this simple language could be extended by a concept of graded belief in order to account for the *intensity* of emotions, that could then depend on the degree of certainty of the beliefs involved in their definition.

[9] To get all useful information, the personal assistant agent may develop several strategies: communicate with the car agent, question the father or the personal assistant of people knowing him, make deductions... but this is not the point here.

Chapter 8
Incorporating Human Aspects in Ambient Intelligence and Smart Environments

Tibor Bosse
Vrije Universiteit Amsterdam, The Netherlands

Mark Hoogendoorn
Vrije Universiteit Amsterdam, The Netherlands

Michel Klein
Vrije Universiteit Amsterdam, The Netherlands

Rianne van Lambalgen
Vrije Universiteit Amsterdam, The Netherlands

Peter-Paul van Maanen
Vrije Universiteit Amsterdam, The Netherlands

Jan Treur
Vrije Universiteit Amsterdam, The Netherlands

ABSTRACT

In this chapter, we propose to outline the scientific area that addresses Ambient Intelligence applications in which not only sensor data, but also knowledge from the human-directed sciences such as biomedical science, neuroscience, and psychological and social sciences is incorporated. This knowledge enables the environment to perform more in-depth, human-like analyses of the functioning of the observed humans, and to come up with better informed actions. A structured approach to embed human knowledge in Ambient Intelligence applications is presented an illustrated using two examples, one on automated visual attention manipulation, and another on the assessment of the behaviour of a car driver.

DOI: 10.4018/978-1-61692-857-5.ch008

HUMAN KNOWLEDGE IN AMBIENT INTELLIGENCE

Ambient Intelligence provides possibilities to contribute to more personal care; e.g., (Aarts, Harwig, Schuurmans, 2001; Aarts, Collier, Loenen, Ruyter, 2003; Riva, Vatalaro, Davide, Alcañiz, 2005). Acquisition of sensor information about humans and their functioning is an important factor, but without adequate knowledge for analysis of this information, the scope of such applications is limited. However, devices in the environment possessing such knowledge can show a more human-like understanding and base personal care on this understanding. For example, this may concern elderly people, patients depending on regular medicine usage, surveillance, penitentiary care, psychotherapeutical / selfhelp communities, but also, for example, humans in highly demanding tasks such as warfare officers, air traffic controllers, crisis and disaster managers, and humans in space missions; e.g., (Green, 2005; Itti and Koch, 2001).

Within human-directed scientific areas, such as cognitive science, psychology, neuroscience and biomedical sciences, models have been and are being developed for a variety of aspects of human functioning. If such models of human processes are represented in a formal and computational format, and incorporated in the human environment in devices that monitor the physical and mental state of the human, then such devices are able to perform a more in-depth analysis of the human's functioning. This can result in an environment that may more effectively affect the state of humans by undertaking actions in a knowledgeable manner that improve their wellbeing and performance. For example, the workspaces of naval officers may include systems that, among others, track their eye movements and characteristics of incoming stimuli (e.g., airplanes on a radar screen), and use this information in a computational model that is able to estimate where their attention is focussed at. When it turns out that an officer neglects parts

of a radar screen, such a system can either indicate this to the person, or arrange on the background that another person or computer system takes care of this neglected part. Note that for a radar screen it would also be possible to make static design changes, for example those that improve situation awareness (e.g. picture of the environment, Wickens, 2002). However, as different circumstances might need a different design, the advantage of a dynamic system is that the environment can be adapted taking both the circumstances and the real-time behaviour of the human into account.

In applications of this type, an ambience is created that has a better understanding of humans, based on computationally formalised knowledge from the human-directed disciplines. The use of knowledge from these disciplines in Ambient Intelligence applications is beneficial, because it allows taking care in a more sophisticated manner of humans in their daily living in medical, psychological and social respects. In more detail, content from the domain of human-directed sciences, among others, can be taken from areas such as medical physiology, health sciences, neuroscience, cognitive psychology, clinical psychology, psychopathology, sociology, criminology, and exercise and sport sciences.

FRAMEWORK FOR REFLECTIVE COUPLED HUMAN-ENVIRONMENT SYSTEMS

One of the challenges is to provide frameworks that cover the class of Ambient Intelligence applications showing human-like understanding and supporting behaviour; see also (Treur, 2008). Here human-like understanding is defined as understanding in the sense of being able to analyse and estimate what is going on in the human's mind (a form of mindreading) and in his or her body (a form of bodyreading). Input for these processes are observed information about the human's state over time, and dynamic models for the human's physical and mental processes. For the mental side such a dynamic model is sometimes called a Theory of Mind (e.g., Baron-Cohen, 1995; Dennett, 1987; Gärdenfors, 2003; Goldman, 2006) and may cover, for example, emotion, attention, intention, and belief. Similarly for the human's physical processes, such a model relates, for example, to skin conditions, heart rates, and levels of blood sugar, insulin, adrenalin, testosterone, serotonin, and specific medicines taken. Note that different types of models are needed: physiological, neurological, cognitive, emotional, social, as well as models of the physical and artificial environment.

A framework can be used as a template for the specific class of Ambient Intelligence applications as described. The structure of such an ambient software and hardware design can be described in an agent-based manner at a conceptual design level and can be given generic facilities built in to represent knowledge, models and analysis methods about humans, for example (see also Figure 1):

- human state and history models

- environment state and history models
- profiles and characteristics models of humans
- ontologies and knowledge from biomedical, neurological, psychological and/or social disciplines
- dynamic process models about human functioning
- dynamic environment process models
- methods for analysis on the basis of such models

Examples of useful analysis methods are voice and skin analysis with respect to emotional states, gesture analysis, and heart rate analysis. The template can include slots where the application-specific content can be filled to get an executable design for a working system.

This specific content together with the generic methods to operate on it, provides a reflective coupled human-environment system, based on a tight cooperation between a human and an ambient system to show human-like understanding of humans and to react from this understanding in a knowledgeable manner. In this sense, 'coupled'

Figure 1. Framework to combine the ingredients

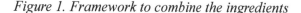

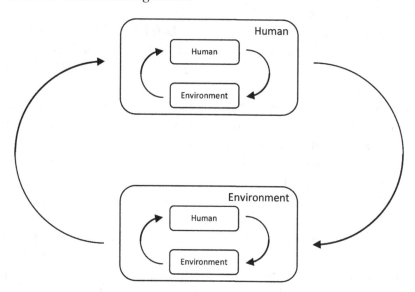

means mutually interacting. For the specific type of applications considered here, however, the coupling takes two different forms; see also Figure 2.

- On the one hand the coupling takes place as interaction between human and environment, as in any Ambient Intelligence application:
  - the environment gets information generated by the human as input, and
  - the human gets information generated by the environment as input.
- In addition, coupling at a more deep, reflective level takes place due to the fact that
  - the environment has and maintains knowledge about the functioning of the human, the environment and their interaction, and
  - the human has and maintains knowledge about functioning of him or herself, the environment, and their interaction

So, in such a more specific human-environment system, being coupled does not only mean that the human and its environment interact, but also that they have knowledge, understanding and awareness of each other, themselves and their interaction. This entails two types of awareness:

- **Human awareness:** awareness by the human about the human and environmental processes and their interaction
- **Technological awareness:** awareness by the environment about the human and environmental processes and their interaction

A general approach for embedding knowledge about the interaction between the environment and the human in Ambient Intelligence applications is to integrate dynamic models of this interaction (i.e. a model of the *domain*) into the application. This integration takes place by embedding domain models in certain ways within agent models of the Ambient Intelligence application. By incorporating domain models within an agent model, the Ambient Intelligence agent gets an understanding of the processes of its surrounding environment, which is a solid basis for knowledgeable intelligent behaviour. Three different ways to integrate domain models within agent models can be distinguished. A most simple way is to use a domain model that specifically models human behaviour in the following manner:

- **domain model directly used as agent model:** In this case a domain model that describes human processes and behaviour is used directly as an agent model, in or-

Figure 2. Reflective coupled human-environment systems

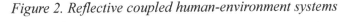
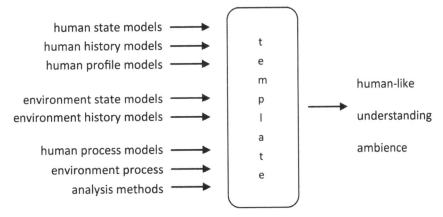

der to simulate human behaviour. Note that here the domain model and agent model refer to the same agent.

Such an agent model can be used in interaction with other agent models, in particular with *ambient agent models* to obtain a test environment for simulations. For this last type of (artificial) agents domain models can be integrated within their agent models in two different ways, in order to obtain one or more of the following (sub)models; see Figure 3. Here the solid arrows indicate information exchange between processes (data flow) and the dotted arrows the integration process of the domain models within the agent models.

- **analysis model:** To perform analysis of the human's states and processes by reasoning based on observations (possibly using specific sensors) and the domain model.
- **support model:** To generate support for the human by reasoning based on the domain model.

Note that here the domain model that is integrated refers to one agent (the human considered), whereas the agent model in which it is integrated refers to a different agent (the ambient software agent).

In the next sections of this chapter, two examples of Ambient Intelligence applications are presented, and the role of human knowledge in these applications is discussed. In the examples, each of the three types of models are discussed. "Automated Visual Attention Manipulation" addresses a case study on automated visual attention manipulation, and "Assessment of Driver Behaviour" addresses assessment of driver behaviour.

AUTOMATED VISUAL ATTENTION MANIPULATION

For people that have to perform complex and demanding tasks, it is very important to have sufficient *attention* for the various subtasks involved. This is particularly true for tasks that involve the continuous inspection of (computer) screens. For example, a person that has to inspect the images

Figure 3. Three ways to integrate a domain model within an agent model

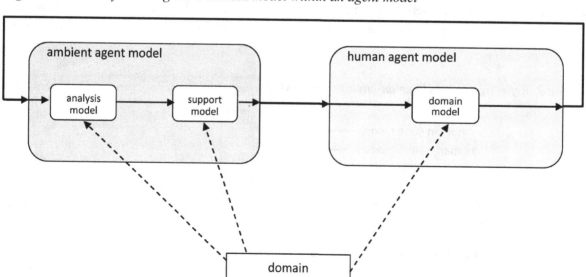

of a surveillance camera can not permit him- or herself to miss part of the events that occur. The same holds for an air traffic controller that inspects the movements of aircraft, or a naval operator that monitors the movements of hostile vessels on a radar screen. In such situations, a person may be supported by an ambient agent that keeps track of where his or her attention is, and provides some assistance in case the attention is not where it should be. To the end, the agent should maintain a *domain model* of the state of attention of the human, as well as an *analysis model* to reason about this domain model, and a *support model* to determine actions based on this analysis. This section introduces such models, and shows how they can be used to perform simulations. To make the models a bit more concrete, they are explained the context of a case study about a naval operator, which is introduced below.

Case Study: A Naval Operator

In the domain of naval warfare, it is crucial for the crew of the vessels involved to be aware of the situation in the field. Examples of important questions that should be addressed continuously are "in which direction are we heading?", "are we currently under attack?", "are there any friendly vessels around?", and so on. To assess such issues, one of the crew members is usually assigned the Tactical Picture Compilation Task (TPCT): the task to identify and classify all entities in the environment (e.g., Heuvelink and Both, 2007). This is done by monitoring several radar screens in the control room for radar contacts, and reasoning with the available information in order to determine the type and intent of the contacts on the screens. For example, available information on the speed and direction of a radar contact can be used to determine the threat of this contact. However, due to the complex and dynamic nature of the environment, the person assigned to the TPCT has to deal with a large number of tasks in parallel. Often the radar contacts are simply too

numerous and dynamic to be adequately monitored by a single human, which compromises the performance of the task.

For these reasons, it may be useful to offer this human operator some support from an intelligent ambient system, consisting of software agents that assist him in the execution of the Tactical Picture Compilation Task. For example, in case the human is directing its attention on the left part of a radar screen, but ignores an important contact that just entered the radar screen from the right, such a system may alert him about the arrival of that new contact. To be able to provide this kind of intelligent support, the system somehow needs to maintain a model of the cognitive state of the human: in this case the human's focus of attention. It should have the capability to attribute mental, and in particular attentional (e.g., Itti and Koch, 2001) states to the human, and to reason about these. In this section, an example of such a system is described, based on (Bosse, Maanen, and Treur, 2006; Bosse, Maanen, and Treur, 2009); see also (Bosse, Lambalgen, Maanen and Treur, 2011). Two types of sensor information are assumed, namely: information about the human's gaze (e.g., measured by an eye tracker), and characteristics of stimuli (e.g., the colour and speed of airplanes on radar screens, or of the persons on surveillance images).

Domain Model

This section introduces the domain model for attentional processes. In "Analysis of the Main Aspects and Their Relations", the main aspects of the model and their relations are introduced. In "The Detailed Model," the detailed model is provided, and "Example Simulation" presents an example simulation trace.

Analysis of the Main Aspects and Their Relations

In this section, a persons "state of attention" is defined as a distribution of values over different locations in the world. Assuming that the person is facing a certain screen, these locations are described by segments of the screen (which, for the case of the naval operator, may contain radar contacts). A high value for a certain location means that the person has much attention for that location. Note that "attention" is not (purely) defined by "visual attention": a person may currently not be looking at a certain location, but may still have attention for it. Rather, our notion of "attention" of an object should be seen as "presence in working memory".

In order to calculate this state of attention, obviously, the person's *gaze direction* is an important factor: where is the person currently looking to? Next, this information should be combined with the *locations of objects*: if the location of an object (e.g., a radar contact) is close to the location of the person's gaze, then the person will have more attention for that object. However, these are not the only factors. Also different *characteristics of objects* (such as brightness and size) are important: usually people have more attention for objects that are bright and large, compared to objects that are dark and small. Together, these three factors can be combined in order to calculate the *current attention contribution* of a certain location: to what extent is the location currently attracting the person's attention? If this current contribution is known for all locations of the screen, then also a *normalised attention contribution* can be calculated for each location, i.e., the relative attention contribution in comparison with all other locations. One step further, this normalised attention contribution of a location can be used to calculate the actual *attention level* that a person has for that location. In order to do this, also the previous attention level plays a role. For example, in case a person recently had much attention for location X, then

currently (s)he will still have some attention, even if the current contribution is very low. Finally, the attention level plays a role in the behaviour of the person: the more attention a person has for a certain location, the more likely (s)he is to act upon that location. In this section, it is assumed that a high attention level for a location leads to a *mouse click* on that location (for example, because clicking on a location helps a naval operator identifying the radar track on that location).

In sum, the following concepts are needed to describe the dynamics of a person's attention:

* gaze direction
* locations of objects
* characteristics of objects
* current attention contribution of locations
* normalised attention contribution of locations
* (old and current) attention level for locations
* mouse clicks on locations

These dynamic relationships that can be identified on the basis of these concepts are visualised in Figure 4. Note that the difference between current attention level and old attention level is not visualised: both are represented by the same node.

The Detailed Model

For the formalization of the conceptual model described above, the LEADSTO language (Bosse et al, 2007) is used. This is a hybrid specification language with the specific purpose to develop simulation models in a declarative manner. It is based on the assumption that dynamics can be described as an evolution of states over time. The notion of state as used here is characterised on the basis of an ontology defining a set of relevant facts (properties) that do or do not hold at a certain point in time. A specific state is characterised by dividing the set of state properties into those that hold, and those that do not hold in the state. In LEADSTO,

Figure 4. Overview of Domain Model for Attentional Processes

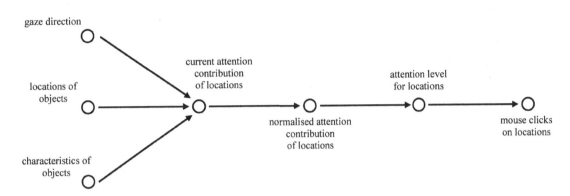

Table 1. Formalisation of concepts used for domain model

concept	formalisation
gaze direction	gaze_at_loc(X:COORD, Y:COORD)
locations of objects	is_at_location(O:OBJECT, loc(X:COORD, Y:COORD))
characteristics of objects	has_value_for(O:OBJECT, V:REAL, A:ATTRIBUTE)
current attention contribution of locations	has_current_attention_contribution(loc(X:COORD, Y:COORD), V:REAL)
normalised attention contribution of locations	has_normalised_attention_contribution(loc(X:COORD, Y:COORD), V:REAL)
(old and current) attention level for locations	has_old_attention_level(loc(X:COORD, Y:COORD), V:REAL) has_attention_level(loc(X:COORD, Y:COORD), V:REAL)
mouse clicks on locations	mouse_click_on(loc(X:COORD, Y:COORD))

direct temporal dependencies between two state properties in successive states are modelled by executable dynamic properties. The LEADSTO format is defined as follows. Let α and β be state properties as defined above. Then, the notation $\alpha_{e, f, g, h} \beta$ means:

If state property α holds for an interval with duration g, then after some delay between e and f, state property β will hold for an interval with duration h.

To formalise the concepts introduced in the previous section for each of them a logical atom is introduced; see Table 1.

Note that some atoms make use of sorts. The specific sorts that are used in the presented model, and the elements that they contain, are shown in Table 2.

In the detailed model, the following LEADS-TO relationships are used: (see Box 1)

This formula determines the current attention contribution of each location. For calculation of the current attention contribution, first the weighted sum of the attribute values of the location is calculated. Here, W1 and W2 (both real numbers between 0 and 1) denote weight factors for the attributes brightness and size, respectively. The weighted sum is divided by $1 + \alpha*r^2$, where r^2 is the *Euclidean distance* between the location loc(X1,Y1) of object O and the current location of a person's gaze loc(X2,Y2), defined by adding the squared value of X1-X2 to the squared value of Y1-Y2. The importance factor α (represented by a real number between 0 and 1) determines the relative impact of distance to the gaze location on the contribution of attention. Note that the formula

Table 2. Sorts used

sort	description	elements
COORD	coordinates of the screen used in the example	{1, 2, ..., max_coord}
REAL	the set of real numbers	the set of real numbers
OBJECTS	objects (e.g. radar contacts) used in the example	{contact05, contact22, contact39, nothing}
ATTRIBUTE	attributes of objects used in the example	{brightness, size}

Box 1. DDR1 (Determining Attention Contributions)

```
If      object O is at location loc(X1,Y1),
  and     the person's gaze is at location loc(X2,Y2),
  and     object O has value V1 for brightness,
  and     object O has value V2 for size,
then     the current attention contribution of location loc(X1,Y1)
is (V1*W1+V2*W2) / (1+ α * (X1-X2)^2+ (Y1-Y2)^2)
is_at_location(O, loc(X1, Y1)) ∧ gaze_at_loc(X2, Y2) ∧
has_value_for(O, V1, brightness) ∧ has_value_for(O, V2, size)
   has_current_attention_contribution(loc(X1, Y1), (V1*W1+V2*W2) / (1+ α * ((X1-
X2)^2+ (Y1-Y2)^2)))
```

Box 2. DDR2 (Normalising Attention Contributions)

```
If      the current attention contribution of location loc(X,Y) is V,
then     the normalised attention contribution of location loc(X,Y)
is V * A divided by the sum of the current attention contributions of all locations
has_current_attention_contribution(loc(X, Y), V) ∧
has_current_attention_contribution(loc(1, 1), V1) ∧
has_current_attention_contribution(loc(1, 2), V2) ∧
has_current_attention_contribution(loc(1, 3), V3) ∧
has_current_attention_contribution(loc(2, 1), V4) ∧
has_current_attention_contribution(loc(2, 2), V5) ∧
has_current_attention_contribution(loc(2, 3), V6) ∧
has_current_attention_contribution(loc(3, 1), V7) ∧
has_current_attention_contribution(loc(3, 2), V8) ∧
has_current_attention_contribution(loc(3, 3), V9)
   has_normalised_attention_contribution(loc(X, Y),V*A / (V1+V2+V3+V4+V5+V6+V7+V8+V9))
```

only takes these two attributes of locations into account. In a more general form, the current attention contribution of a location can be calculated by $\sum_{A:attributes} V(A)*W(A)/(1+\alpha*R^2)$. The formula also assigns a value to empty locations (by using the special element nothing of sort OBJECT, which has value 0 for all attributes). (see Box 2)

This formula determines the normalised attention contribution of each location, by dividing the current attention contribution of the location by the sum of the current attention contributions of all locations. Furthermore, the resulting value is multiplied by a constant a (represented by a real number between 0 and 1) which represents the

Box 3. DDR3 (Determining Attention Levels)

```
If        the normalised attention contribution of location loc(X,Y) is V1,
   and       the old attention level of location loc(X,Y) is V2,
then       the new attention level of location loc(X,Y) is d*V2 + (1-d*V1
has_normalised_attention_contribution(loc(X, Y),V1) ∧
has_old_attention_level(loc(X, Y), V2)
    has_attention_level(loc(X, Y), D*V2 + (1-D)*V1)
```

Box 4. DDR4 (Time Shift)

```
If        the attention level of location loc(X,Y) is V,
then       the old attention level of location loc(X,Y) becomes V
has_attention_level(loc(X, Y), V)
    has_old_attention_level(loc(X, Y), V)
```

Box 5. DDR5 (Determining Mouse Clicks)

```
If        the attention level of location loc(X,Y) is V,
   and       V > th,
then       there will be a mouse click on location loc(X,Y)
has_attention_level(loc(X, Y), V) ∧ V > th
    mouse_click_on(loc(X, Y))
```

total amount of attention that a person can spend. This usually depends on many factors that differ per person and situation (e.g., cognitive abilities, stress, concentration, and so on). (see Box 3)

This formula determines the actual attention level of each location, based on the old attention level and the current (normalised) attention contribution. Here, d is a persistence parameter (represented by a real number between 0 and 1) that determines the relative impact of the old attention level on the new attention level. (see Box 4)

DDR4 is a simple formula that converts the current attention level to the old attention level after each time step. (see Box 5)

This formula determines when the person will click on a certain location. For simplicity, this is assumed to occur in all cases that the person has an attention level V for that location that exceeds a certain threshold th (again represented by a real number between 0 and 1).

Example Simulation

This section presents an example simulation trace that was generated on the basis of the domain model for attentional processes. Figure 5 displays (a selection of) the symbolic concepts involved, and Figure 6 displays (a selection of) the numerical concepts. The settings used for the parameters of the model (see previous section) are as follows: W1=0.8, W2=0.5, α=0.9, A=0.9, d=0.3, th=0.25, max_coord=3.

As shown in Figure 5, there are 9 locations involved (namely loc(1,1), ..., loc(3,3)) and three objects (namely radar contacts contact05, contact22, and contact39). Three rounds of attention generation are shown. During the first round, the person's gaze is at location loc(1,1), later it shifts to loc(2,2) and finally to loc(3,3). Furthermore, contact22 moves from location loc(2,3) to loc(2,2) and to loc(2,1), contact39 is always at location

Figure 5. Example simulation trace for domain model: symbolic concepts

Figure 6. Example simulation trace for domain model: numerical concepts

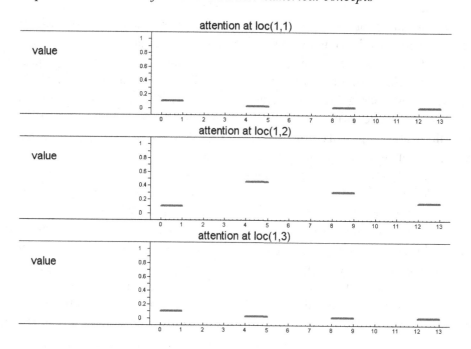

loc(1,2), and contact05 only comes into play during the last round, when it appears at location loc(3,3). The attribute values of the different contacts are shown in the bottom lines of the picture.

The dynamics of the person's attention based on these events are shown in Figure 6 (note that the figure only focuses on three locations, and that intermediate steps that calculate current and normalised attention contribution are not shown). As shown in this picture, initially the person's attention is spread equally over the locations. However, due to the appearance of contact39 at location loc(1,2), the attention for this location strongly increases in the next round. In the rounds that follow, the attention of this location converges back to the initial value.

Model for Analysis

This section introduces the model for analysis of attentional processes. In "Analysis of the Main Aspects and Their Relations," the main aspects of the model and their relations are introduced. In "The Detailed Model," the detailed model is provided, and "Example Simulation" presents an example simulation trace.

Analysis of the Main Aspects and Their Relations

To be able to analyse the dynamics of a human's attention, an ambient agent should, in principle, be equipped with the domain model introduced in "Domain Model," and should be able to reason on the basis of this model. Therefore, the model for analysis should contain more of less the same concepts as the domain model. However, one important difference is that not all concepts that exist in the domain model can actually be observed by the ambient agent. For example, the 'current attention contribution' of a location is not something that is explicitly observable in the world, nor is the 'attention level' of a person

for a location. On the other hand, things like gaze direction, and locations and characteristics of objects can be observed explicitly (e.g., by an eye tracker and a graphical user interface, respectively). As a result, some of the real world concepts from the domain model can directly be incorporated in the analysis model, whereas for some other concepts, the agent should make an estimation: it will form *beliefs* about the state of those concepts. Finally, the analysis model should have information about the human's *desired attention level* for the different locations: where does it want the human to focus at? It is assumed that the agent has tactical domain knowledge that enables it to make such assessments. For example, in the domain of the Tactical Picture Compilation Task, the agent knows the movement information (e.g. speed and direction) of each radar track and uses it to estimate the urgency of identifying the track. This urgency determines the desired attention level at each location. By comparing the desired attention level with the estimated attention level, the analysis model can determine whether there is an *attention discrepancy* at a certain location, i.e., whether the human has too much or too little attention for the location (and, possibly, how big this discrepancy is).

This leads to the following list of concepts that are needed for the attention analysis model:

- gaze direction
- locations of objects
- characteristics of objects
- beliefs on current attention contribution of locations
- beliefs on normalised attention contribution of locations
- beliefs on (old and current) attention level for locations
- desired attention level for locations
- assessments on attention discrepancy for locations

Figure 7. Overview of Analysis Model for Attentional Processes

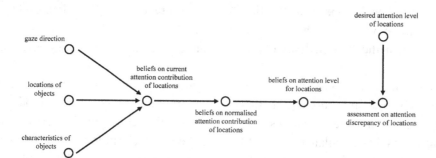

Table 3. Formalisation of concepts used for analysis model

concept	formalisation
gaze direction	gaze_at_loc(X:COORD, Y:COORD)
location of objects	is_at_location(O:OBJECT, loc(X:COORD, Y:COORD))
characteristics of objects	has_value_for(O:OBJECT, V:REAL, A:ATTRIBUTE)
belief on current attention contribution of locations	belief(agent, has_current_attention_contribution(loc(X:COORD, Y:COORD), V:REAL))
belief on normalised attention contribution of locations	belief(agent, has_normalised_attention_contribution(loc(X:COORD, Y:COORD), V:REAL))
belief on (old and current) attention level for locations	belief(agent, has_old_attention_level(loc(X:COORD, Y:COORD), V:REAL)) belief(agent, has_attention_level(loc(X:COORD, Y:COORD), V:REAL))
desired attention level of locations	desire(agent, has_attention_level(loc(X:COORD, Y:COORD), V:REAL))
assessment on attention discrepancy of locations	assessment(agent, has_attention_discrepancy(loc(X:COORD, Y:COORD), V:REAL))

As can be seen, this list is similar to the list of concepts presented in "Analysis of the Main Aspects and Their Relations," where for concept 4, 5, and 6 a belief is introduced, and two new concepts have been added. Furthermore, also most dynamic relationships are similar to those presented in "Analysis of the Main Aspects and Their Relations;" see Figure 7.

The Detailed Model

To formalise the concepts introduced in the previous section, again a number of logical atoms are introduced (most of which are similar to Table 1); see Table 3.

Next, to formalise the dynamic relationships for the analysis model, the following LEADSTO relationships are used: (see Boxes 6, 7, 8, & 9)

Note that dynamic relationship ADR1-ADR4 are very similar to DDR1-DDR4, but that there are two important differences:

1. for some of the concepts, *estimations* (modelled as *beliefs*) are used instead of the real concepts
2. for some of the relationships, the agent uses its own *parameters* (e.g., α_{ag} instead of α, and d_{ag} instead of d), which need not correspond exactly to the actual parameters

Box 6. ADR1 (Determining Beliefs on Attention Contributions)

```
If       object O is at location loc(X1,Y1),
   and      the person's gaze is at location loc(X2,Y2),
   and      object O has value V1 for brightness,
   and      object O has value V2 for size,
then     the agent believes that the current attention contribution of location
loc(X1,Y1)
is (V1*W1_ag +V2*W2_ag) / (1+ α_ag * (X1-X2)^2+ (Y1-Y2)^2)
is_at_location(O, loc(X1, Y1)) ∧ gaze_at_loc(X2, Y2)  ∧
has_value_for(O, V1, brightness) ∧ has_value_for(O, V2, size)
   belief(agent, has_current_attention_contribution(loc(X1, Y1), (V1*W1_ag +V2*W2_ag)/(1+
α_ag * ((X1-X2)^2+ (Y1-Y2)^2))))
```

Box 7. ADR2 (Determining Beliefs on Normalised Attention Contributions)

```
If       the agent believes that the current attention contribution of location
loc(X,Y) is V,
then     the agent believes that the normalised attention contribution of location
loc(X,Y) is
V * A_ag divided by the sum of the current attention contributions of all locations
belief(agent, has_current_attention_contribution(loc(X, Y), V)) ∧
belief(agent, has_current_attention_contribution(loc(1, 1), V1)) ∧
belief(agent, has_current_attention_contribution(loc(1, 2), V2)) ∧
belief(agent, has_current_attention_contribution(loc(1, 3), V3)) ∧
belief(agent, has_current_attention_contribution(loc(2, 1), V4)) ∧
belief(agent, has_current_attention_contribution(loc(2, 2), V5)) ∧
belief(agent, has_current_attention_contribution(loc(2, 3), V6)) ∧
belief(agent, has_current_attention_contribution(loc(3, 1), V7)) ∧
belief(agent, has_current_attention_contribution(loc(3, 2), V8)) ∧
belief(agent, has_current_attention_contribution(loc(3, 3), V9))
   belief(agent, has_normalised_attention_contribution(loc(X, Y),V*A_ag /
(V1+V2+V3+V4+V5+V6+V7+V8+V9)))
```

Box 8. ADR3 (Determining Beliefs on Attention Levels)

```
If       the agent believes that the normalised attention contribution of location
loc(X,Y) is V1,
   and      the agent believes that the old attention level of location loc(X,Y) is V2,
then     the agent believes that the new attention level of location loc(X,Y) is d_ag*V2
+ (1-d_ag)*V1
belief(agent, has_normalised_attention_contribution(loc(X, Y),V1)) ∧
belief(agent, has_old_attention_level(loc(X, Y), V2))
   belief(agent, has_attention_level(loc(X, Y), d_ag*V2 + (1-d_ag)*V1))
```

Box 9. ADR4 (Determining Beliefs on Time Shift)

```
If        the agent believes that the attention level of location loc(X,Y) is V,
then      the agent believes that the old attention level of location loc(X,Y) becomes
V
belief(agent, has_attention_level(loc(X, Y), V))
   belief(agent, has_old_attention_level(loc(X, Y), V))
```

Box 10. ADR5 (Assessing Attention)

```
If        the agent believes that the attention level of location loc(X,Y) is V1,
  and     the agent desires that the attention level of location loc(X,Y) is V2,
  and     V1 < V2,
then      the agent assesses that there is an attention discrepancy at location
loc(X,Y) of size V2-V1
belief(agent, has_attention_level(loc(X, Y), V1)) ∧ desire(agent,
has_attention_level(loc(X, Y), V2)) ∧ V1 < V2
   assessment(agent, has_attention_discrepancy(loc(X, Y), V2-V1))
```

The reason for this second point is that an ambient agent may simply not know what the value of the actual parameter is; therefore it may need to make estimations. However, based on experiences, it may adapt such estimations in an intelligent manner (this topic of parameter adaptation is beyond the scope of this chapter).

Furthermore, one additional LEADSTO relationship is used to determine whether there is an attention discrepancy: (see Box 10)

Note that this is just one example of a (simple) mechanism to assess attention discrepancies. Other, more sophisticated mechanisms would also consider the case that the estimated attention for a location is too high, or would only signal a discrepancy in case the difference between estimated and desired attention is above a certain threshold.

Example Simulation

Also for the analysis model, simulation runs may be performed. An example simulation trace for a situation in which the ambient agent signals a discrepancy is shown in Figure 8 (symbolic concepts) and Figure 9 (numerical concepts). In this trace, a situation is simulated that is similar to the situation shown in "Domain Model." The settings used for the parameters of the domain model are identical to "Domain Model:" W1=0.8, W2=0.5, α=0.9, A=0.9, d=0.3, th=0.25. These same settings are used for the parameters of the analysis model: $W1_{ag}$=0.8, $W2_{ag}$=0.5, α_{ag}=0.9, A_{ag}=0.9, d_{ag}=0.3. Also the characteristics and dynamics of the different objects and the gaze were similar to "Domain Model," with one difference: contact39 is now hardly visible. It is very small, and not very bright (both its size and brightness are 0.01). However, since this is a very urgent contact, the agent wants the human to have at least some attention for its location: see desire(agent, has_attention_level(loc(1, 2), 0.2)) in Figure 8.

As shown in Figure 8, the agent (correctly) believes that the human has very little attention for location loc(1,1), loc(1,2), and loc(1,3). The reason for this is that these locations do not contain any salient objects: there is only a - hardly visible - contact at location loc(1,2). This also explains why the user does not click with the mouse on this location (the user decides to click on locations loc(2,2) and loc(2,3) instead). However, since the agent wants the human to have some attention for location loc(1,2), it derives that

Figure 8. Example simulation trace for analysis model: symbolic concepts

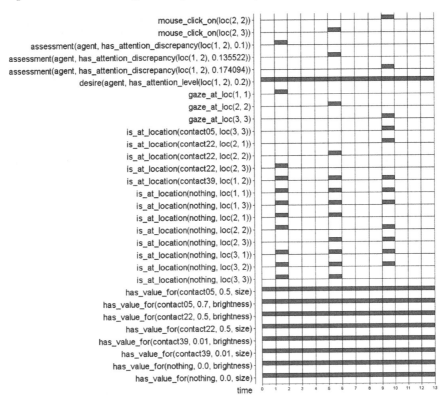

there is a discrepancy at that location: assessment(agent, has_attention_ discrepancy(loc(1, 2), ...)), which increases over time (see Figure 8).

Such discrepancies may be used as input by the support model, in order to determine how the discrepancy should be solved. This is explained in the next section.

Model for Support

This section introduces the model for support based on attentional processes. In "Analysis of the Main Aspects and Their Relations," the main aspects of the model and their relations are introduced. In "The Detailed Model," the detailed model is provided, and "Example Simulation" presents an example simulation trace.

Analysis of the Main Aspects and Their Relations

In case there is a discrepancy between a person's estimated and desired attention level for a certain location, an ambient agent has several alternatives to provide support. To illustrate this, let us assume that there is an urgent radar contact for which the human has no (or not enough) attention. In that case, one possibility is to adapt one (or more) of the characteristics of the object. For instance, the agent could increase the size and/ or brightness of the contact. A second, similar option is to decrease the size and/or brightness of other objects, so that the relative saliency of the contact will increase, or even to make these objects completely invisible. However, such types of support are very intrusive to the activities of the human, and may be disturbing in case (s)he is already very busy with another subtask. Therefore,

Figure 9. Example simulation trace for analysis model: numerical concepts

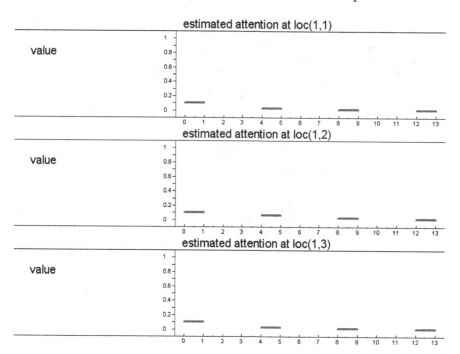

a third alternative is to leave the human alone, and request another party to concentrate on the urgent object. This other party may be another human, but may also be an intelligent software agent (assuming that there exist agents with sufficient domain knowledge to solve the task in a sufficiently satisfactory manner).

In this section, the focus is on the first, relatively simple alternative: a model is introduced that adapts the characteristics of objects that receive too little attention (in this case, by increasing their brightness). To this end, the model needs several types of information as input. First, obviously, it needs to know whether there is an *attention discrepancy* at a certain location. Second, it needs to know which *objects* are at these locations. And third, it needs to know what the *characteristics* of these objects are. Based on that, it will generate *actions to adapt the characteristics of objects*.

In sum, the following concepts are needed for the attention support model:

Figure 10. Overview of support model for attentional processes

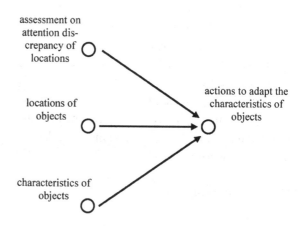

- assessments on attention discrepancy for locations
- locations of objects
- characteristics of objects
- actions to adapt the characteristics of objects

Table 4. Formalisation of concepts used for support model

concept	formalisation
assessment on attention discrepancy of locations	assessment(agent, has_attention_discrepancy(loc(X:COORD, Y:COORD), V:REAL))
locations of objects	is_at_location(O:OBJECT, loc(X:COORD, Y:COORD))
characteristics of objects	has_value_for(O:OBJECT, V:REAL, A:ATTRIBUTE)
actions to adapt characteristics of objects	performed(agent, assign_new_value_for(O:OBJECT, V:REAL), A:ATTRIBUTE)

Box 11. SDR1 (Changing Brightness)

```
If      the agent believes that there is an attention discrepancy at location loc(X,Y)
of size V1,
  and     object O is at location loc(X,Y),
  and     object O has value V2 for brightness,
then    the agent adapts the brightness of object O to value f(V1,V2)
assessment(agent, has_attention_discrepancy(loc(X,Y), V1)) ∧
is_at_location(O, loc(X, Y)) ∧ has_value_for(O, V2, brightness)
  performed(agent, assign_new_value_to(O, f(V1,V2), brightness))
```

The dynamic relationships that can be established between these concepts are visualised in Figure 10.

Note that in reality, a change in characteristics of objects might also result in a change of the human's gaze direction. However, for simplicity, this relationship is not modelled in this chapter; i.e., the gaze movement is assumed to be independent of the objects.

The Detailed Model

To formalise the concepts introduced in the previous section, the following logical atoms are used (of which the first three are taken literally from Table 3); see Table 4.

To formalise the dynamic relationship for the support model, the following LEADSTO relationship is used: (see Box 11)

Here, $f(V1,V2)$ is a function of value $V1$ and $V2$, which can be filled in per domain and per application. A very simple example for $f(V1,V2)$, which is used in the simulation described in the next section, is the following function: $f(V1,V2)$ $= 1$. This function just assigns a maximal brightness to objects at locations with too little attention. However, also more sophisticated function can be used, such as the following: $f(V1,V2) = V2 + V1 * (1 - V2)$. This function (which assumes that both $V1$ and $V2$ are real values between 0 and 1), increases the value of an attribute (in this case brightness) with a fraction of the distance between the current and the maximum value of the attribute. This fraction is taken proportional to the attention discrepancy of the location involved. Furthermore, similar formulae can be used to assign new values to the other attributes.

In addition, the effects of support actions on the world should be modelled. This is not part of the support model itself, but is needed within a simulation, to be able to evaluate whether the support works. This can be done via the following LEADSTO relationships: (see Boxes 12 & 13)

Note that relationship SDR3 assumes the existence of predicates of the form performed(agent, assign_new_value_to(O, A)). To achieve this, these predicates can simply be added to the consequent of SDR1.

Box 12. SDR2 (Assigning New Attribute Values)

```
If       the agent adapts attribute a of object O to value V,
then     attribute a of object O will have value V
performed(agent, assign_new_value_to(O, V, A))
  has_value_for(O, V, A)                        .
```

Box 13. SDR3 (Persistence of Attribute Values)

```
If       attribute A of object O has value V,
  and    the agent does not adapt attribute a of object O,
then     attribute A of object O will keep value v
has_value_for(O, V, A) ∧ not(performed(agent, assign_new_value_to(O, A)))
  has_value_for(O, V, A)
```

Example Simulation

To illustrate the functioning of the attention support model, an example simulation trace is shown in Figure 11 (symbolic concepts) and 12 (numerical concepts). Note that this trace is generated on the basis of the support model in combination with the domain model and the analysis model. The trace addresses the same scenario as addressed in "Model for Analysis." Thus, the settings used for the parameters of the domain model are: $W1=0.8$, $W2=0.5$, $\alpha=0.9$, $A=0.9$, $d=0.3$, $th=0.25$, and the settings used for the parameters of the analysis model are $W1_{ag}=0.8$, $W2_{ag}=0.5$, $\alpha_{ag}=0.9$, $A_{ag}=0.9$, $d_{ag}=0.3$. Also the initial characteristics and dynamics of the different objects and the gaze were similar to "Model for Analysis." Again contact39 is almost invisible at the start of the simulation. However, a difference is that, this time, the agent will adapt the characteristics of this contact.

Figure 12 shows the agent's estimation of the human's attention for the first three locations. Note that, as in "Model for Analysis," this estimation is completely correct (as the agent uses the same parameter settings as the real world). As shown in Figure 12, at the start of the simulation, the agent believes that the human has little attention (about 0.1) for location loc(1,2). Since the

agent wants the human to have more attention (about 0.2) for this location, it derives that there is a discrepancy at that location of 0.1: see assessment(agent, has_attention_ discrepancy(loc(1, 2), 0.1)) at time point 1 of Figure 11. This discrepancy causes the agent to adapt the brightness of contact39 to 1.0, which is effectuated at time point 3. As a result, after a while the human has more attention for location loc(1,2), which makes that (s)he clicks on the location at time point 9. In conclusion, the agent succeeded in attracting the human's attention to the (urgent) location loc(1,2).

Discussion

To support a human in the execution of a complex and demanding task, an ambient agent needs to maintain a domain model of the state of attention of this person. This section introduced such a domain model (in "Domain Model"), as well as an analysis model to reason about this domain model ("Model for Analysis"), and a support model to determine actions based on this analysis ("Model for Support"). Note that, although the example presented here addresses a naval operator who executes the Tactical Picture Compilation Task, in principle the model is generic, and can be ap-

Figure 11. Example simulation trace for support model: symbolic concepts

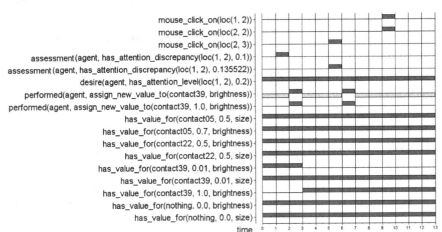

Figure 12. Example simulation trace for support model: numerical concepts

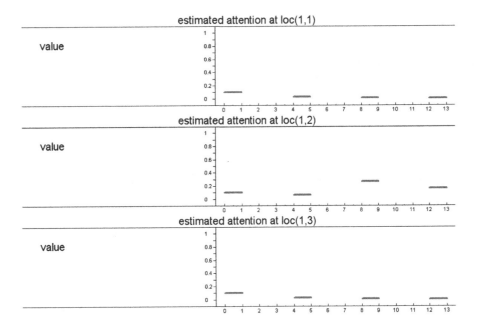

plied to various similar domains (varying from persons that inspect surveillance cameras to air traffic controllers).

The presented model can be extended in various manners. Some possibilities are mentioned below:

- Taking more attributes into account than only brightness and size.

- Addressing not only *bottom-up attention* (i.e., attention influenced by characteristics of the object, such as brightness and size), but also *top-down attention* (i.e., attention influenced by characteristics of the person, such as expectations and goals).

- Modelling not only attention for visible objects, but also attention for invisible objects (e.g., an object that has just disappeared from the screen, or an object that is present on a window that is hidden behind another window).

- Considering different and more complex types of support. The presented model only supports the human by modifying the attributes (in particular the brightness) of urgent objects. Other models could, for example, take over the identification task for those objects in case the human is too busy.
- Considering sophisticated techniques for adaptation of parameter values.

ASSESSMENT OF DRIVER BEHAVIOUR

The second example scenario that illustrates the benefit of human knowledge in ambient intelligence applications is about driving behaviour. Circumstances may occur in which a person's internal state is affected in such a way that driving is no longer safe. For example, when a person has taken drugs, either prescribed by a medical professional, or by own initiative, the driving behaviour may be impaired. For the case of alcohol, specific tests are possible to estimate the alcohol level in the blood, but for many other drugs such tests are not available. Moreover, a bad driver state may have other causes, such as highly emotional events, or being sleepy. Therefore assessment of the driver's state by monitoring the driving behaviour itself and analysing the monitoring results is a wider applicable option. In this section, an ambient agent is presented that assesses a person's driving behaviour, and in case of a negative assessment lets cruise control slow down and stop the car (see Bosse, Hoogendoorn, Klein, and Treur, 2008). The approach was inspired by a system that is currently under development by Toyota. This ambient system that in the near future will be incorporated as a safety support system in Toyota's prestigious Lexus line, uses sensors that

can detect the driver's steering operations, and sensors that can detect the driver's eye movements.

The presented ambient agent makes use of three types of sensor information. First, the driver's steering behaviour is measured, to assess, among others, whether (s)he is making too many unnecessary movements with the steering wheel. Second, a sensor in the steering wheel measures the driver's perspiration. This can be used to assess the percentage of alcohol in the driver's sweat, which correlates linearly with the alcohol percentage in the blood. Third, a camera that is mounted on the front mirror measures the driver's eye movements, which can be used to assess, for instance, whether (s)he is staring too much in the distance, or whether (s)he is blinking much with the eyes.

Domain Model

This section introduces the domain model for driver behaviour. In "Analysis of the Main Aspects and Their Relations", the main aspects of the model and their relations are introduced. In "The Detailed Model," the detailed model is provided, and "Example Simulation" presents an example simulation trace.

Analysis of the Main Aspects and Their Relations

In the literature, various factors are reported that have a (negative) impact on driver behaviour. These factors include high levels of drugs or alcohol, fatigue, specific emotional states, the presence of other persons in the car, loud music, and the use of mobile phones. The system described in this section is designed for measuring three of these factors. However, it can be extended to account for other factors as well.

First, the *level of drugs* in the blood is considered. This is dependent on the amount of *drug intake* over the last hours. Second, and similarly, the *level of alcohol* in the blood (and also in the sweat) is dependent on the amount of *alcohol intake* over the last hours. The third factor is the person's state of *fatigue*, which depends on whether long periods of *non-stop driving* have been performed recently. All of these factors may lead to a state that negatively influences the driving behaviour of the person. In particular, in case a person has taken certain drugs or alcohol, his or her *steering behaviour* may be affected: (s)he will make more sudden movements in the steering wheel, instead of keeping it straight. In addition, the person's eye movements may become abnormal. A person may start to *stare* (i.e., fixating the eyes on one location, without making any saccades) in case (s)he has consumed a lot of alcohol and/or is very tired. In case of the latter, the person's eyes will also *blink* more often than usually. This happens because the state of fatigue makes that the *speed* increases by which the eyes become *dry*, and also because the *basic eye dryness* (the dryness just after a blink) increases. When a person's eyes become too dry (i.e., they are drier than a certain desired *norm*), a person blinks (i.e., rapidly opens and closes the eyelid) in order to spread tears across and remove irritants[1].

In sum, the following concepts are needed to describe the dynamics of a person's driving behaviour:

- drug intake
- alcohol intake
- non-stop driving
- drug level (in blood)
- alcohol level (in blood and sweat)
- fatigue
- steering behaviour
- staring
- eye blinks
- eye dryness
- basic eye dryness
- eye drying speed
- eye dryness norm

These dynamic relationships that can be identified on the basis of these concepts are visualised in Figure 13.

The Detailed Model

To formalise the concepts introduced in the previous section, for each of them a logical atom is introduced; see Table 5. The predicate affects(P:PRODUCT, S:STATE) is not needed per se, but has been introduced in order to create some more generic relationships.

Note that some atoms make use of sorts. The specific sorts that are used in the presented model, and the elements that they contain, are shown in Table 6. Note that, instead of using the qualitative sort LEVEL, also a quantitative sort can be used (e.g., describing alcohol concentration by a real number).

To formalise the dynamic relationships between these concepts, the following relationships are used, expressed in LEADSTO. We both give a semi-formal description in natural language as a formal specification in a logical notation. (see Boxes 14 & 15)

Note that relationship DDR1 and DDR2 can also be replaced by mathematical formulae, stating, for example, how the consumption of alcohol influences the concentration of alcohol in the blood, and how this concentration decreases over time. To this end, more detailed medical knowledge (e.g., about drug and alcohol uptake by the blood) could be exploited. (see Boxes 16, 17, & 18)

Finally, a number of relationships are used to model the processes related to eye blinking, as shown below. Here, d1 and d2 are constants that represent some basic eye dryness in case of, respectively, a high or low state of fatigue. Moreover, s1 and s2 represent the speed by which the eyes become drier in case of, respectively, a high or low state of fatigue. (see Boxes 19, 20, 21, 22, & 23)

Figure 13. Overview of Domain Model for Driver Behaviour

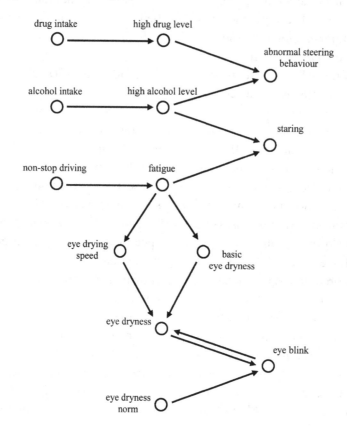

Table 5. Formalisation of concepts used for domain model

concept	formalisation
drug intake	consumed(take_drugs)
alcohol intake	consumed(take_alcohol)
non-stop driving	consumed(non_stop_driving)
drug level	has_level(drug_level, L:LEVEL)
alcohol level	has_level(alcohol_level, L:LEVEL)
fatigue	has_level(fatigue, L:LEVEL)
abnormal steering behaviour	performed(abnormal_steering_behaviour)
staring	performed(staring)
eye blinks	performed(eye_blink)
eye dryness	eye_dryness(I:REAL)
basic eye dryness	basic_eye_dryness(I:REAL)
eye drying speed	eye_drying_speed(I:REAL)
eye dryness norm	eye_dryness_norm(I:REAL)
relations between factors and driver states	affects(P:PRODUCT, S:STATE)

Table 6. Sorts used

sort	description of use	elements
LEVEL	extent to which a certain factor (e.g., drugs, alcohol, or fatigue) is present in a person's body	{low, high}
REAL	the set of real numbers	the set of real numbers
PRODUCT	different products that can be consumed	{drugs, alcohol, non_stop_driving}
STATE	different aspects of the driver's physical state	{drug_level, alcohol_level, fatigue}

Box 14. DDR1 (Consuming a Substance)

```
If      a person consumes a certain product (e.g., drugs),

  and     this product affects a certain aspect of the driver's state (e.g., drug
level in his blood),

then    this aspect will be high (e.g., (s)he will have a high concentration of drugs
in the blood)

consumed(P)   &  affects(P,S)

   has_level(S, high)
```

Box 15. DDR2 (Not Consuming a Substance)

```
If      a person does not consume a certain product,

  and     this product affects a certain aspect of the driver's state,

then    this aspect will be low

not(consumed(P))  &   affects(P,S)

   has_level(S, low)
```

Box 16. DDR3 (From High Drug Level to Abnormal Steering)

```
If      a person has a high drug level,

then    (s)he will perform abnormal steering behaviour

has_level(drug_level, high)

   performed(abnormal_steering_behaviour)
```

Box 17. DDR4 (From High Alcohol Level to Abnormal Steering and Staring)

```
If      a person has a high alcohol level,

then    (s)he will perform abnormal steering behaviour

  and    (s)he will stare

has_level(alcohol_level, high)

   performed(abnormal_steering_behaviour) ∧ performed(staring)
```

Box 18. DDR5 (From Fatigue to Staring)

```
If      a person has a high state of fatigue,

then    (s)he will stare

has_level(fatigue, high)

   performed(staring)
```

Box 19. DDR6 (Effect of Fatique on Eye Dryness)

```
If       a person has a high state of fatigue,
then     the person's basic eye dryness will be d1,
  and      the person's eye drying speed will be s1,
has_level(fatigue, high)
    basic_eye_dryness(d1)    &    eye_drying_speed(s1)
```

Box 20. DDR7 (Basic Eye Dryness)

```
If       a person has a low state of fatigue,
then     the person's basic eye dryness will be d2,
  and      the person's eye drying speed will be s2,
has_level(fatigue, low)
    basic_eye_dryness(d2)    &    eye_drying_speed(s2)
```

Box 21. DDR8 (Eye Blinking)

```
If       a person's eye dryness is X,
  and      the person's eye dryness norm is Y,
  and      X > Y,
then     the person will perform an eye blink
eye_dryness(X)    &
eye_dryness_norm(Y)    &
X > Y
    performed(eye_blink)
```

Box 22. DDR9 (Effect of Blinking on Dryness)

```
If       a person performs an eye blink,
  and      the person's basic eye dryness is d,
then     the person's eye dryness will be d
performed(eye_blink)    &
basic_eye_dryness(D)
    eye_dryness(D)
```

Box 23. DDR10 (Effect of Not Blinking on Dryness)

```
If       a person does not perform an eye blink,
  and      the person's eye dryness is X,
  and      the person's eye drying speed is S,
then     the person's eye dryness will be X+S*(1-X)
not(performed(eye_blink))    &
eye_dryness(X)    &
eye_drying_speed(S)
    eye_dryness(X+S*(1-X))
```

Figure 14. Example simulation trace for domain model

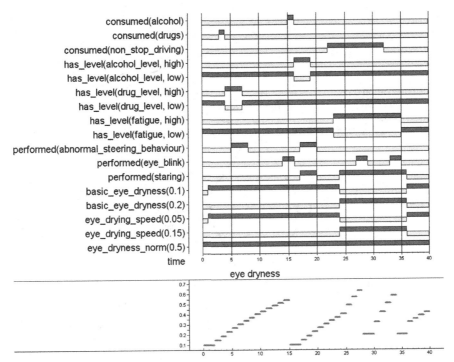

Example Simulation

This section presents an example simulation trace that was generated on the basis of the domain model for driver behaviour. This trace is displayed in Figure 14. The settings for this trace were: d1=0.1, d2=0.2, s1=0.05. s2=0.15, and eye_dryness_norm=0.5. The upper graph denotes the symbolic predicates, and the lower to graphs denote the numerical concepts (i.e., eye dryness). Note that predicates of type affects(…) have been left out, for readability.

As shown in Figure 14, the person involved initially has a low concentration of drugs and alcohol in the blood, and a low state of fatigue. However, at time point 3 (s)he uses some drugs, which leads to an increased drug level between time point 4 and 7 (it is assumed that this lasts 3 time steps), and abnormal steering behaviour. After that, everything is back to normal, until time point 15. At that moment, the person consumes some alcohol, which leads to an increased alcohol

level for a while. As a result of this, the person -again- steers abnormally, and now also stares in the distance. However, until that moment, the person has not been very tired (i.e., a low state of fatigue), which resulted in a basic eye dryness of 0.1 and an eye drying rate of 0.05. As a consequence, the person's eyes have been drying very slowly, and (s)he has blinked only once (at time point 14, when the eyes were drier than the norm of 0.5). However, between time point 22 and 32, the person enters a long period of non-stop driving. As a result, his state of fatigue becomes high, and his values for basic eye dryness and eye drying rate increase. This causes the eyes to dry rather quickly, so that the person blinks twice in a relatively short period. After time point 36, everything is back to normal again.

Model for Analysis

This section introduces the model for analysis of driver behaviour. In "Analysis of the Main

Figure 15. Overview of Analysis Model for Driver Behaviour

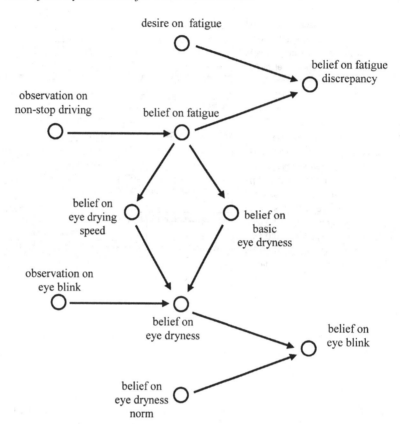

Aspects and Their Relations," the main aspects of the model and their relations are introduced. In "The Detailed Model," the detailed model is provided, and "Example Simulation" presents an example simulation trace.

Analysis of the Main Aspects and Their Relations

To be able to analyse the dynamics of a driver's behaviour, an ambient agent should, in principle, be equipped with the domain model introduced in "Domain Model," and should be able to reason on the basis of this model. Therefore, the model for analysis should contain more of less the same concepts as the domain model. Below, an analysis model to assess the driver's state of fatigue is worked out in detail. This means that the parts of the domain model that address fatigue are re-used

here; to keep the model simple, the parts about drug and alcohol intake are not addressed.

The input that the ambient agent can use to make estimations about the state of fatigue is the amount of *non-stop driving*. The agent can *observe* this, because it knows when the engine is turned on and off. Based on that information, it can derive *beliefs* about the driver's state of *fatigue*. In combination with the agent's *desires* about the driver's state of *fatigue*, this can be used to determine whether there is a *discrepancy* between the driver's believed and desired state of fatigue. However, this is not the only useful information the agent can exploit. Because of the camera on the front mirror, it can also *observe* the driver's *eye blinks*. Therefore, it is useful for the agent to also make predictions (modelled as *beliefs*) about when the driver will perform these *eye blinks*, so that it can compare this with the observed eye

Table 7. Formalisation of concepts used for analysis model

concept	formalisation
observations on non-stop driving	observed(agent, consumed(non_stop_driving))
observations on eye blinks	observed(agent, performed(eye_blink))
beliefs on fatigue	belief(agent, has_level(fatigue, L:LEVEL))
beliefs on eye blinks	belief(agent, performed(eye_blink))
beliefs on eye dryness	belief(agent, eye_dryness(I:REAL))
beliefs on basic eye dryness	belief(agent, basic_eye_dryness(I:REAL))
beliefs on eye drying speed	belief(agent, eye_drying_speed(I:REAL))
beliefs on eye dryness norm	belief(agent, eye_dryness_norm(I:REAL))
desires on fatigue	desire(agent, has_level(fatigue, L:LEVEL))
beliefs on fatigue discrepancy	belief(agent, has_discrepancy(fatigue, L:LEVEL))
beliefs on relations between factors and driver states	belief(agent, affects(P:PRODUCT, S:STATE))

Box 24. ADR1 (Generating Beliefs for Consuming a Substance)

```
if      the agent observes that a person consumes a certain product (e.g., non-stop
driving),

   and     the agent believes that this product affects a certain aspect of the
driver's state (e.g., fatigue),

then      the agent will believe that this aspect will be high

observed(agent, consumed(P)) ∧ belief(agent, affects(P,S))

   belief(agent, has_level(S, high))
```

blinks in order to update its knowledge in case its beliefs (that were generated only on the basis of the information about non-stop driving) are not entirely correct (note that the exact mechanisms to make such updates are beyond the scope of this chapter). In order to make predictions about eye blinks, the agent may use several of the relationships of the domain model (modelled in the form of *beliefs*). These beliefs may involve *eye dryness*, *basic eye dryness*, *eye drying speed*, and *eye dryness norm*.

This leads to the following list of concepts that are needed for the attention analysis model:

- observations on non-stop driving
- observations on eye blinks
- beliefs on fatigue
- beliefs on eye blinks
- beliefs on eye dryness
- beliefs on basic eye dryness
- beliefs on eye drying speed
- beliefs on eye dryness norm
- desires on fatigue
- beliefs on fatigue discrepancy

These dynamic relationships that are formulated based on these concepts are visualised in Figure 15.

The Detailed Model

To formalise the concepts introduced in the previous section, again a number of logical atoms are introduced; see Table 7.

Next, to formalise the dynamic relationships for the analysis model, the following relationships in LEADSTO are used: (see Boxes 24, 25, 26, 27, 28, 29, 30, & 31)

Box 25. ADR2 (Generating Beliefs for Not Consuming a Substance)

```
if       the agent observes that a person does not consume a certain product,
   and      the agent believes that this product affects a certain aspect of the
driver's state,
then     the agent will believe that this aspect will be low
not(observed(agent, consumed(P))) ∧ belief(agent, affects(P,S))
     belief(agent, has_level(S, low))
```

Box 26. ADR3 (Generating Beliefs for Effect of High Fatigue on Eye Dryness)

```
if       the agent believes that a person has a high state of fatigue,
then     the agent believes that the person's basic eye dryness will be d1,
   and      the agent will believe that the person's eye drying speed will be s1,
belief(agent, has_level(fatigue, high))
     belief(agent, basic_eye_dryness(d1)) ∧ belief(agent, eye_drying_speed(s1))
```

Box 27. ADR4 (Generating Beliefs for Effect of Low Fatigue on Eye Dryness)

```
if       the agent believes that a person has a low state of fatigue,
then     the agent believes that the person's basic eye dryness will be d2,
   and      the agent will believe that the person's eye drying speed will be s2,
belief(agent, has_level(fatigue, low))
     belief(agent, basic_eye_dryness(d2)) ∧ belief(agent, eye_drying_speed(s2))
```

Box 28. ADR5 (Generating Beliefs for Eye Blinking)

```
if       the agent believes that a person's eye dryness is X,
   and      the agent believes that the person's eye dryness norm is Y,
   and      X > Y,
then     the agent will believe (predicts) that the person will perform an eye blink
belief(agent, eye_dryness(X)) ∧ belief(agent, eye_dryness_norm(Y)) ∧ x > y
     belief(agent, performed(eye_blink))
```

Box 29. ADR6 (Generating Beliefs for Eye Dryness)

```
if       the agent observes that a person performs an eye blink,
   and      the agent believes that the person's basic eye dryness is d,
then     the agent will believe that the person's eye dryness will be d
observed(agent, performed(eye_blink)) ∧ belief(agent, basic_eye_dryness(D))
     belief(agent, eye_dryness(D))
```

Box 30. ADR7 (Generating Beliefs for Effect of Not Blinking on Dryness)

```
if        the agent observes that a person does not perform an eye blink,
   and       the agent believes that the person's eye dryness is X,
   and       the agent believes that the person's eye drying speed is S,
then      the agent will believe that the person's eye dryness will be X+S*(1-X)

not(observed(agent, performed(eye_blink))) ∧ belief(agent, eye_dryness(X)) ∧
belief(agent, eye_drying_speed(S))
    belief(agent, eye_dryness(X+S*(1-X)))
```

Box 31. ADR8 (Assessing Fatigue)

```
if       the agent desires that the driver has a low state of fatigue,
   and       the agent believes that the driver has a high state of fatigue,
then      the agent will believe that the there is a (high) discrepancy
between the driver's desired and believed state of fatigue

desire(agent, has_level(fatigue, low)) ∧ belief(agent, has_level(fatigue, high))
    belief(agent, has_discrepancy(fatigue, high))
```

Example Simulation

Also for the analysis model, simulation runs may be performed. An example simulation trace for a situation in which the ambient agent signals a discrepancy is shown in Figure 16.

As shown in the figure, this scenario is quite similar to the scenario shown in Figure 14 (with all information related to drug and alcohol intake left out). The driver starts with a low state of fatigue, which means that (s)he performs eye blinks at a low frequency. However, between time point 22 and 32, the person drives a long distance without taking a break, which is observed by the agent. Based on this, the agent believes that (s)he is in a high state of fatigue. Since this is in conflict with the agent's desire (namely that the driver is in a low state of fatigue), the agent derives that there is a fatigue discrepancy (which can be input for the support model introduced in the next section).

Furthermore, notice that, based on its belief that the driver is in a high state of fatigue, the agent also predicts that the driver's eyes will become dry very quickly, and that (s)he will blink at time point 27 and 33. Since the agent's analysis model is 100%

correct, these predictions correspond exactly to reality. However, in cases that the analysis model is not that perfect, the predicted eye blinks will not exactly overlap with the observed eye blinks. In such cases, the parameters used by the model will have to be updated.

Note that the model presented in this section can only be used to assess fatigue based on observations about non-stop driving (and eye blinks). Thus, it covers the bottom part of Figure 15. However, the upper part of that picture has not been worked out here. To this end, an analysis model to assess the driver's drug and alcohol concentration should be developed. One additional difficulty in that case would be that some relevant concepts that exist in the domain model cannot directly be observed by the ambient agent. For example, the amount of alcohol and drug intake cannot easily be measured by an ambient agent. On the other hand, it is possible to measure information about whether the driver shows abnormal steering behaviour or is staring (the concepts on the upper-right hand side of Figure 15). These types of information can be measured, respectively, by the movements of the steering wheel and the camera on the front mirror.

Figure 16. Example simulation trace for analysis model

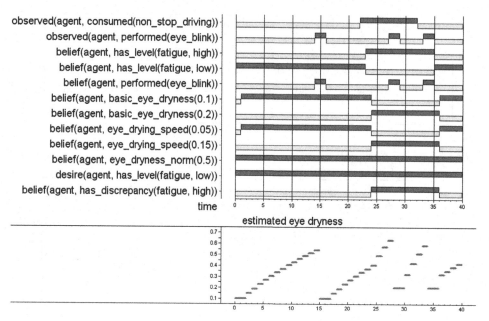

Thus, it is still possible to obtain information about states in these domain models, but these state are only present at the right-hand side of the model. As a result, no simple forward reasoning pattern (of the format "if I believe X and I believe that X leads to Y, then I also believe Y") can be applied. Instead, other reasoning methods should be applied. An example of such a method is *abduction*. In a nutshell, this method has the format "if I believe Y and I believe that X leads to Y, then I also believe that X could have been the case". By applying such methods, on could derive, for example, that the driver has been drinking on the basis of an abnormal steering pattern that is observed (e.g., in case the steering wheel moves more than 3 times from left to right and back within five seconds).

In case different analysis models are developed, which each analyse one particular aspect of the driver's state (e.g., drug concentration, alcohol concentration, and state of fatigue), the results of analyses can also be combined, in order to provide more evidence for a certain hypothesis (e.g., in case the driver has a high blinking frequency and

has not stopped the car for a while, (s)he is very likely to have a high state of fatigue). To this end, it may be useful to develop a multi-agent system, consisting of different (monitoring) agents that each analyse a specific factor separately, and another agent that combines the conclusions of the monitoring agents. Such an approach is taken in (Bosse, Hoogendoorn, Klein, and Treur, 2008). An overview of the multi-agent system used in that paper is shown in Figure 17.

Here, the Steering Monitoring agent receives the sensor input from the Steering Sensoring agent, and uses this to determine whether there is an abnormal steering pattern. Similarly, the Gaze-focus Monitoring agent uses sensor input from the Gaze-focus Sensoring agent to determine whether there is an abnormal gaze-focus pattern. Next, the Driver Assessment agent combines the information from the two monitoring agents to determine whether the driver is in an impaired condition. If this is the case, the Driver Assessment agent sends a signal to the Cruise Control agent, who can support the driver, e.g., by slowing down the car. Development of such a system

Figure 17. Multi-agent system for ambient driver support (Bosse, Hoogendoorn, Klein, and Treur, 2008)

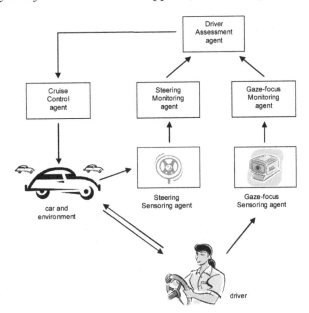

composed of multiple cooperating agents is left as an exercise for the reader.

Model for Support

This section introduces the model for support of a car driver. In "Analysis of the Main Aspects and Their Relations," the main aspects of the model and their relations are introduced. In "The Detailed Model," the detailed model is provided, and "Example Simulation" presents an example simulation trace.

Analysis of the Main Aspects and Their Relations

In case an ambient agent in a car signals that the driver is in an impaired condition, it has several alternatives to provide support. Some possible support actions that are mentioned in the literature are:

- keep a safe distance with the car in front
- slow down the car and safely pull it over

- wake up a sleepy driver (e.g., by making a loud sound)
- give advice to the driver (e.g., tell him or her to take a break)

The choice for which action to perform may depend on the specific state the driver is in. For example, if the agent believes that the driver is only slightly tired, it may advice him or her to take a break. However, when it estimates that the driver is practically asleep, a more urgent measure is necessary.

In sum, the following concepts are needed for the driver support model:

- beliefs about discrepancies between the driver's believed and desired drug level
- beliefs about discrepancies between the driver's believed and desired alcohol
- beliefs about discrepancies between the driver's believed and desired state of fatigue

Figure 18. Overview of Support Model for Driver Behaviour

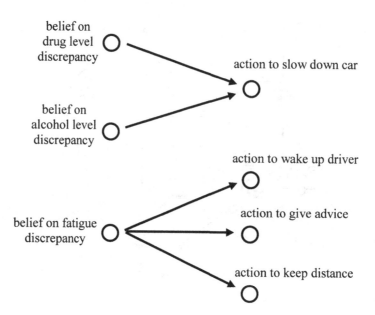

Table 8. Formalisation of concepts used for support model

concept	formalisation
beliefs of discrepancies of the driver's drug level	belief(agent, has_discrepancy(drug_level, L:LEVEL))
beliefs of discrepancies of the driver's alcohol level	belief(agent, has_discrepancy(alcohol, L:LEVEL))
beliefs of discrepancies of the driver's state of fatigue	belief(agent, has_discrepancy(fatigue, L:LEVEL))
actions to keep distance with the car in front	performed(agent, keep_distance)
actions to slow down the car	performed(agent, slow_down_car)
actions to wake up a sleepy driver	performed(agent, wake_up_driver)
actions to give advice	performed(agent, give_advice)

- actions to keep distance with the car in front
- actions to slow down the car
- actions to wake up a sleepy driver
- actions to give advice to the driver

For the dynamic relationships between these concepts, multiple options are possible that connect a specific belief (or a combination of beliefs) to a particular action. In the next section, an example specification is proposed. The relationships used in that example are visualised in Figure 18.

The Detailed Model

To formalise the concepts introduced in the previous section, the following logical atoms are used; see Table 8.

To formalise dynamic relationships that connect beliefs to support actions, the following relationships in LEADSTO are used: (see Boxes 32, 33, 34, & 35)

Note that this knowledge base takes a relatively simple approach, based on a 1:1 mapping between an estimated driver state and a set of support ac-

Box 32. SDR1 (Generating a Wake up Action)

```
if       the agent believes that there is a high discrepancy with respect to the
driver's state of fatigue,
then     the agent will try to wake up the driver
belief(agent, has_discrepancy(fatigue, high))
   performed(agent, wake_up_driver)
```

Box 33. SDR2 (Generating Advice to Keep Distance)

```
if       the agent believes that there is a low discrepancy with respect to the
driver's state of fatigue,
then     the agent will give the driver advice to take a break
belief(agent, has_discrepancy(fatigue, low))
   performed(agent, give_advice) ∧ performed(agent, keep_distance)
```

Box 34. SDR3 (For Alcohol Slowing Down the Car)

```
if       the agent believes that there is a high discrepancy with respect to the
driver's alcohol level,
then     the agent will try to slow down the car
belief(agent, has_discrepancy(alcohol_level, high))
   performed(agent, slow_down_car)
```

Box 35. SDR4 (For Drugs Slowing Down the Car)

```
if       the agent believes that there is a high discrepancy with respect to the
driver's drug level,
then     the agent will try to slow down the car
belief(agent, has_discrepancy(drug_level, high))
   performed(agent, slow_down_car)
```

tions. The reader is encouraged to develop other, more sophisticated methods for support. For example, the agent could first generate a number of support options, and then select the one with the highest feasibility.

Example Simulation

Simulation of the model presented in "The Detailed Model" is straightforward, and therefore not worked out here.

Discussion

In order to assess whether a driver is capable of driving a car, an ambient agent needs to maintain a domain model of the state of alertness this person. This section introduced such a domain model ("Domain Model"), which involved various risk factors such as alcohol intake, drug intake, and periods of non-stop driving. In addition, an analysis model was introduced to reason about this domain model ("Model for Analysis"), as well as a support model to determine actions based on the analysis ("Model for Support").

The presented model can be extended in various manners. Some possibilities are mentioned below:

- Taking more factors into account than alcohol, drugs, and fatigue. For example, emotions, other passengers, music, or mobile phones.
- Modelling the concentration of alcohol and drugs in more detail, using a continuous, numerical approach.
- Developing more complex models for analysis (e.g., a model to analyse steering patterns by taking speed and direction of steering movements into account).
- Designing a multi-agent system (instead of a single agent system) for analysis and support of driver behaviour.
- Considering different and more complex types of support (e.g., a mechanism to compare different options and select the one with the highest feasibility).
- Considering sophisticated techniques for adaptation of parameter values.

CONCLUSION

The scientific area that addresses Ambient Intelligence applications in which knowledge from the human-directed sciences is incorporated, has a high potential to provide nontrivial Ambient Intelligence applications based on human-like understanding. Such understanding can result in better informed actions and will feel more natural for humans. The resulting human-environment systems are coupled not only by their mutual interaction, but also in a reflective manner in the sense that both the human and the ambient system have and/or develop a model of the interactive processes of the human and the environment.

In this chapter, a structured way to realize this reflectiveness in Ambient Intelligence applications has been described. This approach comprises the usage of a *domain* model of the process as a basis for an *analysis* model and a *support* model

within an agent model of an Ambient Intelligence application. The approach has been illustrated in two different examples.

It is possible to also account for automated adaptation of an Ambient Intelligence application. This could be modelled by integrating the domain model in yet another way within an agent model, i.e. a fourth (sub)model in Figure 3:

- *Adaptation model*

To tune parameters in the domain model better to the specific characteristics of the human by reasoning based on the domain model.

By (human and technological) learning, adaptation and development processes for both the human and the environment the human and technological awareness can grow over time.

Reflective coupled human-environment systems can have a positive impact at different aggregation levels, from individual via an organisation within society to the society as a whole:

- *Individual level*
  ◦ more effective functioning
  ◦ stimulating healthy functioning and preventing health problems to occur
  ◦ support of learning and development
- *Organisation level*
  ◦ efficient functioning organisation by wellfunctioning members
  ◦ learning and adaptation of the organisation
- *Society level*
  ◦ limiting costs for illness and inability to work
  ◦ efficient management of environment

Some more specific examples of today's societal challenges, to which reflective coupled human-environment systems can contribute, are elderly care, health management, crime and security. Reflective coupled human-environment systems provide a solid foundation for human-like

Ambient Intelligence applications with significant benefits for individuals, organisations, and the society as a whole. For some further examples of such Ambient Intelligence applications, see (Azizi, Klein and Treur, 2010; Both, Hoogendoorn, Klein, and Treur, 2009) on ambient support of depressed persons.

REFERENCES

Aarts, E., Collier, R., van Loenen, E., & de Ruyter, B. (Eds.). (2003). *Ambient Intelligence. Proc. of the First European Symposium, EUSAI 2003.* Lecture Notes in Computer Science, vol. 2875. Springer Verlag, 2003, pp. 432.

Aarts, E., Harwig, R., & Schuurmans, M. (2001). Ambient Intelligence. In Denning, P. (Ed.), *The Invisible Future* (pp. 235–250). New York: McGraw Hill.

Aziz, A. A., Klein, M. C. A., & Treur, J. (2010). An Integrative Ambient Agent Model for Unipolar Depression Relapse Prevention. *Journal of Ambient Intelligence and Smart Environments, 2*, 5–20.

Baron-Cohen, S. (1995). *Mindblindness.* MIT Press.

Bosse, T., Hoogendoorn, M., Klein, M. C. A., & Treur, J. (2008). A Component-Based Agent Model for Assessment of Driving Behaviour. In: Sandnes, F.E., Burgess, M., and Rong, C. (eds.), *Proceedings of the Fifth International Conference on Ubiquitous Intelligence and Computing, UIC'08.* Lecture Notes in Computer Science, vol. 5061, Springer Verlag, 2008, pp. 229-243.

Bosse, T., Jonker, C. M., van der Meij, L., & Treur, J. (2007). A Language and Environment for Analysis of Dynamics by SimulaTiOn. *International Journal of AI Tools, 16*(issue 3), 435–464. doi:10.1142/S0218213007003357

Bosse, T., van Lambalgen, R., van Maanen, P. P., & Treur, J. (2011). (in press). A System to Support Attention Allocation: Development and Application. *Web Intelligence and Agent Systems Journal.*

Bosse, T., van Maanen, P. P., & Treur, J. (2006). A Cognitive Model for Visual Attention and its Application. In: Nishida, T., Klusch, M., Sycara, K., Yokoo, M., Liu, J., Wah, B., Cheung, W., and Cheung, Y.-M. (eds.), *Proceedings of the Sixth International Conference on Intelligent Agent Technology, IAT'06.* IEEE Computer Society Press, pp. 255-262.

Bosse, T., van Maanen, P.-P., & Treur, J. (2009). Simulation and Formal Analysis of Visual Attention. *Web Intelligence and Agent Systems Journal, 7*, 89–105.

Both, F., Hoogendoorn, M., Klein, M. C. A., & Treur, J. (2009). Design and Analysis of an Ambient Intelligent System Supporting Depression Therapy. In: Luis Azevedo and Ana Rita Londral (Eds), *Proceedings of the Second International Conference on Health Informatics, HEALTHINF'09.* INSTICC Press, pp. 142-148.

Caffier, P. P., Erdmann, U., & Ullsperger, P. (2003). Experimental evaluation of eye-blink parameters as a drowsiness measure. *European Journal of Applied Physiology, 89*(3-4), 319–325. doi:10.1007/s00421-003-0807-5

Dennett, D. C. (1987). *The Intentional Stance.* Cambridge, Mass.: MIT Press.

Gärdenfors, P. (2003), *How Homo Became Sapiens: On The Evolution Of Thinking.* Oxford University Press, 2003.

Goldman, A. I. (2006). *Simulating Minds: The Philosophy, Psychology and Neuroscience of Mind Reading.* Oxford University Press.

Green, D. J. (2005). Realtime Compliance Management Using a Wireless Realtime Pillbottle – A Report on the Pilot Study of SIMPILL. In: *Proc. of the International Conference for eHealth, Telemedicine and Health, Med-e-Tel'05*, 2005, Luxemburg.

Heuvelink, A., & Both, F. (2007). BoA: A cognitive tactical picture compilation agent. In: *Proceedings of the 2007 IEEE/WIC/ACM International Conference on Intelligent Agent Technology, IAT 2007*. IEEE Computer Society Press, 2007, pp. 175-181.

Itti, L., & Koch, C. (2001). Computational Modeling of Visual Attention. *Nature Reviews. Neuroscience, 2*(3), 194–203. doi:10.1038/35058500

Riva, G., & Vatalaro, F. Davide, F., and Alcañiz, M. (eds). (2005). *Ambient Intelligence*. IOS Press, 2001.

Treur, J. (2008). On Human Aspects in Ambient Intelligence. In: *Proceedings of the First International Workshop on Human Aspects in Ambient Intelligence*. Published in: M. Mühlhäuser, A. Ferscha, and E. Aitenbichler (eds.), *Constructing Ambient Intelligence: AmI-07 Workshops Proceedings*. Communications in Computer and Information Science (CCIS), vol. 11, Springer Verlag, pp. 262-267.

Wickens, C. D. (2002). Situation awareness and workload in aviation. *Current Directions in Psychological Science, 11*(4), 128–133. doi:10.1111/1467-8721.00184

KEY TERMS AND DEFINITIONS

Human Awareness: awareness by the human about the human and environmental processes and their interaction

Technological Awareness: awareness by the environment about the human and environmental processes and their interaction

Situation Awareness: the perception of environmental elements within a volume of time and space, the comprehension of their meaning, and the projection of their status in the near future

Tactical Picture Compilation: the task in a military context to identify and classify all entities in the environment by reasoning with the available information, e.g. to determine the threat of a entity.

Visual Attention Manipulation: the task of tracking and steering the visual attention of a human related to the requirements of the task

LEADSTO: a specification language that can be used to formally describe simulation models in a declarative manner, combining numerical and logical statements.

Domain Model: a model of the interaction between the environment and the human in some scenario subject to support via ambient intelligence

Analysis Model: a model that allows for analysis of the human's states and processes by reasoning based on observations (possibly using specific sensors) and a domain model

Support Model: a model that allows for generating support for the human by reasoning based on the domain model

Driver Behaviour: N/A (general term)

ENDNOTES

[1] For a more detailed description of the relation between eye movements and fatigue, see e.g., (Caffier, Erdmann, and Ullsperger, 2003).

Chapter 9
Assistive Technologies in Smart Homes

Tatsuya Yamazaki
National Institute of Information and Communications Technology, Japan

ABSTRACT

This book chapter provides a review of the assistive technologies deployed in smart spaces with a variety of smart home or house examples. In the first place, home networking technologies and sensing technologies are surveyed as fundamental technologies to support smart environment. After reviewing representative smart home projects from across the world, concrete assistive services related with the fundamental technologies in smart environment are deployed not only for the elderly and handicapped but for people in ordinary families as well. Adaptability is one of the key essences in the assistive technologies in smart environment and, for this purpose, human-ware studies including man-machine interfaces, ergonomics and gerontology are needed to be linked with the hardware specific fundamental technologies.

INTRODUCTION

It is an old saying that 'a house becomes a home over time'. From the word 'house', one may imagine only a building. On the other hand, one is likely to feel comfortable and warm at the mention of the word 'home'—a place that residents make livable. A home is formalized by the harmonization of its residents, environments, household appliances,

etc. and is then able to adapt to the lifestyles of its residents. How can we make our homes more adaptable to our lifestyles? Owing to the introduction of ICT (Information and Communication Technology), home functions such as networking, sensing and appliances have become smarter and home adaptability, as a result, has increased greatly. Today, a smart home can be defined as a dwelling that incorporates a communications network which connects the key electrical appliances, sensors and services, and allows them to be remotely

DOI: 10.4018/978-1-61692-857-5.ch009

controlled, monitored or accessed. Such smart homes adapt to their residents autonomously and assist readily in their ways of living, particularly in the case of the handicapped or elderly people. In addition, from the term 'smart', one may expect smart homes to possess the ability to think, predict and take decisions. Such behaviour must be supported by capabilities in the fields of adaptation, communication, pattern recognition and so on.

Speaking of assistive technologies in smart homes, both individual technologies (as outlined above) and total service provisioning technology consisting of individual technologies may be involved. As a result, the area covered by the field of assistive technologies for smart homes is indeed very wide. This chapter samples certain foundational technologies that are related to assistive technologies; further, we overview the smart home research and development area.

The remainder of this chapter is organized as follows. With regard to foundational technologies in smart homes, "Overview" describes home networking technologies while "Sensing Technologies" focuses on sensor technologies. In "Smart Homes in the World", I present representative smart home or house projects from across the world. I pick up several test beds from North America, Europe and Asia and test them in real situations. "Assistive Services in Smart Homes" describes a few examples of assistive technologies as integrated services. Finally, "Conclusion" concludes this chapter.

HOME NETWORKING TECHNOLOGIES

Overview

As mentioned in the previous section, the introduction of ICT into the residential environment has added a new layer of mediation between a house and its residents, and is likely to expedite the adaptation process outlined above. In this section,

we describe networking technologies, as one of the technologies that are integral to smart homes.

Standardized technologies related to home networks are depicted in Figure 1. The technologies are classified into the two axes. The horizontal axis represents home appliance categorization and the three categories are depicted in Figure 1. The first category includes audio/visual appliances such as DVD recorders and digital TVs, which need high-speed connections. The second category is that of information appliances which need medium-speed connections. Lastly, the third category encompasses major appliances such as refrigerators and air-conditioners, which are capable of working on low-speed connections.

The vertical axis in Figure 1 represents the layer structure of the technologies. The upper-most layer corresponds to the applications layer; the lower-most layer lies just above the layer of physical communication. The communication layers in Figure 1 include the general networking technologies standardized in IEEE, ITU, etc. In the application layers in Figure 1, the technologies tend to depend on each application, especially for CODEC.

For the communication lower layer, there are various networking technologies for data link. Each networking technology has its own feature and its usage is differentiated according to purpose, situation and so on. Table 1 briefly summarizes a part of features for representative networking technologies for data link. Since there are several modes or options for each technology and its performance may change according to the environment in which it is used. Therefore the present best specification values are presented in Table 1.

Figure 2 presents a home network architecture, which was standardized as ITU-T J.190 in July, 2002 (ITU-T, 2002), and revised in October, 2007. In the revised version, the home network consists of IP-based and non-IP-based domains; ITU-T J.290, for example, is an application targeted at the home network.

Figure 1. Overview of Standardized Technologies Related to the Home Network

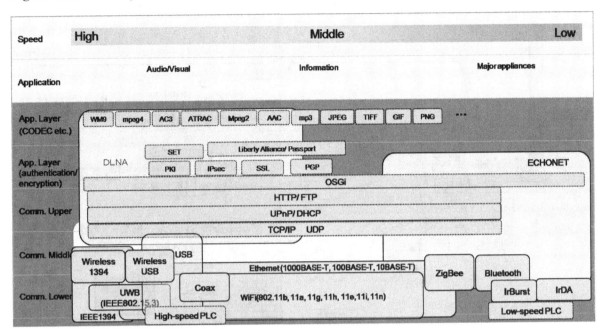

Table 1. Comparison of Networking Technologies for Data Link

	Ethernet	WiFi	UWB	Bluetooth	ZigBee	IrDA
Transmission rate (bps)	1G	54M	320M	3M	250K	16M
Transmission distance	100m	100m	10m	100m	30m	1m
Other features	Stableness, wide popularization	Mobility, wide popularization	Mobility, location detection	Mobility, interference-proof, security	Mobility, low-power	low-cost, security

Each of the next subsections describes an internationally standardized home network technology.

DLNA

The DLNA (Digital Living Network Alliance) was established in 2003 when a collection of companies from around the world agreed that they all made better products when those products were compatible. DLNA published its first set of Interoperability Guidelines in June 2004 and the first set of DLNA Certified products began appearing in market soon thereafter. The latest version of the DLNA Interoperability Guidelines, version 1.5, was published in March 2006, and then expanded in October 2006.

These guidelines enlarge the capabilities of a DLNA-defined network to include more home and mobile devices. They also include specifications for link protection to allow the secure transmission of copyright-protected commercial digital content.

HAVi

HAVi, which stands for Home Audio/Video Interoperability, was developed as a common specification for networking digital home entertainment

167

Figure 2. Overview of Standardized Technologies Related to the Home Network

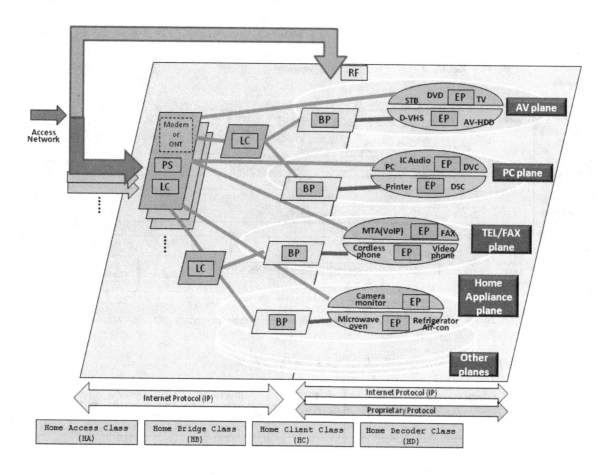

products such as digital audio and digital video by eight of the world's leading manufacturers of audio-visual electronics. The objective of HAVi was to enable the mutual operation of different vender products on the home network by using IEEE 1394.

ECHONET

ECHONET is an abbreviation of Energy Conservation and Homecare NETwork. Several major electric machine manufacturers and associated agencies in Japan established the ECHONET Consortium in 1997 and specified a universal transmission standard for the various services in home network systems that combine home appliances designed by different manufacturers through transmission media.

ECHONET is designed to control home appliances directly and connect home electronics devices through a gateway. This design will enable industry participants to develop a variety of systems having different communication speeds and levels of technological sophistication while maintaining optimum cost performance.

Others

The OSGi (Open Service Gateway Initiative) framework has played an important role as an open standard, and is one of the most promising architectures for building the home gateway which connects the home appliances and creates home networking services (Gouin-Vallerand and Giroux, 2007; West, Newman and Greenhill, 2005).

Sensor networking is another key technology in the area of smart home research. The sensor devices are minute, small enough and to detect environmental information. To make a sensor network by connecting distributed sensor devices, light networking protocols have been developed. ZigBee is one of the most used sensor networking protocols.

The X10 is used for communication among home appliances and electronic devices used within homes. It primarily works by carrying control signals over power lines. The X10 protocol is relatively simple and widely-used.

SENSING TECHNOLOGIES

Overview

For the implementation of assistive technologies, we must consider the interface between systems and humans. Since the interface differs for each service and there is a great variety of services, it is difficult for a single interface to cover every one of the research fields. In particular, the output mechanism of the man–machine interface has a wide range of possibilities. Depending on the service or application, the following questions can be answered.

1. What are the terminals or devices that should be used?
2. What type of media should be used?
3. Who should be the target?

In this section, therefore, the input mechanism of the man–machine interface will be focused on; in addition, sensing technologies will be reviewed briefly, since they are the basis of the assistive technologies.

This aspect of sensing can be separated into ubiquitous sensing and wearable sensing. Very generally speaking, ubiquitous sensors are attached or embedded in the environment and collect the related information from the environmental viewpoint. On the other hand, wearable sensors are attached to the user and collect more human-related contextual information. In Table 2, ubiquitous sensing and wearable sensing are compared qualitatively.

Ubiquitous Sensing

Visual sensors, viz. cameras, are popularly used to capture human activities directly and realistically. It is often that privacy issues become a subject of discussion; moreover, captured images are sometimes difficult to process and understand by the method of computational processing. Nevertheless, in (West, Newman and Greenhill, 2005), West *et al.* used a camera with virtual sensors to determine the activities and events inside a smart house. Their method basically relied on a tradi-

Table 2. Comparison between Ubiquitous Sensing and Wearable Sensing

	Sensing accuracy	User annoyance	System feature
Ubiquitous sensing	Sensor location is stable. Interference is anticipated relatively.	There is no burden by body attachment. Psychological burden may occur because of privacy issues.	Relatively scale is large and architecture is complex. Maintenance cost is usually expensive.
Wearable sensing	Accurate biometric data can be collected. Sensing data are apt to be affected by noise relatively.	Users have to admit physical attachment. Psychological burden depends on usage.	Sensors should be small, light and low-power-driven. Toughness is required.

tional technique of background subtraction. Also Do *et al.* (2006) used camera images to recognize hand gestures when a user wanted to control home appliances by pointing, using hand signs and so on. In (Kim and Kim, 2006), CCD cameras were used to recognize human upper body gestures. As a result of such recognition, curtains and lighting could be controlled.

A microphone is a sensing device that collects human voice as well as various kinds of environmental audio information. In the SAFIR project, they used a speech component as a user interface for natural language interaction. Voice Query Language (Schiller and Gan, 2003) is a solution that allows a user to query or command existing information systems through language. McHugh and Smeaton (2006) tried to detect most events by the use of a single microphone situated inside an actual home. Finally, it was found that the microphone could not function effectively in a normal home environment due to the crosstalk from outside environmental noises.

In the PROSAFE project, a set of infrared motion sensors located on the ceiling of a room were used as a multi-sensory system to estimate the trajectories of humans (Chan *et al.*, 2004). The sensors were linked to a PC, using an RS485 bus. An unsupervised classification method was developed to extract these trajectories.

Owing to the influence of gravity, the measuring weights on the floor can be connected to information on presence (position and general direction of motion) awareness. While Kaddoura, King and Helal (2005) developed a floor-based indoor location tracking system with pressure sensor technologies, Fook *et al.* (2006) used Fiber Bragg Grating (FBG) sensors.

Wearable Sensing

Wearable sensors are very useful to collect vital information in the form of unrestrained and 24-hour monitoring devices. Such devices have been developed for EMGs (electromyograms), ECGs (electrocardiographs), EEGs (electroencephalograms), and for the tracking of body temperature, blood pressure, pulse rate and so on. A wireless ECG sensor was used in the RECATA platform (Zhou and Hou, 2005) — a platform adapted to telemedicine applications, and especially to real-time cardiac arrhythmias tele-assistance and monitoring. It was evaluated by actual patients at a French hospital. Lee, Bien and Kim (2005) have developed a 24-hour continuous health monitoring system that can measure EMG, ECG, EEG, body temperature, and blood pressure. The main purpose of this system was the prediction and on-time detection of strokes and heart attacks. Similarly, O'Donoghue, Herbert and Stack (2006) have developed a wireless sensor device that can monitor ECG, body temperature, blood pressure and pulse rate, non-intrusively.

Despite their application field being prison environments and not homes, portable biomedical devices for health monitoring were developed as a jacket along with wireless Wi-Fi communication (Canelli *et al.*, 2003).

SMART HOMES IN THE WORLD

Using the above-mentioned home network and sensor technologies and combining them with other advanced technologies, smart home projects have been carried out as living laboratories where human-centred services are evaluated in real settings. For example, realistic studies were conducted on how to provide services that were adapted to individual residence situations and on monitoring the daily behavioural patterns of such residents.

In this section, we review major smart house or home projects across the world. The regions covered include North America, Europe and Asia. The smart houses or homes described here are only examples, and there are several projects that are not cited in this chapter. In particular, telecommunication companies around the world have been carrying out projects that are similar in scope.

North America

Adaptive House (Mozer, 2005)

The University of Colorado has constructed a prototype system, the adaptive house, in an actual residence. The adaptive house is equipped with an array of more than 75 sensors that provide information on environmental conditions such as temperature, ambient light levels, sound, motion, door and window openings, and actuators to control the furnace, space heaters, water heater, lighting units, and ceiling fans. The control systems in the adaptive house are based on neural network reinforcement learning and prediction techniques. The adaptive house can, for example, predict when its occupants will return home and determine when to start heating the house so that a comfortable temperature is attained by the time the occupants arrive.

Aware Home Project (Kidd *et al.*, 1999)

The Future Computing Environments (FCE) Group at the Georgia Institute of Technology has constructed a real experimental setting. This living laboratory is known as the Georgia Tech Broadband Institute Residential Laboratory. The site is a three-storey, 5040-square-foot home that functions as a living laboratory for ubiquitous computing in home life. The Aware Home Project is devoted to the multidisciplinary exploration of emerging technologies and services for the home; the development of home services in the fields of health, education, entertainment, security, etc.; and the evaluation of user experiences with such domestic technologies.

MavHome Project (Das *et al.*, 2002; Cook, Youngblood and Das, 2006)

The University of Texas at Arlington has developed the MavHome (Managing an Intelligent Versatile Home) Project. The MavHome is a home environment system that perceives the state of the home through sensors, and intelligently acts upon the environment through controllers. The goal of the MavHome is to create a home that acts as a rational agent in a manner that maximizes the comfort of its inhabitants and minimizes operating costs.

House_n Project (Intille, 2002; Intille, 2006)

At the Massachusetts Institute of Technology (MIT), the House_n group is working towards a vision where computer technology is ever-present, but in a more subtle way than often advocated in popular culture or even in engineering circles. The MIT researchers are seeking sensor-driven pervasive technologies to empower people with information that helps them make decisions, but they do not want to strip people of their sense of control over their environment.

The House_n research is focused on how the design of a home and its related technologies, products, and services should evolve to better meet the opportunities and challenges of the future. The House_n project encompasses a multidisciplinary team that includes researchers in the fields of architecture, computer science, mechanical engineering, nutrition communications, user interface and product design, preventative medicine, social anthropology, and public health. Several key issues related to the building itself and to the services that can be offered are slated for investigation so as to facilitate this home of the future.

EasyLiving Project (Brumitt *et al.*, 2000)

The vision research group at Microsoft Research conducted the EasyLiving Project. They aimed at developing prototype architecture and technologies in order to build intelligent environments. By using the geometric model of a room and taking readings from sensors embedded within, the EasyLiving Project was designed to provide context-aware computing services through video tracking and recognition, and sensor readings in the sensing room.

Gator Tech Smart House

The University of Florida's Mobile and Pervasive Computing Laboratory is developing programmable pervasive spaces in which a smart space exists as both a runtime environment and a software library (Helal *et al.*, 2005). In these spaces, service discovery and gateway protocols automatically integrate system components by using generic middleware that maintains a service definition for every sensor and actuator in the space. The programmers at this facility assemble component services into composite services and applications, which third parties can easily implement or extend to meet their requirements.

One objective of the programmable pervasive spaces project is to design and implement a useful reference architecture and a corresponding middleware for future pervasive computing environments. The other objective is to simplify the complexity of smart environments and make them more programmable and utilizable.

DOMUS Laboratory

The University of Sherbrooke in Canada constructed the DOMUS laboratory, which is a research pole on cognitive assistance in smart homes and mobile computing (Pigot *et al.*, 2005). The DOMUS laboratory is situated at the University of Sherbrooke and consists of an apartment with a living or meeting room, a kitchen, a bedroom, etc. The maintenance of the apartment is supported by the computer server room.

The DOMUS laboratory research is focused on dealing with issues that are part of the computer science plan (distributed systems, personalization, ontology, etc.) as well as those that are related to the multidisciplinary plan (ethics, psychology, healthcare, etc.). Ongoing projects focus on cognitive assistance to people suffering from Alzheimer's type dementia, schizophrenia, cranial trauma, or intellectual deficiencies. This assistance is dispensed through the realization of an intelligent home and associated mobile applications.

Intelligent Home Project

The multi-agent systems lab at the University of Massachusetts in Amherst started a project to develop an intelligent home. They designed and implemented a set of distributed autonomous home control agents and deployed them in a simulated home environment (Lesser *et al.*, 1999). Each agent was associated with a particular appliance to realize resource coordination in the intelligent home environment.

Europe

comHOME (Junestrand, Keijer and Tollmar, 2001)

The comHOME is a smart dwelling of the future and is located at the headquarters of Telia. comHOME consists of three rooms within a lab environment, simulating a smart home. Each space in this home is viewed as a separate situation and not as a part of a complete apartment. The researchers are attempting to connect these spaces by using video-mediated communication.

Gloucester Smart House

The University of Bath in England has been developing devices to support people with dementia. They built the Gloucester Smart House to carry out user-involved evaluations. This project was conducted between 2001 and 2004.

The problem areas addressed included those associated with safety-related tasks such as the use of the cooker, reminder devices for, for instance, the taking of medication, and support systems such as a cookery assistant. The Gloucester Smart House used sensors to monitor the behaviour of its occupants and their interaction with appliances. The house responded by controlling the supporting devices and by providing reminders. Through the evaluations, the researchers demonstrated that the house was capable of detecting gas leaks and turning off the cooker that had caused these leaks.

CUSTODIAN Project
(Davis *et al.*, 2003)

The CUSTODIAN project was carried out from 1999 to 2000 under the coordination of the Robert Gordon University. Its objective was to create a software tool that could be used to facilitate the design of smart homes for people with disabilities.

Subsequent assistive technology research led to the design and installation of a number of smart home systems. In two new housing developments in Aberdeen, for instance, homes were designed for people with severe disabilities. The project involved detailed consultation with various care and housing organizations. The software that had been developed was aimed at facilitating interactions and exchanges among the diverse groups (engineers, designers, care and housing providers, and users) involved in the process of creating the best spatial solutions for these homes.

Philips

Since 1995 or earlier still, the Dutch electronics company Philips has been involved in making homes smarter by employing information and communication technologies such as wireless communications. Their targeted markets include lighting and healthcare systems. In recent times,

the vision of Philips has shifted towards creating energy-efficient homes through improved lighting efficiency, attained by using integrated smart wireless devices.

Siemens

Siemens researchers at Corporate Technology (CT) have a motto—'From smart homes to smart cities'. They have constructed a complete apartment, fully equipped with Siemens household electrical, building, communications, and multimedia technologies, and a test living environment. All these devices can be controlled through a central communications node known as a gateway. Siemens CT has developed software that permits the integration of various wireless protocols and interfaces while building technology systems. Adaptive lighting systems developed by CT, which control natural as well as artificial light, have been installed and tested in their laboratory.

Asia

Welfare Techno Houses
(Tamura *et al.*, 1998)

In 1995, the WTH (Welfare Techno Houses) were constructed in Japan. The concept of these experimental houses was to secure a measure of independence for elderly and disabled persons and to improve their quality of life. People living in the WTH were observed using infrared sensors and sensors to detect the status of doors and windows. In particular, their sleeping and toilet activities were meticulously monitored.

Smart House Ikeda

An experiment was carried out by AIST (Advanced Industrial Science and Technology) at the Smart House Ikeda, wherein a family spent one month living in the smart house (Matsuoka, 2004). Their behaviour was recorded by 167 sensors of 15 types, and was classified automatically into 12 kinds of life actions.

Robotics Room and Sensing Room

As an extension of Robotics Room 2 (Mori *et al.*, 2001), a Sensing Room was constructed at the University of Tokyo (Mori *et al.*, 2004). Although the researchers claim that the Sensing Room can measure details of daily human behaviour over the long term, it appears that there are limitations on the extent of simulation possible of real-life situations, owing to the smallness of the room.

ActiveHome

ActiveHome, a 1200-square-foot apartment-like residential space consisting of a bedroom, a bathroom, a kitchen, and a living room, is located in a university setting. The Information and Communications Department at the University of Korea has conducted several experiments to evaluate context-aware services according to the type of human behaviour (Lee, Park and Hahn, 2005).

ubiHome

Another smart space in Korea is 'ubiHome', which is a test bed for applying ubiquitous computing technologies to smart home environments. The 'ubiHome' provides distributed computing environments wherein contexts are directly exchanged between sensors and services without any centralized context-managing server. The 'ubiHome' also manages user conflicts by exploiting user priorities in terms of context. It provides a personalized service that autonomously takes appropriate action by adapting to an individual's situation; for example, the system automatically monitors the light controls in accordance with the user's preferences (Oh and Woo, 2004).

Intelligent Sweet Home (Park and Bien, 2003)

Intelligent Sweet Home (ISH) is a service-robot-integrated intelligent house platform with a sensor-based home network system. It provides various kinds of daily living assistance to the inhabitants, especially for the elderly and the disabled. The idea was based on the fact that technologies and solutions for smart homes should be human-friendly, and a high level of intelligence should be embedded in the system's controls, actions and interactions with the users.

UKARI Project and Ubiquitous Home (Yamazaki, 2005; Yamazaki *et al.*, 2005; Yamazaki, 2006; Yamazaki, 2007)

Our last example is the UKARI (Universal Knowledgeable Architecture for Real-Life Appliances) Project conducted by NICT (National Institute of Information and Communications Technology), which consists of 6 companies and 10 universities. The UKARI Project researchers built a real-life test bed, termed as the 'Ubiquitous Home', for home context-aware service experiments and real-life experiments to collect several sets of data on of real-life situations. The layout and sensor arrangement of the Ubiquitous Home are shown in Figure 3.

ASSISTIVE SERVICES IN SMART HOMES

As seen in the previous sections, foundational assistive technologies have undergone considerable development, and significant experiments on smart homes have been conducted. Moreover, some coordinated assistive services are on their way to realization. Although assistance is generally considered vital in the case of elderly or (mentally/physically) disabled persons, assis-

Figure 3. Layout and Sensor Arrangement of the Ubiquitous Home

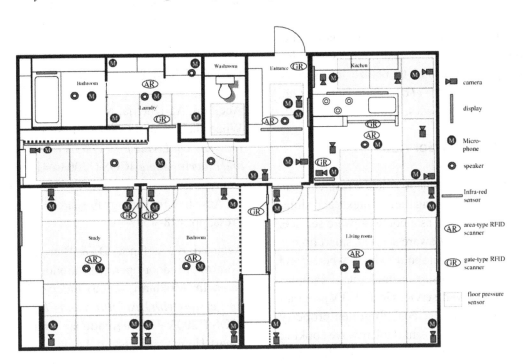

tive technologies can also be applied to ordinary people according to their context and situation. In this section, such representative assistive services are presented.

Accident Avoidance

Even though we may be careful to guard against accidental events in our daily life, there is always the possibility of meeting with an unexpected turn of events. Falls, for instance, are a significant cause of serious injury among the elderly. Using accelerometers and wavelet multi-resolution analysis, Nyan *et al.* (2004) developed a method to detect major incidents such as a fall. Song *et al.* (2008) proposed a similar method using a tri-axial accelerometer and the SVM (Support Vector Machine) algorithm.

Another accident that should be avoided is the misuse of medicine. A small mote-based portable medical system at home can automatically monitor and handle medication to persons with dementia (Fook and Tee, 2007).

Movement Assist

Some disabled persons find it difficult to travel and sometimes use a wheelchair. Some are, however, in need of assistance to control the wheelchair owing to ailments such as trouble in their limbs. ICT can help such people control the wheel chair automatically by programming the wheelchair to recognize signs and instructions emanating from the motion of the head or body of the user (Kvasnica, 2004). Moreover, some researchers have amended the normal wheelchair to add autonomy by using ICT (Avtanski, Stefanov and Bien, 2003; Abellard, 2003; Ernst, 2004).

Health Care

Monitoring Health Status at homes is another assistive service. Sensor and home networking technologies are useful to realize this service (Soede, 2003; Tamura *et al.*, 2004; Boudy *et al.*, 2006; Kukhun and Sedes, 2008).

Context-Aware Service

In general, a service adaptive to the residents can be provided by knowing their contexts and situations. This is known as a context-aware service. To realize a context-aware service, smart homes detect the information about "who," "where," and "when" by using sensing technologies. Examples of the context-aware services include TV program recommendation services, forgotten-property check services and the smart microwave cooking services, all of which have been implemented (Yamazaki, 2006; Russo *et al.*, 2004).

CONCLUSION

This chapter reviewed the assistive technologies deployed in smart homes. Since the range of assistive technologies is wide, some foundational technologies were picked and examples of the entire gamut of assistive services were presented. In addition, the smart home research and development trends from across the world were overviewed.

This review has tended to be a hardware specific survey. However, it is not only hardware technology but software technology as well that is needed for smart home studies. In addition, human-ware studies that deal with man-machine interfaces, ergonomics, gerontology and so on are needed. Developments in software technology and human-ware studies have resulted in making assistive technologies more adaptive and realistic. Assistive technologies are fast becoming ubiquitous and smart homes will, in the near future, assist and favour not only the elderly and handicapped but people in ordinary families as well.

REFERENCES

Abellard, A., Khelifa, M. M. B., et al. (2003) A Wheelchair Neural Control. In Mokhtari M (ed) *Independent Living for Persons with Disabilities and Elderly People*, ICOST'2003 1st International Conference on Smart Homes and Health Telematics, Assistive Technology Research Series 12, IOS Press, pp. 128–133.

Avtanski, A., Stefanov, D., & Bien, Z. Z. (2003) Control of Indoor-operated Autonomous Wheelchair. In: Mokhtari M (ed) *Independent Living for Persons with Disabilities and Elderly People*, ICOST'2003 1st International Conference on Smart Homes and Health Telematics, Assistive Technology Research Series 12, IOS Press, pp. 120–127.

Boudy, J., Baldinger, J.-L., et al. (2006) Telemedicine for Elderly Patient at Home: the TelePat Project. In: Nugent C and Augusto JC (eds) Smart Homes and Beyond, ICOST'2006 4th International Conference on Smart Homes and Health Telematics, Assistive Technology Research Series 19, IOS Press, pp. 74–81.

Brumitt, B., & Meyers, B. (2000). *EasyLiving: Technologies for Intelligent Environments. Handheld and Ubiquitous Computing 2000, LNCS 1927.* Springer-Verlag.

Canelli, N., Pisetta, A., et al. (2003) Development of a "Domotic Cell" for Health Care Delivery in Prison Environments. In: Mokhtari M (ed) *Independent Living for Persons with Disabilities and Elderly People*, ICOST'2003 1st International Conference on Smart Homes and Health Telematics, Assistive Technology Research Series 12, IOS Press, pp. 96–103.

Chan, M., Campo, E., et al. (2004) Monitoring Elderly People Using a Multisensor System. Proc. 2nd International Conference on Smart Homes and Health Telematics (ICOST 2004), Vol. 14, pp. 162–169.

Cook, D. J., Youngblood, M., & Das, S. K. (2006). A Multi-Agent Approach to Controlling a Smart Environment. In Augusto, J. C., & Nugent, C. (Eds.), *Designing Smart Homes The Role of Artificial Intelligence, Lecture Notes in Artificial Intelligence 4008* (pp. 165–182). Springer-Verlag Berlin Heidelberg.

Das, S. K., & Cook, D. J. (2002). The Role of Prediction Algorithms on the MavHome Smart Home Architectures. [Special Issue on Smart Homes]. *IEEE Wireless Communications*, 9(6), 77–84. doi:10.1109/MWC.2002.1160085

Davis, G., Wiratunga, N., et al. (2003) Matching SmartHouse technology to needs of the elderly and disabled. Proc. Workshops of the 5th International Conference on Case-Based Reasoning (ICCBR 03), pp. 29–36.

Do, J.-H., Jung, S. H., et al. (2006) Gesture-Based Interface for Home Appliance Control in Smart Home. In: Nugent C and Augusto JC (eds) Smart Homes and Beyond, ICOST'2006 4th International Conference on Smart Homes and Health Telematics, Assistive Technology Research Series 19, IOS Press, pp. 23–30.

Ernst, T. (2004) E-Wheelchair: A Communication System Based on IPv6 and NEMO. Proc. 2nd International Conference on Smart Homes and Health Telematics (ICOST 2004), Vol. 14, pp. 213–220.

Fook, V. F. S., Hao, S., et al. (2006) Fiber Bragg Grating Sensor System for Monitoring and Handling Bedridden Patients. In: Nugent C and Augusto JC (eds) Smart Homes and Beyond, ICOST'2006 4th International Conference on Smart Homes and Health Telematics, Assistive Technology Research Series 19, IOS Press, pp. 239–246.

Fook, V. F. S., Tee, J. H., et al. (2007) Smart Mote-Based Medical System for Monitoring and Handling Medication among Persons with Dementia. 5th International Conference on Smart Homes and Health Telematics (ICOST 2007), LNCS 4541, Springer, pp. 54–62.

Gouin-Vallerand, C., & Giroux, C. (2007). Managing and Deployment of Applications with OSGi in the Context of Smart Homes. Proc. of the Third IEEE International Conference on Wireless and Mobile Computing, Networking, and Communications (WiMob 2007).

Helal, A., & Mann, W. (2005). *Gator Tech Smart House: A Programmable Pervasive Space* (pp. 64–74). IEEE Computer Magazine.

Intille, S. S. (2002). *Designing a home of the future* (pp. 80–86). IEEE Pervasive Computing, Vol. April-June.

Intille, S. S. (2006) The Goal: Smart People, Not Smart Homes. In: Nugent C and Augusto JC (eds) Smart Homes and Beyond, ICOST'2006 4th International Conference on Smart Homes and Health Telematics, Assistive Technology Research Series 19, IOS Press, pp. 3–6.

ITU-T Recommendation J.190, Architecture of MediaHomeNet that supports cable-based services, 2002.

Jiang, L., Liu, D.-Y., & Yang, B. (2004) Smart Home Research. Proc. of the Third International Conference on Machine Learning and Cybernetics, pp. 659–663.

Junestrand, S., Keijer, U., & Tollmar, K. (2001) Private and Public Digital Domestic Spaces. International Journal of Human-Computer Studies, Vol. 54, No. 5, Academic Press, London, England, pp. 753–778.

Kaddoura, Y., King, J., and Helal, (Sumi) A. (2005). Cost-Precision Tradeoffs in Unencumbered Floor-based Indoor Location Tracking. In: Giroux S and Pigot H (eds) From Smart Homes to Smart Care, ICOST'2005 3rd International Conference on Smart Homes and Health Telematics, Assistive Technology Research Series 15, IOS Press, pp.75–82.

Kidd, C. D., Orr, R. J., et al. (1999) The Aware Home: A Living Laboratory for Ubiquitous Computing Research. Proc. of the Second International Workshop on Cooperative Buildings - CoBuild'99.

Kim, D., & Kim, D. (2006) An Intelligent Smart Home Control Using Body Gestures. Proc. of 2006 International Conference on Hybrid Information Technology (ICHIT 2006), vol.II, pp. 439–446.

Kukhun, D. A., & Sedes, F. (2008) Adaptive Solutions for Access Control within Pervasive Healthcare Systems. Proc. 6th International Conference on Smart Homes and Health Telematics (ICOST 2008), LNCS 5120, Springer, pp. 42–53.

Kvasnica, M. (2004) Six-DoF Force-Torque Wheelchair Control by Means of the Body Motion. Proc. 2nd International Conference on Smart Homes and Health Telematics (ICOST 2004), Vol. 14, pp. 247–252.

Lee, H., Bien, Z., & Kim, Y. (2005). 24-hour Continuous Health Monitoring System for the Disabled and the Elderly in Sensor Network Environment. In: Giroux S and Pigot H (eds) From Smart Homes to Smart Care, ICOST'2005 3rd International Conference on Smart Homes and Health Telematics, Assistive Technology Research Series 15, IOS Press, pp. 116–123.

Lee, J.-S., Park, K.-S., & Hahn, M.-S. (2005) WindowActive: An Interactive House Window On Demand. 1st Korea-Japan Joint Workshop on Ubiquitous Computing and Networking Systems (UbiCNS 2005), pp. 481–484.

Lesser, V., Atighetchi, M., et al. (1999) A Multi-Agent System for Intelligent Environment Control. UMass Computer Science Technical Report 1998-40.

Matsuoka, K. (2004) Aware Home Understanding Life Activities. Proc. 2nd International Conference on Smart Homes and Health Telematics (ICOST 2004), Vol. 14, pp. 186–193.

McHugh, M., & Smeaton, A. F. (2006) Event Detection Using Audio in a Smart Home Context. In: Nugent C and Augusto JC (eds) Smart Homes and Beyond, ICOST'2006 4th International Conference on Smart Homes and Health Telematics, Assistive Technology Research Series 19, IOS Press, pp. 23–30.

Mori, T., et al. (2001) Accumulation and Summarization of Human Daily Action Data in One-Room-Type Sensing System. Proc. IROS (IEEE/RSJ International Conference on Intelligent Robots and Systems) 2001, pp. 2349–2354.

Mori, T., Noguchi, H., et al. (2004) Sensing Room: Distributed Sensor Environment for Measurement of Human Daily Behavior. Proc. of First International Workshop on Networked Sensing Systems (INSS2004), pp. 40–43.

Mozer, M. C. (2005). Lessons from an Adaptive House. In Cook, D., & Das, R. (Eds.), *Smart Environments: Technologies, Protocols, and Applications* (pp. 273–294). Hoboken, NJ: J. Wiley & Sons.

Nyan, M. N., Tay, F. E. H., & Seah, K. H. W. (2004) Segment Extraction Using Wavelet Multi-resolution Analysis for Human Activities and Fall Incidents Monitoring System. Proc. 2nd International Conference on Smart Homes and Health Telematics (ICOST 2004), Vol. 14, pp. 177–185.

O'Donoghue, J., Herbert, J., & Stack, P. (2006) Remote Non-Intrusive Patient Monitoring. In: Nugent C and Augusto JC (eds) Smart Homes and Beyond, ICOST'2006 4th International Conference on Smart Homes and Health Telematics, Assistive Technology Research Series 19, IOS Press, pp. 23–30.

Oh, Y., & Woo, W. (2004) A Unified Application Service Model for ubiHome by Exploiting Intelligent Context-Awareness. Proc. of Second Intern. Symp. on Ubiquitous Computing Systems (UCS2004), pp. 117–122.

Park, K.-H., & Bien, Z. Z. (2003) Intelligent Sweet Home for Assisting the Elderly and the Handicapped. In: Mokhtari M (ed) *Independent Living for Persons with Disabilities and Elderly People*, ICOST'2003 1st International Conference on Smart Homes and Health Telematics, Assistive Technology Research Series 12, IOS Press, pp. 151–158.

Pigot, H., Lefebvre, B., et al. (2003) The Role of Intelligent Habitats in Upholding Elders in Residence. 5th International Conference on Simulations in Biomedicine.

Russo, J., Sukojo, A., et al. (2004) SmartWave – Intelligent Meal Preparation System to Help Older People Live Independently. Proc. 2nd International Conference on Smart Homes and Health Telematics (ICOST 2004), Vol. 14, pp. 122–135.

Schiller, C. A., Gan, Y. M., et al. (2003) Exploring New Ways of Enabling e-Government Services for the Smart Home with Speech Interaction. In: Mokhtari M (ed) *Independent Living for Persons with Disabilities and Elderly People*, ICOST'2003 1st International Conference on Smart Homes and Health Telematics, Assistive Technology Research Series 12, IOS Press, pp.151–158.

Soede, M. (2003) The Home and Care Technology for Chronically Ill and Disabled Persons. In: Mokhtari M (ed) *Independent Living for Persons with Disabilities and Elderly People*, ICOST'2003 1st International Conference on Smart Homes and Health Telematics, Assistive Technology Research Series 12, IOS Press, pp. 3–9.

Song, S.-K., Jang, J., & Park, S. (2008) An Efficient Method for Activity Recognition of the Elderly Using Tilt Signals of Tri-axial Acceleration Sensor. Proc. 6th International Conference on Smart Homes and Health Telematics (ICOST 2008), LNCS 5120, Springer, pp. 99–104.

Tamura, T., Masuda, Y., et al. (2004) Application of Mobile Phone Technology in the Elderly – A Simple Telecare System for Home Rehabilitation. Proc. 2nd International Conference on Smart Homes and Health Telematics (ICOST 2004), Vol. 14, pp. 278–282.

Tamura, T., & Togawa, T. (1998). Fully automated health monitoring system in the home. *Medical Engineering & Physics*, 20, 573–579. doi:10.1016/S1350-4533(98)00064-2

West, G., Newman, K., & Greenhill, S. (2005). Using a Camera to Implement Virtual Sensors in a Smart House. In: Giroux S and Pigot H (eds) From Smart Homes to Smart Care, ICOST'2005 3rd International Conference on Smart Homes and Health Telematics, Assistive Technology Research Series 15, IOS Press, pp. 75–82.

Yamazaki, T. (2005) Ubiquitous Home: Real-Life Testbed for Home Context-Aware Service. Proc. Tridentcom 2005 (First International Conference on Testbeds and Research Infrastructures for the DEvelopment of NeTworks and COMmunities), pp. 54–59.

Yamazaki, T. (2006) Human Action Detection and Context-Aware Service Implementation in a Real-life Living Space Test Bed. Proc. Trident-Com 2006 (Second International Conference on Testbeds and Research Infrastructures for the DEvelopment of NeTworks and COMmunities).

Yamazaki, T. (2007). The Ubiquitous Home. *International Journal on Smart Homes*, *1*(1), 17–22.

Yamazaki, T., Ueda, H., et al. (2005). Networked Appliances Collaboration on the Ubiquitous Home. In: Giroux S and Pigot H (eds) From Smart Homes to Smart Care, ICOST'2005 3rd International Conference on Smart Homes and Health Telematics, Assistive Technology Research Series 15, IOS Press, pp. 135–142.

Zhang, D., Wang, X., et al. (2003) OSGi Based Service Infrastructure for Context Aware Connected Homes. In: Mokhtari M (ed) *Independent Living for Persons with Disabilities and Elderly People*, ICOST'2003 1st International Conference on Smart Homes and Health Telematics, Assistive Technology Research Series 12, IOS Press, pp. 81–88.

Zhou, H., & Hou, K. M. (2005). Real-time Cardiac Arrhythmia Tele-Assistance and Monitoring Platform: RECATA. In: Giroux S and Pigot H (eds) From Smart Homes to Smart Care, ICOST'2005 3rd International Conference on Smart Homes and Health Telematics, Assistive Technology Research Series 15, IOS Press, pp. 99–106.

KEY TERMS AND DEFINITIONS

Smart Homes: Conceptually homes that assist the residents, usually by means of ICT (Information and Communication Technology). From the viewpoint of ICT, fundamental components of smart homes are networking and sensing technologies. Smart homes automatically or autonomously provide services to make the residential life more comfortable, securer or more economical. The residents who need an assistive service are wide-ranging. Therefore universal interfaces for human-machine interaction are desired to be equipped for smart homes.

ITU-T J.190: A home network architecture standardized in ITU-T (International Telecommunication Union-Telecommunication). The title is "Architecture of MediaHomeNet that supports cable-based services". It was standardized in July, 2002 in the first place and revised in October, 2007. The home network was modelled as combination of IP-based and non-IP-based domains and a fundamental architecture of home network was defined.

Ubiquitous sensing: To collect necessary data through sensors attached or embedded in the environment ubiquitously. Although the variation of sensors differs from the targeted service or application, examples are cameras or microphones that may be used for purposes of surveillance and log recording.

Wearable sensing: To collect necessary data through sensors worn by users. Since wearable sensors are in contact with the user, more accurate biometric data can be collected by wearable sensing than by ubiquitous sensing. On the other hand, wearable sensors need to be small, light and low-power-driven.

Assistive services: In general, services that support human activities, usually in a daily life. In the context of smart homes, examples of assistive services are accident avoidance service, movement assist service, health care service, etc.

Ubiquitous Home: A smart home built by NICT (National Institute of Information and Communications Technology), a Japanese research institute. Ubiquitous Home is located inside the building of NICT and perfectly emulates a residential apartment. Ubiquitous Home is equipped with various types of sensors, network infrastructures and networked appliances. Real-life experiments were conducted and a part of collected data are open to be shared for an academic purpose.

Context-aware services: Services which coordinated and provided properly according to the user context, that is for example characteristic, status and situation. It is considered that context-aware services are included or overlapped by assistive services. Although essential contextual information differs from each service, information on place, time and identification is considered as fundamental contextual information.

Chapter 10
System Support for Smart Spaces

Francisco J. Ballestero
Universidad Rey Juan Carlos, Spain

Enrique Soriano
Universidad Rey Juan Carlos, Spain

Gorka Guardiola
Universidad Rey Juan Carlos, Spain

ABSTRACT

There are some important requirements to build effective smart spaces, like human aspects, sensing, activity recognition, context awareness, etc. However, all of them require adequate system support to build systems that work in practice. In this chapter, we discuss system level support services that are necessary to build working smart spaces. We also include a full discussion of system abstractions for pervasive computing taking in account naming, protection, modularity, communication, and programmability issues.

INTRODUCTION

On almost any computer, an operating system is required to build applications effectively. In the same way, it is necessary to consider system issues on their own to build smart spaces and pervasive applications. The main aim of ambient intelligence is to devise a unique computer, *the smart space*, out of the different subsystems in the space (computers, devices, sensors, and so on). In order to correctly achieve such purpose, pervasive services and applications must be provided with system support. Otherwise, we will come across the same pitfalls that were found while implementing applications in the old days, before the adoption of operating systems.

Unless you have appropriate system support for building a smart space, it is very likely that in the end any prototype built will be a mere collection of unrelated services instead of a reasonable and well integrated, operational, system.

The lack of system support leads to implementing the same services multiple times in different parts of the system. In addition, this increases

DOI: 10.4018/978-1-61692-857-5.ch010

the complexity of the environment and makes it unreliable.

In addition, without this support, it is very hard to add new devices to an already deployed system, because of poor modularity. A consequence is that a smart space may become obsolete quickly as new devices and services become available.

At the center of all this is the need for the system (or systems) used to provide appropriate abstractions for building and deploying services for smart spaces (and pervasive computing in general). For example, issues like naming, protection, and resource management can become almost trivial when proper system support is provided. In the absence of system support, these issues may become cumbersome or even impossible to overcome.

Figure 1 presents the top-level view of a smart space, considering it as a system. The key point is that the requirements necessary to build applications on the Smart Space are supported by a homogeneous system layer (which may be built upon an existing platform or not). Any particular list of requirements depends heavily on the particular set of applications considered. Therefore, enumerating such requirements is not feasible as

far as our discussion goes. We will mention some of them as examples later, but a full discussion is outside of the scope of the present chapter.

Exploding the previous diagram, we may identify the following components (depicted in figure 2):

- **Service Discovery:** mechanisms to discover which application services are exported by the subsystems that form the Smart Space.
- **Naming:** mapping methods for objects and services, so that the application can identify and access the resources.
- **Security:** authentication and access control schemes for the resources.
- **Communication and synchronization:** mechanisms for inter-application communication (similar to IPC, but not confined within a single system).

These components are more abstract services than actual components. Depending on the system built, they may be implemented by actual components matching those depicted in the figure or

Figure 1. Top-level view for a Smart Space as a system

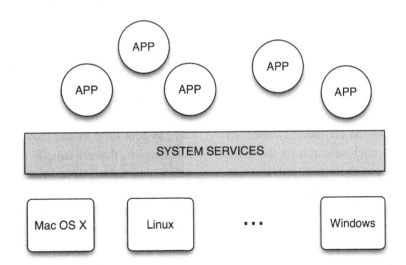

Figure 2. Components and services

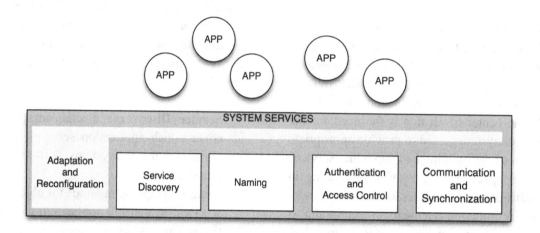

they may be implemented somewhere else. The next section presents several examples.

To guide the exposition, we plan to discuss two different approaches that have worked well in the past: The Plan B (Ballesteros *et al.*, 2007; Ballesteros *et al.*, 2006a; Ballesteros *et al.*, 2006b; Ballesteros *et al.*, 2005) approach (model all resources as synthetic files at the system level, like UNIX did time ago with devices) and the Interactive Workspaces Project (Johanson, Fox and Winograd, 2002; Johanson, Winograd and Fox, 2003; Fox *et al.*, 2000; Johanson and Fox, 2002) model for building smart rooms (mostly using middleware based event delivery systems). Both approaches are quite different. Comparing them regarding the different issues (resource location, naming, protection, etc.) can provide a reasonable view of the different approaches to build systems for these environments. However, this does not mean that the two approaches used to illustrate the discussion are the only ones. In fact, there has been much research on the area. See for example any of the recent publications on IEEE PerCom conference, IEEE Pervasive Computing articles, or Elsevier's Pervasive and Mobile Computing. A full survey of the state of the art for the field would be too extensive for the scope of this chapter.

System Organization and Application Model

To build a smart space or a system to support a smart space it is necessary to have an application model in mind. Depending on this the system shapes how applications are built and deployed on the space. Roughly, there are four different application models to be considered depending on the relationship between the space and applications.

The smart space is itself an application. In this model the smart space software is written solely for the purpose of a particular space. Therefore, system support is negligible other than the set of different modules within the application itself.

Architecting the system as an application without making a distinction between system services and the applications themselves is similar to what happened with computing before the adoption of operating systems. All applications require their own system support and they work using ad-hoc mechanisms that are usually poor because of the effort required to implement good ones just for a single application.

This approach is not very flexible, because it requires to (re)write most, if not all, of the code from scratch for the space. More importantly, it fails to define interfaces or services independent of particular tools and applications to deploy on the space. Even the simplest smart space ends up being more than just one application, because of different facilities and automation required on smart spaces. For example, it is typical to combine the smart space with other (conventional) applications available within a system (Johanson, Fox and Winograd, 2002; Johanson, Winograd and Fox, 2003; Fox *et al.*, 2000; Johanson and Fox, 2002).

The smart space is a new environment for new applications. This is the most popular approach for the development of smart spaces. In most cases, several frameworks are provided including libraries and data models so that applications can be written for the space. The framework is usually portable across heterogeneous hardware and system software, so that different machines can support the smart space, but applications must be written for the framework built.

This framework is normally termed *middleware*, because it sits in the middle, between the operating system and the application. Middleware has the disadvantage of requiring the developer to write the application or at least most parts of it in a particular way to be able to interact with the smart space and other integrated applications. For example, it may be necessary to write the software in Java because the framework is using that platform. This precludes using existing applications to operate on the space. On the other hand, writing new software is usually not a problem. Legacy applications, of interest, may still be completely rewritten or, at least, adapted and integrated into the smart space.

The smart space is a new environment for both old and new applications. It is useful to be able to control traditional applications present on the machines deployed through the space. For example, Gaia (Chan *et al.*, 2005) was able to control productivity applications in a semi-automated way to help the users with their sessions. In general, this requires being able to execute existing applications and/or being able to control them at run time through existing interfaces.

In some cases, all the control that can be exercised on applications is being able to move their interfaces. For example, this is as far as it goes in systems like (Johanson, Fox and Winograd, 2002; Johanson, Winograd and Fox, 2003; Fox *et al.*, 2000; Johanson and Fox, 2002). Nevertheless, it is desirable to be actually able to control external applications found on the space. In order to control an external application and properly integrate it into the smart space, some programmatic interface for the application has to be provided. The interface must be also enough to handle other applications of the same kind. Otherwise, each different instance for a type of application (e.g., different word processors) would require a different program to control it, which would mean that no program could operate on all existing applications of the type considered.

Plan B (Ballesteros *et al.*, 2007; Ballesteros *et al.*, 2006a; Ballesteros *et al.*, 2006b; Ballesteros *et al.*, 2005) provides a way to control external applications by wrapping them inside tiny synthetic file system interfaces, which make application become first class citizens in the smart space. For other software in the space, an application seems to be a little file tree. Operations on the files in the tree become operations on the application supported by the file tree.

A disadvantage of this approach is that external application control has to be less flexible than the one you can provide by rewriting part of the application. However, it makes it possible to integrate existing applications into the space.

The smart space is a set of data services. Although this one is a system model on its own, it is also important for the models discussed so far. A smart space may provide a set of data services to other applications. Therefore, we may think of a smart space as a set of *smart data* services. Applications operate on data provided by the smart space and become *smart* only depending on how *smart* the data actually is.

For example, a space can provide XML data to convey the physical layout of the space and the location of elements in the space. On the other hand, it may provide this data through file interfaces (like in Plan B) or through event channels (like in IWS).

Even if the space is not just the set of data services (as discussed above), this is always an issue as long as applications are able to consume data (e.g., context information) generated by the space.

The main problem is that legacy applications (all software not written just for the space) consume data with legacy formats and understand only particular system interfaces. Plan B addresses this problem by relying on file interfaces (because many applications may operate on files) and by placing the semantics on the user (the system conveys the information but does not try to interpret it). Other systems address this problem by converting data generated by the space to the format required by the application considered.

For example, Plan B relies on simple scripts to convert directories and files into HTML documents to be served by a web browser to provide information about the environment over the web.

ABSTRACTIONS USED TO CONVEY SERVICES

Any abstraction used to convey services in a smart space has to take into account several issues, including: adaptation to a highly dynamic environment, the high likelihood of failure of devices and applications, the use of the context information, the discovery of the services, how they are organized and named, etc. Applications cannot be in charge of dealing with all these factors, so they need the smart space support to handle them. In other words, applications need abstractions to use the smart space services while being decoupled from the smart space infrastructure by an interface. At this point, there are two decisions to make: (i) the level at which the smart space and the application are decoupled, and (ii) the level of

abstraction of the interface provided by the smart space. See figure 3.

Smart Space Support at Middleware Level

Abstractions used to reach the services of the smart spaces are usually based on middleware. Middleware is situated between the operating system and the application can use different models of communication to provide services for the smart space. The most popular models are object-oriented models based on remote method invocation (e.g. Java RMI) and message-oriented models (e.g. Java JMS). The middleware is usually implemented by a framework, that is, a set of libraries and programs that permit applications to access smart space services by invoking methods on local proxy objects.

Let's look at a simplified example: suppose that a program needs to use the audio infrastructure of the smart space to deliver a speech message to a user. In an object-oriented middleware, the program needs to instantiate a new proxy object of the class that abstracts the speech messaging service. The service is abstracted through the object's interface, which is made of methods and attributes. Once it has a reference for this object, the application invokes one or more methods of the corresponding interface to access the smart space service. These invocations are translated to the communication model of this particular middleware, for example, in a remote method invocation that commands the voice messaging service of the smart space to speech the message.

In this case, we can see that, in fact, the abstraction is offered at application level. This means that, in order to access to the remote service, the proxy object has to be integrated with the application.

The more important advantage of this approach is that it is well understood by programmers. In order to make a new application for the smart space, a developer follows a straightforward model: to instantiate an object and then use its interface.

Figure 3. Smart space support at different levels

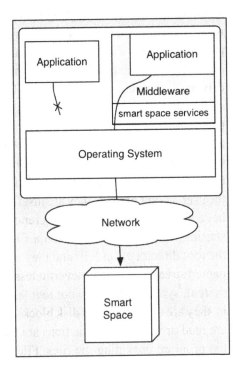
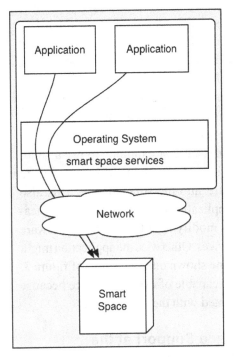

Placing the level of abstraction at the application level leads to serious interoperability issues: the application must be capable of instantiating such a proxy object and of handling it. To do that, it must use the technology imposed by the framework. That would imply using a particular programming language, a specific runtime or virtual machine, and perhaps certain operating system and/or certain dynamic libraries. This issue has a huge impact on the portability of pervasive applications and the heterogeneity of the smart space.

Another problem is that, because of being at application level, users cannot manipulate the abstraction directly from a general-purpose operating system shell (a command line based one or a window system). In order to allow users to do that, a new, specific purpose shell has to be implemented for each operating system used in the smart space.

iROS, the infrastructure used by the Interactive Workspaces (Johanson, Fox and Winograd, 2002; Johanson, Winograd and Fox, 2003; Fox *et al.*, 2000; Johanson and Fox, 2002), is an example of placing the smart space support at the middleware level. In this system, abstractions are provided by a framework that permits applications to instantiate proxy objects to interact with the smart space infrastructure. This infrastructure is based on an event delivery server, called the Event Heap. This central service is a Tuplespace (Carriero and Gelernter, 1989), in which applications can post information in form of tuples of attributes. The applications access to the Event Heap through an Object Oriented framework, by invoking methods of proxy objects that post tuples in the Event Heap.

On systems that rely on middleware, services that need a different computer than the one used to implement the system require adapting the middleware layer to the new platforms and devices. For example, a new operating system or hardware platform requires porting the framework to the new system. When platforms like Java or Inferno are used, this operation is easy to achieve, because the middleware is ready to be executed on the new system. Otherwise, this requires a significant effort.

Integrating new devices with the environment usually requires writing more software to convert the language spoken by the smart space into that used by the device. For example, to use a smart phone to switch off the lights it is necessary to bring software into the smart phone. This also happens to applications. Integrating a new application requires modifying it to use the middleware provided services. Otherwise the application might be like the one shown on the top-left of figure 3, which is not capable of using the space because is not integrated with the middleware.

Smart Space Support at the Operating System Level

As an alternative, abstractions can be placed at the operating system level. This means that the operating system itself provides the smart space abstractions to the applications. This might imply that the operating system must be changed to communicate with the smart space in order to offer these new services, which is a disadvantage of this model.

Suppose the same scenario used before, in which a program needs to send a speech message to a user. In this case, the program has to perform some operation over the local operating system to access the smart space's speech messaging service. In this case, from the application's point of view, there are is difference between accessing a local device (e.g. the audio card of the PC) and accessing a smart space service (e.g. the speech messaging service of the smart space).

The problem here is how to organize the operating system and the kind of interface to access these new devices. Plan B (Ballesteros *et al.*, 2007; Ballesteros *et al.*, 2006a; Ballesteros *et al.*,

2006b; Ballesteros *et al.*, 2005) provides smart space support at the operating system level by using synthetic file systems, following the UNIX (Kerninghan and Pike, 1984) and Plan 9 (Pike *at al.*, 1990) approaches for implementing device drivers. A synthetic file system is a file hierarchy of fake files that are dynamically generated by a program. A synthetic file system is mounted at a point of the namespace of the system (the file tree), and its files are accessed as common files: they can be open, read, written and closed. From the user's point of view, they are just files: (i) they have a path, that is, a name referencing the file starting from an initial point on a file tree (e.g., the root directory), and (ii) and they are untyped, named streams of bytes. Nevertheless, within the system, synthetic files are not regular files. That is, they are not just a set of disk blocks; when they are read or written, some actions are triggered in the program providing the files. Files (common or synthetic) are exported through a network connection using a network file system protocol. Therefore, they offer a way to export services in a distributed system (Pike *et al.*, 1990). Plan B takes advantage of these technologies to provide smart space support at system level.

This approach has an important advantage: abstractions can be accessed by humans and programs using general-purpose tools. A command line suffices to interact with the smart space. Therefore, the smart space becomes highly scriptable.

Regarding programmability, this approach may appear less intuitive to programmers than a common API, especially for those not experienced on UNIX. Nevertheless, the programmer only has to know how to read and write files in the system in order to access smart space services. Besides, it is trivial to implement libraries to wrap the file manipulation and offer a common API to less experienced programmers.

Adding new hardware platforms to the smart space requires porting the operating system used (one of them is several ones are used) to the new platform. Alternatively, perhaps, implementing the changes made to existing systems to let the new one use the services available on the space. This is true in general for services provided at the system level, but note that Plan B exports services as files that can be used remotely. This means that other systems capable of exchanging files over the network can still use Plan B services (without any modification) although they cannot export new services into the space.

To integrate new devices on the space the modification outlined above has to be made to the new device, in general, if the abstractions are provided at the system level. However, as stated above, Plan B enables foreign systems to use its files without any change, which means that devices can be integrated, at least in part, into the system.

Level of the Abstractions Provided by the Smart Space

With independence of the choice made regarding where to place the line separating the application from the smart space support, another choice to make is to decide on the level of the abstractions provided by the smart space.

Note that both choices are orthogonal. For example, support offered at a system level might supply services using a high-level of abstraction. In addition, an abstraction offered at the application level can offer a low level interface. Consider for example a large display in the smart space. The interface can provide operations to draw lines, boxes, and pixels. On the other hand, the interface may provide operations to create UI widgets, such as windows, menus, or buttons. The former is a low level interface. The later is a high level interface. Both interfaces can be offered as OOP methods of a framework (at application level), as well as with synthetic files (at the operating system level).

Both iROS and Plan B are meant to abstract applications through high level interfaces. This helps to improve the interoperability and the heterogeneity of the smart space and to make it more scriptable.

SERVICE DISCOVERY

In general there are two main approaches that can be followed in order to discover services: (i) to use a centralized registry to register the services, in a classical client/server fashion, and (ii) to use a peer-to-peer scheme and an announce/discover protocol between the different peers (the machines serving or/and using the services). There are other schemes, such as mixed approaches and federated approaches. Nevertheless, we will focus on the two main approaches. They are well known, they have been widely discussed in the literature, and both can be applied to a smart space. They have different advantages and issues regarding complexity, coherence, scalability, fault-tolerance, and dependability of the smart space communication infrastructure. The choice depends on the model of the ubiquitous computing environment to be provided.

The centralized approach uses a central service to track pervasive services. This approach is convenient for closed smart spaces, in which the pervasive environment is restricted to an area (e.g. a room, a set of offices, a building, etc.). This approach is also suitable when it is assumed that clients are always able to reach the centralized services of the smart space infrastructure. In this case, the communication infrastructure and the centralized service must be reliable.

On the other hand, in a decentralized scheme each peer announces its own services and discovers other ones using broadcast/multicast communications. Thus, the discovery is based on beacons sent by the different peers that form the ubiquitous system. The decentralized approach is suitable when building an open pervasive environment, in which principals are able to interact among them without depending on the central smart space infrastructure. This approach is also convenient when building pervasive environments over non-reliable, partitionable networks.

iROS (Johanson, Fox and Winograd, 2002; Johanson, Winograd and Fox, 2003; Fox *et al.*, 2000; Johanson and Fox, 2002) uses the Event Heap, a central repository that is based on the tuplespace model of Linda (Carriero and Gelernter, 1989). An application can select events destructively or non-destructively based on a tuple matching algebra. These events are persistent, but they may have an expiration time. The Event Heap decouples clients and servers: all the communication between clients and servers is performed through this service. Service discovery and invocation and, in general, all services in this system are based on the Event Heap.

Plan B (Ballesteros *et al.*, 2007; Ballesteros *et al.*, 2006a; Ballesteros *et al.*, 2006b; Ballesteros *et al.*, 2005) on the other side follows a decentralized scheme. The different machines, or peers, announce services they provide through periodic broadcasted messages. These messages include attributes for the services, such as the type (e.g. audio devices), the location, and the owner. Each peer runs a daemon that keeps track of discovered services and informs the operating system. The operating system is able to mount the discovered synthetic file systems to provide the service to users and programs. Discovered file systems are mounted depending on some restrictions as set by the user (who for example, might want to mount any audio volume in the office, no matter who owns it). These restrictions are expressed along with the name of the service requested. Once a resource matching the adequate restrictions is discovered, it is mounted automatically. Restrictions in the Plan B service announce protocol play the role of tuples used in the IWS.

NAMING

Once the system is able to discover and get a reference (in the broadest sense of the word) to a service, it has to make the service available to the application. Therefore, some kind of organization and a naming convention is needed in order to allow applications to locate an imported service.

In iROS, there is no difference between discovering and using a service. The Event Heap decouples the client application and the server application: when an application needs a service, it sends a request (in the form of an RPC interaction expressed as a tuple) to the Event Heap. The Event Heap eventually delivers the event to the service provider, depending on the attributes expressed in the posted tuple. Thus, it is not necessary to keep any reference to known services in the client. This is possible because the system relies in a centralized naming service.

Decentralized systems need to keep references to known services. In order to send a service request, the machine needs to know where is the server.

There is a plethora of data structures suitable for building a naming system. One of the most popular naming schemes is the tree. There are different ways to implement such a scheme. For example, it can be implemented using markup languages (e.g. XML) in a system's configuration file. As Plan B is based on the file system abstraction, it uses the file system tree itself to provide a uniform, regular approach to naming. Once a file system in mounted inside the namespace, there is a well-known method for accessing it. The path to the file provides the naming mechanism. The set of names can be configured using the mount command to import file systems from the network and bind them to desired names in the local namespace. In Plan B, namespaces are per-process, so different application within the system can have different configurations and import different resources depending on their needs, like in Plan 9 (Pike *et al.*, 1990). Mount points for services follow a naming convention, in order to provide a standard location for each service. For example, the convention requires the audio device to be always mounted on the same point (i.e. /mnt/audio). Therefore, when an application needs to play a sound, it writes audio data to the synthetic file system found at the conventional place. Plan B provides the mechanism to make a service available at any part of the namespace of the process (the file tree) but does not dictate any policy.

PROTECTION

The system should provide some kind of access control to reach the services in the pervasive environment. Access control mechanisms grant or deny access for a principal to use a service, according to a security policy.

Most access control schemes are based on the identity of the principal. Other access control schemes, such as capability-based schemes, are based on authorizations given to principals. Both iROS and Plan B perform identity-based access control for pervasive services.

Both identity-based and authorization-based access control must rely on an authentication scheme: To perform access control or to provide some capabilities, the system must verify that the principal is who he claims to be. In distributed systems, authentication is usually provided by some kind of authentication server using an authentication protocol similar to any of Kerberos (Neumann, 1994) or Sesame (Ashley, 1997).

In iROS, the protection model is provided by a framework named iSecurity (Song *et al.*, 2004). This framework depends on a dedicated authentication service called iSign that provides both client and server applications with authentication certificates. These certificates authenticate users that are executing applications against the Event Heap. All events are delivered to applications together with the corresponding identity. Thus, applications are able to manage their own access control policies for different users of the smart space. The access control is based on Access Control Lists (ACL). ACLs are stored in the server application itself. An ACL holds permissions for users and groups of users (Trust Groups). Trust Groups are lists of users created to simplify the ACL. They are stored in the Event Heap and managed by the server applications.

In Plan B, principals are authenticated by a Plan 9 authentication server (Cox *et al.*, 2002) that uses a protocol similar to Kerberos (Neumann, 1994). The protection model comes for free from the file system model. Plan B uses traditional UNIX permissions for files. The UNIX access control scheme expresses permissions to read, write, and execute a file for its owner, an assigned group of users, and the rest of the world. These permissions suffice to provide reasonable access control for a smart space. However, note that any other file system protection scheme would also work. It all depends on what ACLs are supported by the program that implements the file interface.

There is another authentication and access control scheme for Plan B, called SHAD (Soriano, Ballesteros and Guardiola, 2007a). SHAD allows users to authenticate and share resources in a safe way without depending on centralized authentication services. In order to allow that, SHAD is based on personal mobile security servers and provides Role Based Access Control (Soriano, Ballesteros and Guardiola, 2007b). RBAC allows users to set richer policies and to configure authorization in a more convenient way than using UNIX permissions.

COMMUNICATION AND SYNCHRONIZATION

Communication between different programs is an important topic, because programs must cooperate to make the entire space become intelligent, or to make it appear to be so. There are different communication channels that can be used but which one is preferred is usually a function of the system used to architect the space.

At one end of the spectrum we can find spaces where different software modules communicate via ad-hoc mechanisms, establishing their own channels using TCP or a similar protocol. That is usually the case with spaces built as an application on their own.

In most cases, the framework used to build the space includes some communication facility. For example, event channels used by iROS are also a communication mechanism that can be used to convey information (in the form of events) to software entities. This follows from the design borrowed to some extent from systems like Linda.

In such cases, applications can establish communication with a high level of abstraction, which is usually convenient. However, as an aside, it also means that applications must usually employ such communication mechanism or they will be kept apart from the rest of the system.

Systems like Plan B use a distributed file system interface as the primary communication mechanism. Therefore, communication happens at the system level and is handled by the underlying operating systems deployed through the space. For the most part, applications may remain unaware of communication and rely on files to share information. This is a well-known technique that has been proven to work well in practice.

Heterogeneity is also of relevance because data exchanged through the communication mechanism must remain meaningful no matter the platform sending the data and the one receiving it.

Middleware systems rely on usual marshalling and unmarshaling of native data, on the one hand. On the other, it is also popular to exchange data in the form of XML trees represented in a textual format.

Plan B took the UNIX approach in this respect. Most data is exchanged by placing plain text on files, using a set of conventions known by applications and users on the space. Because text is portable, applications may remain unaware of the data representation.

While it is true that data must adhere to conventions (for example, an X10 power switch must use the string on to represent that the switch is active), there is no way to escape from this. At some level there has to be a convention regarding how data is represented. The popularity of ontologies is simply a reminder of this fact (although in many cases they complicate things by trying to formalize things that in many cases cannot be formalized).

Closely related to communication is synchronization. Different parts of the system must be able to synchronize to work together. In almost all the cases, some form of event mechanism is used for synchronization. It has to be noted that shared memory is not an option because of the distributed nature of devices on the space.

For systems like iROS using events is natural. For systems like Plan B a general purpose event mechanism is built and supplied as a tiny file system capable of creating and exporting event channels.

Polling for changes is also a popular alternative. Especially when gathering context information. Location, user activity, and other related pieces of data are retrieved by inspecting sensors and by asking the underlying platform for such information. In many cases, the only way to achieve this is by polling the source of data in one way or another. For example, Plan B devices extracting context do poll various devices and update a context file tree with context information. When events are preferred, the generic event mechanism is used after polling to deliver updates. Even systems where the primary mechanism is event delivery do poll at some point, perhaps just to deliver events.

A shared whiteboard mechanism is also popular to publish information. However, both event channels and file trees can be used as a whiteboard. It is a matter of registering the relevant information at one place known by applications on the space. On Plan B, this comes for free, because the file trees used as interfaces to services are seen as places where information is kept. In other cases, events have to be listened to update a central repository.

SYSTEM SUPPORT FOR ADAPTATION AND RECONFIGURABILITY

On a pervasive computing environment resources come and go and some services may be deployed at different times on top of different (heterogeneous) platforms and networks. Applications must adapt to changes (if only to keep on working) and how to do that depends on existing system support. That is, applications may find either trivial or impossible to adapt to changes depending on the system and abstractions used.

It is important to note that a smart space may remain in service for a long time. For example, ours is still working (although with different services) after several years. During the time of operation, new devices and services will be deployed and old ones will be retired. It is important for the system to provide support for adapting to changes, so that applications could be left undisturbed.

Another kind of adaptation is a short-term adaptation needed during small time periods. In this case, what matters is for the system to be able to discover changes on available services (see the section on naming above) and to provide a way to customize (perhaps automatically) applications after a change.

Systems like the Interactive Workspaces are more static in this respect, although they consider devices being brought into the space. On the other hand, Plan B relies on automatic reconfiguration of the name space built out of the forest of file trees available on the space. This means that upon a change in resources, the file tree as seen by applications changes to reflect the environmental change.

Thus for example, when a new terminal is used the system is able to start using some of the resources (e.g., audio) provided by the terminal

if the properties of the new devices match the requirements specified by the user (or the user's profile).

It is also important to consider what the system does when a resource is powered down or removed from the space. Applications using the resource must adapt to changes as well. Systems like Gaia keep a user's profile so that the session may be recreated at a different place on the space. Plan B on the other hand lets the application deal with errors resulting from interrupted I/O on the removed device. However, Plan B keeps the file tree operational, although some files may disappear.

The Plan B approach of letting the application receive an error indication is useful to let the software adapt, because in the end only the application knows how to adapt (or the user). For example, on the web users simply hit reload if there is a problem. If a Plan B music player ceases working because the speaker device is gone, the user would hit the Play button again. (Actually, the Plan B player does recover on its own from this kind of error by reopening the file representing the speaker device, which usually opens a different speaker if the previous one is gone. But the example is representative of what other applications do).

EXAMPLES

In this section, we are going to make some examples of use of system provided abstractions. We will not describe any example of IWS here because it has been extensively published (Johanson, Fox and Winograd, 2002; Johanson, Winograd and Fox, 2003; Fox *et al.*, 2000; Johanson and Fox, 2002) how that system is used. Instead, we introduce some Plan B examples because it is a system we implemented (Ballesteros *et al.*, 2007; Ballesteros *et al.*, 2006a; Ballesteros *et al.*, 2006b; Ballesteros *et al.*, 2005).

Figure 4 shows a screenshot of two windows. The left one is the interface for the Plan B context service that determines who is using the facilities on the space. Because our smart space is made of different physical spaces, it makes sense to let the user know who is using the space.

To implement this application, a C program was made that polls for changes a tiny file tree located at /who. It then selects appropriate images for each user and shows who is at the space. For the application, things are not different than they are for a conventional application using files within a single UNIX machine. Therefore, it pays to offer system support for all the services it requires.

On the right of figure 4, a shell session is shown that also uses the context file tree. The first command shown tries to determine who is away from the system. This is achieved by reading all files /who/*/status, which, by convention,

contain strings descriptive of the status for the respective users.

Figure 5 shows other applications used to control lights and power sources and to monitor presence on the space. The window on the top-left part of the figure shows a mosaic with the status for all the devices of interest. Mouse actions on the mosaic switches on/off the device represented by the little rectangle.

Devices that include dim abilities can be tuned by sliders (like the one shown on the top-right part of the figure). Adjusting the gauge value with the mouse also adjust the device level. For example, the slider shown is for a light with dim control.

In both cases, the application is unaware of the mechanisms used to control the sensors and

Figure 4. Context information panel and programmability of the system

Figure 5. Light control and presence detection integrated using scripts

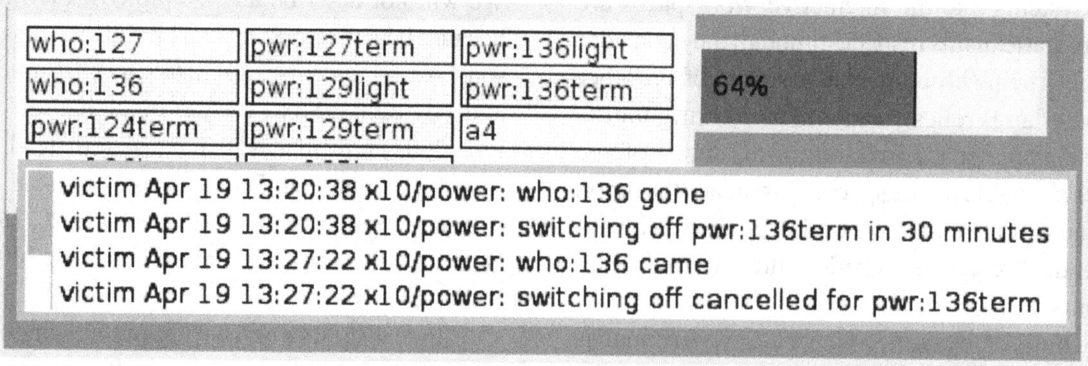

actuators involved, and unaware of other synchronization, protection, naming, and discovery issues. All that is provided by existing system support. The application relies on files available at conventional places to determine which sensors/actuators are installed on the space and then it provides a user interface for them. The names of the files are the names used in the mosaic, for example. In the same way, the percentage shown in the slider is the actual content of the file representing the dim control.

The lower part of figure 5 shows log entries resulting for scripts that rely on files to control the system. In this case, the scripts used to switch on/off the lights without human intervention depending on the activity of the space.

The actual applications shown do not really matter. The point is that by segregating system support into a separate component kept apart from the applications it is easy to write software for the space (at least as easy as it is to write it for the OS where it runs).

CONCLUSION

It is important to consider services in the smart space as system services, and to build upon them to deploy actual applications. Otherwise, the space may become hard to modify and hard to adapt to changes on the long term. The system on the space plays a similar role to the operating system on a computer, relieving applications from the burden of managing resources needed for their operation. Although in some cases the system is a native system (like for example Plan B), in other cases the system is a meta operating system (like for example on Gaia or the Octopus). Either way, it is important to make the distinction between the system and the applications. This has worked well on the computers we use and it is likely to work well also for smart spaces.

REFERENCES

Ashley, P. *Authorization for a Large Heterogeneous Multi-Domain System*. In Proceedings of the AUUG

Ballesteros, F. J., Guardiola, G., Soriano, E., & Leal, K. *Traditional Systems can Work Well for Pervasive Applications. A Case Study: Plan 9 from Bell Labs Becomes Ubiquitous*. IEEE Intl. Conf. on Pervasive Computing and Communications 2005.

Ballesteros, F. J., Soriano, E., Algara, K. L., & Guardiola, G. (2007, September). Plan B: Using Files instead of Middleware Abstractions for Pervasive Computing Environments. *IEEE Pervasive Computing / IEEE Computer Society [and] IEEE Communications Society*, *6*(Issue 3), 58–65. doi:10.1109/MPRV.2007.65

F. J. Ballesteros, E. Soriano, G. Guardiola, K. Leal. *The Plan B OS for Ubiquitous Computing. Voice, Security, and Terminals as Case Studies*. Elsevier Pervasive and Mobile Computing Journal. 2(2006), pp. 472-488. 2006a.

Ballesteros, F. J., Soriano, E., Leal, K., & Guardiola, G. *Plan B: An Operating System for Ubiquitous Computing Environments*. IEEE Intl. Conf. on Pervasive Computing and Communications 2006b.

Carriero, N., & Gelernter, D. (1989, April). Linda in context. *Communications of the ACM*, *32*(issue 4), 444–458. doi:10.1145/63334.63337

Chan, E., Bresler, J., Al-Muhtadi, J., & Campbell, R. *Gaia Microserver: An Extendable Mobile Middleware Platform*. Proceedings of 3rd IEEE Intl. Conf. on Pervasive Computing and Communications, 2005, p. 309-313.

Clifford Neuman, B., & Ts'o, T. (1994, September). Kerberos: An Authentication Service for Computer Networks. *IEEE Communications*, *329*, 33–38. doi:10.1109/35.312841

Cox, R., Grosse, E., Pike, R., Presotto, D., & Quinlan, S. *Security in Plan 9*. In Proceedings of the USENIX 2002 Security Symposium, p. 3-16, 2002.

Fox, A., Johanson, B., Hanrahan, P., & Winograd, T. (2000, May-June). Integrating information appliances into an interactive workspace. *IEEE Computer Graphics and Applications, 20*(Issue 3), 54–65. doi:10.1109/38.844373

Johanson, B., & Fox, A. *The Event Heap: a coordination infrastructure for interactive workspaces*. Mobile Computing Systems and Applications, 2002. Proceedings Fourth IEEE Workshop, p. 83 – 93, June 2002.

Johanson, B., Fox, A., & Winograd, T. (2002, April). The Interactive Workspaces Project: Experiences with Ubiquitous Computing Rooms. *IEEE Pervasive Computing / IEEE Computer Society [and] IEEE Communications Society, 1*(Issue 2), 67–74. doi:10.1109/MPRV.2002.1012339

Johanson, B., Winograd, T., & Fox, A. (2003, April). Interactive Workspaces. *IEEE Computer, 36*(Issue 4), 99–101.

Kernighan, B. W., & Pike, R. (1984). *The UNIX Programming Environment*. Prentice-Hall.

National Conference, Brisbane, Qld., September 1997.

Pike, R., Presotto, D., Thompson, K., & Trickey, H. (1990). Plan 9 from Bell Labs. *EUUG Newsletter, 10*, 3.

Song, Y. J., Tobagus, W., Leong, D. Y., Johanson, B., & Fox, A. *iSecurity: A Security Framework for Interactive Workspaces*. Tech Report, Stanford University, 2004.

Soriano, E., Ballesteros, F. J., & Guardiola, G. (2007a). Human-to-Human Authorization for Resource Sharing in SHAD: Roles and Protocols. *Elsevier Pervasive and Mobile Computing., 3*(6), 607–738.

Soriano, E., Ballesteros, F. J., & Guardiola, G. *SHAD: A Human Centered Security Architecture for the Plan B Operating System*. IEEE Intl. Conf. on Pervasive Computing and Communications 2007b.

KEY TERMS AND DEFINITIONS

Middleware: Software layer in the middle of application and system software.

Smart Space: Physical space enhanced with computer software enhancing the behavior of the space for particular users or activities.

Synthetic File System: An interface to an entity looking like a file tree. Operations on the entity accessed through the interface happen upon reading, writing, creating, and/or removing files.

RPC: (Remote Procedure Call). Mechanism used to operate on remote services similar to a procedure call from the client perspective (would be equal to a procedure call if argument references and error handling is not considered).

ACL: (Access Control List). A mechanism to control access to operations on a resource. It lists which clients may perform which operations on a resource.

Adaptability: Capability of a system to change itself (or at least its behavior) in response to changes in the environment.

Process: Running program.

IPC: Inter-process communication. Mechanisms to permit communication among different processes on a system.

Chapter 11
MEDUSA:
Middleware for End-User Composition of Ubiquitous Applications

Oleg Davidyuk
ARLES Research Team, INRIA, France and University of Oulu, Finland

Nikolaos Georgantas
ARLES Research Team, INRIA, France

Valérie Issarny
ARLES Research Team, INRIA, France

Jukka Riekki
University of Oulu, Finland

ABSTRACT

Activity-oriented computing (AOC) is a paradigm promoting the run-time realization of applications by composing ubiquitous services in the user's surroundings according to abstract specifications of user activities. The paradigm is particularly well-suited for enacting ubiquitous applications. However, there is still a need for end-users to create and control the ubiquitous applications because they are better aware of their own needs and activities than any existing context-aware system could ever be. In this chapter, we give an overview of state of the art ubiquitous application composition, present the architecture of the MEDUSA middleware and demonstrate its realization, which is based on existing open-source solutions. On the basis of our discussion on state of the art ubiquitous application composition, we argue that current implementations of the AOC paradigm are lacking in end-user support. Our solution, the MEDUSA middleware, allows end-users to explicitly compose applications from networked services, while building on an activity-oriented computing infrastructure to dynamically realize the composition.

INTRODUCTION

Ubiquitous computing as envisioned by Mark Weiser (1991) emphasizes the interaction of users with smart spaces composed of humans and multiple networked computing devices. This vision has been empowered by continuing progress in various relevant fields of research, such as wire-

DOI: 10.4018/978-1-61692-857-5.ch011

less communication, mobile computing, mobile sensing and human-computer interaction (HCI). Firstly, wireless communication standards (e.g., WiFi and Bluetooth) have enabled users to access smart spaces using their mobile terminals and portable devices. Secondly, efficient low-powered CPUs make mobile terminals capable of running software platforms which support advanced interoperability between different devices. In addition, mobile sensing technologies, like RFID and GPS, make ubiquitous applications context-aware and they also enable alternative HCI interfaces based on, e.g., physical contact (Nokia, 2009). However, despite all these technical developments, the user-centric nature of ubiquitous computing should not be neglected.

An important new trend in ubiquitous computing research is activity-oriented computing (AOC) (Masuoka, Parsia & Labrou, 2003; Ben Mokhtar, Georgantas & Issarny, 2007; Sousa, Schmerl, Steenkiste & Garlan, 2008a). This paradigm adopts a user-centric perspective and assumes that smart spaces are aware of user needs and activities and reactively, or even proactively, satisfy user demands by composing and deploying the appropriate services and resources. User activities consist of the users' everyday tasks and they can be abstractly described in terms of (i) the situation (context) where the tasks take place, (ii) the system functionalities required for accomplishing the activities, and (iii) user preferences relating to QoS, privacy, security or other non-functional requirements. The system then dynamically realizes the activities by composing the applications according to the activity descriptions.

Under the AOC paradigm, the research community concentrates on introducing autonomic systems, which are self-manageable, self-configurable and adaptive, and therefore do not require user involvement. However, as pointed out by Hardian, Indulska & Henricksen (2008) and confirmed through user evaluation and usability tests (Davidyuk, Sánches, Duran & Riekki, 2008a; Vastenburg, Keyson & de Ridder, 2007),

involving users in application control is essential for ensuring the user acceptance of autonomous products, especially in home or office automation domains. The AOC approach also has another drawback which limits its applicability. Application composition in AOC systems is performed by matching user activity descriptions with the networked services and resources discovered in the vicinity of the user according to the chosen criteria. Although some prototypes allow end-users to adjust the criteria at run-time (Davidyuk, Selek, Duran & Riekki, 2008b; Sousa et al, 2008a), the composition is limited to predefined activity descriptions (i.e. templates), which are complex structures and are not supposed to be understood or modified by end-users. Thus, users are neither able to customize the existing applications nor able to create their own smart space applications.

To address the aforementioned limitations of AOC systems, we present MEDUSA, a middleware solution which enables user-driven application composition in smart spaces. MEDUSA enables end-users to create simple applications from available ubiquitous resources discovered in the vicinity of the users. The applications are composed on the users' handheld devices. MEDUSA employs an end-user approach to programming applying it to the Ambient Intelligence (AmI) and ubiquitous computing domains (Kawsar, Nakajima & Fujinami, 2008; Mavrommati & Darzentas, 2007; Mavrommati & Kameas, 2003; Sousa, Schmerl, Steenkiste & Garlan, 2008b). The design of the MEDUSA middleware builds on our years of experience in developing middleware interoperability solutions, e.g., Amigo (2009) and PLASTIC (2009) as well as RFID-based user interfaces for smart spaces (Davidyuk et al, 2008a; Davidyuk, Sanches, Duran & Riekki, 2009; Riekki, 2007).

This chapter describes the design rationale for the MEDUSA middleware architecture and it organized as follows. "Related Work" surveys the background of ubiquitous application composition focusing on developments in AOC research, thus providing an overview of the key aspects relating to the specification of ubiquitous applications, the middleware support of dynamic application realization and end-user involvement. "MEDUSA Middleware" then introduces MEDUSA, which empowers end-users to participate in application development and further leverages existing open-source middleware solutions for the actual realization of application composition. Finally, "Conclusion" concludes with a summary of the contribution of this study and our plans for future research.

RELATED WORK

Before presenting related work, we first introduce the notion of application composition, which here refers to the process of building applications by assembling services offered by computing devices embedded in the environment.

As shown in Figure 1, the application composition process generally involves two separate actors, the end-user and the service provider. The role of the service provider is to develop and publish services which provide some functionalities through a well-defined interface. The end-user's role is to

create applications and utilize (i.e., interact with) applications. Some authors further separate the roles of application users and application developers, thus assuming that applications are developed and used by different persons (Masuoka et al, 2003; Sousa, Poladian, Garlan, Schmerl & Shaw, 2006; Sousa et al, 2008a). Others do not consider the end-user role at all and suggest a view in which only application developers are involved in the application composition process (Paluska, Pham, Saif, Chau, Terman & Ward, 2008).

The application composition process relies on descriptions of applications and services. An application description specifies the abstract services the application is composed of and the relationships between them (i.e. control and data flows). The application composer is responsible for decision-making in the application composition process. The application composer uses application descriptions and possible composition criteria (preferences, fidelity constraints, costs and so on) provided by the end-user as inputs and chooses available services which satisfy the given criteria. The selection of services is supported by a service discovery protocol, which is responsible for the matchmaking functionality (i.e., for matching service discovery requests

Figure 1. Generic application composition process

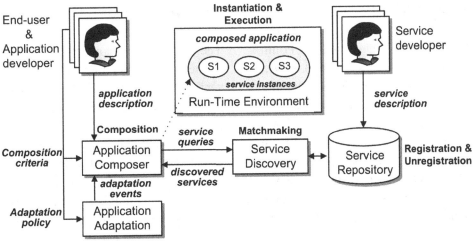

against the service descriptions stored in the service registry). Since the discovered set of services may potentially contain redundant instances, the application composer optimizes (i.e., further reduces) the service set and produces an application configuration (i.e., application composition plan), which satisfies the criteria given earlier. Different application configurations may be produced depending on the criteria and the situation in the environment. After this, the application is instantiated and executed by the runtime environment. During execution, the application can be adapted (i.e. recomposed) according to user defined adaptation policies. These policies are formal rules which trigger predefined actions when the application or user context changes.

To date, several studies supporting ubiquitous application composition have been presented. They focus on service provisioning issues (Chantzara, Anagnostou & Sykas, 2006; Nakano, Takemoto, Yamato & Sunaga, 2006; Takemoto, Ohishi, Iwata, Yamato, Tanaka, Shinno, Tokumoto & Shimamoto, 2004), context-aware adaptation (Preuveneers & Berbers, 2005a; Hesselman, Tokmakoff, Pawar & Iacob, 2006; Jianqi & Lalanda, 2008; Bottaro, Gerodolle & Lalanda, 2007; Bottaro, Bourcier, Escofier & Lalanda, 2007; Handte Herrmann, Schiele & Becker, 2007; Rouvoy, Eliassen, Floch, Hallsteinsen & Stav, 2008; Rouvoy, Barone, Ding, Eliassen, Hallsteinsen, Lorenzo, Mamelli & Scholz, 2009), service validation and trust (Bertolino, De Angelis, Frantzen & Polini, 2008; Bertolino, De Angelis & Polini, 2009; Buford, Kumar & Perkins, 2006), service communication path optimization (Kalasapur, Kumar & Shirazi, 2005), automatic application code generation (Nakazawa, Yura & Tokuda, 2004) and distributed user interface deployment (Rigole, Vandervelpen, Luyten, Berbers, Vandewoude & Coninx, 2005). Several design styles for developing adaptive ubiquitous applications through composition have also been suggested (Paluska et

al, 2008; Saif, Pham, Paluska, Waterman, Terman & Ward, 2003; Sousa et al, 2008b).

Some of the most promising research deals with the activity-oriented computing approach (Masuoka et al, 2003; Ben Mokhtar et al 2007; Sousa et al, 2008a). These solutions focus on decoupling the end-user tasks from their system-level realization, i.e., they let the end-users concentrate on the tasks or activities they need, rather than ask them to specify how they want these tasks to be performed by the system. User tasks are simple everyday activities (e.g., in the home or office environment), which can be achieved by compositing an application. AOC solutions take a user-centric approach to application composition by suggesting that users explicitly provide descriptions to the system via dedicated task composition and control interfaces (Davidyuk et al, 2008a; Messer, Kunjithapatham, Sheshagiri, Song, Kumar, Nguyen & Yi, 2006; Sousa et al, 2006). Other approaches assume that the descriptions are provided to the system implicitly through user context recognition facilities (Ranganathan & Campbell, 2004) or that user task descriptions are developed by application developers (Beauche & Poizat, 2008; Ben Mokhtar et al, 2007).

Next, we discuss related work in the context of three key issues, namely, the specification language, middleware support and end-user involvement.

Specification Language

The specification language is the cornerstone of the application composition process as it essentially serves the following purposes: (i) it enables service providers to advertise the properties of their services, both the functional and the non-functional properties; (ii) it enables the discovery of networked services by matching the service specifications with query requests sent by consumers; and (iii) it helps in arranging the discovered services in order of priority according to their non-functional properties and optimization criteria.

The solution presented by Paluska et al (2008) suggests a design style called "goal-oriented programming" and a proprietary scripting language for describing composite applications. The solution is centered around application developers who create applications by describing the goals and the techniques corresponding to these goals. The goals are abstract decision points which determine the application's structure and behavior by describing the functionalities required by certain parts of the application. Each goal is described in terms of quality parameters, which are used to evaluate whether or not the goal has been reached. The techniques are specified as programming scripts which describe ways of achieving goals, e.g., by using certain hardware instances. The scripts introduced by the techniques do not directly implement application functionalities. Instead, what they provide is rather an abstraction of the existing component modules and ubiquitous devices required by the application. The structure of resulting application resembles trees; this is due to the planning algorithm which is used to optimize the set of techniques.

Most of the work on dynamic composition uses proprietary XML-based specification languages bound to particular application domains (e.g., office automation, web services and mobile services). The structure of these specifications has two features: it is fixed (i.e., adding a new property requires redesigning the language) and it resembles a tree because XML is used. An XML-based specification is used, for instance, by Sousa et al (2006; 2008a) for describing user activities. In their paper, Sousa et al (2008a) present an example activity that describes the services and the service properties required to review a movie clip. Their example consists of two services, namely "play Video" and "edit Text". The properties of these services are "material" (i.e., required files) and "service state" (i.e., video playback position, video dimensions etc.). Sousa's specification supports two types of attributes: numeric (integer) values and enumerations (i.e., sets of integer or string values). Although XML-based languages allow the tailoring of specifications to certain problems (or application domains), they themselves do not contain any interpretation of the concepts (i.e., semantic meaning of attributes, values, etc), instead they leave this task to the system or the application. This means that the interpretation of XML-based specifications depends on the application or system logic, which can potentially create ambiguity if two different systems interpret the same specification differently.

In addition to XML-based models, multiple service specification standards exist, such as Web Services Description Language (WSDL)[1] and YAWL[2]. Although they have been successfully in many existing systems, none of these languages has been accepted as a global standard in the application composition domain. Therefore, systems using different description languages may be incompatible with each other due to the diversity of their service and application descriptions. This is also known as semantic heterogeneity (Halevy, 2005) and it occurs because service description languages in general support different concepts (i.e., the service descriptions vary content-wise) and they may specify the same concepts in different ways. For example, service behavior can be modeled using multiple techniques, such as process modeling (e.g., BPEL[3]) and conversations (e.g., WSCL[4]).

Another alternative to the XML-based approach to specifications is the ontology-based approach which also solves the problem of semantic heterogeneity. This approach models applications, services, their properties and possible relationships between them using a common theory (also called the "upper ontology"), thus enabling the participating parties to reason and match service concepts, even if their descriptions do not comply in syntax.

Several ontology-based languages have been suggested for service descriptions, such as Web

Service Ontology (OWL-S)[5] and Web Services Modeling Framework (WSMF)[6]. These languages have been used frequently, especially in dynamic application composition systems (Hesselman et al, 2006; Lee, Chun & Geller, 2004; Ben Mokhtar et al, 2007; Preuveneers & Berbers, 2005a; Preuveneers & Berbers, 2005b; Ranganathan & Campbell, 2004). For example, iCOCOA (Ben Mokhtar et al, 2007) uses an OWL-S based semantic language for specifying user activities, services and their properties. iCOCOA particularly focuses on dynamic service properties and models service behavior using workflows. Thus, each service is modeled as a set of service operations which are interconnected with control and data relationships. In addition, iCOCOA also describes service QoS properties using qualitative and quantitative attributes.

Another OWL-S based specification language is used by CODAMOS middleware (Preuveneers & Berbers, 2005a; Preuveneers & Berbers, 2005b). The main focus of the CODAMOS middleware is the hierarchical composition of service-based mobile systems, in which each service can consist of a sequence (i.e., a hierarchy) of other services. The CODAMOS specification defines the functional and non-functional properties of the services and relationships between them, i.e., connectors. The connectors link services and provide communication channels within the composed structures of the services. The non-functional properties of the services include contracts specifying user requirements and context and service control interfaces.

The language used for the application and service specification has a significant impact on middleware support in dynamic application composition. This point is discussed further in the next section.

Middleware Support

We distinguish between two important middleware functionalities in the dynamic composition of ubiquitous applications: (i) the application composer, which realizes the application by implementing a matching or a planning algorithm to choose the necessary services and (ii) the functionality that enables the interoperability of the service discovery, the service descriptions and the service communication.

Application composer. The application composer implements an algorithm to select service instances which realize the application. The algorithm performs either a matching or a planning function. Matching algorithms select appropriate services simply by matching the attributes of the services. In contrast, planning algorithms perform optimization and select the set of services which best satisfies a certain criteria. Planning algorithms are usually applied to systems in which finding a solution requires significant time and computing resources.

Application composition by matching is used in the iCOCOA (Ben Mokhtar et al, 2007), Inter-Play (Messer et al, 2006), PCOM (Handte et al, 2005), CASE (Hesselman et al, 2006), USON (Takemoto et al, 2004), Galaxy (Nakazawa et al, 2004) and SesCo (Kalaspur et al, 2005) projects. For example, iCOCOA suggests an application composition engine that uses semantic reasoning as well as a QoS attribute- and a conversation-based matching algorithm. The algorithm dynamically integrates the available service instances into the application according to the service behavior and application QoS constraints.

Several solutions use planning algorithms for application composition (Beauche & Poizat, 2008; Chantzara et al, 2006; Preuveneers & Berbers, 2005a; Ranganathan & Campbell, 2004; Rouvoy et al, 2009; Sousa et al, 2006; Sousa et al 2008a). For example, Sousa et al (2006, 2008a) and Ranganathan & Campbell (2004) use similar planning approaches to compose applications and particularly to address fault-tolerance issues. Their application composition engines take the goal description of an abstract user into account, in addition to the user's current context and preferences, in order to find a sequence of actions which leads to the best realization of the user activity. The resulting sequence (or plan) has to be executed by the application framework to ensure that none of the executions fail because of resource unavailability. This is done by dynamically monitoring the execution of the plan and the resources.

The composition mechanism of MUSIC (Rouvoy et al, 2008; Rouvoy et al, 2009) uses a utility-based planning algorithm, which relies on the normalized utility function determined by the required properties of the application and its current execution context. The utility function defines the relevancy of the QoS properties and reflects the application state (i.e. deployed or running), which affects the way particular QoS properties are estimated. This planning algorithm is also capable of negotiating QoS values directly with service providers during planning.

Another project, CODAMOS (Preuveneers & Berbers, 2005a), uses the Protégé reasoning tool to take the capacities of the client devices into consideration during planning. The algorithm estimates the resource capacities of the client devices and composes the applications accordingly. The algorithm uses the backtracking approach to optimization, i.e., it cuts down on the user preferences if it does not find any suitable solutions which fit the required device set. CODAMOS is a particularly interesting project, as their algorithm optimizes the structure and the functionality of the application according to the available devices, instead of optimizing the set of services to meet the application QoS requirements.

Middleware interoperability. According to Ben Mokhtar (2007), dynamic application composition requires two types of middleware interoperability, that is, among service discovery and among service communication protocols. As discussed in 2.1, the former relates to overcoming semantic heterogeneity and can be addressed through the semantic description of services and composite applications. The latter requires adequate mapping among heterogeneous middleware protocols.

A number of solutions have been proposed recently to address the interoperability issues of the service discovery and communication functionalities. For example, MUSDAC (Raverdy, Issarny, Chibout & de la Chapelle, 2006) and ubiSOAP (Caporuscio, Raverdy & Issarny, 2009) use auxiliary service components to translate messages sent between different service discovery networks. An instance of such a service is added to each service discovery network enabling the clients to use multiple protocols at the same time. Siebert, Cao, Zhou, Wang & Raychoudhury (2007) and Bromberg & Issarny (2005), on the other hand, suggest a universal adaptor approach which implements both client and server-side functionalities for discovering services and for mapping the primitives used by the universal adaptor with the primitives used by various service discovery systems.

Similarly, the ANSO architecture (Bottaro et al, 2007; Jianqi & Lalanda, 2008) suggests using adaptive adaptors to target service discovery heterogeneity issues. ANSO uses the UPnP service discovery protocol and provides explicit mappings of other protocols for UPnP in order to integrate sensors, computing devices and web services into one coherent and manageable network. Unlike other solutions, the ANSO service discovery engine generates a separate proxy component for each service instance which needs to use some other protocol than UPnP. Each time the service providers register a new service, ANSO dynamically generates a proxy using Java reflection (i.e. the bytecode generation technique) which implements the service discovery protocol required by the service instance. The proxies also provide access to the service functionalities.

Application composition also needs another kind of interoperability which is related to service communication protocols. Service communication interoperability is required because service instances may use incompatible communication protocols, such as Bluetooth and Wi-Fi. This kind of interoperability has been addressed in particular by the AmIi (Georgantas, Issarny, Ben Mokhtar, Bromberg, Bianco, Thomson, Ravedy, Urbeita & Cardoso, 2009) and ubiSOAP (Caporuscio et al, 2009) solutions. Although AmIi is a middleware solution which focuses on semantic interoperability, it also provides interoperability among heterogeneous RPC-protocols. For this reason AmIi introduces a proprietary AmIi-COM communication mechanism, which is based on runtime protocol translation. Similar to AmIi, the ubiSOAP middleware addresses communication interoperability. It implements a custom SOAP protocol to enable service communication in wireless networks. We discuss these two solutions in "MEDUSA Middleware Interoperability."

End-User Involvement

In ubiquitous computing, end-users rarely play an active role. This is best demonstrated by research in context-aware application composition in which users do not explicitly interact with the system (i.e., through a user interface), but rather influence the decisions made by the system passively, i.e., through sensors which autonomously capture the users' preferences, their behavior, current needs and other parameters. For this reason, end-user involvement in ubiquitous computing is very limited and often non-existing.

However, end-user involvement has been more extensively studied in AOC research. The AOC systems assume that application composition is performed on the basis of predefined activity templates that are developed by application programmers, thus restricting the role of the end-users in composing applications (or activities) to simply matching the templates. This is simultaneously, both a major drawback and a contradiction in the AOC approach, because, on one side, the AOC promises to support end-user activities, but on the other side it restricts the user choice to activity templates that are predefined by the system. Thus, studies on end-user involvement in AOC have

mainly focused on interfaces used to customize user activity templates.

For example, Sousa et al (2006, 2008a) present a set of user interfaces for customizing activity templates and specifying end-user preferences. These preferences define the constraints and requirements that are taken into account by the application composer, which chooses the service instances for the corresponding user activity. The example shows a user activity template, which includes the following tasks: editing text, browsing the Web and editing a spreadsheet. Users can associate each of these single tasks with specific material (filename or address). Sousa's user interfaces also support multiple dimensions of application QoS requirements and the particular value of each dimension is represented by the slider position. The user chooses the QoS dimension (e.g., latency) and then adjusts the position of the slider to define the value intervals "bad", "moderate", "good" and "excellent".

In our previous study, we suggested another approach to collect user preferences based on physical interfaces (Davidyuk et al, 2008a). In this approach, users specify their preferences by touching the appropriate RFID tags. An example of such an interface, the interface of a ubiquitous movie player application, is shown in Figure 2. This interface allows the users to choose the quality of the video they want to play with the application, which can be "very low", "low", "medium" or "high". For example, in order to choose the maximum quality, users need to touch (i.e. read the tag using a mobile terminal) the corresponding RFID tag labeled "high". Although we find that Sousa's interfaces are more flexible in terms of the value ranges of the captured preferences, RFID-based interfaces demonstrate a higher usability and also require less learning effort.

The InterPlay middleware (Messer et al, 2006) provides several user interfaces for querying device and content information and for obtaining the status of user activities at home. This interface set also includes a work interface for activity

Figure 2. Physical user interface for providing user preferences used by Davidyuk et al (2008a)

composition similar to the one in Sousa's study, which uses a verb-subject-device template. In order to compose an activity, users perform three steps in this interface: (i) they choose an action from the "play", "print" and "show" options, then (ii) they choose a material (i.e., content type) from the "movie", "music" and "photo" options and finally (iii) they choose the target device they want to use to watch the content. However, the template only allows the user to choose one device instance per user activity. If a user needs to compose an activity from two device instances, then this kind of template will not support it.

Summary

Our review of dynamic application composition solutions can be summarized by stating that in the related work the XML-based service specification languages are primarily used. These languages are easier to design than, e.g. ontology-based alternatives, but they neither allow reasoning nor encoding interpretation (i.e., meaning) of specification concepts. Thus, XML-based specifications may cause semantic heterogeneity (i.e., ambiguity) issues. Therefore, we consider the ontology-based

specification approach as the most promising solution for supporting application composition. Such an approach does not require global agreement among service providers and consumers on a specification standard, thus different legacy specification standards can be supported by mapping them to a common ontology. In addition, this approach offers greater flexibility and expressiveness compared to XML-based specification languages. Still, there is one argument against using ontologies in specifications: they increase the overall latencies, because the ontologies need to be processed.

As can be seen from Table 1, only a few of the discussed ubiquitous application composition solutions deal with middleware interoperability. Supporting interoperability is the cornerstone functionality, which allows an application composition system to utilize services that are specified in different service description languages and use various communication protocols. Since the large number of existing (and well-established) service discovery protocols, service description languages and service communication protocols make the adoption of one unique solution a rather unrealistic scenario, then supporting middleware interoperability is an essential requirement for making an application composition system truly ubiquitous.

End-user involvement has been studied in the context of ubiquitous application composition by Sousa et al (2006, 2008a) and Messer et al (2006). However, these two approaches assume that application composition is performed on the basis of predefined activity templates (i.e., application descriptions), which are developed by professional programmers. As a result, end-users are not able to create their own applications and activities according to their own needs. Ideally, applications should be created by the end-users themselves, as they are the ones with in-depth knowledge of their own needs and activities. Involving users in the process of creating applications would result in a better understanding of how the applications should be created and what services should be used. In addition, it would give the users a feeling of having more control over the environment, which is an important factor in

ensuring user acceptability of prototypes as demonstrated by Davidyuk et al (2008a).

MEDUSA MIDDLEWARE

In this section we discuss our earlier research on application composition and explain how our findings motivated the development of the MEDUSA middleware.

In our previous work on ubiquitous application composition we have developed two system prototypes which were used in building our example applications (Davidyuk et al, 2008a; Davidyuk et al, 2008b; Davidyuk et al 2009). These prototypes include a proprietary service discovery protocol, an application composition algorithm and support composite multimedia applications using the application deployment and messaging facilities of the REACHES framework (Sánchez, Cortés & Riekki, 2007). The prototypes use a mobile terminal as a remote control unit which allows the

Table 1. Comparison of related work

Solution	Specifi-cation	Composer	Interope-rability	End-User Involvement
Paluska et al (2008)	Script-based	Planning	-	-
Sousa et al (2006,2008a)	XML	Planning	-	Yes
iCOCOA (Ben Mokhtar et al, 2007)	OWL-S	Matching	Semantic	-
Gaia (Ranganathan & Campbell, 2004)	DAML	Planning	Semantic	-
PerSo (Beauche & Poizat, 2008)	YAWL	Planning	-	-
InterPlay (Messer et al, 2006)	RDF	Matching	-	Yes
PCOM (Handte et al, 2007)	XML	Matching	-	-
CODAMOS (Preuveneers & Berbers, 2005a)	OWL-S	Planning	Semantic	-
ANSO (Bottaro et al, 2007a)	XML	-	Semantic	-
MUSIC (Rouvoy et al, 2009)	XML	Planning	-	-
CASE (Hesselman et al, 2006)	OWL-S	Matching	Semantic	-
USON (Takemoto et al, 2004)	XML	Matching	-	-
Galaxy (Nakazawa et al, 2004)	XML	Matching	-	-
SesCo (Kalaspur et al, 2005)	XML	Matching	-	-
IST-Context (Chantzara et al, 2006)	XML	Planning	-	-
DRACO (Rigole et al, 2005)	XML	-	-	-

users to create audio/video playlists by touching physical objects associated with certain multimedia files (Davidyuk et al, 2008a; Davidyuk et al, 2009). The applications in both prototypes have fixed structures, i.e. composition is performed using predefined application templates.

We also conducted a user study to evaluate the feasibility of the first prototype, which relied on an autonomic algorithm to compose applications. We reported this study in Davidyuk et al (2008a). The end-user involvement in that prototype was very limited and focused on application-related issues, such as choosing multimedia content and specifying preferences over it. As the result, we observed that the user acceptance of the autonomous application composition was very low, because the users were bound by the decisions made by the algorithm. Thus, a key finding was that end-user control in application composition is necessary.

We used the results from the feasibility test as a base for designing our second prototype, CADEAU (Davidyuk et al, 2009) which focuses on end-user control in application composition. Unlike the first prototype, CADEAU allows the user to choose service instances (more specifically, services that represent real devices) manually by touching or interactively. In other words, the users are able to choose the most appropriate means of interaction according to their needs and the situation at hand. However, the prototype restricts application composition to simply matching the predefined application descriptions with the services discovered according to a given criteria. Therefore, the matching approach to composition forces the end-users to rely on applications designed by application developers, instead of giving them the possibility to create their own applications in a do-it-yourself fashion.

The main goals of the MEDUSA middleware are related to providing end-user support for the creation and customization of applications and for controlling the composition process according to user needs. To achieve these goals, MEDUSA utilizes a composition tool for encoding user intent

into applications and a set of control interfaces. These interfaces are based on our previous work (Davidyuk et al, 2009). Another important issue addressed by MEDUSA is interoperability between heterogeneous devices, networks and platforms. Achieving interoperability is a prerequisite for building an open application composition system, as the services constituting an application need to be able to discover each other, exchange information and indeed interpret this information meaningfully. This is especially important if the environment, in which the application composition system operates, consists of services implemented and deployed by independent providers (Ben Mokhtar, Raverdy, Urbeita & Speicys Cardoso, 2008).

The following section introduces the overall architecture of the MEDUSA middleware ("MEDUSA Middleware Architecture") paying special attention to end-user support ("End-User Support") and interoperability ("MEDUSA Middleware Interoperability").

MEDUSA Middleware Architecture

The architecture of MEDUSA is decomposed into the following layers; including end-user support and communication management in ubiquitous computing environment (see Figure 3):

1. The communication interoperability layer uses a common network interface to both integrate and hide the underlying multi-radio networking technologies. The layer contains two entities, namely multi-radio networking and multi-protocol network management (Caporuscio et al, 2009). The first one effectively manages the nodes' multi-radio facilities using a cross-network addressing scheme and provides point-to-point and multicast communication primitives. Whereas, the second entity is responsible for multi-protocol network management, communication mobility and multi-network

Figure 3. MEDUSA conceptual architecture

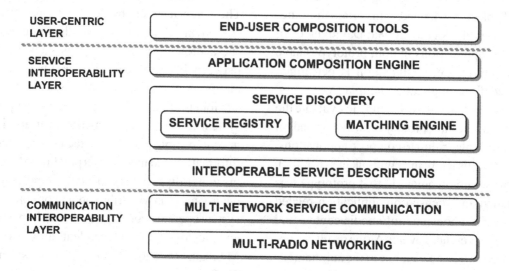

routing. In other words, it handles the mobility of the nodes to the upper layers in a transparent manner and also enables messages to be routed across nodes physically located in different networks. The realization of this layer is further discussed in "MEDUSA Middleware Interoperability."

2. The key role of the service interoperability layer is to enable semantic and syntactic interoperability between different service providers and consumers without imposing them to use a specific standard. Interoperability is achieved using a common service description model, which specifies the mapping function between the service concepts and the functionalities provided by the platform nodes (Ben Mokhtar, 2007). In addition to this, the layer is also responsible for the service discovery, which stores descriptions of available services and performs matchmaking (i.e. searches for service descriptions in the repository matching the service query). The MEDUSA service discovery supports various legacy service discovery protocols through pluggins. This enables interoperability as presented by Georgantas et al (2009). The application composition engine

employs multiple composition algorithms to produce application configurations, which are optimized using QoS requirements and user-defined criteria (Davidyuk et al, 2008b). We describe the realization of this layer in "MEDUSA Middleware Interoperability."

3. The user-centric layer provides functionality for creating and customizing applications and interfaces. The functionality can be used to control the composition of applications at run-time, and it can be used by end-users to encode their intents into the applications before they are composed by the application composition engine. In addition, end-users can utilize this functionality to control the composition process and to adapt applications at run-time by providing information on their preferences and by choosing the service instances that constitute the application. The MEDUSA end-user support is explained in further detail in the following section.

End-User Support

The functionality of the user-centric layer is provided by the end-user application composition tool and the application control interfaces. The composition tool helps users to arrange services into applications according to their own needs. We assume that each ubiquitous environment provides a set of cards which are associated with the service instances available in the environment. These cards can be issued, e.g., by the administrator who is responsible for maintaining and installing the actual services. Thus, each service is represented with a square paper card with a graphical icon on one side and an RFID tag attached the other side. A sticker with an RFID tag used in our prototype and an example set of cards representing a file server, a display, a remote controller, and an audio service are shown in Figure 4. Each of these RFID tags contains a web link to its service description which can be read by touching the card with an RFID-equipped mobile phone, as shown in Figure 4.B. Users can arrange the cards into different structures or sequences (an example sequence is shown in Figure 4.C), which are then read by touching them with a remote controller. In addition to sequences, other structures are also supported, however, they require connection cards to combine different parts of application structures. The main advantage of using a physical RFID-based interface for application composition is that it enables user cooperation and collaboration in designing ubiquitous applications, which would be difficult to realize using a traditional desktop-based user interface.

Once the application structure is provided to the system by touching service cards, end-users have to specify control and data dependencies between the services they have chosen through the mobile phone UI. This step can be supported using a mobile phone-based assistant, e.g. the office assistant in MS Word, which will guide the users through the application composition process. So, for example, if the users designed an application which is not complete structure-wise, the assistant would suggest how to complete the application using e.g., information from a database with existing application descriptions. Similar approach is used by Wisner & Kalofonos (2007) for programming smart homes.

We have tested our approach with a prototype composition tool and an initial set of cards. Figure 5 shows test subjects jointly implementing an application using a mobile phone. Two users participated in the preliminary experiment which lasted 1.5 hours. The users were asked to design abstract applications using the given service cards. In addition, they were allowed to use service cards they thought up themselves, if necessary. Altogether six applications were designed for multiple domains and several additional services were suggested during the experiment. The application domains included home, office, hospital and learning environments. We learned three lessons from this experiment. First, the graphical designs of the tags

Figure 4. Sticker with RFID tag (A), RFID reader (B), and set of service cards (C)

(i.e. icons) have to be self-explanatory and very intuitive for the users, and match their technical background. Secondly, we found that our set of service cards for application composition should be further extended. Thirdly, our composition tool needs a mechanism to motivate users to build applications, because motivation is necessary to balance the effort needed to learn to use the tool.

When the users develop applications using the composition tool, the applications only exists as abstract descriptions which have to be realized by service instances. In other words, the services constituting an application have to be connected at run-time to the service instances available in the environment. This task is performed by the application composer and it can be controlled by users through a set of user interfaces, which have two functions: (i) they allow users to choose among the possible application configurations (i.e. they are able to map the application descriptions to the service instances) suggested by the application composition engine; (ii) they permit

users to directly choose service instances which are physically available in the environment. ME-DUSA offers four different control interfaces, which enable different degrees of user involvement in controlling the application composition.

The MEDUSA middleware supports the following interfaces: manual, semi-manual, mixed-initiative and automatic. The interfaces are shown in Figure 6 where they are arranged according to the level of user involvement and system autonomy they provide.

The manual interface assumes that the users have full control over application composition. Thus, the application composition engine is not utilized at all. The users can choose service instances by touching them with their mobile phone, as shown in Figure 7 (left). The interface is based on RFID technology and each service instance has an RFID tag attached to it. Each occasion when the tag is touched it uniquely identifies the service instance associated with the tag. Non-visual resources, i.e., abstract services or ser-

Figure 5. Two MEDUSA users jointly developing an application

vices located in a hard to reach places (e.g., on the ceiling) are represented with RFID-based control panels. Figure 7 (right) shows an example of such a control panel for a multimedia projector service. The application is started once the user has chosen all the necessary service instances.

The semi-manual interface does not require users to choose manually all the services. The users can select some services by touching while the application composition engine will complete the application configuration by assigning the missing services automatically. The advantage of this interface is that in the fact only a part of the application configuration will be realized by the system. Thus, the users can choose the most important services in their opinion, and leave the less important decisions to be made automatically by the system.

The mixed-initiative interface restricts user control to the options suggested by the application composer. However, the users can always switch to another control interface if they are not satisfied with the options suggested by the system. The options are dynamically produced and ordered according to user defined criteria into a list of application configurations starting with the most attractive one. The list also shows the service instances required by each application. Before the application starts, the user can browse through the application configuration list and identify the services used by clicking the phone's soft buttons. This feature is essential if users need to validate the configuration of the application before starting it.

The autonomic interface uses the application composer to start and run an application configuration without distracting the user. This interface assumes that the user does not want to control the application composition.

MEDUSA Middleware Interoperability

The MEDUSA communication interoperability layer is realized using ubiSOAP (Caporuscio et

Figure 6. MEDUSA end-user control interfaces

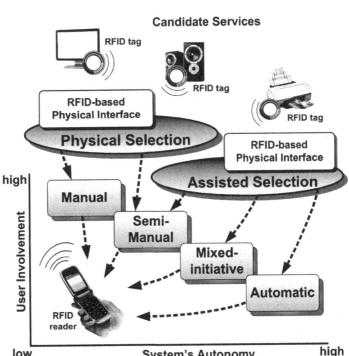

Figure 7. User selecting a service instance (left), and example service control panel (right)

al, 2009). The ubiSOAP communication middleware is specifically designed for resource-limited portable ubiquitous devices which can be interconnected through multiple wireless links (i.e. Bluetooth, Wi-Fi, GPRS, etc). This feature is essential for MEDUSA, because the front-end functionality of the middleware is intended for mobile phones. In addition, ubiSOAP adopts Web Service standards as a baseline for implementing ubiquitous services, extends the standard SOAP protocol with group messaging connectivity and supports node mobility. ubiSOAP has been implemented into two SOAP engines, Axis2[7] and CSOAP[8], which demonstrates that ubiSOAP allows legacy applications to communicate using the standard SOAP protocol. This feature guarantees the compatibility of MEDUSA with thousands of existing online services.

Figure 8 shows the two-layered architecture of ubiSOAP (for details please refer to Caporuscio et al (2009). The lower layer is the ubiSOAP connectivity layer which selects the network based on user policies, as users may require the utilization of a certain type of network for personal reasons. This layer also identifies and addresses the applications in the networking environment. The upper layer is the ubiSOAP communication layer which extends the use of the standard SOAP protocol

for messaging between the participating services by introducing SOAP multi-network multi-radio point-to-point transport and group (multicast) transport protocols.

The upper MEDUSA service interoperability layer is realized by adopting the AmIi (Ambient Intelligence interoperability) service description model (Georgantas et al, 2009) and the interoperable ambient service discovery (Ben Mokhtar, 2007). AmIi is a semantic-based solution which achieves conceptual interoperability between heterogeneous service platforms. This type of interoperability is required from service providers wanting to register services using different service discovery protocols, which in turn use different service description languages. AmIi relies on the interoperable service description model and the multi-protocol service discovery engine. The former enables mapping between the heterogeneous service description languages, thus providing conformance on both a syntactical and a semantic level. As a result, the service descriptions, which were originally written in languages such as UPnP or WSDL, can be translated into corresponding interoperable service descriptions. As shown in Figure 9, the AmIi service description model captures both functional (i.e. interface-related) and non-functional (e.g., QoS) service properties in addition to conveying information on service behavior and service grounding. The latter specifies how the service can be accessed using a legacy communication protocol. The service behavior is modeled using a workflow language e.g. BPEL. However, the model also supports other alternatives. The functional and non-functional properties are either specified with a reference to an existing ontology or they may contain embedded semantic descriptions.

The interoperable service discovery function provided by AmIi is achieved using distributed service repositories which support legacy service discovery protocols. This means that each repository runs a set of plugins (i.e., proxies) associated with various service discovery protocols. The legacy service discovery protocols are then able to exchange service descriptions with the repositories and answer query requests. However, such a mechanism requires the translation of all the service descriptions into one interoperable format, which may potentially increase the latency of the discovery protocol. Therefore, the interoperable ambient service discovery also supports a plugin for registering and querying the service descriptions directly in the interoperable format.

The application composition engine in MEDU-SA is realized using the composition algorithms initially introduced in Davidyuk et al (2008b). These optimization algorithms are based on the theories of evolutionary and genetic computing and they are used to optimize application configurations (i.e. set of services that constitute the application). The algorithms perform the optimization on the basis of user specified criteria (the nearest, the fastest or the cheapest option) and user preferences (fidelity and QoS requirements). For example, an application configuration can be optimized in order to minimize the overall application bandwidth consumption and to maximize the QoS properties of interest. These algorithms are generic and support (i) customizable criteria which may include multiple simultaneous optimization goals and (ii) various application QoS property types which can be added and removed from the service descriptions as needed at run-time.

CONCLUSION

The vision of activity-oriented computing advocates the development of systems which are explicitly aware of user needs and activities. These

Figure 8. ubiSOAP architecture, adapted from Caporuscio et al (2009)

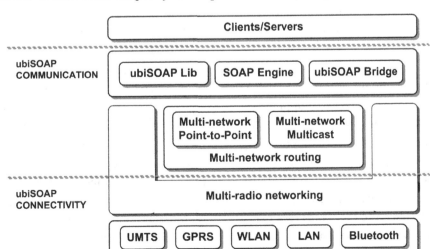

Figure 9. AmIi Service Description Model (AMSD), adapted from Ben Mokhtar (2007)

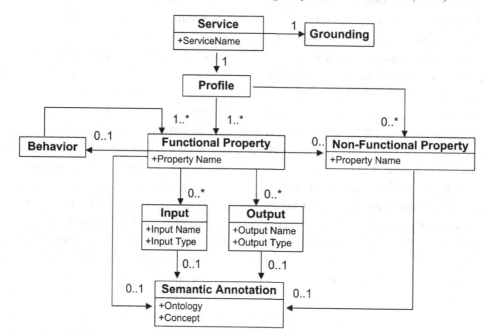

activities consist of the users' everyday tasks (e.g., in a home or office environment), which are abstractly described in terms of the situation in which the activities take place, the required system functionalities for accomplishing the activities, and the user preferences such as fidelity, privacy, or using other non-functional requirements. These activities are dynamically realized by the system composing the applications according to the users' activity descriptions. Although existing AOC systems are autonomous and rarely involve users in the actual application composition, we argue that end-user support in such systems should not be neglected.

To address this issue, we present MEDUSA, our middleware solution for end-user application composition with a two-fold goal: (i) enabling end-user application development and controlling the composition of applications at run-time; and (ii) solving issues relating to the interoperability of service communication and service discovery, thus promising to integrate thousands of already existing services. We discussed the middleware

architecture, its functionality and also presented realizations of MEDUSA utilizing the open-source solutions ubiSOAP and AmIi. MEDUSA end-user support helps avoid predefined activity descriptions, which are used in most of the related work. The end-user composition tool supports users who do not have advanced technical skills. The tool relies on mobile terminals and RFID-based physical interfaces. These technologies were found very promising in terms of usability and user acceptance (Davidyuk et al, 2008a).

One lesson learned from this work is that end-users have a need to create their own services, thus the role of the end-user has to expand into the role of a service developer. Our plan is to achieve this change iteratively, in a few steps. For example, at first our system will allow an exporting functionality which enables user-composed applications to be exported as services, so that these applications can be used to compose other applications.

In the future we are planning on experimenting with the middleware and the end-user composition tool. In addition to conducting performance measurements, we will evaluate our solution with a series of user experiments in order to assess factors relating to user acceptance and usability. It would also be interesting to study what kinds of applications the end-users usually compose and what kinds of services they typically need in different situations.

We are also planning on making the service cards from thick cardboard or pieces of wood, thus the cards would resemble pieces of a jigsaw puzzle. By making different cutout interfaces we will restrict the ways in which the services can be combined together. This is necessary because some services may not be permitted to combine directly, but they can be combined using an auxiliary service in between.

ACKNOWLEDGMENT

This work has been funded by Tekes (National Technology Agency of Finland) under UBICOM technology program, GETA (Finnish Graduate School in Electronics, Telecommunications and Automations) and Nokia Foundation.

The authors would like to thank personnel of ARLES team in INRIA, especially Animesh Pathak, Pushpendra Singh, Elena Kuznetsova and Sneha Godbole for their valuable comments and ideas regarding the functionality of the MEDUSA middleware and the content of this chapter.

REFERENCES

Amigo. (2009). Project web site. Retrieved on November 16, 2009, from http://www.hitech-projects.com/ euprojects/ amigo

Beauche, S., & Poizat, P. (2008). Automated Service Composition with Adaptive Planning. In Bouguettaya, A., Krueger, I. & Margaria, T. (Eds.), *Proceedings of the 6th International Conference on Service-Oriented Computing* (pp. 530-537). Berlin, Heidelberg: Springer-Verlag.

Ben Mokhtar, S., Georgantas, N., & Issarny, V. (2007). COCOA: COnversation-based Service Composition in PervAsive Computing Environments with QoS Support. *Journal of Systems and Software, 80*(12), 1941–1955. doi:10.1016/j.jss.2007.03.002

Ben Mokhtar, S., Raverdy, P.-G., Urbieta, A., & Speicys Cardoso, R. (2008). Interoperable Semantic and Syntactic Service Matching for Ambient Computing Environments. *Proceedings of the 1st International Workshop on Ad-hoc Ambient Computing*, France.

Bertolino, A., De Angelis, G., Frantzen, L., & Polini, A. (2008). The PLASTIC Framework and Tools for Testing Service-Oriented Applications. *In Software Engineering. International Summer Schools, ISSSE, 2006-2008*, 106–139.

Bertolino, A., De Angelis, G., & Polini, A. (2009). Online Validation of Service Oriented Systems in the European Project TAS3. *Proceedings of the ICSE Workshop on Principles of Engineering Service Oriented Systems, PESOS* (pp. 107-110). Washington DC: IEEE Computer Society.

Bottaro, A., Bourcier, J., Escofier, C., & Lalanda, P. (2007a). Context-Aware Service Composition in a Home Control Gateway. *Proceedings of the IEEE International Conference on Pervasive Services.* Washington DC: IEEE Computer Society.

Bottaro, A., Gerodolle, A., & Lalanda, P. (2007b). Pervasive Service Composition in the Home Network. *Proceedings of the 21th International Conference on Advanced Networking and Applications AINA'07* (pp. 596-603), Washington, DC: IEEE Computer Society.

Bromberg, Y.-D., & Issarny, V. (2005). INDISS: Interoperable Discovery System for Networked Services. In Alonso, G. (Ed.) *Proceedings of the of ACM/IFIP/USENIX 5th International Conference on Middleware Middleware'05* (pp. 164-183), New York, NY: Springer-Verlag New York.

Buford, J., Kumar, R., & Perkins, G. (2006). Composition Trust Bindings in Pervasive Computing Service Composition. *Proceedings of the 4th IEEE International Conference on Pervasive Computing and Communications Workshops,* Washington DC: IEEE Computer Society.

Caporuscio, M., Raverdy, P.-G., & Issarny, V. (2009). (To appear). *ubi*SOAP: A Service Oriented Middleware for Ubiquitous Networking. *Journal of Transactions on Service Computing.*

Chantzara, M., Anagnostou, M., & Sykas, E. (2006). Designing a Quality-Aware Discovery Mechanism for Acquiring Context Information. *Proceedings of the 20th International Conference on Advanced Information Networking and Applications, 1(6), AINA '06.* Washington DC: IEEE Computer Society.

Davidyuk, O., Sánchez, I., Duran, J. I., & Riekki, J. (2008a). Autonomic Composition of Ubiquitous Multimedia Applications in REACHES. *Proceedings of the 7th International ACM Conference on Mobile and Ubiquitous Multimedia, MUM '08* (pp. 105-108). New York, NY: ACM.

Davidyuk, O., Sánchez, I., Duran, J. I., & Riekki, J. (2009). CADEAU: Collecting and Delivering Multimedia Information in Ubiquitous Environments. In Kamei K. (Ed.) *Adjunct Proceedings of the 7th International Conference on Pervasive Computing PERVASIVE '09* (pp. 283-287).

Davidyuk, O., Selek, I., Duran, J. I., & Riekki, J. (2008b). Algorithms for Composing Pervasive Applications. *International Journal of Software Engineering and Its Applications, 2*(2), 71–94.

Georgantas, N., Issarny, V., Ben Mokhtar, S., Bromberg, Y.-D., Bianco, S., & Thomson, G. (2009To appear). Middleware Architecture for Ambient Intelligence in the Networked Home. In Nakashima, H., Augusto, J. C., & Aghajan, H. (Eds.), *Handbook of Ambient Intelligence and Smart Environments.* Springer.

Halevy, A. (2005). Why Your Data Wont Mix. *ACM Queue; Tomorrow's Computing Today, 3*(8), 50–58. doi:10.1145/1103822.1103836

Handte, M., Herrmann, K., Schiele, G., & Becker, C. (2007). Supporting Pluggable Configuration Algorithms in PCOM. *Proceedings of International Workshop on Pervasive Computing and Communications* (pp. 472-476).

Hardian, B., Indulska, J., & Henricksen, K. (2008). Exposing Contextual Information for Balancing Software Autonomy and User Control in Context-Aware Systems, *Proceedings of the Workshop on Context-Aware Pervasive Communities: Infrastructures, Services and Applications CAPC '08.* (Sydney, May, 2008).

Hesselman, C., Tokmakoff, A., Pawar, P., & Iacob, S. (2006). Discovery and Composition of Services for Context-Aware Systems. *Proceedings of the 1st IEEE European Conference on Smart Sensing and Context* (pp. 67-81). Berlin: Springer-Verlag.

Jianqi, Y., & Lalanda, P. (2008). Integrating UPnP in a Development Environment for Service-Oriented Applications. *Proceedings of the IEEE International Conference on Industrial Technology ICIT, 08,* 1–5.

Kalasapur, S., Kumar, M., & Shirazi, B. A. (2005). Personalized Service Composition for Ubiquitous Multimedia Delivery. *Proceedings of the 6th IEEE International Symposium on a World of Wireless Mobile and Multimedia Networks WoWMoM '05* (pp. 258-263), Washington, DC: IEEE Computer Society.

Kawsar, F., Nakajima, T., & Fujinami, K. (2008). Deploy Spontaneously: Supporting End-Users in Building and Enhancing a Smart Home. *Proceedings of the 6th International Conference on Ubiquitous Computing, UbiComp '08* (pp. 282-291). Vol. 344. New York, NY: ACM.

Lee, Y., Chun, S., & Geller, J. (2004). Web-Based Semantic Pervasive Computing Services. *Proceedings of the IEEE Intelligent Informatics Bulletin, 4*(2).

Masuoka, R., Parsia, B., & Labrou, Y. (2003). Task Computing - the Semantic Web meets Pervasive Computing. In Fensel D. et al (Eds.) *Proceedings of the 2nd International Semantic Web Conference ISWC'03* (pp. 866-881), Lecture Notes In Computer Science, Vol. 2870. Berlin, Heidelberg: Springer-Verlag.

Mavrommati, I., & Darzentas, J. (2007). End User Tools for Ambient Intelligence Environments: An Overview. In Jacko, J. (Ed.), *Human-Computer Interaction Part II* (pp. 864–872). Springer-Verlag Berlin Heidelberg.

Mavrommati, I., & Kameas, A. (2003). End-User Programming Tools in Ubiquitous Computing Applications. In Stephanidis C. (Ed.), *Proceedings of International Conference on Human-Computer Interaction* (pp. 864-872). London, UK: Lawrence Erlbaum Associates.

Messer, A., Kunjithapatham, A., Sheshagiri, M., Song, H., Kumar, P., Nguyen, P., & Yi, K. H. (2006). InterPlay: A Middleware for Seamless Device Integration and Task Orchestration in a Networked Home. *Proceedings of the Annual IEEE International Conference on Pervasive Computing PerCom'06* (pp. 296-307), Washington DC: IEEE Computer Society. Ben Mokhtar, S. (2007). *Semantic Middleware for Service-Oriented Pervasive Computing*. Doctoral dissertation, University of Paris 6, Paris, France.

Nakano, Y., Takemoto, M., Yamato, Y., & Sunaga, H. (2006). Effective Web-Service Creation Mechanism for Ubiquitous Service Oriented Architecture. *Proceedings of the 8th IEEE International Conference on E-Commerce Technology and the 3rd IEEE International Conference on Enterprise Computing, E-Commerce, and E-Services CEC/EEE'06* (p. 85). Washington DC: IEEE Computer Society.

Nakazawa, J., Yura, J., & Tokuda, H. (2004). Galaxy: a Service Shaping Approach for Addressing the Hidden Service Problem. *Proceedings of the 2nd IEEE Workshop on Software Technologies for Future Embedded and Ubiquitous Systems* (pp. 35-39).

Nokia. (2009). *Point&Find API for mobile phones.* Retrieved on November 16, 2009, from http://pointandfind.nokia.com

Paluska, J. M., Pham, H., Saif, U., Chau, G., Terman, C., & Ward, S. (2008). Structured decomposition of adaptive applications. *International Journal of Pervasive and Mobile Computing, 4*(6), 791–806. doi:10.1016/j.pmcj.2008.04.006

PLASTIC. (2009). *Project's web site.* Retrieved on November 16, 2009, from http://www-c.inria.fr/ plastic/ the-plastic-middleware

Preuveneers, D., & Berbers, Y. (2005a). Automated Context-Driven Composition of Pervasive Services to Alleviate Non-Functional Concerns. *International Journal of Computing and Information Sciences, 3*(2), 19–28.

Preuveneers, D., & Berbers, Y. (2005b). Semantic and Syntactic Modeling of Component-Based Services for Context-Aware Pervasive Systems Using OWL-S. *Proceedings of the 1st International Workshop on Managing Context Information in Mobile and Pervasive Environments* (pp. 30-39).

Ranganathan, A., & Campbell, R. H. (2004). Pervasive Autonomic Computing Based on Planning. [Washington, DC: IEEE Computer Society.]. *Proceedings of the IEEE International Conference on Autonomic Computing ICAC, 04*, 80–87. doi:10.1109/ICAC.2004.1301350

Raverdy, P.-G., Issarny, V., Chibout, R., & de La Chapelle, A. (2006). A Multi-Protocol Approach to Service Discovery and Access in Pervasive Environments. *Proceedings the 3rd Annual International Conference on Mobile and Ubiquitous Systems: Networks and Services MOBIQUITOUS'06* (pp. 1-9), Washington DC: IEEE Computer Society.

Riekki, J. (2007). RFID and Smart Spaces. *International Journal of Internet Protocol Technology, 2*(3-4), 143–152. doi:10.1504/IJIPT.2007.016216

Rigole, P., Vandervelpen, C., Luyten, K., Berbers, Y., Vandewoude, Y., & Coninx, K. (2005). A Component-Based Infrastructure for Pervasive User Interaction. *Proceedings of the International Conference on Software Techniques for Embedded and Pervasive Systems* (pp. 1-16).

Rouvoy, R., Barone, P., Ding, Y., Eliassen, F., Hallsteinsen, S., Lorenzo, J., et al. (2009). MUSIC: Middleware Support for Self-Adaptation in Ubiquitous and Service-Oriented Environments. In Cheng, B. H. et al (Eds.) *Software Engineering For Self-Adaptive Systems* (pp. 164-182), Lecture Notes In Computer Science, Vol. 5525. Berlin, Heidelberg: Springer-Verlag.

Rouvoy, R., Eliassen, F., Floch, J., Hallsteinsen, S., & Stav, E. (2008). Composing Components and Services Using a Planning-Based Adaptation Middleware. *Proceedings of the 7th Symposium on Software Composition SC'08* (pp. 52-67).

Saif, U., Pham, H., Paluska, J. M., Waterman, J., Terman, C., & Ward, S. (2003). A Case for Goal-oriented Programming Semantics. *Proceedings of the System Support for Ubiquitous Computing Workshop at Ubicomp'03*.

Sánchez, I., Cortés, M., & Riekki, J. (2007). Controlling Multimedia Players using NFC Enabled mobile phones. *Proceedings of the 6th International Conference on Mobile and Ubiquitous Multimedia MUM'07*, Vol. 284. (pp.118-124). New York, NY: ACM.

Siebert, J., Cao, J., Zhou, Y., Wang, M., & Raychoudhury, V. (2007). Universal Adaptor: A Novel Approach to Supporting Multi-Protocol Service Discovery in Pervasive Computing. In Kuo T.-W. et al (Eds.), *Proceedings of the International Conference on Embedded and Ubiquitous Computing EUC'07* (pp. 683-693), Lecture Notes In Computer Science, Vol. 4808. Berlin, Heidelberg: Springer-Verlag.

Sousa, J. P., Poladian, V., Garlan, D., Schmerl, B., & Shaw, M. (2006). *Task-Based Adaptation for Ubiquitous Computing, (Tech. Rep.)*. Pittsburgh, PA: Carnegie Mellon University, School of Computer Science.

Sousa, J. P., Schmerl, B., Steenkiste, P., & Garlan, D. (2008a). Activity-Oriented Computing. In Mostefaoui, S., Maamar, Z., & Giaglis, G. M. (Eds.), *Advances in Ubiquitous Computing: Future Paradigms and Directions* (pp. 280–315). Hershey, PA: IGI Publishing. doi:10.4018/978-1-59904-840-6.ch011

Sousa, J. P., Schmerl, B., Steenkiste, P., & Garlan, D. (2008b). uDesign: End-User Design Applied to Monitoring and Control Applications for Smart Spaces. *Proceedings of the 7th IEEE/IFIP Conference on Software Architecture* (pp. 72-80). Washington, DC: IEEE Computer Society.

Takemoto, M., Oh-ishi, T., Iwata, T., Yamato, Y., Tanaka, Y., & Shinno, K. (2004). A Service-Composition and Service-Emergence Framework for Ubiquitous-Computing Environments. [Washington DC: IEEE Computer Society.]. *Proceedings of International Symposium on Applications and the Internet, SAINT, 04-W*, 313–318.

Vastenburg, M., Keyson, D., & de Ridder, H. (2007). Measuring User Experiences of Prototypical Autonomous Products in a Simulated Home Environment. *International Journal of Human-Computer Interaction*, (2): 998–1007.

Weiser, M. (1991). The Computer for the 21st Century. *Scientific American, 265*(3), 94–104. doi:10.1038/scientificamerican0991-94

Wisner, P., & Kalofonos, D. N. (2007). A Framework for End-User Programming of Smart Homes Using Mobile Devices. *Proceedings of the 4th IEEE Consumer Communications and Networking Conference CCNC'07* (pp. 716-721), Washington DC: IEEE Computer Society.

Yang, Y., & Mahon, F. Williams. M.H. & Pfeifer, T. (2006). Context-Aware Dynamic Personalized Service Re-composition in a Pervasive Service Environment. In Ma J. et al (Eds.) *Proceedings of the 3rd International Conference on Ubiquitous Intelligence and Computing UIC'06* (pp. 724-735). Berlin, Heidelberg: Springer-Verlag.

KEY TERMS AND DEFINITIONS

Service-oriented computing: A paradigm that promotes building applications by assembling together independent networking services.

Activity-oriented computing: Promotes the idea of supporting everyday user activities through composing and deploying appropriate services and resources.

End-user application development: Studies tools and programming environments that allow end-users, instead of professional developers, customizing existing or even creating their own applications according to users' needs and knowledge.

Interaction design: A discipline that studies the relationship between humans and interactive products (i.e. devices) they use.

Physical user interface design: A discipline that studies user interfaces in which users interact with the digital would using real (i.e. physical) objects.

Ambient Intelligence (AmI): Refers to computer environments that are populated with tiny networking devices which support people in carrying out their everyday tasks using non-intrusive intelligent technology.

ENDNOTES

1 http://www.w3.org/TR/wsdl
2 http://www.yawl-system.com/
3 http://docs.oasis-open.org/wsbpel/
4 http://www.w3.org/TR/wscl10/
5 http://www.daml.org/services/owl-s/1.0/
6 http://www.wsmo.org
7 available from http://ws.apache.org/axis2/
8 available from http://www-rocq.inria.fr/arles/download/ozone/

Chapter 12
Home Service Engineering for Sensor Networks

Jochen Meis
Fraunhofer Institute for Software and System Engineering, Germany

Manfred Wojciechowski
Fraunhofer Institute for Software and System Engineering, Germany

ABSTRACT

This Chapter deals with the important process related to smart environments engineering, with a specific emphasis on the software infrastructure. In particular, the Chapter focuses on the whole process, from the initial definition of functional requirements to the identification of possible implementation strategies. On the basis of this analysis, a context model as well as the possible choice of relevant sensor types is carried out.

MOTIVATION FOR IT-BASED HOME SERVICES

Demographic factors, including age, income level, marital status, birth and death rate and the average family size, influence the development of populations and available human resources. Taking these factors into account, there will be less human resources to provide home and care services to people. This leads to a demographic distribution that is ill-equipped to provide home and care services since there is a smaller workforce providing for the elderly. The percentage of single households among all ages is rising, leading to a higher demand for services. Single households still require the same quantity of home services as all other households. Finally, there is and will continue to be a higher demand of individual home services across all ages. One possibility to fulfill the higher demand is to use IT as an active part of home services.

"Smart home", "intelligent home", "automated home" and "smart living" are popular terms used to describe home process automation. Of interest

DOI: 10.4018/978-1-61692-857-5.ch012

is how to integrate this automation into a service process realized by a local service provider. To define such integrated services, different aspects thereof must be discussed. These aspects are:

- Focusing service engineering for IT-based services
- Defining and modeling IT-based home services
- Exhibiting context model

In order to model and (automatically) execute IT-based home services, they must be given special treatment, with sensors and actuators taking active roles within the service package as agents supporting or triggering human actions. Sensors and actuators are local components inside the home and can be bundled under the term domotic. For example, an emergency detected by a system leads to a phone call from an assistant or a detected device failure is followed by a technical inspection. The infrastructure to enable such interaction is based upon information and communication technology (ICT). To set up and facilitate such IT-based home services, a methodology and notation for modeling the service process and information pool is needed. Work is currently underway in the research community of service engineering to deliver such a solution. Service engineering, however, does not currently consider local components as an active part of an IT-based home service. The focus lies instead on human tasks done by employers or customers. Till now such details of IT-based home services are not defined and classified - they do not exist. Therefore standard process models for service engineering are examined, assessed and extended to support IT-based home services. In other words, existing methodologies and notations are verified for practice. If they are not sufficient, they must be extended.

HOME SERVICES ENGINEERING IN FOCUS

For modeling IT-based home services, a service blueprint is used to structure service tasks on different levels according to their distance from the customer. It structures process activities on two sides - customer and service provider. The analysis reveals important communication points to customers and shows potential failure points. Data can be presented visually with different notations such as flow charts, extended Event-Driven Process Chain (eEPC) or Business Process Modeling Notation (BPMN). There are much more notations existing, but none of them are capable of modeling IT-based home services without extension. For the purposes, BPMN is used to model the IT-based home service within the service blueprint structure. BPMN is extended by special features to address the entire range of IT-based home services provided within a smart home environment.

Service and Home Service Definition

Currently, there is no concrete scientific definition for service. Through its usage, we are able to extrapolate the following aspects of a service (for more details see Meffert and Bruhn, 2000):

- Activity-oriented
- Process-oriented
- Result-oriented
- Potential-oriented

These four characteristics of services involve different approaches. As a consequence, services are rendered directly to activities within a process to fulfill individual needs or market-oriented results or rather are viewed by human or system realized potentials. A combined view of the three characteristics (potential-oriented, process-oriented, and result-oriented) can be organized within a phase-oriented model. HILKE combined these characteristics to a abstract model (see Figure 1) (Hilke, 1989). This phase-oriented model integrates the aspect "activity-oriented" because the physical and psychological manpower flows into the process and potential (Schüller, 1967).

Following the phase-oriented model, Meffert and Bruhn define services as self-acting marketable efforts which are connected to performance. The service provider's combination of potentials is used to achieve beneficial impact for humans or their objects (Meffert and Bruhn, 2000).

An important consideration for reflection upon services is cooperation with customers (Hilke, 1989; Berekoven, 1974; Meyer, 1992; Engelhardt *et al.*, 1993). The customer is the most important external factor which must be taken into account within every service process. Internal factors like business premises, employees, qualifications, etc. can be altered by the service provider. External factors such as customer needs, development of the market, price trends, etc. can not be influenced by the service provider (Meffert and Bruhn, 2000). These internal and external factors points out the problem when dealing with service properties.

Services do not have, compared to industrial goods, a physical representative. They have their own scope for design and arrangements for interpretation. The development and provision of services are not comparable to development and production of physical goods (Meffert and Bruhn, 2000). Special properties of services are (more details in Meffert and Bruhn, 2000; Meyer, 1992; Zollner, 1995; Pepels, 2004.

- Immateriality
- Less A-priori assessment
- Intangibility
- Uno-Actu-Principle
- Non suitability of storage and transport
- Integration of external factor
- Heterogeneity
- Ex-ante quality assessment

These properties make it difficult to formulate a standard for services. In fact the three characteristics (potential, process, result) will be standardized separately (Meffert and Bruhn, 2000). The combination of standardized characteristics, defined in sub processes, allow a faster development of services and value performance for external factors. The advantages of standardized service sub

Figure 1. Phase-oriented combination of the three characteristics (process, result, potential) of services (see Meffert and Bruhn, 2000)

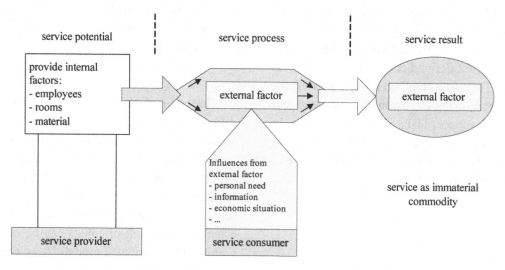

processes are an easier arrangement, organization, participation, commercialization and realization.

Home services are specialized services for customers and homes. The services are consumed directly in customers private surrounding. They are provided within a regional range, because different activities must be done by a qualified person. Currently a uniform definition for home services is not present (Scharp *et al.*, 2004). Typical well known home services are gardening, cleaning or cooking. From a caregiver perspective it's for example good to know, that someone, he is responsible for, is active. For this scenario an active button exists at a cradle of an in-house emergency call system. The person has to push it at least once a day and if he forgot to press the activity button, an emergency call is initiated. This service will called VitalCheck within this chapter.

Definition for IT-Based Home Services

An IT infrastructure is needed to set up IT-based home services. The IT supports service processes for optimization and thus partially automated service organization and delivery. For semi-automatic realization the service process must be synchronized with customer needs. One important characteristic for synchronization is the degree of individuality in the service delivery process. The higher the degree of individuality within service delivery process is, the more is the extensive participation and integration of customer within the service process. The integration again is dependent on the service itself. A classification for such services is (Dahlke and Kergaßner, 1996):

- Off-the-shelf
- Mass customization
- One-of-a-kind

Services classified as off-the-shelf or mass customization are very interesting for integrating IT. Due to the higher potential of standardization,

it is possible to partially, or in some cases even fully, automate the service process. To support service automation a semi-formal language is needed. A semi-formal language allows clear service interpretation and service execution by an IT-system and steps over to IT-services.

IT-services have another focus like IT-based services. To define IT-based services a detailed look at three characteristics has to be made (cp. "Service & Home Service Definition" section). The evaluation for IT-services are as follows (Böhmann, 2004):

- **Service potential:** The IT is used to hold and fetch service potential. The higher the IT-quota for conservation or maintenance is, the higher is the relation to IT-services.
- **Service process:** The IT is used to reproduce and manage the whole service process. The service process includes IT-system integration and IT-activities of the external factor.
- **Service result:** The IT is part of the service result. The service result focuses to IT-system and connected IT-activities to them of the service promises.

Therefore increased commitment of IT leads to IT-services. But IT-services are different than IT-based services. For differentiation a closer look at the three mentioned characteristics is made. IT-services as well as IT-based services utilize IT for realization of service potential and service process. For example to organize and to train the manpower or using a cradle within VitalCheck example. The cradle represents IT used to support the service process. Service result, on the other hand, is different. IT-services explicitly include IT within service result but IT-based services not. This context is represented in Figure 2. IT-based care services generally depend on ICT within the service process and service potential contexts. The service process is driven while the service potential is managed by IT like getting activity notifications by using a cradle. IT-based care services are using ICT for potential representation and / or process organization. In contrast, the service result is independent of ICT like the cared person themselves. When an activity is missing a responsible will visit the person and this is a task without focusing IT.

Special requirements for IT-based home services have to be taken into account for further treatments. The challenge lies in matching sensor tasks with human tasks during execution of the service process. The cradle within VitalCheck can be seen as one possible sensor. Up to this point, tasks were started, stepped into or executed by humans. IT-based home services overtake selected human tasks by including home automation system as part of a service. Within the VitalCheck the activity notification can be signaled by motion detectors in combination with contact sensors. To offer these possibilities a process model, methodology, notation and sensor context model is needed.

Process Models for Service Engineering

The discipline service engineering is currently more or less underrepresented in many industrial nations, but it is growing quickly. The life cycle of services comprises multiple phases. As part of service engineering, two major phases can be differentiated: the service creation and the service delivery. The existing process models for service engineering consider the two phases but also divide additional subjects (Hohm *et al.*, 2004; Schneider *et al.*, 2003; Meiren *et al.*, 1998). To establish high quality services the models, methods and tools have to have domain focus and the service quality has a special meaning for service providers (Schneider *et al.*, 2003). The quality is strongly influenced by the service process quality (Ehrlenspiel, 1995). Compared to software engineering, where overall engineering approaches are placed, a process model supports a general process for development (Balzert, 2000). Process modeling, when compared to the success of software engineering, is useful for developing services. The Deutsche Institut für Normung (DIN) definition for service engineering specifies the

Figure 2. Delimitation of IT-services to IT-based services (see Böhmann, 2004)

	IT-service		IT-based service		
service result	●	●	-	-	
service process	●	○	●	○	
service potential	●	○	●	○	
examples	Application hosting	IT-strategy consulting	e-reservation	e-order and e-delivery	

Impact of ICT	
●	high
○	normal
-	low

methodologies, construction and development of service products and systems (DIN, 1998). To support methodological construction and development process models, methods and tools have to be implemented (Bullinger and Meiren, 2001; Goecke and Stein, 1998). In general, three different types of process models can be identified (see Figure 3) (Schneider *et al.*, 2003; Meiren *et al.*, 1998):

- Linear process models
- Iterative process models
- Prototype process models

A linear process model describes a sequence of activities which illustrate the general process of service development. An activity starts when its preceding activity has finished (Grob and Seufert, 1996). Representatives of linear process models for service engineering are models defined by Edvardsson and Olsson (1996), Scheuing and Johnson (1989), Ramaswamy (1996), the DIN-process model (1998) and the BMBF 4 phase process model (Hohm *et al.*, 2004). Due to default of flexibility by linear process models a step back is integrated within iterative process models. Within service engineering for example the model defined by Shostack (1982) is considered

to be an iterative process model. The prototype process models enable an early prefiguration of service for testing functionality, properties and characteristics. Within the service engineering prototype process models are very effective, but combined with many intermediate development steps consuming time and money.

BMBF 4 Phase Process Model

A research project called "Service engineering für die Wohnungswirtschaft" (SeWoWi, Service engineering for the housing industry) developed a specific process model to engineer new services especially for the housing industry. This process model is named "BMBF 4 phase process model" (Hohm *et al.*, 2004). The process model keeps the typical steps of the service engineering approach in mind. It includes four major steps, which are subdivided into sections. The steps are alternating between design and validation section (see Figure 4).

Starting with situation analysis the surrounding of the housing industry is addressed. This includes questions like how new services will fit to business objectives or do existing services have to be customized. The situation analysis delivers first ideas as input for the following step – the

Figure 3. Overview of different types of process models (Schneider et al., 2003; Meiren et al., 1998)

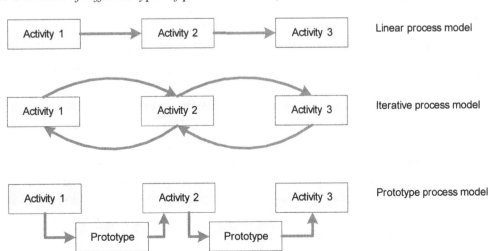

Figure 4. SeWoWi process model (figure follows diagram in Hohm et al., 2004)

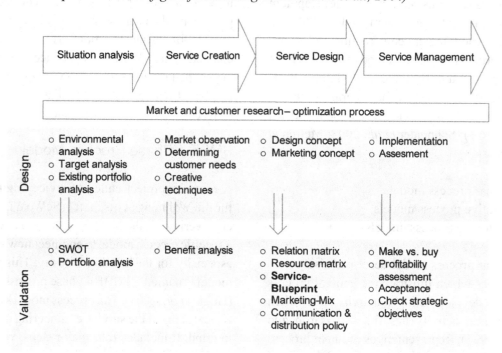

service creation. By using creative techniques, like brainstorming, brainwriting or improvisation, additional ideas are developed. Scenario and interview techniques can also be used to generate and to discuss innovative ideas. All gathered ideas are evaluated with selected criteria to get an idea priority list. High listed ideas will be experimentally designed within the step of service design. The **service design** is very important for service engineering, because it delivers an impression of selected service and helps to improve the service. The service design starts with the design concept, which includes planning service potential, developing service process and describing service result of the selected service. The marketing concept includes marketing goals, strategy and marketing-mix. The validation of the design is done by bringing together the service process – visualized as service blueprint -, the marketing-mix and the communication and distribution policy. The service management establishes the new or modified service. The service potential (staff member, rooms, cooperation partners, etc.) must be prepared and instructed. All activities connected to the service have to be integrated into existing business processes. The marketing concept will be implemented. The quality and success of the service has to be proved with appropriate assessments. If variations become clear, service optimizations can be inserted afterwards.

As seen above various process models exist within service engineering and they are structured into different phases or steps. One important step for service engineering is the service design. The service design includes specifics for blueprinting services in advanced for getting a concrete idea of the service process. During service design the methodology of service blueprint describes the service process. The customer or service provider will initiate activities or will participate in the service process. The challenge within the service processes of IT-based home services is the use and integration of local components.

Special View on Service Blueprint

In smart living scenarios, local components will be a part of customers' activities, thus requiring such components to be installed in the customer's home. The local components produce information and context-dependent information will be sent back to local components. Service blueprint is a process analysis method based on flow charts to systematize the description, documentation and analysis of service processes (Shostack, 1982; Kingman-Brundage, 1989; Shostack, 1984; Fließ, Lasshof and Mekel, 2004; Fließ, 2006; Fließ and Kleinaltenkamp, 2004). The blueprint structures the process activities on either side – customer and provider. Shostack (1982) defined the service blueprint with three activity layers divided by the "line of visibility" and the "line of interaction". The first extension, done by Kingman-Brundage (1989), contains more activity layer. It differs between customer activities, onstage activities, backstage activities, support activities and management activities. These different kinds of activities show the organized process of a service blueprint from customer to service provider including support and management. The second increment, processed by Fließ (2006) and Fließ and Kleinaltenkamp (2004), includes the "line of order penetration". It divides the management activity into preparation and facility activities. The main part, the "line of visibility", didn't change through the increments of the service blueprinting. Figure 5 shows the different structures of increments in service blueprint.

Currently, the service blueprint specific options allow integrating different notations and visualizing different kinds of information such as time, costs and resources within the model. An analysis reflects the important communication points to customers and reveals potential points of failure. It is impossible to visualize automatically generated information from home automation system into the service blueprint. This information is unavailable during the service process and must be assimilated into the IT-based home service process afterwards. Modeling the service process will assist the housing industry in integrating required local components into a building.

Analyzing the collected information and placing it within the service processes to raise the quality of service to customers is the main focus of blueprinting (Schneider, Wagner and Behrens, 2003; Edvardsson and Olsson, 1996). Different service process modeling methods and notations can be used for analyzing customer expectations and designing customer services. The "flow chart" for service design and management by Ramaswamy (1996) adopts different modeling constructions such as decision branch and process owner. It models the steps of functions to deliver services to customers.

Figure 5. Increment overview of Service Blueprint

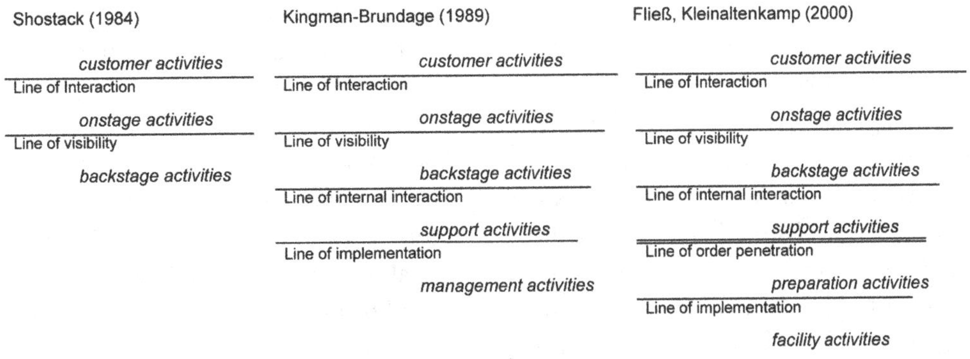

227

In addition to modeling and graphic presentation of service processes, a process model exists for the generation of a service blueprint. This process model outlines the following six steps needed to generate a service blueprint (Fließ, 2006; Fließ and Kleinaltenkamp, 2004):

1. Determine blueprint process
2. Define target audience
3. Understand client's perception of the process
4. Incorporate employee activities and equipment
5. Include internal strategy decisions
6. Identify added value

These six steps describe the standard approach to build up a service blueprint. Taking IT-based home service specialties like home automation system, self acting local components, etc. into account, it is obvious, that extensions to these six steps have to be made. Therefore extensions are made and will be described in "Engineering Focus" section.

Notation Used for Service Blueprint

Service blueprint is visualizing the activities between the customer and service provider. Different notations are already used to fill up the service blueprint layers. Service blueprint uses easy symbols of flow charts. In (Fließ, Lasshof and Mekel, 2004) additional notations (like eEPC, Gantt-diagram) are used to visualize the same process. The visualizations are easy to understand the modeled service process. The BPMN can also be used in combination with service blueprint. BPMN is a favorite modeling language by managers. It includes easy diagrams and concepts and can be expanded by expressive graphical representation. The focus is set to processes, events and activities, modeled with different abstraction levels. On the one hand, the process can be modeled on a high abstraction level to give an overall view of the service. On the other hand many specific details

can be included into the service diagram. A high detail level is necessary for process automation (Siegel, 2008). The Object Management Group (OMG) says: "BPMN is targeted at a high level for business users and at a lower level for process implementers." (OMG, 2004). BPMN is more and more used by business specialists as well as IT-specialists (camunda service GmbH, 2008). The core elements are (see White, 2007; OMG, 2009):

- Flow objects (events, activities, gateways)
- Connecting objects (sequence flow, message flow, association)
- Swimlanes (pools, lanes)
- Artifacts (data objects, group, annotation)

The BPMN is integrated in more than 40 modeling tools. Formally BPMN was designed for Business-Analysts, working with flow charts. BPMN was integrated into the OMG landscape, after the Business Process Modeling Initiative become part of the OMG in 2005. Current version of BPMN is Version 1.2 released in January 2009 and it includes some updates from Version 1.1. A simple example using BPMN is given in Figure 6.

New is that user tasks can be modeled as a special task. The attribute "TaskType" allows "User" and "Manual", so that simple service tasks can be taken into account for human tasks and tasks by local components. The problem is how human tasks can be transferred to system assistance. To model the human tasks is not enough. Although service potential, like staff member, material, time estimation, room etc., has to be integrated into the model as well as human interaction. The additional information will be part of the tasks attribute.

Also the active part of the home automation system, like motion detectors or contact sensors, has to be integrated. Furthermore the additional information in combination with handling time, probabilities, costs and defined states a simulation should be possible.

Table 1. Overview BPMN elements (see OMG, 2009)

Element	Notation
Events (start, intermediate, end)	
Activitiy	
Gateway	
Sequence flow	
Message flow	
Association	
Pool	
2 Lanes inside Pool	
Data object	
Group	
Annotation	

Figure 6. Simple BPMN Example (White, 2007)

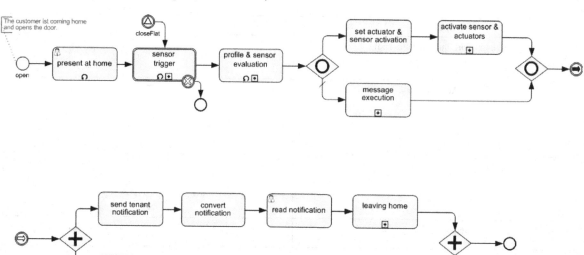

ENGINEERING FOCUS

Service Blueprint Extension

The amount of information available at customer service interactions is a major factor in the service delivery quality. Further information collected from customer profiles and additional local components (e.g. motion detectors thermostat, heating systems) should be integrated in the service design and service management process. Situations and activities triggered by a customer and identified by home automation (e.g. activity gaps, leaving home, leakage, and consumption) assist services like in-house emergency service, relocation service, security service, delivery service etc. With minor modifications and extensions, service blueprint allows dealing with automatically collected information from local sensors and actuators. Customers' activities can be swayed by local components and their individual profiles.

For the purpose of dealing with local components, directly or indirectly influenced by customers, a new line will be added to extend service

blueprint (see Figure 7). The "line of crossbench interaction" separates the activities undertaken directly by customers from those done indirectly by customers through their home automation system (specific for smart homes) (Meis and Schöpe, 2006).

In smart homes, customers will interact more and more with smart local components. Independent acting IT-based home services need to distinguish between direct and indirect customer activities. Two objectives can be developed from the extended structure of the service blueprint. First the extended structure will identify local components needed for the value added service, because they are modeled as self-acting customer activities. Second, it will obviously point out the potential of a service within a smart home. The potential of new innovative services in the area of living offered as a product will include a large quantity of providing a better service, a higher quality, lower prices or a combination of them assisted by local components.

As an example for IT-based home service the VitalCheck will be shown in Figure 8. An outpa-

Figure 7. Extended service blueprint for IT-based home services

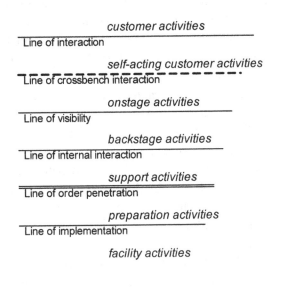

tient care provider is involved to look after their patient. The needed information from a smart home is defined in advance, so afterwards the responsible sensors and actuators can be installed within the home. Therefore, it is helpful to define all offered services and their needed information content. IT-based home services will perform on a service platform, which establish the connection between the home automation system and the employees of the service provider. The service platform will act independently and offers various service interfaces. From the layer perspective the service platform will be part of the backstage activities.

If a critical context is reached, the service platform will notify, depending on escalation breakpoints, relatives, neighbors, friends, persons responsible or caregivers. The caregiver will get the notification from the service platform and afterwards he can check whether he should contact or visit the person or alert emergency services. His decision will depend on the information he has. The service platform provides him actual and prior component values, to confirm his decision. In addition to the component values, it is possible to provide more information about the person. Maybe he was outside and the alarm activates or the last action (like moving, oven

Figure 8. Service blueprint with home automation based on care service example "VitalCheck"

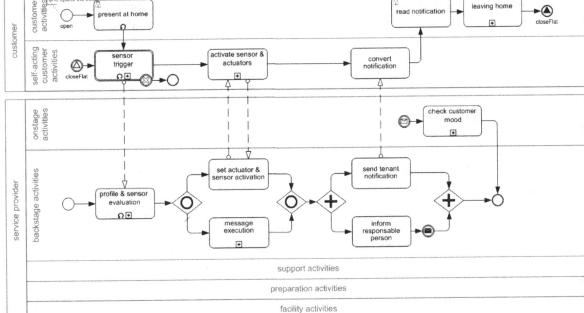

using …) of the customer. The emergency will know where the person is because the service platform provides all collected information.

In order to account for the specific needs of IT-based home services, additional steps for generating a service blueprint are required. First, the potential of the automation system used as the point of interaction between the client and the system must be defined and understood. Next, the dependencies among the individual components of the sensor devices as well as the (intermediate) activities of the service processes must be determined. Furthermore, in addition to the identified added value, a dynamic added value is presented in the final step. This is necessary because the building's automation system introduces the possibility of automatically generated events, thus allowing intervention in the service process. The steps needed to create a service blueprint in the application domain are summarized by the following (italics denote extensions):

1. Determine blueprint process
2. Define target audience
3. Understand client's perception of the process
4. Define and engage domotic context
5. Incorporate employee activities and equipment
6. Domotic enhancement definition
7. Include internal strategy decisions
8. Identify *dynamic* added value

Notation Extension for IT-Based Home Services

The important part within the blueprint, especially for the housing industry, will be the usage of smart home activities and situation detection. Additional graphical representatives allow specifying sensors and actuators within a service process. For integration BPMN allows extension, which shall not change the specified shape of a defined graphical element or marker (e.g. changing a square into a triangle) (OMG, 2009). A new shape representing

a kind of artifact may be added without conflicting to existing shapes. Local components influencing IT-based home services can be compared to a data object in BPMN. Therefore it is obvious, that sensors and actuators are used like data object, but with their own graphical representation and properties. Selected graphical representatives for local components are shown in Table 2.

Each local component includes an own set of properties. Depending of the component, the needed properties can be integrated within the service process. In Table 3 selected properties are listed for components used as indirect customer interaction within a service blueprint. The defined properties derive from the needed hardware plugged into the home. If needed a keyless entry system for an IT-based home service can be defined and is directly effecting the requirements of the infrastructure. Selected properties for components are listed and described in Table 3.

Figure 9 shows a detailed look at the IT-based home service "VitalCheck" including local components and properties. Starting with the attendance of a person, different actions are done inside the home. The person is using the toilet – reached through the floor – and is open windows and doors. Motion detectors and contact sensors will provide their event for evaluation. These actions will be influence the sensor query task. All modeled components have their own property set, their sensor specification (see Figure 9). The sensor specification will be analyzed depending on customer's individualized profile. If no action taken inside the home within a defined period of time, a notification is generated by the service-platform. The notification allows – if needed – triggering additional actuators like optical detectors or reminders, inside the home. Finally the notification will send to a contact person, which includes relatives, neighbors, friends and employees of the service provider.

The service process and its corresponding service tasks are represented directly by the service provider so that sensors and actuators become a

Table 2. Selected graphical representatives of local components

Sensor	Notation	Actuator	Notation
Motion detector		Light actuator	
Glass breackage detector		Camera	
Smoke detector		Optical detector	
Contact sensor		Multimedia Plug	
Water sensor		Electronic locker	
Keyless entry		...	

Table 3. Overview of selected properties for components

Property	Description	Property	Description
ComponentCategory	A Component belongs to "sensor", "actuator" or "mix".	Time	Defines time windows for starting and stopping the component. Subproperties are "startTime", "endTime", "duration", ...
ComponentType	Gives the Component Type: e.g. - MotionDetector - ContactSensor - LightActuator ...	Event	Defines the events, which is thrown and defined as in input in service task
Status	Gives the actual status of the sensor depending of the sensor type.	Behavior	Describes the behavior of the component.
Place	Defines the place, where the component is stored	...	
...			

Figure 9. Reduced example with components details of the IT-based home service "VitalCheck"

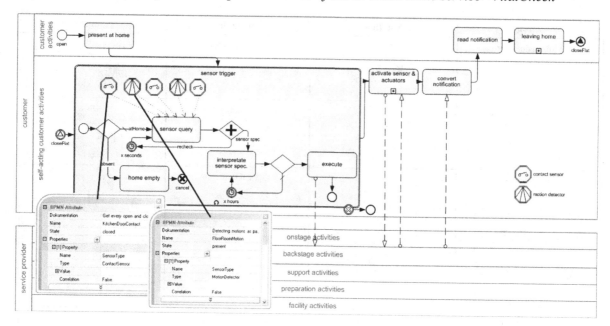

part of a customer's activities. The local components produce and consume information and combine it with human tasks as part of the service process. The human tasks are now supported and accompanied by the individual home sensor network. Basis for home sensor networks will be a comprehensive sensor network model. This sensor network model allows extracting defined contexts of a person – the individual context model.

The dependencies between the home activities will be defined by using the property "reference". This property describes the need of an event – trigged by another home activity – before the home activity starts. The property description will be transferred to a service process compatible XML-structure. Standardized XML-structure carries over to an automatically instantiated and interpreted home automation system.

Exhibiting Context Model

Sensors as part of a home automation system will deliver different kinds of information. This

information can be integrated into the service process as an activator, a decider or a piece of information. The value of sensor information depends strongly on service, infrastructure and service process integration (Heusinger, 2004). For home integration, different home automation systems are available. A home automation system is a system which links components together by using an infrastructure. The infrastructure assures functionality and operability. System intelligence is distributed over components and information exchange is done directly between components (Zentralverband Elektrotechnik- und Elektroindustrie, Handbuch, 1997). The available home automation systems are different in structure, protocol and field of application. They use sensors and actuators to collect information, control components and offer services. Major home automation systems are KNX/EIB, LON, LCN or BacNet (see Heusinger, 2004; Tränkler, 2001; Weinzierl and Schneider, 2001). All of them are using their own description model to represent domotic and functionality. The representations of different home automation systems can be abstracted to a sensor network model including categories. On top of the sensor network model, new services are possible.

Sensors provide different kind of context information regarding the home or its customer. According to Dey (2000) context is "any information that can be used to characterize the situation of an entity. An entity is a person, place, or object that is considered relevant to the interaction between a user and an application, including the user and application themselves". A context model is a formal description of the relevant aspects of the real world. Usually it is used to abstract from the technical details of context sensing and to connect the real world to the technical view of a context adaptive application (Becker and Nicklas, 2004). There are already a number of approaches for context modeling introduced into the context awareness community. In Strang and Linnhoff-Popien (2004) an overview on actual approaches is given. From a software engineering point of view the abstraction of the usage of sensors is a good architectural pattern. From our service engineering point of view such an abstraction cannot be applied.

The choice of sensors is an important aspect in the service design process. A reason is the individual level of acceptance of different types of sensors for each customer. The acceptance of a sensor can depend on different aspects. One important aspect is the sensor's intrusiveness. A camera for example can be used as a sensor for different kind of context information, but a customer might feel uncomfortable being watched by a device. These can be partially be replaced by less intrusive sensors, e.g. a microphone (Kranz *et al.*, 2007). Such an alternative sensor might be less reliable in regard to the correctness of the sensed context information. The aspect of uncertainty therefore is also an important criteria for the selection of a context sensor. Sometimes there is a tradeoff between the intrusiveness and the reliability which has to be regarded. Also the quality of the sensed context information, e.g. the location resolution, can be selection criteria. Another criteria for the acceptance of a context sensor by the customer is his involvement in the context sensing process. For example a RFID-based sensor might require the customer to always carry an RFID-chip. Another example is sensors where the customer is directly asked to give feedback. Such involvement of the inhabitant should be minimized. Because of these characteristics the selection of a sensor is an important part of the service design process.

\Therefore we need to supplement the context model for the sensor network with information on potential context sensors and their characteristics. This part is then needed to discuss on usage of a concrete context sensor and its impact on the service quality. In Table 4 an example for the classification of different location sensors is given.

Table 4. Cut-off of categories and information potential for sensors and actuators

Context model	Sensor description				
Entity and feature		**Intrusiveness**	**Reliability**	**Quality**	**User involvement**
Person.location	RFID	Low	High	Low	High
	Passive IR	Low	Middle	Low	High
	Camera	High	High	High	Low
	Door contacts	Low	Middle	Middle	Low
...	...				

CONCLUSION AND TRANSPORTATION

All additional techniques for modeling service processes must focus on customer. A service model does not guarantee the success of a service, but it helps to understand the customer actions and home automation integration and to reveal, and thus prevent, failure points within the service process. In order for a service to be successful, customers' expectations and experiences need to be integrated into the development process. The gained expectations and experiences must be transferred into the service blueprint and take part in the service design and service management decisions. Simple homes are extended to smart homes by using a service platform and adding smart local components for realizing IT-based home services.

IT-based home services offer a new perspective to provide services directly to the customers within their own environment. The close dependency between the customer and the service provider allows reacting as fast as possible. In the instance of IT-based home services, the reaction is fast because there is no longer a change of media. But currently there is no standard for modeling IT-based home services. Furthermore, the modeled services cannot be transferred directly to an interpreting language performed on an execution platform. The extensions for service blueprint, BPMN and service context provide the first steps to the automatic execution of modeled IT-based home services. Additional research and standardization is needed.

The service blueprint is an element of the design phase of the service engineering process. The extension of service blueprint showed the impact of developing innovative services, especially in the application domain smart living. Further research will focus on developing an adequate interpretation including different local components. Additionally, different aspects of existing home automation systems have to be treated as

properties for components. BPMN is a good example of a service process modeling notation since it addresses business managers as well as process implementers. Extensions allow BPMN modeling IT-based home services. In addition, a mapping from BPMN to BPEL4WS exists and in the BPMN 2.0 there will is a BPMN interchange format specified, which can be used for direct execution. This does not consider the specifics of IT-based home services. If modeling and automatic execution of IT-based home services is available, the housing industry can easily extend their existing services. Using techniques of service engineering, new services can be developed and existing services can be adapted.

REFERENCES

Balzert, H. Lehrbuch der Software-Technik - Software Entwicklung, 2. Auflage ed. Heidelberg; Berlin: Spektrum, Akademischer Verlag, 2000.

Becker C. and Nicklas D., "Where do spatial context-models end and where do ontologies start? A proposal of a combined approach," 2004.

Berekoven, L. (1974). *Dienstleistungsbetrieb*. Wiesbaden: Wesen - Struktur - Bedeutung.

Böhmann, T. Modularisierung von IT-Dienstleistungen, 1 ed. Wiesbaden: 2004.

Bullinger, H.-J., & Meiren, T. (2001). In Bruhn, M., & Meffert, H. (Eds.), *"Service Engineering. Entwicklung und Gestaltung von Dienstleistungen,"* in Handbuch Dienstleistungsmanagement (pp. 149–177). Wiesbaden: Gabler.

camunda service GmbH, "BPM-Pakete haben Nachholbedarf," Berlin: camunda service GmbH, 2008.

Dahlke, B., & Kergaßner, R. (1996). In Kleinaltenkamp, M., Fließ, S., & Jacob, F. (Eds.), *"Customer Integration und die Gestaltung von Geschäftsbeziehungen," in Customer Integration. Von der Kundenorientierung zur Kundenintegration* (pp. 177–192). Wiesbaden.

Dey, A. (2000). *Providing Architectural Support for Building Context-Aware Applications.* Georgia Institute of Technology.

DIN. "Service Engineering - Entwicklungsbegleitende Normund und Dienstleistungen,", DIN-Fachbericht 75 ed Berlin, Wien, Zürich: DIN - Deutsches Institut für Normung, 1998.

Edvardsson, B., & Olsson, J. (1996). Key Concepts for New Service Development. *The Service Industries Journal, 16*(2), 140–164. doi:10.1080/02642069600000019

Ehrlenspiel, K. (1995). *Integrierte Produktentwicklung - Methoden für Prozessorganisation, Produktherstellung und Konstruktion.* München: Carl Hanser Verlag.

Engelhardt, W., Kleinaltenkamp, M., & Reckenfelderbäumer, M. "Leistungsbündel als Absatzobjekte. Ein Ansatz zur Überwindung der Dichotomie von Sach- und Dienstleistungen,", 45 ed 1993, pp. 395-426.

Fließ, S. Prozessorganisation in Dienstleistungsunternehmen. Stuttgart: 2006.

Fließ, S., & Kleinaltenkamp, M. (2004). Blueprinting the service company: Managing service processes efficiently. *Journal of Business Research, 57*(4), 392–404. doi:10.1016/S0148-2963(02)00273-4

Fließ, S., Lasshof, B., & Mekel, M. (2004). *Möglichkeiten der Integration eines Zeitmanagements in das Blueprinting von Dienstleistungsprozessen.* Douglas-Stiftungslehrstuhl für Dienstleistungsmanagement.

Goecke, R., & Stein, S. "Marktführerschaft durch Leistungsbündelung und kundenorientiertes Service Engineering,", 13 ed 1998, pp. 11-13.

Grob H.L and Seufert S., "Vorgehensmodelle bei der Entwicklung von CAL-Software,", Arbeitsbericht 5 ed Münster: Universität Münster, 1996.

Heusinger, W. Das Intelligente Haus, 1 ed. Frankfurt am Main, Berlin, Bern, Bruxlles, New York, Oxford, Wien: Peter Lang GmbH - Europäischer Verlag der Wissenschaft, 2004.

Hilke, W. (1989). In Hilke, W. (Ed.), *"Grundprobleme und Entwicklungstendenzen des Dienstleistungs-Marketing," in Dienstleistungs-Marketing* (pp. 5–44). Wiesbaden.

Hohm D., Jonuschat H., Scharp M., Scheer D., and Scholl G., "Innovative Dienstleistungen "rund um das Wohnen" professionell entwickeln - Service Engineering in der Wohnungswirtschaft," GdW Bundesverband deutscher Wohnungsunternehmen e.V., 2004.

Kingman-Brundage, J. "The ABC's of Service System Blueprinting," Chicago: Bitner M.J.; Crosby L.A., 1989, pp. 30-33.

Kranz M., Schmidt A., Bogdan Rusu R., Maldoado A., Beetz M., Hörnler B., and Rigoll G., "Sensing Technologies and the Player-Middleware for Context-Awareness in Kitchen Environments," 2007.

Meffert, H., & Bruhn, M. (2000). *Dienstleistungsmarketing.* Wiesbaden: Grundlagen - Konzepte - Methoden.

Meiren, T., Hofmann, H. R., & Klein, L. "Vorgehensmodelle für das Service Engineering,", 13 ed 1998, pp. 20-25.

Meis, J., & Schöpe, L. "ServiceDesign with the ServiceBlueprint,". Knowledge Systems Institute Graduate School, Ed. San Francisco, Calif.: Knowledge Systems Institute Graduate School, 2006, pp. 708-713.

Meyer, A. Dienstleistungs-Marketing: Erkenntnisse und praktische Ergebnisse. München, Augsburg: 1992.

Object Management Group. (2004). *Business Process Modeling Notation (BPMN)*. Information.

Object Management Group, "Business Process Modeling Notation Specification - Version 1.2," Object Management Group, 9 A.D..

Pepels W., Marketing, 4 ed Oldenbourg Wissenschaftsverlag, 2004.

Ramaswamy R., "Design and Management of Service Process," 1996.

Scharp, M., Halme, M., & Jonuschat, H. Nachhaltige Dienstleistungen der Wohnungswirtschaft, Arbeitsbericht 9 ed. Berlin: IZT - Institut für Zukunftsstudien und Technologiebewertung, 2004.

Scheer, A.-W., & Thomas, O. "Geschäftsprozessmodellierung mit der ereignisgesteuerten Prozesskette,", 8 ed 2005, pp. 1069-1079.

Scheuing, E. E., & Johnson, E. M. (1989). A proposed model for new service development. *Journal of Services Marketing, 3*(2), 25–34. doi:10.1108/EUM0000000002484

Schneider, K., Wagner, D., & Behrens, H. "Vorgehensmodelle zum Service Engineering," in Service Engineering - Entwicklung und Gestaltung innovativer Dienstleistungen. Bullinger H.-J. and Scheer A.-W., Eds. Berlin, Heidelberg, New York, Barcelona, Hongkong, London, Mailand, Paris, Tokio: Springer, 2003, pp. 117-141.

Schüller, A. Dienstleistungsmärkte in der Bundesrepublik Deutschland. Köln: 1967.

"Service Engineering: Entwicklungsbegleitende Normung für Dienstleistungen," Beuth-Verlag, Berlin, 1998.

Shostack, G. L. (1982). How to Design a Service. *European Journal of Marketing, 16*(January/February), 49–63. doi:10.1108/EUM0000000004799

Shostack, G. L. "Designing Services that deliver,", 62 ed 1984, pp. 133-139.

Siegel J., "BPMN und BPDM - Die OMG-Spezifikationen zur Modellierung von Geschäftsprozessen," 2007.

Siegel J., "BPMN unter der Lupe," 2008.

Strang T. and Linnhoff-Popien C., "A Context Modeling Survey," 2004.

Tränkler, H.-R. (2001). In Tränkler, H.-R., & Schneider, F. (Eds.), *"Zukunftsmarkt intelligentes Haus,"* in Das intelligente Haus - Wohnen und Arbeiten mit zukunftsweisender Technik (pp. 17–34). Pflaum Verlag.

Weinzierl, T., & Schneider, F. (2001). In Tränkler, H.-R., & Schneider, F. (Eds.), *"Gebäudesystemtechnik,"* in Das intelligente Haus - Wohnen und Arbeiten mit zukunftsweisender Technik (pp. 349–361). Pflaum Verlag.

White, S. A. (2007). *Introduction to BPMN*. IBM Cooperation.

Zentralverband Elektrotechnik- und Elektroindustrie. Handbuch: Gebäudesystemtechnik. Grundlagen, 4 überarbeitete Auflage ed. Frankfurt am Main: Zentralverband Elektrotechnik- und Elektronikindustrie/Zentralverband der Deutschen Elektrohandwerke, 1997.

Zollner, G. (1995). *Kundennähe in Dienstleistungsunternehmen*. Wiesbaden: Empirische Analyse von Banken.

Chapter 13
Dynamic Ambient Networks with Middleware

Rebecca L. Willard
Waynesburg University, USA

Baoying Wang
Waynesburg University, USA

ABSTRACT

Ambient Intelligence is the concept that technology will become a part of everyday living and assist users in multiple tasks. It is a combination and further development of ubiquitous computation, pervasive computation, and multimedia.The technology is sensitive to the actions of humans and it can interact with the human or adjust the surroundings to suit the needs of the users dynamically. All of this is made possible by embedding sensors and computing components inconspicuously into human surroundings. This paper discusses the middleware needed for dynamic ambient intelligence networks and the ambient intelligence network architecture. The bottom-up middleware approach for ambient intelligence is important so the lower layers of all ambient intelligence networks are interchangeable and compatible with other ambient intelligence components. This approach also allows components to be programmed to be compatible with multiple ambient intelligence networks. The network architecture discussed in this paper allows for dynamic networking capabilities for minimal interruptions with changes in computer components.

INTRODUCTION

Ambient intelligence is ubiquitous, flexible, human-centered computer systems. The components are embedded into human environments

DOI: 10.4018/978-1-61692-857-5.ch013

to enhance life in a way natural to humans (Anastasopoulos *et al.*, 2005). Ambient intelligence is "a condition of an environment that we are in the process of creating" (Riva *et al.*, 2005). The technology becomes a part of everyday living and can assist the users in many everyday tasks, such as temperature control, kitchen inventory, smart shopping, or security alarms and controls. Technology is supposed to be hidden from the human's eyes and intuitively to use.

Ambient intelligence networks can interact with humans or adjust the surroundings to suit the needs of the user dynamicaclly. For example, an ambient network in a home can be programmed to lower the temperature of the house during the winter to save energy. However, the network will recognize when a user enters the network and will adjust the temperature settings to make the house more comfortable.

In order for dynamic use of ambient networks, the hardware and middleware structures must be compatible so devices can be found and used by different networks. The basic network architecture and middleware services must be universally compatible and dynamic so devices can leave and enter the network freely and use the services provided by the network when needed. The basic structure of the services provided by the network must be similar so the devices can be programmed to utilize the services of any ambient network.

This chapter discusses a bottom-up structure for middleware and hardware architecture for ambient networks. The bottom-up middleware approach for ambient intelligence is important so that the lower layers of all ambient intelligence networks are interchangeable and compatible with other ambient intelligence components. This approach also allows components to be programmed to be compatible with multiple ambient intelligence networks.

The middleware structure provides a logical view of the services an ambient intelligence network is supposed to provide (Grace, Blair and Samuel, 2003). The middleware services must be used universally to ensure compatibility across ambient intelligence networks and various hardware devices. The middleware services must be able to communicate, distribute events or actions to other components, lookup or search for components, find similar components, log sensor data, and find the location of components and users in a network.

This chapter also discusses general requirements and hardware topology needed to make ambient intelligence networks universally compatible. The four categories of hardware topology are discussed, which must be accounted for in ambient networks. The four major categories of hardware are fixed nodes, portable nodes, sensor/actuator nodes, and device nodes. The network architecture discussed in this chapter allows for dynamic networking capabilities for minimal interruptions with changes in computer components.

BACKGROUND

Ambient intelligence is a further development of the field of ubiquitous computation, pervasive computation, and multimedia. Ambient intelligence research covers complete platforms, development methods, and many other features (Vasilakos and Pedrycz, 2006; Bartelt *et al.*, 2005; Funk, Kuhmunch and Niedermeier, 2005). The research field of ambient intelligence starts to spread rapidly and first applications are on their way (Noguchi, Mori and Sato, 2003; Jammes and Smit, 2005; Kahn, Katz and Pister, 2000; Kirk and Newmarch, 2005). For example, an intelligent refridgerator communicates to a consumer in the supermarket on which food is needed to cook a certain dish. The ambient intelligient system consists of many services: an online shop which provides cooking recipes, networking which enables communication between different components, RFID (Radio-frequency identification) readers in the refridgerator which are used to search for the right food and guide the consumer to the corresponding locations. The consumer can simply interact with a monitor on his shopping car (Lugmayr, 2006) to accomplish the smarting shopping process.

There are many different definition of ambient intelligence based on the different perspectives.

Wikipedia defines ambient intelligence as following: "In computing, ambient intelligence (AmI) refers to electronic environments that are sensitive and responsive to the presence of people. Ambient intelligence is a vision on the future of consumer electronics, telecommunications and computing that was originally developed in the late 1990s for the time frame 2010–2020. In an ambient intelligence world, devices work in concert to support people in carrying out their everyday life activities, tasks and rituals in easy, natural way using information and intelligence that is hidden in the network connecting these devices. As these devices grow smaller, more connected and more integrated into our environment, the technology disappears into our surroundings until only the user interface remains perceivable by users" (WIKIPEDIA, "Ambient Intelligence").

According to ISTAG (Riva *et al.*, 2005), ambient intelligent system consist of two essential components: the ambient component addresses hardware technology as part of the natural environment of people; and the intelligence component defining software technology and the algorithmic component. The ambient component is "a system/environment/background view, including sensor technology obtaining data about the physical world, real-time embedded systems, ubiquitous communication ranging from pico-networks to mobile broadband access, I/O devices supporting touch-less control, and adaptive software" (Riva *et al.*, 2005). The intelligence component is "user/person/foreground view, that is including media management and handling, natural interaction, computational intelligence, contextual awareness, and emotional computation" (Riva *et al.*, 2005).

The topology of ambient intelligence is distribution, rather than a central database system (Hall and Cervantes, 2004). It is service oriented system. Many service oriented architecture specifications address the issue of service discovery and the issue of distant communication in distributed environments. In ambient intelligence, the analogue world mixes with the digital world.

The primary goal is that the technology of ambient intelligience systems is to be more and more hidden and naturally to use (Raverdy and Issarny, 2005; Lugmayr, 2006). Such a goal is achieved by layered approaches so that detailed networking, data manipulation and communication are located at lower layers and hidden to users.

AMBIENT MIDDLEWARE

The basis for ambient intelligent networks described in this section is a layered approach. The lower layers of a network are the less sophisticated layers that simply collect raw data. The middle and upper layers of a system provide varying degrees of computation and processing of the data (Ahlgren *et al.*, 2005).

A dynamic network is imperative to the purpose of an ambient intelligence (Bottaro and Anne, 2006). It is essential to create universally accepted and universally used control functions for ambient networks (Niebert *et al.*, 2004). To realize the goals and purposes of ambient networks, the middleware of different networks must be similar so both the lower layers and upper layers can be designed and programmed in a way that can utilize almost any device that may enter the network. One important goal of ambient network middleware services should be to program devices to sense changes in the ambient network environment and to adapt to the changes (Acampora and Vincenzo, 2008). However, due to the wide variety of uses and components of ambient networks, it will be hard to determine what changes each device should detect and it will be hard to update devices if the devices need to detect other changes.

This section describes a logical view of the middleware technical services needed for an ambient network. If this logical view is followed universally, networks can be designed to be compatible.

Communication Services

The first technical service of ambient middleware is called communication services (Anastasopoulos *et al.*, 2005). This service is the most basic service provided by the network. The communication service is the link or connection between the middleware network as a whole and the individual services or applications that the network uses. The network system must be able to communicate synchronously and asynchronously with each network function (Anastasopoulos *et al.*, 2005). Synchronous communication is direct communication, where all parties involved in the communication are present at the same time/event, while asynchronous communication does not require that all parties involved in the communication need to be present and available at the same time.

Event Services

The second middleware service is event services (Anastasopoulos *et al.*, 2005). An event is "the occurrence of some facts that can be perceived by or communicated to the ambient intelligence environment" (Acampora and Vincenzo, 2008). The event service is very similar to the communication services. This technical service sends a message to other selected ambient intelligence components within the network when an event occurs (Anastasopoulos *et al.*, 2005). If the status of one component changes, the event service notifies all other components that need this information (Anastasopoulos *et al.*, 2005). This service maintains the integrity of the information across the network. However, the messages are sent only to selective devices so the network does not use unnecessary resources, such as extra bandwidth. For example, in most ambient intelligence networks, changes in temperature control would not need to be sent to a Personal Digital Assistant (PDA), because the PDA would not store or process that information.

Each ambient intelligence network would need to be programmed independently to determine what events need to be sent to which component. All networks would need to determine what information should be sent to components that are new to the network. Different networks collect and process different information and that information needs to be sent to different places. The details of this service could not be universally programmed, but the basic service must be offered.

Lookup Service

The third service needed for ambient networks is a lookup service (Anastasopoulos *et al.*, 2005). This service is needed because of the dynamic nature of these networks. Devices can randomly join or leave the network, new devices may be installed, and old devices can be updated (Bottaro and Anne, 2006). Each of these devices may provide a different quality of service and specific properties (Bottaro and Anne, 2006). The lookup service has many functions. The main function is to provide a mechanism for the components connected to an ambient network to find each other (Anastasopoulos *et al.*, 2005).

Components register with the lookup service when they enter the ambient intelligence network (Bottaro and Anne, 2006). Applications or components have the capability to ask the lookup service what other services are available that suit their needs (Anastasopoulos *et al.*, 2005). The lookup service will find the possible services that fit the needs and it will determine which one will give the best service quality and is best suited for the inquiring application (Anastasopoulos *et al.*, 2005). The lookup service even goes one step further and notify an application or component if a different service that is more suitable becomes available (Anastasopoulos *et al.*, 2005). For example, if a PDA is the most suitable device for a notification, but there are no PDAs in the network, the application will start by using a different device. If a PDA enters the network, the lookup service will notify the application that a PDA became available.

Similarity Search Services

The fourth middleware service is the similarity search service. The similarity search service is comparable to the lookup service. Sometimes in ambient networks, an application wants to use a specific type of device. Due to the dynamic nature of ambient networks, however, the type of device may not be available. The similarity search service is only used in these situations. This service will determine what component is available that is the most similar or will work the best for the desired task (Anastasopoulos *et al.*, 2005). It is a lookup service that is more sophisticated and can compare the functions and quality of service for different devices.

One issue with this service that needs to be addressed is what would happen in a situation where there are no components to complete the task. In dynamic networks, it could be possible that none of the available devices will work for the desired task. Since the bottom-up approach does not discuss this situation, each individual network will need to address and plan for this problem independently, because it is a problem that will have many different answers and will need to be based on personal preferences and needs.

Logging Service

The fifth service is a logging service. The logging service is designed to be used for data received from sensors (Anastasopoulos *et al.*, 2005). In some ambient networks, such as in an assisted-living home or store, an analysis of past data can be helpful. The logging service stores sensed data over time. This data is entered into a file of the system's history so other services or users can perform evaluations or calculations of the data (Anastasopoulos *et al.*, 2005). This service is similar to decision support systems or data warehouses

in other systems. It helps ambient networks learn from the experience and the usage pattern over the time in order to provide more personal services.

The data collected for a logging service would need to be sent and stored on a fixed node such as a stationary desktop or server. These devices have the greatest storage and computing capacity and would be most able to collect and process the data efficiently without interfering in other services.

Context Services

The final middleware aspect for ambient intelligence networks is called context service. Ambient network applications must be aware of the location of users and hardware components (Anastasopoulos *et al.*, 2005). For example, the network and applications must know if the user is within the range of an alarm so it knows whether it can use that alarm to notify you (Anastasopoulos *et al.*, 2005).

The context service must determine two types of locations – absolute and relative. Absolute localization determines the physical location of a person or component (Anastasopoulos *et al.*, 2005). For example, it determines the city, building, floor, and room where a person is located. Relative localization determines what ambient network components or users are nearby (Anastasopoulos *et al.*, 2005). For example, if a signal bell or alarm needs to be used, the context service might determine what alarms or alert lights are the closest to the user.

AMBIENT NETWORK ARCHITECTURE

Ambient intelligence networks must be created to recognize and use any device that may enter the network. Therefore, the networks and devices must be programmed similarly to be universally compatible. This section identifies and describes some of the architectural requirements needed to create a universal platform for ambient intelligence networks and it also describes the hardware topology that is present in ambient intelligence networks.

General Architecture Requirements

Ambient intelligence networks must all follow a universal framework to ensure congruency (Nieber *et al.*, 2004). Having a framework that is followed in every ambient intelligence network will allow for "building a unified communication environment out of the resources of individual networks" (Nieber *et al.*, 2004). This will allow for devices, services, and applications to be compatible on any ambient network.

The composition of ambient intelligence networks must also be able to operate around technology constraints that may arise (Nieber *et al.*, 2004). For example, if a network is programmed to use a specific device, it must be able to continue to work in spite of that specific device failing or leaving the range of the network.

Ambient intelligence networks should also be designed to allow new functions to be added (Nieber *et al.*, 2004). Programming networks to allow for new functions or changes to be added is important because it allows users to take existing networks and update them to meet their needs. Networks also needs to be flexible and changeable because of the rapid changes in technology. If networks are programmed to allow adding new functions, networks can be updated without having to re-write large amounts of code. If this is followed, technology companies can also build a general network and allows the users or business to customize their network based on their own flexible needs.

Hardware Topology

A hardware device that can be used in an ambient intelligence network is a node. As in other networks, each node is connected over the network by a link and the links form a path through which information passes between two nodes (Ahlgren *et al.*, 2005). Each node in a specific group is able to communicate with other nodes within the network range from that group (Anastasopoulos *et al.*, 2005). This section identifies and describes the four main kinds of node groups available in an ambient intelligence network.

Fixed Nodes Group

The first hardware device group type is called the fixed nodes group or FNG (Anastasopoulos *et al.*, 2005). Devices that belong to this group are in a fixed location and, therefore, form the fixed infrastructure of the ambient network. Fixed nodes are equipped with exceptionally high powers of computation and a large bandwidth for network communication (Anastasopoulos *et al.*, 2005). However, these nodes are in a fixed location due to power constraints or network wires (Anastasopoulos *et al.*, 2005).

This group must form the basis of the ambient network because they are the only stable group. FNG devices are unlikely to leave the network but devices in the other groups are dynamic and can enter and leave the network without restraint or are only used in conjunction with another device. The majority of the devices in this group are desktop computers (Anastasopoulos *et al.*, 2005).

Portable Nodes Group

The second type of hardware device group that is included in the ambient network architecture is the portable nodes group or PNG (Anastasopoulos *et al.*, 2005). In some ways, these are similar to the fixed node group devices. Portable node group devices have greater power of computation than the two groups described below. These devices also have a large bandwidth for network communication (Anastasopoulos *et al.*, 2005). The major difference in computing capabilities between portable node group devices and fixed node group is that portable devices have fewer functions or uses than fixed devices (Anastasopoulos *et al.*, 2005). Portable node group devices are limited in the number of utilities or applications.

However, unlike fixed node group devices, portable node group devices are mobile and can enter and leave the network at anytime (Anastasopoulos *et al.*, 2005). These devices are not a stable part of ambient network architecture, so the network cannot depend on their presence to function. Some common portable node group devices are notebook computers, pocket personal computers, PDAs, or other handheld devices (Anastasopoulos *et al.*, 2005).

Sensor/Actuator Nodes Group

The third ambient network architecture group is called the sensor/actuator nodes group or S/ANG. Sensor/actuator nodes have very specific and limited computing functions (Anastasopoulos *et al.*, 2005). Sensor/actuator node group devices either collect or process data from sensors within the network and communicate by sending this data to another network component, such as a desktop computer, that can further process the information (Anastasopoulos *et al.*, 2005).

These devices can be fixed or mobile (Anastasopoulos *et al.*, 2005). An example of a sensor/actuator node would be a device that collects information from a sensor monitoring the temperature of a house.

Device Group

The fourth group of ambient network architecture is called the device group or DG (Anastasopoulos *et al.*, 2005). Devices in this group have no computing power. There are two main functions of device group components: to collect context information and to replicate another application's output. The most unusual characteristic of this group is that these devices must be used in

conjunction with or incorporated through another component (Anastasopoulos *et al.*, 2005).

Some examples of devices in this group are Bluetooth headsets, sensors, or a video screen (Anastasopoulos *et al.*, 2005). None of these components can do anything without using a component from another group. For example, a video screen cannot display information without being connected to a computer.

This is a very general view of the hardware topology. In order to create ambient networks, programmers must look at the differences and similarities among different devices and brands. The small differences in devices and brands can make a big difference in writing programs that will work on all devices.

The middleware and hardware groups described in this paper are needed to create dynamic, compatible ambient intelligence networks. Each network needs to be programmed and designed to work with devices from each hardware group and every ambient network needs to provide the same middleware services. Only with universal middleware services and guidelines can the purpose of dynamic ambient networks be successful. Without a basic structure, ambient networks cannot be compatible and devices used in one network may be useless in another network.

CONCLUSION

In this paper, a logical view of a bottom-up approach to ambient intelligence middleware was presented. The middleware services must be used universally to ensure compatibility across ambient intelligence networks and various hardware devices. The middleware services must be able to communicate, distribute events or actions to other components, lookup or search for components, find similar components, log sensor data, and find the location of components and users in a network.

This logical view of ambient intelligence in middleware is one of very few that has been defined. The view appears to cover all of the important services needed for every ambient network, but the structure must be tested to determine if this bottom-up, layered structure will work as it needs to.

This paper also presented the general requirements needed in every ambient intelligence network to allow for changes and compatibility. The four categories of hardware topology are also discussed, which must be accounted for in ambient networks. The four major categories of hardware are fixed nodes, portable nodes, sensor/actuator nodes, and device nodes.

The biggest challenge of creating ambient networks will be defining a standard that will be accepted and followed throughout the technology world. The implementation of ambient intelligence networks in everyday environments will take many years because even after a standard is established, it will take time for all the companies to start to follow the standard.

REFERENCES

Acampora, G., & Vincenzo, L. (2008). A Proposal of Ubiquitous Fuzzy Computing for Ambient Intelligence. *Information Sciences*, *178*(Issue 3), 631–646. doi:10.1016/j.ins.2007.08.023

Ahlgren, B., et al. "Ambient Networks: Bridging Heterogeneous Network Domains." Personal, Indoor and Mobile Radio Communications, 2005. PIMRC 2005. IEEE 16th International Symposium on (2005). Page(s): 937-941

Anastasopoulos, M., et al. "Towards a Reference Middleware Architecture for Ambient Intelligence." 20th ACM Conference on Object-Oriented Programming Systems, Languages and Applications. San Diego, 2005. Pages: 12-17.

Bartelt, C., Fischer, T., Niebuhr, D., Rausch, A., Seidl, F., & Trapp, M. "Dynamic Integration of Heterogeneous Mobile Devices." the Workshop in Design and Evolution of Autonomic Application Software (DEAS 2005), ICSE 2005, May 21, 2005, St. Louis, Missouri, USA. Pages: 1 - 7

Bottaro, A., & Anne, G. "Extended Service Binder: Dynamic Service Availability Management in Ambient Intelligence." First Workshop on Future Research Challenges for Software and Service. Vienna, Austria, April 2006. Pages: 383-388

Funk, C., Kuhmünch, C., & Niedermeier, C. "A Model of Pervasive Services for Service Composition", CAMS05: OTM 2005 Workshop on Context-Aware Mobile Systems, Agia Napa, Cyprus, October 2005. Pages 215-224.

Grace, P., Blair, G. S., & Samuel, S. "ReMMoC, A Reflective Middleware to Support Mobile Client Interoperability", International Symposium on Distributed Objects and Applications (DOA), Catania, Sicily, Italy, November 2003. pp. 1170-1187.

Hall, R. S., & Cervantes, H. (2004). Challenges in Building Service-Oriented Applications for OSGi. *IEEE Communications Magazine*, (May): 144–149. doi:10.1109/MCOM.2004.1299359

http://en.wikipedia.org/ wiki/ Ambient_intelligence retrieved on May 27, 2009.

Jammes, F., & Smit, H. Service-oriented paradigms in industrial automation. IEEE Transactions on Industrial Informatics. v1. 2005. pp. 62-70.

Kahn, J., Katz, R. and Pister K., Emerging Challenges: Mobile Networking for 'Smart Dust', J. Comm. Networks, Sept. 2000, pp. 188-196.

Kirk, R., & Newmarch, J. "A Location-aware, Service-based Audio System", IEEE Consumer Communications and Networking Conference, 2005. pp. 343-347.

Lugmayr, A. (2006). "The Future is 'Ambient'", Proceedings- Spie The International Society. *Optical Engineering (Redondo Beach, Calif.)*, *6074*, 60740J.

Niebert, N. (2004). Ambient Networks - An Architecture for Communication Networks Beyond 3G. *IEEE Wireless Communications*, *11*(Issue 2), 14–22. doi:10.1109/MWC.2004.1295733

Noguchi, H. Mori, T. and Sato, T. Network middleware for utilization of sensors in room, in Proceedings of IEEE/RSJ International Conference on Intelligent Robots and Systems, vol. 2, 2003, pp. 1832-1838.

Raverdy, P., & Issarny, V. "Context-aware Service Discovery in Heterogeneous Networks", IEEE International Symposium on a World of Wireless, Mobile and Multimedia Networks (WoWMoM 2005), June 2005. Pages: 478 – 480.

Riva, G., Vatalaro, F., Davide, F. & M. Alcañiz (2005). *Ambient Intelligence: From Vision to Reality* (pp. 45–48). IOS Press.

Vasilakos, A., & Pedrycz, W. (2006). *Ambient Intelligence, Wireless Networking, and Ubiquitous Computing*, (pp. 12–45). Norwood, MA: Artech House, Inc.

KEY TERMS AND DEFINITIONS

Hardware architecture: The identification of the physical components and their interrelationships, which allows hardware designers to understand how their components fit into a system architecture and provides software component designers important information needed for software development and integration.

ISTAG: Information Society Technologies Advisory Group is an advisory body to the European Commission in the field of Information and Communication Technology. It developed the ambient intelligence concept. The members are appointed as individuals rather than to explicitly represent the interests of a single group, although it could still be argued that its members largely represent big companies and research institutions.

Middleware architecture: The organization of the software layer that lies between the operating system and the applications on each site of the system.

Network architecture: The entire framework for the network's physical components and their functional configuration, including operational principles and procedures.

Node: A device attached to a computer network or other telecommunications network, such as a computer or router, or a point in a network topology at which processes intersect or branch.

PDA: Personal Digital Assistant is a mobile device, also known as a palmtop computer. An important function of PDAs is synchronizing data with a computer system. Today the vast majority of all PDAs are smartphones.

RFID: Radio-frequency identification is the use of an object (typically referred to as an RFID tag) applied to or incorporated into a product, animal, or person for the purpose of identification and tracking using radio waves.

Chapter 14
Using Context Awareness for Self–Management in Pervasive Service Middleware

Weishan Zhang
China University of Petroleum, P.R. China

Klaus Marius Hansen
University of Copenhagen, Denmark

ABSTRACT

Context-awareness is an important feature in Ambient Intelligence environments including in pervasive middleware. In addition, there is a growing trend and demand on self-management capabilities for a pervasive middleware in order to provide high-level dependability for services. In this chapter, we propose to make use of context-awareness features to facilitate self-management. To achieve self-management, dynamic contexts for example device and service statuses, are critical to take self-management actions. Therefore, we consider dynamic contexts in context modeling, specifically as a set of OWL/SWRL ontologies, called the Self-Management for Pervasive Services (SeMaPS) ontologies. Self-management rules can be developed based on the SeMaPS ontologies to achieve self-management goals. Our approach is demonstrated within the LinkSmart pervasive middleware. Finally, our experiments with performance, extensibility, and scalability in the context of LinkSmart show that the SeMaPS-based self-management approach is effective.

INTRODUCTION TO CONTEXT AWARENESS MIDDLEWARE

Pervasive computing (Henricksen & Indulska, 2006) is becoming more of a reality in everyday life. As a consequence, we face the question of how to handle the diversity of related computing facilities, ranging from powerful computers to small mobile phones and sensors, and how to tackle the diversity of network technologies. It is becoming

DOI: 10.4018/978-1-61692-857-5.ch014

urgent that we can develop and deploy various personalized pervasive service applications efficiently. Pervasive middleware is an underlying technology to address these challenges.

Context awareness (Satyanarayanan, 2001) is one of the key features in pervasive middleware and it is used to achieve openness and dynamism of pervasive systems, to promote knowledge sharing between pervasive systems, and to provide secured and personalized services for end users. These all depend on the knowledge on when and where a service can happen, what triggers a service provision and how to provide a service to whom. That is, to be aware of the current context and to take actions based on this context. This awareness of contexts depends heavily on the modeling approach adopted in the context-awareness implementation as context models decide their potentials to provide reasoning capabilities for achieving the awareness of contexts.

It has been discussed that Web Ontology Language (OWL) ontologies (OWL, 2009) have stronger context modeling capabilities than OWL's predecessor DAML+OIL and other counter-parts such as XML, key-value pairs, object-oriented models, and RDF (Strang & Linnhoff, 2004; Zhang & Hansen, 2008a). The main reason is that OWL ontologies are using the Open World Assumption (OWA), and openness is the nature of pervasive computing system, therefore OWL ontologies can provide better reasoning potentials and allows applications to intelligently react on a range of different contexts changes (Strang & Linnhoff, 2004; Zhang & Hansen, 2008a). In practice a choice of context modeling approach also depends on whether an approach is easy to use and efficient for resource-constrained devices, and therefore a hybrid of different context modeling approaches can be used, e.g., in the way we are doing in the Hydra EU project (Zhang & Hansen, 2008a), where a combination of key-value pair and OWL ontologies are used for context modeling as will be detailed later. The LinkSmart middleware is a product of the Hydra project, and is a pervasive

service middleware for networked and embedded devices based on service oriented architecture using pervasive web services.

The rest of the chapter is organized as follows: In "Middleware for Self Management in AmI," we will discuss the importance of context-awareness for self-management, and will survey current self-management pervasive middleware to compare their features. "Combining Context Awareness and Self Management in Pervasive Service Middleware" presents the design of the SeMaPS ontologies and our hybrid context-awareness framework. We next discuss self-management rule specification in "Self-Management Specification with the SeMaPS Ontologies using SWRL Rules." Following this, we show the LinkSmart self-management architecture, design and implementation, based on the SeMaPS ontologies in "Architecture and Implementation of Self-Management in LinkSmart." We demonstrate our approach with an example scenario in "A Self-Management Scenario with the LinkSmart Middleware," together with evaluations and discussions. Finally, we conclude the chapter.

MIDDLEWARE FOR SELF MANAGEMENT IN AMI

Recently, there has been a growing demand for high dependability of pervasive systems. When pervasive middleware is concerned, its manageability depends on whether it has some degree of self-management, as also is the case for Ambient Intelligence (AmI) (Kephart & Chess, 1996).

Self-Management in Pervasive Middleware

Self-management is an enabler of dependability, potentially leading to higher quality of pervasive systems. Self-management (Kephart & Chess, 1996) features constitute a broad list, including self-configuration, self-adaptation, self-optimization, self-protection, self-optimization, self-diagnosis, and self-healing. These are all important in achieving dependability for pervasive systems towards the vision of Ambient Intelligence.

Context-awareness can play a critical role in realizing self-management features. For example, self-protection can be aided by taking into consideration the current context a user is in and then choosing an appropriate security mechanism to protect information, considering the quality of service (QoS) of the security protocols and the corresponding security requirements. Therefore, it is natural and reasonable to extend context-awareness to support self-management.

In the next section, we will first present a survey on self-management features of existing pervasive middleware, and based on this, we will discuss shortcomings of the existing research work. Then we will motivate our contributions of this chapter.

Survey on Self-Management Capabilities for Pervasive Middleware

There are a number of existing pervasive middleware implementations that can provide a degree of self-management. We now briefly look at these middleware implementations and compare their self-management capabilities, including context models used, and whether the full spectrum of self-management is supported. As self-management is complex and sometimes difficult to achieve, self-management is often goal-based with a corresponding planning mechanism to help to make decisions on how to achieve that goal. Therefore, we will also investigate which planning algorithms are used in the surveyed middleware. We compare context model capabilities according to the work in (Strang & Linnhoff, 2004; Zhang & Hansen, 2008a).

GLOSS (GLObal Smart Space) (Coutaz et al., 2003) is an infrastructure to support people's interaction with their surroundings in a smart way, in which location, person, time and activity

are modeled with pure XML mark up, which has very limited capabilities for expressing complex contexts and deriving new contexts (Strang & Linnhoff, 2004; Zhang & Hansen, 2008a). The Context Toolkit (Dey, 2000) applies key-value pairs to model contexts, which has low formality for contexts in a precise and traceable manner, and has low context reasoning capabilities. Furthermore, the specific format makes knowledge sharing in Internet-scale computing difficult. CASS (Context-awareness sub-structure) (Fahy & Clarke, 2004) uses the relational model for context modeling. The relational model used by CASS has limited reasoning capabilities when compared to the requirements of AmI.

Hydrogen (Hofer et al., 2003) adopts object-oriented context models, and uses a peer-to-peer style for context information sharing and exchange. The object-oriented context model is not intrinsically consistent with the nature of openness of pervasive systems. The Context Managing Framework (Korpipaa at al., 2003) used RDF as a context model, which has limited reasoning capabilities and expressive power compared with OWL. In Gaia (Ranganathan et al., 2003), DAML+OIL, a predecessor of OWL, was used to model contexts. Among its weakness compared to OWL is that DAML+OIL can not define symmetry of properties. Gaia did not explicitly model dynamic contexts, e.g the status of devices and components, which are vital to conduct self-management activities. It has STRIPS-based planning which decides how goals are to be achieved (Ranganathan & Campbell, 2004).

The CARISMA middleware (Capra et al., 2003) applies a sealed-bid auction approach used in micro-economics to dynamically resolve policy conflicts during context changes. It adopts a kind of object-oriented model for context modeling: it defines wrappers to sensors and monitors encompassing event messages that deliver context information according to triggers/policies set by the applications. This object-oriented context model limits its reasoning capability for awareness of contexts, and henceforth for the achievement of self-management.

Utility functions are used for planning in Aura (Sousa et al., 2006) and they are composed of three parts: configuration preferences, supplier preferences, and QoS preferences. An object-oriented context model is used: a context observer collects context, and reports context changes to task managers. ReMMoC (Grace et al., 2005) uses architecture specifications for reconfigurations, and the ReMMoC service discovery component reconfigures itself at run time. However, ReMMoC does not support planning and there is no explicit context model in ReMMoC either. The modeling of dynamism was not covered by Aura and ReMMoC, which is important to achieve self-management goals.

CoBrA (Context Broker Architecture) (Chen et al., 2003) is implemented using software agents, and OWL ontologies are used for context modeling, but CoBrA does not consider the dynamism of pervasive systems. SOCAM (Service-oriented Context-Aware Middleware) (Gu et al., 2005) has a distributed structure where OWL ontologies are used to hierarchically model contexts. Amigo (Mokhtar et al., 2006) is dedicated to smart home applications, where context ontologies defined with OWL are used for achieving personalized service provisioning and adaption. Dynamic contexts for devices and self-management are not considered in CoBrA, SOCAM, or Amigo.

The MUSIC project is applying utility functions for planning in self-adaption (Alia et al., 2007; Rouvoy et al., 2008). The overall utility function is a weighted sum of a set of one-dimensional utility functions, where the weight corresponds to the importance of each QoS dimension preferred by a user. We have also found that it is hard to précising give weight for different dimensions, which may limit the usability of this utility based approach. MUSIC is concentrating on self-adaption/configuration, but not other self-management properties such as self-protection.

Now we summarize our survey in Table 1, in which we characterize a pervasive middleware based on its self-management capabilities, the planning algorithms used, whether that middleware is considering the dynamism of devices, and its reasoning capabilities (based on the reasoning on context models of the middleware). By dynamism of devices, we mean that the status of a device (device aliveness and other device runtime status), the corresponding events associated with a device, and also messages exchanged between these devices and services running on these devices. From the introduction in "Introduction" and this survey, we conclude that:

- Dynamism should be explicitly modeled and handled for achieving self-management capabilities as pervasive systems are intrinsically dynamic
- Existing pervasive middleware implementations are not addressing the full spectrum of self-management capabilities: only a few are focusing on one or two aspects of self-management
- To improve the reasoning capabilities for achieving context-awareness and other self-management capabilities, OWL context models are useful

Therefore in this chapter, we will present our work on addressing the above problems. Our main contributions are:

- Proposing the extension of context-awareness for supporting self-management
- Presenting the design of the SeMaPS context ontologies to support context-awareness-based self-management that features the modeling of dynamism of pervasive service systems
- Proposing a hybrid context-awareness framework based on SeMaPS
- Presenting Semantic web-enabled self-management using SWRL (Semantic

Table 1. Comparing self-management pervasive middleware

	Self-management capabilities	Planning algorithms	Device dynamism modeling	Context model
SOCAM	NA (Not Available)	No	No	OWL
CASS	NA	No	No	Relational data
Context Toolkit	NA	No	No	Key-value pair
Hydrogen	NA	No	No	Object-oriented
CoBRA	NA	No	No	OWL
GLOSS	NA	No	No	XML
ReMMoC	Self-configuration	No	No	No
Gaia	NA	STRIPS-based planning	No	DAML+OIL
CARISMA	NA	Sealed-bid auction	No	Object-oriented
Aura	NA	Utility function	No	Object-oriented
Amigo	NA	No	No	OWL
MUSIC	Self-adaption/self-configuration	Utility function, Genetic algorithms (under investigation)	No	OWL

web Rule Language)[1] rules based on the SeMaPS context ontologies

- Illustrating the above ideas with a usage scenario and evaluating our approach.

COMBINING CONTEXT AWARENESS AND SELF MANAGEMENT IN PERVASIVE SERVICE MIDDLEWARE

In this section, we present the SeMaPS context ontologies that are modeling the dynamism of pervasive systems, and provide extensible features to build the support for the full spectrum of self-management capabilities. A hybrid approach for context-awareness is furthermore proposed.

Context Modeling with the SeMaPS Ontologies

Context can be static or dynamic (Zhang, Hansen & Kunz, 2009). For example, a person has a birth date, which is static and a device has CPU information that is also static. But the available device memory and the remaining battery are dynamically changing. These dynamic contexts are critical to take self-management actions and make decisions for self-management planning. For example, when device battery is low, some measures may need to be taken to save power, an example of which can be switching communication protocol from TCP to Bluetooth.

SeMaPS Ontologies Structure and Relationships

There are quite a few context ontologies proposed by researchers, for example the SOUPA (Chen et al., 2003) and Amigo (Mokhtar et al., 2006) ontologies. But these ontologies have not considered the necessary dynamic context as these ontologies were not designed to support self-management. Thus, there are no specific concepts in them to support self-diagnosis and other self-management capabilities. In the SeMaPS ontologies, special ontologies are designed to cater for dynamic context, and different context types are considered, including user context, network context, and environment context. These are used further to develop more complex context and diagnosis/

monitoring rules using SWRL. In practice, we restrict our ontologies to OWL-DL to guarantee tractable reasoning within acceptable performance limits. The structure and relationships between ontologies are shown in Figure 1.

SMaPS Ontologies Design

To model devices in pervasive systems, we designed a set of ontologies. First, we use a Device ontology to model high-level device information such as Device (as a concept) type classification (e.g. mobile phone, PDA, or thermometer) as in the Amigo project ontologies (Mokhtar et al., 2006) and manufacturer information. To facilitate self-diagnosis, we model a system with a System concept as composed of devices that provide services. A corresponding object property hasDevice is added, which has the domain System and range Device. The Device concept has a data-type property currentMalfunction which is used to

store the inferred device malfunction (including device error) diagnosis information at runtime. To model device capabilities including software platform and hardware platform that come with a device, we have developed a SoftwarePlatform and a HardwarePlatform, in which virtual machines, CPU, and other software/hardware information is modeled.

To model where events can happen, we designed a set of Location ontologies, in which different ways of location modeling are modeled in separate ontologies, namely SetBased, SemanticBased, CoordinateBased and GraphBased inspired by (Flurry, 2004). To model when things can happen, a Time ontology is used to model time- and date-related knowledge, for example a time zone concept. The Time ontology is developed based on OWL-Time [2].

A Person ontology is used to model a user in pervasive systems. It primarily encodes information about a person's preferences. Depending on

Figure 1. SeMaPS ontologies structure

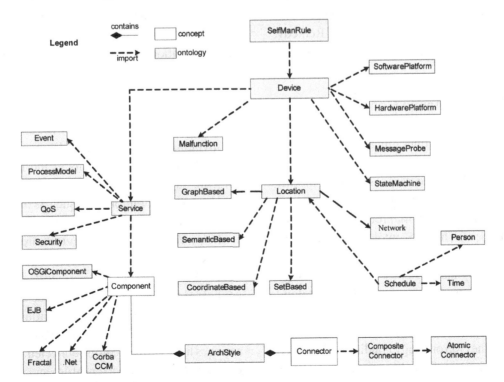

the domain to be used (Health care, Agriculture, Smart building as targeted by LinkSmart), different set of properties are defined. For example, the hasHome property is defined for Smart building domain, which is then used to monitor the distance of a person from his home in order to adopt different security policies at home.

In the Security ontology (Zhang et al., 2009), protection goals are defined with instances of the SecurityObjective class in order to describe which configuration of the system is suited to achieve certain protection goals. The concept Security-Protocol represents the actual configuration that could be applied to the system. This concept is the only one whose instance refers to specific software modules or settings. The resource consumption of each instance is represented by the properties requiresComputingPower and requiresMemory.

To facilitate self-diagnosis and self-healing, knowledge of malfunctions and recovery resolutions is encoded in a Malfunction ontology (called Error ontology historically). It classifies device malfunctions into two categories: Error (including complete device failure) and Warning (including function scale-down, and plain warning). Two further concepts, Cause and Remedy, are used to describe the origin of a malfunction and its resolution.

Modeling of Dynamic Contexts

The QoS ontology defines important QoS parameters, such as goodput, round trip time (RTT), reliability, and service response time. Another important QoS parameter is the power and memory consumption for a network protocol and for a security protocol. The QoS ontology also contains properties for these parameters, for example their nature (dynamic, static). The QoS ontology is developed based on the Amigo QoS ontology (Mokhtar et al., 2006). QoS is associated with a profile of a service that is provided by a device. Therefore, the QoS ontology is indirectly imported by the Device ontology. Moreover, a service is

realized by one or some software components, possibly following a software architecture style. Therefore, all these ontologies are linked with the Device ontology, directly or indirectly imported.

Commonly, embedded devices are designed and run as state machines. To model device state changes, a StateMachine ontology has been developed based on (Dolog, 2004). A State concept has data-type property isCurrent to indicate whether a state is current or not for device status monitoring, a doActivity object property is used to specify the corresponding activity in a state, and also a data-type property hasResult is defined to check a corresponding execution result at runtime of that activity. Additionally, three extra data-type properties are used to model historical action results, which are then use to conduct history-based self-management (the most current results are also usually used).

To model invocations of services, a MessageProbe ontology is developed and used in the monitoring of the liveness of a device/component/service. The SocketProcess concept is used to model a process running in a client or service, and SocketMessage models a message sent between a client and a service when TCP/UDP communication is used. There is also a concept called IPAddress, which is related to Device with a property hasIPAddress in the Device ontology. Some object properties including invoke, messageSourceIP, and messageTargetIP are used to build the invoking relationships, and data type property initiatingTime is used to model the timestamp for a message. The MessageProbe ontology may also be used to monitor QoS, such as the request/response time of a corresponding service call.

In pervasive systems, services are dynamic and they can join and leave anytime. This makes it necessary to ensure that services/components connected are matching each other and that overall constraints are followed, e.g. exhibiting a certain architecture style, for a new configuration after services have connected/disconnected. A Service ontology (using a service profile) and a Process-

Model ontology (using a service process model) are used to model services and service dynamisms, similar to the idea in the OWL-S[3] ontologies for Semantic web services. The Service ontology and ProcessModel ontology can be used at runtime to conduct self-configuration and self-adaptation, based on quality-of-service requirements (supported by the QoS ontology).

To validate software architecture at runtime, SeMaPS includes a set of software architecture asset ontologies, including an ontology for software architecture styles, an ontology for the OSGi[4] component model and ontologies for other popular component models, and ontologies for software connector ontologies. These ontologies are used to validate an architecture style at run time, and to identify interface mismatches during service composition.

To model events that can be sent and received at runtime, we have designed an Event ontology. An Event is composed of one Topic and several Key-Value pairs. The main type of events is StateChange events that represent the state changes of devices and components. Properties like the frequency, key name, and the corresponding value are defined to know the details of an event. The Event ontology is also used at development time to help developers know what kind of events that a device can trigger and process.

The OSGi component model is a widely used component models, also in pervasive computing. Our OSGiComponent ontology is based on OSGi DS (Declarative Services) specification (OSGi, 2007). It specifies the Component's dynamic status, for example whether it is enabled, and also static characteristics such as its reference to other service, its implementation interface, and services provided.

In summary, the dynamic contexts are modeled with runtime concepts and properties in the related ontologies, mainly the StateMachine ontology, MessageProbe ontology, Event ontology, Malfunction ontology, Component ontologies, QoS ontology, Service and ProcessModel ontology,

and other concepts and properties in the Device ontology, such as currentMalfunction and System. The currentMalfunction is used to store the current diagnosis information for the malfunction.

Figure 2 shows partial details of the Device ontology, StateMachine ontology, MessageProbe ontology, Malfunction ontology, OSGiComponent ontology in order to facilitate the understanding of rules shown in "Self-Management Specification with the SeMaPS Ontologies using SWRL Rules."

In the next section, we will describe the architecture and design of our context-awareness framework. A hybrid context modeling approach will be used considering the capabilities of pervasive devices, and a hybrid approach for context provision. Key-value pairs can be used for resource-constrained devices, and OWL/SWRL ontologies for powerful devices. Both SPARQL and SWRL query are used for push-based context provision in the hybrid context framework.

A Hybrid Context-Awareness Framework

As a pre-requisite for context-awareness, context data acquisition is essential. First, we clarify what data is going to be collected, how this data is going to be used. This data includes:

- **Sensor data:** dynamically sensed sensor data, e.g., GPS data and static meta-data, e.g., unit and precision.
- **Resource data:** hardware and software of a device, both dynamic (e.g., available memory) and static (e.g., CPU information)
- **Network data:** dynamic data including latency and other dynamic network properties, static properties such as the capacity of the network
- **Property data:** characteristics of services and components such as platform information, dependencies, etc. and QoS properties of them

Figure 2. Partial details of the SeMaPS ontologies

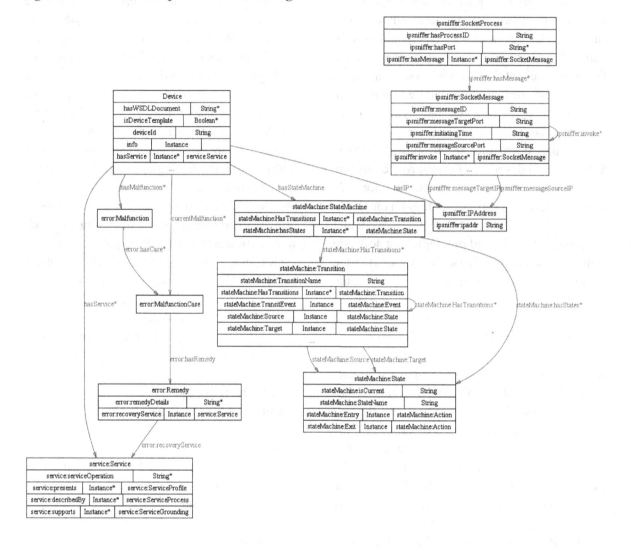

- **User data:** the profile of a user including preferences
- **Security data:** (virtualized) identification of a user or device and security protocols
- **State data of devices:** for example measuring, stop, or start

Corresponding to the above data acquisition list and the requirements from self-management components, we propose the context-aware self-management framework shown in Figure 3. First, contexts should be acquired (sensed), from context sources (including devices, services, sensors,

components). If necessary, this sensed raw data should first be normalized for uniformity. After the context data is normalized, this data should be formed as context and related context models updated (in the form of OWL/SWRL individuals using the ContextOntologyUpdate component). These context changes should trigger events, including internal events for ontology update (e.g., instance changes), external events published on a specific topic (e.g., "LocationChange"). These external events can then be listened to by context consumers to be used to trigger actions and/or further events. The internal events can be used

Figure 3. Architecture of a hybrid Context-awareness framework

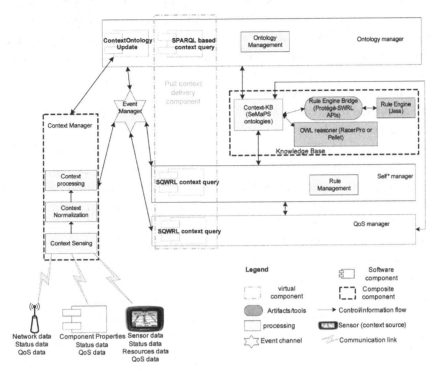

to trigger execution of SWRL reasoning as in self-management, and then this inferred information will be interpreted to check whether further actions are needed.

The approach requires the existence of a component for context management ("Context Manager") in the figure, but is decoupled from a specific component through eventing. The design rationale of the Context Manager and the preliminary context awareness approach as implemented in the LinkSmart middleware is described in (Badii et al., 2008).

The awareness of contexts for self-management in LinkSmart is using the SeMaPS ontologies-based approach, which is realized either as queries to the context knowledge base (which are the SeMaPS ontologies), which can be considered as pull approach for context provision, or as parsing events that contains key-value pair contexts, which can be considered as push approach for context

provision. We can have multiple approaches for the pull mode of context provision.

- SPARQL queries for context provision that are realized in LinkSmart by both the Protege-OWL APIs and the Jena[5] APIs.
- SQWRL[6] queries implemented using the Protege-OWL/SWRL APIs

SQWRL is used as the primary way of context provision and has the advantage over SPARQL that is has queries with OWL-specific semantics, while SPARQL is operating over RDF triples.

SELF-MANAGEMENT SPECIFICATION WITH THE SEMAPS ONTOLOGIES USING SWRL RULES

A self-management scheme is used to manage self-management activities. The scheme can usually be concretized from a general case: if certain condition is true, then a consequence can happen (i.e. deriving a result). This consequence can then be related to a certain event that triggers corresponding actions to meet self-management goals. In the cases that have been investigated, the conditions are all related the contexts where self-management scheme works in. As we have modeled all contexts in the SeMaPS ontologies, we will choose an approach that is related to the Semantic web, and is capable of specifying such self-management cases in a flexible way.

SWRL is a W3C recommendation for the rule language of the Semantic web, which can be used to write rules to reason about OWL individuals and to infer new knowledge about those individuals. An SWRL rule is composed of an antecedent part (body), and a consequent part (head). Both the body and head consist of positive conjunctions of atoms. A SWRL rule states that if all the atoms in the antecedent (body) are true, then the consequent (head) must also be true. SWRL is built on OWL-DL and shares its formal semantics: conclusions reached by SWRL rules have the same formal guarantees as the conclusions reached using standard OWL constructs[7]. Therefore, SWRL is an ideal candidate for us to choose.

In our work, all variables in SWRL rules bind only to known individuals in an ontology in order to develop DL-Safe rules that are decidable. SWRL provides built-ins such as math, string, and comparisons that can be used to specify extra constraints, which are not possible or very hard

to achieve by OWL itself. In an SWRL rule, the symbol "∧" means conjunction, "?x" stands for a variable, "→" means implication, and if there is no "?" in the variable, then it refers to a specific OWL instance.

Self-Diagnosis Rules

We realize two types of self-diagnosis rules (Zhang & Hansen, 2008b): device-level rules and system/application-level rules. Device-level rules are used for a certain type of devices and are usually generic for that type of devices, for example, if the battery level of a mobile phone is lower than 10%, a "low battery" event could be sent. The device-level rule can also be application specific in some cases. System/application-level rules can consider multiple devices, services, and subsystems to make an overall diagnosis of an underlying system. For example from a trend analysis, a diagnosis can be made that some devices are working correctly.

Now we show a device-level rule and a system level rule (Box 1) that are parts of a pig farm monitoring system (cf. "A Self-Management Scenario"). This rule is used to monitor the most recent value measured by a flow meter. If is the value is suddenly greater than 16 gallons per second, then it means that the pump is broken:

Now we present an application-level rule (Box 2).

Here, thermometers named PicoTh03_Outdoor and PicoTh03_Indoor respectively are used to

Box 1.

```
                Rule: PigFarmFlowmeter
device:FlowMeter(?device) ∧
device:hasStateMachine(?device, ?statemachine) ∧
statemachine:hasStates(?statemachine, ?state) ∧
statemachine:doActivity(?state, ?action) ∧
statemachine:actionResult(?action, ?result) ∧
abox:isNumeric(?result) ∧
swrlb:greaterThan(?result, 16.0)
  → sqwrl:select(?result) ∧
error:currentMalfunction(device:Flowmeter, error:PumpBroken)
```

Box 2.

```
Rule: PigFarmVentilatingDown
device:hasStateMachine(device:PicoTh03_Outdoor, ?statemachine) ∧
statemachine:hasStates(?statemachine, ?state) ∧
statemachine:doActivity(?state, ?action) ∧
statemachine:actionResult(?action, ?result) ∧
statemachine:historicalResult1(?action, ?result1) ∧
statemachine:historicalResult2(?action, ?result2) ∧
statemachine:historicalResult3(?action, ?result3) ∧
swrlb:add(?tempaverage, ?result1, ?result2, ?result3) ∧
swrlb:divide(?average, ?tempaverage, 3) ∧
swrlb:subtract(?temp1, ?result, ?result1) ∧
swrlb:subtract(?temp2, ?result1, ?result2) ∧
swrlb:subtract(?temp3, ?result2, ?result3) ∧
swrlb:add(?temp, ?temp1, ?temp2, ?temp3) ∧
swrlb:greaterThan(?average, 15.0) ∧
swrlb:lessThan(?average, 30.0) ∧
swrlb:lessThan(?temp, 0) ∧
device:hasStateMachine(device:PicoTh03_Indoor, ?statemachine_b) ∧
statemachine:hasStates(?statemachine_b, ?state_b) /\
statemachine:doActivity(?state_b, ?action_b) ∧
statemachine:actionResult(?action_b, ?result_b) ∧
statemachine:historicalResult1(?action_b, ?result1_b) ∧
statemachine:historicalResult2(?action_b, ?result2_b) ∧
statemachine:historicalResult3(?action_b, ?result3_b) ∧
swrlb:subtract(?temp1_b, ?result_b, ?result1_b) ∧
swrlb:subtract(?temp2_b, ?result1_b, ?result2_b) ∧
swrlb:subtract(?temp3_b, ?result2_b, ?result3_b) ∧
swrlb:add(?temp_b, ?temp1_b, ?temp2_b, ?temp3_b) ∧
swrlb:greaterThan(?temp_b, 0)
 → error:currentMalfunction(device:VentilatorMY0193, error:PowerDown)
```

measure both an indoor and an outdoor temperature. In the season that has conformable temperature, i.e., when the outdoor temperature is between 15 and 30 degree, the indoor temperature should follow the same trend as the outdoor temperature; otherwise, we can infer that the ventilator is down. The processing of this rule will obtain the trends via the difference of continuous temperature measurements of both the indoor and outdoor temperatures. If the trends are different, an instance of the currentMalfunction property ("VentilatorDown") of concept MalDevice (which is VentilatorMY0193) will be inferred. Then the Malfunction ontology will be checked for the resolution of the problem based on the malfunction cause. In our case, the Malfunction ontology would give us the suggestion: power supply off because a fuse is blown.

When conducting diagnosis, it is not possible to determine precisely that an error has a certain cause; it is a probabilistic situation in essence. Existing Semantic web languages are based on classical logic that cannot easily represent uncertainty. We are currently using a simple solution for building probability directly into SWRL rule by a data type property currentMalProbability if needed.

Self-Configuration Rules

The runtime architecture of pervasive service systems can be dynamic when new services are composed, and/or services join or leave the system. Therefore, it may be important that for these configurations, it can be validated that certain architecture styles can be followed, and interfaces

Box 3.

```
           Rule: Arch_InterfaceMismatch
archstyle:CurrentConfiguration(?con) ∧
archstyle:hasArchitecturePart(?con, ?comp1) ∧
osgi:componentName(?comp1, ?compname1) ∧
osgi:reference(?comp1, ?ref1) ∧
osgi:cardinality(?ref1, ?car1) ∧
swrlb:containsIgnoreCase(?car1, "1.") ∧
osgi:interface(?ref1, ?inter1) ∧
osgi:interfaceName(?inter1, ?name1) ∧
component:componentServiceDetails(?comp1, ?pr1) ∧
service:presents(?pr1, ?prservice1) ∧
profile:has_process(?prservice1, ?process1) ∧
process:realizedBy(?process1, ?aprocess1) ∧
process:hasInput(?aprocess1, ?input1) ∧
process:parameterValue(?input1, ?ivalue1) ∧
archstyle:hasArchitecturePart(?con, ?comp2) ∧
architectureRole(?comp2, ?role2) ∧
archstyle:archPartName(?role2, ?rolename) ∧
swrlb:equal(?rolename, "Repository")  ∧
osgi:service(?comp2, ?ser2) ∧
osgi:provide(?ser2, ?inter2) ∧
osgi:interfaceName(?inter2, ?name2) ∧
osgi:componentName(?comp2, ?compname2) ∧
component:componentServiceDetails(?comp2, ?pr2) ∧
service:presents(?pr2, ?prservice2) ∧
profile:has_process(?prservice2, ?process2) ∧
process:realizedBy(?process2, ?aprocess2) ∧
process:hasInput(?aprocess2, ?input2) ∧
process:name(?aprocess2, ?proname2) ∧
process:hasOutput(?aprocess2, ?proout2) ∧
process:parameterValue(?input2, ?inputtype2) ∧
process:parameterValue(?proout2, ?ovalue2) ∧
swrlb:equal(?name1, ?name2) ∧
swrlb:substringBefore(?temp1, ?ivalue1, "+")  ∧
swrlb:equal(?temp1, ?compname2) ∧
swrlb:substringAfter(?temp, ?ivalue1, "+")  ∧
swrlb:substringBefore(?op, ?temp, "#")  ∧
swrlb:equal(?op, ?proname2) ∧
swrlb:substringAfter(?temp2, ?ivalue1, "#")  ∧
swrlb:substringBefore(?inputtype, ?temp2, "$")  ∧
swrlb:notEqual(?inputtype, ?inputtype2) ∧
swrlb:substringAfter(?returntype, ?ivalue1, "$")  ∧
swrlb:equal(?ovalue2, ?returntype)
   → sqwrl:selectDistinct(?comp1, ?comp2, ?inputtype, ?inputtype2) ∧
sqwrl:select("Input type mismatch: invalid references")  ∧
sqwrl:select("add an adaptor connector as a bridge") ∧
 hasComponent(?con; adaptConnector1)
```

of services are matched. The following rule (Box 3) gives an example of this approach:

For an OSGi component, with a reference which has a cardinality of the form "1." (at least one reference to other services), there must be a component providing that required service. In a referencing component, the references to another component is modeled in the hasInput datatype property in the ProcessModel ontology, in the format "component name (including package name)+operation name#input types with orders$return type" (strings are used to make the rule development and processing easier). Then this information is compared with that from the referenced component with respect to a specific interface. If they are not exactly matched, then

Box 4

```
            Rule: Security_ResourceQuery
sec:SecurityProtocol(?protocol) ∧
sec:requiresComputingPower(?protocol, ?power) ∧
sec:requiresMemory(?protocol, ?memory) ∧
sec:authenticityObj(?protocol, ?auth) ∧
sec:hasStrength(?auth, ?value)
  → sqwrl:select(?protocol, ?memory, ?power, ?auth, ?value)
```

the references are invalid. Additionally, if a connector resolving the interface mismatch exists (its instance is adaptConnector1 as in the OSGiComponent ontology), we can then infer that an adaptor connector needs to be added to the current configuration in order to correctly match the references, shown in the last line of the rule. This inferred result can then be published, and bind the "AdaptConnetor1" into the current configuration. An upcoming feature of SQWRL is set operations (O'Connor, 2009) including the standard set operator sqwrl:intersection. This operator can be used to decide whether two sets of operations in the two to-be-matched interfaces match, which can simplify the rules for matching a large interface. In general this will allow for a form of quantification over sets.

Self-Protection Rules

An important set of SWRL rules contains queries for contexts using SQWRL. A simple example, shown next (Box 4), is used for the awareness of security contexts:

This rule will retrieve the security protocols and their memory consumption, CPU power consumption, authenticity level, and strength from the security ontologies. Subsequently, these values are used in the planning layer (Figure 4) to calculate the best options for the security protocols to be used.

Self-Adaptation Rules

In the LinkSmart middleware, we rely on web service communication using SOAP. SOAP transport over different communication protocols (e.g., TCP, UDP, or Bluetooth) has different round trip time, throughput and power consumptions (see, e.g., (Sperandio et al., 2008) for values). Some aspects of these differences include (Sperandio et al., 2008):

- To save battery, Bluetooth should be used as the communication protocol, but this is at the expense of low round trip time and throughput.
- To increase reliability, TCP should be used
- To increase throughput, UDP should be used.

Based on this, we have developed SWRL rules that help choose the appropriate transportation protocols. An example is given next (Box 5).

This rule is the realization of the first conclusion we listed above: if the highest priority is to save battery, then Bluetooth is set as the current connection.

We can see that these kinds of rules consider mostly a local or small scope of QoS requirements. To globally optimize for QoS (as well as other parameters), we use a Genetic Algorithms-based approach in our Goal Management Layer as reported in Zhang & Hansen (2009).

Figure 4. LinkSmart 3 layered and the Semantic web-enabled self-management architecture

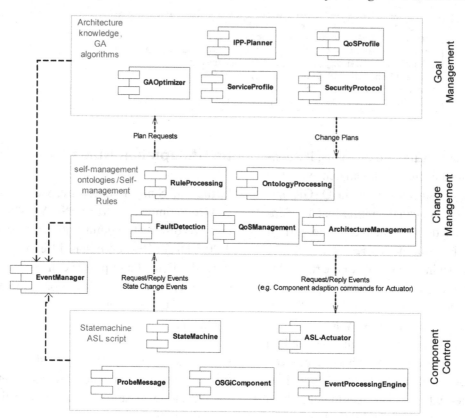

Box 5.

```
Rule:BluetoothConnection
qos:CurrentQoSRequirement(?req) ∧
qos:hasQoSParameter(?req, ?para) ∧
qos:priority (?para, ?prio) ∧
swrlb:containsIgnoreCase (?prio, "highest")  ∧
qos:PowerDrain(?para)
  → network:NetworkConnection (network:BT) ∧
       network:isCurrrent (network:BT, "true")
```

QoS Monitoring and Calculation Rules

An example of QoS monitoring shown in the next rule (Box 6) is used to calculate the latency of the network, providing that the same time clock is used for both client and service, i.e. the time is synchronized for clients and servers.

This rule will build service invocation relationships among the messages using the "invoke"

property. A complete invocation relationship is built based on information from probes inserted into web service clients and servers: one probe message for client invocation, one for service execution start, one for service execution finish, and finally one for the client receiving a response. The time stamps of these four probe messages can then be utilized to calculate the related QoS properties.

Box 6.

```
                   Rule:QoSMonitoring
probe:messageID(?message1, ?messageid) ∧
probe:messageID(?message2, ?messageid) ∧
probe:messageID(?message3, ?messageid) ∧
probe:messageID(?message4, ?messageid) ∧
abox:hasURI(?message1, ?u1) ∧
abox:hasURI(?message2, ?u2) ∧
abox:hasURI(?message3, ?u3) ∧
abox:hasURI(?message4, ?u4) ∧
swrlb:containsIgnoreCase(?u1, "clientbegin")  ∧
swrlb:containsIgnoreCase(?u2, "servicebegin")  ∧
swrlb:containsIgnoreCase(?u3, "serviceend")  ∧
swrlb:containsIgnoreCase(?u4, "clientend")  ∧
probe:messageSourceIP(?message1, ?ip1) ∧
probe:ipaddr(?ip1, ?ipa1) ∧
probe:ipaddr(?ip2, ?ipa2) ∧
probe:hasMessage(?process1, ?message1) ∧
probe:hasProcessID(?process1, ?pid1) ∧
probe:messageTargetIP(?message1, ?ip2) ∧
probe:initiatingTime(?message1, ?time1) ∧
probe:messageSourceIP(?message2, ?ip3) ∧
probe:messageTargetIP(?message2, ?ip4) ∧
probe:ipaddr(?ip3, ?ipa3) ∧
probe:ipaddr(?ip4, ?ipa4) ∧
probe:messageTargetPort(?message2, ?port2) ∧
probe:hasMessage(?process2, ?message2) ∧
probe:hasProcessID(?process2, ?pid2) ∧
probe:initiatingTime(?message2, ?time2) ∧
probe:messageSourceIP(?message3, ?ip5) ∧
probe:messageTargetIP(?message3, ?ip6) ∧
probe:ipaddr(?ip5, ?ipa5) ∧
probe:ipaddr(?ip6, ?ipa6) ∧
probe:messageTargetPort(?message3, ?port3) ∧
probe:hasMessage(?process3, ?message3) ∧
probe:hasProcessID(?process3, ?pid3) ∧
probe:initiatingTime(?message3, ?time3) ∧
probe:messageSourceIP(?message4, ?ip7) ∧
probe:messageTargetIP(?message4, ?ip8) ∧
probe:ipaddr(?ip7, ?ipa7) ∧
probe:ipaddr(?ip8, ?ipa8) ∧
probe:messageTargetPort(?message4, ?port4) ∧
probe:hasMessage(?process4, ?message4) ∧
probe:hasProcessID(?process4, ?pid4) ∧
probe:initiatingTime(?message4, ?time4) ∧
temporal:duration(?d1, ?time1, ?time4, temporal:Milliseconds) ∧
temporal:duration(?d2, ?time2, ?time3, temporal:Milliseconds)
  → probe:invoke(?message1, ?message2) ∧
sqwrl:select(?ip1, ?ipa1, ?pid1, ?ipa2, ?port2, ?pid2, ?d1, ?d2)
```

Discussion

The SeMaPS ontologies cover a relatively complete set of knowledge for self-management, including dynamic information about pervasive device states, service invocation relationships, and events exchanged between services. SWRL is flexible and powerful for developing self-management rules based on the SeMaPS ontologies. As SWRL is limited to unary or binary relationships, a SWRL rule may therefore become too long to be readable. However, SWRL is mainly used for machine processing. The development of OWL/SWRL ontologies is supported by the Protégé ontology development environment. Our experiences with this tool show that a developer will find it relatively easy to develop SWRL rules as the user interface (as a tab page in Protégé) for SWRL rule development facilitates the development of SWRL rules well. At the same time, SWRL rules can be parameterized and dynamically generated according to different requirements, which further improves its usability.

The disadvantage of OWL-DL and SWRL is that operations (such as queries) are computationally complex which may cause a performance bottleneck and we will show the performance later for our prototype. Furthermore, SWRL (and OWL) support monotonic inference only, which does not support negation as failure as it would lead to non-monotonicity. SWRL does not support disjunction either.

In the next section, we will present the LinkSmart self-management architecture, which is supported by the SeMaPS ontologies and the self-management rules presented in this section.

ARCHITECTURE AND IMPLEMENTATION OF SELF-MANAGEMENT IN LINKSMART

In LinkSmart, self-management features should cover a spectrum of self-management requirements. This includes:

- [Req 1.] Self-diagnosis for devices/services, systems, and applications, including device/service status monitoring, global resource consumption monitoring, suggestions for malfunction recovery.
- [Req 2.] Self-adaption, including quality of service (QoS) based adaptation, for example switching communication protocols to achieve different level of reliability and performance, and energy awareness for adaptation

- [Req 3.] Self-configuration, including QoS-based configuration for components, energy awareness for configuration
- [Req 4.] Self-protection, based on QoS requirements, including choosing the most suitable security strategies for component and service communication in a global manner
- [Req 5.] Self-management planning, including optimization of service selection based on multiple QoS requirements, and multiple planning algorithm support

Separation of concerns is a general guideline followed by us to choose architecture styles, and to design self-management components. As can be seen from the requirements, we need to monitor dynamic changes of services/components/devices, and then be able to respond to these changes. Finally we need to plan how to obtain the best candidates from a number (potentially a big number) of services (can be security services) following QoS requirements. This naturally motivates us to choose a layered architecture style, to separate these different concerns. More specifically, we employ the three-layer architecture proposed by Kramer & Magee (2007) in which there is a Component Control Layer, a Change Management Layer, and a Goal Management Layer.

Architecture of the 3L LinkSmart Self-Management

The LinkSmart self-management architecture is shown in Figure 4. The three layers are interacting through events, through a publish-subscribe (Eugster et al., 2003) component called LinkSmart Event Manager, which means the interaction of components in the architecture is independent of how they are distributed.

Now we detail the design of these three layers supported by SeMaPS ontologies.

Component Control Layer

The Component Control Layer is used to monitor low-level events, and to actuate changes to configure and re-configure a system.

In order to monitor and diagnose devices based on their status, as required in ***Req1***, state machines for devices are used to handle the dynamism of devices, and are deployed on devices. Additionally, to monitor service response time and network quality, a message probe has been designed and implemented to monitor service invocations, in which performance of round trip service calls, and network latency can be calculated. To achieve this, we use the Limbo web service compiler (Hansen et al., 2008) to generate the state machine code and message probe code for services. These message probe code is deployed with the corresponding services and its clients.

To actuate changes, an architectural scripting language (ASL; (Ingstrup & Hansen, 2009)) component is used to load/unload and enable/disable components and services. This component is also responsible for the execution of plans generated by an IPP planner (Koehler et al., 1997) in the Goal Management Layer.

Change Management Layer

The Change Management Layer executes predefined schemes for managing changes that require self-management actions. As discussed earlier, an SWRL rule-based approach is adopted as SWRL can harmoniously work with the underlying SeMaPS context ontologies. These schemes can be used for various self-management purposes, including self-diagnosis (Zhang & Hansen, 2008b), self-configuration, self-adaptation, self-protection, self-optimization (Zhang & Hansen, 2008c; Zhang & Hansen, 2009; Zhang et al. 2009). A "scheme" in this approach is composed with one or a set of rules and corresponding ontologies in which these rules are expressed.

Requirement 2 to 5 imply that our considered self-management features are related QoS modeling and optimization, e.g., energy-awareness. QoS properties are encoded in a set of QoS ontologies. How to apply the SeMaPS ontologies and SWRL self-management rules to these ontologies is presented in "Self-Management Specification with the SeMaPS Ontologies using SWRL Rules." Additional components are needed to provide support functions such as processing rules and ontologies. A rule engine processes the rules and produces output that needs to be parsed in order to, e.g.:

- Generate the results for fault detection, for example issuing an event for the diagnosed results for a device that has no response. Therefore a fault detection component is need in this layer.
- Generate the results for software architecture issues, like interface mismatch. Therefore, a software architecture management component is needed in this layer.
- Generate the results for QoS management, for example, network latency calculation and issuing the corresponding events for this.

The Change Management Layer is also responsible for invoking the planning layer to generate a new management scheme in case none of the existing ones can be applied to a detected problem. An example is that a new device is joining the network in a public area, which then needs to find appropriate security protocols following the restrictions on memory consumption, security levels, and latency. The Goal Management Layer will then be responsible for finding an optimized solution for this case.

Goal Management Layer

From requirement 4 and 5, we can see that finding optimized solutions from a large number of alternatives (e.g. choosing the most suitable con-

Figure 5. Processing sequence of the self-management components

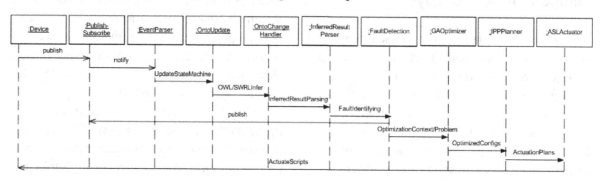

figuration of security protocols for communication channels meeting requirements for memory and CPU consumption) is an important task in self-management for pervasive service computing. Therefore in the Goal Management Layer, there should be components to conduct these optimizations. Genetic Algorithms (GAs) is one of the most successful computational intelligence approaches. They are effective in solving multi-objective optimization problems, and are chosen as one of the optimization algorithms used in the Goal Management Layer, as reported in (Zhang & Hansen, 2009).

The usage of GAs will generate an optimal solution/configuration for a multi-objective optimization problem as we usually meet in self-management, and based on a desired configuration, a change management plan that transfers the system from its current state to the optimized one is generated by the IPP planner. The idea is that an ASL script is generated, and the execution of the script will take the system architecture from its current to the desired state.

Implementation of the Semantic Web-Enabled Self-Management

Figure 5 shows the high level processing sequence of the Semantic web-enabled self-management involving all the three layers, but with self-diagnosis as an instance for illustration. When there are state updates, the diagnosis module, as

a subscriber to this topic, will update the device state machine. These updates are observed by the registered listeners, and at the same time OWL/SWRL rule inference will be started. The new inferred information will be added to an inferred result queue, which is also observed. Next, the observer for inferred result will parse the new inferred result, and publish the corresponding diagnosis result. The fault is then identified. If there is a need to optimize the configuration, the Genetic algorithms based optimizer will be invoked to optimize based on QoS requirements. Then the IPP planner will be called to generate an activation plan to be executed by the ASL actuator, to execute the enabling/loading service or the other way around.

The SWRL APIs from Protege-OWL are used for the implementation and execution of SWRL rules, as these APIs are the only available for SWRL. We adopted a mix of the Blackboard architecture style and the Layered architecture style (Shaw & Garlan, 1996), and used the Observer pattern in both the updating of ontologies and inferred result parsing, where all ontologies are loaded to memory in advance for performance reasons as loading of OWL ontologies is an expensive operation. For the implementation of the Genetic Algorithms-based optimizations, we are using JMetal GA framework[8], which has a good separation of concerns in terms of its ability to support different GAs after a problem is abstracted.

A SELF-MANAGEMENT SCENARIO WITH THE LINKSMART MIDDLEWARE

In this section, we will demonstrate the SeMaPS ontologies based self-management with a pig farm monitoring scenario to exemplify how to use our approach, which is based on the scenario presented in our former publications (Zhang & Hansen, 2008b; Zhang & Hansen, 2008c).

A Self-Management Scenario

Bjarne is an agricultural worker at a large pig farm in Denmark. He is using a PDA to help his daily routines. As he walks through the pens, he checks to see if some pigs need care and whether they are provided the correct amount of food. Bjarne takes out his PDA to make notes. However, his work is interrupted by a sound from his PDA, indicating that a high priority alarm has arrived. Apparently, the ventilation system in the pig stable has malfunctioned. Being high priority, the context-aware alarm system has decided that what Bjarne is doing now has lower priority, and sends him the alarm. At this time, the battery of his PDA is running out and the pig farm managing client on his PDA automatically switches to using Bluetooth as the underlying communication protocols to connect with the nearby devices. After sending an acknowledgement message via Bluetooth, the system begins diagnosis and soon it decides that the cause of the problem is 'power supply off because of fuse blown'. Then he can prepare a fuse and repair the ventilator. After repairing it, he signs off the alarm, and chooses one of the predefined log messages, describing what he has done.

In this scenario, the pig farm system needs to monitor the pumps, ventilators to make sure they are working normally, and is using thermometers to monitor the temperature of the pigsty. The rules introduced before are then used for achieve self-diagnosis, and also to switch communication protocols as needed when battery of PDA is low, i.e., for self-configuration.

Implementation and Evaluation

Rule PigFarmFlowmeter and PigFarmVentilating-Down introduced in "Self-Diagnosis Rules" are among the rules used for the scenario introduced above. The flow meter and thermometers are used for monitoring, for example knowing whether a pump is down which is used to water pigs, and knowing whether the farm ventilation is working. The high priority alarm is the one for notifying that the ventilator is down. This alarm is the execution result of rule PigFarmVentilatingDown.

The self-management component uses 266M bytes (minimum 64M bytes) Java heap memory for running Protege-OWL/SWRL APIs (Protege 3.4 formal release), JVM 1.6.02-b06 (JDK1.5 or later is sufficient to run our prototype). We have tested the self-management component with Windows XP SP3, but should be compatible with any platform that supports Java SE. The hardware platform used to test the performance was a Think-pad T61P T7500 2.2G CPU, 7200rpm hard disk, and 2G DDR2 RAM. The time measurements are in milliseconds. The size of the SelfManRule ontology is 405,515 bytes, and contains 26 rules, including six rules for the PigFarm system.

As the Semantic web and reasoning mechanisms is currently a heavyweight solution, it is crucial to measure the performance of reasoning at runtime. We here show the results of performance tests with two rule groups: a Flowmeter rule group that has three rules and a Ventilator rule group that has one rule. We can see from Table 2 that it takes less than 280ms in average for running each rule group. It takes around 300ms from receiving flow meter monitoring and thermometer monitoring events until the end of rule group inference. The ontology update time is measuring the time for update the StateMachine ontologies for thermometers and flow meters. The update time is around 600ms in average.

Table 2. Performance tests for running two rule groups at the same time

	Ontology update time	Flowmeter rule group		Ventilator rule group	
		Inferring time	Receiving event to inferring finished	Inferring time	Receiving event to inferring finished
First run	656	297	359	266	266
Second run	656	265	266	297	297
Third run	547	265	297	266	281
Fourth run	577	266	312	266	282
Fifth run	547	265	282	250	250
Average	596.6	271.6	303.2	269	275.2

We have furthermore measured the scalability of our solution as reported in (Zhang & Hansen, 2008b). Measuring from publishing of a diagnosis event until the end of inferring and publishing of related inferring result, the time taken is linear in the number of events that need to be processed for our measurements. This is reasonable as the inferring need to be executed accordingly as a corresponding event is received. We have also tested the performance of the Genetic Algorithms based optimization taking into consideration QoS requirements and restrictions. It takes between 300 and 500ms to find optimized solutions for self-configuration (Zhang & Hansen, 2009), and self-protection (Zhang et al., 2009) according to our recommendation of parameter setting for the used Genetic Algorithms. This performance is acceptable for planning work in self-management in our context.

Discussion

We started the self-management implementation with the self-diagnosis module with the rule for temperature trend monitoring. Next, we added the MessageProbe ontology and service invocation dynamics handling to detect whether a service is available or not. A generic SWRL rule processing (including rule group processing) component was then developed. The later addition of the Goal Management Layer did not affect the design and implementation of the two bottom layers. The optimization for self-protection and self-configuration are making use of the rule processing and rule grouping features, with separate components to interpret the reasoning results. In summary, the rule processing and rule grouping features are generic and have good extensibility for adding new features. The design shows good overall separation of concerns.

The performance shown above is acceptable for self-management, taking consideration of the reasoning potentials brought with the Semantic web, and also the optimization capabilities from the Genetic Algorithms in the Goal Management Layer.

CONCLUSION

To improve the dependability of pervasive systems, it is important to equip them with self-management capabilities by making use of the context-awareness capabilities. To achieve self-management, the underlying context models should consider the dynamism of pervasive devices, services, and networks. In this chapter, we have presented a set of OWL/SWRL ontologies, called the SeMaPS ontologies, which have been developed for self-management. We have presented a Semantic web-enabled approach to self-management that makes use of the SeMaPS ontologies and a hybrid context-awareness framework. In the hybrid context-awareness framework, a key-value pair approach of context modeling and an OWL ontologies approach are combines. The context-awareness-based self-management is demonstrated with an example scenario, where performance is evaluated to show it performs adequately.

ACKNOWLEDGMENT

The research reported in this chapter has been supported by the Hydra EU project (IST-2005-034891). Thanks to our colleagues Mads Ingstrup and Joao Fernandes at Aarhus University for their efforts to realize parts of the self-management features. Thanks to Julian Schütte from Fraunhofer Institute for Secure Information Technology for the design of the security ontology, and Peter Kostelnik from Technical University of Košice for his work on the event ontology, device ontology design and other ontology related work.

REFERENCES

Alia, M., Eide, V., Paspallis, N., Eliassen, F., Hallsteinsen, S., & Papadopoulos, G. (2007). A Utility-based Adaptivity Model for Mobile Applications. In Proceedings of the 21st International Conference on Advanced Information Networking and Applications Workshops-Volume 02, IEEE Computer Society Washington, DC, USA, 556–563.

Badii, A., Adetoye, A., Thiemert, D., & Hoffman, M. (2008). A Framework Architecture for Semantic Resolution of Security for AMI. Proceedings of the Hydra Consortium Internet of Things & Services (IOTS) Workshop at the TrustAMI08 Conference 18-19 September 2008, Sophia Antipolis, http://www.strategiestm.com/conferences/internet-of-things/ 08/ proceedings.htm

Capra, L., Emmerich, W., & Mascolo, C. (2003). CARISMA: Context-Aware Reflective mIddleware System for Mobile Applications. *IEEE Transactions on Software Engineering, 29*(10), 921–945. doi:10.1109/TSE.2003.1237173

Chen, H., Finin, T., & Joshi, A. (2003). An ontology for context-aware pervasive computing environments. Special Issue on Ontologies for Distributed Systems. *The Knowledge Engineering Review, 18*(3), 197–207. doi:10.1017/S0269888904000025

Coutaz, J., Dearle, A., Dupuy-Chessa, S., Kirby, G., Lachenal, C., Morrison, R., et al. (2003). Working document on GlOSS ontology. Technical Report IST-2000-26070 Report D9.2, Global Smart Spaces Project.

Dey, A. K. (2000, December). Providing Architectural Support for Building Context-Aware Applications. PhD thesis, College of Computing, Georgia Institute of Technology.

Dolog, P. (2004). *Model-driven navigation design for semantic web applications with the UML-guide*. Engineering Advanced Web Applications.

Eugster, P., Felber, P., Guerraoui, R., & Kermarrec, A. (2003). The Many Faces of Publish/Subscribe. *ACM Computing Surveys, 35*(2), 114–131. doi:10.1145/857076.857078

Fahy, P., & Clarke, S. (2004). CASS: a middleware for mobile context-aware applications. In Proceedings of the Workshop on Context Awareness, MobiSys 2004.

Flury, T., Privat, G., & Ramparany, F. (2004). OWL-based location ontology for context-aware services. Proceedings of the Artificial Intelligence in Mobile Systems (AIMS 2004), 52-57.

Grace, P., Blair, G. S., & Samuel, S. (2005). A reflective framework for discovery and interaction in heterogeneous mobile environments. *ACM SIGMOBILE Mobile Computing and Communications Review*, *9*(1), 2–14. doi:10.1145/1055959.1055962

Gu, T., Pung, H. K., & Zhang, D. Q. (2005, January). A service-oriented middleware for building context-aware services. *Journal of Network and Computer Applications*, *28*(1), 1–18. doi:10.1016/j.jnca.2004.06.002

Hansen, K. M., Zhang, W., & Fernandes, J. (2008). Flexible Generation of Pervasive Web Services Using OSGi Declarative Services and OWL Ontologies. 15th Asia-Pacific Software Engineering Conference (APSEC 2008), 3-5 December 2008, Beijing, China. IEEE, 135-142.

Henricksen, K., & Indulska, J. (2006, Feb.). Developing context-aware pervasive computing applications: Models and approach. *Journal of Pervasive and Mobile Computing*, *2*(1), 37–64. doi:10.1016/j.pmcj.2005.07.003

Hofer, T., Schwinger, W., Pichler, M., Leonhartsberger, G., & Altmann, J. (2003). Context-awareness on mobile devices – the hydrogen approach, in: Proceedings of the 36th Annual Hawaii International Conference on System Sciences (HICSS'2003), Big Island, Hawaii, USA.

Ingstrup, M., & Hansen, K. M. (2009). Modeling Architectural Change: Architectural scripting and its applications to reconfiguration. In Proceedings of WICSA/ECSA 2009, 337-340.

Kephart, J., & Chess, D. (1996). *The Vision of Autonomic Computing. 2003. Mitchell, M.: An Introduction to Genetic Algorithms*. Bradford Books.

Koehler, J., & Nebel, B. Ho_mann, J., & Dimopoulos, Y. (1997): Extending planning graphs to an adl subset. In: Recent Advances in AI Planning. Volume 1348 of Lecture Notes in Computer Science. Springer, 273-285

Korpipaa, P., Mantyjarvi, J., Kela, J., Keranen, H., & Malm, E. (2003). Managing context information in mobile devices. *Pervasive Computing, IEEE*, *2*(3), 42–51. doi:10.1109/MPRV.2003.1228526

Kramer, J., & Magee, J. (2007). Self-Managed Systems: an Architectural Challenge. International Conference on Software Engineering, 259-268

Mokhtar, S. B., Bromberg, Y.-D., & Georgantas, N. (2006, March). Amigo middleware core: Prototype implementation and documentation. Deliverable 3.2, European Amigo Project.

O'Connor, M. J., & Das, A. K. (2009). OWL: Experiences and Directions (OWLED), Fifth International Workshop, Chantilly, VA. In Press.

OSGi Alliance. (2007). OSGi service platform - service compendium. Technical Report Release 4, Version 4.1, OSGi.

Ranganathan, A., & Campbell, R. H. (2004). Autonomic pervasive computing based on planning. Autonomic Computing, 2004. Proceedings. International Conference on In Proceedings of International Conference on Autonomic Computing. 80-87.

Ranganathan, A., McGrath, R. E., Campbell, R. H., & Mickunas, M. D. (2003). Use of ontologies in a pervasive computing environment. *The Knowledge Engineering Review*, *18*(3), 209–220. doi:10.1017/S0269888904000037

Rouvoy, R., Eliassen, F., Floch, J., Hallsteinsen, S., & Stav, E. (2008). Composing Components and Services using a Planning-based Adaptation Middleware. In Pautasso, C., & Tanter, É. (Eds.), *SC 2008. LNCS* (*Vol. 4954*, pp. 52–67). Heidelberg: Springer.

Satyanarayanan, M. (2001). Pervasive computing: vision and challenges. [see also IEEE Wireless Communications]. *Personal Communications, IEEE, 8*(4), 10–17. doi:10.1109/98.943998

Shaw, M., & Garlan, D. (1996). *Software Architecture: Perspectives on an Emerging Discipline*. Prentice Hall.

Sousa, J. P., Poladian, V., Garlan, D., Schmerl, B., & Shaw, M. (2006). Task-based adaptation for ubiquitous computing. Systems, Man and Cybernetics, Part C: Applications and Reviews. *IEEE Transactions on, 36*(3), 328–340.

Sperandio, P., Bublitz, S., & Fernandes, J. (2008). Wireless Device Discovery and Testing Environment. Technical Report D5.9, Hydra Consortium. IST 2005-034891.

Strang, T., & Linnhoff, P. C. (2004). A context modeling survey. In First International Workshop on Advanced Context Modelling, Reasoning And Management, UbiComp.

Web Ontology Language (OWL) homepage (2009-11-3). http://www.w3.org/ 2004/ OWL/.

Zhang, W., & Hansen, K. M. (2008a) Semantic web based Self-management for a Pervasive Service Middleware. Second IEEE International Conference on Self-Adaptive and Self-Organizing Systems. IEEE Computer Society. Venice, Italy. 245-254.

Zhang, W., & Hansen, K. M. (2008b). An OWL/ SWRL based Diagnosis Approach in a Web Service-based Middleware for Embedded and Networked Systems. Proceedings of The 20th International Conference on Software Engineering and Knowledge Engineering (SEKE 2008). Redwood City, San Francisco Bay, USA, 893-898.

Zhang, W., & Hansen, K. M. (2008c). Towards Self-managed Pervasive Middleware using OWL/ SWRL ontologies. Fifth International Workshop on Modeling and Reasoning in Context (MRC 2008), Held together with HCP 08, Delft, The Netherlands, 9-12 June 2008, 1-12.

Zhang, W., & Hansen, K. M. (2009). Evaluation of NSGA-II and MOCell Genetic Algorithms for Self-management Planning in a Pervasive Service Middleware. 14th IEEE International Conference on Engineering of Complex Computer Systems (ICECCS 2009), Potsdam. Germany. June 2-4 *2009*, 192-201.

Zhang, W., Hansen, K. M., & Kunz, T. (2009). Enhancing Intelligence of a Product Line Enabled Pervasive Middleware. Pervasive and Mobile Computing Journal. Elsevier. June, 2009. http:// dx.doi.org/ 10.1016/ j.pmcj.2009.07.002.

Zhang, W., Schütte, J., Ingstrup, M., & Hansen, K. M. (2009). A Genetic Algorithms-based Approach for Optimized Self-protection in a Pervasive Service Middleware. The Seventh International Conference on Service Oriented Computing (ICSoC 2009), Stockholm. Sweden Nov 24-27 *2009*. Springer LNCS 5900, 404-419.

ENDNOTES

1. http://www.w3.org/Submission/SWRL/
2. http://www.w3.org/TR/owl-time/
3. http://www.w3.org/Submission/OWL-S/
4. http://www.osgi.org
5. http://jena.sourceforge.net
6. http://protege.cim3.net/cgi-bin/wiki.pl?SQWRL
7. http://protege.cim3.net/cgi-bin/wiki.pl?SWRLLanguageFAQ
8. jmetal.sourceforge.net/

Chapter 15
An Ontology–Based Context–Aware Infrastructure for Smart Homes

Bin Guo
Institut TELECOM SudParis, France & Keio University, Japan

Daqing Zhang
Institut TELECOM SudParis, France

Michita Imai
Keio University, Japan

ABSTRACT

A general infrastructure that can facilitate the development of context-aware applications in smart homes is proposed. Unlike previous systems, our system builds on semantic web technologies, and it particularly concerns the contexts from human-artifact interaction. A multi-levels' design of our ontology (called SS-ONT) makes it possible to realize context sharing and end-user-oriented customization. Using this infrastructure as a basis, we address some of the principles involved in performing context querying and context reasoning. The performance of our system is evaluated through a series of experiments.

INTRODUCTION

Computing is moving from traditional computers towards everyday artifacts (e.g., appliances, furniture, cups) to make them "smart" and "intelligent". By making use of the perceived contexts, smart artifacts can support a variety of human-centric applications. This paper proposes a general infrastructure which can ease the developers' efforts in building context-aware systems.

Designing and developing context-aware applications in smart environments have been drawing much attention from researchers in the recent years. However, context-aware services

DOI: 10.4018/978-1-61692-857-5.ch015

have never been widely available to home users. Recent research results show that building and maintaining context-aware systems is still a complex and time-consuming task due to the lack of an adequate infrastructure support. We believe that such an infrastructure requires the following supports:

- *A formal context model that can facilitate context sharing and reuse among different smart environments and context-aware services.* Raw context data obtained from various sources comes in heterogeneous formats, and applications that do not have prior knowledge of the context representation cannot use the data directly. Therefore, a single method for explicitly representing context semantics is crucial for sharing common understandings among independently developed context-aware systems. A unified knowledge-representation scheme also makes it easy to implement knowledge-reuse operations, such as extending or merging existing context definitions. Without such a common context model, most context-aware systems have to be written from scratch at a high cost.

- *It can easily integrate with generic reasoning and querying engines.* Context reasoning is a key aspect of context-aware systems because high-level contexts (e.g., *is there anyone near the book? what is the person doing?*) cannot be directly provided by low-level sensor contexts (e.g., *the lighting level of a room, the 3D coordinate data of a book*). On the contrary, they have to be derived by reasoning. A context querying engine is also important for context-aware systems, which allows applications to selectively access the contexts they concern by writing expressive queries.

- *It can be easily customized by users from different families.* As a platform that is intended to work for users from different families, the knowledge infrastructure

should be "sharable" and "easily-customizable", because different families usually have different domestic settings, daily routines, and user considerations.

To facilitate rapid prototyping of context-aware applications, we proposed a new system infrastructure called *Sixth-Sense*. Unlike similar studies that mainly use ad-hoc data structures to represent contexts, *Sixth-Sense* explores the Semantic Web (Berners-Lee, Hendler, & Lassila, 2001) technology to define a common ontology that can assist the development of context-aware systems. The *Sixth-Sense Ontology* (*SS-ONT*), expressed by the Semantic Web language OWL (Web Ontology Language) (Smith, Welty, & McGuinness, 2003), reflects portion of contexts that typically exist in smart home environments, such as sensors, locations, smart objects, humans, as well as interactions between humans and objects.

Benefiting from the hierarchical definition structure and semantic sharing natures of the ontology, *SS-ONT* enables home-knowledge sharing and customization among different families. By exploring this formal context modeling method, *Sixth-Sense* also integrates several standardized approaches and tools that support expressive querying and reasoning of defined facts and contexts. In a word, the *Sixth-Sense* infrastructure builds a good foundation to support the development of a wide variety of context-aware applications in smart home environments. A series of experiments have been conducted to evaluate the performance as well as the effectiveness of our knowledge infrastructure.

In the following sections, we will firstly give a brief survey on existing context modeling methods, and then describe how to explicitly represent contexts by using the Semantic Web language – OWL. Readers can also learn about the methods on ontology-based context querying and reasoning from this chapter. Samples demonstrating how to use this infrastructure to quickly create context-aware applications are also presented.

BACKGROUND ON CONTEXT-AWARE SYSTEM DEVELOPMENT

According to Dey's definition (Dey & Abowd, 2000), context is *"any information that can be used to characterize the situation of an entity"*. An entity is a person, or an object that is considered relevant to the interaction between a user and an application, including the user and application themselves. A general context infrastructure acts as middleware between context-aware applications and context producers, which can provide a formal method for context modeling, support context collection, including reasoning mechanisms for inferring high-level contexts, from various context producers, and provide query mechanisms for applications to selectively access to the contexts they need.

Context Modeling Methods

Numerous approaches have been explored for context modeling. Most studies are based on informal context models, such as using simple data structures (e.g., key-value pairs) (Beigl, 2001; Philipose, 2004) and data formats (e.g., pictures) (Siio, Rowan, Mima, & Mynatt, 2003), relational database schemes (Yap, Srinivasan, & Motani, 2005), XML (Lampe & Strassner, 2003), and programming objects (e.g., Java classes). As these systems rely on ad hoc representations of contexts, independently developed applications cannot interoperate based on a common understanding of context semantics. Furthermore, because the expressive power involved in using such ad hoc schemes is quite low, it is difficult for them to perform context reasoning by introducing generic inference engines. As a result, the reasoning tasks of these systems are mostly implemented as programming procedures, which makes the overall systems inflexible and difficult to maintain.

Recent research work has focused on providing toolkit or infrastructure support to tackle the above issues. The pioneering work of Context Toolkit provided an object-oriented architecture for rapid prototyping of context-aware applications (Salber, Dey, & Abowd, 1999). It gives developers a set of programming abstractions which separate context acquisition from actual context usage and reuse sensing and processing functionality. The Smart-Its project proposed a generic layered architecture for sensor-based context computation, including separate layers for feature extraction and abstract context derivation (Beigl & Gellersen, 2003). Henricksen et al. (02) model context using both ER and UML models, where contexts can be easily managed with relational databases. A logic-programming based platform for context-aware applications is described in LogicCAP (Loke, 2004).

To serve as a general context modeling method for smart homes, it must meet several requirements: First, it can easily represent the heterogeneous, semantically-rich relationships among physical (like smart artifacts and humans) and logical entities (like software agents); Second, it should be *"sharable"* among different families and be *"easily-extendable"* to meet an evolving environment (e.g., a new kind of sensor is deployed or a new relationship between two object classes is needed to be added); Third, it can work easily with an inference engine to interpret collected contexts; Finally, it should provide an *"easy-to-use"* tool for users to customize the knowledge definition of their home. However, the context management studies mentioned above cannot fulfill these requirements: (1) the expressive power of the methods they used are still low (e.g., semantics like a symmetry property *"isNear"* and taxonomy information cannot be easily represented by them); (2) they don't provide adequate support for organizing contexts in a formal structure format, which makes it difficult to realize knowledge sharing and reuse among different families; (3) they provide no generic mechanism for context querying and reasoning. Though LogicCAP introduced a formal logic-based model for context reasoning, it didn't support knowledge sharing and was difficult to be maintained by end users; (4) existing methods are mainly designed for experts, and they don't provide any interface for end users. To address these issues, recently there emerges another way —an ontology-based approach, to model contexts, as surveyed in the next section.

Ontology-Based Context Modeling

The term "ontology" has a long history in philosophy, in which it refers to the subject of existence. In the context of knowledge management, ontology is referred as the shared understanding of some domains, which is often conceived as a set of entities, relations, axioms, facts and instances (Uschold & Gruninger, 1996).

Table 1 gives a comparison of database schemes and ontology schemes. In contrast to database-based knowledge infrastructure, context modeling based on ontology is more normalized and promising (Strang & Linnhoff-Popien, 2004), which can at least offer the following four advantages. First, ontology is based on a set of normalized standards and it can explicitly represent the knowledge of a domain. Second, an ontology provides a shared vocabulary base, which enables the sharing of context semantics, and its hierarchical structure facilitates developers to reuse and extend domain ontologies. Third, using ontology as a knowledge infrastructure allows the underlying system implementations to be better integrated with various existing logical reasoning mechanisms. Finally, there are free, mature and user-friendly ontology editors, like Protégé-OWL editor (available at: "http://protege. stanford.edu/"). These four characters just meet the requirements mentioned in last subsection, so our system explored the ontology-based method to build our context model.

Semantic Web is an effort that has been going on in the W3C to provide rich and explicit descriptions of Web resources. The essence of it is a set of standards for exchanging machine-understandable information. Among these standards, Re-

Table 1. Database vs. Ontology

Axis of Comparison	Database Schemes	Ontology Schemes
Modeling Perspective	Intended to model data being used by one or more applications	Intended to model a domain
Structure vs. Semantics	Emphasis while modeling is on structure of the tables	Emphasis while modeling is on the semantics of the domain - emphasis on relationships, also facts/knowledge.
What concerns	Efficiency and effectiveness on data processing Expressiveness	Expressiveness, semantic sharing, logical reasoning

source Description Framework (RDF) provides data model specifications (Brickley & Guha, 2000), and OWL enables the definition and sharing of domain ontologies. Using OWL, one can (1) formalize a domain by defining classes and the properties of those classes, (2) define individuals and assert their properties, and (3) reason about these classes and individuals. From a formal point of view, OWL is rooted in description logic (DL), which allows OWL to exploit DL reasoning tasks such as class consistency and consumption.

Web ontology and other Semantic Web technologies have been recently employed in modeling and reasoning contexts in different ubiquitous computing domains. Ranganathan and Compell (2003) developed a middleware for context awareness, where they represented context ontology written in DAML+OIL. Chen et al. (2004a) proposed an agent-oriented Context Broker Architecture (CoBrA) infrastructure for semantic context representation and privacy control. Semantic Space is a pervasive computing infrastructure that exploits Semantic Web technologies to support explicit representation and flexible reasoning of contexts in smart spaces (Wang et al., 2004). Yamada, Iijima, and Yamaguchi (2005) explores ontology to describe the knowledge of digital information appliances in home networks. Ontology has also be employed in sensor network systems to deal with some particular issues, such as relation-based data search (Lewis, Cameron, Xie, & Arpinar, 2006) and the adaptive problem of sensor networks (Avancha, Patel, & Joshi, 2004)

The above studies explored ontology to cope with different problems in different domains, such as context modeling in context-aware systems and problem description in sensor networks. However, though the ontology-based method has been employed into many domains for context modeling, there lacks an ontology definition for human-artifact interaction systems. Because so many interactions in smart homes occur between people and everyday artifacts (such as books, toothbrushes, etc.), ignoring such rich and valuable contexts will limit the reasoning tasks that a knowledge infrastructure can support. For example, inferring whether a teacup is filled with tea, or recognizing an artifact-related behavior such as a person drinks, cannot be implemented by existing studies. Moreover, all ontologies in above systems are defined by experts, and there lacks a study to investigate whether end users can exert control over ontology definitions. Because home users have the most intimate knowledge about their living environments, they should be allowed to manage the ontology definition of their home. Following we will present a general context infrastructure called *Sixth-Sense* to deal with all these problems.

ONTOLOGY-BASED CONTEXT MODELING

Our *Sixth-Sense* infrastructure explores the Semantic Web technology to model contexts. The ontology, called *SS-ONT*, is described using the OWL language.

Ontology Design: Three Levels

As shown in Figure 1, we define *SS-ONT* at three levels. Definition of the first two levels, namely *upper ontology* and *domain ontology*, follows the general ontology design strategy as reported by Wang et al. (2004). The upper ontology is a high-level ontology that provides a set of basic concepts common to different smart spaces (e.g., homes and offices). The domain-specific ontology is an extension of the upper ontology created by adding detailed concepts and their features in different application domains (e.g., "*dinner table*" class and "*cabinet*" class in home domain, as illustrated in Figure 1). These two levels make *SS-ONT* more extensible, and enable developers to customize particular domain ontologies by inheriting the classes defined in the upper ontology. Different from previous systems, as a user-oriented system, we also introduce a third level, *instance definition* or *user-customization* level, which allows home users to insert particular static contexts according to their domestic settings (e.g., *there is a light called 'light-22'* and *'light22' is installed in the living room*) to a shared domain ontology. In this way, each family can share a common knowledge structure and only adding the static contexts of their home, which helps to realize application sharing and customization.

There have been many attempts in the ontology research field that try to bridge the gap between real world and computational world, such as SOUPA (Chen et al., 2004b) and ULCO (Wang et al., 2004). In light of the advantage of knowledge reuse provided by using an ontology, we referenced several such consensus ontologies and standard specifications about home devices, such as ECHONET (please refer to "http://www.echonet.gr.jp/" for details). Different from previous ontology studies, our *SS-ONT* ontology is an attempt to model portion of knowledge in the domain of human-artifact interaction. From the various contexts, we selected 14 top-level classes, and a set of sub-classes, to form the skeleton of *SS-ONT*,

Figure 1. Three levels' design of SS-ONT

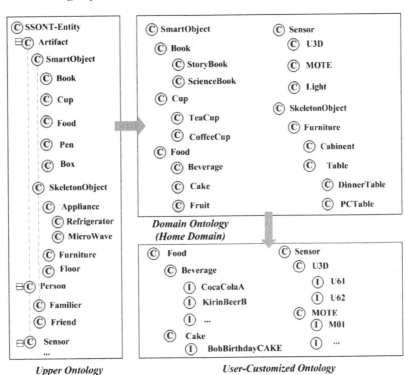

Upper Ontology
(Smart Spaces)

User-Customized Ontology
(A Specific Home)

i.e., the *SS-ONT* upper ontology (see Figure 1). In next subsection, we will describe an extension of this upper ontology to the smart home domain.

A Domain Specific Ontology for Smart Homes

Part of an extension of the *SS-ONT* upper ontology for the smart-home domain is shown in Figure 2. This specific ontology defines a number of concrete subclasses, properties, and individuals in addition to the upper classes. The current version of the *SS-ONT* (ver-1.5) domain ontology consists of 108 classes and 114 properties, the whole definition of which is available at: "http://www.ayu. ics.eio.ac.jp/members/bingo/SS-ONT-v1.5.owl"). It models a portion of the general relationships and properties associated with artifacts, people, sensors, behaviors as well as applications (e.g., games) in smart environments. In the following subsections we explain how to achieve this using OWL.

Person

The "*Person*" class defines typical vocabularies for describing a person. Besides the profile description, there are specific predicates for representing human locations, human behaviors, and activities.

Location: The location of a human is modeled in several ways in response to varied references. For example, the predicate "*isLocatedIn*" denotes the spatial relation between a person and a room, the range of "*isLocatedAt*" is a skeleton object (such as a bed, explained later), and the predicate "*isLocatedNear*" represents a person's location by nearby *SmartObject* individuals. Different representation formats deliver different contexts, which makes the reasoning task more effective.

Figure 2. SS-ONT domain ontology model (partial)

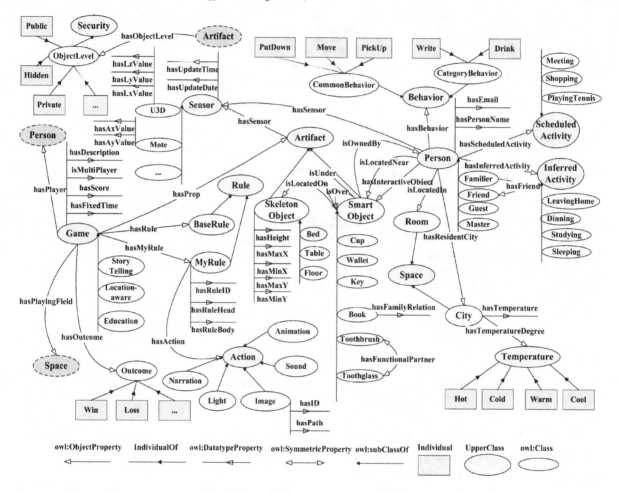

Human Behavior: In the course of human-artifact interactions, ongoing human behaviors are reflected by changes in smart-artifact states. As shown in Figure 2, the class "*Behavior*" has two subclasses, "*CommonBehavior*" and "*CategoryBehavior*". The former denotes behavior types that exist among all object categories, for instance, *pick up*, *put down*, and *move*, while the latter indicates category-specific behaviors, such as the "*drink*" behavior to the cup category. In human-artifact interactions, each human behavior also implies that a potential object is being interacted with, and we use "*hasInteractiveObject*" to reflect this information.

Human Activity: The OWL class "*Activity*" also has two subclasses: "*InferredActivity*" and "*ScheduledActivity*". The former denotes activities derived from context reasoning, while the latter refers to activities retrieved from software services, such as *Google Calendar API* (available at: "http://code.google.com/apis/calendar/").

Artifacts

Explicit definitions of physical artifacts are important for context-aware applications. Defined artifact properties can be classified into four types: object profiles and status information, location description, relationships with humans and with other objects.

Profile and Status: These describe, respectively, the static (e.g., *hasName, hasShape*) and dynamic properties (e.g., *hasTiltAngle, isMoving*) of an artifact.

Location: We use a symbolic expression method to represent the location of objects. For example, the location of a key can be represented as "on (key, table)", which means the key is on a table. Obviously, a symbolic expression provides an efficient way for humans to locate things than using an absolute-location expression (i.e. raw coordinate data). To achieve this, we borrowed two subclasses from the *Artifact* class: *SkeletonObject* and *SmartObject* (see Figure 2). Several large and mostly static artifacts (e.g., furniture) are selected to serve as skeleton objects, while other small, easily movable objects (e.g., keys, books) are categorized into the *SmartObject* class. In *SS-ONT*, the property "*isLocatedOn*" is defined to represent the spatial relation between smart object *A* and skeleton object *B* (meaning *A* is located on *B*). There are also two OWL inverse properties terms, *isOver* and *isUnder*, defined to reflect the spatial relations between two *SmartObject* instances (e.g., a pen is placed on a book).

For the *SkeletonObject* instances, because they may act as reference objects for other smart objects, particular properties are needed to specify their coverage area. For example, the coverage area of a skeleton object with a rectangular surface can be expressed by four OWL data-type properties, *hasMaxX, hasMinX, hasMaxY, hasMinY*, which involve the maximum and minimum horizontal coordinate values. There is also another datatype property "*hasHeight*" to denote the height of this skeleton object, and a *Boolean* property "*hasLoadedObject*" to express whether there is any object located on this skeleton object. An OWL object property "*isLocatedIn*" is defined to specify in which room the skeleton object is placed.

Relationships with humans: This mainly reflects the relations between artifacts and humans. *isOwnedBy* and *isInteractedBy* are two such properties.

Relationships with objects: In addition to physical relations, there are also logical relations among artifacts (e.g., the functional relation between a toothbrush and a glass). Compared to physical relations, logical relations among objects can sometimes provide more precise and direct information (e.g. based on the known functional relationship, we can guess that the object placed in a glass is a toothbrush and not a pen). To the best of our knowledge, the logical relations between physical objects have not been discussed in previous smart-space studies. The current *SS-ONT* defines four types of logical relations.

Definition 1 (Combinational Relation): Some objects act as components of, or accessories for, other objects, e.g., the relation between a mouse and a computer (reflected by the predicate "*hasComponent*" in SS-ONT).

Definition 2 (Partner Relation): Several objects can cooperate as partners to perform human-centric tasks and are always arranged in the same way. Possible instances include tables and chairs, and foreign language books and electronic dictionaries. The partner relation between objects is symmetric, and a symmetric property "*hasWorkPartner*" is used to represent this relation.

Definition 3 (Family Relation): Objects that belong to the same category sometimes have this relationship because, in daily life, similar operational modes exist when humans deal with objects in the same category (e.g., people deposit all their books on the bookshelf, or place all their umbrellas in the umbrella stand near the door). We use a *Boolean* property "*hasFamilyRelation*" to denote if the object class has such a relation.

Definition 4 (Functional Relation): Some objects are designed for a specific functional purpose, which is to serve the work of another object (e.g. a toothbrush and its supporting glass). We use "*hasFunctionalPartner*" to express this relation.

Security Filter

As reported in Meyer et al. (2003), if people prefer to live in a context-aware home for extra convenience, every measure should be taken to prevent information about their private life being accessed by anybody else. Otherwise, they will not find life in a smart house pleasant. In an artifact-rich environment, residents should have the authority to decide which artifacts can be publicly viewed and which cannot, and also the authority to allow access to different users, such as friends.

To this end, we classify the "*Person*" class into four subclasses: *Master*, *Familier* (denotes a family member), *Friend*, and *Guest*. In addition, artifacts are classified into five types: *Public* (only available to guests, e.g., magazines), *Protected-Family* (objects shared by the whole family, such as a shoe cabinet), *ProtectedFriend* (an object that is privately owned but set by its owner to friend level), *Private* (personally owned objects, such as diaries), and *Hidden* (objects whose sensors do not work well). A *Familier* user can manage and view the private objects that belong to him (defined by the "*isOwnedBy*" property) and all the other three object types. As a specific *Familier* user, a *Master* user has intimate knowledge about home computers and is responsible for maintaining *ProtectedFamily* objects (e.g., insert an item about a new chair, or modify information about an existing object). If a *Familier* user *A* has a friend *B*, and *A* sets the level of a private object *O* to *ProtectedFriend*, then *O* will be accessible to *B*. In addition, for the sake of security, objects at the *Hidden* level can only be tracked by *Master* users. With this protection policy, if *B* wants to view an object owned by *Familier* user *C*, his access will be denied.

Sensor

Our system uses various sensors, including U3D (ultrasonic 3D location, which can provide 3D coordinate data) sensors, MICA2 mote sensors (see "http://www.xbow.com/"), light sensors and Kinotex pressure sensors (see "http://www.tactex.com/"). Such sensors are either attached to smart artifacts or placed on skeleton objects. To reflect the various sensor types, we borrowed several subclasses from the top class "*Sensor*", like *U3D* and *Mote* (see Figure 2). A number of properties that represent sensor values are also defined. For instance, "*hasLxValue*", "*hasLyValue*" and "*hasLzValue*" denote the real-time coordinate values derived from U3D, whereas "*hasOldLx-Value*", "*hasOldLyValue*", and "*hasOldLzValue*" represent the sensor data for the last update time (i.e., a historical data record). "*hasLight*" denotes the lighting value of the ambient environment. The relationship that an object equips a sensor is represented by the "*hasSensor*" property.

Category Standard

It is known that people always use the concept of category to distinguish between objects. Each category should have one or more particular attributes that distinguish it from other object categories. A definition of this artifact category principle is given here.

Definition 5 (Category Standard): Each artifact category has its product or design standards, such as shape, size, color or weight, which are all endowed with specific or unified definitions (e.g., the size of a sheet of paper can be A4, B5; while the shape of the bottom of a cup is usually circular). We call this category standard.

Use of category standard helps recognize objects and *SS-ONT* has a top class "*Standard*" to reflect this concept. The class *Standard* has a number of individuals (e.g., *Cup-Standard*, *Key-Standard*) that denote different category

Figure 3. Context Representation Samples

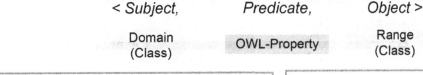

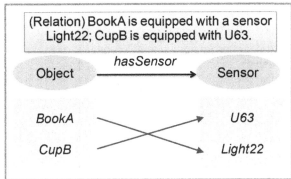

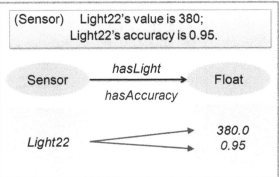

standards. A set of properties is defined to describe the standards. OWL datatype properties "*hasMaxLength*", "*hasMinLength*", "*hasMaxWidth*", "*hasMinWidth*", "*hasMaxHeight*", and "*hasMinHeight*" are used to express the general dimension range of this object category. Properties such as "*hasWeight*" and "*hasShape*" are used to describe the common attributes of the same kind of object as well.

Context Representation and User Customization

Context in an OWL ontology-based model is represented as a triple pattern, namely (<*subject, predicate, object*>). According to OWL axioms, each OWL property (the predicate element of a context) defined in "A Domain Specific Ontology for Smart Homes" has a domain and range restriction, which represents the subject and the object element of a context.

Some context examples are shown in Figure 3. If we want to represent a context that describes the relation between two objects, such as "an object is equipped with a sensor", we can simply add a "*hasSensor*" property between the related

objects (e.g., *BookA* and *Light 22* in Figure 3). If we want to represent a data-type context such as the accuracy or sensory value of a sensor, we can also assign the values to related properties of the sensor. The example shown in Figure 3 includes a description that *Light22*'s lighting value is 380.

Since different homes have different context descriptions, some contexts should be customized by end users who have intimate knowledge about their homes and daily routines, which leads to the work of the third ontology design level mentioned in "Ontology Design: Three Levels." As shown in Figure 4, the tool we chose to support this user-oriented task is the Protégé-OWL editor, because it provides well-designed documents, user-friendly interfaces and error-checking mechanisms, which make it a good tool for home computer users. User-oriented ontology customization in our system is mainly about individuals, i.e., the ABox definition of *SS-ONT*, which reflects things that are different among different homes. Following is the description of two such tasks.

1. **Creating individuals.** It is not possible for an ontology developer to predefine the instances (e.g., *a story book about Harry*

Figure 4. A screenshot of the Protégé-OWL editor

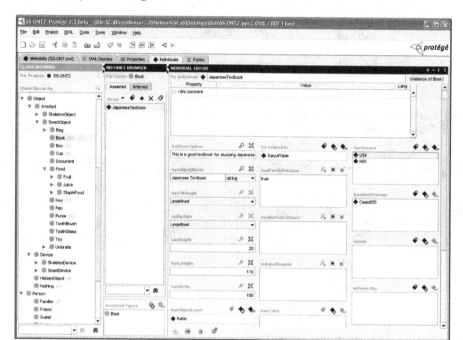

Potter) of the defined classes within an individual home. Instead, this should be defined by end users. It is not difficult to achieve using Protégé. For example, to create an instance for a certain class, a user can first choose the targeted class (e.g., *Book*) from the '*classes tree*' shown in the left frame of Figure 4, and then press the '*create instance*' button in the '*instance browser*' frame (middle part of Figure 4) to create an new instance (with a desired name) for the selected class. Users can repeat the above steps to create more individuals that exist in their home. Table 2 summarizes the types of individuals needed to be created by users.

2. **Setting properties.** Another task for ontology customization is to assign static-property values (or *static contexts*) for new created individuals. As shown in right part of Figure 4, when an individual is created, all its properties will be listed in the '*individual editor*' frame. For an OWL data-type property like *hasEmail*, users can directly fill its

Table 2. Concepts needed to be configured by users

Tasks	Concepts to be Configured
Creating individuals	Person, Artifact (Smart artifact, Skeleton artifact), Device, Sensor, Space (City, Room)
Asserting properties	Human or object profiles (e.g., hasName, hasSize, hasRecidentCity)
	Sensor deployment information (e.g., hasSensor)
	Static human-artifact relations (e.g., isOwnedBy)
	Static artifact-artifact relations (e.g., hasFunctionalPartner, hasComponent)
	Static human-human relations (e.g., hasParent, hasFriend)
	Space settings (e.g., hasPart, isPartOf)

value in the relevant form, while for an OWL object property like *hasSensor*, users should press the diamond-shaped buttons on the '*resource widget*' of this property, and specify its value in the pop-upped resource dialog, where all available individuals within the range of this property are listed (see

Figure 5. A screenshot of the "hasSensor" property-setting dialog

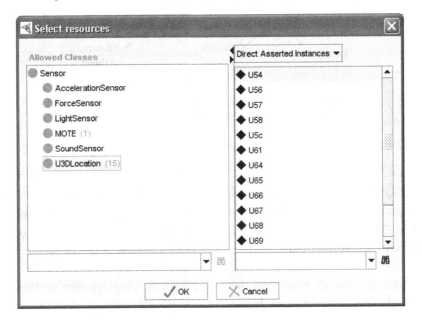

Figure 5). A set of user-controlled properties are also listed in Table 2.

CONTEXT INFRASTRUCTURE IMPLEMENTATION

Context Querying Engine

The query engine provides an abstract interface that enables applications or agents to extract desired context information from *SS-ONT*. We provide two query mechanisms, *Protégé-OWL API based query* and *SPARQL mbased query*.

Two Query Methods

For experts and programmers, a simple way to query *SS-ONT* is to use the Protégé-OWL API, which allows them to query the content of user-defined classes, properties or individuals from an OWL file by writing Java codes. However, the efficiency and expressive power of this kind of query is not high. To support advanced queries, we also adopted the SPARQL language (Prud'hommeaux & Seaborne, 2006), the Semantic Web query language developed by W3C, as the context query language. A SPARQL query statement consists of a series of triple patterns and modifiers. Different form traditional SQL queries, as the query language for the Semantic Web, SPARQL-based query approach allows users to perform "semantic querying" of defined concepts. Here we give an example to manifest the differences of different query methods.

Figure 6 shows a simple model that expresses the relations between several classes (extracted from *SS-ONT*). A simple query over this model, says, "*List all the objects that are placed on the DinnerTable*", is listed in Table 3. This example clearly indicates that writing SPARQL query statements is easier and more intuitive than writing Java codes (i.e., API-based method), especially for non-programmers. But for expert users, the Protégé-OWL API based query method is sometimes more flexible because it can be seamlessly integrated into their Java-based projects. From this example we can also find that SQL-based query is mainly based on integer/string comparisons, which ignores any relationships between

Figure 6. A partial graph model of SS-ONT

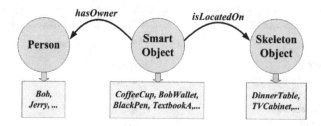

physical entities (i.e., objects, sensors, humans). In contrast, SPARQL-based query is founded on semantic mapping, and its WHERE clause consists of several "triple-formatted" semantic descriptions. For example, the triple pattern in this example (i.e.,<?object ssont: isLocatedOn ssont: DinnerTable>) denotes *"what are placed on the dinner table"*. It's obvious that, for smart home systems, semantic querying is more intuitive and expressiveness, and it may traverse some hops in SQL querying.

SPARQL-based Query Mechanism

As mentioned previously, the WHERE clause in a SPARQL query consists of several triple patterns. These triples together comprise what is known as a *graph pattern*, which can be used to identify the shape of the graph that we want to match against. That's to say, a query in our system attempts to match the triples of the graph pattern to the *SS-ONT* context model (as shown in Figure 2). Each matching binding of the graph pattern's variables to the model's nodes becomes a query solution, and the values of the variables named in the SELECT clause become part of the query results. The example mentioned previously used one triple pattern to match the two nodes (i.e., *SmartObject* and *SkeletonObject*) of the sample model given in Figure 6. For a relatively complex query, says, *"Whose objects are placed on the DinnerTable"*, which relates to all three nodes in the sample model, we need to use a two-triple-formed graph pattern for its matching, as shown

in Equation (1). Therefore, for users who want to perform SPARQL-based query in our system, they have to firstly examine the graphical context model to identify which nodes (or what shape) are to be covered in the target query, and then they can combine the triples related to these nodes and create the graph pattern for this query.

$$PREFIX \ ssont :< http : // \sim /SSONT.owl\# >$$
$$SELECT \ ?person$$
$$WHERE\{?object \ ssont : isLocatedOn \ ssont : DinnerTable \ .$$
$$?object \ ssont : hasOwner \ ?person \ .\}$$

$$(1)$$

Context Reasoning

As discussed previously, *Sixth-Sense* infrastructure supports two reasoning levels: ontology reasoning based on OWL axioms and logic inference via user-defined rules. Below, we will explain these two reasoning levels in detail.

Ontology Reasoning

Since OWL is rooted in description logic (DL), in the design stage of *SS-ONT*, we used a well-known DL reasoner *Racer* (Haarslev & Moller, 2001) to execute class subsumption (i.e., to compute the inferred class hierarchy) and consistency checking. *Racer* is capable of detecting inconsistencies as demonstrated in the following example: if a property like *"isLocatedIn"* is defined as an OWL functional property, and in the definition of the *"Person"* class, a minimum cardinality restriction "2" is added to this property, there will be a

Table 3. A sample query using three different query methods

Query Method	Query Expressions

Protégé-OWL API Query	JenaOWLModel owlModel = ProtegeOWL.createJenaOWLModelFromInputStream(SS-ONT_Path);
	OWLNamedClass cls = owlModel.getOWLNamedClass("SmartObject");
	Collection instances = cls.getInstances(true); //*retrieve all instances of the SmartObject class*
	for (Iterator jt = instances.iterator(); jt.hasNext();) {
	    OWLIndividual object = (OWLIndividual) jt.next();
	    RDFProperty isLocatedOn = owlModel.getRDFProperty("isLocatedOn");
	    RDFIndividual ske_Obj = (RDFIndividual) individual.getPropertyValue(isLocatedOn);
	    if(ske_Obj.getBrowerText() = = "DinnerTable")
	        System.out.println(object.getBrowerText());
	}

SPARQL Query	PREFIX ssont: <http://~/SSONT.owl#> SELECT ?object WHERE{?object ssont: isLocatedOn ssont: DinnerTable .}
SQL Query	Select Object_Name From SmartObject, LocationRelation Where SmartObject.ObjectID=LocationRelation.EntityID and LocationRelation.Place=DinnerTable

conflict between these two definitions according to the principles of OWL language. If we run *Racer*, it detects this inconsistency. In the course of *SS-ONT*'s extension, users are also suggested to use *Racer* to keep their ontology's consistency.

In current implementation of *Sixth-Sense*, we also provide a mechanism to perform rule-based ontology reasoning among the defined ontology concepts (i.e., classes, properties, and instances). It is implemented by combining the Protégé-OWL API and Jess inference engine (Friedman-Hill 07). Our ontology reasoning mechanism supports both RDF Scheme (RDF-S) and OWL Lite (please refer to Smith et al. (2003) for more details). The RDF-S reasoner supports all the RDF-S specifications (e.g., sub-class, sub-property), while the OWL reasoner supports simple constraints on classes and properties (e.g., symmetry, cardinality and transitivity). All these rules are needed to be pre-defined, and some examples are listed in Table 4.

Ontology reasoning is useful for context-aware systems, by which some implicit relationships among defined ontology concepts can be derived. For example, the *Transitivity* rule can be used in spatial reasoning. In *SS-ONT*, we use

an OWL transitive property "*isPartOf*" to define relations between different "*Place*" instances. If two "*isPartOf*" relations between "*Room412*" and building "*25#*", and between "*25#*" and "*Yagami_Campus*" are known, an implicit context that "*Room412*" is part of the "*Yagami_Campus*" can then be deduced. This is because the spatial relation "*isPartOf*" is transitive.

User-Defined Rule-Based Reasoning

In the logical reasoning level, the creation of user-defined inference rules is allowed, which makes the reasoning more flexible. A wide range of high-level contextual information, such as "*what is the user going to do*", "*how should the agent react to the current situation*", can be deduced at this level.

Currently, *SS-ONT* supports rules in the form of the *Semantic Web Rule Language* (SWRL) (Horrocks et al., 2004) and Jess. SWRL is based on a combination of the OWL DL and OWL Lite sublanguages, and it enables users to write Horn-like rules to reason about OWL individuals and to infer new knowledge about these individuals.

Table 4. Instances of OWL ontology reasoning rules

Name	Rule
SubClassOf	(?a rdfs:subClassOf ?b) ∧ (?b rdfs:subClassOf ?c) ⇒ (?a rdfs:subClassOf ?c)
SubPropertyOf	(?a rdfs:subPropertyOf ?b) ∧ (?b rdfs:subPropertyOf ?c) ⇒ (?a rdfs:subPropertyOf ?c)
Transitivity	(?p rdf:type owl:TransitivityProperty) ∧ (?a ?p ?b) ∧ (?b ?p ?c) ⇒ (?a ?p ?c)
InverseOf	(?p1 owl:inerseOf ?p2) ∧ (?a ?p1 ?b) ⇒ (?b ?p2 ?a)
Functional Property	(?p rdf:type owl:FunctionalProperty) ∧ (?a ?p ?m) ∧ (?a ?p ?n) ⇒ (?m owl:sameAs?n)
Symmetry	(?p rdf:type owl:SymmetricProperty) ∧ (?a ?p ?b) ⇒ (?b ?p ?a)

Box 1.

$$< entity(?x) > \ \land \ \lfloor existed\_objectproperty(?x, \ ?y)\rfloor \land$$
$$< datatypeproperty(?x, \ ?num1) \ \{>, \ <, \ =, \ >=, \ <=\} \ num2 >$$
$$\rightarrow < inferred\_objectproperty(?x, \ ?value1) > \qquad (2)$$

We chose to use SWRL for several reasons: (1) SWRL is a standardized language to realize rule interoperation on the Web, that's to say, to share rules among different rule-based systems, and it is designed to be the rule language of the Semantic Web. (2) The Protégé team from Stanford University has developed a full-featured editor for SWRL (O'Connor, Knublauch, & Tu, 2005). The SWRL editor tightly integrates with Protégé-OWL (the editor we used to define *SS-ONT*), and also supports inference with SWRL rules using the Jess rule engine. The highly-interactive interfaces, well-designed documentation and error-checking support make it an ideal tool for experienced users to create, edit and test their rules. (3) One of the most powerful features of SWRL is its ability to support a range of built-ins. A built-in is a predicate that takes one or more arguments and evaluates them as true if the arguments satisfy the predicate. For example, an *"equal"* built-in is defined to accept two arguments and return true if the arguments are the same. Using the built-ins, users can create more flexible rules and more interesting applications. As an emerging rule language, there does not exist a standard rule engine

that can directly execute SWRL rules. Therefore, we integrated a famous rule engine, **Jess**, into our infrastructure. Jess has its own rule language, so current implementation of *Sixth-Sense* also supports rules written in the Jess rule language.

Defining Human-Artifact Interaction Rules
Note that inference rules are used to deduce high-level, implicit contextual knowledge from low-level raw data, and that the premises of some inference rules might involve facts that are newly derived from other inference rules. That is, inference rules are designed and implemented at different levels, as illustrated in Figure 7.

Basic Attribute Reasoning: The lowest reasoning level, the task of which is to read in the updated sensory data (or software data) to derive some state, attribute, or location information about the relevant entity (the entity might be an artifact, a person, or an abstract concept like the weather). Inference rules at this level usually follow the form given in Equation (2): (see Box 1)

Where: $< x_1, x_2, ..., x_n >$ denotes that the formula can involve *"1"* or *several* such terms defined

Figure 7. Different reasoning levels

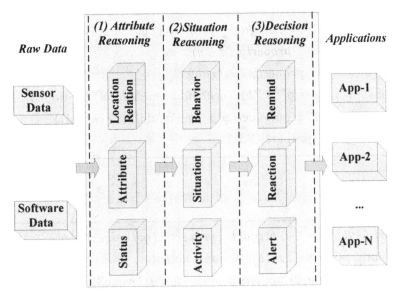

Table 5. Instances of inference rules for human-artifact interaction

Rule Type Name(Level)	Inference rules represented by SWRL
Weather R1 (L-1)	$City(?x) \land hasTemperature(?x, ?y) \land swrlb : greaterThanOrEqual(?y, 25) \land$ $\rightarrow hasTemperatureDegree(?x, Hot)$
Location Near R2 (L-1)	$Person(?m) \land hasSensor(?m, ?s1) \land U3D(?s1) \land SmartObject(?n) \land$ $hasSensor(?n, ?s2) \land U3D(?s2) \land hasLxValue(?s1, ?x1) \land$ $hasLxValue(?s2, ?x2) \land swrlb : subtract(?x3, ?x1, ?x2) \land$ $swrlb : abs(?x4, ?x3) \land swrlb : lessThanOrEqual(?x4, 500) \land$ $hasLyValue(?s1, ?y1) \land hasLyValue(?s2, ?y2) \land swrlb : subtract(?y3, ?y1, ?y2) \land$ $swrlb : abs(?y4, ?y3) \land swrlb : lessThanOrEqual(?y4, 500) \rightarrow isLocatedNear(?m, ?n)$
Behavior Pick-Up R3 (L-2)	$SmartObject(?A) \land hasSensor(?A, ?s) \land U3D(?s) \land Person(?B) \land$ $hasLzValue(?s, ?z1) \land hasOldLzValue(?s, ?z2) \land$ $swrlb : greaterThan(?z1, ?z2) \land swrlb : subtract(?z3, ?z1, ?z2) \land$ $swrlb : greaterThan(?z3, 120) \land isLocatedNear(?B, ?A)$ $\rightarrow hasBehavior(?B, PickUp) \land hasInteractiveObject(?B, ?A)$

in this set; $[x_1, x_2, ..., x_n]$ denotes that *"0"* or *several* such terms defined in this set can be used in this formula; and $\{x_1, x_2, ..., x_n\}$ denotes that the formula can only select one term from the set. Two instances of rules (*R1* and *R2*) at this level are given in Table 5. Note that some SWRL built-ins are used in these rules, preceded by the namespace qualifier: *swrlb*).

Situation Reasoning: By reflecting changes in information relating to the state or location of smart artifacts, some higher level contexts, such as a human's behavior or current activity in a room, can be deduced. A general form for this

Box 2.

$$
\begin{aligned}
&human(?\,x)\wedge\, <artifact(?\,y)>\,\wedge\, <artifact\_property(?\,y,?\,value1)>\,\wedge\\
&<human\_artifact\_location\_property(?\,x,?\,y)>\,\wedge\\
&[existed\_human\_behavior\_property(?\,x,?\,value2)]\\
&\rightarrow\{human\_behavior\_property(?\,x,?\,b)\wedge behavior\_metadata(?\,b,?\,value3),\\
&human\_activity\_property(?\,x,?\,a)\wedge activity\_metadata(?\,a,?\,value4)\}\\
&\wedge[other\_human\_artifact\_property(?\,x,?\,y)]
\end{aligned} \tag{3}
$$

Box 3.

$$
\begin{aligned}
&human(?\,x)\wedge\, <artifact(?\,y)>\,\wedge\, <artifact\_property(?\,y,?\,value1)>\,\wedge\\
&<human\_behavior\_property(?\,x,?\,b),human\_activity\_property(?\,x,?\,a)>\,\wedge\\
&[behavior\_property(?\,b,?\,value2),activity\_property(?\,a,?\,value3)]\\
&\rightarrow reaction\_property(?\,x,?\,value4)
\end{aligned} \tag{4}
$$

kind of rule is given in Equation (3), and an instance (*R3*) is listed in Table 5. (see Box 2)

Decision Making: How the application or agent reacts to the current situation or environment change is dealt with at this reasoning level. For instance, when a robot detects the resident is going to leave home, and also finds that his room-key is still on the table, the robot can generate an alert to the human, reminding him to take the key. A general form for this kind of rule is given in Equation (4). (see Box 3)

Integration with an Inference Engine

To allow Jess to integrate with SWRL rules, our *Sixth-Sense* infrastructure performs the following four steps: (1) represents relevant OWL knowledge defined in *SS-ONT* as Jess facts; (2) represents SWRL rules as Jess rules; (3) performs inference using those rules; and (4) reflects the resulting Jess facts back to the *SS-ONT* knowledge base. In our system, the interaction between SWRL rules and the Jess rule engine is implemented through the SWRL-Jess Bridge API (O'Connor, Knublauch, & Tu, 2005).

The interaction between Jess and SWRL is data-driven in our system. That is, when the system receives updated data from a sensor node, the *four*-step interaction process will be triggered to see if some new facts can be deduced. An example of the overall inference process is shown in Figure 8. In this example, if a smart book, called *BookA*, was previously on the table and, in the next instant, picked up by a nearby person (called *Bob*), the sensed location value (read from U67) for *BookA* will change. Rule *R3* (see Table 5) will be triggered then and a new fact – *Bob* picks up *BookA* – will be generated.

DEVELOPING APPLICATIONS

To demonstrate the feasibility of our programming platform, we prototyped a smart-home application platform called *Home-Explorer*, which involves various applications related to human-artifact interaction in smart homes. In this section, we will describe two such services: an object locating service and a reminding service.

Figure 8. Rule-based reasoning process

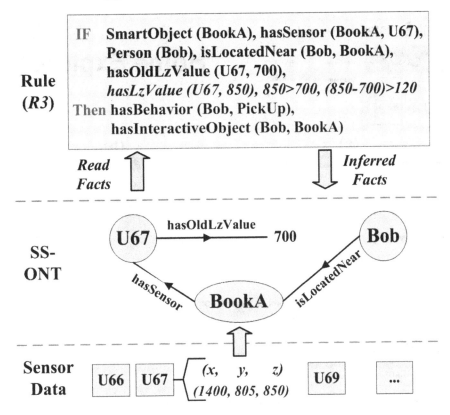

Box 4.

$$SmartObject(?\,A) \wedge SkeletonObject(?\,B) \wedge hasSensor(?\,A,?\,h) \wedge U3D(?\,h) \wedge hasLxValue$$
$$(?\,h,?\,x) \wedge hasMaxX(?\,B,?\,x1) \wedge hasMinX(?\,B,?\,x2) \wedge swrlb:lessThanOrEqual(?\,x,?\,x1) \wedge$$
$$swrlb:greaterThanOrEqual(?\,x,?\,x2) \wedge hasLyValue(?\,h,?\,y) \wedge$$
$$hasMaxY(?\,B,?\,y1) \wedge hasMinY(?\,B,?\,y2) \wedge swrlb:lessThanOrEqual(?\,y,?\,y1) \wedge \qquad (5)$$
$$swrlb:greaterThanOrEqual(?\,y,?\,y2) \wedge hasLzValue(?\,h,?\,z) \wedge$$
$$hasHeight(?\,B,?\,z1) \wedge swrlb:greaterThanOrEqual(?\,z,?\,z1) \rightarrow isLocatedOn(?\,A,?\,B)$$

Object Locator Application

We use the following scenario to describe the functionality of this service: *Assume it is 7:50 in the morning, and Bob is going to work and must catch the bus at 8:00. He always wears his watch to work, but has forgotten where he put it last night. As a result, he takes too much time to find it and misses the bus.*

Situations like this are common and cause us considerable inconvenience in our daily lives. However, a real-world search system will make it possible to solve this problem. Using the U3D (see "A Domain Specific Ontology for Smart Homes") sensor data from smart artifacts and the

Figure 9. Real-world search interfaces

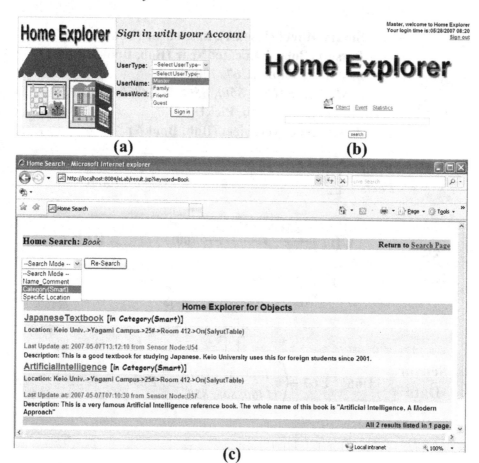

(a)

(b)

(c)

following rule R4 (formulated by Equation (5)), we developed an indoor-object search system. (see Box 4)

This rule can be explained as this: For smart object A and skeleton object B, if the downward projection of A is within the range of B, we conclude that A is placed on B.

The main interfaces of this service are given in Figure 9. To ensure security, a user must first login to our system via an authorized account, including user type (e.g., *Master* or *Friend*), user name, and password (see Figure 9 (a)). In terms of the security policy defined in our ontology, different user types can only search objects in the relevant level (e.g., a *Friend* user cannot search a *FamilyProtected* level object). This policy is reflected

in the search results. The main search interface (see Figure 9 (b)) is presented after the user logs in. The user can then input a certain keyword to search for objects. We provide three search modes: (1) *Search by object names*: an individual object (e.g., Bob's wallet) can be searched for using this mode; (2) *Search by category*: in some cases, people may want to find a series of objects that belong to the same category (e.g. Bob may want to choose an interesting book to read from among his books). From the ontology's point of view, this mode can be interpreted as returning all the OWL individuals that belong to a targeted OWL class; and (3) *Search by particular locations*: this can list all the smart objects placed on a specific skeleton object (e.g. what objects are placed on the dinner

Box 5.

$$SmartObject\ (?x) \wedge U3D(?y) \wedge hasSensor\ (?x, ?y) \wedge hasUpdateTime\ (?y, ?t) \wedge$$
$$swrlb: intervalGreaterThan\ (?t, 2) \rightarrow notUpdate\ (?x, "true") \tag{6}$$

$$Butter\ (?x) \wedge notUpdate\ (?x,\ "true") \rightarrow InRefrigerator\ (?x,\ "true") \tag{7}$$

table). Figure 9 (c) illustrates the search results for books obtained via the category search mode, where the relative locations of these searched books are listed.

Out-Refrigerator Application

A possible scenario for this application is as follows: *Assume Bob's mother, Jane, wants to make butter bread for her family. However, after using the butter, she forgets to put it back in the refrigerator. As a result, the unused butter melts before she notices it after cooking.*

People are prone to forget things, and situations like this are common in everyone's life. However, *Home-Explorer* can reduce the occurrence of such problems. The Out-Refrigerator Alert component is designed to send alerts when objects that should be kept in a refrigerator are left outside it for a considerable period of time (e.g., 15 minutes). To implement this application, we must first find out whether the object being monitored is in the refrigerator, and if not, what time it was taken out. Because we use U3D sensors to track object locations, and the ultrasonic sensor signal is blocked when U3D sensors are placed in a closed container (e.g., a refrigerator) (Guo, Satake, & Imai, 2006), we investigate whether a monitored object, such as the butter, is in the refrigerator in terms of the update time property of its U3D sensor. If the U3D sensor it is equipped with is not updated during a predefined period, we consider that this object is in the refrigerator. *R5* and *R6*, formalized by Equation (6) and Equation (7), are used to represent this reasoning process. As the Protégé version (Protégé 3.3_bata) we used does not provide for

the implementation of time-related built-ins, we define two new built-ins, *intervalGreaterThan* and *intervalLessThan*, to measure whether the interval between the current time and the latest sensor update time is greater than (or less than) an assigned length of time (refer to "http://protege.cim3.net/cgi-bin/wiki.pl?SWRLTab" to see how to define new build-ins). For instance, in *R5*, the *intervalGreaterThan* built-in returns 'true' when the sensor has not been updated for two minutes or more. (see Box 5)

If the prior state of the butter is in a refrigerator, but at a certain time t_i, its U3D sensor value is updated (measured by the *intervalLessThan* built-in), we consider that the butter has been taken out of the refrigerator, and at the same time we record t_i as the starting time that the butter left the refrigerator. This inference process is formalized by *R7* in Equation 8. In addition, if the butter is out of the refrigerator for more than 15 minutes, an alert message will be sent to Jane. This decision-making rule, *R8*, is given in Equation 9. (see Box 6)

The contexts used in this application, i.e., the butter is in (or out of) the refrigerator, fit in the category of artifact-artifact relationships. The relationships between artifacts are significant for most artifact-managing or home-monitoring systems. Our infrastructure can be used to implement other similar applications in these two fields. For example, Smart Toolbox (Lampe & Strassner, 2003), which sends alerts if an operator forgets to put the tools he has used back in a toolbox; an application that can send a message to a mother that her children have not put their toys in order

Box 6.

$$Butter(?\,x) \wedge inRefrigerator(?\,x,\ "true") \wedge U3D(?\,y) \wedge$$
$$hasSensor\ (?\,x,?\,y) \wedge hasUpdateTime\ (?\,y,?\,t) \wedge swrlb: intervalLessThan\ (?\,t,1) \tag{8}$$
$$\rightarrow inRefrigerator\ (?\,x,\ "false") \wedge outStoragePlaceSince(?\,x,?\,t)$$

$$Butter(?\,x) \wedge inRefrigerator(?\,x," false") \wedge outStoragePlaceSince(?\,x,?\,t)$$
$$\wedge Person(Jane) \wedge swrlb: intervalGreaterThan\ (?\,t,15) \tag{9}$$
$$\rightarrow ourStoragePlaceAlert\ (Jane,?\,x)$$

after playing with them; and a monitoring application that can generate a warning if it detects that a cup of tea is placed near a notebook PC or on a book. The logical relations between artifacts are also crucial to smart home systems, as reported in our previous work (Guo, Satake, & Imai, 2006).

EVALUATION AND DISCUSSION

This section evaluates the performance our *Sixth-Sense* infrastructure described in "Context Infrastructure Implementation," which involves two aspects: (1) its runtime performance; (2) the effectiveness of our rule-based reasoning mechanism.

Performance of Context Reasoning

We identified context reasoning (the reasoning mechanisms are presented in "Context Reasoning") as a potential performance bottleneck of our infrastructure, so we did a series of experiments to evaluate its runtime performance. These experiments were conducted on two Windows workstations with different hardware configurations (1.0 GB RAM with P4/1.0GHz, and P4/2.6GHz). We used three context data sets to evaluate our system's scalability. These three test data sets, including the real one we used in our test environment, i.e., *SS-ONT-v1.5*, which can be parsed into about 2000 RDF triples, and two other simplified versions of *SS-ONT* (with fewer classes

and instances), amount to about 1000 triples and 500 triples, respectively. We used four rule sets to test the performance of our context reasoner. The smallest rule set includes only one rule, and the biggest set has twenty rules.

The results of the experiments are illustrated in Figure 10 (results are calculated as the average of five runs.). From these results, it is not difficult to conclude that logic-based context reasoning depends on four major factors: *size of ontology*, *CPU speed*, *number of rules applied*, and *complexity of rules*. More concretely, the size of ontology is determined by its *TBox* definition - definitions about OWL classes and properties, and *ABox* definition - definitions about instances, i.e., the number of smart objects and persons in a smart home. The prior part is almost the same in different homes while the latter part changes among different families. The complexity of a rule is in essence determined by the number of atoms in this rule. From the above observations, we further derive an estimation model that can be used to anticipate the reasoning time for different scales of smart spaces, as shown in formula Equation (10). (see Box 7)

Where: $Size_{TBox}$ denotes size (in triples) of the *TBox* definition of *SS-ONT*; M and N denote, respectively, the number of individuals (e.g., smart objects and persons) and the number of rules; $Class\_Num_i$ and $ProNum\_Obj_i$ denote the number of parent classes and the number of properties for a specific individual i; Complexity of a

Figure 10. Performance of reasoning with the changes of ontology scale (left) and CPU speed (right)

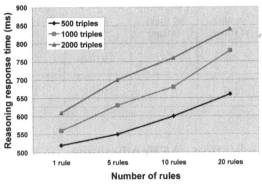 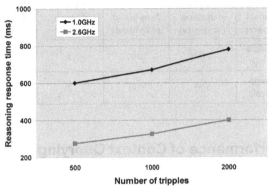

Box 7.

$$ReasoningTime \approx \mu \frac{(Size_{TBox} + \sum_{i=1}^{M}(Class\_Num_i + ProNum\_Obj_i))^{\exp 1} * (\sum_{i=1}^{N} Comp\_Rule_i)^{\exp 2}}{Speed_{cpu}}$$

(10)

rule i is denoted by $Comp\_Rule_i$; *exp1* and *exp2* are two exponents whose values are between 0 and 1 (according to the results shown in Figure 10, their values cannot be bigger than 1), which reflect the rate of change of the reasoning time with the growth of ontology size and complexity of rules; μ is a positive coefficient that relates the change in reasoning time to the change in its influence factors (e.g., ontology size, number of rules, etc.).

Let *exp1 = 1* and *exp2 = 1*, we can get the upper-limit value of reasoning time, i.e., the anticipated maximum reasoning time. Suppose there is a 1000 triple-sized ontology, and 10 rules with a mean complexity of 8, we can read from Figure 10 that its reasoning time is about 680 ms (P4/1.0 GHz). Using these values to replace the related variables in Equation (10), we can derive an approximate value for μ: $\mu = 0.085$. The proposed estimation model is useful for us to anticipate the computational overhead for a given-sized problem.

As listed in Table 6, for a middle-scaled smart space (a space with 50 smart objects and 50 rules), the anticipated maximum reasoning time is about 1.7s; whereas for a large-scale one (100 smart objects and 100 rules involved), this value increases to 2.8s.

The performance study results reveal that rule-based context reasoning is a computation-intensive task, and the reasoning time will be human-perceivable if the scale of the smart space increases. However, for most non-time-critical applications (e.g., searching for smart objects), as their real-time requirement is not likely to be critical, a perceivable delay (one or two seconds) caused by context reasoning is acceptable. The evaluation results also suggest that, with a suitably scaled ontology size and rule-set complexity, our rule-based context reasoning mechanism can also work in some time-critical applications, for example, security and emergency situations.

Table 6. Anticipated maximum reasoning time in different scales of smart spaces

Smart Spaces	Size of TBox definition (in triples)	Number of Individuals	Parent Classes (Mean)	Number of Properties (Mean)	Number of Rules	Complexity of Rules (Mean)	CPU Speed	Maximum Reasoning Time
Middle-Scale	500	50	3	7	50	8	2.0 GHz	1.7s
Large-Scale	500	100	2	6	100	7	3.0 GHz	2.8s

Performance of Context Querying

We used SPARQL query statements like Equation (1) to test the performance of context querying (the querying mechanism is described in "Context Querying Engine"). The experiment result, shown in Figure 11, clearly demonstrates that the increase of matched facts results in a corresponding increase of the querying response time. This is mainly because a bigger number of matched facts imply that there are more contexts (in *SS-ONT*) defined in the graph pattern of the query statement applied, which cost more matching time. Size of the ontology applied becomes another factor that influences the querying performance of our system. We also measured the loading time (time cost from OWL files to the main memory) of different-scaled ontology data sets. The result shown in Figure 11 indicates that the loading time of an ontology data set is, to some extent, proportional to its size.

Above experiment results and the proposed computational model also suggest several possible ways to improve the performance of our system. Controlling the scale of context dataset seems to be one good way to significantly reduce the context reasoning time. For example, we can separate the *SS-ONT* ontology into several sub-domain ontologies (e.g., kitchen domain and bathroom domain), and provide several resource-rich devices to process them respectively. Designing optimized rules and utilizing high-performance processors are two other effective ways to achieve a better reasoning performance.

Evaluation of Effectiveness

Because our context reasoning mechanism is layered on a series of inference rules, the effectiveness of the defined rules will directly influence the performance of our system. Eight typical inference rules are reported in this chapter, and we tested them one by one. The experiments were performed as follows. For each inference rule R_i, we set the test time as three days, during which a situation similar to the scenario for R_i recurred 30 times. R5~R8, which belong to the same application, were evaluated together, but with one difference in that we used a large carton with a cover to replace the refrigerator because there was no actual refrigerator in the test area. We carried out thirty tests and put the object being monitored in the carton for 15 of the tests and took it out for the other 15.

Definition 6 (Diverse Test Types): In our system, an *accurate* test is one in which an object property (e.g., location) or a hidden object is detected at the right time by the right rule, it scores a true positive (TP). An incorrect claim scores a false positive (FP), and it's an *error* test. A *failed* test is one in which an inference rule should have been triggered but was not, which scores a false negative (FN).

We then used two standard metrics to summarize our system's effectiveness. *Precision* is the probability that a given inference about an object property is correct. *Recall* is the probability that our system will correctly infer a given true event. They can be expressed in formula Equation (11):

Figure 11. Performance of querying

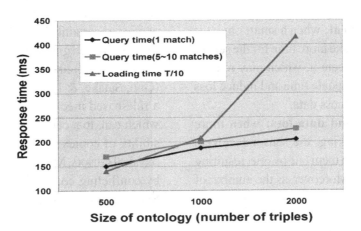

Table 7. Experiment results of effectiveness

Rule Name (Rule ID)	True positives	False positives	False negatives	Precision (%)	Recall (%)
R1	30	0	0	100	100
R2	26	3	4	90	87
R3	24	1	6	96	80
R4	29	1	0	97	100
R5	15	3	0	83	100
R6	15	3	0	83	100
R7	13	0	2	100	87
R8	12	0	3	100	80
Total	164	11	15	94	92

$$Precision = \frac{TP}{TP + FP}; \quad Recall = \frac{TP}{TP + FN} \quad (11)$$

The results are shown in Table 7, where it can be seen that *Sixth-Sense* correctly inferred that an event occurs 94 percent of the time. Of the totally 180 tests that actually happened, *Sixth-Sense* detected 89 percent correctly. For the three reasoning levels discussed in "User-Defined Rule-Based Reasoning," the lower the level of the rule, the better it performed (in terms of the "recall" measurement results). This is mainly because higher-level rules mostly rely on the inference results of lower-level rules. For example, to use *R3*, we must first find out whether the person and the object referred to in this rule are near each other; this information would be previously deduced by a lower-level rule *R2*. The results indicate that our rule-based detection mechanism is technically sound. By analyzing the real-time testing data, we generalized the following reasons that a failed or error test may occur.

1. **Sensor noise:** sometimes sensor noise exceeds the threshold value we set in the rules,

which may trigger an inference rule, causing an error test.

2. **Object movement:** when a smart object is moving, it will take more time for the sensor receivers to relocate it, which may require a longer sensor update time and induce loss of important process data.

3. **Sensor delay and data loss:** when more sensors are working, the sensor receivers have to cope with them one by one, resulting in more delays. Moreover, as the number of sensors increases, the processing capacity of the sensor system decreases. According to our experimental statistics, when 15 U3D sensors are working simultaneously, about 4% of the sensor data will be lost.

4. **Inference rule defects:** Other failures may be due to the inference rules themselves. The imprecise nature of the common-sense knowledge background sometimes makes it difficult to create a faultless inference rule.

In summary of the above evaluations, to further improve our system's performance, better sensor technologies, transmission protocols, underlying common sense knowledge, and reasoning technology still need to be used.

Discussion about Uncertainty

In the current implementation of our infrastructure, we assume that the sensed or inferred context information is complete and accurate. However, this assumption is sometimes unjustified, as the acquired contexts are sometimes inconsistent for the following reasons: (1) the raw data collected from physical sensors is inaccurate or unavailable as a result of noise or sensor failure (e.g., sensor is broken or out of buttery); (2) contexts acquired from distinct sources may be inconsistent with each other; (3) the inference rule has flawed or incomplete domain heuristics (e.g., for *R6*, when the butter is placed in a drawer, the sensor attached to it also does not update). Keeping the robustness

of ubiquitous computing applications is important for end users to accept them and use them. Though context uncertainty is not the main concern of this chapter, we still give an overlook for handling this problem under our infrastructure. In our prior work (Guo, Satake, & Imai, 2006), we have proposed a rule-based mechanism to detect sensor faults, which can, to a certain extent, decrease the possibility of context inaccuracy caused by fallible physical sensors. More inconsistencies are induced by conflicting conclusions from different rules. Since the SWRL rule language we used is based on a strictly monotonic interpretation, contradictory conclusions are allowed to be drawn under its current semantics. To deal with this issue, we can introduce well-founded non-monotonic semantics, such as the defeasible logic (Antoniou & Wagner, 2003) and assumption-based logic (Chen et al., 2004a), into ontology-based context-aware reasoning.

SUMMARY

We have described our efforts to incorporate semantic web technologies into the development of context-aware systems. By providing explicit ontology representation, expressive context querying, and flexible reasoning mechanisms, the *Sixth-Sense* infrastructure facilitates the building of context-aware applications in smart homes. The *Home-Explorer* application shows how applications based on our infrastructure can be easily implemented. The runtime performance and the effectiveness of our system are validated through a series of experiments.

In our preliminary implementation, the *SS-ONT* ontology is stored in OWL files. Operations related to this ontology, such as query and update, are implemented by loading, updating, and saving these files, which reduces the execution efficiency to lower than that achieved using traditional databases. We therefore need to investigate more efficient methods of OWL storage and retrieval, such as Sesame (Broekstra, Kampman, & Harmelen, 2002). The context infrastructure discussed in this chapter mainly handles the contexts generated in a single smart domain (e.g., gathering contexts produced in a smart home and allowing the access of contexts by local applications). We are now exploring mechanisms to manage and exchange contexts in multi-domain ubicomp environments (Guo, Sun, & Zhang, 2010), by which applications can retrieve the needed contexts from remote domains. Furthermore, the infrastructure presented here is designed to facilitate the development of context-aware applications by developers (or experts), an extended version that also allows end users (or non-experts) to design/customize context-aware applications (in terms of their preferences, particular needs, distinct environment settings) is leveraged by the OPEN platform (Guo, Zhang, & Imai, 2011).

REFERENCES

Antoniou, G., & Wagner, G. (2003). Rules and defeasible reasoning on the semantic web. In Proc. of RuleML Workshop 2003 (pp. 111-120).

Avancha, S., Patel, C., & Joshi, A. (2004). Ontology-driven Adaptive Sensor Networks. In Proc of MobiQuitous'04 (pp. 194-202).

Beigl, M., & Gellersen, H. W. (2003). SmartIts: An embedded platform for smart objects. In Proc. of the Smart Objects Conference, Grenoble, France.

Beigl, M., Gellersen, H. W., & Schmidt, A. (2001). Mediacups: Experience with design and use of computer-augmented everyday objects. *Computer Networks, 35*(4), 401–409. doi:10.1016/S1389-1286(00)00180-8

Berners-Lee, T., Hendler, J., & Lassila, O. (2001). The Semantic Web. *Scientific American, 284*(5), 34–43. doi:10.1038/scientificamerican0501-34

Brickley, D., & Guha, R. (2000). Resource Description Framework (RDF) schema specification, from http://www.w3.org/ TR/ RDF-schema.

Broekstra, J., Kampman, A., & Harmelen, F. (2002). Sesame: A Generic Architecture for Storing and Querying RDF and RDF Schema. In Proc. of the 1st International Semantic Web Conference (pp. 54-68), Sardinia, Italy.

Chen, H., Finin, T., Joshi, A., Perich, F., Chakraborty, D., & Kagal, L. (2004a). Intelligent agents meet the semantic web in smart spaces. *IEEE Internet Computing, 19*(5), 69–79. doi:10.1109/MIC.2004.66

Chen, H., Perich, F., Finin, T., & Joshi, A. (2004b). Standard ontology for ubiquitous and pervasive applications. In *Proc. of the MobiQuitous'04.* SOUPA.

Dey, A. K., & Abowd, G. D. (2000). Towards a better understanding of context and context-awareness. In Proc. of CHI'2000.

Friedman-Hill, E. (2007). Jess, the Rule Engine for the Java Platform, from http://www.jessrules.com/ jess index.shtml.

Guo, B., Satake, S., & Imai, M. (2006). Sixth-Sense: Context Reasoning for Potential Objects Detection in Smart Sensor Rich Environment. In Proc. of the IAT-06 (pp. 191-194), Hong Kong, China.

Guo, B., Sun, L., & Zhang, D. (2010). *The Architecture Design of a Cross-Domain Context Management System* (pp. 499–504). Germany: In Proc. of Percom Workshops.

Guo, B., Zhang, D., & Imai, M. (2011). Towards a Cooperative Programming Framework for Context-Aware Applications. *Personal and Ubiquitous Computing, 15*(3), 221–233. doi:10.1007/s00779-010-0329-1

Haarslev, V., & Moller, R. (2001). Racer system description. In Proc. of the International Joint Conference on Automated Reasoning (pp. 701-705).

Henricksen, K. (2002). *02*. Modeling Context Information in Pervasive Computing Systems. In Proc. of Pervasive.

Horrocks, I., Patel-Schneider, P. F., Boley, H., Tabet, S., Grosof, B., & Dean, M. (2004). SWRL: A Semantic Web Rule Language Combining OWL and RuleML, from http://www.daml.org/2004 /04/ swrl.

Lampe, M., & Strassner, M. (2003). *The potential of RFID for moveable asset management*. Seattle: In Proc. of UbiComp.

Lewis, M., Cameron, D., Xie, S., & Arpinar, B. (2006). ES3N: A Semantic Approach to Data Management in Sensor Networks. In Proc. of 5th International Semantic Web Conference (ISWC'06).

Loke, S. W. (2004). Logic programming for context-aware pervasive computing: Language support, characterizing situations, and integration with the Web. In Proc. of the IEEE/WIC/ACM Conference on Web Intelligence (WI'04).

Meyer, S., & Rakotonirainy, A. (2003). A survey of research on context-aware homes. In Proc. of the ACSW frontiers 2003 (pp. 159-168), Adelaide, Australia.

O'Connor, M. J., Knublauch, H., & Tu, S. W. (2005). Supporting rule system interoperability on the semantic web with SWRL. In Proc. of the 4th International Semantic Web Conference.

Philipose, M., Fishkin, K., Perkowitz, M., Patterson, D., Kautz, H., & Hahnel, D. (2004). Inferring activities from interactions with objects. *IEEE Pervasive Computing / IEEE Computer Society [and] IEEE Communications Society, 3*(4), 50–57. doi:10.1109/MPRV.2004.7

Prud'hommeaux, E., & Seaborne, A. (2006). SPARQL Query Language for RDF, from http://www.w3.org/TR/rdf-sparql-query/.

Ranganathan, A., & Campbell, R. H. (2003). A Middleware for Context-Aware Agents in Ubiquitous Computing Environments. In Proc. of ACM/IFIP/USENIX International Middleware Conference, Brazil.

Salber, D., Dey, A. K., & Abowd, G. D. (1999). The Context Toolkit: Aiding the development of context-enabled applications. In Proc. of CHI'99 (pp. 434–441).

Siio, I., Rowan, J., Mima, N., & Mynatt, E. (2003). Augmented everyday things. In *Proc. of the Canadian annual conference Graphics Interface*. Digital Decor.

Smith, M., Welty, C., & McGuinness, D. (2003). Web ontology language (OWL) guide Version 1, from http://www.w3.org/ TR/ owl-guide/.

Strang, T., & Linnhoff-Popien, L. (2004). A Context Modeling Survey. In Proc. of the First International Workshop on Advanced Context Modeling, Reasoning and Management (pp. 33-40).

Uschold, M., & Gruninger, M. (1996). Ontologies: Principles, Methods and Applications. *The Knowledge Engineering Review, 11*(2), 93–136. doi:10.1017/S0269888900007797

Wang, X. H., Zhang, D. Q., Dong, J. S., Chin, C. Y., & Hettiarachchi, S. (2004). Semantic Space: An infrastructure for smart spaces. *IEEE Pervasive Computing / IEEE Computer Society [and] IEEE Communications Society, 3*(3), 32–39. doi:10.1109/MPRV.2004.1321026

Yamada, T., Iijima, T., & Yamaguchi, T. (2005). Architecture for Cooperating Information Appliances Using Ontology. In Proc. of 19th Annual Conference of the Japanese Society for Artificial Intelligence.

Yap, K. K., Srinivasan, V., & Motani, M. (2005). Max: Human-centric search of the physical world. In Proc. of SenSys'05 (pp. 166–179).

KEY TERMS AND DEFINITIONS

Smart Artifacts: Sensor-enhanced everyday artifacts, which can perceive contexts from ambient environment and act on behalf of user needs.

Context-Aware Systems: refer to a general class of intelligent systems that can sense their physical environment, i.e., their context of use, and adapt their behavior accordingly.

Context: any information that can be used to characterize the situation of an entity. An entity is a person, or an object that is considered relevant to the interaction between a user and an application, including the user and application themselves.

Context-Aware Infrastructure: it acts as middleware between context-aware applications and context producers, which can collect contexts from various context providers and provide query mechanisms for applications to selectively access to the contexts they need.

Context Reasoning Engine: An engine in the context-aware infrastructure that can infer high-level contexts (e.g., user behavior) from low-level contexts (e.g., raw sensory data).

Ontology: In the context of knowledge management, ontology is referred as the shared understanding of some domains, which is often conceived as a set of entities, relations, axioms, facts and instances

Semantic Web: The Semantic Web is an evolving development of the World Wide Web in which the meaning (semantics) of information and services on the web is defined, making it possible for the web to understand and satisfy the requests of people and machines to use the web content. It derives from World Wide Web Consortium director Sir Tim Berners-Lee's vision of the Web as a universal medium for data, information, and knowledge exchange.

OWL: The Web Ontology Language (OWL) is a family of knowledge representation languages for authoring ontologies, and is endorsed by the World Wide Web Consortium. OWL can be used to explicitly represent the meaning of terms in vocabularies and the relationships between those terms.

SWRL: it is a proposal for a Semantic Web Rule Language, combining sublanguages of the OWL Web Ontology Language (OWL DL and Lite) with those of the Rule Markup Language (Unary/Binary Datalog).

Chapter 16
State–of–the–Art Assistive Technology for People with Dementia

Clifton Phua
Institute for Infocomm Research, Singapore

Aung-Phyo-Wai Aung
Institute for Infocomm Research, Singapore

Patrice Claude Roy
Sherbrooke University, Canada

Weimin Huang
Institute for Infocomm Research, Singapore

Hamdi Aloulou
Institute for Infocomm Research, Singapore

Mohamed Ali Feki
Alcatel-Lucent Bell-Labs, Belgium

Jit Biswas
Institute for Infocomm Research, Singapore

Jayachandran Maniyeri
Institute for Infocomm Research, Singapore

Andrei Tolstikov
Institute for Infocomm Research, Singapore

Alvin Kok-Weng Chu
Temasek Polytechnic, Singapore

Victor Siang-Fook Foo
Institute for Infocomm Research, Singapore

Duangui Xu
Temasek Polytechnic, Singapore

ABSTRACT

The work is motivated by the expanding demand and limited supply of long-term personal care for People with Dementia (PwD), and assistive technology as an alternative. Telecare allows PwD to live in the comfort of their homes for a longer time. It is challenging to have remote care in smart homes with ambient intelligence, using devices, networks, and activity and plan recognition. Our scope is limited to mostly related work on existing execution environments in smart homes, and activity and plan recognition algorithms which can be applied to PwD living in smart homes. PwD and caregiver needs are addressed in a more holistic healthcare approach, domain challenges include doctor valida-tion and erroneous behaviour, and technical challenges include high maintenance and low accuracy. State-of-the-art devices, networks, activity and plan recognition for physical health are presented; ideas for developing mental training for mental health and social networking for social health are explored.

DOI: 10.4018/978-1-61692-857-5.ch016

There are two implications of this work: more needs to be done for assistive technology to improve PwD's mental and social health, and assistive software is not highly accurate and persuasive yet. Our work applies not only to PwD, but also the elderly without dementia and people with intellectual disabilities.

INTRODUCTION

Long-term personal care or hands-on assistance is getting more expensive and difficult in many developed countries. This is directly related to:

- Demographic trends of the population living longer and having fewer children
- Higher economic costs of professional healthcare and lesser government subsidies
- Increased workload for healthcare professionals and insufficient residential care facilities
- Dementia is a common, multiple-stage, and long-duration disease

Dementia is caused mainly by Alzheimer's disease and Dementia with Lewy Bodies, and is significantly more common in the elderly population over 65 years old. People with Dementia (PwD) gradually lose memory and cognition, meaning that they lose their ability to carry out Activities of Daily Living (ADLs are familiar tasks for day-to-day function), learn, solve problems, and communicate.

Worldwide, there are more than 20 million PwDs, with more than 4 million new cases every year. PwDs projected to increase by 100% in developed countries and 300% in Asia by 2040 (Ferri et al., 2005). In Europe, about 2% of the elderly population have mild dementia, which is around 2 million PwD (Information Society Technologies (IST) Program, 2009).

It is generally acknowledged that PwDs have common patterns of symptom progression. There are 7 stages with loose classifications of mild (Stages 2 and 3), moderate (Stages 4 and 5), and

severe (Stages 6 and 7). In summary, PwD will exhibit no symptoms in Stage 1, and start to have some mental and social health problems in Stages 2 and 3. PwD will exhibit severe mental and social health problems and some physical health problems in Stages 4 and 5, and will completely lose all physical, mental, and social abilities in Stages 6 and 7. After diagnosis, PwD die within an average of 4 to 6 years and can vary from 3 to 20 years (Reisberg et al., 1984). Dementia can be mitigated with medication, more specifically, moderate-to-severe Alzheimer's disease can treated with consumption of memantine (Reisberg et al., 2003). If the PwD is personally cared for at home, most caregivers (usually PwD's spouse or children) are known to be under extreme stress and require some form of assistance (Belle et al., 2006).

The current practice is to place PwD with severe dementia in residential care: community-based group homes and supervised apartments. Healthcare professionals will be assigned to look after PwD and has to be physically present in order to prompt PwD step-by-step in ADLs. An alternative to personal care is assistive technology, more specifically cognitive assistance. With the right devices, networks, activity and plan recognition, assistive technology can possibly accommodate the aging populations needs with minimal primary caregivers stress and at an affordable healthcare cost. In addition, many elderly has substantial amount of disposable income and are not adverse to using technology (Brewster, & Goodman, 2008). Therefore, only assistive technology can help to scale long-term care to many PwD and it seems that assistive technology which monitors human behaviour will eventually become more robust for common use (Philipose, 2009).

Assistive technology touches across broad fields such as health science (ADLs and cognitive errors), electrical and electronics engineering (sensors and networking protocols), and computer science (activity and plan recognition). It can refer to terms such as ambient intelligence, healthcare informatics, telemedicine, telehealth, gerontechnology, and telecare. Ambient Intelligence (AmI) is the use of sensitive and responsive sensors, capable of working together to support PwD and caregivers in improving their physical, mental, and social health. Healthcare informatics is the study of use and processing of data, information, and knowledge applied to medicine, healthcare and public health.

The following assistive technology terms are part of healthcare informatics. Telemedicine is the transfer of medical information over the telephone, Internet, or other telecommunications technologies for consulting, remote medical procedures or examinations. Likewise, telehealth is the delivery of health-related services and information. The field of gerontechnology is the to bring technological environments to health, housing, mobility, communication, leisure and work of elderly people.

The most relevant form of assistive technology for PwD is telecare, which allows PwD to live in the comfort of their homes for a longer time. Telecare is remote care in smart homes with ambient intelligence, using devices, networks, and activity and plan recognition. It automatically logs and detects when PwD needs help or makes errors, and timely prompts or coaches PwD (for example, through audio or visual prompts), notifies caregivers (for example, through in-house or phone alerts) or healthcare professionals (for example, through a 24-hour monitoring centre) to take some corrective action.

In a smart home, telecare can be as basic as using webcams, microphones, speakers, and display monitors for PwD to interact with caregivers

and healthcare professionals; and having safety devices such as smoke detectors, carbon monoxide detectors, and intrusion detectors (Darling, February 2006). However, in this basic form of telecare, there is no activity or plan recognition to sense all PwD's indoor physical activities and plans, and automatically detect when PwD needs physical help and respond accordingly.

Assistive technology for PwD aims to:

- Reduce PwD distress by helping them stay more independent and mentally active during the mild-moderate stages of dementia (Stages 2 to 5), to postpone the movement to residential care
- Reduce caregivers' and healthcare professionals' burden by supporting them to reduce the workload
- Provide early detection, and monitor and indicate the severity of dementia in order to contain it

Here, our focus is more on long-term care rather than safety-related assistive technology (for example, fall detection). We want to find out if the PwD is adhering to essential regimens (for example, consumption of medication and having sufficient sleep), basic ADLs (for example, eating) and instrumental ADLs (for example, preparing meals) over a period of time.

While safety enhancement is of higher perceived priority and value than long-term care, the drawback is that safety enhancement has higher liability than long-term care as the former needs to be highly accurate (if not accurate, the PwD's life could be in danger) and requires immediate action by caregivers or healthcare professionals (if not accurate, the caregivers or healthcare professionals will be inconvenienced) (Philipose, 2009).

Our scope is limited to related work (from published papers and patents) on existing execution environments in smart homes, and activity and plan recognition algorithms that can be applied to PwD living in smart homes. This book chapter is primarily written from an assistive technology researcher point-of-view, for researchers who are working on or interested in state-of-the-art assistive technology for PwD. It may be of interest to PwD's caregivers, healthcare and medical professionals.

"Challenges" is on PwD and caregiver needs, domain and technical challenges. "Assistive Hardware" explores relevant assistive hardware (grouped by devices, networks, environments, and standards) to address the needs and challenges. Running on top of the assistive hardware, "Assistive Software" examines assistive software (for physical, mental, and social health). "Discussion" discusses about how well does state-of-the-art meet the needs and challenges and what are some critical challenges, and "Conclusion" concludes.

CHALLENGES

PwD and Caregiver Needs

The ability of PwD to perform Activities of Daily Living (ADLs) is often used to determine his or her functional status and type of healthcare services required. ADLs can be classified into two categories: 7 basic ADLs and 6 instrumental ADLs. Basic ADLs involve self-care tasks such as eating, bathing, and dressing and undressing (McDowell, & Newell, 1996). Instrumental ADLs involve more tool-usage and consist of preparing meals, using the telephone, and light housework (Bookman, Harrington, Pass, & Reisner, 2007). In addition, exercise and active lifestyles reduces dementia risk or slows down dementia deterioration (Abbott et al., 2004; Verghese et al., 2003). Most current state-of-the-art assistive technology is aimed at improving PwD's physical health (and to some extent, mental health) by increasing their ability to carry out ADLs, exercise, and lead active lifestyles.

However, we argue that a more holistic healthcare approach can be taken towards assistive technology for PwD. It is possible that PwD can have physical health but not mental or social health (and vice versa), and mental and social problems are more common in mild-moderate dementia stages than physical problems. Assistive technology can be used to sense, respond, and improve PwD's mental and social health; can also possibly delay the decline in physical health. We use and modify the health triangle (Presley, 2009) as a guide for making healthy choices and it is made up of (with key factors):

- Physical health (exercise, nutrition, sleep, and hygiene)
- Mental health (learning and stress management)
- Social health (family relationships and peer relationships, especially problems of spousal death and elder abuse)

The technology wish-list for PwD (Sixsmith, Gibson, Orpwood, & Torrington, 2007) and Cogknow project's objectives identified by PwD (Bengtsson, & Savenstedt, 2007) also supports our argument that mental and social health of PwD's should not be neglected. Out of about 20 items, the top-6 wish-list items wanted by PwD are to improve mental and social health; only the 6th item is to improve physical health:

- Reminiscencing of personal histories
- Stimulating social participation
- Helping in conversation prompting
- Encouraging use of music
- Improving community of relationships
- Supporting sequence of activities

Similarly, Cogknow addresses four areas of remembering, social contact, activities of pleasure, and enhanced feelings of safety.

Since PwD's everyday life is shaped by a variety of diverse factors, in designing assistive technology, we have to know as much about the PwD's profile (for example, life and health history, cultural background, and social network) and needs to cater for (for example, dementia stage, additional medical conditions, and language ability). Given their advanced age, PwD are often associated with changes in levels of eyesight, hearing, mobility and other abilities that affect the use of assistive technology.

We also have to know about caregivers' profile (for example, relationship to PwD, and contact and employment information) and needs (for example, medical condition and financial health); and the living environment (for example, type and layout of rooms, furniture, and fittings).

In particular, PwD at different dementia stages may require very different assistive technologies (Sixsmith, Gibson, Orpwood, & Torrington, 2007). For PwD with mild-moderate levels of impairment, assistive technology should focus on encouraging physical activity, helping PwD's with ADLs, aiding communication, memory, stress reduction, and social interaction. For PwD with severe levels of impairment, assistive technology should change to areas involving close personal contact with a single caregiver.

Domain Challenges

In our particular domain for PwD, assistive technology requires doctor validation, field trials, and is within ethical guidelines. Doctors have to be supportive and familiar with assistive technology in order to educate PwD and their caregivers. Also, they need to be able to remotely communicate, configure, and monitor some home-bound PwD.

A relevant field trial is conducted on physical and mental health of elderly (not specifically for PwD) in 40 smart homes (Blackburn, Brownsell, & Hawley, 2006). Of the elderly who entered hospital, a retrospective manual review of data collected from their bed, chair, motion, door, and electrical usage sensors show that some admissions can be predicted beforehand. Of the elderly

who attended a computer course, they found the Internet useful for communication, information, and shopping. Another field trial is conducted on social health of PwD using Web-based Alzheimer's Carer Internet Support System (ACISS) (Vehvilainen, Zielstorff, Gertman, Tzeng, & Estey, 2002). ACISS is designed to provide 42 caregivers with clinical, decision-making and emotional support over 6 months, was demonstrated to be beneficial to PwD's caregivers.

Unlike telemedicine that is subject to medical-device standards and approval from government health standards organisation, there is no regulatory standard for telecare. Ethical guidelines for assistive technology for PwD include humanistic concerns (respect for PwD and caregivers), research needs and concerns, technology promises and concerns in societal context (Mahoney et al., 2007).

The domain challenges in assistive software for physical health include erroneous behaviour, multiple people, interleaved plans, and ineffective prompts. PwD's behaviour will be more random and erroneous than a healthy person's, and can deviate from correct plans in 6 different ways (Baum, & Edwards, 1993) with many possibilities. It is interesting to note that other than errors in human behaviour, there are also errors in device, network, and software errors. There will be multiple people in a smart home, and it is necessary to distinguish PwD's behaviour from one another, caregivers, or other people. PwD can frequently multi-task, performing multiple ADLs or plans in parallel, and it is important to track each task individually to detect errors in human behaviour. In the event of behavioural errors, assistive technology can prompt PwDs. Text and audio prompts are language-specific and one-way communication, therefore image and interactive prompts can be useful.

Technical Challenges

The technical challenges in assistive hardware and software include high maintenance, low accuracy and persuasiveness (Philipose, 2009), and lack of standards. The vision is to create a behaviour monitoring system that is simple enough for PwD and caregivers to easily install and maintain by themselves. Currently, this is very challenging as some devices have short battery life and can malfunction easily; behaviour monitoring has to be conducted in controlled environments (for example, type and layout of rooms, furniture, and fittings); and not robust to real-life contingencies (for example, network or power failure). A patent describes detection of device, network, and software errors in a smart home for PwD (Son, Ku, Park, & Moon, June 2008). It allows PwD or caregivers to configure error handling, and recognises the type of error in order to handle it.

Another vision is to have a behaviour monitoring system that is highly accurate and persuasive for a long period of time. Currently, given that the accuracy of a system is highly dependent on many hardware and software components, the focus of most related research is usually on a specific ADL or task for a short time. Even if the system is highly accurate and when there are errors in PwD behaviour, it is another technical challenge to prompt PwD until the ADL is completed.

There are existing and evolving standards for assistive hardware, such as smart home devices and networks, but this seems to be lacking for assistive software. For example, plan recognition is an important part of behaviour monitoring. But what consitutes a plan is subjective in terms of context (PwD and home environment), breadth (type of activities), depth (granularity of activities), and importance (some plans are more important than others).

ASSISTIVE HARDWARE

Devices

Various devices can be configured to work together as sensing modalities of PwD physical behaviour. The following ambient ones are widely available and relatively cheap. These include pressure sensors that measure resistance changes when physical pressure is applied (for example, for detecting when PwD is sitting on a chair or bed). Near-field Radio-Frequency IDentification (RFID) antennas and tags that identify and track objects using radio waves. Mid-range RFID antennas and tags can be read from a few metres. Near-field RFID tags can be read from a range of few inches (for example, for detecting when PwD has placed cup and plate on table). Pyroelectric InfraRed (PIR) sensors which are electronic motion detectors that measures InfraRed (IR) light radiating from objects which are moving in its field of view (for example, for detecting when the PwD has walked into the dining area). Reed switch sensors are electrical sensors operated by an applied magnetic field (for example, for detecting when the PwD has opened and closed the cupboard).

Other devices are widely available but are more expensive. Accelerometers are wearable electromechanical sensors that measure acceleration forces (for example, for detecting PwD's hand movements). Video sensors capture and evaluate scenes recorded by a video camera (for example, for detecting the number of people, and to annotate and audit the PwD's plans). Ultra-wideband (UWB) transmitters and sensors uses radio technology for precision locating and tracking (for example, for detecting multiple people and their locations).

Some ambient devices are research prototypes and not readily available. They include battery-powered, reader-free wireless shake sensors which are a better alternative than RFID tags for larger objects; and battery-less, midrange Wireless Identification and Sensing Platform (WISP) tags which can also do sensing and computing for small objects (Philipose, 2009).

Many commercial off-the-shelf smart home devices are stand-alone, some of which are useful for PwD and caregivers. Some ambient device categories include cameras and surveillance, intercoms and phones, and timers and automation[1]. Assistive ambient devices can also be categorised into those for ADLs and safety support (physical health), memory support (mental health), and social support (social health) (Castellot, 2007a). For ADL support is an automatic bedroom light that lights up at night when PwD gets out of bed. Pressure sensors are placed under the bed legs to detect bed occupancy. For safey support is a sounder beacon which uses visual indication and low siren to alert PwD with hearing difficulties to a ringing telephone. For memory support is an automatic pill dispenser which can be programmed to beep and partially open during medication time. For social support, there is a telephone that has photos of family and close friends for one-touch dialing.

Some everyday-life, wearable, or often-carried devices can be used as sensing modalities of PwD physical behaviour. These devices include Global Positioning System (GPS), smart mobile phones, watches, and cameras with projectors. PwD carries indoor personal GPS devices in order for caregivers to locate them. Smart mobile phones can be specially configured for PwD to give them access to functionalities such as Internet with email/chat/talk/sms/conferencing, calendar/organiser, touchscreen, keyboard, camera/video recorder, GPS, and music player (Armstrong, Nugent, Moore, & Finlay, 2009). Watches can provide continuous telemetric monitoring of PwD sleep and exercise activity, and changes in behaviour. In addition, watches have been used as ID transmitters, with sensors at multiple hazard locations to prompt PwD and can be used for multiple people detection (Power, June 2004). To augment the physical world with digital information and use natural hand gestures to interact with that information,

a camera and a small projector (mobile phones can have camera and projector functionalities) is mounted on a hat or coupled in a pendant like mobile wearable device (Mistry, Maes, & Chang, 2009).

Networks, Environments, and Standards

Stationary devices can be integrated using over the Internet or intranet, such as Service-Oriented Architecture (Aiello, Marchese, Busetta, & Calabrese, 2004) and Universal Plug-and-Play (Yang, Schilit, & McDonald, April 2008) for smart home networking.

Mobile devices can be integrated using a wireless network communication model. These protocols or specifications include WiFi, Zigbee, X10, and INSTEON. WiFi is widely supported in laptops and mobile phones, although WiFi connectivity has to be enabled and may not very reliable when moving around the smart home. Zigbee is targeted at radio-frequency (RF) applications, such as home automation, which require a low data rate, long battery life, and secure networking. X10 and INSTEON are RF protocols for communication among devices for home automation. Unlike X10, INSTEON does not require a master controller or complex routing software.

There are smart home environments, which usually includes a kitchen, dining, living, and bedrooms, which can be used for PwD and have made their benchmark data available to validate algorithms (Intille et al., 2006; Yamazaki, & Toyomura, 2008; Cook, Schmitter-Edgecombe, Crandall, Sanders, & Thomas, 2009; Phua et al., 2009).

The PlaceLab datasets (Intille et al., 2006), available over the Web, were collected in a complex and naturalistic environment with comprehensive sensing that took 2 years to build. There are close to 300 ambient sensors, categorised into 6 main groups: 80 wired switch sensors, more than 50 interior condition sensors, 125 object movement sensors, 9 PIR sensors, 9 video sensors, and 18 microphones throughout the smart home. The participant wore 2 accelerometers on his wrist and upper thigh and a wireless heart-rate monitor. He performed four hours of unscripted, intensive, common household activities such as doing laundry, preparing a meal, baking, and performing light cleaning tasks. The participant also integrated additional activities, such as making telephone calls, watching TV, using the computer, grooming, and eating. These datasets are manually and fully-annotated using a hierarchy of 89 activity, body posture, and room/context types.

The NICT ambient datasets (Yamazaki, & Toyomura, 2008), available from the authors, were collected from 17 video sensors, 30 microphones, 1774 floor pressure sensors, 28 PIR sensors, 5 cushion, and 2 sleep sensors. The participants were an elderly couple who performed ADLs for 16 days.

The CASAS ambient datasets (Cook, Schmitter-Edgecombe, Crandall, Sanders, & Thomas, 2009), available over the Web, were specifically targeted at PwD in smart homes. The simulated datasets were collected from PIR sensors, temperature, and switch sensors on 5 activities such as telephone use, hand washing, meal preparation, eating and medication use, and cleaning. Some datasets were deliberately collected with specific or general errors, interleaved plans, or multiple people.

The AIHEC datasets (Phua et al., 2009), available from the authors, simulates various errors by PwD in the kitchen and dining rooms during meal-time. It encompasses 1 basic ADL (eating) which is important to long-term well-being and 3 instrumental ADLs (preparing meals, using the telephone, and light housework). Each dataset is performed by 1 of 6 actors, and is annotated with 4 correct plans and 3 erroneous plans.

The Continua Health Alliance[2] sets the standard to enable organisations to build interoperable devices, smart home networks, and telehealth platforms. The Cogknow project (Information Society Technologies (IST) Program, 2009) has strong involvement of worldclass medical/clinical experts who work closely with PwD. This allows the project to build devices and environments to help PwD navigate through their day. The TRIL center (Intel Research Lab, 2009) conducts multi-disciplinary research including ethnographic and anthropological research which addresses elderly needs, clinical modeling which identifies specific behaviour which is correlated with certain risks (for example, PwD forgetting spouse's name is a sign of severe dementia) technological platform which monitors and analyses behaviour to provide feedback to the elderly, caregivers, or healthcare professionals.

ASSISTIVE SOFTWARE

Dependent on assistive hardware, assistive software refers to the techniques or algorithms that monitor PwD physical, mental, and social health.

Physical Health

An activity is a task and a plan is the intention to carry out a task. Activity Recognition (AR) and Plan Recognition (PR) are two main types of behaviour recognition assistive software for physical health. In a smart home environment, AR and PR involve real-time, round-the-clock, unobtrusive monitoring of PwD's ADLs. One form of definition:

a. AR is to directly recognise a task from sensor values
b. PR is to indirectly recognise a task from reasoning and interpretation of multiple types of sensors and their sensor values

Our form of definition:

c. AR is both (a) and (b)

d. PR is to detect a logical sequence of activities recognised by (c), in our case, a plan is the PwD's intention to carry out an ADL. Erroneous-Plan Recognition (EPR) aims to detect errors in the PwD's execution of ADLs

Physical Activity Recognition

Physical AR is challenging because of the need to develop and maintain many types of sensors, refine sensor models, and reason about system behaviour (Vurgun, Philipose, & Pavel, 2007). There are 3 types of physical AR techniques (suitable for different types of sensors) (Philipose, 2009):

1. Handwritten deterministic rules which are easy to understand (for example, it is easy to write rules for object use from RFID, place from location, activity from object use, and limited activity from place)
2. Handwritten statistical rules, learned parameters which are easy to understand, and handle uncertainty and are more accurate (for example, it is not easy to write rules, and is not enough labeled data for activity from video, object use from video, face from video, and transportation routines from location)
3. Learned statistical mapping which does not require rules (for example, there is sufficient labeled data for location from WiFi, walking/running from accelerometers, and ID from speech)

The Dynamic Bayesian Network (DBN) is a learned statistical Bayesian network that represents sequences of activities. It is frequently used in physical AR, for example, to detect and analyse the activities or sequence of objects being manipulated by the PwD. Using RFID and video sensors with DBNs (Park, & Kautz, 2008), the accuracy is more than 80% of 16 household activities involving 33 objects with significant

noise (Wu, Osuntogun, Choudhury, Philipose, & Rehg, 2007). Using pressure, PIR, and reed switch sensors together with pillbox and activity beacon as prompting devices, DBNs can improve medication adherence to more than 70% (Vurgun, Philipose, & Pavel, 2007). Using handwritten rules for AR, DBNs are used to probabilistically estimate the activities and sequential Monte Carlo approximation to probabilistically solve for the most likely activities (Philipose et al., 2004). Using accelerometer, DBNs are used to detect eating activity (Tolstikov, Biswas, Tham, & Yap, 2008).

Using PIR, reed switch, and RFID sensors, supervised learning is used to create binary classifiers for each activity. Binary classifiers for PIR sensors outperformed other sensors (Logan, Healey, Philipose, Tapia, & Intille, 2007). Using large, publicly available, handcontributed commonsense databases as joint models with parameters mined from the Web, semi-supervised learning is used (Pentney, Philipose, Bilmes, & Kautz, 2007).

The Partially Observable Markov Decision Process (POMDP) is a generalisation of a Markov Decision Process (MDP) where the underlying state cannot be directly observed and have to be inferred. POMDP is used to represent sequential decision processes. Using only video sensors, POMDP is used recognise PwD's handwashing activity (Mihailidis, Boger, Craig, & Hoey, 2008; Mihailidis, Boger, Canido, & Hoey, 2007; Hoey, von Bertoldi, Poupart, & Mihailidis, 2007; Boger, Poupart, Hoey, Boutilier, Fernie, & Mihailidis, 2006, 2005; Poupart, 2005; Mihailidis, Carmichael, & Boger, 2004). For PwDs in moderate-to-severe stages, the COACH system tracks, monitors state, matches MDP policy, and prompts PwD. It has a display monitor in front of the basin that can give minimal or maximum verbal/video assistance (for example, it has about 20 prompts, mostly detailed ones).

COACH combines a Bayesian sequential estimation framework for tracking hands and towel, with a POMDP sequential and decision-theoretic framework for computing policies of action. In addition, they use colour tracking without markers and movement detection. They demonstrate that POMDPs can account for the specific needs and preferences of PwD and can scale up to millions of states.

There are other relevant AR techniques. Using video sensors, neural networks are used to detect emotions (Benta, Cremene, & Todica, 2009). Probabilistic Content Free Grammar (pCFG) considers sensor data as abstract activity events and performs sequence mining. Using body sensors, pCFG is used to detect elderly dietary behaviour in smart homes, such as eating/drinking intake cycles and identifying food categories (Amft, Kusserow, & Troster, 2007). Interleaved Hidden Markov Models (HMM) are used for a portable reminder system (Modayil, Levinson, Harman, Halper, & Kautz, 2008; Modayil, Bai, & Kautz, 2008). Evidence theory is used to model the effects of sensor failure on accuracy of AR (Hong et al., 2008). UWB is used for hierarchical representation of sensor values and object location (Bauchet, Giroux, Pigot, LussierDesrochers, & Lachapelle, 2008).

Physical Plan Recognition

A hybrid recognition model based on probabilistic description logic (Roy, Bouchard, Bouzouane, & Giroux, 2007) and a keyhole PR model based on lattice theory and action description logic (Bouchard, Giroux, & Bouzouane, 2006, 2007) are used to model interleaved plans in PwD's kitchen scenario ADLs (for example, making pasta).

Usability, landmarks, and MDP are applied to wayfinding using mobile phone (Liu et al., 2009). Using GPS for outdoor location tracking and active badges for indoor location tracking, state estimation, PR, and machine learning are detect spatial disorientation (Kautz, Etzioni, Fox, & Weld, March 2003; Kautz, Arnstein, Borriello, Etzioni, & Fox, 2002; Kautz, 1987).

Other relevant PR techniques include erroneous-plan recognition (Feki, Biswas, & Tolstikov, 2009), event calculus (Chen et al., 2008a), and first order logic (Chen et al., 2008b). Polya's urn is used to detect sudden changes using PIR (Fouquet, Vuillerme, & Demongeot, 2009) and profile patterns are generated with agent monitoring for multiple people detection (Rammal, & Trouilhet, 2008). Intertransaction association rule mining (Luhr, West, & Venkatesh, 2007), client modeler, plan manager, and reminder generation (Pollack et al., 2003) have been used. There are other nondomain-related, but relevant PR techniques, such as Abstract HMM Model for plan recognition (Bui, 2003), DBN in gaming (Albrecht, Zukerman, & Nicholson, 1998), and keyhole and intended plan recognition (Carberry, 2001).

We introduce "Dementia Assistance" which is an application for doctors to remotely monitor home-bound PwD's plans and detect errors.

Figure 1 shows that PwD's activities will be tracked, and doctors can configure events, existing or new plans. There is a plan library for each PwD in a specified scene.

"Dementia Assistance" can detect new activities that are not contained in existing plans. Also, it gives the doctor the possibility to add or remove these new plans according to each PwD case. When PwD enters a scene, the scene plans graph is created. The PwD activities graph is formed progressively every time the PwD carries out an activity.

Figure 2 explains the sequence diagram. After the creation of the scene graph by "create scene graph" method, the fragment "loop" represents the repetition of PwD activities. There is verification of activity in the scene graph. If this is not true, then a process of detection of new activities plan is launched. Otherwise, there is a set of message exchanges between different devices to detect errors and to generate prompts for the PwD.

These message exchanges are described below:

- If activity exists in the scene graph, then application verifies the existence of this activity in the PwD activities graph and checks the following steps:
- If activity is in this graph, then the application recommends to the PwD to perform the activity following the previous activity accepted by the application. If the activity is not in the PwD activities graph, then the application informs the PwD that the plan is achieved.
- If activity is not contained in the PwD activities graph, then the application verifies the existence in the PwD activities graph using one of activities that precedes this activity in the scene graph.
- If it is the case, then the application adds the activity performed to the PwD activities graph and triggers a timer with the duration that separates current activity and activities that follow it. When the timer timeouts, the application prompts the PwD to perform the activity that follows the activity for which the timer was launched and restart the timer. If the activity is performed before the end of the timer, then it is stopped.

- If the PwD has skipped some activities in his plan, the application will search for the last activity performed in this activities plan and prompt PwD to perform the appropriate activity that follows it.
- If the PwD leaves a branch of the graph to a new different one, then a timer is launched with a set duration. When the timer timeouts, the application prompts PwD to perform the appropriate activity that follow the activity and restarts the timer. If PwD performs this activity before the end of the timer, it is stopped.

Figure 3 show a class represents each scene and every scene class inherits methods to manipulate a scene graph from the "Scene Graph" class. "Assistance" class allows monitoring of plans, "Timer" class detects errors and "Movements Graph" manipulates the PwD graph. "Learning" and "Learning Timer" classes are used when a new plan is detected by the application. "Event"

Figure 1. Use case diagram that shows software functions and interaction with external actors

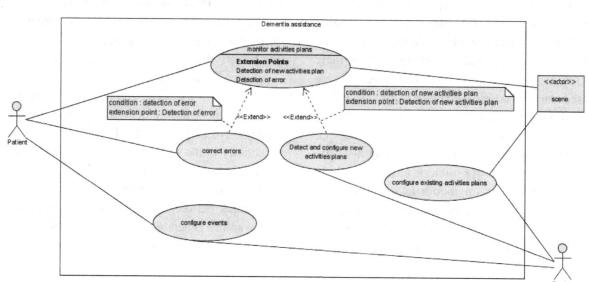

Figure 2. Sequence diagram

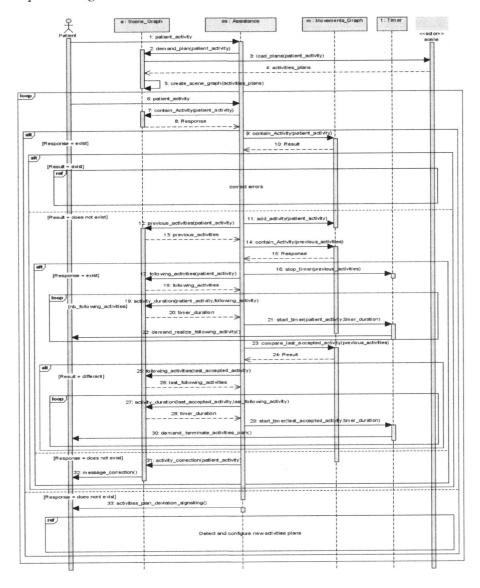

class and "Event timer" class are used to configure and follow activities.

Figure 4 presents user interface of the application. The entire graph consists of different activities of the current scene occupied by the PwD, the orange-colored nodes contain unrealised activities, and black-colored nodes represent realised activities. The blue node is the current activity being carried out by the PwD. Relying on possible plans defined in the specified scene, the application can detect errors performed by PwD and generate prompts for the PwD.

Mental and Social Health

Mental Training

To encourage learning and improve memory, there are generic brain training games that can be used by home-bound PwD as thinking tools (Bengtsson,

Figure 3.Class diagram

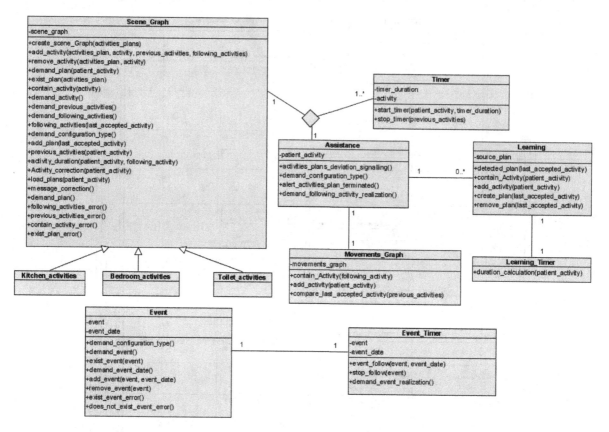

& Savenstedt, 2007). Smartbrain[3] is an interactive multimedia website which allows PwD to engage in 19 different stimulation programs with 15 pro/regressive difficulty levels. Brain-Training[4] provides mental stimulation by allowing PwD to solve simple maths problems, count people going in and out of a house simultaneously, drawing pictures on a touch screen, and reading classic literature out aloud.

Using PwD's profile and behavioural data, social network and Web-based general knowledge data, depending on context, recommendation algorithms can help PwD "think" by narrowing down to the most relevant items. For example, recommendation algorithms can pre-select the TV channels that PwD will like to watch, suggest online news and games, or remind them about some important facts in their personal life.

To reduce stress from memory loss, life-logging systems can help PwD to remember their recent experiences by capturing images passively throughout the day and as a by-product of PwD's activities (Sellen, Fogg, Hodges, & Wood, 2007). SenseCam (Harper et al., 2007) reports that, through reviewing images, life-logging systems can present different perspectives to PwD's own recollections and enrich their memories.

Using video sensors or PwD's computer usage, computer vision or opinion mining algorithms can detect stress in PwD and respond by playing their favourite music or videos.

Mental training and social networking can be further developed for:

- Early detection of dementia symptoms (often unrecognised and undocumented in

Figure 4. User interface

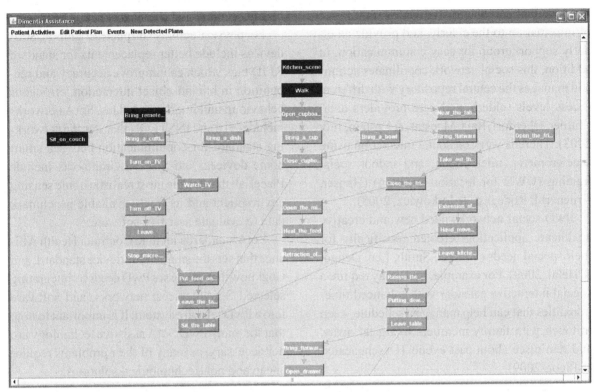

private practice settings (Valcour, Masaki, Curb, & Blanchette, 2000))

- Better assessment of PwD's dementia stage (other than using verbal questions (Teng et al., 1994), computer games (Jimison, Pavel, McKanna, & Pavel, 2004), and kitchen task assessments with 6 types of cognitive errors (Baum, & Edwards, 1993))
- Accurate measurement of PwD's quality-of-life (other than using PwD's and their caregivers' written reports (Logsdon, Gibbons, McCurry, & Teri, 2002))

Social Networking

Dementia-related online information exchanges will help PwD and caregivers get relevant advice and share similar experiences, learn more dementia, and about dementia treatment and care. These healthcare social networks, in the form of forums,

messageboards, and chats provided by Alzheimer associations[5] support these information exchanges (Castellot, 2007b). Social network analysis can be used to identify who are the PwD and caregivers who provide the most important, timely, possibly life-enhancing information.

To reduce home-bound PwD isolation and depression, there are benefits for computer-based social network activities and games. PwD can communicate, engage, participate, and continue to be part of virtual communities. To increase monitoring with more human-touch, for example using video sensors, social networking can help doctors, caregivers, other family members, and friends to remotely monitor and communicate with PwD. Since PwD are less likely to use social networking, one idea is to have a sensor-based microblogging system where the PwD recognised activities and plans are uploaded to the PwD's social network.

There is a private social network for the elderly in residential care that disseminates wide range of information to the elderly, and provide an elderly support group for easy communication. In addition, this social network coordinates actions and manages the central repository with different access levels (elderly and care providers data) (Turner, Oberdorf, Raja, Maestas, & Epstein, July 2003). There is work on social interaction using accelerometer, microphone, and indoor social training (UWB for location tracking) (Hanser, Gruenerbl, Rodegast, & Lukowicz, 2008).

PwD social networks need new and creative healthcare applications created specifically for their special needs (Weitzel, Smith, Lee, Deugd, & Helal, 2009). For example, PwD may require a special interactive calendar with enhanced functionalities that can help manage a schedule, keep in touch with family members living far away, and reminisce about past events (Descheneaux, & Pigot, 2009).

There are relevant studies which examine features, threats, and design directions to improve social connectedness (Morris, Lundell, & Dishman, 2004), and highlight four main ubiquitious computing principles: early assessment, adoptable technology, social connectedness, and human-computer interaction (Morris, Lundell, Dishman, & Needham, 2003).

DISCUSSION

In this section, we address two questions critically:

1. How well does State-of-the-Art (SotA) assistive technology meet the challenges?
2. What are some critical challenges for physical activity and plan recognition?

For Question 1, SotA assistive technology meets some challenges to improve physical health, but little has been done to improve mental and social health. In addition, assistive software is not highly accurate and persuasive yet.

For physical activity and plan recognition, SotA devices include better replacements for standard RFID tags which can improve accuracy and recognition in human-object interaction, erroneous behaviour, and interleaved plans. SotA networks include X10 and INSTEON that sets the networking standard for communication between smart home devices, and SotA environments include PlaceLab that has the most realistic home sensing environment and its freely available benchmark data to evaluate assistive software.

SotA standards include Continua Health Alliance that sets the smart home device standard, and Cogknow that addresses PwD needs by integrating selected SotA devices, networks, and software into a PwD telecare system. It is important to note that for many PwD SotA assistive technology are not necessary, as many of their problems require cheap and non-technological solutions.

For Question 2, some critical challenges are in physical activity and plan recognition include collaborative work, data labeling, and commercialisation.

Assistive technology for PwD requires a large team effort with diverse expertise and resources, and geographically separated researchers should work together to integrate the more successful hardware and software, carry out more trials together, and set standards.

For physical AR, data labeling was time consuming but was necessary to train DBNs (Vurgun, Philipose, & Pavel, 2007), and subjective and may require custom annotation tools (for example, quick-action body postures and activities which involve multiple devices are hard to label correctly) (Intille et al. 2006). For our definition of physical PR, data labeling is simpler than AR because there are much fewer plans than activities.

The main commercial market is professional health care organisations, working closely with doctors and healthcare professionals. The secondary market is for independent use by PwD, working closely with caregivers (Bengtsson, & Savenstedt, 2007). There are several business fears about assistive technology (Philipose, 2009). They include device infeasibility as many believe sensors do not "really" work yet, wearable devices as PwD can neither tolerate nor manage wearable sensors, partnerships with truck roll, call centre, telecommunication, sales and marketing companies which will reduce the profit.

CONCLUSION

In conclusion, assistive technology has been defined and its goals stated for PwD. We described the PwD and caregiver needs using the health triangle, the need for doctor validation and field trials, and other domain and technical challenges. Next, we explored relevant assistive hardware, such as ambient sensors, wireless communication protocols, smart home testbeds, and standards organisations, to address the needs and challenges. We examined assistive software for ADLs, mental training, and social networking. Finally, we discussed how well does state-of-the-art meet the needs and challenges and what are some critical challenges for physical activity and plan recognition. The assistive technology explored here applies not only to PwD, but also the elderly without dementia and people with intellectual disabilities.

REFERENCES

Abbott, R., White, L., Ross, W., Masaki, K., Curb, D., & Petrovitch, H. (2004). Walking and dementia in physically capable elderly men. *Journal of the American Medical Association*, *292*(12), 1447–1453.

Aiello, M., Marchese, M., Busetta, P., & Calabrese, G. (2004). Opening the home: a web service approach to domotics. Technical Report DIT-04-109, University of Trento.

Albrecht, D., Zukerman, I., & Nicholson, A. (1998). Bayesian models for keyhole plan recognition in an adventure game. *User Modeling and User-Adapted Interaction*, *8*(1-2), 5–47.

Amft, O., Kusserow, M., & Troster, G. (2007). Probabilistic parsing of dietary activity events. In *Proc. of BSN07, 13*, 242–7.

Armstrong, N., Nugent, C., Moore, G., & Finlay, D. (2009). Mapping user needs to smartphone services for persons with chronic disease. In *Proc. of ICOST09*.

Bauchet, J., Giroux, S., & Pigot, H. LussierDesrochers, D., & Lachapelle, Y. (2008). Pervasive assistance in smart homes for people with intellectual disabilities: A case study on meal preparation. In *International Journal of ARM, 9*.

Baum, C., & Edwards, D. (1993). Cognitive performance in senile dementia of the alzheimer's type: the kitchen task assessment. *The American Journal of Occupational Therapy.*, *47*(5), 431–436.

Belle, S., Burgio, L., Burns, R., Coon, D., Czaja, S., & Gallagher-Thompson, D. (2006). Enhancing the quality of life of dementia caregivers from different ethnic or racial groups: A randomized, controlled trial. *Annals of Internal Medicine*, *145*(10), 727–738.

Bengtsson, J., & Savenstedt, S. (2007). *Evaluation of field trial 1. Technical report*. COGKNOW Consortium.

Benta, K., Cremene, M., & Todica, V. (2009). Towards an affective aware home. In *Proc. of ICOST09*.

Blackburn, S., Brownsell, S., & Hawley, M. (2006). Assistive technology for independence (at4i) executive summary.

Boger, J., Poupart, P., Hoey, J., Boutilier, C., Fernie, G., & Mihailidis, A. (2005). A decision-theoretic approach to task assistance for persons with dementia. In *Proc. of IJCAI05*, 1293–9.

Boger, J., Poupart, P., Hoey, J., Boutilier, C., Fernie, G., & Mihailidis, A. (2006). A planning system based on markov decision processes to guide people with dementia through activities of daily living. *IEEE Transactions on Information Technology in Biomedicine, 10*(2), 323–333.

Bookman, A., Harrington, M., Pass, L., & Reisner, E. (2007). *Family Caregiver Handbook.* Cambridge, MA: Massachusetts Institute of Technology.

Bouchard, B., Giroux, S., & Bouzouane, A. (2006). A smart home agent for plan recognition of cognitively-impaired patients. *Journal of Computers, 1*(5), 53–62.

Bouchard, B., Giroux, S., & Bouzouane, A. (2007). A keyhole plan recognition model for alzheimer's patients: First results. *Applied Artificial Intelligence, 21*(7), 623–658.

Brewster, S., & Goodman, J. (2008). Hci and the older population.

Bui, H. (2003). A general model for online probabilistic plan recognition. In *Proc. of IJCAI03*, 1309–15.

Carberry, S. (2001). Techniques for plan recognition. *User Modeling and User-Adapted Interaction, 11*(1-2), 31–48.

Castellot, R. (2007a). *Report detailing state-of-the-art in cognitive reminder devices. Technical report.* COGKNOW Consortium.

Castellot, R. (2007b). *Report detailing state-of-the-art in services supporting cognitive disabilities. Technical report.* COGKNOW Consortium.

Chen, L., Nugent, C., Mulvenna, M., Finlay, D., Hong, X., & Poland, M. (2008a). Using event calculus for behaviour reasoning and assistance in a smart home. In *Proc. of ICOST08*, 81–9.

Chen, L., Nugent, C., Mulvenna, M., Finlay, D., Hong, X., & Poland, M. (2008b). A logical framework for behaviour reasoning and assistance in a smart home. In *International Journal of ARM, 9.*

Cook, D., Schmitter-Edgecombe, M., Crandall, A., Sanders, C., & Thomas, B. (2009). Collecting and disseminating smart home sensor data in the casas project. In *Proc. of CHI09 Workshop on Developing Shared Home Behavior Datasets to Advance HCI and Ubiquitous Computing Research.*

Darling, J. (February 2006). Remote supervision system and method.

Descheneaux, C., & Pigot, H. (2009). Interactive calendar to help maintain social interactions for elderly people and people with mild cognitive impairments. In *Proc. of ICOST2009.*

Feki, M., Biswas, J., & Tolstikov, A. (2009). Model and algorithmic framework for detection and correction of cognitive errors. *Technology and Health Care Journal, 17*(3), 203–219.

Ferri, C., Prince, M., Brayne, C., Brodaty, H., Fratiglioni, L., & Ganguli, M. (2005). Global prevalence of dementia: a delphi consensus study. *Lancet, 366*, 2112–2117.

Fouquet, Y., Vuillerme, N., & Demongeot, J. (2009). Pervasive informatics and persistent actimetric information in health smart homes. In *Proc. of ICOST09.*

Hanser, F., Gruenerbl, A., Rodegast, C., & Lukowicz, P. (2008). *Design and real life deployment of a pervasive monitoring system for dementia patients* (pp. 279–280). Proc. of Pervasive Computing Technologies for Healthcare.

Harper, R., Randall, D., Smyth, N., & Evans, C. (2007). Thanks for the memory . In *Proc. of HCI07*. Heledd. L., & Moore. R.

Hoey, J., von Bertoldi, A., Poupart, P., & Mihailidis, A. (2007). Assisting persons with dementia during handwashing using a partially observable markov decision process. In *Proc. of ICVS07*.

Hong, X., Nugent, C., Mulvenna, M., Mcclean, S., Scotney, B., & Devlin, S. (2008). Assessment of the impact of sensor failure in the recognition of activities of daily living. In *Proc. of ICOST08*, 136–144.

Intille, S., Larson, K., Tapia, E., Beaudin, J., Kaushik, P., Nawyn, J., & Rockinson, R. (2006). Using a live-in laboratory for ubiquitous computing research. In *Proc. of PERVASIVE06*, 349–365.

Jimison, H., Pavel, M., McKanna, J., & Pavel, J. (2004). Unobtrusive monitoring of computer interactions to detect cognitive status in elders. *IEEE Transactions on Information Technology in Biomedicine*, *8*, 248–252.

Kautz, H. (1987). A Formal Theory of Plan Recognition. PhD thesis, Dept. of Computer Science, University of Rochester.

Kautz, H., Arnstein, L., Borriello, G., Etzioni, O., & Fox, D. (2002). An overview of the assisted cognition project. In *Proc. of AAAI02 Workshop on Automation as Caregiver: The Role of Intelligent Technology in Elder Care*.

Kautz, H., Etzioni, O., Fox, D., & Weld, D. (March 2003). Foundations of assisted cognition systems. Technical report, University of Washington. Intel Research Lab. (2009). Technology for independent living.

Liu, A., Hile, H., Borriello, G., Kautz, H., Brown, P., Harniss, M., & Johnson, K. (2009). *Informing the design of an automated wayfinding system for individuals with cognitive impairments*. Proc. of Pervasive Computing Technologies for Healthcare.

Logan, B., Healey, J., Philipose, M., Tapia, E., & Intille, S. (2007). A long-term evaluation of sensing modalities for activity recognition. In *Proc. of Ubicomp07*, 483–500.

Logsdon, R., Gibbons, L., McCurry, S., & Teri, L. (2002). Assessing quality of life in older adults with cognitive impairment. *Psychosomatic Medicine*, *64*, 510–519.

Lu"hr, S., West, G., & Venkatesh, S. (2007). Recognition of emergent human behaviour in a smart home: A data mining approach. *Pervasive and Mobile Computing*, *3*(2), 95–116.

Mahoney, D., Purtilo, R., Webbe, F., Alwan, M., Bharucha, A., & Adlam, T. (2007). In-home monitoring of persons with dementia: Ethical guidelines for technology research and development. *Alzheimer's & Dementia*, *3*, 217–226.

McDowell, I., & Newell, C. (1996). *Measuring Health: A Guide to Rating Scales and Questionnaires* (2nd ed.). New York: Oxford University Press.

Mihailidis, A., Boger, J., Canido, M., & Hoey, J. (2007). The use of an intelligent prompting system for people with dementia: A case study. *ACM Interactions (Special issue on Designing for seniors: innovations for graying times)*, *14*(4):34–7.

Mihailidis, A., Boger, J., Craig, T., & Hoey, J. (2008). The coach prompting system to assist older adults with dementia through handwashing: An efficacy study. *BMC Geriatrics*, *8*(28).

Mihailidis, A., Carmichael, B., & Boger, J. (2004). The use of computer vision in an intelligent environment to support aging-in-place, safety, and independence in the home. *IEEE Transactions on Information Technology in Biomedicine*, *8*, 238–247.

Mistry, P., Maes, P., & Chang, L. (2009). Wuw wear ur world: a wearable gestural interface. In *Proc. of Human factors in Computing Systems*, 4111–6.

Modayil, J., Bai, T., & Kautz, H. (2008). Improving the recognition of interleaved activities. In *Proc. of UbiComp08*, 40–3.

Modayil, J., Levinson, R., Harman, C., Halper, D., & Kautz, H. (2008). Integrating sensing and cueing for more effective activity reminders . In *AI in Eldercare*. New Solutions to Old Problems.

Morris, M., Lundell, J., & Dishman, E. (2004). Catalyzing social interaction with ubiquitous computing: a needs assessment of elders coping with cognitive decline. In *Proc. of CHI04*, 1151–4.

Morris, M., Lundell, J., Dishman, E., & Needham, B. (2003). New perspectives on ubiquitous computing from ethnographic study of elders with cognitive decline. In *Proc. of UbiComp03*, 227–242.

Park, S., & Kautz, H. (2008). Hierarchical recognition of activities of daily living using multi-scale, multi-perspective vision and rfid. In *Proc. of IET Conference*.

Pentney, W., Philipose, M., Bilmes, J., & Kautz, H. (2007). Learning large scale common sense models of everyday life. In *Proc. of AAAI07*, 465–470. Christopher Presley. (2009). Health triangle. Information Society Technologies (IST) Program. (2009). Helping people with mild dementia to navigate their day.

Philipose, M. (2009). Technology for long-term care: Scaling eldercare to the next billion. In *Proc. of ICOST09*.

Philipose, M., Fishkin, K., Perkowitz, M., Patterson, D., Fox, D., Kautz, H., & Hahnel, D. (2004). Inferring activities from interactions with objects. *IEEE Pervasive Computing / IEEE Computer Society [and] IEEE Communications Society, 3*(4), 50–57.

Phua, C., Foo, V., Biswas, J., Tolstikov, A., Aung, A., Maniyeri, J., et al. (2009). 2-layer erroneous-plan recognition for dementia patients in smart homes. In *Proc. of HealthCom09*.

Pollack, M., Brown, L., Colbry, D., McCarthy, C., Orosz, C., Peintner, B., Ramakrishnan, S., & Tsamardinos, I. (2003). Autominder: An intelligent cognitive orthotic system for people with memory impairment.

Poupart, P. (2005). Exploiting Structure To Efficiently Solve Large Scale Partially Observable Markov Decision Processes. PhD thesis, University of Toronto, Toronto, Canada.

Power, M. (June 2004). System for monitoring patients with alzheimer's disease or related dementia.

Rammal, A., & Trouilhet, S. (2008). Keeping elderly people at home: A multi-agent classification of monitoring data. In *Proc. of ICOST08*, 145–152.

Reisberg, B., Doody, R., Stffler, A., Schmitt, F., Ferris, S., & Mbius, H. (2003). Memantine in moderate-to-severe alzheimer's disease. *The New England Journal of Medicine, 348*(14), 1333–1341.

Reisberg, B., Ferris, S., Anand, R., Leon, M., Schneck, M., Buttinger, C., & Borenstein, J. (1984). Functional staging of dementia of the alzheimer type. *Annals of the New York Academy of Sciences, 435*, 481–483.

Roy, P., Bouchard, B., Bouzouane, A., & Giroux, S. (2007). A hybrid plan recognition model for alzheimer's patients: Interleaved-erroneous dilemma. In *Proc. of IAT07*, 131–7.

Sellen, A., Fogg, A., Hodges, S., & Wood, K. (2007). Do life-logging technologies support memory for the past? an experimental study using sensecam. In *Proc. of CHI07*.

Sixsmith, A., Gibson, G., Orpwood, R., & Torrington, J. (2007). Developing a technology wish-list to enhance the quality of life of people with dementia. *Gerontechnology (Valkenswaard), 6*, 2–19.

Son, Y., Ku, T., Park, J., & Moon, K. (June 2008). Inference-based home network error handling system and method.

Teng, E., Hasegawa, K., Homma, A., Imai, Y., Larson, E., & Graves, A. (1994). The cognitive abilities screening instrument (casi): a practical test for cross-cultural epidemiological studies of dementia. *International Psychogeriatrics, 6*(1), 45–58.

Tolstikov, A., Biswas, J., Tham, C., & Yap, P. (2008). Eating activity primitives detection a step towards adl recognition. In *Proc. of Health-Com08*, 35–41.

Turner, K., Oberdorf, V., Raja, G., Maestas, G., & Epstein, H. (July 2003). System and method for facilitating the care of an individual and dissemination of information.

Valcour, V., Masaki, K., Curb, J., & Blanchette, P. (2000). The detection of dementia in the primary care setting. *Archives of Internal Medicine, 160*, 2964–2968.

Vehvilainen, L., Zielstorff, R., Gertman, P., Tzeng, M., & Estey, G. (2002). Alzheimer's caregiver internet support system (aciss): Evaluating the feasibility and effectiveness of supporting family caregivers virtually. In *American Medical Informatics Association 2002 Symposium*.

Verghese, J., Lipton, R., Katz, M., Hall, C., Derby, C., & Kuslansky, G. (2003). Leisure activities and the risk of dementia in the elderly. *The New England Journal of Medicine, 348*, 2508–2616.

Vurgun, S., Philipose, M., & Pavel, M. (2007). A statistical reasoning system for medication prompting. In *Proc. of Ubicomp07*, 1–18.

Weitzel, M., Smith, A., Lee, D., Deugd, S., & Helal, S. (2009). Participatory medicine: Leveraging social networks in telehealth solutions. In *Proc. of ICOST2009*.

Wu, J., Osuntogun, A., Choudhury, T., Philipose, M., & Rehg, J. (2007). A scalable approach to activity recognition based on object use. In *Proc. of ICCV07*, 1–8.

Yamazaki, T., & Toyomura, T. (2008). Real-life experimental data acquisition in smart home and data analysis tool development. *In International Journal of ARM, 9*.

Yang, J., Schilit, B., & McDonald, D. (2008, April). Activity recognition for the digital home. *Computer, 41*(4), 102–104.

KEY TERMS AND DEFINITIONS

Smart Home: A residential dwelling with increased automation of household appliances and features.

Ambient Intelligence: A smart home that is sensitive and responsive to the presence of people.

Dementia: The progressive decline in cognitive function due to damage or disease in the body.

Plan Recognition: Behavior monitoring that determines the intention to carry out a task or tasks.

Assistive Technology: A generic term that includes assistive, adaptive, and rehabilitative devices for people with disabilities.

Mental Training: A set of brain-related exercises that aims to improve cognitive function.

Healthcare Social Network: A set of positive relationships between people that can provide physical and mental support.

ENDNOTES

[1] http://www.smarthome.com
[2] http://www.continuaalliance.org
[3] http://www.smartbrain.net
[4] http://www.braintraining.com.au
[5] http://www.alz.org, http://www.alzheimer.ca, http://alzheimers.org.uk

Chapter 17
Challenging Issues of Ambient Activity Recognition for Cognitive Assistance

Patrice C. Roy
Université de Sherbrooke, Canada

Bruno Bouchard
Université du Québec à Chicoutimi, Canada

Abdenour Bouzouane
Université du Québec à Chicoutimi, Canada

Sylvain Giroux
Université de Sherbrooke, Canada

ABSTRACT

In order to provide adequate assistance to cognitively impaired people when they carry out their activities of daily living (ADLs) at home, new technologies based on the emerging concept of Ambient Intelligence (AmI) must be developed. The main application of the AmI concept is the development of Smart Homes, which can provide advanced assistance services to its occupant when he performs his ADLs. The main difficulty inherent to this kind of assistance services is to be able to identify the on-going inhabitant ADL from the observed basic actions and from the sensors events produced by these actions. This chapter will investigate in details the challenging issues that emerge from activity recognition in order to provide cognitive assistance in Smart Homes, by identifying gaps in the capabilities of current approaches. This will allow to raise numerous research issues and challenges that need to be addressed for understanding this research field and enabling ambient recognition systems for cognitive assistance to operate effectively.

INTRODUCTION

People with a cognitive deficit often have difficulty carrying out their activities of everyday life (Baum & Edwards, 1993). These people, mostly elderly, wish to stay at home, where they feel comfortable and safe, as long as possible. Governments aim to help them for social reasons as well as economical ones. However, keeping cognitively impaired

DOI: 10.4018/978-1-61692-857-5.ch017

people at home involves many risks that must be controlled. In order to do that, the physical and human environment must be specifically designed to compensate for the cognitive impairments and the loss of autonomy, thus constituting an economically viable alternative to the exhaustion of caregivers (Pigot *et al.*, 2003). This is why a growing worldwide community of scientists now works on the development of new technologies based on the emerging concept of Ambient Intelligence (AmI), which can be considered as the key to solve the challenge of maintaining semi-autonomous people at home safely.

Ambient intelligence (Capezio *et al.*, 2007; Ramos *et al.*, 2008) refers to a multidisciplinary approach that consists of enriching a common environment (room, building, car ...) with technology (sensors, identification tags ...), in order to build a system that makes decisions to benefit the users of this environment, based on real-time information and historical data. In this way, technology merges with the environment, becoming non-intrusive, but stands ready to react to the occupant needs and to provide assistance. The main application of this AmI concept concerns the development of Smart Homes (Augusto & Nugent, 2006), which can be seen as houses equipped with ambient agents able to bring advanced assistive services to a resident, for the performance of his Activities of Daily Living (ADL). The main difficulty inherent in this kind of assistance is to identify the on-going inhabitant ADL from observed basic actions and from the events produced by these actions. This difficulty is yet another instance of the so-called plan recognition problem (Carberry, 2001) studied for many years in the field of Artificial Intelligence (AI), in many and varied applicative contexts. A plan, in AI, can be defined as the formal description of an activity. It corresponds to a chain of actions linked by different ordering/temporal relations, which describe the causes, the effects and the goal of a particular activity.

From the recent AmI point of view, which constitutes our focus in this chapter, the activities

(plan) recognition problem can be summarized as the process of interpreting low-level actions (which are detected by sensors placed in environment's objects) in order to infer the goal (the on-going activities) pursued by a person (Patterson *et al.*, 2003). One of the main objectives of this recognition process is to identify errors in the performance of activities, in order to target the right moment when assistance is needed and to choose one of the various ways a smart home (observer agent) may help its occupant (a cognitively-impaired patient) (Bouchard *et al.*, 2007). Hence, due to our context specificity, the challenging issues related to activities recognition tend to become much more complex and require dealing with a high possibility of observing errors emanating from the weakening cognitive functions of the patient. For instance, a distraction (e.g. phone call, unfamiliar noise, etc.) or a memory lapse can lead him to perform actions in the wrong order, to skip some steps of his activity, or to perform actions that are not even related to his original goal.

The AI community produced, in the last two decades, many fundamental works addressing the activities recognition issue, which can be divided into three major trends of research. The first one comprises works based on logical approaches (Kautz, 1991; Christensen, 2002; Wobcke, 2002), which consist in developing a theory, using first order logic, for formalizing the recognition activity into a deduction or an abduction process. The second trend of literature is related to probabilistic models (Albrecht *et al.*, 1998; Boger *et al.*, 2006; Charniak & Goldman, 1993; Patterson *et al.*, 2005), which define activity recognition in terms of probabilistic reasoning, based primarily on some Markovian models or on Bayesian networks. The third trend covers emerging hybrid approaches (Avrahami-Zilberbrand & Kaminka, 2007; Geib & Goldman, 2005; Roy *et al.*, 2009a) that try to combine the two previous avenues of research. It considers activities recognition as the result of a probabilistic quantification on the hypotheses obtained from a symbolic (qualitative) algorithm. A significant amount of the fundamental works of these three trends were concerned with applicative contexts different from AmI, supposing most of the time that the observed entity always acts in a coherent way, avoiding many important issues such as recognizing erroneous activities. Presently, there are many research groups (Giroux *et al.*, 2009; Haigh et al., 2006; Patterson *et al.*, 2007; Pollack, 2005) working on assistive technologies that try to exploit and enhance those activity recognition models in order to concretely address the activities recognition problem in a AmI context for cognitive assistance in a smart environment.

This chapter will investigate in detail the challenging key issues that emerge from research in the field of ambient agents in smart homes, under the context of activity recognition. We will clearly describe the specific functional needs inherent in cognitive assistance for effective activity recognition, and then we will present the fundamental research that addresses this problem in such a context. This chapter is more of a survey and an analysis of existing works that have been studied for potential integration into our laboratories, rather than a focused evaluation report. Our objective is to identify gaps in the capabilities of current techniques and to suggest the most productive lines of research to address this complex issue. As such, the contribution is of both theoretical and practical significance. We raise numerous research issues and challenges that need to be addressed for understanding the domain and enabling ambient multi-agent systems for cognitive assistance to operate effectively. The chapter is organized as follows. "Foundations of Activity Recognition" presents the foundations that characterize the problem of activity recognition and position it in an AmI context. "Related Works on Activity Recognition" draws a detailed portrait of the three types of activity recognition models that exist, presents their forces and weaknesses in our specific context, illustrates their application with small examples, and presents several AmI

projects that exploit those models to concretely deploy cognitive assistance solutions. Finally, we conclude this chapter with perspectives and future challenges facing this fascinating issue.

FOUNDATIONS OF ACTIVITY RECOGNITION

Activity recognition has been an active topic of artificial intelligence for the last three decades. However, it is only recently, with the emergence of the concept of ubiquitous computing and the development of smart environments, that this issue has become a master piece in the grand scheme of ambient intelligence. Hence, to clearly understand the multiple facets that characterize the problem of activity recognition, we must first clarify the basic notions related to the topic and show how this issue evolved in 30 years to its actual form in a context of AmI. The first thing that needs to be clarified is: what are exactly these ADLs that we want to recognize in a smart environment?

What is an Activity of Daily Living (ADL)

The notion of Activities of Daily Living (ADL) has been defined by Dr. Sidney Katz in 1963 (Katz *et al.*, 1963) as the set of activities that an individual performs as a routine for self-care, for instance dressing, personal hygiene, preparing meals, etc. Health professionals often refer to the ability or inability to perform ADL as a measurement of the functional status of a person. This measurement is useful for assessing cognitively impaired patients; in order to evaluate what type of health care services they may need (Giovannetti *et al.*, 2002). Therefore, ADL represent a set of common activities that a normal person is minimally supposed to be able to do by himself to pretend to be autonomous. We can distinguish three types of ADL: basic, instrumental, and enhanced (Rogers *et al.*, 1998).

Basic Activities of Daily Living (ADL) consist of core fundamental self-care tasks, such as bathing, dressing and undressing, eating, transferring from bed to chair, voluntarily controlling urinary and fecal discharge, using the toilet, walking, etc. These activities are composed of only a few steps and do not require a real planning. One needs to be able to carry out these activities by himself, even to be considered as a semi-autonomous person.

Instrumental Activities of Daily Living (IADL) are not related to fundamental functioning, but enable a person to be relatively autonomous at home and within a community. These activities are, for instance, preparing meals, taking medications, shopping for groceries or clothes, using a telephone, managing money, etc. Instrumental ADL are more complex and require a higher level of planning and a better judgment than basic ADL.

Enhanced Activities of Daily Living (EADL) correspond to tasks requiring adaptation or learning on behalf of the person because of the nature of the environment. For instance, calling back a person from a new phone that keeps a record of the ten last callers is a task that requires an adaptation (to the new phone).

In the literature on ambient assistive technologies (Giroux *et al.*, 2009; Haigh et al., 2006; Patterson *et al.*, 2007; Pollack, 2005), researchers often refer to these three types under the general term of ADL. However, most of the recognition systems developed for assisting people at home focus only on basic and instrumental ADL.

The Origins of Activity Recognition

The problem of recognizing activities constituted a very active topic in AI in the last 30 years. At the beginning, this issue was described in general terms by the need "… to take as input a sequence of actions performed by an actor and to infer the goal pursued by the actor and also to organize the action sequence in terms of an plan (activity) structure" (Schmidt et al., 1978). From that definition, the expression *activity recognition*

refers to the fact that one supposes the existence of activity structure (i.e. a set of organized actions in time) chosen at the start by the acting entity (for instance a patient), and constituting the result that the observing agent (for instance a smart environment) seeks to recognize. By referring to the literature on the subject, one can realize that this vision of the problem is a historical heritage from the first expert systems, which was originally used to resolve *planning* issues (Waern *et al.*, 1995). The *planning* problem also constitutes a well-known challenge in the AI community, which can be considered as the inverse of activity recognition (Russell & Norvig, 2003). This difficulty consists, for an agent, to identify a sequence of actions (a plan) which, at the end of its execution, will allow the agent to achieve a pursued goal (Georgeff, 1987). Therefore, the problem of recognizing an activity can be seen as the inverse operation, in which another agent (observer) that does not know the planned objective has the task of identifying the goal of the first agent by inferring, from the observed actions, the possible on-going activities set.

Classification of Different Types of Activity Recognition

According to Cohen (Cohen *et al.*, 1982) and Geib (Geib, 2007), activity recognition can be characterized by the relationship that exists between the observed agent and its observer. This relationship allows dividing activity recognition into three different categories: *intended recognition, adversarial recognition*, and *keyhole recognition*. Each of these recognition types makes a particular assumption about the existing relationship between the observing agent and the observed entity, which will guide the recognition process. This categorization is often used by authors to clarify their point of view on the recognition context and constitutes a fundamental element that we must take into account when we want to build a suitable recognition model for a specific context such as AmI.

Intended activity recognition: in a context of *intended recognition*, one supposes that the agent in action explicitly knows that he is being observed and therefore, that he adapts deliberately his behaviour by performing his activity in a way that will facilitate the recognition process of the observer (Kautz, 1991). Consequently, this form of recognition presumes a direct cooperative effort on behalf of the observed entity. This category of recognition was introduced by Cohen (Cohen *et al.*, 1982). His goal was to position the activity recognition problem within the particular context of natural language processing, in which he worked at the time. Some researchers supposed, based on the cooperative assumption of this category, that the observer agent could directly ask the actor in case of uncertainty, in order to get clarifications on the on-going task (Lesh *et al.*, 1998). From an AmI point of view, it is unrealistic to position the problem of recognizing activities in a context of intended recognition. For instance, in the case of a smart environment trying to assist an Alzheimer's patient in his everyday tasks, it is clearly unrealistic to make any assumption about the capacity of the patient to sustain a cooperative effort by adapting his behaviour in order to positively influence the recognition process. Moreover, we cannot let the system ask an observed patient for clarifications in case of uncertainty, because this action would result in an increased cognitive charge on behalf of the patient, which is unsuitable in a context of cognitive assistance.

Adversarial activity recognition: this second category has been added by Geib (Geib & Goldman, 2001) as a counterpart of the first category, the intended recognition. In this type of recognition, rather than presuming that the observed agent will positively contribute to the recognition process, one supposes that the acting agent will likely attempt to spoil the process in a competitive gesture on its behalf (Mao & Gratch, 2004). In other terms, the observer agent is considered as an enemy by the actor agent and therefore, it will voluntarily try to perform actions that are incoherent with its goals in order to lead the observer to infer false conclusions about its behaviour. This type of recognition is better suited to a military context (Heinze *et al.*, 1999) or to a context where competitive agents are deployed inside video games to challenge a human player (Albrecht *et al.*, 1998). However, this type of recognition does not fit well in an AmI context of cognitive assistance. Even if we need to take into account the fact that a cognitively impaired patient will likely perform activities in an erroneous way, as is commonly the case in an adversarial context to misguide the observer, these errors are not performed deliberately. They are rather the consequence of the patient's symptoms and not the result of a well planned strategy to compromise the recognition process.

Keyhole activity recognition: this last category characterizes the assumption that the acting agent does not really know that it is been observed and thus, it will not attempt to influence the recognition process or to misguide the observer conclusions (Cohen *et al.*, 1982). With this category, Cohen wanted to define a generic frame for activity recognition, in which one make a supposition of a neutral cooperative effort on behalf of the acting agent. Hence, Cohen made an analogy, from which the name of the category emanate, with a person being observed, inside a room, through the keyhole of the door, so that this person acts in the room naturally without worrying about being observed and without trying to help or compromise the on-going recognition process. We can position the activity recognition problem, from an AmI point of view, in this category that suit better to the context of cognitive assistance, in which we certainly cannot presume the collaboration nor that the deliberate nuisance of the patient (Bouchard, 2006).

From Activity Recognition to Ambient Recognition in a Smart Environment

Since the arrival of Cohen's and Geib's classifications, the definition of keyhole activity recognition has been reformulated many times, notably by (Goldman *et al.*, 1999; Boger *et al.*, 2005; Bouchard *et al.*, 2007; Patterson *et al.*, 2003), in order to clarify its scope in a context of ambient recognition in a smart environment. In these new refined definitions, we note a tendency to try making explicit the notion of environment, which refers to the recognition context, and to clearly link this notion with the challenges related to the recognition issue. For instance, Goldman (Goldman *et al.*, 1999) presents the recognition of activities as "… a process of inferring the ongoing plans of an agent from the observations of the results of its actions". This definition may seem very similar to that of Smith. However, there is a subtle, but important clarification in this definition that makes explicit the distinction between the notion of actions performed by the acting agent, who is not directly observable, and the observations (perceptions) that can be directly sensed by the recognition (observing) agent. Patterson, in his definition (Patterson *et al.*, 2003), further clarifies the issue by describing the recognition problem as "… the task of inferring the activity performed by observed entity from data provided by low-level sensors". This new vision of the recognition problem is now shared by most researchers working in the AmI community and adheres to the recent paradigm of ubiquitous computing (Weiser, 1991), which seeks to create augmented environments with miniaturized processors, software (agents) communicating between each other, and multi-modal sensors that are embedded in any common everyday objects, making them effectively invisible to the resident, in order to give services and assistance when needed. From this point of view, activity recognition is situated in a precise applicative context where

the observed actor evolves inside an environment with low-level sensors that are triggered by the results of the interaction between the actor and his environment, and where the observing agent can perceive these triggers.

Starting from these recent visions of activity recognition and in the light of our work in ambient assisting environments, we propose to clarify the problem by distinguishing four main issues that aim to characterize the entire process of recognizing activities in this particular context of AmI. These issues are: (1) gathering sensor inputs, (2) interpreting and merging sensor data to get useful information, (3) identifying basic actions performed, and (4) interpreting high-level behaviour. Each of these issues poses different challenges that should be addressed separately but complementarily if one wants to deploy an effective recognition process. From our experience, we propose to attack these four issues by deploying a 4-layer recognition architecture, as shown in Figure 1, where each layer is in charge of one specific issue, but works the other layers to give the desired output. Operations that must be done at each layer can be executed by a single agent, but it is more profitable to organize them in a multi-agents architecture (Bouchard *et al.*, 2005), where one agent is responsible for processing one layer and where other agents can communicate with any agents to get the results at the end of the processing phase.

Issue 1 (Gathering sensor inputs). The first issue concerns inputs gathered directly from sensors. When a resident interacts with a smart environment equipped with sensors, they generate low-level triggers that must be caught by some kind of software agent with an event-trigger system. This agent should provide an abstraction layer that allows the higher level agents to register to a common source in order to get the information coming from the sensors in a unified form. The main challenge related to this first issue is to deal with the heterogeneity of the sensors, which communicate with different protocols and differ-

Figure 1. Architecture of activity recognition inside a smart environment composed of 4 layers

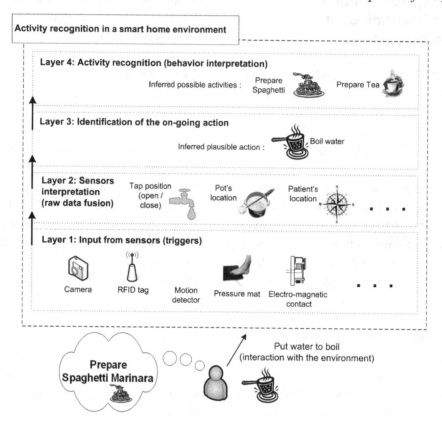

ent means, for instance Wi-Fi, electrical network, etc. In Figure 1, we show an illustrative example of this first step of the recognition process, where a resident wants to cook some spaghetti marinara. To do that, he interacts with his environment by performing the action of putting water on to boil. This action triggered some sensors: Radio Frequency Identification (RFID) tags on the pot, pressure mat in front of the oven, movement detector, etc. All these triggers must be caught and converted into events that will be placed, for instance, inside a database that will be accessible by the higher level agents. An important challenge related to this operation concerns the precision of the sensors and the type of sensors used. It has been shown that using complex sensors, such as video cameras and microphones, is not very effective. Although in principle the data captured by these sensors should be as useful as that cap-

tured by the key human senses of sight and hearing, in practice the task of extracting features from rich low-level representations has proved to be very challenging in an unstructured environment (Choudhury *et al.*, 2006). A more effective approach is to use targeted simple sensors, such as RFID tags or accelerometers, which give precise information about a particular small set of features. Finally, another important aspect of this issue concerns the detection of sensors' failures. One must be able to cross-validate information given by a single sensor in order to detect incoherencies that may be caused by a defective sensor. This challenge must be addressed in order to ensure the efficiency and the safety of any assistive system.

Issue 2 (Interpretation-fusion of sensor data). The second issue concerns the process of merging the data gathered from various sensors by addressing the first issue, in order to extract bounded, concise and useful information from it. The objective is to refine the raw data provided by sensors in order to infer from it useful information that we seek to know. A real living environment is vast and has too many objects. Hence, a smart environment is usually composed of only several key object properties that one wants to monitor in order to identify the desired on-going activities. For instance, we might only want to check if a senior takes his medication correctly. Therefore we will only need to monitor few objects and not the entire house. If we take our example in Figure 1, we can suppose that we wanted to monitor three object properties, which are the tap state (open or close), the pot's location, and the resident's location. The action of putting water on to boil, executed by the resident, has triggered, for instance, the movement detector and the pressure mat in front of the oven. These two pieces of information are, in this second layer, merged and use to infer the approximate patient's location, which is one of the key properties that will allow us to conduct further deductions about on-going activities. The challenge related to this part of the recognition process is mainly a fusion issue: how can we aggregate all the data from sensors to extract useful and targeted information about the recognition process?

Issue 3 (Identification of basic actions). The third issue concerns the identification of the basic actions performed by the resident during his activities. For instance, in our example of Figure 1, we want to be able to infer that the resident performed the basic action "put water on to boil" in the course of preparing spaghetti marinara. This problem is complex and is related to a classification issue. More precisely, the challenge is to take as input the state S_{t-1} containing all properties of the monitored environment's objects at time *t-1*, and to take all the changes that happened in the state S_t of these objects at the actual time *t*, and to identify from a taxonomy of action structures the most adequate action type that can correspond to

these observed changes (Bouzouane *et al.*, 2005). In our example, the action structure of "put water on to boil" could have been described by a set of preconditions specifying that, prior to the action, the tap should have been closed and the pot should have been in a cabinet. The set of postconditions of the action could have specified that some of the monitored environment's properties should change, such as, the pot is now located on the oven, the patient is located near the oven, and the tap should have been open. Therefore, the observations of those changes in the monitored object's properties would have allowed an agent, by using a classification algorithm, to identify the resident's low-level action.

Issue 4 (Activity recognition: behaviour interpretation). The fourth and last issue that we distinguish concerns the interpretation of the behaviour of the resident, assuming that we identified correctly the basic actions he performed in the smart environment. It is usually based on an abduction principle for the construction of hypotheses about the possible activities, and on a matching process linking the observed actions with some models of activities included in a knowledge base related to the application domain. This knowledge base describes the structure of all activities (plans) that the observed agent can potentially carry out. At each observation of an action occurrence, the recognition agent tries to build hypotheses based on knowledge included in this library. In our cooking example, the identified basic action "put water on to boil" could lead to two different explanations: (i) the resident is preparing spaghetti or (ii) the resident is making tea. As you can see, there can be many possible activities that can explain the observations, and thus the behaviour of the resident in a smart environment. Therefore, one of the main challenges related to this issue is to differentiate among the concurrent possible hypotheses. In a context of cognitive assistance, another important challenge concerns the dilemma existing between identifying the performance of multiple interleaved activities versus erroneous ones (Roy *et al.*, 2009a). More precisely, this dilemma occurs when we need to interpret a new observed action, different from the expected one and which is not related to current on-going activity. This action can constitute an erroneous continuation from the started activity or the beginning of a new coherent activity resulting of the temporary interruption of the first one; the challenge is to make the distinction between these two possibilities. This facet of the issue must be taken into account if one wishes to establish an effective recognition model capable of distinguishing correct or incorrect behaviour. The behaviour interpretation constitutes the most important and complex part of the recognition process; this is why our research focused mostly on this issue. Thus, the section on related research will focus on analyzing recognition systems with respect to this particular issue.

Architectural Issues: Practices and Guidelines for Easy Deployment

From an architectural perspective, one can ask questions about how implementing different solutions to activity recognition may have an impact on a complete Ambient Intelligence system, as well as on specific parts such as sensing. For instance, is implementing a recognition system using a probabilistic approach, such as Bayesian network, will change the physical and software architectural needs of the system? Our answer to this question is that, in a four layers architecture as we proposed, the impact of choosing a solution over another one will be strictly limited to the architecture of the concerned layer. For example, using a Bayesian inference system for implementing layer 4 will only impact the data structures and algorithms used in this layer. The layer 3 could use a logical approach to identify the basic action and send the result to the layer 4, without having to concretely know how this one works. In fact, the whole point of implementing the smart environment with four separated modules

is precisely to create abstraction layers that will isolate each part of the system, allowing them to use each other as black boxes.

Ambient Intelligence applications and real-world smart homes are starting to appear outside research centres. As a consequence, part of the focus is switching from fundamental research activities to practical best practices and guidelines for the easy deploy of such systems in real environment. At the light of our knowledge and experience, and based on the successful experiments conducted by several researchers (Bouchard *et al.*, 2007; Haigh *et al.*, 2006; Mihailidis *et al.*, 2008, Patterson *et al.*, 2005), we can draw some recommendations to help design rapidly efficient and concrete smart environments:

1. Use only a limited number of targeted basic and robust sensors
2. Restraint the recognition to at most several important activities
3. Implement a learning mechanism to improve the recognition process
4. Do not try to be to precise, focus on recognizing main steps of the activity.

First, if one want to shortly develop a fully functional smart environment, he is better to use simple and robust sensors, such as pressure mat and electromagnetic contacts, than video cameras and microphone (Choudhury *et al.*, 2006). These kind of rich sensors are complicated to interpret and it need more research to be really efficient, even if some works has recently shown encouraging progress (Hoey *et al.*, 2009). Secondly, modeling a library containing only few or several important activities, such as taking medication, eating, etc., is also a good practice in order to build a concrete system. The agent will be limited in its ability to help to person, so the point is to focus the recognition process on the issues that the assistive agent is able to manage, while ignoring the other activities. Thirdly, the implementation of a good learning mechanism that improves the recognition process has proven to be a very effective solution (Patterson *et al.*, 2005; Mihailidis *et al.*, 2008), in particular while using a probabilistic approach. It allows adjusting the recognition to a specific patient's profile. Finally, the way you model and try to recognize activities will have an important impact on the quality of your results. Trying to be to precise by modeling activities with too many steps will only lead you to troubles and complexity and will not bring you any good. However, modeling activities with only several key steps, for instance 8-10 steps, has also proven to be a good match of complexity and effectiveness (Boger *et al.*, 2005; Bouchard *et al.*, 2007). The question is: which steps are really important for the assistance process and which ones can be regrouped? Following these four simple guidelines will greatly improve the design of effective smart environments.

Ambient Activity Recognition and Context Awareness

When humans interact with each others, they are able to use implicit situational information (context) in order to increase the efficiency of their interactions. In order to improve the effectiveness of computational services offered by smart home technologies, this notion of *context* must be taken into account by the computing devices. In AmI, the term *context* refers to a set of both human and physical factors (Lin *et al.*, 2007). Human factors related to the context are divided into three categories: information on the user (knowledge of habits, emotional state, biophysiological conditions, etc.), the user's social environment (co-location of others, social interaction, group dynamics, etc.), and the user's tasks (spontaneous activity, engaged tasks, general goals, etc.). Likewise, context factors related to physical environment are divided into three categories: location (absolute position, relative position, co-location, etc.), infrastructure (surrounding resources for computation, communication, task performance...), and physical conditions (noise, light, pressure, etc.). In classical

applicative fields, such as in video games, one usually supposes that the recognition context is static, and thus, that the observed agent plans in advance the performance of an activity and then sticks to its plan until the end or until something goes wrong. In a smart environment, the recognition context becomes reactive (Rao, 1993), and thus, one presumes that the acting agent works with a high-level (meta) plan and reconsiders each future planned action after executing one in order to seize opportunities, reacting to the evolving state of its environment and to its stimulus. Therefore, recognizing activities of people in a smart environment implies that the observed person may coherently change his strategy in the middle of an activity, deviating from the initial set of planned actions, but keeping the same objective in mind. This assumption amplifies the problem because the recognition system must be *aware* of the context and react accordingly in order to make a clear distinction between an erroneous deviation of a planned activity due, for instance, to a cognitive deficit, and a correct deviation justified by the evolving context (arising opportunities) coherently with the person's inferred goals. This reality causes a great deal of confusion in the AmI community between the notions of activity recognition and the one of *context awareness*. In simple terms, context awareness refers to the capacity, for a system, to take several aspects and preferences of a user (human and physical factors) into account and react accordingly (Mastrogiovanni *et al.*, 2009; Rosemann & Recker, 2006). These reactions are not necessarily complex; they can be simple reflexes or automation. Activity recognition is much more than that and refers to a high level interpretation of an observed agent behaviour while taking into account the context's elements. The objective of this recognition is to predict the behaviour in order to identify the various ways a smart environment may help its occupant. Hence, activity recognition is, from our point of view, naturally composed of context awareness abilities, but not the inverse. Therefore, activity recognition is richer and constitutes the key issue in order to be able to deploy ambient agents capable to help maintaining people at home safely (Augusto & Nugent, 2008).

Recognition Based on Temporal Constraints Propagation

Mining temporal patterns in multiple streams of sensor data is extremely useful for prediction of human behaviours and habits in the ambient environment. A large work for modeling temporal patterns deals with applications of Hidden Markov Models (HMM) or Dynamic Bayesian Networks (DBN) based on time series. For instance, The Barista system (Patterson *et al.*, 2007) is a fine-grained ADL recognition system that uses object IDs to determine which activities are currently executed. It uses radiofrequency-identification (RFID) tags on objects and two RFID gloves that the user wears in order to recognize activities in a smart home. The system is composed of a set of sensors (RFID tags and gloves) that detects object interactions, a probabilistic engine that infers activities with observations from the sensors, and a model creator that allows creating probabilistic models of activities from, for instance, written recipes. The activities are represented as sequences of activity stages. Each stage is composed of the objects involved, the probability of their involve-ment, and a time to completion modeled as a Gaussian probability distribution. The activities are converted into DBN by the probabilistic engine. By using the current sub-activity as a hidden variable and the set of objects seen and time elapsed as observed variables, the engine is able to probabilistically estimate the activities from sensor data. The engine was also tested with an HMM in order to evaluate the accuracy precision of activity recognition. These models were trained with a set of examples where a user performs a set of interleaved activities. However, some HMM models perform poorly, and the DBN model was able to identify the specific on-going activity with

recognition accuracy higher than 80%, which is very impressive. This approach is able to identify the currently carried out ADL in a context where activities can be interleaved. However, this ap-proach does not take into account the erroneous realization of activities, because the result of the activity recognition is the most plausible on-going ADLs. Moreover, one of the limitations is the standard estimation algorithms which assume that the topology of the structure in terms of states and transitions is known. Clearly, when it comes to data mining, finding the structure is the crux of the problem where incorporating temporal pat-terns across different timescales, is difficult as mentioned in (Pauwels *et al.*, 2007). Using Allen's temporal logic as alternative method of predicting user actions, and being employed in the MavHome smart home project (Jakkula & Cook, 2007), is very interesting but it present some limitation such as the ambiguities of the 13 relations proposed, where from each couple of actions we can extract more than one relations. Moreover, all the most approaches ignore the home scenario where there are multiple inhabitants performing multiple tasks at the same or different times.

RELATED WORKS ON ACTIVITY RECOGNITION

The literature related to activity recognition in AI can be divided into three major trends of research: logically based theories (Kautz, 1991; Nerzic, 1996; Py, 1991; Wobcke, 2002), probabilistic methods (Albrecht *et al.*, 1998; Boger *et al.*, 2006; Charniak & Goldman, 1993; Patterson *et al.*, 2005), and hybrid models (Avrahami-Zilberbrand & Kaminka, 2007; Geib & Goldman, 2005; Roy *et al.*, 2009a), which are an emerging alternative that combines the two prior avenues of research in order to disambiguate more efficiently the possible hypotheses that can explain the observations. In this section, we present the most significant ap-proaches in activity recognition. We discuss the advantages and limitations of each of them in a context of AmI and cognitive assistance, where one of the main goals is to recognize erroneous behaviour. Finally, we describe several AmI applicative systems that used those models to concretely address the problem of assisting impaired patients in a smart environment.

Logical Activity Recognition Approaches

Logical approaches to activity recognition (Kautz, 1991; Nerzic, 1996; Py, 1991; Wobcke, 2002) consist of developing a theory using, for instance, first-order logic, in order to formalize the recognition activity by an inference process. This approach was initiated by the innovative and well-known work of Kautz (Kautz, 1991), in which a recognition agent uses a collection of plan schemes (action types), as shown in Figure 2, which are formalized by first-order logical axioms,

thus constituting an abstraction/decomposition hierarchy. In Kautz's model, the abstraction and composition relations of the hierarchy are encoded with first order axioms of type (ABS) and (DEC), specified as follows:

$$(\text{ABS}) \ \forall x(\alpha(x) \rightarrow \alpha'(x)),$$

$$(\text{DEC}) \ \forall x(\alpha(x) \rightarrow (\alpha_1(step_1(x)) \wedge \ldots \wedge (\alpha_n(step_n(x)) \wedge \kappa),$$

where α corresponds to an event (action or plan), α' corresponds to a direct abstraction of the specified event α (ABS), and α_i corresponds to the ith step of a certain plan (event) α (DEC). For instance, the axiom (ABS) $\forall x(\text{Make\_Pasta\_Dish}(x) \rightarrow \text{PrepareMeal}(x))$ would formally describe the abstraction relation that exists between "Make pasta dish" and "Prepare Meal" in the example of Figure 2. The *End* event characterizes the root of the hierarchy, and is used to control the

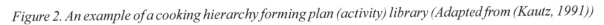

Figure 2. An example of a cooking hierarchy forming plan (activity) library (Adapted from (Kautz, 1991))

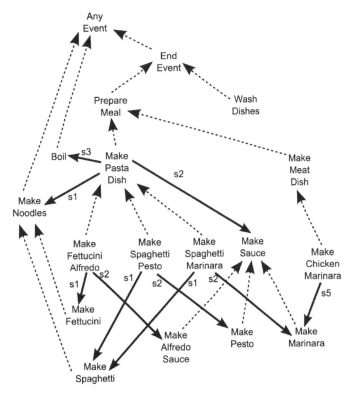

inferential recognition process. The symbol κ denotes a conjunction of constraints (preconditions, temporal relations, etc.). This event hierarchy is presumed sound (i.e. there are no errors in the defined relations between events) and complete (i.e. it contains all the events that the agent can possibly observe).

With this plan library, the activity recognition process uses a set of inference rules in order to extract a minimal interpretation model, which consists of a disjunction of hypotheses (possible plans), which cover the set of observed actions. In fact, this corresponds to applying the circumscription theory proposed by McCarthy (McCarthy, 1980) to the plan library based on observations. These inferential rules are described as follows:

(EXA) $\forall x(\alpha(x) \rightarrow \alpha_1(x) \vee ... \vee \alpha_n(x))$

(DJA) $\forall x(\neg\alpha_1(x) \vee \neg\alpha_2(x))$

(CUA) $\forall x(\alpha(x) \rightarrow END(x) \vee (... \vee$
$\exists y(\alpha^j(x) \wedge (x = step_{ij}(y))) \vee ...))$

(MCA$_n$) $\forall x_1 ... \forall x_n (END(x_1) ... END(x_n) \rightarrow$
$\vee i_j(x_i = x_j))$

The exhaustivity (EXA) inferential rule specifies that an observed event x, of type α, is necessary also an instance of one of the specialized events $\{\alpha_1 ... \alpha_v\}$. For instance, if an observed event x is known to be of "Make pasta dish" type, we can infer that it also belong to one of its specialized types: "Make fettuccini alfredo", "Make spaghetti pesto" of "Make spaghetti marinara". In addition, the disjunction rule (DJA) specifies that this event must be the instance of only one specialisation. So, x cannot be of type "Make fettuccini alfredo", and of type "Make spaghetti pesto" at the same time. These two rules are limited, however, by the fact that they lose their validity if the plan library is incomplete or if multiple inheritances are admitted in the hierarchy. The composition rule (CUA) implies that an observed event is

necessarily related to an intention (i.e. an *End* event) or is a component of one plan linked to an intention. For instance, the observed event x can be of "Make meat dish" type, which is a subtype of *End*. If x does not belong to a subtype of *End*, then it corresponds to a component (for instance "Make marinara") of one its specialisations. As we can see, this rule also assumes that the library is complete. Finally, the last rule concerns the minimum cardinality (MCA) and is based on the assumption that if two observations can be interpreted by the same *End* event, then these observations necessarily emanate from the same intention. In other terms, in case of doubt, we always keep the simplest explanation.

One problem with most of the logical recognition models is that the result of the activity recognition process consists of a disjunction (a set) of possible plans which are equiprobable, so that a hypothesis cannot be privileged within this set. In a context of AmI for cognitive support, where assistive actions must be taken quickly, purely logically-oriented recognition approaches are rarely used, because one should possess a way to quantify relevant hypotheses in order to make educated guesses about the person's on-going activities. Another important issue with most of the existing logical approaches is that they do not consider that an activity can be carried out in an erroneous way, because of the common assumption made about observed agent coherency. Consequently, the observation of an action different from the expected one will be only interpreted as an interruption of the current plan, in order to carry out another goal. This assumption is not realistic in our context because an unexpected action can result from an error by a patient suffering from cognitive disorder or by a healthy but fallible person. This uncertainty on the interpretation to be given to an unexpected action reveals the dilemma previously mentioned: erroneous realization versus interleaved realization. Yet another problem consists of the assumption that the plan library of the recognition agent is complete, in which all the possible activities which an observed agent can carry out are known. This assumption is also unrealistic in a context where an observed agent is erratic and can carry out errors, which cannot all be included in the library.

Some logical approaches, like those of Nerzic (Nerzic, 1996) and Py (Py, 1991), proposed an extension to the Kautz (Kautz, 1991) model in order to specifically recognize erroneous behaviour. In Py (Py, 1991), this extension consists of adding an event type called *Error* to the plan hierarchy defined by Kautz, where his intention is to explain all the observations that cannot be explained by normal events. This approach was used to implement a computer-assisted teaching system for geometry, where students plan the use of theorems in order to prove some geometrical properties. Hence, the majority of mistakes made by students, which consisted of using a theorem when its antecedents are not all true, are recognized and corrected by the system. However, as Kautz (Kautz, 1991) pointed out, this kind of approach is limited because it requires predefining a modified version of each plan for each error type, which can be costly when the system deals with several plans. In Nerzic (Nerzic, 1996), the extension consists of a second order mechanism to merge two observed actions that seem to be made for different purposes, in order to interpret the second observed action as an erroneous continuation of the first one. The main limitation of this approach is due to second-order logic, which does not admit a complete proof theory. Therefore, this approach has not yet been formalized or implemented.

Probabilistic Activity Recognition Models

The second kind of activity recognition approaches is based on probabilistic models (Albrecht *et al.*, 1998; Boger *et al.*, 2006; Charniak & Goldman, 1993; Patterson *et al.*, 2005), with or without a learning mechanism. These approaches are mainly based on Bayesian networks (Charniak

& Goldman, 1993; Albrecht *et al.*, 1998) and on Markovian models (Boger et al., 2006; Patterson *et al.*, 2005) and try to define activity recognition in terms of probabilistic reasoning. This allows us to resolve the hypotheses equiprobability problem that results from logical approaches. The general idea consists of attributing a probability to each plan in the library and defining a stochastic model that allows for revising these probabilities according to the observed actions and the known state of the environment. Consequently, the result of the recognition process consists of plans (hypotheses) having the highest probability.

One of the precursors of this kind of approaches is Charniak (Charniak & Goldman, 1993), which proposed the first recognition process based on Bayesian networks. A Bayesian network can be defined as a directed acyclic graph (DAG) structure, where nodes represent random variables and arcs represent causal dependencies between variables (Russell & Norvig, 2003). In Charniak & Goldman (1993), possible activities currently carried out by the observed person are defined in the networks with variables that correspond to high-level hypotheses, each one possessing a handcrafted *a priori* probability. Figure 3 shows a simple example of that kind of Bayesian networks, where there are three possible activities h_1 (Make fettucini marinara) = 0.35, h_2 (Make spaghetti marinara) = 0.3, h_3 (Prepare tea) = 0.35. Primitive actions, which are considered directly observable, are also represented with variables with a causality link defined between them and their corresponding activities. For instance, in Figure 3, the link between h_1 and e_2 specifies the fact that putting water on to boil is a step in the achievement of making spaghetti marinara. The probabilities $P(e_2/h_1) = 0.5$ of this link characterizes the fact that there is a 50% chance that if e_1 is observed, it is because h_1 is the on-going activity performed by the observed agent.

The inferential recognition process of this model consists of evaluating the posterior probability of each hypothesis (activity) by propagat-

Figure 3. An example of a simple Bayesian network for plan recognition

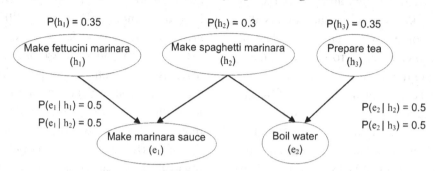

ing the conditional influences through the network using the Bayes' theorem, following an observed event. Consequently, the probabilities update in the network allows us to select the most likely interpretation (hypothesis) for a given set of observed actions. To illustrate that process, let's return to the example of Figure 3 and suppose that we just observed e_2 (Boil water). The Bayes' theorem allows us to calculate the posterior probability $P(h_1/e_2)$ by dividing $P(e_2/h_1) * P(h_1)$ by the sum of the multiplication of all other events ($P(e_2/h_1) * P(h_1)) + (P(e_2/h_2) * P(h_2)) + (P(e_2/h_3) * P(h_3))$. Therefore, the new probability is P(Prepare tea/ Boil water) = $(0.5 * 0.35) / ((0.5 * 0.35) + (0.5 * 0.3) + (0 * 0.35)) = 0.54$, meaning that there is approximately now 54% chance that the on-going activity is "Prepare tea", knowing that "Boil water" has been observed.

Following the trace of work on Bayesian approaches, another group of scientists (Boger *et al.*, 2006; Patterson *et al.*, 2005) recently proposed to define the activity recognition process, in a context of AmI, using hidden Markov models (HMM) (Rabiner, 1990). This type of approach tries to represent the library of the observing agent with a set of discrete possible states. These states aim to characterize all the possible configurations of the smart environment, assuming that these configurations are not directly observable but can only be evaluated from inputs given by sensors. A stochastic model specifies and quantifies the dynamics that bounds the transition between those states. The structure of the HMM is defined as a tuple <S, Obs, A, B, π>, where S is the set of all possible states of the environment, Obs is the set of observable inputs given by sensors, A (actions) is the probability transition matrix between states, B is the probability matrix that links inputs from sensors with environment states, and π defines the *a priori* probability of each state at the beginning of the recognition process, when there is no input. Figure 4 shows an example of such an HMM, where S = {"Boil water", "Make noodle", "Make sauce"}, indicating that the environment can be in three possible states: (1) the person is putting water on to boil, (2) the person is making noodles, (3) the person is preparing sauce. In this example, we have three sensors: one on the tap, one on the noodle box, and one on the stove. The probability of getting an input from the tap while being in the state "Boil water" could be, for instance, very high. Similarly with the noodle box sensor and "Make noodle", and the stove sensor and "Make sauce". Knowing the *a priori* probability of each state defined in π, and the probability matrices A and B, the Viterbi algorithm (Forney, 1973) is used to estimate the most likely sequence of states that fits a sequence of observations. For instance, if the tap sensor came on at time *t*, the noodle box RFID tag indicate it is moving at time *t+1*, and the stove sensor came on at time *t+2*, the most likely inferred states sequence would be "Boil water", "Make noodle", "Make sauce".

Figure 4. An example of a simple HMM for activity recognition in a smart environment

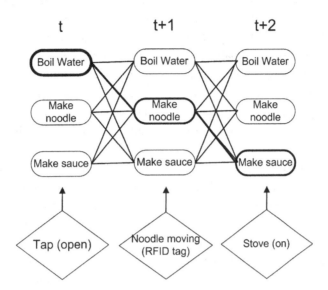

This sequence could then be linked with the corresponding activity.

Some Markovian recognition approaches also use a learning technique, mostly based on the Expectation-maximization algorithm (Blimes, 1998) and on some modified version of the Viterbi Algorithm (Wilson & Philipose, 2005), to improve the efficiency of the recognition model over time. The procedure consists in adjusting/updating probabilities in π, A and B based on the result of each episode of observation/recognition. The disadvantage of this kind of learning techniques is that they need a very large amount of training data to be efficient, and they are also limited in the recognition of unusual or novel events that did not occur beforehand. Moreover, learning erroneous patterns as correct ones may be problematic.

At the end, it is clear that several limitations restrain the use of probabilistic models for recognizing activities in a context of AmI. The first problem is the complexity of the calculations, notably with Bayesian networks, at the time of the propagation in the network, mainly observed when the network's size is large. Most of these calculations are somewhat blind because the prob-

abilities of all possible hypotheses are evaluated, even the one of irrelevant activities. Moreover, the estimation of the prior probabilities is a difficult and important task, because the probabilistic inference's precision depends on this estimation. Also, the probabilistic distribution in the plan library must remain uniform. Therefore, when a new activity of state is added to the library, the probabilistic distribution must be re-evaluated in order to be uniform. This constitutes a clear limitation in a context where one wishes to learn new behaviour coming from a patient. Furthermore, the result of the recognition consists of the most likely plan (activity) to be carried out, which constitutes a problem in a context where multiple activities can be performed in an interleaved or an erroneous way.

Hybrid Activity Recognition Approaches

A promising new avenue of research in activity recognition, which we are exploring at our labs, consists of combining logical and probabilistic plan recognition approaches in order to disambiguate more efficiently the hypotheses explaining the observations (Avrahami-Zilberbrand & Kaminka, 2007; Geib & Goldman, 2005; Roy *et al.*, 2009a). It consists of combining a symbolic approach for hypothesis construction with a probabilistic inferential process. The symbolic recognition layer filters the hypotheses by passing only a relevant recognition space to the probabilistic inference engine, instead of considering the whole set of plans included in the library, as the classical probabilistic approaches usually do.

As an example, the hybrid logical-probabilistic activity recognition that we proposed in the (Roy *et al.*, 2009a) model tries to address the recognition issue by using lattice theory and Description Logics (DL) (Baader *et al.*, 2007), which transform the activity recognition problem into a classification issue. Description logics are a well-known family of knowledge representation formalisms that may be viewed as fragments of first-order logic. The main strength of DL is that they offer considerable expressive power going

far beyond propositional logic, although reasoning is still decidable. The proposed model is made to provide an adequate basis to define algebraic tools used to formalize the inferential process of activity recognition for Alzheimer's patients. To summarize, our approach consists of developing a model of minimal interpretation for a set of observed actions, by building a plan lattice structure as shown in Figure 5.

In this model, the uncertainty related to the anticipated patient's behaviour is characterized by an intention schema. This schema corresponds to the lower bound of the lattice and is used to extract the anticipated incoherent plans, which are not pre-established in the knowledge base that the patient may potentially carry out as a result of the symptoms of his disease. However, it is not sufficient to be able to disambiguate the relevant hypotheses. Therefore, the addition of a probabilistic quantification on the lattice structure is an interesting and effective alternative. The proba-

Figure 5. A plan lattice structure that models two plans: "Cooking pasta" and "Preparing tea"

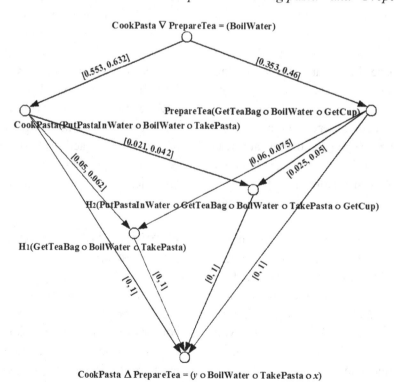

bilistic quantification that we proposed is based on samples of observation frequencies obtained at the end of a training period while the system learns the usual routines of the patient. This knowledge allows us to create a profile of the patient that offers a relevant basis to accurately estimate the probabilities of possible ongoing plans. This approach was implemented and tested in the DOMUS experimental infrastructures, where we have simulated different scenarios based on 40 low-level actions and 10 activities of daily living. Each of these activities corresponds to a common kitchen task (cooking cake, cooking pasta, making tea, etc.) sharing several actions with some other activities, in order to create a realistic context where plans can be interleaved and can lead to many different kinds of planning errors (realization, initiation, sequence, completion, etc). The observation's frequencies of the erroneous and coherent behaviours are based on the frequencies described in the study of (Giovannetti *et al.*, 2002), done on 51 patients suffering from neurodegenerative diseases, which include Alzheimer disease. The results clearly show very good performance in recognizing certain types of errors. However, the computational complexity of this approach limits its large scale application. Moreover, the rigidity of the proposed activity description formalism that, for instance, does not take temporal relations between actions into account, limits the capacity of the approach to recognize all error types.

To address those issues, we very recently proposed a complete new version of this model (Roy et al., 2009b) based on possibilistic description logic (Qi *et al.*, 2007) instead of probability theory. We propose to quantify the uncertainty related to each plan by a measurement of necessity and possibility of the activity, which will depend on the correlation between the data collected through the various distributed sensors of the smart home. From these measurements, we will deduce the confidence interval of the normal and abnormal behaviours. We can primarily justify the use of a

confidence interval rather than a point probability, as previous approaches traditionally used, since all is possible in the case of Alzheimer's behaviour. Hence, unusual behaviour is not considered as the residual result of normal behaviour. More precisely, measurements of possibilities of the erroneous and normal activities can converge respectively towards the maximum value of the confidence interval. This is one of the aspects that the theory of probability is not able to take into account.

There are two other important hybrid approaches that can be distinguished (Avrahami-Zilberbrand & Kaminka, 2005; Geib & Goldman, 2005). In Geib (Geib & Goldman, 2005), activity recognition is based on probabilistic abductive reasoning to generate agent goal hypotheses, while taking into account observation loss uncertainty, when the system is unable to observe an action because, for instance, of a hardware error. These inherent difficulties with the recognition of low-level actions are not related to high-level behavioural errors, which occur when an observed agent carries out, in an erroneous way, an activity planned in advance. The Geib (Geib & Goldman, 2005) approach uses a Hierarchical Task Network (HTN) to represent the plan library. There are several methods to achieve possible goals. Each method consists of a set of primitive actions (which can be partially ordered) that an agent must carry out to achieve a specific goal. Also, this approach is centered on a model of plan execution in which the recognition agent takes as input the set of possible plans (activities) of an observed agent at the start of the recognition process. This set of plans determines the set of pending actions, which represents all the actions that the observed agent can carry out according to the partial order of the selected plans. When this agent carries out one of the pending actions, a new set of pending actions is generated by removing the executed action and by adding the set of newly enabled actions. An action is enabled when its predecessor is carried out. The set of pending actions allows considering the realization of multiple plans in an interleaved way. This execution model is formalized with logical rules in the notation of Poole's logic of probabilistic Horn abduction (PHA) (Poole, 1993). After each observation, the probability of each hypothesis (a selected plan) is evaluated. Also, for each action in the pending set, the probability that the action will be the next one executed is evaluated. In a situation where the recognition process is unable to recognize or to observe an action, due to a system error or to deficient sensors, the recognition system evaluates the most plausible action that occurs. This action is chosen from the pending set, which limits the system in a context where the observed entity can carry out an activity in an erroneous way, in which some actions are not in the pending set. To be in the pending set, an action in an activity must be directly executable and its predecessors in the activity must be carried out. An extension of this model was used in an agent-based monitoring and support system for elderly people (Haigh *et al.*, 2006). Geib's approach suffers from several limitations. Firstly, the recognition agent must know all the observed agent's possible plans in order to be able to use the model of plan execution. In a context of AmI for cognitive assistance, the recognition agent does not know all the possible activities that may be performed, and hence it must consider all the plans from the library, thus increasing the complexity of the recognition process. Finally, this model considers that the observed agent generally acts in a coherent way. The only exception is the case of goal abandonment, which can be induced by coherent behaviour, such as a person that reacts to the dynamics of his environment where a goal cannot be carried out, or by incoherent behaviour, such as an Alzheimer's patient that forgets what he is doing.

In (Avrahami-Zilberbrand & Kaminka, 2005), the hybrid activity recognition is based on a symbolic algorithm that filters valid hypotheses from the plan library and on a probabilistic reasoner that ranks these hypotheses based on their likelihood.

The symbolic algorithm uses a single root directed acyclic graph to represent the plan library. In this graph, each node represents a plan step, a vertical edge decomposes a plan step into sub-steps, and a sequential edge specifies the temporal order of execution of the sub-steps. For each observation period *t*, the algorithm uses a decision tree, like those used in the learning program C4.5 (Quinlan, 1993), to evaluate the plan step which corresponds to the set of the environment's observed characteristics which can be partially observable. In order to consider partial observability of the environment, the decision tree uses branches that represent missing values of the environment's properties. The identified plan step is tagged with a timestamp *t* in the plan library, and this timestamp is propagated to the parent node and sub-steps of the identified plan step. The use of timestamps in the plan library allows one to consider temporal consistent plans only, and to take into account the fact that a plan can be interrupted in order to execute another one, and return later to the remaining plan steps in the first plan. The symbolic algorithm, at a time *t*, selects in the plan library the hypotheses that are temporally consistent with the timestamp *t* and its predecessors. The probabilistic reasoner uses a Hierarchical Hidden Markov Model (HHMM) (Rabiner, 1990) version of the plan library to evaluate the probability of each hypothesis, in order to rank them. However, it is difficult to correctly evaluate the probability for all transitions in the plan library, in or to have a good probabilistic quantification of the hypotheses. Recently, (Avrahami-Zilberbrand & Kaminka, 2007) extends this model with a utility function in order to allow the incorporation of the observer's biases and preference. Hence, the recognition process ranks the hypotheses based on their expected utility to the observer, which is obtained by multiplying the posterior probability of a hypothesis by its utility to the observer. Nevertheless, a hypothesis utility is difficult to measure in our cognitive assistance context, contrary to economic and strategic worlds. Also, this approach considers that an observed agent cannot carry out erroneous plans because it is assumed that he has a coherent behaviour which is temporally consistent.

Ambient Intelligent Applications of Activity Recognition Models

There are presently many research groups working on assistive technologies that exploit activity recognition models in order to concretely address the problem of cognitive assistance in a smart environment (Augusto & Shapiro, 2007). This section presents, as examples, an overview of some of these projects and shows how they use activity recognition techniques.

The DOMUS and LIARA labs (Giroux *et al.*, 2009) explore together how smart environment technologies can be used to provide pervasive assistance to people suffering from cognitive deficits (Alzheimer disease, head traumas, schizophrenia, etc.). The DOMUS lab has an actual apartment on the campus of the University of Sherbrooke, which is basically composed of a kitchen, a living room, a dining hall, a bedroom, and a bathroom, augmented with sensors such as smart tags (RFID), location and identification systems for objects and people, audio and video devices, etc. This infrastructure has three layers. The lowest layer is made of hardware, embedded processors, and networks. Middleware is deployed on top of the hardware layer. At the highest level, smart care services are deployed. Prototypes are thus experimented, validated and progressively transferred to real life. In the long term, the goal of this project is to create a net of intelligent assistive agents forming together a system of ambient intelligence able to efficiently decide: (i) if help is needed and the right moment to provide assistance, (ii) how to assist (simply giving advices or performing required actions in the user's place), (ii) which devices to use to assist among the available effectors (light, TVs, micros, speakers, PDAs, etc.), (iii) how to take into account

the user profile, in particular the user's specific cognitive deficits and abilities. In this context, activity recognition becomes a key issue that needs to be addressed. It is why the researchers at the DOMUS and LIARA labs have explored several representation models to attack this issue. The first avenue that we explored for recognizing activities described the user's daily task using a hierarchical structure (Bauchet & Mayers, 2005). The model included two types of task nodes: the goal of the occupant and the method to attain it. Leaves are methods of terminal tasks, which mean an atomic way to attain a concrete goal. To monitor the proper completion of activities, temporal information is introduced for task nodes. Despite the good results that have been shown in real case assistance scenarios, the system appears to be somewhat limited, owing to the facts that it is only able to monitor one specific ADL and that the assistance agent reacts after the user error. To address this limitation, we decided to explore a second avenue of research by developing a hybrid recognition approach, which took the form of a lattice-based model in DL (Roy *et al.*, 2009a) that we described in previous section. This model has shown good results in simulation with real case scenarios, especially in recognizing typical errors of Alzheimer's patients. A new version of this model (Roy *et al.*, 2009b), based on probabilistic description logic (Qi *et al.*, 2007) was recently proposed to address the main limitations of the model, which were computational complexity and recognition of certain types of errors.

The Autominder system (Pollack, 2005) is a well-known application of cognitive assistance in smart homes. It is a planning assistant developed at the University of Michigan. This system provides reminders about the daily activities of its user. It tracks the behaviour of the user in order to reason about the need to provide reminders and when it is most appropriate to issue them. The Autominder system has three main components: a plan manager, a client modeller, and a personal cognitive orthotic. The plan manager stores the user's activity in the client plan, updates it, and identifies possible conflict in it. The client modeller tracks the execution of the plans given information about the user's observable activities, and stores its belief about the execution status in the client model. The personal cognitive orthotic reasons about inconsistencies between what the user is supposed to do and what he is currently doing, and determines when to issue reminders. In order to take into account several kinds of temporal constraints, the user's plans are modeled as Disjunctive Temporal Problems (DTPs). The client manager uses sensor information to infer activities performed by the user. If the likelihood that a planned activity has been executed is high, the client manager sends this information to the plan manager, which updates the client plan by recording the time of execution and by propagating all the affected constraints. The client model is represented by a Quantitative Temporal Bayes Net (QTBN), which was developed for the reasoning about fluents and probabilistic temporal constraints (Pollack, 2005). A QTBN is composed of a dynamic Bayes network that contains all causal relationships and a standard Bayes net that represents all temporal relationships. The personal cognitive orthotic uses the client plan and the client model to determine which reminders should be issued and when. The Autominder system is able to take into account situations where the user performs multiple activities and to prompt reminders when some erroneous behaviour occurs (mainly temporally related). The handcrafted temporal constraints defined on the model of the on-going activity are compared to the observed actions sequence, and if they are not satisfied by the observations, the patient is considered to be in error. This approach is limited by the fact that the system does not know the type of the committed error. For instance, the patient may have performed actions in the wrong orders or may have forgotten to carry out an action, both cases breaking the temporal constraints. However, the first case corresponds to a sequence error and the second one to an omission. In order to correctly adapt the form of assistance to the patient's profile, one must know the type of error that he committed. Moreover, cognitive errors are not always related to temporal constraints.

The Independent LifeStyle Assistant™ (I.L.S.A.) (Haigh *et al.*, 2006) is a monitoring and support system for elderly people developed at Honeywell Laboratories. The purpose of this system is to allow elderly people to live longer in their homes by reducing the caregivers' burden. The I.L.S.A. concept consists of a multi-agent system that incorporates different home automation devices to detect and monitor activities and behaviour. It uses Geib's hybrid plan recognition model (Geib & Goldman, 2005) that we described in the previous section for its task tracking component. This component generates hypotheses concerning the person's goals, while taking into account uncertainty concerning observation loss when the system is unable to observe an action due to sensor errors. Therefore, the task-tracking component addresses only the problem of recognizing low-level errors. The high-level behavioural errors, which occur when an observed person carries out, in an erroneous way, an activity planned in advance, are not taken into account. A field study of the I.L.S.A. system was made to complete an end-to-end proof-of-concept and consists mainly of monitoring two ADLs (medication and mobility) and issuing alerts and information to family caregivers. However, the hybrid recognition component was eventually removed from the I.L.S.A system field study because of experimental code, scalability, and memory management issues and to reduce the deployment cost induced by the size of the set of sensors.

CONCLUSION: CHALLENGES AND PERSPECTIVES

The new development towards pervasive assisted living will stimulate research in many fields of artificial intelligence, such as multi-agent approaches as a development paradigm for this open and hardly dynamic environment. For forty years, artificial intelligence has not ceased being used on a large scale through expert system applications, web search agents, etc. If the internet indicated the advent of conventional planetary networks, the next evolution, that will support the development of the artificial intelligence, relates to new challenging issues concerning how a network of agents will be deployed within our natural living environment to enable ambient intelligence and cognitive support. As a contribution to this emerging field, this chapter investigated in detail the challenging key issues related to the development of ambient agents in a smart environment, under the context of activity recognition. First, we clearly defined the notion of activity recognition by positioning it in the context of AmI. Then, we clarified the specific functional needs inherent in cognitive assistance for effective activity recognition and distinguished four main issues related to the recognition process. In order to guide further work, we suggested following a 4-layer multi-agent architecture that aims to address each of these issues. Thereafter, we presented a detailed survey and an analysis of existing work on activity recognition that has been studied for potential integration into our laboratories. This survey allowed us to identify gaps in the capabilities of current techniques and to give hints on the most productive lines of research to address this complex issue. Finally, we presented several projects in the AmI community that exploited activity recognition models to concretely deploy cognitive assistance solutions. As such, the contribution of this chapter is of both theoretical and practical significance. We raised numerous research issues and challenges that need to be addressed for understanding the field and enabling ambient multi-agent recognition systems for cognitive assistance to operate effectively.

REFERENCES

Albrecht, D. W., Zukerman, D. W., & Nicholson, A. E. (1998). Bayesian Models for Keyhole Plan Recognition in an Adventure Game. *User Modeling and User-Adapted Interaction*, 8(1-2), 5–47.

Augusto, J. C., & Nugent, C. D. (2006). Designing Smart Homes: The Role of Artificial Intelligence. Vol. 4008 of LNAI, Springer-Verlag, Berlin, Germany.

Augusto, J. C., & Shapiro, D. (2007). *Advances in Ambient Intelligence. In the Frontiers of Artificial Intelligence and Application (FAIA) Series* (*Vol. 164*). IOS Press.

Avrahami-Zilberbrand, D., & Kaminka, G. A. (2005). Fast and complete symbolic plan recognition. Proc. of the 19th Int. Joint Conference on Artificial Intelligence (IJCAI'05), International Joint Conference on Artificial Intelligence, pp.653–658.

Avrahami-Zilberbrand, D., & Kaminka, G. A. (2007). Incorporating Observer Biases in Keyhole Plan Recognition (Efficiently!), in: Proc. of the Twenty-Second AAAI Conference on Artificial Intelligence, AAAI Press, Menlo Park, CA, pp.944–949.

Baader, F., Calvanese, D., McGuinness, D. L., Nardi, D., & Patel-Schneider, P. F. (2007). *The Description Logic Handbook: Theory, Implementation, and Applications* (2nd ed.). Cambridge University Press.

Bauchet, J., & Mayers, A. (2005). Modelisation of ADLs in its Environment for Cognitive Assistance, In: Proc. of the 3rd International Conference on Smart homes and health Telematic, ICOST'05, Sherbrooke, Canada, pp. 3-10.

Baum, C., & Edwards, D. F. (1993). Cognitive Performance in Senile Dementia of the Alzheimer's type: The Kitchen Task Assessment. *The American Journal of Occupational Therapy.*, *47*, 431–436.

Bilmes, J. (1998). A gentle tutorial on the EM algorithm and its application to parameter estimation for Gaussian mixture and hidden Markov models. Technical Report ICSI-TR-97-021. Technical report, University of Berkeley.

Boger, J., Hoey, J., Poupart, P., Boutilier, C., Fernie, G., & Mihailidis, A. (2006). A Planning System Based on Markov Decision Processes to Guide People with Dementia Through Activities of Daily Living. *IEEE Transactions on Information Technology in Biomedicine, 10*(2), 323–333.

Boger, J., Poupart, P., Hoey, J., Boutilier, C., Fernie, G., & Mihailidis, A. (2005). A decision theoretic approach to task assistance for persons with dementia. In Proceedings of the 19th International Joint Conference on Artificial Intelligence (IJCAI05), pp. 1293–1299.

Bouchard, B. (2006). Un modèle de reconnaissance de plan pour les personnes atteintes de la maladie d'Alzheimer basé sur la théorie des treillis et sur un modèle d'action en logique de description, Ph.D. Thesis, University of Sherbrooke, Canada, 268 pages.

Bouchard, B., Bouzouane, A., & Giroux, S. (2005). Plan recognition in Smart Homes: An approach based on the Lattice Theory [HWRS]. *Journal of Human-friendly Welfare Robotic Systems, 6*(4), 29–45.

Bouchard, B., Bouzouane, A., & Giroux, S. (2007). A Keyhole Plan Recognition Model for Alzheimer's Patients: First Results, Journal of Applied Artificial Intelligence (AAI), Taylor & Francis publisher, Vol. 22 (7), July 2007, pp. 623–658.

Bouzouane, A., Bouchard, B., & Giroux, S. (2005). Action Description Logic for Smart Home Agent Recognition, in Proc. of the International conference on Human-Computer Interaction (HCI) of the The International Association of Science and Technology for Development (IASTED), Phoenix, USA, pp 185-190.

Capezio, F., Giuni, A., Mastrogiovanni, F., Sgorbissa, A., Vernazza, P., Vernazza, T., & Zaccaria, R. (2007). Perspectives of Ambient Intelligence. In *Journal of the Italian AEIT Association* (pp. 42–49). Sweet Home.

Carberry, S. (2001). Techniques for plan recognition. *User Modeling and User-Adapted Interaction, 11*(1–2), 31–48.

Charniak, E., & Goldman, R. (1993). A Bayesian Model of Plan Recognition. *Artificial Intelligence Journal, 64*, 53–79.

Choudhury, T., Philipose, M., Wyatt, D., & Lester, J. (2006). Towards activity databases: Using sensors and statistical models to summarize people's lives, in IEEE Data Engineering Bulletin, pp. 49-58.

Christensen, H. B. (2002), "Using Logic Programming to Detect Activities in Pervasive Healthcare", Proc. Int'l Conf. Logic Programming, Springer, pp. 185–199.

Cohen, P. R., Perrault, C. R., & Allen, J. F. (1982). Beyond question answering, in Strategies for Natural Language Processing, W.G. Lehnert and M.H. Ringle eds., Lawrence Erlbaum Associates, Hillsdale, NJ.

Forney, G. D. (1973). The Viterbi algorithm. *Proceedings of the IEEE, 61*, 268–278.

Geib, C. (2007). Plan recognition. In Kott, A., & McEneaney, W. M. (Eds.), *Adversarial Reasoning: Computational Approaches to Reading the Opponent's Mind* (pp. 77–100). Chapman & Hall/CRC.

Geib, C., & Goldman, R. (2001). Plan Recognition in Intrusion Detection Systems. In DARPA Information Survivability Conference and Exposition (DISCEX).

Geib, C., & Goldman, R. (2005) Partial Observability and Probabilistic Plan/Goal Recognition, in: Proc. of the IJCAI-05 workshop on Modeling Others from Observations (MOO-05), Int. Joint Conference on Artificial Intelligence, pp. 1–6.

Georgeff, M. P. (1987), Planning, in Annual Reviews, Palo Alto, California.

Giovannetti, T., Libon, D. J., Buxbaum, L. J., & Schwartz, M. F. (2002). Naturalistic action impairments in dementia. *Neuropsychologia, 40*(8), 1220–1232.

Giroux, S., Leblanc, T., Bouzouane, A., Bouchard, B., Pigot, H., & Bauchet, J. (2009) The Praxis of Cognitive Assistance in Smart Homes, IOS Press book chapter, pp. 1-30, (to appear).

Goldman, R. P., Geib, C. W., & Miller, C. A. (1999). A new model of plan recognition. In Proceedings of the Conference on Uncertainty in Artificial Intelligence, San Francisco, CA (USA), 245–254.

Haigh, K., Kiff, L., & Ho, G. (2006). The Independent LifeStyle Assistant: Lessons Learned. *Assistive Technology, 18*(1), 87–106.

Heinze, C., Goss, S., Lloyd, I., & Pearce, A. (1999). Plan Recognition in Military Simulation: Incorporating Machine Learning with Intelligent Agents. In Proc of IJCAI-99 Workshop on Team Behaviour and Plan Recognition, pp. 53–64.

Hoey, J., Von Bertoldi, A., Craig, T., Poupart, P. & Mihailidis, A. (in press, 2009). Automated Handwashing Assistance For Persons With Dementia Using Video and A Partially Observable Markov Decision Process. Computer Vision and Image Understanding (Special Issue on Computer Vision Systems).

Jakkula, V. R., & Cook, D. J. (2007). Using temporal relations in Smart Environment Data for Activity Prediction, in: Proc. of the 24th International Conference on Machine Learning, Corvallis, Oregon, pp. 1–4.

Katz, S., Ford, A. B., Moskowitz, R. D., Jackson, B. A., & Jaffe, M. W. (1963). Studies of illness in the aged. The index of ADL: A standardized measure of biological and psychosocial function. *Journal of the American Medical Association, 185*(12), 914–919.

Kautz, H. (1991) A Formal Theory of Plan Recognition and its Implementation, Reasoning About Plans, Allen J., Pelavin R. and Tenenberg J. eds., Morgan Kaufmann, San Mateo, C.A., 69-125.

Lesh, N., Rich, C., & Sidner, C. L. (1998). *Using plan recognition in Human-Computer Collaboration, TR-98-23, Mitsubishi Electric Research Lab*. MERL.

Lin, F., Yang, C., & Yang, Y. (2007). Supporting Learning Context-aware and Auto-notification Mechanism on an Examination System. In T. Bastiaens & S. Carliner (Eds.), Proceedings of World Conference on E-Learning in Corporate, Government, Healthcare, and Higher Education 2007, pp. 352-359.

Mao, W., & Gratch, J. (2004). Decision-Theoric Approach to Plan Recognition. Technical Report ICT-TR-01-2004, Institute for Creative Technologies (ICT).

Mastrogiovanni, F., Sgorbissa, A., & Zaccaria, R. (2009). On the Problem of Describing Activities in Context-Aware Environments. In The International Journal of Assistive Robotics and Mechatronics, vol. 9, no. 4.

McCarthy, J. (1980). Circumscription – A Form of Non-Monotonic Reasoning. *Artificial Intelligence, 13*, 27–39.

Mihailidis, A., Boger, J., Craig, T., & Hoey, J. (2008). The COACH prompting system to assist older adults with dementia through handwashing: An efficacy study. *BMC Geriatrics, 8*(28), 1–18.

Nerzic, P. (1996). Two Methods for Recognizing Erroneous Plans in Human-Machine Dialogue. In AAAI'96 Workshop: Detecting, Repairing and Preventing Human-Machine Miscommunication.

Patterson, D. J., Fox, D., Kautz, H. A., & Phillipose, M. (2005). Fine-Grained Activity Recognition by Aggregating Abstract Object Usage, in: Proc. of the IEEE International Symposium on Wearable Computers, Osaka, Japan, pp.1–8.

Patterson, D. J., Kautz, H. A., Fox, D., & Liao, L. (2007). Pervasive computing in the home and community. In Bardram, J. E., Mihailidis, A., & Wan, D. (Eds.), *Pervasive Computing in Healthcare* (pp. 79–103). CRC Press.

Patterson, D. J., Liao, L., Fox, D., & Kautz, H. (2003). Inferring High-Level Behavior from Low-Level Sensors. In Proceeding of UBICOMP 2003: The 5th International Conference on Ubiquitous Computing, Anind Dey, Albrecht Schmidt, et Joseph F. McCarthy, editors, volume LNCS 2864, 73–89. Springer-Verlag.

Pauwels, E. J., Salah, A. A., & Tavenard, R. (2007). Sensor Networks for Ambient Intelligence, in: Proc. of the 9th IEEE workshop on MultiMedia Signal Processing, Chania, Crete, Greece pp. 13–16.

Philipose, M., Fishkin, K. P., Perkowitz, M., Patterson, D. J., Fox, D., Kautz, H., & Hähnel, D. (2004). Inferring Activities from Interactions with Objects, IEEE Pervasive Computing: mobile and Ubiquitous Systems, 3(4), pp. 50–57.

Pigot, H., Mayers, A., & Giroux, S. (2003). The intelligent habitat and everyday life activity support, in: Proc. Int. Conf. in Simulations in Biomedicine, Slovenia, pp.507–516.

Pollack, M. E. (2005). Intelligent Technology for an Aging Population: The Use of AI to Assist Elders with Cognitive Impairment. *AI Magazine*, *26*(2), 9–24.

Poole, D. (1993). Probabilistic Horn abduction and Bayesian networks. *Artificial Intelligence*, *64*(1), 81–129.

Py, D. (1992). Reconnaissance de plan pour l'aide à la démonstration dans un tuteur intelligent de la géométrie. Thèse de Doctorat, Université de Rennes 1.

Qi, G., Pan, J. Z., & Ji, Q. (2007). Extending description logics with uncertainty reasoning in possibilistic logic. In Proc of the 9th ECSQARU, Hammamet, Tunisia, pp. 828-839.

Quinlan, J. R. (1993). *C4.5: Programs For Machine Learning*. San Mateo, CA: Morgan Kaufmann.

Rabiner, L. R. (1990). A tutorial on hidden Markov models and selected applications in speech recognition. IEEE 77(2), pp. 267–296.

Ramos, C., Augusto, J. C., & Shapiro, D. (2008). Ambient Intelligence: the Next Step for Artificial Intelligence. *IEEE Intelligent Systems*, *23*(2), 5–18.

Rao, A. S. (1993) Reactive plan recognition. Technical Report 46, Australian Artificial Intelligence Institute, Carlton, Australia.

Rogers, W. A., Meyer, B., Walker, N., & Fisk, A. D. (1998). Functional limitations to daily living tasks in the aged: focus groups analysis. *Human Factors*, *40*, 111–125.

Rosemann, M., & Recker, J. (2006). *Context-aware process design: Exploring the extrinsic drivers for process flexibility. T. Latour & M. Petit, 18th international conference on advanced information systems engineering. Proceedings of workshops and doctoral consortium* (pp. 149–158). Luxembourg: Namur University Press.

Roy, P. C., Bouchard, B., Bouzouane, A., & Giroux, S. (2009a). A hybrid plan recognition model for Alzheimer's patients: interleaved-erroneous dilemma, Web Intelligence and Agent Systems (WIAS): An Int. *Journal, IOS Press*, 7(4), 375–397.

Roy, P. C., Bouchard, B., Bouzouane, A., & Giroux, S. (2009b): Ambient Activity Recognition: A Possibilistic Approach, Proc. of the IADIS International Conference on Intelligent Systems and Agents 2009 (ISA'09), Algarve, Portugal, 21 - 23 June 2009, pp. 1-5.

Russell, S., & Norvig, P. (2003). *Artificial Intelligence: A Modern Approach* (2nd ed.). Englewood Cliffs, NJ: Prentice-Hall.

Schmidt, C., Sridharan, N., & Goodson, J. (1978). The plan recognition problem: an intersection of psychology and artificial intelligence. *Artificial Intelligence*, 45–83.

Waern, A., & Stenborg, O. (1995). Recognizing the plans of a replanning user. In Proc. of the IJCAI-95 Workshop on The Next Generation of Plan Recognition Systems, Montréal, Canada, pp. 113–118.

Weiser, M. (1991). The Computer for the 21st Century, *Scientific American*.

Wilson, D., & Philipose, M. (2005). Maximum A Posteriori Path Estimation with Input Trace Perturbation: Algorithms and Application to Credible Rating of Human Routines. In IJCAI-05, Proceedings of the Nineteenth International Joint Conference on Artificial Intelligence, Edinburgh, Scotland, UK, pp. 895–901.

Wobcke, W. (2002). Two Logical Theories of Plan Recognition. *Journal of Logic Computation*, *12*(3), 371–412.

KEY TERMS AND DEFINITIONS

Activity of Daily Living (ADL): Set of common activities that an individual carries out as a routine for self-care (dressing, personal hygiene, preparing meal …). It is often used as a measurement in order to assess the functional status of a cognitively impaired patient.

Activity Recognition: To infer the goals of an observed entity and the action sequence that was planned to achieve those specific goals from the observation of the environmental conditions from the sensor inputs. In order to achieve the goals of an activity, a sequence of actions must be planned. So, activity recognition is also known as plan recognition or behaviour recognition.

Knowledge Representation: Artificial intelligence area that is concerned with how to use a symbol system to represent a domain of discourse and how to use functions that allow inference about individuals/objects within this domain of discourse.

Multi-agent System: System composed of multiple intelligent agents that interact with each others. An intelligent agent is an autonomous entity (software, person …) that observes his environment and acts in it according to a set of goals (rational).

Context-Awareness System: System that uses context in order to provide relevant information and/or services to the user, where relevancy depends on the user's task. A context is any information that can be used to characterize the current state/situation of an entity (person, place, object …).

Erroneous Behaviour: Behaviour that represents the realization of one or multiple activities, where some of them are erroneous. Multiple error types could occur: initiation (incapacity to begin an activity), organization (incapacity to use the good tools in order to realize an activity), realization (forgot some steps or add irrelevant actions), sequence (the activity's steps are carried out in an unordered way), judgment (the activity is carried out in an unsafe way), and completion (incapacity to determine the activity realization end).

Interleaved Activities: Behaviour that represents the realization of multiple activities, where their actions can be interleaved. Some activities can share some actions.

Chapter 18
Logic–Based Approaches to Intention Recognition

Fariba Sadri
Imperial College London, UK

ABSTRACT

In this chapter we discuss intention recognition in general, and the use of logic-based formalisms, and deduction and abduction in particular. We consider the relationship between causal theories used for planning and the knowledge representation and reasoning used for intention recognition. We look at the challenges and the issues, and we explore eight case studies.

1 INTRODUCTION

Intention recognition, also called *goal recognition*, is the task of recognizing the intentions of an agent by analyzing some or all of their actions and/or analyzing the changes in the state (environment) resulting from their actions. *Plan recognition* is closely related to intention recognition, and extends it to recognizing the plan (i.e. the sequence of actions, including future actions) the observed agent is following in order to achieve his intention. Throughout this paper we will use *goal* and *intention* interchangeably.

Work on intention recognition has been going on for about 30 years. Examples of early work are attributed to Schmidt et al. [1978], Wilensky [1983], and Kautz and Allen [1986]. Much of the early work has been in the context of language and story understanding and automatic response generation, for example in Unix help facilities. However, new applications, such as assisted living and ambient intelligence, increasingly sophisticated computer games, intrusion and terrorism detection, and the military have brought new and exciting challenges to the field. For example assisted living applications require recognizing the intentions of residents in domestic environments in order to anticipate and assist with their needs.

DOI: 10.4018/978-1-61692-857-5.ch018

Applications in computer systems intrusion or terrorism detection require recognizing the intentions of the would-be-attackers in order to prevent them. Military applications need to recognize the intentions of the enemy maneuvers in order to plan counter-measures and react appropriately. Programs that make moral decisions (e.g. Pereira and Saptawijaya 2009) need to reason about intentions, in particular to decide whether untoward consequences of some actions were intended by the agent that performed them or were merely unintended side-effects.

Logic has been a powerful tool in intention recognition. Since the early days, for example in the work of Charniak and McDermott [1985], abduction has been used as the underlying reasoning mechanism in providing hypotheses about intentions. Also, conceptually, intention recognition is directly related to planning, and logic has been the basis of many causal theories describing the relationship between actions and effects.

Intention recognition is a rich and challenging field. Often multiple competing hypotheses are possible regarding the intentions of an observed agent. The choice between these hypotheses is one challenge, but there are many others. One, for example, is that circumstances, including the adversarial nature of the observed agent, may afford only partial observability of the actions. Geib and Goldman [2001] make a contribution in this respect, as do Sindlar et al [2008], described in Section 4 in this article. Furthermore, would-be-intruders and would-be attackers may even deliberatively execute misleading actions.

Another challenge is the case where the acting agent may have multiple intentions and may interleave the execution of his plans of actions for achieving them, or the case where the actor is concurrently trying alternative plans for achieving the same intention. Intention recognition becomes more difficult when we attempt to interpret the actions of cognitively impaired individuals who may be executing actions in error and confusion, for example in the case of Alzheimer patients (Roy

et. al. 2007). Similar complications arise and are magnified when attempting to analyze the actions and intentions of multiple (co-operating) agents (e.g. Sukthankar and Sycara 2008).

In this article we will explore the field of intention recognition, and in particular we will focus on single agent cases, as opposed to multi-agents, and on logic-based approaches. We will explore the logical basis of intention recognition, and provide an analysis of eight case studies. The case studies are chosen from the literature for their variety of methodologies and applications. We will not consider many probabilistic approaches to intention recognition, except in two of the case studies, Pereira and Anh [2009b] and Demolombe and Frenandez [2006], both of which combine logic with probabilities.

In Section 2 we look at the background and issues involved in intention recognition. In Section 3 we look at the possible relationships between logic-based causal theories and knowledge representation and reasoning for intention recognition. In Section 4 we describe and analyze the eight case studies. Finally, in Section 5, we conclude with a further discussion of the challenges.

2 BACKGROUND AND ISSUES

2.1 Applications

Much of the early work on intention recognition has been in the context of language and story understanding and automatic response generation, for example in Unix help facilities. Other early application areas include interfaces for computer-aided design [for example Goodman and Litman 1992] and collaborative problem-solving [Lesh, Rich, Sidner 1999]. More recent applications have been in diverse areas. We mention some of these below.

Assisted technologies, in general, and in the care of the elderly at home, in particular, are popular application areas for intention recognition. Giroux et al. [2008] address intention recognition for the purpose of providing assistance for cognitively impaired individuals. Pereira and Anh [2009b] and Geib and Goldman [2005], for example, focus on intention recognition in the care of the elderly. Roy et al. [2007] address complications in intention recognition when tracking the behaviour of Alzheimer patients. In the case of such patients it cannot always be assumed that their actions are based on an organized rational plan. So if the next action they execute is not what is expected it can be due to an interleaving of two plans for two different goals or it can be due to error.

Canberry and Elzer [2007] consider a more unusual, but related, application, namely the recognition of intention behind (bar chart) graphics, in order to convey the "messages" of bar charts to sight-impaired individuals. The application involves recognizing the intention of the designer of a bar chart, by analyzing an XML representation of the chart that provides information about the heights, colours and labels of the bars.

Another application area is interactive storytelling. LOGTELL [Karlsson et al. 2007], for example, is a logic-based tool for interactive storytelling. During story creation, the user can intervene by inserting events chosen from a pre-specified list. Plan recognition is then used to find plans from a plan library that subsume these events. The user chooses from amongst them and the story unfolds accordingly.

Other major application areas include computer system intrusion detection [Geib and Goldman 2001], terrorism intrusion detection [Jarvis et al. 2004], and computer games, such as real-time strategy games [Cheng and Thawonmas 2004]. The military and the civil policing also provide major applications and challenges to the field. Mao and Gratch [2004], for example, address intention recognition from observing military movements. Suzić and Svenson [2006] consider the application of riot control in urban environments. Observations include movements of groups of people.

From these and contextual information, such as location (how close they are to which buildings) and resources (what weapons they have) their intention is hypothesized.

2.2 Classification

Cohen, Perrault, Allen [1981] classify intention recognition as either *intended* or *keyhole*. In the *intended* case the agent which is being observed wants his intentions to be identified and intentionally gives signals to be sensed by the other (observing) agent. This would apply, for example, in the case of language understanding where the speaker wants to convey his intentions. In the *keyhole* case the agent which is being observed either does not intend for his intentions to be identified, or does not care; he is focused on his own activities, which may provide only partial observability to the other agent. This might be the case with help systems that provide unsolicited guidance, for example in an ambient intelligence system at home.

A third class, identified by Geib et al [2001], is *adversarial*, where the actor is hostile to his actions being observed, for example where the actions are aimed at intrusion in a network system. In fact we can take the classification further, to *diversionary,* where the actor is in fact attempting to conceal his intentions by performing misleading actions. Much of the work on intention recognition concentrates on the first two classes, namely *intended* and *keyhole*. Pereira and Anh [2009a] is a recent attempt to deal with diversionary intentions.

2.3 Components, Formalisms, Methodologies

Typically there are at least three components in an intention recognition system: (1) a set of intentions from which the system chooses, (2) some form of knowledge about how actions and plans achieve goals, and (3) a sequence of observed actions executed by the agent whose intention is being recognized. A possible additional component may be a set of hypotheses currently held about the agent's intention. For example, in a home setting, if it is late at night the recognition system may hypothesize that the resident normally prepares a hot drink or prepares a hot water bottle for bed.

There is often an assumption that the actions the observed agent executes are aimed at achieving a goal, i.e. are part of the execution of a plan formed by the agent. Of course, in difficult cases, for example in the case of Alzheimer patients, or students learning a new skill, the actions may be erroneous. There is very little work, if any, that is directed at recognizing *reactive* behaviour, namely recognizing that actions may be done in reaction to external events and stimuli, and not part of a *proactive* plan.

The intention recognition problem has been cast in different formalisms and methodologies. Prominent amongst these are logic-based formalisms, case-based approaches, and probabilistic approaches. Accordingly, component (2) of the system, i.e. the knowledge about the relationship between actions and goals, may be, for example, in the form of logic-based specifications of macro-actions [Demolombe and Fernandez 2006], or in the form of cases [for example Cox and Kerkez 2006], or in the form of plan libraries specified as Hierarchical Task Networks (HTNs) [Geib and Goldman 2005]. Geib and Steedman [2003] cast intention recognition as a problem in parsing, much as in natural language processing. Accordingly, they map Hierarchical Task Networks into context-free grammars, and the parsing is used to group together individual observations into structures that are meaningful according to the grammars.

A common assumption is that the observer agent has (full) knowledge of the planning rules (sometimes called behaviour rules) of the acting agent. For intention recognition, the observer agent may use the same representation of the planning rules, or more commonly, some transformed form of those rules which are not useful for planning, but

are useful specifically for intention/plan recognition. This transformation may be, for example, in the form of HTNs augmented with probabilities and/or utilities, or the only-if direction of if-then planning rules. This will be discussed further in Section 3.

Depending on how the knowledge is cast in component (2), component (1), i.e. the set of all possible intentions at the disposal of the system, may be explicitly represented or may remain implicit. If it is explicit it will be represented as a list of possible intentions, or a set of logical facts naming each possible intention. If it remains implicit then it can be identified as, for example, the topmost operator in the HTN used in component (2), or as (ground instances of) the predicates in the conclusion of certain sets of rules used in (2).

Component (3) of an intention recognition system is the sequence of observed actions. The assumption that the acting agent's actions can be observed unambiguously is a strong one which may be justified only in virtual environments such as simulations or games. To provide the intention recognition system with this component, *activity recognition* may be used in conjunction with intention recognition.

2.4 Activity Recognition

Activity recognition, typically, uses data from cameras and RFID (Radio Frequency Identification) readers and tags to track movements of humans and to identify the objects they handle. In the home environment, for example, RFID tags may be attached to household objects. Park and Kautz [2008a,b], for example, describe a prototype human activity recognition system built in the Laboratory for Assisted Cognitive Environments (LACE). The system uses a combination of distributed multi-view visions system and RFID readers and tags to provide information about activities, at both coarse (e.g. walking around) and fine (e.g. opened cupboard door, took out cereal box) levels.

Box 1.

```
building(X, public) ∧ door-open(X)^0.1 → may-enter(X)
building(X, private) ∧ door-open(X)^0.9 → may-enter(X)
```

Mihailidis, Fernie, Barbenel [2001] focus on recognizing very specific activities such as hand washing, similarly to Barger et al. [2002] who focus on meal preparation. PROACT [Philipose et al. 2005] employs a Dynamic Bayesian Network representing daily activities such as making tea. The user wears special gloves that can read the RFID tags of objects such as cups. Making tea is modeled as a three stage process, with high probabilities of using the kettle in the first stage and the box of tea-bags in the second stage, and medium probability of using milk, sugar or lemon in the third stage. PROACT then uses information about the objects being used and time elapsed between their usages to hypothesise possible activities. Sanchez, Tentori, Favela [2008] describe work on activity recognition in a hospital. Here information about the context, such as location, time, and the roles of the people present, and RFID-tagged artifacts is used to determine what an actor is doing, for example the activities of a ward nurse and doctor during the morning hours, handling reports and case files, may be interpreted as patient care or clinical case assessment.

Whatever the formalism and methodology, the result or output of intention recognition is a hypothesis or a set of hypotheses about the intention of the observed agent. The output of plan recognition, in addition includes hypotheses about the plan of the observed agent, including the rest of his actions yet to come. The process of generating such results may be iterative whereby hypotheses are generated, predictions are made and tested, and hypotheses are refined.

2.5 Pruning the Space of Hypotheses

A substantial issue in intention (and plan) recognition is the problem of narrowing down the space of possible hypotheses. Various techniques have been used to accomplish this. Early approaches impose minimality or simplicity constraints, for example via circumscription [Kautz and Allen 1986]. Here circumscription is used, in effect, to characterize the assumption that all the actions observed are being executed towards a minimum number of (top-level) intentions.

Appelt and Pollock [1992] use weighted abduction where weights are attached to conditions of the rules used for intention recognition. The cost of proving a conclusion G is the sum of the costs of proving the conditions of a rule whose conclusion matches G. The cost of proving each condition depends on whether it is assumed via abduction, or is true or is proved using other rules. The weights are intended to capture domain-specific information. For example, consider the two rules in Box 1.

They both state that you may enter a building if its door is open, but the cost of assuming that the door of the building is open is much higher if it is a private building than if it is a public one. Put another way, intuitively, more evidence is required to believe that the door of a private building is open than to believe that the door of a public building is open.

More recently, Jarvis, Lunt and Myers [2005] couple a form of abductive reasoning with domain information regarding frequency of certain actions (for example *high* for renting a car and *low* for filing a crime report). Then in the application of terrorism intention recognition they use this in-

Box 2.

```
pa_rule(lw(T), (9,10)) ← time(T),schedule(T, football)
pa_rule(lw(T), (1,10)) ← time(T), (T>23∨ T<5)
pa_rule(lw(T), (3,10)) ← temp(T, TM), TM>30
```

formation to impose a maximum on the frequency of observed actions that are to be used in the recognition system. Thus, given a set of observed actions the system focuses on a subset of these that have a maximum threshold frequency, in effect ignoring the more "common" actions and focusing on the more "rare" or "unusual" ones. Another approach is to make use of ordering constraints. For example, Avrahami-Zilberbrand and Kaminka [2005], reject hypotheses of plans which have some matching observed actions but also have actions that should have been executed earlier but have not been observed.

Associating weights with conditions of rules as in Appelt and Pollock [1992] and *a priori* frequencies to actions as in Jarvis, Lunt and Myers [2005] may be thought of as forms of probabilistic reasoning. Other more explicit forms of probabilistic reasoning to prune the search space have also been used [e.g. Geib and Goldman 2001, 2005, Geib and Steedman, 2003]. The probabilistic approaches may be based on Baysian reasoning or the hidden Markov model [Bui 2003]. Another approach is situation-sensitive Causal Bayes Nets [Pereira and Anh 2009b] where logic programming clauses are used to specify probabilities of intentions given information about the current state, including time of day and temperature. For example, modifying the authors' notation, the rules in Box 2 state that the probability of (the observed agent) liking to watch TV (*lw*) is 90% when football is on, 10% when it is between 23:00 and 5:00 hours, and 30% if the temperature is higher than 30 degrees.

Mao and Gratch [2004] combine probabilities with utilities in plan recognition to choose from amongst competing hypotheses. The domain they consider is that of military maneuvers. Intentions

are given pre-specified associated utilities as well as probabilities. The utilities are from the point of view of the observed agent. Thus if the observed actions lead to two equally probable hypotheses they are ranked according to their utilities, preferring the hypothesis which has higher utility, i.e. the one believed to be more profitable for the actor. Avrahami-Zilberbrand and Kamonka [2007], on the other hand, exploit utility from the point of view of the observer, for example ordering the hypothesized intentions according to how dangerous they may be to the observing agent. In their work this is particularly useful when there is uncertainty about the observations. For example, in CCTV monitoring at an airport, someone is observed putting down their luggage and it is uncertain if they have picked it up and taken it with them or not. The hypothesis that they have left the luggage with criminal intent is preferred as it is more dangerous from the point of view of the observers.

Once a hypothesis is chosen from amongst the possible ones, how it is used depends on the application of intention recognition. For example, two contrasting applications are identifying terrorist activity and providing assistance at home. In the first [e.g. Jarvis et al. 2004], the objective is to prevent the terrorists achieving their intentions by first identifying the intentions. In the second, for example the care of the elderly at home in an ambient intelligence setting, the objective is to help and guide the elder towards achieving his intention, by first identifying the intention. Notice that these two applications correspond to the adversarial (and possibly diversionary) and keyhole (and possibly intended) classes of problems, respectively.

To our knowledge, there is no work studying the relative effectiveness of the different approaches to intention recognition. Mayfield [2000] proposes three criteria for evaluating the effectiveness of the outcome of plan recognition systems, *applicability*, *grounding* and *completeness*. *Applicability* refers to how useful the explanation generated by the plan recognition system is to the program (or person) which is to use it in terms of its content, granularity and level of detail. *Grounding* refers to how well the plan recognition system takes account of all that is known about the actor and context (apart from actions that are observed). *Completeness* refers to how well the explanation that is produced covers all of the observations. But the three notions remain informal and anecdotal in Mayfield's work and his focus is on dialogue understanding particularly in the context of Unix help facility.

3 LOGIC-BASED APPROACHES TO INTENTION RECOGNITION

3.1 Abductive Approaches

Abduction is a prominent methodology used in intention recognition and forms the basis of several of the papers reviewed in the case studies in Section 4. Abduction [Peirce 1958] is a form of defeasible reasoning, often used to provide explanations for observations. For example given a logic rule

```
room-is-hot ← heating-is-on
```

deduction allows deriving *room-is-hot* from the knowledge that *heating-is-on*, and abduction allows abducing *heating-is-on* to explain the observation of *room-is-hot*. In the abductive framework [e.g. Kakas, et al. for abductive logic programming], in general, given a background theory T, and an observation (or goal) Q, an abductive answer to Q is a set Δ, such that Δ

consists of special pre-specified abducible atoms, $T \cup \Delta \vdash Q$ and $T \cup \Delta$ is consistent. In addition, in particular in abductive logic programming, an extra requirement may be that $T \cup \Delta$ satisfies a given set of integrity constraints. For example an integrity constraint

```
heating-is-on ∧¬boiler-working ⇒
false
```

will result in disregarding *heating-is-on* as an explanation for *room-is-hot* if it is believed that the boiler is not working.

Charniak and McDermott [1985] were possibly the first to suggest that intention/plan recognition could be framed as an abductive problem. Their focus was on intention recognition, or *motivation analysis*, as they called it, in the context of story comprehension. They identified intention recognition as the reverse of planning. In the latter, given a task, reasoning is employed to determine what actions would achieve it. In the former, given an action, reasoning is employed to determine what tasks it could help achieve, either directly, or in conjunction with other possible actions. This reversal of the reasoning employed for planning gives the flavour of abductive reasoning, but their actual formalization did not strictly conform to the abductive framework, as described above. For the purpose of intention recognition plan schemas were compiled in the form *todo(G, A)* denoting that action A achieves goal G. Thus observing an instance of A, the same instance of G could be hypothesized as a possible intention.

Abduction can provide multiple hypotheses explaining an observation. Charniak and McDermott [1985] suggested a number of criteria for choosing between multiple hypotheses. One is to prefer a hypothesis that uses the most specific characteristics of the observed action. For example, if we observe that Tom picks up a newspaper, there may be two possible explanations, he intends to read it or he intends to swat a fly with it. According to the specific characteristics criteria, the first

Box 3.

```
Goal ← Conjunction of actions and preconditions of actions and other proper-
ties and temporal constraints.
```

is preferred because it uses the characteristic of the newspaper as a readable object, whereas the second uses the characteristic of the newspaper just as an object. Another criterion suggested is to prefer a hypothesis which requires fewer additional assumptions (similar to the *global* criteria mentioned below). For example the explanation of swatting a fly requires an additional assumption that there is a fly, and may thus be less preferred to the explanation of reading if that requires no additional assumptions.

More recently, two broad types of criteria, *global* and *local*, are often used for ranking, and thus choosing from amongst the explanations. The global criteria may, for example, prefer explanations that are minimal in some sense, for example syntactically in terms of the number of facts. The local criteria, on the other hand, may associate some form of evaluation metric with each rule in the background theory, and provide an evaluation metric for a set of hypotheses by combining the metrics of the rules from which the hypotheses originated.

In intention recognition the background theory, T, is a characterization of the relationships between actions and intentions, the observations, Q, are the actions of the observed agent, and the explanations, Δ, are hypotheses about the agent's intentions. In general, as explained in Section 2, whatever form of reasoning is employed, whether abductive, deductive, probabilistic or a mixture, intention recognition requires some knowledge of how actions achieve goals. As observed by Charniak and McDermott, such a theory is conceptually closely related to causal theories used for planning.

3.2 Causal Theories for Planning and Theories for Intention Recognition

A common premise of intention recognition is that the observed agent is rational (even though forgetful and chaotic in some cases), and is pursuing a course of actions he believes will help him achieve a goal. This course of action must be the result of reasoning with some causal theory for planning. A further assumption is that the observer agent has some knowledge of this causal theory, although he may not use that same theory for intention recognition. The theory that he uses for intention recognition may have some direct logical relationship to the observed agent's causal theory, or may have some loose and informal relationship to it. The theory and the representation that the observer agent uses lends itself naturally to the form of reasoning that the agent needs to employ.

Logic-based causal theories essentially include rules of the form (Box 3).

Causal theories are often general purpose and can be used for different applications, such as planning and prediction. Here, we are only interested in their use for planning. In particular there are two approaches, planning from first principles and planning from second principles or plan libraries. Both can be formalized, for example in the situation calculus [Reiter 2001] or the event calculus [Kowalski and Sergot 1986], or some other formalism of actions and effects. So for example an abductive theory of the event calculus for first principle planning may include the rules in Box 4.

The rules state that a property P holds at a time T if an event E happens earlier which initiates P and P persists (at least) from the occurrence of E until T. P persists between two times if it is not

Box 4.

```
holds-at(P,T) ← happens(E, T1) ∧ initiates(E,P) ∧ T1<T ∧ persists(T1,P,T)
persists(T1,P,T) ← not clipped(T1,P,T)
clipped(T1,P,T) ←happens(E, T2) ∧ terminates(E,P) ∧ not out(T2,T1,T)
out(T2,T1,T) ← T=T2          out(T2,T1,T) ← T<T2          out(T2,T1,T) ←
T2<T1
Integrity constraint:          happens(A, T), precondition(A,P) ⇒ holds(P,
T).
```

Box 5.

```
initiates(unlock-door(R), gain-entry(R))
terminates(lock-door(R), gain-entry(R))
precondition(unlock-door(R), have-key(R))
precondition(unlock-door(R), in-front-of(R))
```

Box 6.

```
gain-entry(laboratory, T+3) ← goto(reception, T) ∧ pick-up-key(laboratory,
T+1) ∧ goto(laboratory, T+2) ∧ unlock-door(laboratory, T+3). ¹
```

clipped in that interval. *P* is clipped in an interval if an event *E* happens that terminates *P* and it cannot be shown that *E* occurred outside that interval. Here *not* can be thought of negation as failure. The integrity constraint states that an action can be done only if its preconditions hold.

Domain dependent information is used to specify rules defining *initiates* and *terminates*, and *preconditions* of actions, for an example see Box 5.

These state that unlocking the door to a room initiates gaining entry to that room, and has preconditions having the key to and being in front of that room, and locking the door terminates gaining entry. In this abductive framework the set of abducible predicates will consist of *happens* and the ordering relations, = and <.

A theory for planning from second principles, based on the above theory, and further information about preconditions and effects of actions, may have rules such as Box 6.

Thus a **logic**-based plan library is typically of the form:

$$G \leftarrow A_1; \ldots; A_n$$

where ";" denotes (conjunction and) sequencing.

Very little work uses causal theories from first principles for intention recognition (one exception is Quaresma and Lopes [1995] reviewed in Section 4), although an obvious major advantage of it would be to increase the chances of recognizing the intention behind unusual and unpredicted clusters of actions, and also lending itself well to recognizing short and medium term intentions as well as the ultimate intention.

Much of intention recognition work assumes a plan library, at least used by the observed agent. If the observer agent uses the same representation, i.e.

$$G \leftarrow A_1; \ldots; A_n$$

Box 7.

$$G\sigma \leftarrow (A_{1;} \; ...; \; A_{i-1;} \; A_{i\;+1}; \; ....; \; A_{j-1;} \; A_{j+1}; \; ....; \; A_{k-1;} \; A_{k+1}; ...; \; A_n) \; \sigma$$

Box 8.

$$G \rightarrow (A_{11;} \; ...; \; A_{1n}) \; \lor \; (A_{21;} \; ...; \; A_{2m}) \; \lor \; .... \; \lor \; (A_{p1;} \; ...; \; A_{pq})$$

then the reasoning employed for intention recognition can be deductive, reasoning from the observed actions to the goal they establish, i.e. given instances of $A_{1;} \; ...; \; A_n$ deduce the appropriate instance of G. To make the approach more flexible this can be combined with probabilities, in the sense of increasing the probability of (an instance of) G being the intention as an increasing number of the (appropriate instances of the) actions A_i are observed. This is essentially the basis of the work of Demolombe and Fernandez [2006], described in Section 4. The reasoning may also be a combination of deductive and abductive, reasoning deductively from the occurrence of some actions $c_{i;} \; c_{j;} ...; c_k$ matching actions $A_{i;} \; A_{j;} ...; \; A_k$, with a most general unifier σ, to deduce a residue (resolvent) (see Box 7) and then abducing (possibly a ground instance $\sigma\phi$ of) the remaining actions ($A_{1;} \; ..., \; A_{i-1,} \; A_{i+1,} \; ....., \; A_{j-1,} \; A_{j+1,} \; ...., \; A_{k-1,} \; A_{k+1,} \; ...., \; A_n$) σ (and any conditions, such as the ordering), and thus hypothesizing the goal $G\sigma$ (or $G\sigma\phi$). This is essentially the approach used by Quaresma and Lopes [1995], also described in Section 4.

Another approach is for the observer agent to employ a theory of rules of the form

$$G \rightarrow A_{1;} \; ...; A_n$$

or more generally see Box 8 which is the only-if half of a completed plan library. Then the reasoning employed is abductive, finding hypotheses G that would explain observations of actions A_{ij}. This is essentially the approach used by several of the studies described in Section 4, in particular Sindlar, Dastani et al. [2008], Myers [1997], Jarvis,

Lunt, and Myers [2005], and Dragoni, Giorgini and Serafini [2002]. However, Sindlar, Dastani et al. [2008] use a meta-level representation of the form (see Box 9).

Myers [1997], Jarvis, Lunt, and Myers [2005] employ an HTN (similar to a production rule type) representation of the form (see Box 10).

In Dragoni, Giorgini and Serafini [2002], which is perhaps one of the most complex approaches because of its use of bridge rules, a crucial part of the knowledge remains an implicit assumption, in particular the knowledge that links certain actions (in their case utterances of speech acts) and the intention behind them. Note that none of these representations is intended to be used for planning, thus although the declarative reading of the representations of Sindlar et al, Myers, and Jarvis, Lunt, and Myers are at some variant with the only-if part of completed planning rules, their usage, can, in effect, be interpreted as using abductive reasoning with such only-if representations.

In the next section we will look at a number of studies that essentially employ some form of logic-based approach for intention recognition. They cover diverse methodologies and applications. Most work described here falls in the **keyhole** or *intended* classification of intention recognition. Mulder and Voorbraak [2003] deal with enemy intention recognition, and thus conceptually fall in the *adversarial* classification.

We start with case studies (4.1, 4.2, 4.3) that use simpler forms of knowledge representation, primarily in the form of simple plan libraries, and consequently use simpler forms of reasoning. We

Box 9.

$$
\begin{aligned}
&\texttt{goal (G)} \rightarrow \texttt{plan(A}_{11;} \texttt{ ...; A}_{1n}) \qquad\qquad \texttt{goal (G)} \rightarrow \texttt{plan(A}_{21;} \texttt{ ...; A}_{2m}) \qquad\qquad \texttt{.......} \\
&\texttt{goal (G)} \rightarrow \texttt{plan(A}_{p1;} \texttt{ ...; A}_{pq}) \texttt{ .}
\end{aligned}
$$

Box 10.

$$
G \rightarrow A_{11;} \ ...; \ A_{1n} \qquad\qquad G \rightarrow A_{21;} \ ...; \ A_{2m} \qquad ...... \qquad G \rightarrow A_{p1;} \ ...; \ A_{pq}.
$$

then consider an approach based on BDI-agents (4.4), and move on to two approaches (4.5 and 4.6) that combine logic with probabilities. The second of these (4.6) also incorporates the notion of states as in the situation calculus [Reiter 2001]. We finish with two approaches (4.7 and 4.8), possibly the most complex of the eight, which incorporate some concept of theory of the mind, as bridge rules in the first, and epistemic operators in the second. This latter approach also incorporates an extension of the event calculus [Kowalski and Sergot 1986] as the background theory used for intention recognition.

4 CASE STUDIES OF LOGIC-BASED APPROACHES

4.1 Mulder and Voorbraak [2003]

This paper addresses *tactical* intention recognition, which the authors define as the recognition of enemy plans. In general such a task will have various identifying characteristics, for example the specialized domain of the military, the tendency of the enemy to attempt to mislead the observers, or at the very least to try and avoid detection and prevent recognition of their plans and intentions. The paper makes a contribution related to the last of these features, by way of allowing observation of actions that do not convey all the information about the actions. In other words, and in logical terms, where other work below, and in the majority of the literature, assumes fully grounded

observations, such as the action *land(jet101, airbase1)*, here the action may be non-ground and existentially quantified, for example $\exists X \ land(X, airbase1)$, namely that it has been observed that something has landed at *airbase1*.

The work assumes fairly simple plan libraries, with rules of the form:

$$
G \leftrightarrow A_1, \ ...., A_n,
$$

where the A_i are actions, not defined by any other rules (i.e. they are simple rather than macro-actions). The "," denotes conjunction. The reasoning for intention recognition is abductive. So given a plan library P, and set of observations O, the reasoning seeks to find abductive explanations for all observations in O, i.e. it seeks to find sets of hypotheses Δ, made up of ground atoms in the predicates that appear in the goal side (the Gs) in the rules in P such that

$$
P \cup \Delta \vdash O.
$$

The example in Box 11 helps illustrate the contribution:

The first rule in P specifies that an attack is planned from *airbase1* if something lands at *airbase1* and is loaded with *missiles*. The second rule specifies that an aid flight is planned if what is loaded is *aid-supplies*, and the carrier is covered with the Red Cross symbol.

Now an observation *load(jet1, missiles, airbase1)* leads to one hypothesis, namely *attackplan(airbase1)*. However, an observation $\exists Z$

Box 11.

```
P:    attack-plan(airbase1)  ↔ land(X, airbase1), load(X, missiles, airbase1)
    aid-plan(airbase1)   ↔ land(X, airbase1), load(X, aid-supplies, airbase1),
                         cover-with-red-cross(X, airbase1)
```

load(jet1, Z, airbase1) leads to two potential hypotheses, namely *attack-plan(airbase1)* and *aid-plan(airbase1)*. A further observation $\exists X$ *cover-with-red-cross(X, airbase1)* still maintains two hypotheses, one consisting of *aid-plan(airbase1)* explaining both observations, and the other consisting of both *attack-plan(airbase1)* and *aid-plan(airbase1)*, each explaining one of the observations.

No proof procedures are suggested for the abduction, but it is worth noting that most, if not all, proof procedures for abductive logic programming will capture this reasoning (for example, the iff-proof procedure of Fung and Kowalski [1997] and the CIFF proof procedure of Mancarella et al. [2009]).

Note that with a slightly more elaborate representation of plan libraries it may well be possible to avoid reasoning with existentially quantified representations of observations. One such formalization could make use of binary representations of the known information about the observed actions, such as, for example for an event e_1 of landing:

$$act(e_1, land)\ destination(e_1, airbase1),$$

ignoring what is not known, here the aircraft involved in e_1, and for an event e_2 of loading:

$$act(e_2, load)\ base(e_2, airbase1)\ carrier(e_2, jet1),$$

ignoring what is not known, here the cargo that is loaded. To accommodate such a representation the *attack-plan* rule, above, for example, can be written as the example in Box 12 ignoring any

temporal constraints and persistence requirements between events *E1* and *E2*.

4.2 K. Myers [1997]

Myers' [1997] work is motivated by making planning technology more accessible and easier to use. Here the two parties, the observer and the actor are, respectively, a planning system and its user. The planning system observes the user's inputs and attempts to guess his intentions. Conventionally, in a planning system the user inputs a goal and the system provides a plan of actions that, if executed, would achieve the goal. In Myers' work the planning system is expected to do more than this. In effect, the planning is to be done co-operatively, where, as well as (optionally) inputting a goal, the user can participate in the planning process by providing the planning system with a partial plan which may consist of a (partial) list of tasks (actions or subgoals).

The system then attempts to identify from this partial plan any higher level goals the user may have, and then to complete the partial plan to achieve these higher level goals. For example let us consider the domain of travel planning. A traveler may provide a partial list of tasks, for example visiting the Swiss embassy and visiting a ski shop. A co-operative planner, in principle, may fill in the gaps, by deducing or guessing (abducing) that the user wishes to take a ski holiday trip to Switzerland, and can then complete the plan by including the actions of booking the flight and booking accommodation in Switzerland.

The top level goals come from a predefined set, either explicitly pre-defined by the programmer, or otherwise available to the system. The planner

Box 12.

```
attack-plan(airbase1) ↔ act(E1,land), destination(E1,airbase1),
carrier(E1,X), act(E2, load)                    base(E2, airbase1),
cargo(E2, missiles),  carrier(E2, X),
```

uses plan libraries, and in fact the same libraries, both for planning and for intention recognition. The plan libraries are based on the Hierarchical Task Network (HTN) model of planning [Erol, Hendler, Nau, 1994]. HTN defines operators for reducing goals to subgoals. For example:

O1: B → C, D
O2: C → K, L, M
O3: C → P, Z
O4: D → W

We comment later, at the end of this section, on the logic of HTNs and the above representation.

Here *O2*, for example, is an operator that reduces goal *C* to subgoals *K, L,* and *M*. So, given this HTN, for example if the user provides a goal *C* two plans are possible, one consisting of subgoals *K, L, M,* and the other of *P, Z*. On the other hand, if instead of providing a goal the user provides a partial plan consisting of *P*, the intention recognition system guesses that the intention is *C*, and ultimately B, and provides a plan *P, Z, D*, further refined to *P, Z, W*, compatibly with the user-given partial plan.

The planning is done conventionally, using the operators to reduce goals to subgoals. On the other hand, the determination of the user goals from the actions or subgoals they input is based on a form of abduction. For example, in the case above, *B* is an abductive explanation, according to Myers, for the user input subgoal *P*. This is determined using the notion of an *abductive chain* which in this case is:

$P \rightarrow^{O3} C \rightarrow^{O1} B$.

It is assumed that the operator definitions are non-recursive. If there are alternative top level goals possible via such abductive chains then a subset is chosen. The choice is not addressed in the paper. Once a set of goals is identified then the planning proceeds as in HTN with the constraint that the final plan should accommodate the user-given tasks. A brief analysis of the approach provides an exponential upper-bound for the process of constructing the abductive chains, but argues that this is dependent on the length of the chain and in practice this length is small, while the user-given partial plans can reduce the search space of the planning phase.

Note that the use of abduction here calls for a logical interpretation of the HTN representation. The operator decompositions are similar to goal-reduction rules in production systems. Kowalski and Sadri [2009] argue the difficulties of giving such rules model-theoretic semantics and provide an alternative framework that does provide such semantics. However, it seems relatively straightforward to put Myers' work in the context of the discussion of the different logic-based approaches in Section 3, and to relate the abduction done here to the formal notion of abduction defined there. This can be done by representing the operators in a logically meaningful way. For example, if we interpret operator definitions O1-O4, above as goal-reduction rules for goals *B*, *C* and *D*, we can have:

$B \leftrightarrow C \wedge D$
$C \leftrightarrow (K \wedge L \wedge M) \vee (P \wedge Z)$
$D \leftrightarrow W$.

Box 13.

```
template Physical_Attack(?group, ?target)
            purpose destroy(?group, ?target)
            tasks   1. reconnaissance(?group, ?target)
                    2. prepare_attack(?group, ?target)
                    3. attack(?group, ?target);
            Ordering 1-->3, 2-->3;
```

Now the abductive reasoning required in the construction of the *abductive chains* is classical abduction using the rules above in the only-if direction ($\rightarrow$), as in Sindlar et al. [2008], described below. The planning can also be seen as classical planning using the *if* direction ($\leftarrow$) of the rules.

4.3 Jarvis, Lunt, and Myers [2004, 2005]

This work essentially uses the approach of Myers [1997] in the application of terrorist intention recognition. This more recent work uses an architecture called CAPRe (Computer-Aided Plan Recognition). Here the plan libraries are in the form of templates, and they contain more information than the earlier work [Myers 1997]. They are non-ground (i.e. contain non-ground parameters) and have additional information, such as ordering and preconditions of tasks, and *frequency* and *accuracy* of observations, described later.

An example, in a slightly simplified notation, is Box 13.

The template specifies a macro-action *Physical_Attack* of a target by a group. The purpose (effect) of the action is the destruction of the target by the group. The action is decomposed into three partially ordered actions, reconnaissance of the target and preparation of the attack, followed by the attack.

Ignoring the ordering and the variables this template corresponds to the HTN representation (see Box 14) in the earlier notation (Subsection 4.2). A more elaborate abstract template example, ignoring variables, may be:

```
template TaskName
        purpose P
        tasks            T₁, ...., Tₙ
        Ordering set of Tᵢ -->Tⱼ
        Effects
                E₁ at T₁, frequency F₁
                ....
                Eₙ at Tₙ, frequency Fₙ;
```

The task *TaskName*, itself may be a subtask of another task, and so on. Ignoring the ordering and the *frequency* information the template above corresponds to the HTN representation:

```
P → TaskName
TaskName → E₁,..., Eₙ
E₁ → T₁
....
Eₙ → Tₙ.
```

The *frequency* information represents the frequency of a given action occurring in *normal* circumstances. It is given values *high, medium* or *low*. For example car renting has a *high* frequency, whereas filing reports of missing persons has a *low* frequency. The templates may also include information about *accuracy* of normal observation, again given as *high, medium* or *low*. For example observing a missing person report has higher accuracy than recollecting a license plate. *Frequency* and *accuracy* can be thought of as qualitative

Box 14.

```
destroy → Physical_Attack
Physical_Attack→   reconnaissance, prepare_attack, attack
```

measures used as a means of introducing probabilities in reasoning about choices.

The intention recognition is done in two phases. The first is mostly identical to the approach of Myers [1997], with so-called *task seedlings* generated from the templates in a way that is the similar to the generation of *abductive chains* from Hierarchical Task Networks. The variables are treated by unification, along the chain. The templates allow constructing chains starting from observation of actions T_i or effects E_i. Any additional *frequency* information can be used to specify maximum frequencies of observations for which abductive explanations are sought, thus, for example allowing the system to ignore commonly occurring (supposedly mundane) actions and changes in the environment. Similarly the *accuracy* information can also be used to impose a threshold on the observations that are taken into account.

The result of the first phase of intention recognition consists of explanations for each chosen observation. The second phase attempts to combine these to provide compatible explanations for clusters of observations. It does so by considering sets of increasing size of the explanations generated by the first phase, and removing from consideration incompatible sets. A set is incompatible if it has incompatible variable bindings, or incompatible orderings or other constraints within the templates.

As in the case of Myers [1997] the performance of the system is dependent on the length and number of the *task seedlings* (*abductive chains*). It is also dependent on the number of sets of explanations that have to be considered in the second phase. As might be expected, experimental results report a degradation in performance with increasing number of observations to be explained and

increasing noise, i.e. activities that are unrelated to any attack plan.

It can be noted that the templates can be formalized in logic along the lines sketched in Subsection 4.2. For example the *Physical_Attack* template can be formalized as seen in Box 15.

It would be interesting to explore if such a logic-based formalization, together with more recent abductive proof procedures with constraint handling, such as Mancarella et al. [2009] would have any impact on the performance of the system. Such an abductive proof procedure allows merging into one process the two phases of generation of explanations for each action and then finding compatible clusters. This merging may prevent generation of hypotheses and clusters that would in the end prove incompatible.

4.4 Sindlar, Dastani et al. [2008]

In this work the acting agent is assumed to be a BDI-type agent [Rao and Georgeff 1995], with the particular feature of interest being the planning rules that govern its behaviour. These rules are assumed to be of the form

$$G \leftarrow B | \pi$$

to be interpreted in logical terms as stating that goal G holds if B is believed and plan π is executed.

A plan may consist of sequences of actions and non-deterministic choice. Actions include test actions, that do not bring about any changes but simply check a condition. For example the planning rule shown in Box 16 specifies that g is achieved if b is believed and actions $a1$ and $a2$ are executed and, if $\varnothing 1$ holds then if $\varnothing 2$ holds

Box 15.

```
destroy(Group, Target, Time) ↔ physical-attack(Group, Target, Time)
physical-attack(Group, Target, Time) ↔ reconnaissance(Group, Target, Time1),
prepare-          attack(Group, Target, Time2), attack(Group, Target, Time),
Time1<Time, Time2<Time
```

Box 16.

```
g ← b | a1; a2; (Ø1?; ((Ø2?; (a3; a4) + (¬Ø2?; a5))) + (¬Ø1?)); a6; a7
```

a3 and *a4* are executed, otherwise *a5* is executed, and then *a6* and *a7* are executed.

The observer agent may also be a BDI-type agent with a belief base, goals and behaviour rules. Those details are not relevant for the purposes of intention recognition. What is relevant and important here about the observer agent is that, in addition to any rules that govern its own (planning and action execution) behaviour, it has knowledge of the planning rules of the observed agent. This knowledge is in the form of rules of the form shown in Box 17.

The agent is assumed to have one such rule for every rule of the form

$$G \leftarrow B | \pi$$

in the observed agent's knowledge base. Rules of the form RD are not used for planning purposes, and are not particularly appropriate for such purposes, but they are used specifically for intention recognition. They are used abductively to provide explanations for observed actions in terms of the goals and beliefs of the observed agent. The abducible predicates are *goal* and *belief*. Thus although the observer agent is a BDI-type agent, the rules it uses for intention recognition have logical semantics, and they lend themselves to abductive reasoning.

The work is based on several assumptions: One is that the observed agents execute one plan at a time, so there are no interleaved plan executions.

Another is that the observer is able to determine ahead of time whether they have *complete* observability or whether their observations will be *late* or *partial*. *Complete* observation means that all the non-test actions of the actor will be observed. *Late* observation means that the observer may miss some of the initial actions of the plan executed but from the first observation it will observe all the non-test actions executed henceforth. *Partial* observation means that the observer may miss some of the actions, anywhere in the plan. In all three cases all the actions that are observed are assumed to be observed in the correct sequence.

In each of the three cases a match is sought between the sequence, S, of actions that have been observed and the sequence in π in the conclusion of rules of the form RD. The test actions in π are ignored in this matching process. In the case of *complete* observability the match sought is between S and any initial segments of π (for example *a1;a2* or *a1;a2;a3;a4* or any other initial segment of *a1;a2; a3;a4;a6;a7* or *a1;a2; a5;a6;a7* or *a1;a2; a6;a7* in the above example). In the case of *late* observation the match sought is between S and any sub-segments of π (for example *a1;a2* or *a2;a3;a4* or *a2;a5* or …). In the case of *partial* observation the match sought is between S and any *dilution* of any sub-segment of π, i.e. any sub-segment with any elements removed (for example *a2;a4* or *a1;a4;a7* or …).

When such a match is found the conditions *goal(G)* ∧ *belief(B)* of the RD rule provide the

Box 17.

$$goal(G) \land belief(B) \rightarrow plan(\pi) \qquad\qquad\qquad\qquad\qquad\qquad\qquad\text{RD}$$

Box 18.

$$goal(G) \land belief(B) \rightarrow (a_1;a_2;...;a_n) \lor ...\lor (a'_1; a'_2; ..; a'_m)$$

necessary abductions and thus the hypothesis regarding the intention of the acting agent.[2] We can interpret what is done here in the general abductive framework outlined in Section 3 by considering the theory used for abduction to consist of the rules of the form in Box 18 where the a_i and a'_i are non-test actions, and ";" can be thought of, as is common, as representing conjunctions of actions and their sequential ordering. The three cases of observability can be thought of as providing additional integrity constraints. *Complete* and *late* observability both impose the constraint that there is no action in between any two actions observed where one immediately follows the other. *Complete* observability additionally imposes the constraint that there is no action before the first one observed. *Partial* observability imposes neither constraint.

Two quite intuitive propositions are proved. One states that the number of possible explanations (abductive hypotheses) generated in case of *complete* observability is less or the same as that generated by *late* observation which in turn is less or the same as *partial* observation. The other is that in each case the number of possible explanations decreases or at most stays the same as the number of observations increases.

4.5 Demolombe and Fernandez [2006]

This paper proposes a framework that combines probabilities with a situation calculus-like formalization of actions. The situation calculus [Reiter 2001] and its further extension GOLOG [Levesque

et al 1997] are logical formalisms for specifying actions and their effects. They allow specifying macro-actions or, in the authors' terminology, procedures. For example, the procedure for dealing with a fire on board a plane may be represented as Box 19 to express the sequence of actions *turn fuel off, turn full throttle* and *turn mixture off.* Note the similarity between this notation and the HTN concept and representation.

Each single atomic action maps one state to its successor, and correspondingly macro-actions map one state to another. The assumption of the paper is that the actor and the observer, essentially, have the same "knowledge" about such procedures. However, a central aim of the paper is to allow intention recognition of human actors who may interleave procedures. To this end for each procedure that is in the knowledge base of the actor a modified one is included in the knowledge base of the observer. The modification is done by explicitly adding arbitrary actions, and any constraints, to the procedure definition. For example the definition above is modified to Box 20.

This modified definition represents the same procedure for tackling fire on board, but explicitly allows any sequence of actions σ, except *turn-fuel-on* in between turning the fuel off and turning full throttle, and between turning full throttle and turning the mixture off. Here, in the *fire-on-board* procedure, the three actions *turn-fuel-off, turn-full-throttle, turn-mixture-off* are *explicit* actions, and the action *turn-fuel-on* is said to be *prohibited* in between any two of the *explicit* actions. The other actions in σ are said to

Box 19.

```
tackle-fire-on-board =def turn-fuel-off; turn-full-throttle; turn-mixture-off
```

Box 20.

```
tackle-fire-on-board =def turn-fuel-off; (σ/turn-fuel-on); turn-full-throttle;
(σ/turn-fuel-on);                          turn-mixture-off.              OD
```

be *tolerated*. The set of all defined procedures provides the set of all possible intentions.

Those familiar with formalisms such as the situation calculus or the event calculus [Kowalski and Sergot 1986] may relate the observer definition OD to a formalization that makes explicit the required persistence of properties between actions or states. An example of such formalization, letting the *Si* to represent states or time points, is shown in Box 21 where action *turn-fuel-off* is specified to have the effect *fuel-off*.

The paper assumes that there is full visibility of actions performed, or rather that the intention recognition is performed purely on the basis of the observed actions. In the initial state before any action is observed all possible intentions are given pre-specified probabilities. Then as actions are observed they are matched with the procedure definitions. The matching includes *explicit* actions in procedure definitions, such as *turn-fuel-off* in the example above, as well as any *prohibited* or *tolerated* actions.

With each match probabilities are updated for the possible intentions. The probability of an intention is increased, if the next expected *explicit* action is observed, it is decreased if a *tolerated* action is observed, and decreased to a greater extent if a *prohibited* action is observed. All the probability increments and decrements are by pre-specified amounts. The computation cost of evaluating the probabilities is said to be linear with respect to the number of observations for a given procedure.

The following is a modified version of an example in the paper. Consider three procedures

P1 = a; (σ/g); b; c
P2 = d; σ; e
P3 = a; σ; f

where, σ denotes a sequence of arbitrary actions. If in the initial state *s0* action *f* is observed then the probabilities of the agent having the intention *P1*, *P2* or *P3* in state *s0* are equal and low. So $P(Int(P1, s0))=P(Int(P2, s0))= P(Int(P3, s0))$, where $P(Int(Q,S))$ denotes the probability of intention *Q* in state *S*. Then in the resulting state, $s1=do([f],s0)$ in modified situation calculus notation, if action *a* is observed, the probabilities of *P1* and *P3* are increased equally in state *s1*. Now if an action *m* is observed in state $s2=do([f,a],s0)$ there is still a match with *P1* and *P3*, but action *m* lowers the probability of both *P1* and P3, thus, for example $P(Int(P1, s0))< P(Int(P1, s1))$ and $P(Int(P1, s2))< P(Int(P1, s1))$. If now an action *g* is observed in state *s3*, where $s3= do([f,a, m],s0)$, it reduces the probability of *P1*, because *g* is a prohibited action for *P1* in sate s3. But the observation of *g* does not affect the probability of *P3*, thus $P(Int(P1, s3))< P(Int(P3, s3))$.

4.6 Pereira and Anh [2009b]

This paper describes an implemented logic programming framework incorporating *situation-sensitive Causal Bayes Nets* (CBNs) for intention recognition in the care of the elderly. The CBNs

Box 21.

```
happen(tackle-fire-on-board, S3) ← happen(turn-fuel-off, S1), happen(turn-
full-throttle, S2),                                          happen(turn-
mixture-off, S3), S1<S2, S2<S3, persists(S1, fuel-off,, S3)
    persists(S1, fuel-off,, S3)← ¬(happens(turn-fuel-on, S), S1<S, S<S3),³
```

provide a graphical representation and are translated into a declarative language called P-log which represents the same information in logical terms. P-log combines logical and probabilistic reasoning, the logical part based on Answer Set Programming (ASP) [Hu et al. 2007] and the probabilistic part based on CBNs.

The CBNs consist of nodes representing causes, intentions, actions and effects of actions. Causes give rise to intentions, somewhat like reactive production rules (e.g. *if you are thirsty then you (intend to) drink*). Intentions give rise to actions, somewhat like goal reduction production rules (e.g. *if you intend to drink then you look for a drink*), as in Figure 1, where we have ignored effects of actions.

In the running example in the paper there is one action, which the elder is observed to be doing, namely looking for something in the sitting room. The possible intentions from which the system can choose from are looking for a book or for a drink or for the TV remote or the light switch. The causes include that he is thirsty, likes to read, or likes to watch TV.

The paper makes no assumptions about the planning methodology or representation of the observed agent or about its relationship to the knowledge and representation used by the observer. The approach is entirely based on the knowledge representation the observer uses specifically for intention recognition. The observer's statistical knowledge is compiled as CBNs. The causes in the resulting CBNs are either attributes that can be observed, for example *the light is on* or *the TV is on*, or are attributes that can be surmised, for example *(the elder is) thirsty*. The causes have

pre-specified probabilities, either unconditional probabilities represented as facts (in P-log), for example:

the probability of being thirsty is 50/100,

or situation-sensitive probabilities represented as rules, for example:

the probability of being thirsty is 70/100 ← temperature is higher than 30.

Similarly the probability distributions of intentions conditional on causes are given as rules, for example:

the probability of the intention to look for a drink is 9/10 ← the light is on ∧ thirsty,

as are the probability distributions of actions conditional on intentions, for example see Box 22.

(Note the conjunction of possible intentions in the conditions of these rules, especially in the first rule, rather than the disjunction). All the probability distributions, conditional and unconditional, as well as the list of all possible intentions, causes and actions are pre-specified by the designer of the system.

Given this formalization, intended specifically for a P-log implementation, the probabilities of each of the possible intentions can be obtained conditional on observations regarding the status of the light and TV (on/off), temperature (all related to the causes), and the observation that the elder is looking for something.

Figure 1. An abstract CBN

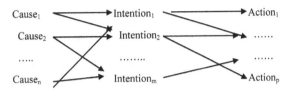

Box 22.

```
    the probability of looking for something is 99/100 ←  the intention to
look for a book is true ∧ the          intention to look for a drink is true
∧ the intention to look for the remote is true
    the probability of looking for something  is 30/100 ← the intention to
look for a book is false ∧ the          intention to look for a drink is true
∧ the intention to look for the remote is false.
```

It is difficult to frame this approach in the methodologies seen in Section 3. Whilst the reasoning from actions to possible intentions has the flavour of abduction, here the explanation is a conjunction of possible intentions rather than a disjunction. This is because of the use of Bayesian networks. Each possible intention is exclusive of the others, and thus the choice of one intention excludes the others.

Notice that there is some similarity between this work and that of Mulder and Voorbraak, described in Subsection 4.1. In both the observer does not have all the information about the action it observes (here he observes that the elder is looking for *something*, for example). Here, however, the intentions are designed to provide hypotheses as to what the missing information may be (the *something* is a book, or the T.V. remote, for example).

4.7 Dragoni, Giorgini and Serafini [2002]

This paper explores the use of abductive reasoning in intention recognition in the context of agent communication. Here the two parties are the hearer (*h*) and the speaker (*s*) of a speech act. The hearer attempts to recognise the beliefs and intentions of the speaker that led to the speaker producing the communication. To this end a *multi-context* system of beliefs and intentions is proposed where the contexts are connected via bridge rules. The contexts correspond to agents' beliefs and intentions, and nested beliefs and intentions about other agents' beliefs and intentions. For example $B_i I_j$ is a context, representing the beliefs of agent *i* about the intentions of agent j. The formula $\varnothing$ in this context, denoted $B_i I_j : \varnothing$, states that agent *i* believes that agent *j* intends $\varnothing$. $B_i B_j I_i$ is another context, representing the beliefs of agent *i* about the beliefs of agent *j* about the intentions of agent *i*. The formula $\varnothing$ in this context, denoted $B_i B_j I_i : \varnothing$, states that agent *i* believes that agent *j* believes that *i* intends $\varnothing$.

There are three kinds of bridge rules: *reflection down*, *reflection up*, and *belief-to-intention* rules. *Reflection down* allows an agent *i* to reason with its image of the mental state of another agent *j* (for example eliminating B_j from $B_j \varnothing$). *Reflection up* allows agent *i* to lift up the result of such reasoning to ascribe it to its beliefs about agent *j* (see Box 23).

Box 23.

> If B_i: $B_j P$ and B_i: $B_j (P \rightarrow Q)$ then by *reflection down* $B_i B_j$: P and $B_i B_j$: $(P \rightarrow Q)$. Then within the context $B_i B_j$, by modus ponens we can derive Q. Thus $B_i B_j$: Q, and by *reflection* up we obtain B_i: $B_j Q$.

Box 24.

> B_i: *raining*
> on ---------------------   *or*   --------------------------------------
> I_i: *bring-umbtrella*
>
> B_i: *temp-higher-20° ∧ conditioning-*
> I_i: *stop-working*

The *belief-to-intention* rules connect an agent's intention to its belief (see Box 24).

The first rule, above, states that if i believes it is raining then i will have the intention of bringing an umbrella. The second rule states that if i believes that the temperature is higher than 20° and the conditioning is on then i will have the intention to stop working. The *belief-to-intention* rules have the flavour of simple reactive production rules. In particular they capture the relationship between the beliefs and intentions of the hearer. On the other hand, what the hearer believes about the reactive behaviour rules of the speaker are formalized as ordinary rules. For example in the context B_h the rule in Box 25 represents what h believes about s's intention when s believes the temperature is higher than 20° and the conditioning is on. The underlying language of the multi-context system is propositional logic, and the inference within contexts is based on natural deduction.

The work assumes a plan-based model of speech acts, in the sense that speech acts, as any other actions, have pre-conditions and post-conditions. Thus, an agent might utter certain speech acts as part of a plan to achieve an intention. The core assumption in the paper is thus that there is a causal relationship between an agent's mental state and his utterances.

Two speech acts are considered, *inform(s,h,∅)* and *request(s,h,∅)*, where s represents the speaker and h, the hearer. *inform(s,h,∅)* represents s telling h that ∅ holds, and *request(s,h,∅)* represents s asking h whether or not ∅ holds. A pre-condition and a post-condition of *inform(s,h,∅)* are, respectively, that s believes ∅, and h believes that s believes ∅. Similarly, a pre-condition and a post-condition of *request(s,h,∅)* are, respectively, that s believes neither ∅ nor ¬∅, and s believes that h believes that s intends to believe ∅.

The hearing agent does not necessarily have any representation of the planning rules of the speaker, in contrast, say, with the work of Sardar, et al [2008]. The hearing agent may have some snippets of the behaviour rules of the speaker, such as the rule in Box 26, but he does not have any explicit representation of any planning rules of the speaker that link the speaker's intentions to his utterances of speech acts. The absence from the knowledge of the hearer of any explicit representation of any link between the intentions of the speaker and the actions that the hearer can observe (i.e. utterances of speech acts) is one crucial difference between this work and all others in this study, and, in fact in the majority of work in the literature on intention recognition.

In the absence of any such explicit information, the hearing agent works with the assumption that

Box 25.

$$B_s(\text{temp-higher-20}° \wedge \text{conditioning-on}) \rightarrow I_s(\text{stop-working})$$

Box 26.

$$B_s(\text{temp-higher-20}° \wedge \text{conditioning-on}) \rightarrow I_s(\text{stop-working}),$$

the intention behind a speech act *inform(s,h,∅)* is to bring about in *h* a mental state in which *h* believes or intends some formula ψ, and the intention behind a speech act *request(s,h,∅)* is to bring about in *s* a mental state in which *s* believes or intends some formula ψ. Also in both cases ∅ is expected to be useful for the derivation of the belief or intention ψ. So the intention recognition task with respect to both utterances is to determine what the ψ is, and whether the intention is to believe ψ or to intend ψ.

So, to attempt to put this in the frameworks discussed in Section 3, conceptually, it is as if the hearer has rules of the form in Box 27 with which it performs abduction.

When *h* receives a speech act *inform(s,h,∅)*, to find out what such a ψ may be, and to determine if the intention is for him to believe or intend ψ, he reasons about the changes that ∅ brings to its beliefs and intentions. It can reason about the consequences of ∅ within any context available to it. For example, an appropriate context would be what *h* believes *s* believes about *h*'s beliefs, i.e. the context $B_h B_s B_h$. The new consequences of ∅, i.e. the consequences that were not derivable before the addition of ∅, but are derivable afterwards, are candidate hypotheses for the intention of the speaker.

A similar process is adapted with the speech act *request(s,h,∅)*. As an example suppose *h* receives a communication *request(s, h, conditioning-on)*, to be interpreted as *s* asking *h* if the conditioning is on. Then if *h* can prove within its image of the mental state of *s* that *temp-higher-20°* and the image also includes or allows the derivation of

the rule in Box 28 then *h* can hypothesise that *s*'s intention behind the speech act is to stop working.

It is interesting to contrast this approach with the abductive approach of Quaresma and Lopes [1995], described below, which also focuses on recognizing the intentions behind the same speech acts. One of the main differences between the two is that the plan-based model of the speech acts, in terms of their preconditions and effects are made explicit in Quaresma and Lopes. Another main difference is that Quaresma and Lopes avoid using bridge rules.

4.8 Quaresma and Lopes [1995]

This work uses an abductive version of the event calculus [Kowalski and Sergot 1986] and a modified version of the action language of Gelfond and Lifschitz [1992] for recognizing the intention behind speech acts. As with Dragoni et al., it focuses on the *request* and *inform* speech acts. This work differs from the majority of the papers on intention recognition in that it uses a theory of planning from first principles as the background theory for intention recognition. This background theory would presumably be used by the speaker, *s*, to reduce goals to subgoals, and thus plan, and is used by the hearer, *h*, with a mixture of deductive and abductive reasoning, for intention recognition.

The background theory is a modification of the event calculus which is a causal theory formalized in Horn clause logic augmented with negation as failure. Box 29 contains their core event calculus rules, which have minor differences with those we presented in Section 3.

Box 27.

```
If IₛIₕψ and "∅ is relevant to ψ" then inform(s,h,∅)
If IₛBₕψ and "∅ is relevant to ψ" then inform(s,h,∅)
If IₛIₛψ and "∅ is relevant to ψ" then request(s,h,∅)
If  IₛBₛψ and "∅ is relevant to ψ" then request(s,h,∅).
```

Box 28.

```
temp-higher-20° ∧ conditioning-on → stop-working,
```

They state that a property P holds at time T if an earlier event E happens which initiates P and E succeeds and P persists (at least) from the occurrence of E until T. P persists between two times, (the time of) E and T, if it is not clipped in that interval. It is clipped in that interval if an event C happens and succeeds and it cannot be shown that C occurred outside that interval. Here *not* is negation as failure.

An event E succeeds if its preconditions hold at the time of its occurrence, Box 30 where $P_1, ...,$ P_n and the preconditions of the action operator A of event E.

These core rules are augmented by the following rules that enable abductions:

$$F \leftarrow \text{not } \neg F \quad R1$$

$$\neg F \leftarrow \text{not } F \quad R2$$

where $\neg$ is classical negation, and F is an abducible predicate. These rules model abductive reasoning and state that if it is not possible to prove $\neg F$ then F should hold, and vice versa. In the intention recognition framework the abducible predicates are *happens/1, act/2* and *</2*. The semantics assumed is that of Well Founded Semantics of Extended Logic Programs [Pereira et al. 1992].

Epistemic operators are used to describe the agents' mental state, for an example see Box 31.

Event-calculus-type rules describe the relationships between the components of the mental state

(as ramifications), for example, the rule in Box 32, where *to(α,P)* is a term representing the plan of performing α to make P true. The rule specifies that if an agent A believes that by doing α P will become true and A intends to do α, and A believes P will be achieved, then A intends to do α in order to make P true. Furthermore, such rules are augmented by an integrity constraint (see Box 33) stating that at no time does the agent believe both that P is true and P will be achieved.

Speech acts are specified as actions within the event calculus framework, for example Box 34 state that a speech act event of S informing H some information P succeeds if S believes P and intends to inform H about it, and it initiates H believing that S believes P. Similarly for the *request* speech act, Box 35 is stating that the preconditions of S requesting H to do an action A are that S believes H can do the action and S intends to make the request, and the action has the effect that H believes S believes it wants action A done.*

Further rules can provide a relationship between the speaker and hearer, for example trust (see Box 36).

Now, in principle, the hearer, h, receiving, for example, a *request* speech act to do an action A will deductively entail the initiation of a belief in h that s intends A (R5). This, in turn, through deduction and also abductions of *</2, happens,* and *act* atoms, leads to hypotheses about what s intends to achieve (using rules including the core rules, R1, R2, R3, R4).

Box 29.

```
holds-at(P,T) ← happens(E), initiates(E,P), succeeds(E), E<T, persists(E,P,T)
persists(E,P,T) ← not clipped(E,P,T)
clipped(E,P,T) ←happens(C), terminates(C,P), succeeds(C), not out(C,E,T)
out(C,E,T) ← T=C          out(C,E,T) ← T<C          out(C,E,T) ← C<E.
```

Box 30.

```
succeeds(E) ← act(E, A), hold-at(P₁,E), ….., holds-at(pₙ,E)
```

Box 31.

```
int(a, actn)      to specify that agent a wants action actn to be done
bel(a, p)         to specify that agent a believes that p is true
ach(a, p)         to specify that agent a believes p will be achieved as a con-
sequence of the actions of          some agent (itself or others).
```

5 CONCLUSION AND CHALLENGES

In this paper we discussed logic-based approaches to intention recognition. We looked at different knowledge representation and reasoning mechanisms and we looked at the relationship between the two. We then described and analysed a number of concrete contributions in this field. Table 1 summarises some of the features of these contributions.

Intention recognition has been a long-standing research area. Not surprisingly, its application areas have changed during the intervening years, moving from Unix help facilities and language understanding to broader areas of ambient intelligence and intrusion and terrorism detection. Advances in activity recognition, sensor technology and RFID tags and readers allow further developments towards realistic and difficult applications.

Many challenges remain. Much of the current work assumes the existence of plan libraries. This requires much human effort in predicting and formalizing plans, and may be unrealistic in many cases. It may also be unrealistic to assume that the observer agent has knowledge of the plan libraries

of the observed agent. Furthermore, the intention recognition system is hampered in cases where an agent attempts novel ways of achieving a goal.

Much work on intention and plan recognition assumes one single observed agent, with one single intention and plan. Consequently it attempts to find hypotheses consisting of one intention or one plan that explains all the observed actions of that agent, with the possible exception of what it considers as "insignificant" actions. However, terrorism detection and other applications increase the demand for multi-agent intention recognition, where several agents co-operate on the same or on several related intentions. This brings with it many challenges, for example to identify related clusters of agents, to identify which agents are contributing to which intentions and to identify intentions and plans from interleaved actions. The challenges become harder if agents deliberately act to mislead the observers.

Dealing with partial observability of actions, whether in a single agent case or a multi-agent one remains an issue. Current solutions include attempting to detect that an action has been executed from changes in the environment (Geib and

Box 32.

$$holds\_at(int(A,to(\alpha,P)),\ T) \leftarrow holds\_at(bel(A,\ to(\alpha,P),\ T)\ \wedge\ holds\_$$
$$at(int(A,\ \alpha),\ T)\ \wedge \qquad\qquad holds\_at(ach(A,P),\ T) \qquad\qquad R3$$

Box 33.

```
holds_at(bel(A,P),  T),  holds_at(ach(A,P),  T)  ⇒ false,
```

Box 34.

```
succeeds(E) ← act(E, inform(S, H,P)), hold-at(bel(S,P),E), holds-at(bel(S,
int(S, inform(S,H,P))),E)
initiates(E, bel(H, bel(S,P))) ← act(E, inform(S, H,P))
```

Box 35.

$$succeeds(E) \leftarrow act(E,\ request(S,\ H,\ A)),\ hold\text{-}at(bel(S,cando(H,$$
$$A)),E),\ holds\text{-}at(bel(S,\ int(S,\ request(S,\ H,\ A))),E) \qquad R4$$
$$initiates(E,\ bel(H,\ bel(S,int(S,A)))) \leftarrow act(E,\ request(S,\ H,$$
$$A)) \qquad R5$$

Box 36.

$$holds\text{-}at(bel(H,P),\ T) \leftarrow holds\text{-}at(bel(H,\ bel(S,P)),\ T).$$

Table 1. A summary of some of the features of the reviewed works

	Formalism	Reasoning	Application	Requires full observability?
Mulder & Voorbraak	Simple plan libraries	Abductive	Tactical/Military	No, allows existentially quantified variables in observations
Myers	HTN plan libraries	Abductive	Co-operative planning	Yes
Jarvis et al.	Extended HTN plan libraries	Abductive	Terrorist intention recognition	No, allows varying degrees of accuracy in observations
Sindlar et al.	Transformed BDI-type plan libraries	Abductive	Generic	No, allows partial observability
Demolombe & Fernandez	Situation calculus and GOLOG	Probabilistic	Generic, but with focus on recognizing intentions of airplane pilots	Yes
Pereira & Anh	Causal Bayes nets	Baysian	Generic, but with focus on elder care	No, allows partial observability
Dragoni et al.	Multi-context logical theories with bridge rules	Abductive	Recognising intentions behind speech act utterances	Yes
Quaresma & Lopes	Event calculus	Abductive	Recognising intentions behind speech act utterances	Yes

Goldman 2001), even though the action itself has not been observed, and attaching probabilities to non-observation of actions (Geib and Goldman 2005). However, a general, efficient and extendible solution remains a challenge. This may be an area where abduction can make a further contribution in intention recognition. Abductive reasoning can be used to generate hypotheses about what actions have been executed, unobserved, compatibly with actions that have been observed, so that together they can account for the changes in the environment.

Most systems in keyhole and intended recognition cases assume that the actor is rational and is following a correct plan. This may not be the case in some applications. For example in tutoring systems the student may make mistakes, and in ambient intelligence systems the actor, maybe an Alzheimer patient, may execute actions in confusion.

ACKNOWLEDGMENT

I am grateful to Bob Kowalski for discussions about the topic of intention recognition, and to him and Luis Pereira for reading and commenting on earlier drafts of this paper. I am also grateful to the anonymous reviewers.

REFERENCES

Appelt, D. E., & Pollack, M. E. (1992, March). Weighted Abduction for Plan Ascription. *Journal of User Modeling and User-Adapted Interaction, 2*(1-2), 1–25.

Avrahami-Zilberbrand, D., & Kaminka, G. A. (2005). Fast and complete symbolic plan recognition. In Proceedings of the International Joint Conference on Artificial Intelligence (IJCAI) 2005, 653-658.

Avrahami-Zilberbrand, D., & Kaminka, G. A. (2007). Incorporating observer biases in keyhole plan recognition (efficiently!). In Proceedings of the Twenty-Second National Conference on Artificial Intelligence (AAAI-07), 2007.

Barger, T., Alwan, M., Kell, S., Turner, B., Wood, S., & Naidu, A. (2002) Objective remote assessment of activities of daily living: analysis of meal preparation patterns. Poster presentation, http://marc.med.virginia.edu /pdfs/ library/ ADL.pdf, 2002.

Bui, H. H. (2003). A general model for online probabilistic plan recognition. In Proceedings of the International Joint Conference on Artificial Intelligence (IJCAI) 2003.

Carberry, S., & Elzer, S. (2007). In Paliouras, G. (Ed.), *Exploiting evidence analysis in plan recognition. UM 2007, LNAI 4511 C. Conati, K. McCoy* (pp. 7–16). Springer-Verlag Berlin Heidelberg.

Charniak, E., & McDermott, D. (1985). *Introduction to artificial intelligence*. Reading, MA: Addison Wesley.

Cheng, D. C., & Thawonmas, R. (2004). Case-based plan recognition for real-time strategy games. In Proceedings of the 5th Game-On International Conference (CGAIDE) 2004, Reading, UK, Nov. 2004, 36-40.

Cohen, P. R., Perrault, C. R., & Allen, J. F. (1981). Beyond question answering. In Lehnert, W., & Ringle, M. (Eds.), *Strategies for Natural Language Processing* (pp. 245–274). Hillsdale, NJ: Lawrence Erlbaum Associates.

Cox, M.T., Kerkez, B. (2006).Case-Based Plan Recognition with Novel Input. *International Journal of Control and Intelligent Systems* 34(2), 2006, 96-104.

Demolombe, R., & Fernandez, A. M. O. (2006). Intention recognition in the situation calculus and probability theory frameworks. In *Proceedings of Computational Logic in Multi-agent Systems*. CLIMA.

Dragoni, A. F., Giorgini, P., & Serafini, L. (2002). Mental States Recognition from Communication. *Journal of Logic Computation, 12*(1), 119–136.

Erol, K., Hendler, J., & Nau, D. (1994). Semantics for hierarchical task-network planning. CS-TR-3239, University of Maryland, 1994.

Fung, T. H., & Kowalski, R. (1997). The IFF Proof Procedure for Abductive Logic Programming. *The Journal of Logic Programming*, 1997.

Geib, C. W., & Goldman, R. P. (2001). Plan recognition in intrusion detection systems. In the Proceedings of the DARPA Information Survivability Conference and Exposition (DISCEX), June, 2001.

Geib, C. W., & Goldman, R. P. (2005). Partial observability and probabilistic plan/goal recognition. In Proceedings of the 2005 International Workshop on Modeling Others from Observations (MOO-2005), July 2005.

Geib, C. W., & Steedman, M. (2003). On natural language processing and plan recognition. In Proceedings of the International Joint Conference on Artificial Intelligence (IJCAI) 2003, 1612- 1617.

Gelfond, M., & Lifschitz, V. (1992). Representing actions in extended logic programs. In Proceedings of the International Symposium on Logic Programming, 1992.

Giroux, S., Bauchet, J., Pigot, H., Lusser-Desrochers, D., & Lachappelle, Y. (2008). Pervasive behavior tracking for cognitive assistance. The 3rd International Conference on Pervasive Technologies Related to Assistive Environments, Petra'08, Greece, July 15-19, 2008.

Goodman, B. A., Litman, D. J. (1992). On the interaction between plan recognition and intelligent interfaces. User Modeling and User-Adapted Interactions 2(1-2), 1992, 83-115.

Hu, P. H., Son, T. C., & Baral, C. (2007, July). Reasoning and planning with sensing actions, incomplete information, and static laws using Answer Set Programming. *Theory and Practice of Logic Programming, 7*(4), 377–450.

Jarvis, P., Lunt, T., & Myers, K. (2004). Identifying terrorist activity with AI plan recognition technology. In the Sixteenth Innovative Applications of Artificial Intelligence Conference (IAAI 04), AAAI Press, 2004.

Jarvis, P. A., Lunt, T. F., & Myers, K. L. (2005). Identifying terrorist activity with AI plan-recognition technology. *AI Magazine, 26*(3), 73–81.

Kakas, A., Kowalski, R., & Toni, F. (1998). The role of logic programming in abduction. In: Gabbay, D., Hogger, C.J., Robinson, J.A. (Eds.): Handbook of Logic in Artificial Intelligence and Programming 5, Oxford University Press, 1998, 235-324.

Karlsson, B., Ciarlini, A. E. M., Feijo, B., & Furtado, A. L. (2006). Applying a plan-recognition/plan-generation paradigm to interactive storytelling, The LOGTELL Case Study. Monografias em Ciência da Computação Series (MCC 24/07), ISSN 0103-9741, Informatics Department/PUC-Rio, Rio de Janeiro, Brazil, September 2007. Also in the Proceedings of the ICAPS06 Workshop on AI Planning for Computer Games and Synthetic Characters, Lake District, UK, June, 2006, 31-40.

Kautz, H., & Allen, J. F. (1986). Generalized plan recognition. In Proceedings of the Fifth National Conference on Artificial Intelligence (AAAI-86), 1986, 32-38.

Konolige, K., & Pollack, M. E. (1989). Ascribing plans to agents: preliminary report. In 11th International Joint Conference on Artificial Intelligence, Detroit, MI, 1989, 924-930.

Kowalski, R., & Sergot, M. (1986). A logic-based calculus of events. In *New Generation Computing*, Vol. 4, No.1, February 1986, 67-95.

Kowalski, R. A., & Sadri, F. (2009). Integrating logic programming and production systems in abductive logic programming agents. Invited papers, the 3rd International Conference on Web Reasoning and Rule Systems, October 25-26, Chantilly, Virginia, USA, 2009.

Lesh, N., Rich, C., & Sidner, C. L. (1999). Using plan recognition in human-computer collaboration. In Proceedings of the Seventh International Conference on User Modelling, Canada, July 1999, 23-32.

Levesque, H., Reiter, R., Lesperance, Y., Lin, F., & Scherl, R. (1997). GOLOG: A Logic Programming Language for Dynamic Domains. *The Journal of Logic Programming, 31*, 59–84.

Mancarella, P., Terreni, G., Sadri, F., Toni, F., & Endriss, U. (2009). The CIFF proof procedure for abductive logic programming with constraints: theory, implementation and experiments. *Theory and Practice of Logic Programming, 9*, 2009.

Mao, W., & Gratch, J. (2004). A utility-based approach to intention recognition. AAMAS 2004 Workshop on Agent Tracking: Modelling Other Agents from Observations.

Mayfiled, J. (2000). Evaluating plan recognition systems: three properties of a good explanation. *Artificial Intelligence Review, 14*, 351–376.

Mihailidis, A., Fernie, G. R., & Barbenel, J. (2001). The use of artificial intelligence in the design of an intelligent cognitive orthosis for people with dementia. *Assistive Technology, 13*, 23–39.

Modayil, J., Bai, T., & Kautz, H. (2008). Improving the recognition of interleaved activities. In the Proceedings of the 10th International Conference on Ubiquitous Computing (UbiComp), 2008, 40-43.

Mulder, F., & Voorbraak, F. (2003). A formal description of tactical plan recognition. *Information Fusion, 4*, 47–61.

Myers, K. L. (1997). Abductive completion of plan sketches. In Proceedings of the 14th National Conference on Artificial Intelligence, American Association of Artificial Intelligence (AAAI-97), Menlo Park, CA: AAAI Press, 1997.

Myers, K. L., Jarvis, P. A., Tyson, W. M., & Wolverton, M. J. (2003). A mixed-initiative framework for robust plan sketching. In Proceedings of the 13thInternational Conferences on AI Planning and Scheduling, Trento, Italy, June, 2003.

Park, S., & Kautz, H. (2008). Hierarchical recognition of activities of daily living using multi-scale, multi-perspective vision and RFID. The 4th IET International Conference on Intelligent Environments, Seattle, WA, July 2008a.

Park, S., & Kautz, H. (2008). Privacy-preserving recognition of activities in daily living from multi-view silhouettes and RFID-based training, AAAI Fall 2008 Symposium on AI in Eldercare: New Solutions to Old Problems, Washington, DC, November 7 - 9, 2008b.

Peirce, C. S. (1958). *The collected papers of Charles Sanders Peirce* (Burks, A. W., Ed.). *Vol. VII-VIII)*. Cambridge, MA: Harvard.

Pereira, L. M., Alferes, J. J., & Aparicio, J. N. (1992). Contradiction removal semantics with explicit negation. In Proceedings of the Applied Logic Conference, 1992.

Pereira, L. M., & Anh, H. T. (2009a). Intention Recognition via Causal Bayes Networks plus Plan Generation, in: L. Rocha et al. (eds.), Progress in Artificial Intelligence, Procs. 14th Portuguese International Conference on Artificial Intelligence (EPIA'09), Springer LNAI, October 2009.

Pereira, L. M., & Anh, H. T. (2009b). Elder care via intention recognition and evolution prospection, in: S. Abreu, D. Seipel (eds.), Procs. 18th International Conference on Applications of Declarative Programming and Knowledge Management (INAP'09), Évora, Portugal, November 2009.

Pereira, L.M., Saptawijaya, A. (2009). Modelling Morality with Prospective Logic, in International Journal of Reasoning-based Intelligent Systems (IJRIS), 1(3/4): 209-221.

Philipose, M., Fishkin, K. P., Perkowitz, M., Patterson, D. J., Hahnel, D., Fox, D., & Kautz, H. (2005). Inferring ADLs from interactions with objects. *IEEE Pervasive Computing / IEEE Computer Society [and] IEEE Communications Society*, 2005.

Quaresma, P., & Lopes, J. G. (1995). Unified Logic Programming Approach to the Abduction of Plans and Intentions in Information-Seeking Dialogues. *Journal of Logic Programming, Elsevier Science Inc.*, 1995, 103–119.

Rao, M., & Georgeff, P. (1995). BDI-agents: from theory to practice. In the Proceedings of the First International Conference on Multi-agent Systems (ICMAS'95), 1995.

Reiter, R. (2001). *Knowledge in action: logical foundations for specifying and implementing dynamical systems*. MIT Press, 2001.

Roy, P., Bouchard, B., Bouzouane, A., & Giroux, S. (2007). A hybrid plan recognition model for Alzheimer's patients: interleaved-erroneous dilemma. IEEE/WIC/ACM International Conference on Intelligent Agent Technology, 2007, 131- 137.

Sanchez, D., Tentori, M., & Favela, J. (2008). Activity recognition for the smart hospital. *IEEE Intelligent Systems*, (March/April): 50–57.

Scmidt, C., Sridharan, N., & Goodson, J. (1978). The plan recognition problem: an intersection of psychology and artificial intelligence. *Artificial Intelligence, 11*, 45–83.

Sindlar, M. P., Dastani, M. M., Dignum, F., & Meyer, J.-J. Ch. (2008). *Mental state abduction of BDI-based agents. Declarative Agent Languages and Technologies*. Estoril, Portugal: DALT.

Sukthankar, G., & Sycara, K. (2008). Robust and efficient plan recognition for dynamic multi-agent teams (Short Paper). In Proceedings of 7th International Conference on Autonomous Agents and Multi-agent Systems (AAMAS 2008), International Foundation for Autonomous Agents and Multi-agent Systems, May 2008.

Sukthankar, G., & Sycara, K. (2008). Hypothesis Pruning and Ranking for Large Plan Recognition Problems. In Proceedings of the Twenty-Third AAAI Conference on Artificial Intelligence (AAAI-08), June 2008.

Suzić, R., & Svenson, P. (2006). Capabilities-based plan recognition. In Proceedings of the 9th International Conference on Information Fusion, Italy, July 2006.

Wilensky, R. (1983). *Planning and understanding*. Reading, MA: Addison Wesley.

KEY TERMS AND DEFINITIONS

Intention Recognition: The task of recognizing the intentions of an agent by analyzing some or all of their actions and/or analyzing the changes in the state (environment) resulting from their actions.

Keyhole Intention Recognition: Intention recognition in cases where the agent which is being observed either does not intend for his intentions to be identified, or does not care; he is focused on his own activities, which may provide only partial observability to the observing agent.

Intended Intention Recognition: Intention recognition in cases where the agent which is being observed wants his intentions to be identified and intentionally gives signals to be sensed by the observing agent.

Plan Recognition: The task of recognizing not just the intention but also the plan (i.e. the sequence of actions, including future actions) the observed agent is following in order to achieve his intention.

Abductive Reasoning: A form of defeasible reasoning allowing to draw hypothesis to explain some evidence or observations.

The Event Calculus: A causal theory of events, times and time-dependent properties formalized in Horn clause logic augmented with some form of negation.

The Situation Calculus: A causal theory formalized in classical logic specifying how actions change situations.

ENDNOTES

[*] Note that here, slightly differently from Dragoni et al., the *request* is aimed at requesting that an action be done. But, in practice, by requesting that an *inform* action be done, the effect is the same, i.e. to request some information.

[1] Here we are ignoring persistence and we are assuming that no action occurs between any two times T and T+1.

[2] In the paper both the planning rules of the observed agent and rules of the form RD of the observer agent are assumed to be ground.

[3] Alternatively to conform to the notation we introduced in Section 3 we could write:
$persists(T1,P,T) \leftarrow not\ clipped(T1,P,T)$
$clipped(T1,P,T) \leftarrow happens(E,\ T2) \wedge terminates(E,P) \wedge not\ out(T2,T1,T)$
$terminates(turn\text{-}fuel\text{-}on,,\ fuel\text{-}off)$.

Chapter 19
Inference of Human Intentions in Context Aware Systems

Katsunori Oyama
Iowa State University, USA

Carl K. Chang
Iowa State University, USA

Simanta Mitra
Iowa State University, USA

ABSTRACT

Most of context models have limited capability in involving human intention for system evolvability and self-adaptability. Human intention in context aware systems can evolve at any time, however, context aware systems based on these context models can provide only standard services that are often insufficient for specific user needs. Consequently, evolving human intentions result in changes in system requirements. Moreover, an intention must be analyzed from tangled relations with different types of contexts. In the past, this complexity has prevented researchers from using computational methods for analyzing or specifying human intention in context aware system design. The authors investigated the possibility for inferring human intentions from contexts and situations, and deploying appropriate services that users require during system run-time. This chapter first focus on describing an inference ontology to represent stepwise inference tasks to detect an intention change and then discuss how context aware systems can accommodate requirements for the intention change.

A Context Aware (CA) room may know that it is dark outside and that the resident has just entered the home and decide to turn on the lights, whereas if the resident enters the home with guests, it may decide to not only turn on the lights but to also turn down the heat a notch to adjust for more people. To be able to respond in this context aware manner, software systems in the CA room usually incorporate an adaptable and distributed device framework that consists of sensor networks and mobile computing techniques. Here the framework's role is to acquire stimulus data using sensors, match the perceived sensory stimulus to

DOI: 10.4018/978-1-61692-857-5.ch019

a context, and trigger actions based on the recognized context.

It is well known that changes in circumstances under which individuals use a system influence them to change goals and hence intent. For example, someone may have taken a walk in his neighborhood for years without fear, but he may be afraid to do so now due to the recent increases in the crime rate and needs new security measures to allow him to continue on his walks. Thus, these changes in intent result in changes in system requirements. As researchers and software system developers, we must answer the question, "How can the system find out about the changing intentions of users so that our system can respond with appropriate services?" In the past, human intention has been modeled (for example, the BDI model (Bratman, 1987)) and implemented in software systems. Researchers today have found various types of sensor information helpful in designing computational methods to mine human intentions; for instance, the non-invasive Brain Computer Interface (Cichoki et al., 2008) analyzes the behaviors of users equipped with wearable sensors.

This chapter introduces understanding human intention as an important factor for the development of effective CA services. Unfortunately, most context models in pervasive computing have a limited capability for involving human intention in system evolvability; human intention may demonstrate a need for a service to adapt, however, CA systems based on these context models can provide only standard services that are often insufficient for specific user needs. As intention changes may occur at any time, pre-defined system requirements can become obsolete or contrary to user needs. Consequently, such a CA system cannot continue to deliver services as expected.

It is hard to extract human intentions from contexts due to their tangled relations with different types of contexts. In the past, this complexity has prevented researchers from using computational methods for analyzing or specifying human intention in CA system design. Recently, the authors have investigated the inference of human intention from context information (Oyama et al., 2008a), deploying appropriate services that users require on the fly (Chang et al., 2008; Ming et al., 2008), and finding potential future goals as system requirements (Oyama et al., 2008b).

In this chapter, we first focus on describing an inference ontology to represent inference tasks to detect changing human intentions and then discuss how CA systems can identify requirements changes due to changing human intentions. We review the execution of step-by-step inference tasks in a CA system which proceeds in a hierarchical fashion from low level processes where device information is captured to higher level human-machine interaction processes.

1 BACKGROUND

1.1 Contexts and Situations

The definitions and interrelationships of contexts, situations and human intentions have been intensively discussed in pervasive computing (Bihler et al., 2005) and service computing (Rolland et al., 07). These notions were originally studied in different areas and then later were put together with the goal of improving human-centric systems including robots, software agents in smart home, and mobile systems.

One definition of context is "any information that can be used to characterize the situation of an entity" (Dey, 2001) for providing relevant information or services to the user. Typically, context awareness research centers on designing context models or context ontology (Strang & Linnhoff-Popien, 2004; Krummenacher & Strang, 2007), and developing various context reasoning and monitoring methods. Many of the existing approaches synthesize information about locations, surrounding objects, physiological readings of people etc., for use in real world applications. For example, location-based services (Bellavista et al., 2008) provide personalized functions based on the location gathered from context information. As one of the drivers for a pervasive computing paradigm, context awareness creates a new means of human-computer interaction because as it is deployed all over the environment, it changes the way users interact with the environment or the software system. CA systems rely heavily on the information of various sensor data to adapt to changing situations of the user who is using a CA service. Normally the CA system contains three major steps: collecting environmental data through sensor network or control frameworks, understanding contexts from the environmental data based on a context model for the system, and triggering actions to provide services. These three steps constitute a spiral of executions of a CA system; any change in the existing environment could cause changes in the context model and therefore trigger new actions, which in turn again affect the environment and so on.

The definition of situation has evolved over a long period. In the 1980's, situation was originally studied to investigate a mathematically based theory of natural language semantics (Barwise & Perry, 1980). In a later work (Barwise & Perry, 1983), situation was defined in terms of information conveyed in a discourse that can be clearly recognized in common sense and human language. Based on this definition, events and episodes are situations in time. Today, researchers have different perspectives on how to define situation, often depending on research objectives. For instance, some define situation as "histories" (i.e., finite sequences of primitive actions (Levesque et al., 1998)). Situation can also be reasoned by aggregating a set of predicates that explicitly represent information about sensory data (Mastrogiovanni et al., 2008). Sequencing such predicates can form the history of a finite number of instances in a first-in first-out order. For our purposes, generally a situation is a set of contexts in the application over a period of time that affects future system behavior (Yau et al., 2008).

1.2 Domain Ontology and Task Ontology

In Information Science, an ontology can be briefly described as "an explicit specification of a conceptualization" (Gruber, 1993). An ontology constitutes a knowledge base for a domain and formally specifies concepts within the domain and relationships between these concepts. For developing such a knowledge base, each concept is defined as a *class*. In every class, *slots* declare properties and attributes of the concept. Restrictions or constraints on slots are called *facets*. If the classes in the ontology are element of the area of study, then the ontology constitutes domain ontology, whereas, if the class in the ontology is an operational task, then the ontology constitutes a task ontology.

Generally, a domain ontology provides domain specific knowledge to applications, and provides metadata of contexts and situations. There are existing domain ontologies that can be used to share contexts or situations in specific problem domains. For example, CONON (Wang et al., 2007) is a context ontology that uses Web Ontology Language (OWL) for modeling context and supports logic based situation reasoning. CONON divides a context model into two layers, an upper context ontology and a domain-specific ontology, where the upper context ontology defines specification of a general context that is common to a wide range of domains. Other examples include a situation ontology that explicitly specifies situation as a set of contexts over a period of time that affects future system behavior (Yau & Liu, 2006), and another situation ontology that uses first-order logic based reasoning rules and also provides a way to convert the ontology specifications to the first-order logic representations.

Unlike domain ontologies, a task ontology is a system of concepts and terms for explaining problem-solving structures (Mizoguchi et al., 1995), where a problem-solving structure represents a task hierarchy of agents and rules to manipulate instances of domain ontologies. Such a task hierarchy includes necessary elements of domain ontologies to explain how the targeted task is executed, but does not define the semantic structure of the elements, such as the internal structure of contexts and situations. Inference ontology separates the concerns of inference tasks into capturing sensor data, situation reasoning, memorizing situations captured, simulation of possible human intentions, and interaction with users (Oyama et al., 2008a). Each inference task is based on the premise that contexts are continuously captured by sampling sensor data at an adequate sampling rate.

1.3 Human Intentions

Human intentions have been investigated in the areas of Philosophy (Bratman, 1987; Scheer, 2004), Cognitive Science (Tomasello et al., 2005) and Artificial Intelligence (Cohen & Levesque, 1990; Rao Georgeff, 1991). Unfortunately, human intentions are hard to untangle from the variety of contexts that are extracted from sensor data and an individual user's captured intention can often be found to be false positive or false negative due to several reasons. A few of these reasons are:

- **Diversity of human actions:** Correlations between intentions and behavioral actions vary from person to person since different human beings have different knowledge and skills (Oyama et al., 2008a). In other words, even with the same intention, individuals will act in different ways based on their knowledge, skills, and preferences.
- **Uncertainty:** Intention changes can occur due to unpredictable influences such as environmental changes, and thus it is not always true that a person chooses same action in a particular situation. Thus, a person with the same intention may choose a different sequence of actions on observing a change in the environment.

- **Intention hierarchy:** Intentions are organized in a hierarchy with smaller intentions (for example, walking, making a call, and entering a building) at the leaves and higher intentions (for example, staying in a building, attending a conference etc) at higher levels. The higher-level intentions play a role in defining the reason for each of the lower-level intentions.

Definitions of intention often attempt to deal with how human mental states are viewed. Bratman (Bratman, 1987), a philosopher, originally described intention as one of the belief, desire, and intention (BDI) mental states that motivate action. Since then, his BDI model has been widely accepted by computer scientists.

In response to Bratman's BDI model, Scheer created an opposing definition to avoid some of the problems associated with modeling mental states in the BDI fashion by explaining intention as simply a course of action that one has adopted (Scheer, 2004). It is worthy of note that both Bratman's and Scheer's definitions seem to agree that intentions can be analyzed through the observation of an individual's actions. Recently, the observation of human intention has been utilized in sensor-based approach which uses probabilistic models (Kelley et al., 2008; Cichoki et al., 2008). This approach has resulted in capture of human intentions in response to a captured sequence of actions - although these are only the lowest level intentions.

Cohen and Levesque (1990) formalized some philosophical aspects of the BDI model based on temporal logic. In their formalism, intentions are defined in terms of temporal execution plans to achieve a goal. An agent committed to intentions from the planning result will maintain goals until the goals are believed to be achieved or found to be unachievable, whereas an agent that has a relativized commitment to the intentions is similarly committed to the goals, but may drop them when some specified conditions are believed to hold.

Rao and Georgeff (1991) also contributed to the BDI model by defining BDI logic for expressing logical interrelationships between beliefs, goals, and intentions.

X-BDI (Mora et al., 1999; Meneguzzi & Luck, 2008), Jason (Bordini et al., 2005) and other BDI based agents focus on finding an optimal execution plan to reach the declarative goal under a given environment. An application of X-BDI was recently applied to try to infer human emotions. Jaques and Vicari (2007) proposed a BDI logic-based approach, which provided a process of affective diagnosis to infer human emotions from sequences of a subject's actions. This BDI logic-based approach used the X-BDI agent to infer the student's emotion based on the combination of his/her actions and used a decision tree to display the results. Although the purpose of their system interface is to infer human emotions instead of human intentions, it is highly related to human intentions since the BDI logic-based approach attempts to deduce the student's emotion from a certain sequence of actions. X-BDI's graphical user interface provided a convenient means to obtain feedback and consequently their system interface allows simple revisions and frequent modifications of the information about the student. The student model is built dynamically from each interaction in real-time.

AI researchers have used the BDI model to develop agent-planning and robotics technologies because it provides a computationally-efficient architecture for dealing with multiple targets. In other words, the agent can mimic a human intention in order to make an optimal execution plan of actions to reach chosen goals (i.e., desires). This approach tries to characterize agents using anthropomorphic notions, such as mental states and actions. Usually, these notions and their properties are formally defined using logical frameworks that allow theorists to analyze, specify and verify rational agents.

To improve CA systems, it is important to devise ways to capture human intentions precisely.

Intention changes need to be detected quickly in order to find the user's emerging need. Thus, one important question that we will address in the next section is "How can we find changing intention of a user so that our system incorporates appropriate services?" We will also review the characteristics of human intentions and the possibility of inferring human intentions from contexts and situations.

2 THE HUMAN DIMENSIONAL INFERENCE ONTOLOGY

Ontology of human intentions is useful for development of better CA systems. As we show in this section, the DIKW (Data, Information, Knowledge, and Wisdom) hierarchy provides a useful structure for such ontology.

2.1 The DIKW hierarchy

Originally, Ackoff came up with definitions for classification of human thinking using a Data, Information, Knowledge, Understanding, and Wisdom structure (Ackoff, 1989). This classification explains how a person can acquire some Knowledge by memorizing a fact instead of by understanding it; the understanding process is unnecessary to store one's Knowledge but required for Wisdom. Later, Bellinger interpreted Understanding as part of the transition process from Data to Wisdom; specifically, the Understanding process is not only seen as a transition from Knowledge to Wisdom, but also the process where Information is acquired from the meaning of Data (Bellinger et al., 2004). Thus, he re-classified the knowledge hierarchy as Data, Information, Knowledge, and Wisdom (DIKW).

The DIKW hierarchy has been often used to explain human thinking (in the context of knowledge management). Recently, Oyama et al. used this DIKW hierarchy to describe the thought process of an expert system designer by describing a concept from the Data to the Wisdom (Oyama

Figure 1. DIKW hierarchy

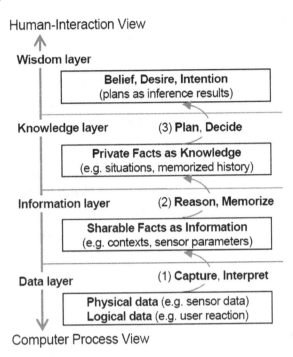

et al., 2006). A system designer's Knowledge is described only in relation to the facts in Information and the predicates in this Knowledge level is used to build up the Wisdom. As human intentions belong to the decision-making process which is within the scope of human knowledge and wisdom, a DIKW hierarchy can provide a comprehensive model for human intentions.

In Context Awareness, the structure of DIKW is often useful to create a domain and task hierarchy from Data, Information, and Knowledge to Wisdom (Oyama et al., 2008a). For example in Figure 1, if an agent of CA system stores a fact from captured contexts and situations in the information layer, the fact will be memorized as history in the knowledge layer. To infer a human intention, the agent makes possible plans based on facts held in the knowledge layer. The plan in the wisdom layer can be compared and analyzed in relation to actions executed by users; this is how the system can confirm the user intention. User interfaces demonstrate the inference result,

allowing the user to react, again confirming (or denying) for the system whether the result is precise enough. By processing Data to Wisdom, each layer of DIKW provides a task hierarchy for the inference of human intentions. This hierarchy is applied to build an inference ontology in the next subsection.

2.2 The HDIO ontology

For modeling the process of inferring human intentions, Oyama et al. used the approach of using an ontology to share knowledge of the required inference tasks and introduced the human-dimensional inference ontology (HDIO) (Oyama, 2008a). This ontology adopted the DIKW structure in order to classify the task hierarchy for a CA agent (see Figure 2).

In the HDIO, at the "Data" level, the computer typically communicates with devices (i.e., sensors and actuators) or other software interfaces. Here, an inference task is a collection of

Figure 2. Task hierarchy in CA agents using HDIO

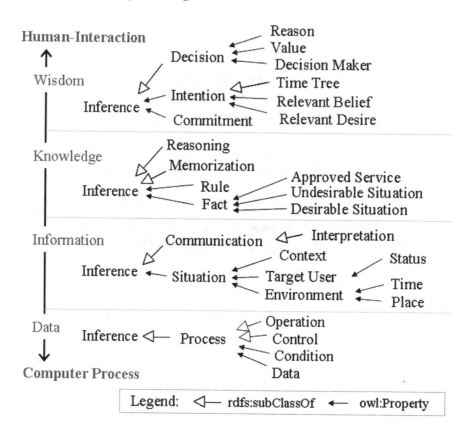

Processes (Operations and Control) that handles Data based on some Conditions. This computer process periodically generates contexts with machine-readable format from physical data or logical data.

At the "Information" level, the inference tasks use the captured data sets and interprets them as contexts. By aggregating contexts over a period, situations are modeled and instantiated with Target User with the user's Status and Environment (Time, Place).

At the "Knowledge" level, captured situations are used for reasoning. This reasoning creates candidates of a human intention using Rules and Facts which are specific situations memorized by an individual CA agent for use in future reasoning. For example, a service approved by the user is memorized as a type of fact, functioning as a preferable service for the Target User reacting to

the specific captured situations. Undesirable and Desirable situations are also stored as Facts. If an Undesirable Situation is detected, the CA agent makes plans for aiming at a Desirable Situation by providing services.

At the "Wisdom level", inference results represent human Intention. Each Intention is deduced using BDI model-based time trees of where one time point is situation of the Target User. If there is more than one result at the same time, the CA agent makes a Decision with help from Reason, Value and Decision Maker. Note that Value prioritizes each decision and Decision Maker can be the Target User or the CA agent which is delegated to automate similar kinds of decisions. The inference task asks the Target User to commit the results of decision in order that the CA agent can learn of changes in the user's intention; this commitment is necessary only if exceptions occur during the

Figure 3. Partial OWL serialization

```
<?xml version="1.0"?>
<rdf:RDF
  xmlns:rdf="http://www.w3.org/1999/02/22-rdf-syntax-ns#"z
  xmlns:rdfs="http://www.w3.org/2000/01/rdf-schema#"
  xmlns:owl="http://www.w3.org/2002/07/owl#"
  xml:base="http://www.sevolution.com/owl/HDIO.xml/"
>

<owl:Class rdf:ID="Inference">
  <rdfs:label>Inference</rdfs:label>
  <rdfs:subClassOf rdf:resource="#Wisdom" />
  <rdfs:subClassOf>
    <owl:Restriction>
      <owl:minCardinality
        rdf:datatype="http://www.w3.org/2001/XMLSchema#int"> 1
      </owl:minCardinality>
      <owl:onProperty rdf:resource="#has_Result"/>
    </owl:Restriction>
  </rdfs:subClassOf>
  <rdfs:subClassOf>
    <owl:Restriction>
      <owl:onProperty rdf:resource="#has_Result"/>
      <owl:allValuesFrom rdf:resource="#Human_Intention" />
    </owl:Restriction>
  </rdfs:subClassOf>
  <rdfs:subClassOf>
    <owl:Restriction>
      <owl:cardinality
        rdf:datatype="http://www.w3.org/2001/XMLSchema#int"> 1
      </owl:cardinality>
      <owl:onProperty rdf:resource="#has_Commitment"/>
    </owl:Restriction>
  </rdfs:subClassOf>
  <rdfs:subClassOf>
    <owl:Restriction>
      <owl:onProperty rdf:resource="#has_Commitment"/>
      <owl:allValuesFrom rdf:resource="#Human_Intention" />
    </owl:Restriction>
  </rdfs:subClassOf>
</owl:Class>

<owl:Class rdf:ID="Time_Tree">
  <rdfs:label>Time Tree</rdfs:label>
  <rdfs:subClassOf>
    <owl:Restriction>
      <owl:minCardinality
        rdf:datatype="http://www.w3.org/2001/XMLSchema#int"> 1
      </owl:minCardinality>
      <owl:onProperty rdf:resource="#has_Time_Point"/>
    </owl:Restriction>
  </rdfs:subClassOf>
  <rdfs:subClassOf>
    <owl:Restriction>
      <owl:onProperty rdf:resource="#has_Time_Point"/>
      <owl:allValuesFrom rdf:resource="#Situation" />
    </owl:Restriction>
  </rdfs:subClassOf>
</owl:Class>

<owl:Class rdf:ID="Decision">
  <rdfs:label>Decision</rdfs:label>
  <rdfs:subClassOf rdf:resource="#Inference" />
  <rdfs:subClassOf>
    <owl:Restriction>
      <owl:minCardinality
        rdf:datatype="http://www.w3.org/2001/XMLSchema#int"> 1
      </owl:minCardinality>
      <owl:onProperty rdf:resource="#has_Reason"/>
    </owl:Restriction>
  </rdfs:subClassOf>
  <rdfs:subClassOf>
    <owl:Restriction>
      <owl:onProperty rdf:resource="#has_Reason"/>
      <owl:allValuesFrom rdf:resource="#Predicate" />
    </owl:Restriction>
  </rdfs:subClassOf>
  <rdfs:subClassOf>
    <owl:Restriction>
      <owl:minCardinality
        rdf:datatype="http://www.w3.org/2001/XMLSchema#int"> 1
      </owl:minCardinality>
      <owl:onProperty rdf:resource="#has_Value"/>
    </owl:Restriction>
  </rdfs:subClassOf>
  <rdfs:subClassOf>
    <owl:Restriction>
      <owl:onProperty rdf:resource="#has_Value"/>
      <owl:allValuesFrom rdf:resource="#Priority" />
    </owl:Restriction>
  </rdfs:subClassOf>
  <rdfs:subClassOf>
    <owl:Restriction>
      <owl:cardinality
        rdf:datatype="http://www.w3.org/2001/XMLSchema#int"> 1
      </owl:cardinality>
      <owl:onProperty rdf:resource="#has_Decision_Maker"/>
    </owl:Restriction>
  </rdfs:subClassOf>
  <rdfs:subClassOf>
    <owl:Restriction>
      <owl:onProperty rdf:resource="#has_Decision_Maker"/>
      <owl:allValuesFrom rdf:resource="#Person" />
    </owl:Restriction>
  </rdfs:subClassOf>
</owl:Class>

</rdf:RDF>
```

service run-time. These concepts of inference, decision, and time tree are defined as classes in an OWL representation (see Figure 3).

In the Knowledge and Wisdom levels, an inference task performs on time trees (see the next subsection) and simulates human belief, desire and intention. In Rao and Georgeff's BDI logic,

Figure 4. Tasks for capturing situations

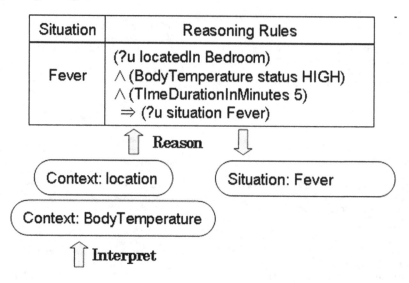

a time tree represents the Belief-Desire-Intention world using a temporal structure with a branching time future and a single past (Rao & Georgeff, 1991). A time point in a particular world is also called situation, and thus paths on the time tree indicate changing situations in the BDI world.

The main contribution of the inference ontology is that it reuses standardized tasks to infer human intentions. This inference ontology includes the following features:

- An agent in the CA system captures situations of a user in a real-time and parallel manner from given context information.
- The agent simulates branching time trees to represent an individual's intention based on the BDI model.
- The Inference result by this agent is intention as a temporal sequence of the user actions.
- To make a step-by-step procedure for the inference, the agent follows the task hierarchy of HDIO in a bottom-up fashion.

2.3 Inference Tasks by BDI Model

We shall illustrate the HDIO inference process by using online healthcare support services as an example, which in recent years have been the focus of attention of researchers who have developed ubiquitous computing infrastructures to deliver personalized healthcare services (Zhang et al., 2004). The major technologies of such systems include a self-sensing mechanism, a context-processing framework, and a service interoperability platform. For example, the U-Healthcare system (Ko et al., 2006) uses wearable devices, which provide physiological data to detect health condition, such as, body temperature, blood pressure, stress level, pulse and respiration. The U-Healthcare system is based on a context awareness model, including the technology and infrastructure for healthcare support services. Purchasers and receiver services can be offered through various user interfaces including different types of medical devices and the Internet portal.

Requirements of user interface to serve medical services are diversified by type of illness, severity of symptoms, and prescription method. To analyze changing requirements during incremental development of such a healthcare system, CA

Figure 5. Tasks for deduction of human intentions

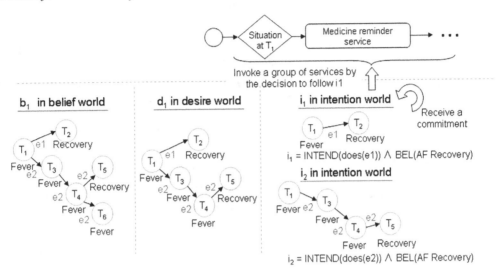

agents running on the user interface will play an important role in speculating the service user's needs in real time as well as providing context information of his/her health condition. Let us assume that the system is serving the user while he/she is sick and that originally only a few services such as "to remind the service user to take medicine" are available.

To capture a situation, the CA agent accesses physiological data from the wearable devices, and interprets them as contexts. As shown in Figure 4, the situation of "Fever" can be reasoned from fulfillment of the following conditions in the first-order logic representation: if the location is bedroom, the body temperature indicates high level, and this status lasts for 5 minutes. Note that reasoning rules involving information of human behavior requires maintenance to fix parameters; each individual user has different optimal parameters such as the level of body temperature depending on his/her health condition.

When a situation is captured, a CA agent responds in real time by invoking a service. However, there may be more than one service related to the captured situation and there may be dependencies among these services. To evaluate the choices, the CA agent employs a reasoning process

where it attempts to simulate a temporal sequence of these situations as a branching time tree and to then formulate possible plans as intentions of the user. Here, plans are represented as time trees and there is a time tree to simulate each of human belief, desire, and intention (as shown in the example depicted in Figure 5).

In the example, the belief time tree has a captured situation placed at the current time point T5. The root time point, at time point T0, provides a starting observation with initial conditions. The other situations at past times are historical data that the CA agent has captured and stored. These situations can include some goals that are set by system designers during the CA development process or those assigned by the users. These goal situations are labeled as desirable situations with the rest of the situations being labeled as undesirable situations. In the figure, the situations of a patient with the status of recovery (time points T2 and T5) are set as desirable situations. The transition from T1 to T2 represents the patient having recovered in 1 day by using some medicine. In contrast, the transition from T1 to T5 represents the patient having recovered in 3 days without having taken any medicine. In this case, the user

Figure 6. Interaction between context aware agent and system user

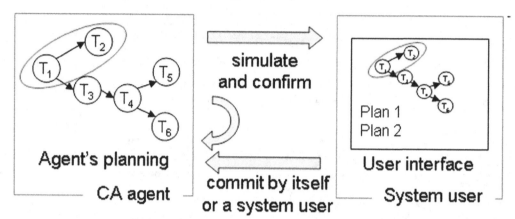

had made this decision to avoid the potential side effects of taking the medicine.

As shown in the figure, the desire time tree consists of only the paths to desirable situations from the belief world. Thus, the path to T6 is eliminated in the desire world because of the undesirable situation in the given period.

By aggregating situations of the service user, the CA agent simulates possible human intentions based on given situations. In the intention time tree, i1 and i2 are choices of paths to desirable situations. In the intention world, i_1 represents that the service user may intend to take medicine, and believes that he/she will recover. On the other hand, i_2 indicates that the service user intends to do nothing because he/she does not wish to take a medicine by considering one of the doctor's advices. If there is any intention change (e.g., the user needs to be reminded of taking medicine, but one day the medicine reminder service was not used, alternative plans in the intention world will be asked to be "confirmed or committed" by the user. The CA agent memorizes the path (i2) for future inferences.

Thus, the CA agent simulates a human intention, and attempts to confirm adaptive services for the service user if his/her intention changes significantly by choices of alternative plans and services. The CA agent also waits periodically

for commitment from the user (Figure 6). This interaction cycle through the simulation of human intention and via the commitment process is also recorded as service usage history in order to elicit potential requirements of the user, especially when an intention change is detected. Human intention change, once monitored, facilitates capture of new requirements. These new requirements trigger the next cycle of system evolution for upgrading the system to a new version.

3 HUMAN INTENTIONS AND GOALS IN REQUIREMENTS ENGINEERING

3.1 Requirements Engineering Process

CA system is typically designed to customize a service to meet user needs based on the context information; however, as CA system is developed for personalized service, the system's requirements will inevitably need to be modified in accordance with the user's intention change (Oyama et al., 2008b). In Requirements Engineering, the notion of goal (which is optative statements similar to human desires) is widely accepted as a means for eliciting, elaborating, structuring, specifying, analyzing, negotiating, documenting, and modifying requirements (Lamsweerde & Letier, 2003). Some goals are decomposable and consistent with each other in a goal tree based on AND and OR relationships. Usually stakeholders in the preliminary material available to requirements engineers explicitly state goals, although sometimes they are initially implicit for system designers and even users themselves (Lamsweerde, 2001). In the ever-changing situations of a modern-day service environment there is a constant demand for evolution of services for users' ultimate satisfaction. For instance, the personalized healthcare service as mentioned in section 2.3 is designed to achieve a goal (say, to follow medical advices correctly for case of hypertension) by providing healthcare services to the user; it is always possible that the goal itself will need to be modified to adapt the healthcare services to the user's evolving intention. Thus, detection of changing intentions and speculation of potential goals in real-time to help guide evolution of services is now an emerging challenge in requirements engineering for CA systems.

The interrelationships between goals, intentions, and contexts have been discussed in service requirements engineering (Rolland et al., 2007). A requirements engineer can capture human intentions of an individual user by observing his/her behavioral patterns; on the other hand, goals are relatively more objective and through explicit knowledge representation can be more easily shared with stakeholders. For example, one of the goals when using a car navigation service is to reach the user's destination. The user may intend to take the shortest way leading to targeted destinations or the easiest way in terms of effort level. Some of the paths can be identified as "the shortest way" by estimation of the distance between two locations, while more specific, personalized paths are captured only through manual input or observing behavioral patterns (e.g. history of chosen paths, tendency of driving speed, etc.). The generally accepted notion is that every target of an intention can be represented as a goal and that

they are related and complementary concepts. Behavioral patterns often indicate human intentions that are temporarily different from that of the service user even with the same goal; more than one human intention can achieve a goal. In other words, humans can generally find different ways to achieve a goal.

Since the requirements engineering process cannot completely elicit all of the possible situations and intentions of the service user, which inevitably evolve over time, we need an evolutionary mechanism to adapt the system to intention changes under targeted goal. Therefore, task ontologies defining the relationship between intentions and the targeted goal are useful to instantiate these relationships as update of requirements so that engineers can incrementally evolve the next generation of the system in a real-time manner. In the case of the HDIO hierarchy shown in Figure 2, although there is no definition about goals, each intention is designed to couple with "Relevant Desire," which depicts a user's simulated desire to represent the desirable situation in the desire world.

3.2 Service Evolution in Ambient Intelligence Environment

Due to new situations, a user may develop new intentions to move forward towards his/her goals. For example, the user may have been using the medicine reminder system to handle his ill health situation. However, due to rapidly deteriorating health, the user may want to use a new call service where he/she can get immediate support. In such a case, the CA system needs to add this new path in the desire, belief, and intention worlds. The desirable situation is now achievable by this new way. The requirements for the system would need to be updated to include both "medicine reminder service" and the new "call service" (see Figure 7). In the figure, the business process is represented by using the Business Process Modeling Notation (BPMN) and shows the plan in the intention

Figure 7. Part of updated business process

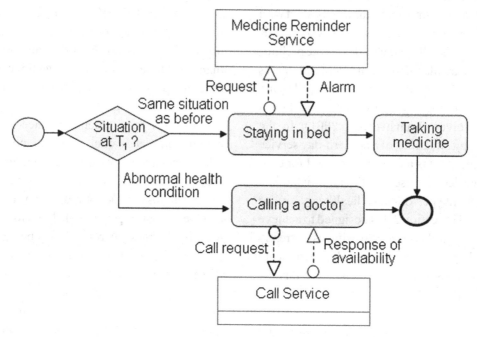

world. The plan is committed after receiving feedback from the user. In the future, when the system recognizes a similar situation based on facts memorized by the CA agent, it sends an enquiry to a doctor for immediate response.

Situ framework (Chang et al., 2009) is the recent approach that illustrates an architecture platform to conduct this service evolution scenario. This situation-theoretic framework defines a computational model of situation to monitor human intention and derive potential goals (i.e., service requirements) in accordance with the intention change at run-time. Observation of intention changes seems to be a critical process to shorten the development cycle of service, which is highly-adaptable to individual cases.

4 CONCLUSION

This chapter reviewed ontology approaches to infer human intentions in context awareness and introduced the HDIO that represents the infer-

ence tasks to bridge the computer process view and the human interaction view. Such inference ontology plays an important role in manipulating situations and human intentions associated with the corresponding contexts. The definitions of context, situation and human intention underlying pervasive technologies and inference mechanisms have been increasingly adopted in many research areas for developing evolvable services that adapt to changing human intentions. The ultimate goal of ontology approach in context awareness would be to build a comprehensive intention model that is applicable to broad areas of common interest, especially pervasive computing, requirements engineering, and run-time software evolution.

For further improving the evolvability of real-world systems, there are two complementary approaches that can be considered: short-term evolution (also called "accommodation") and long-term evolution. Short-term evolution is to handle exceptions and to make correct reactions at runtime; on the other hand, long-term evolution is to monitor a user's behavior and give feedback into new system requirements based on human intentions. Both approaches will improve the techniques for developing evolvable services in a complementary style, as in most cases user participation results in environmental changes and exceptions.

ACKNOWLEDGMENT

We are deeply grateful to Laurel Tweed in the Department of Computer Science, Iowa State University, for her detailed and constructive comments.

REFERENCES

Ackoff, R. L. (1989). From Data to Wisdom. *Journal of Applied Systems Analysis*, *16*, 3–9.

Barwise, J., & Perry, J. (1980). The situation underground. Working Papers in Semantics vol. 1, Stanford University Press.

Barwise, J., & Perry, J. (1983). *Situations and Attitudes. Bradford Books*. MIT Press.

Bellavista, P., Küpper, A., & Helal, S. (2008). Location-Based Services: Back to the Future. *IEEE Pervasive Computing / IEEE Computer Society [and] IEEE Communications Society*, *7*(2), 85–89. doi:10.1109/MPRV.2008.34

Bellinger, G., Castro, D., & Mills, A. (2004) Data, Information, Knowledge, & Wisdom. Retrieved November 10, 2009, from http://www.systems-thinking.org/ dikw/dikw.htm.

Bihler, P., Brunie, L., & Scuturici, V. M. (2005) Modeling User Intention in Pervasive Service Environments. IFIP International Conference on Embedded and Ubiquitous Computing, 977-986.

Bordini, R., Hübner, J., & Vieira, R. (2005) Jason and the Golden Fleece of agent-oriented programming.

Bratman, M. E. (1987). *Intentions, Plans, and Practical Reason*. Cambridge, MA: Harvard University Press.

Chang, C. K., Jiang, H., Ming, H., & Oyama, K. (2009). Situ: A Situation-Theoretic Approach to Context-Aware Service Evolution. *IEEE Transactions on Services Computing*, *2*(3), 261–275. doi:10.1109/TSC.2009.21

Chang, C. K., Oyama, K., Jaygarl, H., & Ming, H. (2008). On distributed run-time software evolution driven by stakeholders of smart home development. Second International Symposium on Universal Communication (ISUC'08), 59-66.

Cichoki, A., Washizawa, Y., Rutkowski, T., Bakardjian, H., Phan, A., & Choi, S. (2008). Noninvasive BCIs: Multiway Signal-Processing Array Decompositions. *Computer*, *41*(10), 34–42. doi:10.1109/MC.2008.431

Cohen, P. R., & Levesque, H. J. (1990). Intention is choice with commitment. *Artificial Intelligence*, *42*(2-3), 213–261. doi:10.1016/0004-3702(90)90055-5

Dennett, D. (1989). *The Intentional Stance*. MIT Press.

Dey, A. K. (2001). Understanding and Using Context. *Personal and Ubiquitous Computing*, *5*(1), 4–7. doi:10.1007/s007790170019

Gruber, T. R. (1993). A translation approach to portable ontologies. *Knowledge Acquisition*, *5*(2), 199–220. doi:10.1006/knac.1993.1008

Jaques, P. A., & Vicari, R. M. (2007). A BDI approach to infer student's emotions in an intelligent learning environment. *Computers & Education*, *49*(2), 360–384. doi:10.1016/j.compedu.2005.09.002

Kelley, R., Tavakkoli, A., King, C., Nicolescu, M., Nicolescu, M., & Bebis, G. (2008), Understanding human intentions via hidden markov models in autonomous mobile robots. Third ACM/IEEE international conference on Human robot interaction (HRI'08), 367-374.

Ko, E. J., Lee, H. J., & Lee, J. W. (2006). Ontology-Based Context Modeling and Reasoning for U-HealthCare. The Institute of Electronics [IEICE]. *Information and Communication Engineers*, *90-D*(8), 1262–1270.

Krummenacher, R., & Strang, T. (2007) Ontology-Based Context Modeling. Third Workshop on Context Awareness for Proactive Systems.

Lamsweerde, A. v. (2001). Goal-oriented Requirements Engineering: A Guided Tour. Fifth IEEE International Symposium on Requirements Engineering (RE'01), 249-262.

Lamsweerde, A. v., & Letier, E., From Object Orientation to Goal Orientation: A Paradigm Shift for Requirements Engineering. Radical Innovations of Software and Systems Engineering, 325-340.

Levesque, H., Pirri, F., & Reiter, R. (1998). Foundations for the situation calculus. Electronic Transactions on Artificial Intelligence, 159-178.

Mastrogiovanni, F., Sgorbissa, A., & Zaccaria, R. (2008). Understanding events relationally and temporally related: Context assessment strategies for a smart home. Second International Symposium on Universal Communication, 217-224.

Meneguzzi, F. R., & Luck, M. Composing high-level plans for declarative programming. Workshop on Declarative Agent Languages and Technologies, 69-85.

Ming, H., Oyama, K., & Chang, C. K. (2008). Human-Intention Driven Self Adaptive Software Evolvability in Distributed Service Environments. Twelfth IEEE International Workshop on Future Trends of Distributed Computing Systems (FT-DCS'08), 51-57.

Mizoguchi, R., Welkenhuysen, J. v., & Ikeda, M. (1995). Task ontology for reuse of problem-solving knowledge. Second International Conference on Knowledge Building and Knowledge Sharing (KB & KS'95), 46-57.

Móra, M., Lopes, J., Vicari, R., & Coelho, H. (1999). BDI models and systems: Bridging the gap, Workshop on Intelligent Agents V, Agent Theories, Architectures, and Languages (ATAL'98), London, UK, 11-27.

Oyama, K., Jaygarl, H., Xia, J., Chang, C. K., Takeuchi, A., & Fujimoto, H. (2008a). A Human-Machine Dimensional Inference Ontology that Weaves Human Intentions and Requirements of Context Awareness Systems. Computer Software and Applications Conference (COMPSAC'08), 287-294.

Oyama, K., Jaygarl, H., Xia, J., Chang, C. K., Takeuchi, A., & Fujimoto, H. (2008b). Requirements Analysis Using Feedback from Context Awareness Systems. Second IEEE International Workshop on Requirements Engineering For Services (REFS '08), 625-630.

Oyama, K., Takeuchi, A., & Fujimoto, H. (2006). CAPIS Model Based Software Design Method for Sharing Experts' Thought Processes. Computer Software and Applications Conference (COMPSAC'06), 307-316.

Rao, A. S., & Georgeff, M. P. (1991). Modeling rational agents within a BDI-architecture. Second International Conference on Principles of Knowledge Representation and Reasoning (KR'91), 473-484.

Rolland, C., Kaabi, R. S., & Kraïem, N. (2007). On ISOA: Intentional Services Oriented Architecture. *International Conference on Advanced information Systems Engineering (CAISE)*, 158-172.

Scheer, R. (2004). The 'Mental State' Theory of Intentions. *Philosophy (London, England)*, *79*(1), 121–131. doi:10.1017/S0031819104000087

Strang, T., & Linnhoff-Popien, C. (2004). A context modeling survey. *UbiComp: First International Workshop on Advanced Context Modelling, Reasoning and Management.*

Tomasello, M., Carpenter, M., Call, J., Behne, T., & Moll, H. (2005). Understanding and Sharing Intentions: The Origins of Cultural Cognition. *The Behavioral and Brain Sciences*, *28*(5), 675–691. doi:10.1017/S0140525X05000129

Wang, X., Zhang, D. Q., Gu, T., & Pung, H. K. (2007). Ontology Based Context Modeling and Reasoning using OWL. *Workshop on Context Modeling and Reasoning at IEEE Fifth International Conference on Pervasive Computing and Communications Workshops (PerComW'07)*, 14-19.

Yau, S. S., Gong, H., Huang, D., Gao, W., & Zhu, L. (2008). Specification, Decomposition and Agent Synthesis for Situation-Aware Service-Based Systems. *Journal of Systems and Software*, *81*(10), 1663–1680. doi:10.1016/j.jss.2008.02.035

Yau, S. S., & Liu, J. (2006). Hierarchical Situation Modeling and Reasoning for Pervasive Computing. *Third Workshop on Software Technologies for Future Embedded & Ubiquitous Systems (SEUS)*, 5-10.

Zhang, D., Yu, Z., & Chin, C. Y. (2004). Context-aware infrastructure for personalized HealthCare. *International Workshop on Personalized Health, IOS Press*, 154-163.

KEY TERMS AND DEFINITIONS

BDI Model: An agent planning model/architecture that mimics Belief-Desire-Intention of human mental state, originally described by Bratman in 1987.

DIKW: Abbreviated name of Data-Information-Knowledge-Wisdom that classifies a human thinking for knowledge management.

Evolution of Services: The development process that creates new generation of a context aware service. There is an emerging challenge in requirements engineering to analyze an individual user's behavior based on context information.

Human Intention: Temporal sequence of user actions in order that the system user can achieve a goal. It belongs to the decision-making process which is within the scope of human knowledge and wisdom.

Human-Dimensional Inference Ontology: A task ontology designed to infer human intentions based on DIKW hierarchy. This ontology models execution of step-by-step inference tasks in a CA system that proceeds in a hierarchical fashion from low level processes where device information is captured to higher level human-machine interaction processes

Inference Ontology: Task ontology in which a set of inference tasks are modeled. In this chapter, the inference tasks are referred to as the tasks for inferring human intentions from context information.

Intention Change: An uncertainty factor that drives system requirements to be obsolete or contrary to user needs. Human intention change, once monitored, facilitates capture of new requirements; these new requirements trigger the next cycle of system evolution for upgrading the system to a new version.

Chapter 20
Ambient Intelligence for Eldercare – the Nuts and Bolts:
Sensor Data Acquisition, Processing and Activity Recognition Under Resource Constraints

Jit Biswas
Institute for Infocomm Research, Singapore

Andrei Tolstikov
Institute for Infocomm Research, Singapore

Aung-Phyo-Wai Aung
Institute for Infocomm Research, Singapore

Victor Siang-Fook Foo
Institute for Infocomm Research, Singapore

Weimin Huang
Institute for Infocomm Research, Singapore

ABSTRACT

This chapter provides examples of sensor data acquisition, processing and activity recognition systems that are necessary for ambient intelligence specifically applied to home care for the elderly. We envision a future where software and algorithms will be tailored and personalized towards the recognition and assistance of Activities of Daily Living (ADLs) of the elderly. In order to meet the needs of the elderly living alone, researchers all around the world are looking to the field of Ambient Intelligence or AmI (see http://www.ambientintelligence.org).

DOI: 10.4018/978-1-61692-857-5.ch020

I. INTRODUCTION

With environmental sensors and wearable sensors gathering data continuously (Figure 1), and with software algorithms working behind the scene, making sense (intelligence) of the data, continuously processing the data, the awareness of the context and situation of elderly can be achieved. In this manner the elderly can be assisted at points of need and thereby helped to live independently at home.

Distributed sensors have long been used to sense the status of environmental parameters such as humidity, temperature and light intensity and aggregate / record these parameters from a distance. Sensors have also long been used for surveillance and military applications, for example in the identification and tracking of targets such as aircrafts and battle tanks. Environmental changes, or aircraft / objects in motion may be thought of as causal agents for phenomena changes in a so-called *phenomena aware system*. The activities and behavioral patterns of the elderly may also be regarded as phenomena in an ambient space which is aware of various phenomena unfolding in the space.

Section II discusses sensor data acquisition and the notion of micro-context as a paradigm for building systems that support diverse applications which build upon knowledge captured from ambient spaces. The notion of activities as phenomena is presented in section III. In section IV is presented one of the most important aspects of any sensor data acquisition system, namely, feature extraction. It is important to note that feature extraction and further operational processes that follow, such as classification and reasoning are essentially independent of sensing modality. We illustrate this with a variety of examples in this chapter, ranging from video camera based agitation detection to pressure sensor based sleeping posture detection (both in section IV), and other modalities such as ultrasound sensors and accelerometers in subsequent sections. Algorithmic techniques are based on micro-context and are essentially agnostic of sensing modality. With the help of an example, section V introduces one of the key challenges of ambient intelligence – the

Figure 1. Activities of elderly as ambient phenomena

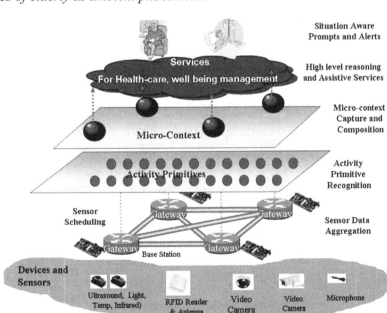

challenge of satisfying the needs of applications through *information quality*. Each of these sections discusses real-life application case studies either from deployment in hospitals or nursing homes or from advanced laboratory prototypes. In order to put these applications in perspective we also include in section VI, a discussion on resources in terms of costs of computation and communication and the inherent benefits of carrying out judicious sensor selection in the presence of resource constraints. The mathematical tool employed to carry out the sensor selection is a dynamic Bayesian network. A few remarks on related work especially pertaining to higher level artificial intelligence techniques are presented in Section VII. The chapter ends with a short summary and conclusions in section VIII.

II. SENSOR DATA ACQUISITION AND MICRO-CONTEXT

To build a platform upon which ambient intelligence applications can be built, some core technologies and capabilities are needed. First,

a wireless sensor network comprising a variety of ambient and wearable sensors must be incorporated into the smart home for the elderly. Data acquisition, filtering, segmentation and classification must be carried out at a low level, in order to extract features from sensor produced data into meaningful *micro-context* information (see definition below) which is stored in appropriate data archives. This data then becomes the basis for activity recognition, behavior understanding and learning. Algorithms must be flexibly mixed and matched, and therefore need to access micro-context stored in a generic format. Our approach relies on multi-modal information fusion and activity primitive recognition at the low level, and context aware reasoning at the high level (Figure 2). Central to our approach is the notion of micro-context which is defined as follows.

We define micro-context as low level information about a person, object or activity that has been established to be accurate at the acceptable level of uncertainty. A piece of information is deemed to be micro-context if it provides a reliable fragment of useful information that contributes to the corpus of knowledge in an ambient

Figure 2. Micro-context as a tool for multi-modal recognition

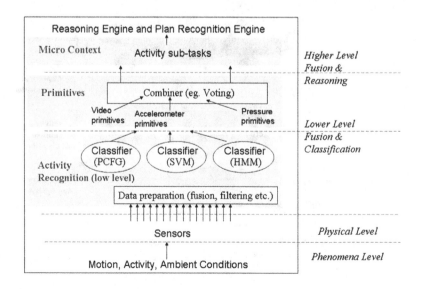

space. Since we are concerned only with activities of daily life of elderly, for our purposes an activity is defined as follows:

An activity is a stipulated and specified pattern that is associated with the performance of normal or routine functions in and around the home. Examples are eating, napping, watching television, ambulating (walking around), etc. An activity consists of a set of tasks, and each task can be inferred from logical conditions met by micro-context information in the ambient space.

Micro-context as a tool for multi-modal recognition. Sensor generated data may be used in its raw form or in various distilled forms. Once sensor data has been transformed into features the data is now information since semantics have been associated with it. In our work we have developed some generic ways of composing fragmented information gleaned through multi-modal sensors, into a form that might be usable by high level agents. Fundamentally there are two approaches for recognizing higher level states from sensor observed data, statistical approaches and value based approaches.

Statistical approaches. Statistical approaches aim to combine raw data from all sensors at their lowest level, using statistical techniques such as Hidden Markov Models (HMM) and other types of statistical models. These techniques may be employed for multiple sensors of the same modality (e.g. multiple video cameras), or for multiple sensors of multiple modalities. Although quite effective, we have found that the calibration of sensors and information overload are the two problems with this approach. Usually this approach is suitable for uni-modal processing.

Value based approaches. Value based approaches are hierarchical and compositional, and work with intermediate states obtained from different sensors. They are able to limit the information flow towards the sink (query source), and are thus suitable for large deployments of multi-modal sensors. They operate on the basis of thresholding and feature extraction. If the value of a sensor reading or an extracted feature has crossed a certain threshold, it may be assumed that a certain type of Activity Primitive has occurred. Based on the discrimination capability of a sensor, a set of features is defined for a particular modality and a particular algorithm.

Figure 2. shows how micro-context, or the aggregation of low uncertainty information is carried out. Sensor information from a particular modality is processed to obtain a set of features. The features are mapped into activity primitives through training algorithms, either on site or remotely. The set of activity primitives however is most often characterized by a high degree of uncertainty due to a variety of reasons such as insufficient data, noise, or unavailability of transmission capacity (Biswas, Naumann and Qiu, 2006). We therefore employ another level of processing (combiner stage), in order to reduce the level of uncertainty in the activity primitives. In section VI we show how in the presence of uncertainty from two modalities, namely accelerometer (with noisy information) and RFID (with insufficient information), we are able to classify eating activity primitives with a high degree of certainty. Combining the results of multiple sensed activity primitives is carried out in a combiner stage, the output of which is reliable (low uncertainty) information about the targeted conditions, activities or behaviors that are being analyzed.

Note that micro-context consists of partial states of the system which are in some sense intermediate states towards the making of a final inference. These partial states may not have significance in and of themselves, nevertheless, in our architecture they are published as micro-context information which exists independently. Note that micro-context is more than just a piece of information from a sensor. It is either partial context, or a portion of a movement or behavior which has been detected. There is a confidence value annotating this information.

III. ACTIVITIES AS PHENOMENA

We must remember that the elderly who are being monitored are usually not cooperative or unable to cooperate appropriately in order to input into a system helpful details about their current activities. This means that one needs to rely entirely on sensor data and traces of events that can be gleaned from such data, in order to reconstruct the history of events in the activities of daily life of the elderly. What are the phenomena of interest in eldercare? This stems from the specific activities of interest and the manifestation of these activities as phenomena. Since the only way we observe phenomena is through sensing them, we view phenomena of interest as manifestations of those activities we are interested in. For example there is ambient light in the environment at all times, however the phenomena of interest to us is the change of ambient lighting when a person enters or leaves the room (assuming that he is switching on or off the lights, as the case may be), since it triggers the important "Person in Room" or "Person outside Room" events. To begin with, let us list a few possible phenomena of interest:

- Whole body movements (ambulant / wheelchair based)
- Limb movements (hand / leg movements made in place)
- Vocalizations (speech such as in a telephone conversation or sounds as in crying or shouting)
- Sounds (familiar) made in carrying out activities (e.g. clinking sound made while stirring tea or coffee)
- Object manipulations that are directly linked to activities (e.g. coffee mug, vacuum cleaner)

It is desirable to have a comprehensive model of phenomena awareness that couples physical / natural phenomena and their observations to the computational infrastructure used to carry out the observation in terms of sensing, interpretation of sensed data, data networking and information extraction, fusion and ultimately understanding of processed data in terms of high-level human semantics. Figure 3 illustrates the overall relationship between ambient phenomena, sensor data acquisition, activity recognition and behavior understanding and eventually sensor selection. Activities of the elderly are a subset of what may be thought of as dynamic phenomena in an ambient space. When multiple sensors are used to monitor phenomena in an ambient space, one major concern is the effective use of the monitoring resources. For example if ultra-sound sensors have been used to monitor movements in a room in a non-intrusive and privacy preserving manner. However, this imposes a load on the networking and computational resources of the smart home environment. In order to address this concern, we consider the selection of sensors which monitor the state of phenomena, resource contention between different applications, resource allocation on individual sensors and their collective performance, and phenomena state distribution.

IV. FEATURE EXTRACTION FROM SENSOR DATA

This section presents a framework and techniques for recognition and automated rating of complex agitation behavior for patients with dementia using video camera. In particular, we present an approach using hierarchical descriptors for video feature extraction for understanding, modeling and recognizing complex agitation behavior in patients.

The application considered is that of monitoring agitation in persons with dementia. Agitation is a common difficult behavioral symptom of dementia that can distress and disable patients and caregivers, prompt institutionalization, and lead to physical and pharmacologic restraints. Agitation occurs at some time in about half of

Figure 3. Sensor data acquisition and information processing for Activity Recognition and Behavior Understanding

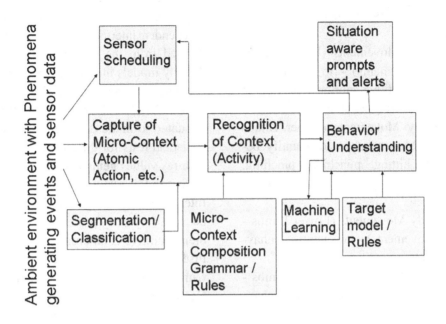

all patients with dementia. Associated behaviors include aggression, combativeness, disinhibition, and hyperactivity. Agitation is a manifestation of a lack of psychological well-being of the subject; indeed measuring agitation exhibited by a dementia patient is an effective proxy to measuring his/her psychological well-being. Once agitation is detected through formal quantification by means of a scale called the SOAPD scale (Hurley *et al.*, 1999), assessment for common systemic causes such as infection, dehydration, constipation or other illnesses, as well as changes in medication could follow.

The Scale for the Observation of Agitation in persons with Dementia (SOAPD) (Hurley *et al.*, 1999) is a unique observational tool that measures agitation by direct patient observation. It is patient focused, conceptually distinct and psychometrically adequate. SOAPD relies on a construct of body motions and vocalizations. The degree of agitation is evaluated by rating the duration of different types of whole and partial body movements and vocalizations; a total of 11 different observation items are to be observed simultaneously and rated every 5 minutes. Obviously this is an ideal candidate for automation, if this can be achieved.

In our work we have investigated the effectiveness of non-intrusive and automated surveillance with video camera, for some aspects of the SOAPD scale. By means of these approaches, not only will technology replace the intensive manpower needed in the use of this tool, it will also extend the capabilities of monitoring for agitation in dementia to the patients' own homes. The remainder of this section presents the highlights of our algorithms that carry out the necessary feature extraction for automated SOAPD observation.

The SOAPD scale characterizes several behavioral traits, and not all of them are amenable to analysis by video. The following SOAPD behavioral traits can be detected by video:

- Whole Body Movements (Not in Place) - to infer behaviors of disturbed pacing
- Whole Body Movements (In Place) - to quantify behaviors of fidgeting, restlessness etc.
- Partial Body Movements (Upper Limbs - Inward Movements) - to quantify behaviors of repetitive motions, picking at clothes, hand wringing etc.
- Partial Body Movements (Upper Limbs - Outward Movements) - to quantify behaviors of hitting, punching, pinching, throwing, grabbing, banging, scratching, squeezing etc.
- Partial Body Movements (Lower Limbs - Inward Movements) - to quantify behaviors of feet tapping
- Partial Body Movements (Lower Limbs - Outward Movements) - to quantify behaviors of kicking

The first challenge is to translate the behavioral indicators into computational descriptors that model the above semantic traits effectively. A standard research methodology consists of mounting a video camera on the ceiling of the observation chamber of ward in the hospital, to capture and measure behavioral parameters of the patient being monitored. This step is followed by obtaining the segmented map of the tracked subject and use human body models to register specific body parts of interest, after which these models will be used for the actual quantification process.

Similar techniques are used in our work but we use a multi-tiered approach. Besides using standard machine learning approaches such as the use of Hidden Markov Models, we adopt a hierarchical representation of features extracted from video processing that is useful for translating the behavioral indicators into computational descriptors. *Primitive feature descriptors* at the lowest level are extracted directly from the temporal segmentation maps of the patient, based on centroid, orientation angle, etc. These primitive

descriptors contain important primitive information for us to derive *higher level feature descriptors* through *fusion* and *reasoning*. It is a very important step for agitation modeling to design independent intermediate-level feature descriptors derived from the primitive descriptors, which directly models individual traits of the agitation behavior. The relevance and usefulness of each intermediate-level feature descriptor will then be validated to make it more exact by conducting statistical tests for differences in mean between features values generated by samples of the behavior and of a closely similar "null behavior". Intermediate-level feature descriptors are used because they carry more semantic meaning than their primitive counterparts. The next stage is to fuse the intermediate-level feature descriptors within a coherent framework that facilitates effective and accurate recognition of the behavior of interest. These can be achieved through many fusion techniques such as Bayesian network, neural network, weighted voting, etc.

Agitation Behaviour Feature Descriptors. Measuring the movement in agitation behaviors is one of the key ways to obtain primitive and intermediate feature descriptors. It can provide valuable information for quantifying both whole body movements and partial body movements in different dimensions. Typically, to observe motions of the object, there are three different techniques for different purposes:

- Motion detection to identify whether the objects are moving or not.
- Motion estimation to identify how the objects are moving.
- Motion segmentation to identify moving objects from other moving points.

We first select motion detection, because we would like to detect whole body movements and partial body movements. Among different motion detection methods, frame subtraction (Gonzales and Woods, 1992) to observe whole body movements and partial body movements involves less computation for real time processing, and the result is comparable to that using other motion detection methods for stable indoor environments. For complex scenes, Bayesian background modeling (Liyuan *et al.*, 2004) can be applied. Though its computational cost is higher it gives better segmentation result. Frame subtraction can detect changes caused by moving objects between two successive frames in the sequence. In frame subtraction, one can use either the *background subtraction* or *temporal differencing* technique. Background subtraction is computing the difference between interest frame image and background image and is commonly used in video surveillance (Zhuang, 2004) as it is very useful to detect continuous motion. Temporal differencing (Lee, 2000) is computing the difference between two consecutive frame images and is useful to detect slow motion of object such as the agitation behaviors in our case. Although there are many variants on the temporal differencing method, the simplest way is to use temporal frame differencing. We will describe the procedure which consists of two stages - preprocessing and thresholding.

Preprocessing. In two consecutive frames, $F_t(x,y)$ and $F_{t+1}(x,y)$ as shown in equation 1, the difference value *DF(x,y)* between the corresponding pixel values is defined as the pixel's changes, which can be caused by the object movement from a position in the current frame to another position in the next frame.

$$DF(x,y) = \left| F_{t+1}(x,y) - F_t(x,y) \right| \qquad (1)$$

Besides the motion of objects, camera channel noise and noises brought by flickering of light could also cause the change of the pixel value at a particular location from frame to frame. However, within a short period, the pixel value's change brought by motion in agitation behavior usually is larger than those brought by noise. To reduce noise, we perform some preprocessing such as filtering in spatial and temporal domains. Although using a simple large threshold can remove most of the noise, we cannot use it because the some movement of the elderly patient may be small. In our case, we use median filter which is one of nonlinear filters to remove noise as it provides excellent noise-reduction capability. Before filtering, RGB color images are first converted to 256 gray level images. When the body is moving, the pixel value changes from background to object or from object to backgrounds. By subtracting two consecutive frames, the difference of the pixel values is obtained and that differential is considered as the body's movement. To prevent negative value after undergoing subtraction, the difference value is converted into absolute value and stored, and the value may be considered as primitive descriptor named *frame primitive descriptor*.

Thresholding. A threshold function is then used to determine that the output image from subtraction contains significant motion or not. A threshold function is defined by the number of changing value pixels. Unless the body is moving, the number of changing value pixel is not obviously differing. If the number of changing value pixels is more than threshold value, there is some motion between two consecutive frames. Threshold value determines the sensitivity which refers to how small a motion. By adjusting the threshold value, we can change the sensitivity, and obtain intermediate descriptors such as the movement index.

DF(x,y) > THRESHOLD → MOVING

DF(x,y) < THRESHOLD → STATIONARY

For constrained environments such as detection of agitation of patient on the bed, we use a simple approach by dividing the bed into six regions which consists the head region, body region and leg region as know the distribution of body parts. Each region has left-right side, for detection of partial body movement. Based on the number of changing value pixels and the region where the change has occurred, we can determine which part of the body has moved, and level of the movement. For each current frame, we find the number of changing value pixels corresponding to each region, and for each region, we set a threshold that determines the level of movement: slight, medium or heavy. According to this data we expect to see how the patient moves as well as the intensity of the movement. Figure 4 shows the GUI for finding the frame subtraction technique used for movement detection.

Through the intermediate feature descriptors, and combined with the prior-knowledge of the constraint environment such as the bed, appropriate algorithms have been developed to infer partial body movements such as the patient's head turning left-right, body turning left-right and leg turning left-right etc. Though frame subtraction is one good way to obtain intermediate descriptors, it will work well generally in constrained environments or with a closed world assumption. However, for unconstrained environments, typically the segmented map of the tracked subject is

first obtained and human body models are used to register specific body parts of interest, after which these models are used for the actual quantification process to produce the intermediate descriptors as described before. In the next section, we present the technique of obtaining primitive and intermediate feature descriptors using skin detection techniques for semi or unconstrained environments.

Skin Color Segmentation. Skin color segmentation is another way of obtaining feature descriptors. For detecting movement in the human body, many primitive descriptors such as color, shading, edges, texture, motion and depth may be obtained. In our case, we use skin color segmentation to evaluate body actions of the extremities that are directed outwardly. The main objective in skin color segmentation is to determine the position of exposed human skin areas which are usually the areas of face, hands and feet. Different color spaces have been used in skin color segmentation including RGB, normalized RGB, HSV/HSI, YCrCb, YUV, etc. In some cases, color classification is done using only pixel chrominance

Figure 4. GUI to show the level of the patient movement

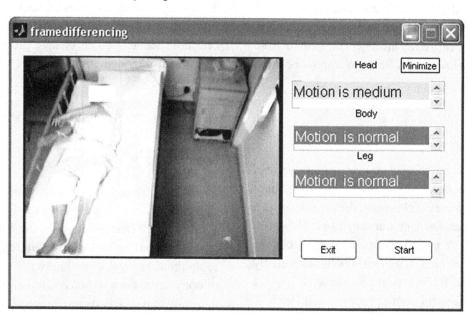

because it is expected that skin segmentation may become more robust to lighting variations if pixel luminance is discarded. The color representation used in this process is HSI.

Several algorithms have also been proposed for skin color segmentation. They are the multilayer perception (Phung, Chai and Bouzerdoum, 2001), the Bayesian classifier with the histogram technique (Jones and Rehg, 2002; Wang and Chang, 1997), Gaussian classifiers (Menser and Wien, 2000; Yang and Ahuja, 1999), and piecewise linear classifiers (Chai and Ngan, 1999; Garcia and Tziritas, 1999; Sobottaka and Pitas, 1998). In our work, we used the Bayesian classifier with the histogram technique. First, we use several images including skin and non-skin regions in different lighting conditions to build the offline histograms for the skin and non-skin distributions. In order to obtain skin and non-skin regions from images, a seed-pixel that has similar skin color to that of the subject has been used. In our case, skin pixels were obtained by manually labeling and masking the skin regions in the set of images. From each labeled image, we achieve two masks, one binary image mask for skin regions and one for non-skin region.

Skin Color Selection tools. To train the system with the skin and non-skin areas, we need to have a set of pictures in which the skin and background are separated. We limit the behavior

to an indoor environment where the ambience does not vary too much and collect the skin and non-skin picture segments from each frame captured by the video camera. To separate the skin from the non-skin area, we develop a simple tool to semi-automatically help to reduce the time required for masking the skin color when we have large numbers of frames upon which the masking of skin needs to be performed. The input to this tool is a frame and outputs are two masks: one for the skin area and the other for non-skin area as shown in Figure 5.

The user selects a pixel as a seed in the skin area, and the tool then automatically finds the similar pixels around this seed recursively. We still need to apply some post-processing techniques to the result as morphological functions in order to obtain the final result.

As shown in Figure 5, from the original frame, the abovementioned tool masks the skin area by black pixels. All the remaining pixels which are in white, are non-skin pixels. From this mask, the non-skin mask is also created by making the skin area a little bigger. This guarantees that no skin pixel will be considered as a non-skin pixel. For a particular frame, two masks are required, namely the skin mask and the non-skin mask. For a particular patient, we found that 20 frames at different positions are sufficient for skin detection.

Figure 5. Sample images for labeling skin and non-skin regions

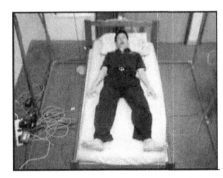

(a) Original image   (b) Skin Area   (c)Non Skin(reversed)

Skin Color Modeling and Detection. For skin color distribution and non-skin color distribution, the HSI histogram models with 32 bins per channel were computed. We then calculated the discrete probability distribution for determining which pixels belong to skin and non-skin regions, by using a 32x32x32 bin resolution histogram. Conditional probability densities were directly computed by dividing the count of pixels in each histogram bin by the total number of pixels in the histogram. The conditional densities may be represented by *P(hsi/fg)* and *P(hsi/bg)*, where *fg* refers to skin pixels, *bg* refers to non-skin pixels.

We classify a pixel as a non-skin pixel or a skin pixel by using a Bayes classifier. Using Bayes' formula, we compute *P(fg/hsi)*, the probability of observing skin and *P(bg/hsi)*, the probability of observing non-skin. The Bayes rule is as given in Equation 2.

$$P(fg/hsi) = \frac{P(hsi/fg)P(fg)}{P(hsi/fg)P(fg) + P(hsi/bg)P(bg)}$$
(2)

Using Equation 2, the decision boundary may be described where the ratio of *P(fg/hsi)* and

P(bg/hsi) exceeds threshold *K* that is based on a risk factor associated with misclassification. For example,

$$\frac{P(fg/hsi)}{P(bg/hsi)} = \frac{P(hsi/fg)P(fg)}{P(hsi/bg)P(bg)} > K$$
(3)

After some manipulations, equation 3 may be rewritten as equation 4.

$$\frac{P(hsi/fg)}{P(hsi/bg)} > K \times \frac{1 - P(fg)}{P(fg)}$$
(4)

Where *P(fg)* is the probability of an arbitrary pixel in an image being skin. Implementation of histogram for training data is shown in Figure 6.

In our work, noise in the result classified as skin region is inevitable. To remove the noise, we use morphological filters such as opening and closing filters and techniques such as the connected pixel method. In our test example (see Figure 5), we expect to see five skin regions in one frame. The number of skin regions in one frame may change because of patient's position and lighting conditions. Through skin segmenta-

Figure 6. The Flow Chart of Offline Training Data

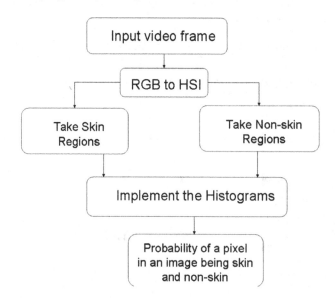

tion, we can obtain other features such as region of interest of the patient and primitive feature descriptors such as the centroid, x and y coordinates, shape, etc.

Complex Automated agitation behaviour Recognition. For accurate agitation behavior recognition, we adopt different strategies for obtaining and fusing the primitive and intermediate feature descriptors from frame subtraction and skin segmentation techniques. Techniques such as statistical, value-based, Bayesian, weighted-voting, etc are used to infer the high level behavior.

Statistical Approach. Statistical pattern classification approach is an intuitive classical way to recognize agitation actions. It can lead to higher recognition rates for those agitation behaviors associated with feature descriptors for which the data is highly spread in its range of values, as compared to the so called "value-based" approach. Agitated behavior such as kicking, hitting etc., can be represented in terms of multiple feature descriptors. Each feature descriptor is either an original observation, i.e. the primitive one, or a transformed or combined observation which is an intermediate descriptor. Intuitively, each agitation action is different from others due to some unique body physical movements associated with that action. For example, kicking action is differentiated from others by a sequence of continuous movements of one or two legs. Therefore, for good discrimination ability, we should try to choose feature descriptors that can describe body movements uniquely associated with each action. For example, the distance variation between two feet allows us to uniquely identify "kicking". The objective of the statistical classifier is to establish decision boundaries in the feature space that separates different agitation actions. Such decision boundaries are determined by the probability distribution of patterns associated with each agitation action, which is learned from training data. Our agitation behavior classifiers consist of two hierarchical layers, Hidden-Markov-Models (HMMs) layer and Support Vector Machine (SVM) layer,

i.e., during the real-time behavior monitoring, all the HMMs are operated in parallel, and the state estimation with maximal likelihood is chosen as inputs to a Support Vector Machine (SVM) built for a larger time scale for the final state estimation.

Value-based Approach. Value-based approach is another simple effective way to detect agitation behaviors for cases in which the behaviors have feature descriptors of data that are of less widely spread value range and have clear patterns of peak or trough with limited noises. It is usually combined with algorithms such as peak detection, averaging, etc. using some rules to infer the agitation behavior, though it is not a general approach and is closely correlated to the particular set of agitation behaviors being identified. We will illustrate two simple value-based approaches using the feature descriptors obtained from frame subtraction and skin segmentation.

From frame subtraction, we obtain the primitive descriptor which is the number of different pixels for detecting agitation behavior such as patient turning left and right. First we track his bed movement on the bed. We then classify the patient position on bed into 3 states - turning left, turning right and idle. Transition among these states can be represented as shown in Figure 7.

Using simple rules we can define when the system should change the states, from the feature descriptor which is the number of different pixels from frame subtraction algorithm. If we call the number of different pixels on the left side is N_{Left}, on the right side is N_{Right} then the four transition rules above can be expresses as shown in Box 1.

Though this approach is simple, it is found to be effective in detecting patient turning left and right movement. In another approach to detect agitation feature descriptors from skin segmentation, it was observed that the occurrence of any agitation behavior such as kicking changes the value of the centroids of the ellipse blobs, called as x and y feature descriptors and produces a peak in the signal as shown in Figure 8. Therefore a peak detection algorithm similar to work done by

Figure 7. Turning Movement Classification

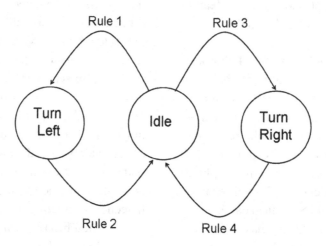

Box 1.

Rule1:	Current State =	IDLE	^ N_{Left}	>	THRESHOLD
Rule2:	Current State =	TURN_LEFT	^ N_{Left}	<	THRESHOLD
Rule3:	Current State =	IDLE	^ N_{Right}	>	THRESHOLD
Rule4:	Current State =	TURN_RIGHT	^ N_{Right}	<	THRESHOLD

Tay (see http://silvertone.princeton.edu) at Princeton University can be implemented that locates possible positive or negative peaks, and controlled by a threshold value.

The threshold value dictates the degree of "peakiness" that is allowed for a local maximum to be considered a possible peak. Based on peak detection algorithm of the peaks of x and y along the rows and columns, we are able to classify the kicking behaviors of the patients. Compared to the statistical approach, both can give comparable reasons but this algorithm is much simpler.

We now briefly describe the inference of agitation behavior based on intermediate feature descriptors using inference combiner such as Bayesian, SVM, weighted voting, etc.

Bayesian Networks. In many situations, the knowledge on the relationship among different feature descriptors in describing patients' agitation behavior is lacking; or the knowledge we feel interested may not directly expressed by the classification results. Bayesian networks entitle us another powerful alternative to obtain the required information from the low level behavior classification results. In our approach, we initially obtain a set of sample feature descriptor data from experiments, and use this set of sample feature descriptor data to train the Bayesian networks. During the real monitoring, such trained Bayesian networks are used to combine the outputs from the HMMs layer.

Support Vector Machine. Support Vector Machine (SVM) is a supervised binary classification algorithm. Support vector machine projects data points into a higher dimensional space, specified by a kernel function, and computes a maximum-margin hyper-plane decision surface that separates the two classes. In the agitation behavior classification for feature descriptors, we adopted the *LIBSVM* implementation, in which

Figure 8. Peak Analysis

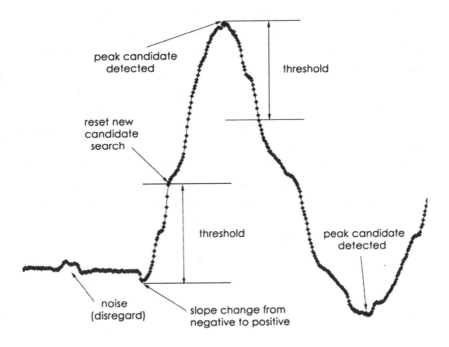

k-class classification are supported by performing C_2^k pair-wise 2-class classification. We choose radial basis function (RBF) kernel:

$$K(x_i, x_j) = \exp(-\gamma |x_i - x_j|^2), \gamma > 0$$

By combining the primitive and intermediate descriptors from HMMs layer, we are able to obtain two major improvements. Firstly, compared with the approach to merely selecting the HMM output with maximal likelihood as the final classification result, SVM combiner can remove some consistent false alarms from certain sensor modality during a particular action period, e.g., the false alarm of hitting action from video modality while actual rolling. Secondly, SVM combiner can further classify those relatively complex actions that can only be recognized with the collaborated observations from multiple intermediate feature descriptors.

Weighted Voting. Weighted voting is a type of system in which some members' votes carry more weight than others. It is very useful to fuse many intermediate feature descriptors from video

camera to inference the agitation behavior as it can reduce the false alarm rate.

We are currently also exploring techniques such as neural networks for agitation behavior recognition. Some approaches have more advantages than the others and it really depends on the specific agitation behaviors.

Frame Subtraction. As explained before, we can use frame subtraction to extract pixel difference index descriptors to detect patient turning left and right. To illustrate, we use one of the agitation behavior of patient as an example. Figure 9 illustrates the detection of patient turning left and right by using pixel index descriptors obtained from frame subtraction technique.

Using the value-based approach with simple rules as explained before, we achieve recognition rate of more than 90%.

Skin Segmentation Results. We use skin segmentation to extect feature descriptors such as x, y cooordinates, shape, etc. Then we perform detection of kicking. Figure 10 illustrates the exposed skin color of the patient that are represented by an ellipse which we record the centroid coordination, in the format of (x,y). There are two different states: the first frame is idle state, and the second frame is kicking state.

When the patient is kicking, we can see that the x coordinate is almost unchanged, while the y coordinate varies a lot as shown in Figure 11.

Based on the y coordinate feature descriptor, we use peak detection algorithm to detect kicking as shown in Figure 12. Each peak detected is considered as one time kicking. In order to achieve processing efficiency and eliminate noisy and false ellipses information, we apply a simple moving average filter to eclipses' coordinate values before feeding into the peak detection algorithm. The filtering of coordinates eliminates the unnecessary processing of small variation of peak changes. In order to eliminate the false detection of kicking on wrong ellipse coordinates,

the algorithm rejects the ellipse information when the number of ellipse readings is less than five.

The main drawback of this approach is that it is highly dependent on the accuracy of skin segmentation algorithm to properly recognize or track the position of one ellipse information from underlying video data acquisition system. The approximate nominal coordinates of the ellipses are also based on different specified regions and there can be misinterpretation of ellipse and overlapping of ellipse. However, with this approach, the kicking behavior is detected at a high recognition rate of above 90%.

Now, we will compare with the statistical approach. Based on the center coordinates of the five ellipses generated from skin segmentation, kicking action is described in terms of variation of the following 5 scalars between adjacent video frames, <two feet position, distance between feet, distance between two feet and head>. We then

Figure 9. Frame Subtraction to detect turning left/right

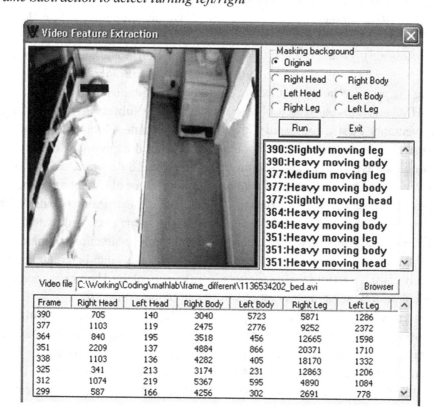

Figure 10. Skin Segmentation to detect kicking and banging

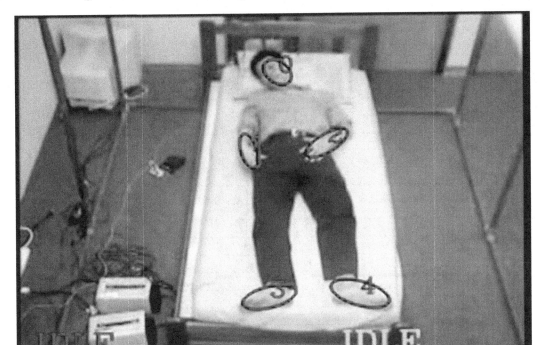

adopt a HMM model to recognize the kicking action based on those chosen features. In this HMM model, we employ mixture multivariate Gaussian distribution for the emission probability, and estimate all the model parameters by the EM algorithm from training data set. The recognition rate for kicking action is above 90%.

Next we will demonstrate the use of inference combiner to reduce the false alarm. By making the results of both the value-based approach and statistical approach as intermediate descriptors, we use weighted voting to infer the agitation behavior. If we know that the value-based approach is giving better recognition rate compared to statistical, we

Figure 11. Idle state and kicking state

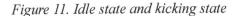

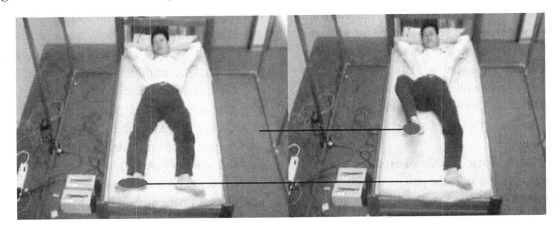

Figure 12. Ellipse Positions and Kicking Detection

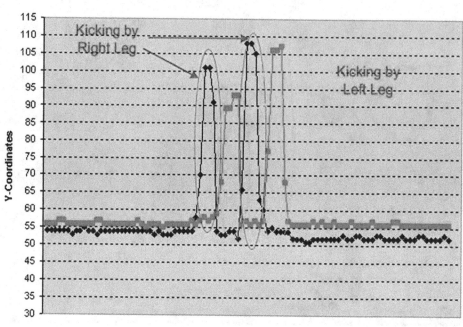

can give more weights to the feature descriptors given by the value-based approach. We found that the use of the approach can greatly reduce the false alarm rate especially in unforeseen circumstances which is very crucial for healthcare applications.

Machine learning techniques may be employed to further enhance the recognition of agitation behavior. This is a huge challenge to do it in an unconstrained environment when dealing with complex human beings' behavior. Hence, while video processing techniques may be good for pattern recognition, using ambient sensors such as pressure, ultrasound, acoustic, etc. will greatly help in the behavior recognition process. In our work many experiments were conducted to simulate patient's behavior and the results were encouraging.

Related work in video-based systems for human behaviour. In the past many video-based systems (Ribeiro and Santos-Victor, 2005; Hong-

seng, Nevatia and Bremond, 2004; Hu, Huang and Surendra, 2006; Liyuan *et al.*, 2004; Ayers *et al.*, 2001) have been developed to support automated recognition of human behavior for surveillance purposes. There are also video systems that specifically address support for monitoring of people and elderly for healthcare purposes (Liu *et al.*, 2007). The CareMedia project (Matsuoka, 2004) by Carnegie Mellon University and University of Pittsburgh used video and audio information to automatically track persons and characterize selected actions and activities. The University of Toronto (see http://www.ot.utoronto.ca/iatsl/index.htm) built an automated computer vision tool for detecting and preventing unsafe use of staircase by older adults. The University of Washington used video data to identify anomalies and actions (see http://grail.cs.washington.edu/projects/). While these works have greatly contributed to the research in human behavior recognition, our work is different from them in that we perform rating of the agitation level. The level of detail required in carrying out agitation detection poses several challenges for the pattern recognition from video data, and is still a very challenging unsolved problem. The use of hierarchical descriptors has been proposed in order to preserve semantics of information in recorded video. This makes it ready for mass customization through personalization, and easy extensibility to meet the new requirements when the healthcare needs change. The approach presented herein, especially the use of micro-context makes it possible to easily integrate the proposed system into more sophisticated behavioral understanding systems using context aware sensing and actuation.

Although feature extraction varies from sensor to sensor and from situation to situation, in this section, we have presented the challenges of automated agitation behavior recognition using video camera. We presented a framework consisting of hierarchical representation of feature descriptors and machine learning techniques to inference agitation behavior. The above approach may be extended to include more sensor modalities for agitation recognition as in many instances using video processing may not be sufficient for successful recognition of complex agitation behavior, or other types of activities and behavior in unconstrained environments and for personalization needed for mass deployment.

As an illustration of the above, we demonstrate how these features extraction methodologies can be successfully applied into sleeping posture and activity recognition of a patient on his bed, with a matrix of pressure sensors under the mattress (called a *pressure sensing map*).

In this paragraph and the next, we discuss details about the framework and techniques for sleeping posture recognition using pressure sensing map. As a first step, the interface pressure distribution acquired from pressure sensors is used to extract features and classify the sleeping posture of the subject. Usually, the pressure sens-

ing map consists of multiple pressure sensors and the direct use of raw pressure measurements into posture classification may not be feasible under resource constrained situations. Moreover, the erroneous readings due to particular sensors may bias the data processing causing incorrect posture classifications. In order alleviate the problems of handling high volume of sensors data and measurements from erroneous sensors, the following 3-level posture recognition framework as shown in Figure 13, is proposed to classify a particular sleeping posture of the subject.

As a first step, the salient and relevant features are extracted from raw pressure readings with respect to particular sleeping postures using Principal Component Analysis (PCA) (Lay, 2005) and descriptive statistic based approaches. PCA transforms the complex high-dimensional data sets into low-dimensional data sets revealing hidden and simplified underlying structures that cannot be observed directly. The spatial features for particular postures can be identified by finding orthogonal linear combinations of principal components (PC)s extracted from original pressure measurements. Similarly, descriptive statistics provide basic features of sleeping posture through statistical summaries based on data dispersion and central tendency of pressure distributions. In order to compute features based on descriptive statistics, the pressure sensing map is divided into a certain number of spatial regions or areas, based on rows and columns of spatially distributed pressure sensors as they appear in the pressure sensing map conFigureuration. Because of spatial variability of different sleeping postures, the applied pressure intensity distributions over mentioned areas, rows and columns are statistically varied. By exploiting these features from pressure readings over different areas, rows and columns of bed, the sleeping posture can be classified. After extracting different pressure features, different types of semi-supervised learning classifiers such as SVM, Bayesian networks, etc are applied to classify the sleeping postures independently. Fi-

Figure 13. Sleeping Posture Recognition using pressure sensing map

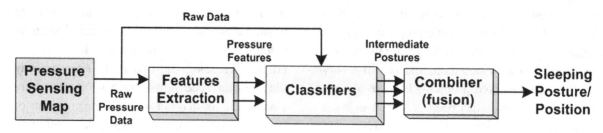

nally, the weight voting fusion combines the intermediate posture outputs from different classifiers in order to eliminate the uncertainty, erroneous or bias features and misclassification rates, resulting from multiple pressure sensors, through feature extraction and classification stages.

From this example, we can conclude that features extraction methodologies used for one sensor modality are easily adapted into processing of measurements from other modalities as well as on different types of applications.

V. DATA FUSION INFORMATION QUALITY

Whilst the previous section characterized the problem of feature extraction for complex actions in place, this section examines tracking of a users movements in space. Several tracking technologies and algorithms are in use. These rely on a variety of sensors ranging from UWB, video cameras, RF signal strength detection, etc. However, not many approaches make use of the notion of information quality to characterize the tradeoff between sensor data acquisition and application need. Thus in this section we study information quality in the context of data fusion.

We develop a system using ultrasonic sensors for trajectory monitoring. We present a methodology to select sensors based on quality of individual sensor. We have devised a trajectory-matching algorithm to classify trajectories. The trajectories

are divided into several routine classes and the current trajectory is compared against the known routine trajectories. In our fusion algorithm, we consider the spatial loss a factor to the application's performance. Our experiment studies the relevant of speed and spatial loss versus the matching accuracy. The initial result shows the potential usage of sonar sensor in monitoring indoor movement of people, in capturing and in classifying trajectories.

Data fusion is the process of combining the data from multiple sources of information to produce a quality data which is useful in making the decisions. There are several techniques for tracking and monitoring of a target (Thang *et al.*, 2006). The ultrasonic sensor used in our experiments gives us only the distance to the target. Two sensors are enough to locate the position of the target given that target is within the range of both the sensors. As we have more than two sensors deployed in the field we need to select the best quality two sensors for data fusion and to monitor the target accurately. Selection of suitable sensors to monitor the moving object and to put unused sensors in the sleep mode to save the energy is discussed later in this section. In elderly tracking application data fusion plays the role of combining the raw data from two sensors and computes the resultant velocity vector. The algorithm we propose is based on simple principle called trilateration. This principle is used to locate the objects on earth by GPS satellites. As we consider 2-D plane we need only 2 sensors.

Figure 14 shows sensors S_1 and S_2 with footprints FT1 and FT2. The movements of the subject in different regions is shown by the letters denoted as ``A'' to ``I''. Let us first consider the areas covered only by a single sensor. Movements A, B, I and J are detected by only sensor S_2, whereas movements D, E, F and G are detected by only sensor S_1. In these areas the readings may be incorrect because the sensors are capable of detecting only movement in the longitudinal direction. Lateral movements (i.e. across the footprint) are not detected. For instance A and I are correctly detected by S_2 but B and H are not detected by S_2. Likewise, G and E are correctly detected by S_1 but F and D are undetected by S_1. Movements C, J, and K are covered by both the sensors and give us a x-component of velocity, v_x and a y-component of velocity, v_y. From these we can calculate the magnitude, or the scalar component (v_{res}^{scalar}) and the angular component (v_{res}^{θ}) of the resultant velocity using the formulae:

$$v_{res}^{scalar} = \sqrt{v_x^2 + v_y^2}$$
$$v_{res}^{\theta} = \tan^{-1}(v_x / v_y)$$

Thus the region covered by both the sensors is called the Critical Coverage Area (CCA). The placement of sensors should be chosen in such a way that the area of movement falls as far as possible within the CCA of S_1 and S_2. This means that the sensors should be placed in such a way that the CCA covers most of elderly person's activities.

Data Fusion- Spatial Loss Compensation. All the above discussion in this section assumes that two sensors detect the same point on the target. However, in most of the cases this is not true. The error cause because of this assumption is called spatial loss. In this section we present a method to compensate for spatial loss. For example, in elderly tracking application when a person is walking each sensor may detect different parts of the body. Consider Figure 15. If O is the point detected by the data fusion algorithm. Then the target can be anywhere in the area made by three points O, A, and B. The simplest way to estimate the location of the target is to compute the center point of the triangle OAB. The steps to compute the center point is 1) determine the two points at

Figure 14. An example of quality of data fusion

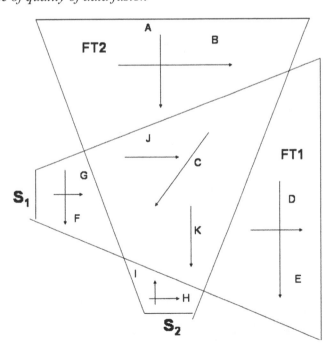

the boundary of one sensor, 2) select the point is which is further from the other sensor, and 3) compute the center point of the triangle *OAB*.

To evaluate the performance of the spatial loss compensation method explained above, a small program was written in Visual Basic. This program communicates with the sensors through the API provided by the manufacturer. The motion detectors are connected to a PC through LabPro. The LabPro is inbuilt with microprocessor which can communicate with PC via RS-232 or COM port and is dedicated for data collection and control of output from sensors. To verify the spatial loss compensation method, two sensors were placed in a room and a person is allowed to stand in the range of two sensors. We collected the data at various distances. The results for these experiments are reported in (Thang *et al.*, 2006). From the results we observe that spatial loss compensation improved the accuracy of the data fusion. The improvement depends on the distance between the two sensors and distance of the target from the sensors. The reason being, when the two sensors are placed near they seem to detect the close points in the body.

Quality-based Sensor Selection. For the purposes of data fusion, we need to select two best quality sensors depending on the target location. In Algorithm 1 we present an algorithm for sensor selection based on the sensor foot-print quality profile.

Elderly Monitoring Application. In this section we explain our elderly monitoring and trajectory identification using ultrasonic sensors. We explain how we represent a trajectory and then a method to classify the trajectories. After that we go ahead and present a method to identify abnormal trajectories.

Trajectory Representation. We divide the living place to be monitor into smaller rectangular regions as shown in Figure 16, the rectangular regions (boxes) are numbered from left to right and bottom to top as shown. With this division living place looks like a grid and a trajectory can be represented as a sequence of boxes. For example Figure 16 shows a trajectory from living room to bath room with the sequence as 4→13→20→28→…68. It is worthwhile to note that, if we represent the trajectory as a sequence of points from the data fusion, we have more

Figure 15. An example to illustrate the spatial loss compensation

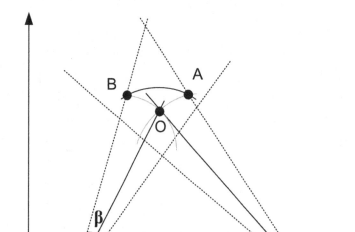

Algorithm 1.

```
Procedure Sensor_Selection
Input: stream of sensor readings, one from each of an array of sensors S₁, S₂,
... Sₙ
Output: stream of fused distance values, d₁, d₂, ... dₖ, where dᵢ is the resul-
tant distance value produced by the fusion of the iᵗʰ entry from each input
stream, i.e. Sⁱ₁, Sⁱ₂, ... Sⁱₙ.
{
Step 1: Initialize the candidate set with all the sensors in the array.
Step 2: Obtain the quality profile information from the quality profile data-
base for each sensor. Apply the distance heuristic. Order the sensors based
on the quality of coverage. Select the two highest ranked sensors.
Step 3: Apply spatial heuristics to remove obviously wrong combinations of
sensors, such as those that are directly interfering with each other. If
there is likelihood of interference, pick one sensor at random and remove it
from the candidate set. Then return to Step 2.
}
```

number of data points at a sampling rate of 10 samples/sec is very high even in a short period of time. So, instead of considering each point we used the center of the rectangular region as a point. This method of representation is less complex and eases the process of trajectory classification and abnormal trajectory identification. In our trajectory representation approach we need to convert the many number of points from fused data to a center point of a box. For this purpose we buffer the fused data from the sensors and find a rectangular box that has maximum number of points. This single box represents the trajectory for that data buffer. In elderly people monitoring application the data points from the fused data are converted to rectangular boxes and trajectory is formed on the fly. Therefore, a trajectory is a sequence of boxes and length of the trajectory is defined as the number of boxes in the sequence.

Trajectory Matching and Classification. Initially we train the system to collect a set of known trajectories. For the purpose of training and experimental evaluation we asked volunteers to walk along the known paths on the floor. We collected this data for each known path. We created three classes of trajectories. Class A: from door to kitchen, Class B: from kitchen to door, and Class C: from living room to bed room and back to living room. The number of trajectories collected for each path is 10-15 and 20-30 in our experimental evaluation. After the training process, we have a set of trajectories $\{T_1, T_2, ... T_n\}$ for each known path. Now, we allowed new users

Figure 16. An example to illustrate a trajectory

to walk in the region covered by the four sensors deployed. Let the trajectory created by the new user is unknown trajectory '*U*'. Then the problem of matching and classification is to match trajectory '*U*' against the known trajectories taken during the training process and if it matches display which trajectories it matches with. If it does not match with any known trajectories we declare this as new trajectory and learn this trajectory and insert in the database if it is repeated certain number of times. The distance from one box '*b*' to a trajectory '*T*' with a sequence of boxes $a_1 \rightarrow a_2 \rightarrow ... \rightarrow a_n$ is defined as: $d(b, T) = min\{d(b, a_j)| i > j\}$. The most matching trajectory to a box '*b*' is the trajectory whose distance to the box is minimum across the distances from all the trajectories and has to be less than a predefined threshold. I.e., $MMT(b) = \{i|d(b, T_i) < d(B, T_j), \forall j \neq I$ and $d(b, T_j) < threshold\}$. The idea of trajectory classification is to find the most matching trajectory from the set of trajectories in different classes for each box in unknown trajectory '*U*'. After this step we compute the maximum appearance trajectory from all the matching trajectories.

VI. ACTIVITY RECOGNITION ALGORITHMS

In this section is presented in brief, a methodology where multi-modal sensors are used to perform activity recognition using various types of classifiers at the lower level, and Bayesian reasoning at the higher level. The use of a higher level reasoning algorithm allows user context to be built into the system in a manner that allows the selection of sensors in a judicious fashion, thereby leading the way for the conservation of resources in the smart home. This is an area of great importance, since energy efficiency is becoming increasingly important in all facets of life, and in all types of applications – mobile as well as fixed, wireless as well as wired.

Activity detection model. For the detection of the current activity we chose the model of the Dynamic Bayesian Network (DBN) (Murphy, 2002). Similar to the Bayesian Network model (Jensen, 2001), it utilizes the hierarchical dependency between variables to provide a compact representation of the joint probability distribution of the variables describing the system. This hierarchical dependency is represented graphically using a directed acyclic graph where directed arcs of the graph represent casual dependency between random variables. The random value of a variable represented by anode *i* statistically depend on the values of the variables represented by the nodes which have an arcs pointing to the node *i* and this dependency is expressed in terms of a conditional probability distribution. Such nodes are called *parents* of a node *i*.

Usually, the top node of the Bayesian network model describes the system state θ, called *hypothesis*, which we want to estimate. The leaf nodes of the network represent measured values z_i, in the case of sensor network - sensor readings or *features* - processed sensor readings mapped into discrete number of states having specific meaning. Given the joint probability distribution, we can estimate the probability that the system is in a particular state based on the sensor data measured. It is done by computing the marginal distribution $Pr(\theta = \theta_k | \mathbf{Z})$ The example of a Bayesian network for eating activity detection is shown on the Figure 17.

In the case of a Bayesian network, the joint probability distribution is given by:

$$P(X)\prod_{i=1}^{n} P(X_i | \pi(X_i)),$$

where $\pi(X_i)$ is a set of parents of node *i*. This joint probability distribution does not in any way include a notion of time and valid for values taken at the same time moment.

Figure 17. The example of the Bayesian Network for recognition of eating in the kitchen. The top node represent the activity we want to detect. Blue nodes represent the features provided by different sensor modalities. Actions node has three possible values: Nobody present, Person in the kitchen and Person eating

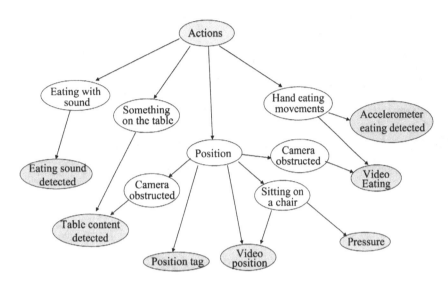

The joint probability distribution for DBN model is given by

$$P(X_{1:T}) = \prod_{t=1}^{T}\prod_{i=1}^{n} P(X_i^t \mid \pi(X_i^t)),$$

which actually means that there is a hierarchical dependency not only in space but also in time. For simplicity, however, it is assumed that the dependency on past exists only for a subset of all variables, called *temporal* variables, and that they only depend on its own state in the previous time moment. Therefore each node of the DBN has either the same set of parents as in BN or the same as in BN plus the same node from the previous time moment. Because of this assumption it is sufficiently to keep two time slices of the network - previous and immediate future and update each time the distribution for temporal nodes of the previous time slice to keep the state of the DBN.

Similar to the Bayesian network, usually the top nodes of the both copies of the DBN represent a hypothesis Θ and leaf nodes represent sensor readings or features z_i. The only difference is that both Θ and z_i have different time indexes in the two copies of the BN, e.g. $z_i^{(t-1)}$ and $z_i^{(t)}$. The estimation of the state in the past is done using evidence readings $\mathbf{Z}^{(t-1)}$ and prediction of the system state at the next time moment $\Theta^{(t)}$ is done by obtaining the distribution $Pr(\Theta^{(t)} = \theta_k)$.

Figure 18 shows an example of the Dynamic Bayesian network corresponding to the Bayesian network from the Figure 17. For activity detection the corresponding DBN has to be learned for a particular set of sensor modalities, actions to be detected and intermediate system states involved. Then by using sensor measurements as evidence we can use the joint probability distribution to compute the probabilities of respective person's actions.

Although it may not be the best model from the point of view of the pattern recognition, the DBN model was chosen because of the three reasons. First, complete probability distribution provided by the BN model allows us to estimate

Figure 18. The Dynamic version of the Bayesian Network from the previous Figure. Yellow nodes are temporal nodes. In this case, the temporal nodes are Activity, Something on the table, Position and Sitting

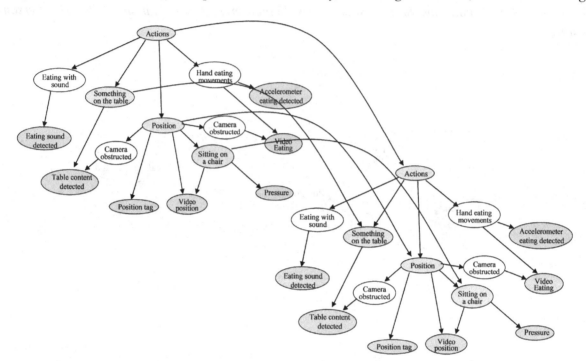

usefulness of the particular sensor modality. Second, state propagation between different time slices of the DBN model allows us to predict which sensor modality will be useful in the future based on the past data. These two properties are used in the our framework of the Phenomena-aware sensor resource management (Tolstikov *et al.*, 2007) based on the notion of the Quality of Information obtained from the sensor network. The third reason was the fact that the temporal dependency of the state in the DBN model also allows us to improve recognition accuracy and certainty level by reusing the past system state information.

Eating activity detection. Here we present our preliminary experimental results of the eating detection. The goal of the setup is to detect the time when a person is eating or drinking while sitting at the table. Currently only two modalities are used: accelerometer attached to a person's wrist (Figure 19) and a short-range RFID reader

installed under the table surface (Figure 20). As a source of accelerometer data we used a data acquisition board of an Intel iMote2 sensor mote from Crossbow (see http://www.xbow.com/). Due to the size of the RFID tag we decided to tag only rather big objects. In the current scenario, RFID tag is attached to a cup. Because of the currently limited number of the sensor modalities we have to assume that the person is already sitting at the table.

The structure of the DBN used in our experiment is presented on the Figure 21. The conditional probability probability distributions for the DBN were learned by first recording training scenario on the video, then by manually annotating all the states of the BN in the training dataset and then using the dataset to derive the probability distributions.

The nodes have the following state space:

Figure 19. Accelerometer on a person's wrist was used to track the hand movements

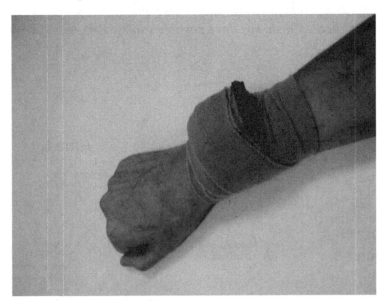

Figure 20. RFID reader installed in the table (right) was used to track the time when cup was not used

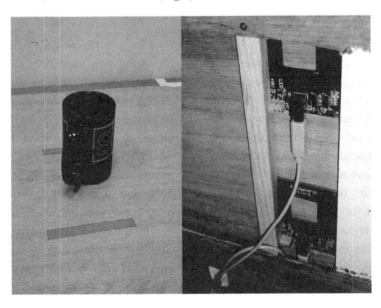

- *Actions* - three states are considered: person is Present, person is Eating, person is Drinking.
- *Cup is being used* - Yes, No
- *Cup detected on the table* - Yes, No
- *Hand movements* - Resting, Picking, Holding, Cup holding, hand near Mouth

- *Accelerometer data* - 50 times a second raw 3 axis accelerometer data

The temporal dependency on the *Actions* node is necessary if we want to improve the accuracy and certainty of the activity detection by reusing past information. Temporal dependency of the

Figure 21. The Dynamic Bayesian Network used for detection of eating activity in the kitchen. The top node represent the activity we want to detect. Blue nodes represent the sensor data provided by the different sensor modalities. Actions node has three possible values: Person is Present, Person is Eating and Person is Drinking

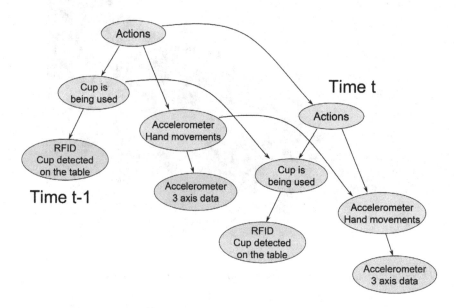

node *Cup is being handled* filters out the noise due to the reader occasionally losing and re-gaining the RFID tag without tag being removed. Temporal dependency of the node *Hand movements* provides a mapping between continuous accelerometer data and discrete hand states. It is currently implemented as a separate Hidden Markov Model recognizer using the HMMPak (see http://www.public.asc.edu/~tmcdani/hmm.htm) HMM implementation. To implement activity detection using DBN we used Bayes Net toolbox (see http://bnt.sourceforge.net/).

We present the results of the activity detection on the Figure 22. Four graphs presented show the most likely activity detected using accelerometer only, RFID only, both accelerometer and RFID, and actual person's actions. Each graph shows three states: top is *Drinking* state, bottom - *Present*. In the case of accelerometer only detection five states of the *Hand Movement* node were used, *Eating* state mapped into *Hand near Mouth* and *Drinking* state mapped to *Cup holding*. Since

RFID tag was used only for the cup, we were unable to detect eating with RFID only. Actual state was taken by a human operator using video recorded during eating. As it can be seen, accelerometer only recognition data was very noisy, while RFID had very high certainty but was incomplete. However, the combination of such noisy and incomplete data using DBN produced quite good recognition accuracy.

Note, that in this experiment we obtained results for eating and drinking which have rather high certainty but are very short in duration. Therefore it is necessary to say a few words on the usability of these results. These or similar results can be used in the following ways. First, straightforward use, is to compute the total time a person eats or drinks during the day and use it as a metric in assessing person's ability to live independently. Second way, is to use these result as a *micro-context* for use by a higher-level pattern recognition systems such as grammar-based in the more general context of *Person is taking food,*

Figure 22. This Figure shows the comparison of the actual state of the system with the estimated state derived using accelerometer and RFID. The bottom-most graph shows the actual state, other states from the top-most down, are obtained using only accelerometer (captures extra, i.e. false events), only RFID (misses several events), both accelerometer and RFID (captures events with 95% accuracy and minimal false alarms albeit with slight time delay)

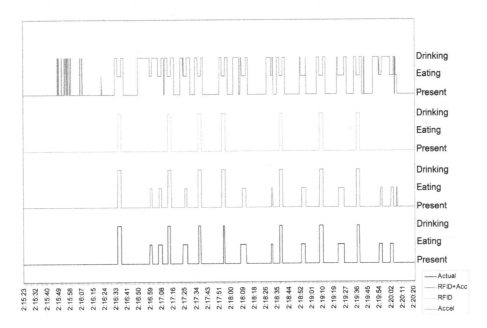

assuming that there are sensors for detection of other types of activities to establish such a more general context.

The states *Resting*, *Holding* and *Picking* of the node *Hand Movements* currently do not make much contribution to the recognition of the eating activity. However, when combined with some other modality they may give enough information to produce a result of sufficient quality to be used as a *micro-context* in the context of, for example, meal preparation or reading a newspaper in the kitchen.

Implementation Issues of Activity Detection Using DBNs

The main problem with using the Dynamic Bayesian Network is the high complexity of the Bayesian Network inference. In fact, the complexity is exponential in the width of the tree of the network,

thus DBN twice as computationally expensive as the BN, simply because it stores two copies of the network. However, there are certain approaches which may help reduce the size of the network thus making it possible to use DBN in real-time applications without requiring too much computational resources. The first approach enables us to reduce the size of the network by using one-way causal dependency between different time slices of the DBN. In particular, the distribution of values at a particular moment depends on the evidence obtained at this moment, and on the distribution of temporal nodes at the previous moment. Therefore, after the evidence of the previous moment is taken into account, we may only keep the temporal nodes in the network for the next step, thus reducing the size of the BN. Note, that this size reduction does not affect the result of the inference.

A second approach makes use of the fact that certain kinds of activities occur, or are of interest only if they occur at certain locations. In this case, the activity state space, as well as sensors may be divided across a number of smaller DBNs, each dealing with a set of activities at a particular location, and incorporating a small set of sensors. This approach does affect the result of the inference, however it may be reasonable for many real systems.

In our experience, it is possible to run in real-time, the activity detection application with the rate of 10 estimations per second, using up to 10 nodes of BN per time slice using usual desktop or notebook PC, and the Bayesian Network inference engine provided by a Microsoft MSBNx toolkit (see http://research.microsoft.com/en-us/um/redmond/groups/adapt/msbnx/).

VII. HIGHER LEVEL INTELLIGENCE

In (Ramos, Augusto and Shapiro, 2008) the authors provide the vision of the next step of research progression that must be made in order to accommodate the requirements of Ambient Intelligence. The paper quotes the reports of the ISTAG, which is the European Commission's Information Society Technologies Advisory Group, and particular stresses the importance of the supporting people in their daily lives. It draws upon the basic infrastructure provided by pervasive computing, ubiquitous computing, context awareness and embedded systems.

As we have purposed in this chapter, to focus on nuts and bolts of activity recognition systems, we have purposely avoided higher level concepts in Ambient Intelligence. However the techniques presented in this chapter may be viewed as being a substrate or a set of enablers for higher level artificial intelligence techniques that employ a rich diversity of algorithms and techniques existing in the body of literature in AI. In (Dubois, 2008) the author provides a bibliography of uncertainty in

knowledge representation and reasoning, covering areas that have been much researched within the AI community for decades. These are, for example, probabilistic and possibilistic reasoning, uncertainty management, plausible reasoning, fuzzy sets and rough sets, imprecise probabilities, random sets, evidence theory and information fusion. The framework and underlying technologies based on micro-context that has been presented in this chapter are to be viewed as sources for information that feeds into such reasoning systems with levels of uncertainty that make it possible to tune, refine or modify the reasoning paradigms and regimes as needed by the application.

VIII. CONCLUSION

This chapter looks specifically at the phenomena in an ambient space characterized by the activities of elderly people who are living alone in their homes. These constraints allow one to put bounds in an otherwise hard problem. The resulting problem, though still challenging, has solutions in specific cases.

To conclude this chapter it may be said that micro-context is a valid and useful paradigm for building systems that support diverse applications that make use of ambient intelligence. Though sensors capture a plethora of information, it is useful to think of the phenomena of interest. In our case activities are the phenomena of interest to be observed. A few applications were considered, the objective being to deal with the topic in depth rather than in breadth, giving the reader a taste of what it takes to put together an activity recognition application using ambient intelligence gathered from sensors. The first detailed example illustrates feature extraction from video frames. A second detailed example discusses data fusion from ultrasound sensors that use reflected sound waves to measure distance of a moving object, and hence put together a simple and unobtrusive tracking system. The third detailed example discusses the integration of multiple modalities of sensors to reason at the higher level, while dealing with resource management issues such as sensor selection to conserve resources based on the notion of information quality. This model used for this is the Dynamic Bayesian Network (DBN) model. We have presented encouraging results of applying an DBN based algorithm for recognizing activities of daily living of elderly. The algorithm reduces the complexity of determining states. This work points to a general algorithmic approach that a) uses multiple modalities of sensors for gathering data, b) detects activity primitives and c) stores detected activity primitives as micro-context for future use.

The pragmatic view of activity recognition adopted in this chapter is critical, we believe, since it insulates the applications from the underlying technologies. The sensors and sensing devices in an ambient space may change in accordance with the prevailing potential of the technology. However the algorithms which depend on abstractions of state and high level reasoning do not change.

ACKNOWLEDGMENT

The authors would like to thank the referee for useful comments that have helped improve the quality of the manuscript and clarify the presentation of the material.

REFERENCES

Ayers, (2001). Monitoring Human Behavior from Video Taken in an Office Environment. *Image and Vision Computing, 19*(23), 833–846. doi:10.1016/S0262-8856(01)00047-6

Biswas, J. F. Naumann F., & Qiu Q. (2006). Assessing the completeness of sensor data, *Proceedings of the Second International Conference on Database Systems for Advanced Applications*, (pp. 717-732)

Chai D. & Ngan K. N. (1999). Face Segmentation Using Skin Color Map in Videophone Applications, *IEEE Trans. Circuits and Systems for Video Technology*, (9) (pp. 551-564)

Dubois, D. Uncertainty in Knowledge Representation and Reasoning: A partial bibliography, Institut de Recherche en Informatique de Toulouse (IRIT) www.irit.fr/~Didier.Dubois

Garcia, C., & Tziritas, G. (1999). Face Detection Using Quantized Skin Color Regions Merging and Wavelet Packet Analysis. *IEEE Transactions on Multimedia*, (1): 264–277. doi:10.1109/6046.784465

Gonzalez, R. C., & Woods, R. E. (1992). *Digital Image Processing, 2nd Ed.* (pp 626-629).

Hongseng, S., Nevatia, R., & Bremond, F. (2004). Video-based event recognition: activity representation and probabilistic recognition methods. *Computer Vision and Image Understanding*, (96): 129–162. doi:10.1016/j.cviu.2004.02.005

Hu, N., Huang, W., & Surendra, R. (2006). Robust Attentive Behavior Detection by Non-Linear Head Pose Embedding And Estimation, *ECCV'06* (LNCS 3953). (pp.356-367)

Hurley A. C., et. al. (1999). Measurement of Observed Agitation in Patients with Dementia of the Alzheimer Type, *Journal of Mental Health and Aging*

Jones M. J. & Rehg J. M. (2002). Statistical Color Models with Application to Skin Detection, *International Journal of Computer Vision*, (46.) (pp. 81-96)

Lay, D. C. (2005). *Linear Algebra and its Applications, 3rd updated Edition.* New York: Addison-Wesley.

Lee, M. S. (2000). Detecting People in cluttered indoor scenes, *Proceedings of the. IEEE International Conference on Computer Vision and Pattern Recognition.* (1) (pp. 804-809)

Liu, C. D., Chung, P. C., Chung, Y. N., & Thonnat, M. (2007). Understanding of human behaviors from videos in nursing care monitoring systems. *Journal of. High Speed Networks, 16*(1), 91–103.

Liyuan, L., Huang, W., Gu, I. Y. H., & Tian, Q. (2004). Statistical Modeling of Complex Backgrounds for Foreground Object Detection. *IEEE Transactions on Image Processing, 13*(11), 1459–1472. doi:10.1109/TIP.2004.836169

Liyuan, L., Huang, W., Tian, Q., Gu, I. Y. H., & Hleung, M. (2004). Human Detection and Behavior Classification for Intelligent Video Surveillance, *International Symposium on Computer Vision, Object Tracking and Recognition*

Matsuoka, K. (2004). Aware Home Understanding Life Activities, *Proceedings of 2nd International Conference On Smart Home and Health Telematics*, IOS press (pp. 186-193)

Menser B. & Wien M. (2000). Segmentation and Tracking of Facial Regions in Color Image Sequences, *SPIE Visual Communication and Image Processing,* (4067), (pp. 731-740).

Murphy, K. P. (2002). Dynamic Bayesian Networks: Representation, Inference and learninng, *Ph.D. Thesis, University of California*, Berkeley Jensen F. V. (2001). Bayesian Networks and Decision Graphs, S*pringer Verlag, New York, Inc., New York*, NY, USA

Phung, S. L., Chai, D., & Bouzerdoum, A. (2001). A Universal and Robust Human Skin Color Model Using Neural Networks, *Proceedings of the INNS-IEEE International Joint Conference on Neural Networks*, (4). (pp. 2844-2849)

Ramos, C., Augusto, J. C., & Shapiro, D. (2008). Ambient Intelligence - the Next Step for Artificial Intelligence. *IEEE Intelligent Systems, 23*(2). doi:10.1109/MIS.2008.19

Ribeiro, P. C., & Jose Santos-Victor, J. (2005). Human Activity Recognition from Video: modeling, feature selection and classification architecture, International Workshop on Human Activity Recognition and modeling, (pp.61-70).

Sobottaka, K., & Pitas, I. (1998). Novel method for Automatic Face Segmentation, Facial Feature Extraction and Tracking. *Signal Processing Image Communication, 12*(3), 263–281. doi:10.1016/S0923-5965(97)00042-8

Thang, V., Qiu, Q., Aung, A. P. W., & Biswas, J. (2006). *Applications of Ultrasound Sensors in a Smart Environment.* Pervasive and Mobile Computing Journal.

Tolstikov, A., Tham, C. K., Wendong, X., & Biswas, J. (2007). Information quality mapping in resource-constrained multi-modal data fusion system over wireless sensor network with losses, *Proceedings of the International Conference on Information, Communications and Signal Processing*, December 2007.

Wang, H., & Chang, S. F. (1997). A Highly Efficient System for Automatic Face Detection in Mpeg Video. *IEEE Transactions on Circuits and Systems for Video Technology*, (7): 615–628. doi:10.1109/76.611173

Yang M. H. & Ahuja N., (1999). Gaussian Mixture Model for Human Skin Color and Its Applications in Image and Video Databases, *SPIE Storage and Retrieval for Image and Video Databases*, (3656). (pp. 45-466).

Zhuang, X. (2004). Vehicle detection and segmentation in dynamic traffic image sequences with scale-rate, *Proceedings of the IEEE International Conference on Intelligent Transportation Systems*, (pp. 570-574).

Chapter 21
Behaviour Monitoring and Interpretation:
Facets of BMI Systems

Björn Gottfried
University of Bremen, Germany

ABSTRACT

This Chapter provides an introduction to the emerging field of Behaviour Monitoring and Interpretation (BMI in short). The study of behaviour encompasses both social and engineering implications: on one hand the scientific goal is to design and represent believable models of human behaviours in different contexts; on the other hand, the engineering goal is to acquire relevant sensory information in real-time, as well as to process all the relevant data. The Chapter provides a number of examples of BMI systems, as well as discussions about possible implications in Smart Environments and Ambient Intelligence in a broad sense.

This chapter gives an introduction to the field of Behaviour Monitoring and Interpretation, BMI for short (Gottfried 2007-2009). The monitoring of behaviours and their interpretation is of interest in different areas; that is, in the natural sciences as well as in engineering disciplines. On the one hand, ethologists are interested in behaviours of humans and animals in order to gain insights into the nature of their character and evolution. On the other hand, engineers aim at automating processes, in particular those which involve behaviours of humans. Taking the technologies available in the first decade of the 21st century, it shows that many possibilities exist in order to support both ethologists in observing behaviours of humans and engineers in devising new tools to evaluate behaviours automatically. In fact, there is a trend which consists in connecting humans to electronic devices in order to record their behaviours. Furthermore, a broad community of cognitive scientists exists who approach problems in interpreting these behaviours in different ways; they involve computer scientists, engineers, psychologists, geographers, ethologists and many others who are all contributing with their specific discipline in making advances in behaviour monitoring and interpretation.

DOI: 10.4018/978-1-61692-857-5.ch021

It is the goal of the present text to outline the research direction of behaviour monitoring and interpretation. This is to consolidate the different disciplines and their different methods in order to arrive at a common view on BMI. This includes the terminology and the components necessary of any BMI application.

The body of this chapter is structured in the following way. An initial example is presented in Section 1. It makes clear what BMI is about from the point of view of an application example. Simultaneously, a framework is outlined that derives from this example. It will be shown how other application examples fit into this very same framework in the following sections. Each of those following examples introduces typical facets of BMI systems: spatial scales are relevant in that each object to be monitored is found at a specific spatial scale (Section 2); direct and indirect observations are to be distinguished since objects are either directly observed or indirectly by means of changes that occur in the environment (Section 3); monitoring processes either occur in reality or in virtual spaces or in mixed reality scenarios (Section 4); behaviours are either purposeful and active, or they reflect typical everyday behaviours (Section 5); it can be distinguished whether a monitoring system works by means of deploying local or allocentric techniques (Section 6); eventually, similar as each application is found at a specific spatial scale, temporal scales are to be distinguished considering both durations and the speed with which observed behaviours are carried out (Section 7). A discussion section closes this chapter with a couple of issues (Section 8): different purposes of BMI applications are identified; their relationships to related areas, namely Ambient Intelligence (AmI) and Smart Environments (SmE) is discussed; the importance of AI methods is pointed out; ethical issues are considered. Finally, an outlook on future work is presented (Section 9).

THE WALKING BEHAVIOURS OF PEDESTRIANS

One of the most fundamental behaviours of both humans and animals consists in getting from one place to another place. In order to monitor such *locomotion behaviours* a whole bunch of technologies are available today (Millonig et al. 2009). Furthermore, several authors propose different methods in order to evaluate sets of positional data, i.e. locomotion behaviours. The purpose of such investigations varies from studying which paths people take (Andrienko et al. 2007) to how their particular walking style looks like (Millonig et al. 2008). We shall initially present a study which is about walking behaviours of pedestrians and about the walking styles that can be distinguished when observing them going shopping.

A Study on Walking Behaviours

In their project, (Millonig et al. 2008) are aiming at the classification of pedestrian walking behaviours. Their study does not rely on positional data alone, but considers further factors that might influence walking styles, e.g. emotional or life style related factors. Their motivation is that mainly in urban areas people are interested in the use of mobile tools for wayfinding combined with location based services in order to coordinate activities. Then, the following problem is identified. What is called an *optimal route* or *useful information* might significantly differ among people. Therefore, a tool which can be judged to work successful, should tailor service information to the individual context of the user. In their study the authors follow the assumption that a comprehensive investigation of walking behaviours will reveal specific patterns which are typical for different individuals. Such patterns can later be employed as essential parameters for determining the individual context of a pedestrian navigation and location based service tool. Since there are endless walking patterns of pedestrians conceivable, they confine their study in an appropriate way by addressing just those pedestrians who are going shopping. It is then the aim to look for whether typical shopping patterns can be distinguished, and hence, whether people can be categorised into different types of shoppers.

In order to analyse the observed patterns they employ several qualitative-interpretative and quantitative-statistical methods. For simplicity, we summarise their methodology as follows. First, people are observed without knowing anything about it; for this purpose a hand-tracking tablet computer is used by the observer who tracks the pedestrians under observation. Second, the people who have been observed get interviewed in order to let them tell a little bit about themselves, their intentions and social background. Finally, they get further tracked, this time equipped with a Bluetooth Smartphone or a GPS logger for indoor and outdoor tracking, respectively. The acquired data is basically analysed by using speed histograms and by clustering the obtained trajectories. As a result there are three types of shoppers identified, namely *swift, convenient,* and *passionate shoppers*. These types of shoppers can be easily employed in a pedestrian navigation and location based service tool.

The Ingredients of BMI Applications

Having a closer look at this *shopping scenario*, a couple of technologies and methods can be identified that are needed for this study. In accordance to their successive application, it shows that a sequence of typical steps is to be performed in order to realise any BMI scenario.

i.   The first step consists in analysing what the actual object of interest is. Here, the assumption is that locomotion patterns will help to tell apart different shopping-characters.

Therefore, the attention of the researchers is drawn to the kinds of walking behaviours pedestrians perform. The research question is which typical walking patterns exist that will provide information about observed pedestrians.

ii.   In order to track pedestrians, that is in order to record their positions, such technologies as Bluetooth Smartphones and GPS loggers are deployed. Besides such automatic tracking tools a hand-tracking tablet computer is used by an observer. The purpose of all these tools is to capture specific aspects of behaviours the investigator is interested in. Measurement tools transform the real being of those behaviours into data sets which can be evaluated automatically. This transformation of specific aspects which are manifested in behaviours of humans constitutes the second step which abstracts from the actual object of observation.

iii. The third step is about the representation of the captured data. Depending on the ultimate goal of the BMI application, necessary and sufficient aspects can be determined. Since types of walking patterns derive from trajectories of the observed pedestrians, space-time coordinates will be required. Additionally, this step includes pre-processing procedures that might ease the later evaluation of the data. Here, trajectories are smoothed to some extent since small deviations from actual walking patterns are due to imperfect measurement tools.

iv.   The next step is about the employment of methods in order to infer properties of the given data. In our example, in particular speed histograms and clustering techniques are used. They enable the classification of the different trajectories. In general, this analysis step is the most complex one because the analysis of the data can involve every kind of data analysis method. By contrast, all previous steps are confined to the determination of specific intermediate results: the identification of the object of interest, the choice of appropriate measurement tools, and the employment of an adequate data representation for the acquired data.

v. The final step concerns the interpretation of the data, in our example of the observed pedestrians. This step is consequently concerned with the semantic level. It states how the observed phenomena are to be interpreted. In the example scenario, the researchers found out that three dominant patterns arise which are to be distinguished and to which we can attach meanings such as *swift, convenient, and passionate shopper*.

The described steps form a sequence within which these steps build on top of each other, as shown in Figure 1. Such a layered architecture illustrates that each layer introduces another abstraction on top of another layer: (i) from all observable phenomena researchers analyse which of them are of interest for the study at hand; (ii) measurement tools focus on specific aspects and neglect everything else which might be captured with other measurement tools; (iii) the chosen representation abstracts from the raw data and transforms the data into some useful representa-

tion; this representation restricts the data to be considered and organises the data in a specific way; (iv) properties of the data are extracted and these properties summarise an often huge amount of data while leaving out details which are not relevant anymore; (v) eventually, the extracted properties are to be mapped to conceptual terms which are comprehensible for the application at hand; this last step even abstracts from everything that could still be related syntactically to the acquired data. In this way, actual behaviours of humans (i) are related to conceptual terms (v). This is in fact a non-trivial task.

The BMI Framework

From the introductory example we learn that the process of comprehending observed behaviours can be decomposed into a number of sub-processes that build on top of each other. This enables the comparison of different BMI applications with respect to the different abstraction layers. As a consequence, research can focus on single layers and sometimes the implementation of specific layers might even be useful for different BMI applications. This would indeed be an advantage the BMI community could make much of a profit from. Looking at the second layer, it shows that

Figure 1. The BMI Framework adapted from (Gottfried et al. 2009, Gottfried 2009 (a))

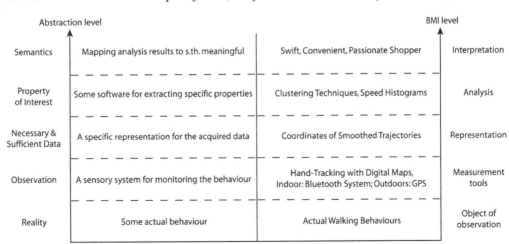

Abstraction level			BMI level
Semantics	Mapping analysis results to s.th. meaningful	Swift, Convenient, Passionate Shopper	Interpretation
Property of Interest	Some software for extracting specific properties	Clustering Techniques, Speed Histograms	Analysis
Necessary & Sufficient Data	A specific representation for the acquired data	Coordinates of Smoothed Trajectories	Representation
Observation	A sensory system for monitoring the behaviour	Hand-Tracking with Digital Maps, Indoor: Bluetooth System; Outdoors: GPS	Measurement tools
Reality	Some actual behaviour	Actual Walking Behaviours	Object of observation

measurement tools have in fact been developed independent of specific applications; the very same tools are used in different contexts; devices could have been developed which record positions of humans and objects without the need to focus on a specific application. But regarding other layers the generality of methods applied at those layers, that is their application independence, is sometimes less obvious.

It shows that in fact each of the five layers is indispensable. It is impossible to leave out a layer when implementing a complete BMI application. Measurement tools are necessary in order to transform observations into data sets. Representations are also necessary, since analysis methods work on such representations. On the other hand, this might be less obvious as representations can be quite simple, whereas the necessity of measurement tools is more obvious. Similarly, analysis methods can in principal be simple, for example, when solely transforming them into a diagram for presentation purposes. But still, no layer can be left out entirely, and thus, the layers can be conceived of as the skeleton of any BMI application. Therefore, we shall henceforth conceive the five-layer architecture as the *BMI framework*. That other BMI applications fit into this framework will be shown in the following sections.

In addition to the generality of the BMI framework, it will be shown which other facets determine BMI applications. These other facets include several spatial and temporal characteristics under which the object of interest acts. Furthermore, different modes of observation can be distinguished, as can be different modes of reality in which observations take place. Then, the interaction between the object of interest and the deployed technology matters. Eventually, specific constraints of the analysis process might be of special relevance. Figure 2 summarises these facets.

SPATIAL SCALES

In the introductory example pedestrians have been observed in a shopping mall. In the second example we shall look at another spatial scale; this scenario is about the monitoring of behaviours of mice living in a cage. This example shows both how it fits into the very same framework as the first scenario and that any BMI scenario is in particular distinguished by the spatial scale at which behaviours are monitored. This is mainly of concern for the second layer which is about the choice of adequate measurement tools. Similar measurement tools will be used at similar spatial scales, while different spatial scales require often different measurement tools; mice in an indoor cage that measures less than 2 metre in each direction can hardly be tracked by GPS. On the other hand, specific sensors are applicable at quite different spatial scales, as will be shown below.

Movement Behaviours of Mice

Behaviour patterns of mice living in a semi-natural environment are investigated by (Kritzler et al. 2007). They compare motion behaviours of mice, which have a genetic predisposition to develop Alzheimer's disease, with their wild-type conspecifics. Aim of this project is the systematic support of behavioural observations made by humans. For this purpose an RFID system is installed in the cage where the mice live in order

Figure 2. Some fundamental facets of BMI applications

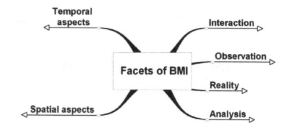

Figure 3. The role of the spatial scale in BMI applications

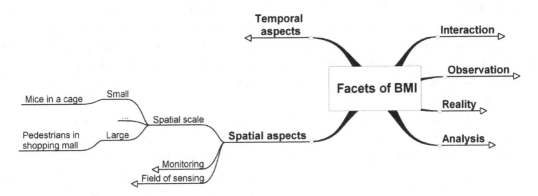

to capture their positions; the cage comprises a number of different levels. At the highest level a video monitoring system is additionally installed capturing more precisely motion behaviours. This scenario also partitions into different abstraction layers with the bottom layer representing the behaviours of interest. By contrast, the top layer shows the interpretation of the observed behaviours. All intermediate layers mediate between observation and interpretation; this mediation is the purpose of BMI.

BMI Integration

The bottom layer of this *mice scenario* is similar to the bottom layer of the *shopping scenario* in that the object of interest in both cases is the same, namely locomotion behaviours. However, different measurement tools are employed in the mice scenario: RFID tags allow different mice to get distinguished at specific places, a scale measures their weights, and a video tracking system is installed at the top most level of the cage. Apart from this video monitoring system the mice are tracked less precisely than the humans in the shopping scenario; only RFID readers at specific places enable the determination of where a mouse currently is, and hence, to determine positions at a rather coarse level. While in the shopping scenario the chosen representation

mainly involves space-time points for the positions of pedestrians, in the mice scenario positions are represented for those locations where RFID readers are installed; additional data are to be represented in both scenarios, information about the social background of tracked pedestrians and information such as weights of the tracked mice. While the shopping scenario looks at the speed histograms and applies clustering techniques for identifying typical patterns, in the mice scenario directions of movements along tubes, speeds, contacts with specific devices, and the duration of stay at specific places are determined. Interpretation results include dominance patterns of mice, their home range as well as drinking and emigration behaviours (See Figure 3).

Altogether it shows that the mice scenario can be equally well described with the BMI framework as the shopping scenario. Although the scenarios are quite different concerning the objects to be tracked and the environments where this tracking is to be performed, they have obviously something in common which is covered by the BMI framework. It concerns the structure every BMI application adheres to, from the observation part up to the interpretation of the observed phenomena. But it is not only this structure how those scenarios relate. One could easily think of RFID scenarios which do determine positions of pedestrians similarly as in the mice scenario. That is, though the object of interest is different already the layer of measurement tools might provide equal means in different scenarios. Things are even more interchangeable at the representation layer; here, the abstraction from the observed object is quite strong: we can easily think of common representations which are useful and applicable in very different scenarios. This holds equally well for the analysis layer which is based upon the representation employed.

DIRECT VERSUS INDIRECT OBSERVATIONS, STATIONARY VERSUS LOCATION INDEPENDENT OBSERVATIONS, ANYTIME BEHAVIOUR VERSUS POST HOC ANALYSIS

The previous two scenarios directly monitor the object of interest, i.e. locomotion behaviours of pedestrians and mice by such techniques as GPS, Bluetooth, and RFID. The measurement tools directly determine coordinates of the tracked objects and we speak of *direct observations*. By contrast, *indirect observations* occur whenever measuring objects or environments with which the actual object of interest interacts. For example, it can be observed that light is switched on at 9 p.m. This information is not so much a direct behaviour of the person under observation but a consequence of his behaviour. That is, behaviours are inferred by looking at changes that occur in the environment of the object under observation.

The Handwashing Scenario

An example is described in the following which uses indirect observations. Additionally, the purpose of its discussion is to better comprehend the generality of the BMI framework since this example is not about locomotion behaviours like the previous scenarios. Also, two other BMI facets are explained with this scenario.

In (Adlam 2009) several studies are reported which are about supporting people with dementia in their activities of daily living (ADLs). The impaired memory functioning of people suffering from dementia makes it difficult even for mildly affected people to successively carry out activities of daily living; they get sometimes disoriented part way through an ADL and might have difficulties to remember what to do next. As an example (Adlam 2009) discuss the ADL of handwashing, as this activity must be completed several times every day. Also, it shows that sometimes cognitively impaired people leave out specific steps when washing their hands. Consequently, an assistant for handwashing would be very useful.

The system (Mihailidis 2008) have developed is called COACH which is the acronym for Cognitive Orthosis for Assisting aCtivites in the Home. It has been developed to assist people to complete ADLs independently from a human caregiver. On the one hand, this takes the caregivers one of the many burdens he has to deal with. On the other hand, it enables greater independence of the person with dementia. Also, the current solution in which a caregiver supervises the person with dementia as they go about their daily activities can be exhausting and humiliating for everyone involved.

COACH does both it monitors a person who is completing an ADL and it provides prompts on a display if the person appears to be having trouble with her current activity. Thereby, COACH tries to provide support but in a way that the person using COACH still remains in control over the activities she is doing. Prompts of the system need to be accurate, clear, delivered just in time, but also sensitive to the individual capabilities of the person. For this purpose, the system has to take into account the context of its environment, enabling it to react appropriately to the individual needs of the monitored person. This includes such things as where the person is in the activity, what her preferred ordering of ADL steps is, and how responsive the person is regarding different kinds of prompts. That is, the support of COACH is not part of its BMI subsystem, but this support determines constraints on what BMI has to monitor.

The monitoring part of the handwashing scenario consists of a computer and an overhead camera placed over the sink. The support component consists of a flat-screen monitor and speakers. Images from the camera are relayed to the computer, which analyses them for hand, towel, and soap positions in each incoming frame. Then, for analysis, handwashing can be broken into a couple of steps with some variances among them; steps include turning on the water, taking the soap, using the soap, and so on. The system interprets images from the camera in order to determine where in the activity the person is and whether she needs assistance. The latter is the case when she has missed a step, is completing steps in the incorrect order, stops the activity before it is completed, or has been inactive for some period of time. If assistance is required, the system plays an audio or audio/visual prompt to guide the person to the next appropriate step. For more details on the assistance part and an evaluation with people using this system we refer to (Adlam 2009) (See Figure 4).

BMI Integration

While in the shopper and mice scenarios positions of moving objects are considered, in this handwashing scenario positions of hand, towel, and soap are considered. That is, instead of tracking a single object, a couple of objects are to be tracked. Moreover, the handwashing scenario is much more restricted in terms of the environment and context which is to be taken into account (*stationary monitoring*). Pedestrians and mice might move along ever changing paths in complex environments, each time differently (*location independent monitoring*). By contrast, handwashing procedures are much more similar to each other than behaviours monitored in the other scenarios. This indeed might be the reason why the video processing has a real chance here, namely because the lighting conditions are always similar, as are the appearances and positions of the objects to be detected and tracked.

What the handwashing scenario additionally distinguishes from the other scenarios is that not only the person under observation is monitored, but also appliances such as the water tap. That the water is turned on, or the occurrence of other changes in the environment, are mainly due to activities of the person under observation. From

Figure 4. The role of the mode of monitoring in BMI applications

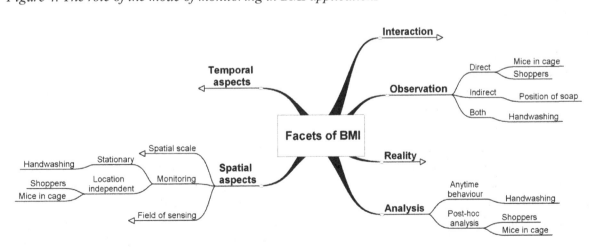

such changes conclusions can be drawn about the behaviour of the person. The monitoring of environmental changes for the purpose of inferring behaviours of a person is what we call indirect observations as opposed to the direct observation of how the person behaves.

Yet another distinction to the former two scenarios is that the monitoring and interpretation in the handwashing scenario requires a kind of *anytime behaviour* in order to let the system react at any time. As soon as a step during the handwashing procedure is left out, or if some other irregularities occur, the system has to react promptly. That is, while the first two scenarios carry out a post hoc analysis of data acquired over a period of time, there is an online monitoring and interpretation of behaviours in the case of the COACH system. This generalises to other assistant systems which would immediately have to react in accordance to observed behaviours.

REALITY VERSUS VIRTUALITY VERSUS MIXED REALITY

Another aspect concerns the mode of reality in which behaviours are observed. In the discussed scenarios the mode of reality is always the same. Each of these scenarios takes place in reality. But there are also behaviours we might want to observe which take place in a virtual space. For example, in the context of human-computer interfaces (HCI) it might be of interest to investigate how well people manage a specific computer desktop interface. Such an analysis might in particular concern *mouse movements* as the next scenario will show. But when distinguishing reality and the virtual space within a computer we should also mention augmented and mixed reality scenarios which might get more important in future. Here, it will be necessary to assess how activities of humans can actually be supported by devices that make use of mixed realities.

The Desktop Mouse

In (Aoidh 2007) the authors are interested in determining the implicit level of interest of users in information arranged on a desktop screen. Their analysis is based on mouse movements. The user interacts with spatial objects through a GIS (Geographic Information System) interface. Meanwhile, the system monitors the actions of the user. In particular, it is looking at the information which is displayed under the mouse positions. This is used for inferring interests and disinterests of the displayed information. Aim is to automatically generate a user profile that captures the interests of the user (See Figure 5).

BMI Integration

While the user works with his computer-mouse, he is moving his arm. However, these arm movements are not particularly informative. It is just the cursor of his mouse on the desktop screen which is of interest. Or more precisely, mouse movements, mouse clicks, interface button clicks as well as their location and duration are considered; also, zoom and re-centre operations on maps are recorded. These observations require quite different means than when monitoring behaviours in real spaces. Additionally, it shows that these mouse movement behaviours are much simpler to monitor than real space behaviours, since the positions of the cursor are already available by the system, which just displays at those positions the mouse pointer. Similarly, the representation of the raw data is almost for free. The real challenge is confined to the upper two BMI layers in order to make sense of the observed mouse movements.

A fundamental assumption is made within this scenario: that there is a correspondence between the visual perception of the user and the position of her mouse. The authors support this assumption with a large body of work that shows this correspondence between eye movements and mouse movements. Accordingly, an object which

Figure 5. The role of the mode of reality in BMI applications

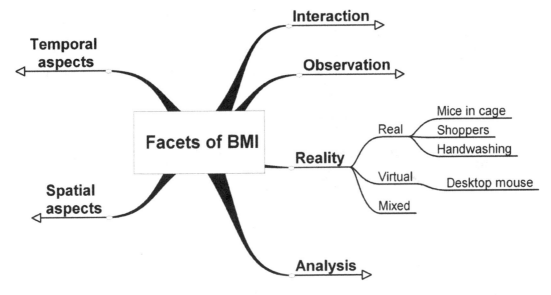

is closer to the mouse position seems to be of greater interest than an object far away from the mouse cursor. Then, some of the interest measures are simple to obtain. The position of the mouse, its duration of stay and zoom in operations are all indicators for the significance to the user of the displayed information.

ACTIVE VERSUS PASSIVE BEHAVIOURS

Yet another aspect concerns the distinction between active and passive behaviours. Most behaviours which are monitored and evaluated are passive in the sense that people are not aware of being observed. They behave as usual and do not have to artificially adjust their behaviours according to the monitoring system. Active behaviours are different in that they are performed consciously as a mode of interaction. People are aware of a monitoring system with which they interact purposefully.

Hand Gesture Behaviours

Hand gesture behaviours are investigated by (Goshorn 2008) in order to provide a non-verbal mode of interaction to command an ambient assisted living environment. The motivation is that many appliances are to be controlled in a household. But handicapped people have difficulties with even simple everyday life activities, as opening curtains, windows, or doors, switching devices on and off etc. If remote controls fell under the bed where a person is bedridden, they would need some way to communicate; hands free communications would be the easiest.

In their system (Goshorn 2008) distinguish a set of six atomic hand postures. In order to monitor hand postures of the user, cameras are mounted at the ceiling of the rooms. Hand gestures are made of sequences of postures, and hence, in principal an arbitrary large vocabulary of gestures, or gesture words, can be defined on the alphabet of postures. Each gesture is interpreted as a command to the ambient assistant system (See Figure 6).

Figure 6. The role of the mode of interaction in BMI applications

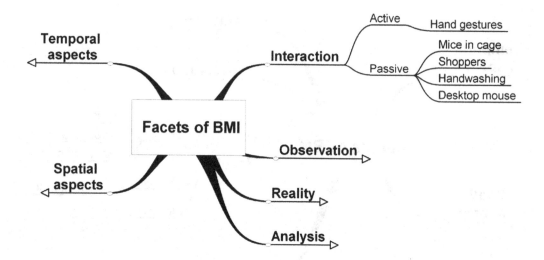

BMI Integration

Using video technologies in order to obtain atomic postures of the user commands, the monitoring part of this scenario is much more complex than in all previous scenarios: the shopper, mice, and desktop mouse scenarios all employ positioning techniques which are easier to analyse than complex video sequences. The handwashing scenario also uses a camera. But in this case the visible context is much more restricted than in the hand gesture scenario; hand postures can be formed in different angular positions with respect to the ceiling-mounted cameras; also, the different hand postures are to be distinguished by the system. By contrast, in the handwashing scenario it suffices to tell apart hands, soap and towel which significantly differ with respect to their sizes, colours, and shapes; moreover, their positions are to be determined under confined conditions.

ALLOCENTRIC VERSUS LOCAL MONITORING

A mobile, such as a person, can be monitored locally, for example by wearable technologies worn by the observed person. But the term *local* might also denote an environment that measures locally at some place the actions of the inhabitant. Then mechanisms are conceivable that propagate monitored data through a network. A specific case which is found at the other end of locality would propagate all data through the whole network of an environment so that every place in this environment gets to know what is going on everywhere else. Such a broadcasting of information through the environment leads to what we refer to as the *allocentric monitoring* technique. Obviously, one should be careful when employing such a broadcasting technique in order to avoid over-stressing the network. But when carefully applied, the following scenario shows how effective this technique can be.

Figure 7. The role of the field of sensing in BMI applications

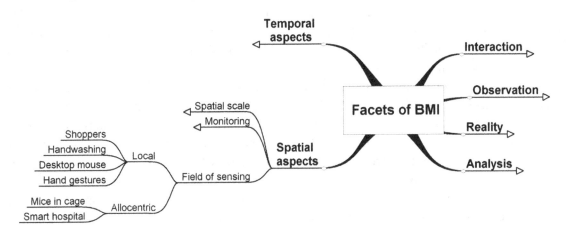

The Smart Hospital

The idea of smart hospitals as described in (Heine 2004; Mea 2004) concentrates on increasing the efficiency of clinical processes and knowledge management; that is, these authors are confined to the monitoring at the document level. By contrast, in (Gottfried 2009b) the spatiotemporal behaviours of patients, doctors, and visitors in a smart hospital site are analysed. For this purpose, allocentric monitoring techniques are employed, as shown in the following.

A nurse has to take care of several patients. For deciding whom to visit next, urgency and space constraints as well as probably other constraints are to be taken into account. In order to realise this scenario the smart environment should provide means for tracking individuals through the entire hospital site. It shows that it is sufficient to determine positions at the level of single rooms. The smart environment can then tell the nurse where she has to go next. More formally, the whereabouts of a nurse as well as the rooms of the patients she has to visit are represented by regions. A nurse is contained in a region and regions represent disjoint places (different rooms and hallways). The smart environment further represents the specific arrangement of regions

at the hospital site, i.e. its spatial topology. With this representation, the smart environment can provide the nurse an efficient plan about where to go next when she is finished with a patient. Then, describing the path the nurse has to take amounts to consider an ordered list of regions she has to cross. The ubiquitous hospital system always knows everything about the current whereabouts of the nurse, her patients, and doctors, if required, and it knows everything about urgent cases to which it can promptly adapt the plan for the nurse (See Figure 7).

BMI Integration

Using allocentric monitoring techniques a fundamental distinction to all other scenarios is made, concerning the second BMI layer. All previous scenarios employ local monitoring techniques alone. Although the object of interest in the hospital scenario is the same as in the shopping scenario in that in both cases locomotion activities of individual people are considered, the monitoring techniques significantly differ. This shows that the object of interest per se does not determine how the second BMI layer looks like. By contrast, it might sometimes be useful to combine different monitoring techniques in order to compensate for difficulties single techniques suffer from. What the hospital scenario distinguishes is that it makes immediately available the positions of individuals at all locations in the hospital site.

The upper layers are influenced when employing an allocentric view on the given environment in that an adequate representation of the whole environment is necessary to make use of allocentric information. The analysis layer might eminently benefit from the allocentric view since it provides comprehensive information about the whereabouts of everybody being tracked within the whole hospital site. Further scenarios that make use of allocentric monitoring techniques in a smart hospital are presented in (Gottfried 2009b).

TEMPORAL ASPECTS

Similar as each scenario is found at one specific spatial scale, all behaviours are realised at specific temporal scales. This concerns both the speed with which behaviours are performed and the duration of monitoring.

In the shopping scenario the duration of observation depends on several factors of the individual shopper; monitoring durations last from a few minutes up to more than one hour. By contrast, in the mice scenario the idea is the systematic observation over long periods of time, i.e. 24 hours a day, seven days a week. The speed of the change of positions is in both cases moderate and can be adequately captured by the mentioned measurement tools. However, there might be differences which are characteristic for the different objects under observation; a mouse might suddenly run very fast to and fro in its cage which is hardly the case of pedestrians going shopping.

In the *desktop mouse scenario* temporal information is automatically captured by the operating system and can be used for evaluating objects of interest that relate to the duration with which the user places the mouse cursor near an information item. In the handwashing scenario the different

steps are also performed in a moderate speed. Here, the temporal information is also used in order to decide whether a step has been carefully carried out, for example, in order to test whether the person is inactive for some period of time although having not completed the current step yet. Also, the order of steps can be read off their temporal order. In the gesture scenario, the user should learn to perform postures in a way that the system has a real chance for interpretation; probably they should not change too fast. The hospital scenario tracks people moving with moderate speed; a challenge is to make available the whereabouts of each one as fast as possible at each place within the whole smart hospital to let the system make use of this information as accurate as possible and without delay (See Figure 8).

DISCUSSION

The following Table 1 summarises the discussed aspects and their realisation in the different scenarios. While the BMI framework allows the comparison with respect to the different layers, there exist further aspects which might be of interest and which are discussed in this section.

The distinction between *stationary* and *location dependent* monitoring systems is left out here, since all scenarios, but the handwashing scenario, require mobile monitoring systems. Similarly, only the mice scenario is restricted to a post hoc analysis; all other scenarios require a kind of anytime behaviour in order to provide assistance at any time. Though, at this stage of *learning typical patterns*, the shopper scenario also carries out a post hoc analysis; but as soon when employed in a pedestrian navigation tool, an anytime behaviour would also be required for this navigation tool (See Figure 9).

Figure 8. The role of temporal characteristics in BMI applications

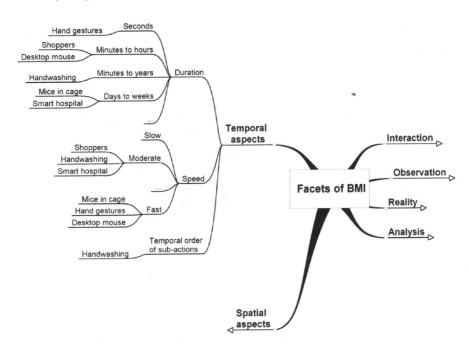

Purposes of BMI Applications

Looking at the different scenarios we learn that there are several different purposes for BMI. In order to be useful for the individual, assistant systems need to take the particular behaviours of their users into account. Mobile tools like the pedestrian navigation and location based service tool of the first scenario would *optionally benefit* from the system being able to determine the walking style of its users; similar is the situation for the desktop mouse scenario where the monitoring

enables the system to automatically recognise information items which are of interest for the user; in both cases the system can provide better support in case the additional information is available (walking style or specific information items of interest). The handwashing scenario is different: it *needs* to determine user behaviours in order to be of any value. For the shopping scenario BMI is an option, for the handwashing scenario BMI is indispensable. This generalises to all applications in the field of Ambient Assisted Living (AAL); this includes the hand gesture scenario as

Table 1. The Scenarios and how they compare with respect to the different BMI aspects

Scenario	Spatial Scale	Observation	Reality	Speed	Duration	Act. / Pas.	Field of Sensing
Shopper	Shopping Mall	Direct	Real	Moderate	Minutes to hours	Passive	Local
Mice	Cage	Direct	Real	Fast	Days to weeks	Passive	Allocentric
Handwashing	Home / Basin	Both	Real	Moderate	Minutes to years	Passive	Local
Mouse	Desktop	Indirect	Virtual	Fast	Minutes to hours	Passive	Allocentric
Gestures	Hand	Direct	Real	Fast	Seconds	Active	Local
Hospital	Hospital area	Both	Real	Moderate	Days to weeks	Passive	Allocentric

Figure 9. Some of the most fundamental BMI facets and typical categories

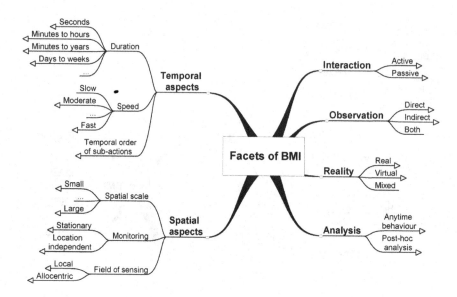

well as the smart hospital. For the mice scenario the situation is quite different. Here, BMI is not the prerequisite of an assistant system but *fulfils a more direct purpose*: a tool is provided that aids ethologists to systematically monitor animal behaviours.

Several areas are conceivable which would also benefit from BMI techniques. Among others, marketing, i.e. to analyse how people behave when being confronted with commercials in a shopping mall; security at stations and airports or other public places and buildings; BMI as sub-modules in other broader disciplines, like Ambient Intelligence and Smart Environments, i.e. whenever it is necessary to analyse the behaviours of users. For the latter, we should have a look at how BMI relates to those other areas.

Related Areas

How does BMI relate to other areas in the context of computing paradigms? The first paradigm that veered away from the *desktop computer*, and thus, which changed the mode of interaction between the user and the system, is the notion of *mobile*

computing. The idea is to use computers like PDAs while moving - as opposed to *portable computers* which can be taken to another place, too, but which are practical of use only when deployed stationary. *Wearable computing* is a specific sub-concept of mobile computing that requires the technology being integrated, for example in the clothing; in this way, the technology can be employed easier (Boronowsky 2008). The conception of *ubiquitous computing* is the next step which focuses on the omnipresence of mobile computing (Weiser 1991) and emphasises simultaneously its unobtrusiveness, in particular invisibility. Closely related to ubiquitous computing is the notion of *pervasive computing* that puts more emphasis on the interoperability of devices, software services, and seamless interconnectivity. The latter is mainly coined by industry which introduces ever more terms; one of the latest being *everyware* (Greenfield 2006). More relevant is the area called *ambient intelligence* which adds to ubiquitous computing the requirement of intelligence, reflecting that the environment should be able to recognise people, adapt to individual preferences and even acts upon the behalf of the users (Aarts 2009). What all these conceptions have in common is that their goal is to introduce new computing paradigms in the most general way.

BMI aims at something different. It is not another computing paradigm but rather about making use of the aforementioned paradigms. It should be seen as an area that exploits these technologies in order to advance specific applications, in particular those applications which are about monitoring and interpreting behaviours. In this way, it has been proposed to introduce the *computational theory of ethology* (Gottfried 2009a). Advantages of this research direction have been extensively discussed within the previous sections.

In the context of mobile computing, BMI enables the monitoring of people or objects that act in space and time. The field of wearable computing has the advantage that the monitoring requires no specific environment that needs to be equipped in a particular way. The ubiquity of monitoring technologies, in particular for non-wearable computing, is a necessity for a monitoring that is free of interruptions; it requires a pervasive network infrastructure. But just as soon as the monitoring techniques adhere to principles of ambient intelligence an unobtrusive monitoring becomes possible that is adapted to the individual and that enables the intelligent analysis of the acquired behaviour patterns. BMI exploits all these infrastructures extensively. However, BMI can especially be seen as a subfield of ambient intelligence, namely a subfield that observers and interprets observations so as to make them available for smart environments and other applications in the context of ambient intelligence.

The Value of AI Methods

From the point of view of data analysis a great body of work has been investigated within the Artificial Intelligence community. BMI investigations which make use of specific AI methods are presented in (Gottfried 2009a). In this paper, it is also shown how BMI can be conceived of as a computational approach to ethology in that it automates processes ethologists of different disciplines are faced with. In particular it is shown how BMI introduces with AI techniques new means for improving the situation for ethologists significantly; namely, long lasting standardised observations are enabled (Kritzler 2007); temporally disjoint behaviours can be appropriately related (Kiefer 2007); typical patterns which are observed can be learned (Aztiria 2008); and, a complex bottom-up ↔ top-down analysis is possible (Aghajan 2007, Terzic 2007).

A more fundamental issue arising within the BMI framework relates to the classical *symbol grounding problem* in Artificial Intelligence research, as discussed by (Harnad 1990). It addresses the question as to how the semantic interpretation of a formal symbol system is made possible; in particular when this meaning is supposed to be intrinsic to the system, as opposed to just being parasitic in the head of the human being who is interpreting what the formal system derives. As far as a BMI application intends to support an ethologist with evaluating her observations, an intrinsic interpretation of the system is in fact not necessary; the interpretation is made by the ethologist herself. But when the BMI system is employed in a smart environment which has to interpret the sensory data so as to enable an automatic assistance system to support the inhabitant, the situation is different. In this case, the system intrinsic interpretation should basically be seen as a relation between what the BMI sub-system recognises and how the smart environment should react in order to provide according support. Then it does not help to ask for what the intrinsic system interpretation is, without taking into account how the system reacts after having made specific observations – that is to say, in a way, the understanding of the system is its reaction to its observations. This is similar to what the Austrian philosopher Wittgenstein said about what the meaning of a symbol is, namely that it is defined by how that symbol is used (Wittgenstein 1953).

Privacy and Ethical Issues

One important aspect is that in the discussed scenarios privacy concerns and ethical issues arise which should be carefully taken into account. For instance, (Alder 1998) argues for an approach which emphasises communication in the design and implementation of monitoring systems, enabling an acceptable balance between potential abuses and benefits.

The question is what it takes for people to accept that their environment is monitoring every move of them in order to take care of them (Aarts 2009). Much of this acceptance, the authors argue, will depend on the perceived functional benefit of ambient intelligence environments and on the availability of mechanisms that enable participants to make their own choices in a way that is understandable, transparent, and independent of their comprehension level. Such mechanisms should comply with the basic agreements that have been formulated within the European Charter of Fundamental Rights, which are concerned with properties such as autonomy, freedom of choice and self-determination. Solutions to these problems are often communicated as embedded ethics, trying to handle the fundamental rights already at the level of the design of the corresponding embedded systems (Aarts 2009).

(Panina 2005) investigates electronic performance monitoring (EPM) as an effective tool in ensuring production quality standards in distant operations. They discuss the use of EPM on a global scale that may lead to an array of questions regarding appropriateness and effectiveness of this procedure in different cultural contexts. Defining BMI in a broad sense, EPM would be another example for a monitoring scenario taking place in a virtual space. But in the current context, the work by (Panina 2005) is of interest because privacy concerns in the aforementioned BMI scenarios would similarly have to consider the appropriateness of specific mechanisms with respect to the given cultural background.

FUTURE PROSPECTS

What will BMI provide in future? Supporting people in their everyday life requires smart environments to recognise activities of their inhabitants. Activity recognition is a research area that is still in its infancy. There is much research necessary in order to let systems recognise arbitrary complex activities of both single people as well as of groups of people. Future will provide BMI systems which enable an understanding of what is going on in a much more precise, accurate, systematic, and reliable way than what humans are able to observe. The correct interpretation of observed behaviours is probably the most difficult challenge. But ever growing knowledge bases and effective reasoning mechanisms will sooner or later let the systems get successful even in their interpretation abilities.

In particular the following areas might benefit from BMI in the future:

- the monitoring of public spaces, in particular airports and stations, for security purposes;
- the control of traffic networks that aim at a higher safeness for all traffic participants;
- the support of handicapped people in the context of Ambient Assisted Living applications;
- the industrial application of new machines that adapt to the humans working with them for more efficient and effective production cycles;
- the systematic and more comprehensive investigation in empirical sciences, such as ethology;
- the advancement of smart living spaces which are in charge of our well-being and a more comfortable life.

As soon as these prospects and previous visions, as those by Marc Weiser, become a reality, BMI systems will have reached indeed a sophisticated level of maturity.

REFERENCES

Aarts, E., & de Ruyter, B. New research perspectives on Ambient Intelligence. In Journal of Ambient Intelligence and Smart Environments, 1, 5-14, IOS Press, 2009.

Adlam, T., Carey-Smith, B., Evans, N., Orpwood, R., Boger, J., & Mihailidis, A. (2009). Implementing Monitoring and Technological Interventions in Smart Homes for People with Dementia, Case Studies. In Gottfried, B., & Aghajan, H. (Eds.), *Behaviour Monitoring and Interpretation, Smart Environments*. IOS Press.

Aghajan, H., & Wu, C. (2007). From Distributed Vision Networks to Human Behaviour Interpretation. In Gottfried, B. (Ed.), *BMI'07* (*Vol. 296*, pp. 129–143). CEUR.

Alder, G. S. (1998). Ethical Issues in Electronic Performance Monitoring: A Consideration of Deontological and Teleological Perspectives. *Journal of Business Ethics*, *17*, 729–743. doi:10.1023/A:1005776615072

Andrienko, N., & Andrienko, G. Extracting patterns of individual movement behaviour from a massive collection of tracked positions. In: BMI 2007, vol. 296, 1–16. CEUR (2007)

Aoidh, E. M., & Bertolotto, M. (2007). Improving Spatial Data Usability by Capturing User Interactions. In Fabrikant, S. I., & Wachowicz, M. (Eds.), *The European Information Society, Leading the Way with Geo-information, Lecture Notes in Geoinformation and Cartography*. Springer.

Aztiria, A., Izaguirre, A., Basagoiti, R., & Augusto, J. C. (2008). Autonomous Learning of User's Preferences improved through User Feedback. In Gottfried, B., & Aghajan, H. (Eds.), *BMI'08* (*Vol. 396*, pp. 87–101). CEUR.

M. Boronowsky, O. Herzog, M. Lawo. Wearable Computing: Information and Commun-ication Technology Supporting Mobile Workers (Wearable Computing: Informations- und Kommunikationstechnologie zur Unterstützung mobiler Arbeiter). it - Information Technology 50(1), 30-39, 2008.

Goshorn, R., Goshorn, D., & Kölsch, M. The Enhancement of Low-Level Classifications for Ambient Assisted Living. In B. Gottfried and H. Aghajan, editors, *2nd Workshop on Behaviour Monitoring and Interpretation (BMI'08)*, volume 396, pages 87–101. CEUR Proceedings, 2008.

Gottfried, B. (Ed.). *Behaviour Monitoring and Interpretation*, volume 296. CEUR Proceedings, 2007.

Gottfried, B. (a). Behaviour Monitoring and Interpretation: A Computational Approach to Ethology. In B. Mertsching et al. (eds.). *The German Conference on Artificial Intelligence*. LNAI 5803, 572-580, Springer, 2009.

Gottfried, B. (2009). (b). Locomotion Activities in Smart Environments. In Nakashima, H., Augusto, J. C., & Aghajan, H. (Eds.), *Handbook of Ambient Intelligence and Smart Environments. To appear at Springer*.

Gottfried, B., & Aghajan, H. (Eds.). *Behaviour Monitoring and Interpretation*, volume 396. CEUR Proceedings, 2008.

Gottfried, B., & Aghajan, H. (Eds.). (2009). *Behaviour Monitoring and Interpretation – Smart Environments*. IOS Press.

Greenfield, A. (2006). *Everyware: the dawning age of ubiquitous computing*. New Riders. 11–12.

Harnad, S. (1990). The Symbol Grounding Problem. *Physica D. Nonlinear Phenomena*, *42*, 335–346. doi:10.1016/0167-2789(90)90087-6

Heine, C., & Kirn, S. Adapt at agent.hospital - agent based support of clinical processes. In Proceedings of the 13th European Conference on Information Systems, The European IS profession in the Global Networking Environment, ECIS, Turku, Finland, June 14–16, p 14, 2004.

Kiefer, P., & Schlieder, C. (2007). Exploring Context Sensitivity in Spatial Intention Recognition. In Gottfried, B. (Ed.), *BMI'07* (*Vol. 296*, pp. 102–116). CEUR.

Kritzler, M., Lewejohann, L., & Krüger, A. (2007). Analysing Movement and Behavioural Patterns of Laboratory Mice in a Semi Natural Environment Based on Data collected via RFID-Technology. In Gottfried, B. (Ed.), *BMI'07* (*Vol. 296*, pp. 17–28). CEUR.

Mea, V. D., Pittaro, M., & Roberto, V. Knowledge management and modelling in health care organizations: The standard operating procedures. In: M. Wimmer (ed), Knowledge Management in Electronic Government, 5th IFIP International Working Conference, KMGov Krems, Austria, May 17–19, 2004, Proceedings, Springer, LNCS, vol 3035, 136–146, 2004.

Mihailidis, A., Boger, J., Craig, T., & Hoey, J. (2008). The COACH prompting system to assist older adults with dementia through handwashing: An efficacy study. *BMC Geriatrics, 8*(28).

Millonig, A., Brändle, N., Ray, M., Bauer, D., & van der Spek, S. (2009). Pedestrian Behaviour Monitoring: Methods and Experiences. In Gottfried, B., & Aghajan, H. (Eds.), *Behaviour Monitoring and Interpretation, Smart Environments*. IOS Press.

Millonig, A., & Gartner, G. (2007). Shadowing, Tracking, Interviewing: How to Explore Human ST-Behaviour Patterns. In Gottfried, B. (Ed.), *BMI'07* (*Vol. 296*, pp. 29–42). CEUR.

Panina, D., & Aiello, J. R. (2005). Acceptance of electronic monitoring and its consequences in different cultural contexts: A conceptual model. *Journal of International Management, 11*, 269–292. doi:10.1016/j.intman.2005.03.009

Terzic, K., Hotz, L., & Neumann, B. (2007). Division of Work During Behaviour Recognition. In Gottfried, B. (Ed.), *BMI'07* (*Vol. 296*, pp. 144–159). CEUR.

Thirde, D., Borg, M., Ferryman, J., Fusier, F., Valentin, V., Brémond, F., & Thonnat, M. A Real-Time Scene Understanding System for Airport Apron Monitoring. *IEEE International Conference on Computer Vision Systems (ICVS 2006)* in New York City, USA, 2006.

Weiser, M. (1991, September). The Computer for the Twenty-First Century. *Scientific American*, 94–10. doi:10.1038/scientificamerican0991-94

Wittgenstein, L. Philosophische Untersuchungen, Oxford 1953.

Chapter 22
Recognising Human Behaviour in a Spatio–Temporal Context

Hans W. Guesgen
Massey University, New Zealand

Stephen Marsland
Massey University, New Zealand

ABSTRACT

Identifying human behaviours in smart homes from sensor observations is an important research problem. The addition of contextual information about environmental circumstances and prior activities, as well as spatial and temporal data, can assist in both recognising particular behaviours and detecting abnormalities in these behaviours. In this chapter, we describe a novel method of representing this data and discuss a wide variety of possible implementation strategies.

INTRODUCTION

Over recent years, research in ambient intelligence and smart homes has experienced a significant push, with applications in a wide range of areas. One field where smart homes are of particular interest, and therefore reoccur frequently as a research topic, is those homes that aim to improve the quality of life of the inhabitant in one way or another. This could be by providing a safer environment through security measures such as intruder alerts, adjusting environmental conditions such as temperature to the inhabitants' activities, or supporting the inhabitants in their daily activities. In particular, the last of these has gained a significant amount of attention. This does not come as a surprise, since more and more individuals wish to live independently in their own homes into old age, but are not always able to do so, due to diminishing physical or mental capabilities caused by the aging process or diseases like Alzheimer's. An ambient intelligence system can monitor individuals in a non-obtrusive way in their own homes, and provide assistance whenever necessary. This assistance can range from assurance (making sure that the individual is safe and performing routine

DOI: 10.4018/978-1-61692-857-5.ch022

activities) through support (helping individuals to compensate for impairment by providing reminders) to assessment (determining the physical or cognitive status of the inhabitant and alerting a caregiver if necessary).

To offer effective assistance, the ambient intelligence has to analyse the activities performed by the individuals in the home, and to infer the behaviours from those activities, as they are observed through the sensors. We use the term 'behaviour' to mean a particular set of activities performed with a particular aim in mind, such as doing the laundry, cooking dinner, making breakfast, or watching TV. These are sometimes known as Activities of Daily Living (ADLs). One of the main challenges in behaviour recognition is that the exact activities are not directly observed: the only information provided are the sensor observations, which could be that the kitchen light is on, the oven is turned on and the burner is on; the inference that therefore somebody is cooking is left to the intelligent part of the system. This is particularly true where cameras are not used, something that we tend to assume since cameras are intrusive and the images can be difficult to analyse.

AN AI PERSPECTIVE ON THE SMART HOME

The smart home can be separated into the sensory system and the ambient intelligence that works to interpret the sensor observations. In this chapter, we do not explicitly consider what types of sensor are (or could be) available in the home, but assume that the sensory stream is available in the form of a sequence of 'tokens', i.e., there has been some preprocessing of the sensor readings (and possibly the fusion of different sensor data) into a sequence of observations that can be used by some form of ambient intelligence system. For our purposes, we consider that the task of the smart home is to segment this token stream into

different behaviours, and to identify whether or not this behaviour is typical of the user and, if not, whether it is sufficiently abnormal to warrant any action (such as calling a carer, or interacting with the inhabitant, for example in the form of reminders: 'did you remember to turn the gas off?').

Our interpretation of the smart home problem from the point of view of artificial intelligence is that it requires solutions to at least some of the following problems:

- **behaviour recognition and segmentation** Assuming that data from the sensors is represented by tokens, the home needs to process that data into individual behaviours and classify them. The challenges in this are that the data is potentially noisy, certainly provide only a partial snapshot of the person's activities, and may well be difficult to segment with complete accuracy. There is lots of work going on in this area around the world, with most of the more successful approaches being based on probabilistic methods such as Hidden Markov Models, e.g., (Nguyen *et al.*, 2005; Duong *et al.*, 2005).

- **novelty detection** Once the inhabitant's behaviour has been recognised, the system needs to decide whether or not the behaviour is novel (i.e., the inhabitant is doing something that they have not done before, so that the classification failed) or if they are doing it in an odd way (either the activities that they are performing are not quite correct, or the time or place are unusual). We believe that it is not generally possible to do this by standard classification, and so novelty detection systems are the most promising approach (Rivera-Illingworth, Callaghan and Hagras, 2007; Jakkula and Cook, 2008; Marsland, 2003). For a smart home that performs monitoring, this is the part of the system that is likely to produce warnings for carers, or some form of interaction with the inhabitant.

- **the ability to utilise temporal, spatial, and contextual data** This chapter is primarily focussed in this area. Using this data can improve the results of both behaviour recognition and novelty detection; indeed, it may be essential for detecting novelty in many cases.
- **ontologies and background knowledge** While a purely evidence-based system would obviously be nice, there are things that are not possible using such a system, such as dealing with new appliances. For example, an ontology of household appliances could be used so that when a new kitchen appliance is bought, the system can theorise tentatively about how the inhabitant could expect to use it. Equally, background knowledge about what is expected may speed up the learning required in the system. While our philosophical preference is for a system that does not require prior knowledge to be installed, this may not be practical in many cases.
- **lifelong learning** Things change within houses and in people's behaviours all the time. The system needs to be able to learn about these things so that it does not annoy the inhabitants and carers by providing multiple false alarms. In addition, not all behaviours may appear during a training phase, and so the system has to be able to receive feedback and update itself.
- **the ability to communicate with other smart homes** One way to avoid the problem of not all behaviours being seen during training could be to use data that other smart homes have seen, and update itself when the inhabitant performs those behaviours. Additionally, if one house detects a potential danger, it can warn other houses that they may see the same thing.

A different way to see some of the same features is to use the classification of abnormal behaviour suggested by Russell Dewey (2009):

- **statistical abnormality** Based on deviation from what statistically is considered usual behaviour.
- **violation of socially-accepted standards** Based on judgment made by religious, cultural, or social groups.
- **theoretical approaches** Based on a theory of personality development.
- **subjective abnormality** Based on a personal assessment of normality.
- **biological injury** Based on the impact that abnormal biological processes such as disease or injury have on the individual.

The first of these categories is the typical one that machine learning systems can deal with. It takes only statistical evidence into consideration and refrains from any judgments made by the individual living in the house or people related to that individual. The idea is that an individual starts off with behaviours that are considered normal (regardless of how strange they might appear when imposing cultural, social, or religious values). These behaviours are used to initialise the ambient intelligence without any prior knowledge of what to expect, and then learns to identify the particular individual's behaviours, for example through machine learning techniques or some logic-based approach. Once trained, the ambient intelligence is used to detect behaviours that deviate from normal behaviours, and thus are considered abnormal.

Although, in principle, this seems to be a straightforward process, there is a major problem with this approach. Behaviours might change over time, but only some of the changes indicate abnormal behaviour. To be able to distinguish between the changes that are normal and those that are abnormal, the behaviours have to be interpreted in context. In other words, the ambient intelligence not only needs observational information, but also has to be aware of the context in which the observations were made.

The second of the five types of abnormality – and to some extend the third – could be approached using an ontology-based method that had codified knowledge of societal norms, and applied them, while the fifth type could be identified through the integration of knowledge from many different smart homes, so that behaviours that caused accidents, or disease diagnoses that changed behaviours, could be identified.

In this chapter, we focus on how to create a mapping between sensor streams and human behaviours, the problems that this task brings with it, and possible ways to overcome these problems. We will do this by considering the properties of the sensor stream with regard to space, time, and context.

USING CONTEXTUAL DATA IN A SMART HOME

Considering just the token level, contextual information can take two different forms: the input of additional sensors that provide background information either about the environment (e.g., temperature, humidity level) or the user (e.g., emotion detection), and previous outputs from the behaviour recognition system, so that the system has some idea of what the person has been doing recently.

Before going further it is worth considering two different ways in which all of this data can be used. The first is as part of the behaviour recognition, while the second is as part of the 'normality' detection unit. To see the difference, consider the following two statements:

1. it is morning and the toaster is being used, so the inhabitant is probably making breakfast
2. the inhabitant is making breakfast and it is morning, so everything is normal

The first uses the temporal data, together with the sensor data, to decide on a behaviour, while the second uses the behaviour and the temporal data to ensure that the behaviour should not cause concern. Both of these interpretations have implicitly used spatial data, since the sensors that identify use of the toaster are presumably in the kitchen. So is one of these two ways to think about the problem more useful, or are both valid? It seems to us that there is no reason to limit the use of any potentially useful data to just one task. Note, however, that the second one allows for the possibility of negation: if it is morning and the person does not make breakfast, then this could be identified as abnormal. This may well be particularly important for a smart home, where things not done could be important either for safety (e.g., gas not turned off) or as a sign of increasing forgetfulness. Detecting that things are not done can be particularly challenging for probabilistic machine learning methods, and this could be one place where logic-based intelligence is required.

The other challenge is that much of the data is not useful for a particular decision. Statements like 'it is 22.5 degrees and the toaster is being used, so they are probably making breakfast' are unlikely to be correct very often. Working out which correlations are useful is a tricky data mining problem, and one that could require a lot of data. If we need to collect a very large dataset before the smart home can be used, then people may not be prepared to use the system. For these reasons, we prefer to separate out the behaviour recognition system from the environmental data, although ideally the system should be able to take advantage of that data if it is available.

THE DIFFERENCE BETWEEN SPACE AND TIME FOR SMART HOMES

Considering the sequence of sensor-based tokens, the first thing to realise is that spatial and temporal properties of the data are to some degree different:

sensors are physically located in one particular location, and are (in the main) unlikely to move. There are exceptions to this, from furniture rearrangement to sensors on objects such as laptop computers that could be used in different rooms, but in general the location of sensors is fixed, and so if the fridge sensor emits the token that says that the fridge has been opened, then it seems reasonable to assume that there is somebody in the kitchen. For this reason, in general there is no need to tag tokens with the spatial location of the sensor. However, the same thing is obviously not true about time – the same sequence of tokens at two different times of day could well be labelled differently, from trivial differences (cooking in the morning is breakfast, while cooking in the evening is dinner) to more important ones: cooking in the middle of the night could well be a sign of confusion. For this reason, it is usual to consider the sequence of tokens as being tagged with time stamps, but not with spatial locations.

There are a couple of corollaries to this difference between space and time: token activations can occur simultaneously in time, and tokens from different sensors can be interleaved, but if there is only one person in the house, then sensors in two different rooms, or even two parts of the same room, cannot be simultaneously triggered (assuming that the areas covered by the sensors are not overlapping). Another is that we have different resolutions of space and time: the pre-defined set of sensors may well miss out several locations, so that we have no idea what the person is doing at various locations – if the fridge door sensor goes on, and then an hour later goes off, this does not necessarily mean that the person stood at the fridge for an hour, more likely they forgot to close the door properly. However, if they did not trigger any other sensors in that hour, then the system will not know what they were doing. Temporally, there is information in the fact that other sensors were not triggered, but not spatially.

This leads us to the problems of temporal and spatial persistence (Guesgen and Hertzberg,

1996), which can roughly be characterised as follows. Assuming that we do not have complete knowledge about what is and is not true at every individual time point, we need to identify those facts whose truth at some time point will persist until some later time point. For example, if the fridge sensor emits the token that says that the fridge has been opened and we then assume that there is somebody in the kitchen, for how long would we maintain this assumption? If we get a token from the living room sensor after a while, we know that the assumption is no longer true. On the other hand we cannot guarantee the persistence of the assumption if no tokens are received, since it may be that the person left the kitchen, but did not trigger any other sensors.

In addition to the temporal persistence problem, we have to deal with the spatial persistence problem. Assuming that we do not have complete knowledge about what is and is not true at every location in the home, we wish to identify those facts known to be true at some location that persist to other locations of the home. For example, if we know that a particular spot on the sofa has enough light to read a book, would we be able to make this assumption for all locations on the sofa or even the whole living room? At which points in space do we drop the assumption of persistence? In answering this question, we face the problem that space does not have an intrinsic direction, as time does. There are many possible trajectories that we can take to move away from a particular location. To get some order into this large set of possible movements through space, we usually employ concepts such as topology, orientation, and distance; something we discuss briefly later on.

Returning to temporal information, there are a number of ways to specify a particular point in time: either absolute (15:34:34 GMT on 24/06/09) or relative (15 minutes after I get home from work, or the first Sunday after the first full moon after the vernal equinox). Implicit in those three different descriptions is another interesting feature, which is the 'granularity', i.e., the resolution at which the time is represented. This is very different in the three cases, and choosing the appropriate resolution is certainly non-trivial. This becomes even more important when you consider how behaviours could be classified in time. It is common for working people to have pretty much two modes of behaviour: week days and the weekend. Within these two times, the pattern is often fairly constant, but there is a marked difference between the two. Things that would be normal on a weekend (lying in bed reading the newspaper, for example) would be a definite sign of something being wrong on a weekday. A smart home that treated every day the same would annoy people very quickly!

When looking at spatial information, the issue becomes even more complex in that it is not clear which space we are referring to. One could argue that normal three-dimensional space is natural to describe movement within the smart home, but this space does not directly correspond to the space that the sensors cover. For example, the sensor at the entry door has a particular coverage area, and if this sensor is activated, then we can assume that a person has entered this area. If immediately after that another sensor is activated, we might conclude that the person has moved from the entry-door area to the area covered by the other sensor, but only if the areas do not overlap. In summary, the space we are really looking at is a space of more or less abstract three-dimensional regions that are related to each other in some way. The question is how the regions are related to each other.

Three types of spatial information are commonly distinguished: topological information, orientation information, and distance information. Children acquire a sense for these types of spatial information in that order (Piaget and Inhelder, 1948), which might suggest that topological information is the most essential one among the three. It is indeed the case that by using this type of information, we can draw conclusions that can significantly improve the behaviour recognition in smart homes. For example, suppose that a sensor

is activated in the kitchen and then another sensor is activated in the living room. If we know that the corridor is between the kitchen and the living room and that the sensor in the corridor has not been activated, then we can conclude that the sensors are activated by different people and therefore belong to different activities. This assumes that the topological relations between the areas are such that they do not allow activation of the sensors in the kitchen and the living room by one person without activating the sensor in the corridor. Note that this does not necessarily mean that the areas covered by the three sensors have to be pairwise disjoint; they can partially overlap. We provide more details on topological relations in a later section of this chapter.

Orientation information is the second type of spatial information that can be used to put sensor data into context. It can be viewed as a ternary relation of the primary object, the reference object, and a frame of reference. For example, the TV (primary object) might be in front of the chair (reference object), where the edge of the seat of the chair is used as frame of reference. The role of the frame of reference here is to define the front side of the object; without it, a given orientation relation between two objects might have several interpretations. Three different types of reference frame (Levinson, 2003) are usually distinguished:

- **intrinsic** The orientation is given by some inherent properties of the reference object.
- **relative** The orientation is imposed by the point of view from which the reference object is seen.
- **absolute** The orientation is given by fixed directions such as the cardinal directions (north, south, east, west) or the direction provided by gravity.

For example, a statement like 'the banana peel is in front of the person' can either put the banana peel close to the person's toes (intrinsic reference frame) or at the person's side (relative reference frame with the observer watching a person passing by). In an absolute reference frame, we would rather use a statement like 'the banana peel is north of the person'.

In principle, we can map one reference frame into another, and therefore we are not concerned here about which particular reference frame has been chosen to represent orientation in the ambient intelligence of our smart home. What is important to us is the fact that orientation can add an additional quality to the reasoning process. For example, if the chair is not facing the TV and the sensor in the chair is activated, then it is unlikely that the person is watching TV.

The third type of spatial information is distance information. There are numerous studies of how humans make subjective judgements regarding distances, in particular how humans determine how close objects are to each other. According to (Gahegan, 1995), the human perception of closeness or proximity is influenced by the following:

- In the absence of other objects, humans reason about proximity in a geometric fashion. Furthermore, the relationship between distance and proximity can be approximated by a simple linear relationship.
- When other objects of the same type are introduced, proximity is judged in part by relative distance, i.e., the distance between a primary object and a reference object.
- Distance is affected by the size of the area being considered, i.e., the frame of reference.

Although it is tempting to use Euclidian distance to determine proximity, this is not always the most intuitive way. Humans often have a qualitative rather than a quantitative perception of distance (Guesgen, 2002). For example, if the person sits down in front of the fire and we know that there is a 'safe' distance between the two, then we can conclude that there is no danger of the person getting burned.

CONTEXT AWARENESS

There are a great many forms that context awareness can take within a smart home. The way that it is most commonly understood is that individual activities of people within the house are not interpreted in isolation, but that additional data concerning, for example, the temperature and what other activities they have been engaging in are included. This can be useful since behaviours that appear to be perfectly normal in isolation can look odd with further inspection: sitting by a log fire is perfectly normal in winter, but when the temperature is 30 degrees (Celcius) in the house, it might be an indication of mental instability. Likewise, cooking dinner is a perfectly normal activity, unless it is the third one that the person has cooked today. Considered from this point of view, context awareness is concerned with adding information to assist in either the classification of the activity, or the analysis of whether or not this behaviour is 'normal'.

However, there are many other ways in which context awareness can be included in a smart home, which go beyond the individual home. The ability of the house to communicate with other houses can be very useful. From the public health point-of-view the detection of pandemics and other spreads of illness could be greatly improved by each smart home reporting on the health status of its inhabitants based on its normal monitoring of their behaviour. Considering the same thing in reverse, once houses were informed that there was disease prevalent within their locality, they could warn the inhabitants to be vigilant, or to stock up on food, etc. Additionally, there is a large pool of data that can be accessed in this way to enable houses to learn about behaviours that could be demonstrated by their own inhabitants in the future.

There is another aspect of context awareness, which is that individual behaviours can act to prime (make more likely) or inhibit others. If the house inhabitant has just had breakfast, then behaviours such as leaving the house (to go to work) or have another cup of coffee (at the weekend) may be quite likely, whereas going to bed is not. By increasing or decreasing the probability of other behaviours being seen after the current one based on experience, the system could improve the accuracy of the behaviour recognition. The length of time for which the priming remains active could also be tuned, so that some behaviours have long-term effects, while others are rather shorter. There are a wide variety of mechanisms that could be used to implement this kind of behaviour, from weights to a Hierarchical Hidden Markov Model (Fine, Singer and Tishby, 1998). A schematic of this idea is shown in Figure 1.

A SYSTEM USING SPATIAL, TEMPORAL, AND CONTEXTUAL DATA FOR BEHAVIOUR RECOGNITION

Figure 2 shows a possible way to visualise our proposed system. Tokens are used to identify possible behaviours using some system, such as Hidden Markov Models. The decision about which behaviour is currently seen may be made using just the token stream, as has been considered in a variety of papers in the literature, e.g., (Fine, Singer and Tishby, 1998). However, if there is other data available, then it could be used to weight the decision. In the figure we have separated the three groupings that we have been discussing, considering different representations for spatial, temporal, and contextual information.

Figure 1. The use of priming between behaviours can increase or decrease the probability of a particular behaviour being recognised next, based on the current behaviour

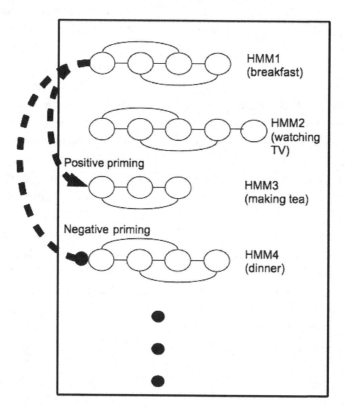

Figure 2. Our conceptual system uses different map layers to represent spatial, temporal, and contextual data

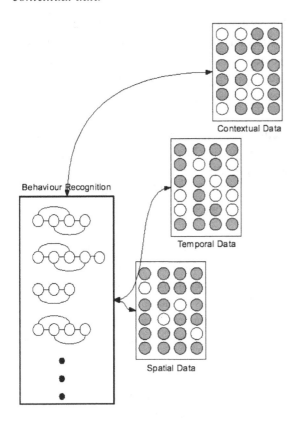

A basic version of the behaviour recognition system could be trained first, to identify behaviours based solely on their token pattern. The spatial, temporal, and contextual data concerning these tokens could then be used as input for each of these different modules. One problem that would have to be solved is to identify suitable granularities for the spatial and temporal representations. Our current thinking is that this should be possible to do by adding many different granularities into the map and allowing the ambient intelligence to work out for itself which are useful. Figure 3 shows two different representations of time, considering it as an absolute measure (left) and as a relative measure (right). Note that there is considerable overlap between different granularities: for example, seasons can be mapped to months without too much difficulty.

If we consider the absolute time map shown on the left of Figure 3 as representing the time that an activity begins (or at least, when the system first detects it), then times at machine precision are unlikely to be useful, since humans do not work at this scale. Each behaviour that occurs will cause a pattern of activation in the various

Figure 3. Two temporal maps based on absolute time (left) and relative time (right). The concepts in white are those that identify the current time, so it is a weekend afternoon in March, which is (Southern hemisphere) summer

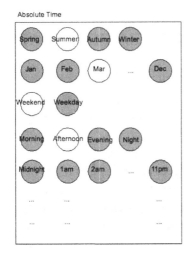

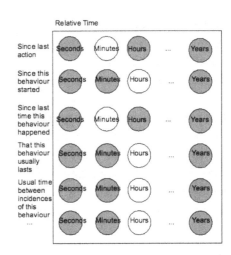

'nodes', and as each behaviour is seen repeatedly, so the commonalities of the pattern should be identified and correlated with the particular behaviour. A number of different ways in which this could be done are discussed later.

Similarly, the relative time map should produce patterns of how long behaviours take, and how frequently they occur. This can include historical information, such as the normal length of time that this behaviour takes, and the common time between incidences of this behaviour. As more than one of the timescale options can be switched on simultaneously (to represent, for example, that incidences of a certain behaviour happen between daily and weekly) this is not too restrictive in the variation in time that it allows – this could be considered as something like a fuzzy logic system, where at midday, both the morning and afternoon representations would be on, not a crisp cut-off between the two.

One problem with our representation using this map is that only the time from the one previous behaviour is considered, and there may well be longer-range correlations. However, even using just one previous behaviour can lead to the system requiring massive amounts of data in order to identify patterns, something that will obviously get even worse with more inputs; the so-called curse of dimensionality. A possible way to avoid this problem is to include additional information and reasoning methods, for example by using logical representations of some parts of the data space. For now, we mention just one other thing that might be useful, which is that some actions are more likely than others to be used as a reference in this kind of relative timing – you are more likely to say that you had a cup of tea 10 minutes after you got in from work than that you had the tea 2 minutes after you fed the cat. If the actions that are used in practice for relative timing are possible to identify, then it could partially alleviate this problem.

POSSIBLE IMPLEMENTATION STRATEGIES

The system that we have shown in Figure 2 does not give any indication at all about implementation methods, and during the previous discussion we have highlighted some places where probabilistic methods would be useful, and others where they would not. There are certainly lots of possible methods, and we do not aim to describe a possible implementation that is sure to work; rather, we wish to produce an overview of some technologies that might be useful, particularly in light of the discussion of spatial and temporal data, and context.

We reiterate that the idea behind the different maps of the data is that what is interesting is 'patterns' of data in the maps, not the activation of individual nodes in each of the maps. If the latter were the case, then a set of pure logic rules of the form 'if it is winter and the temperature is less than 20 degrees then the inhabitant should turn the heater on' would be more suitable than our system. There are, of course, ways to learn these types of rules from data, using methods such as inductive logic programming, but that is not our focus here. However, there are plenty of places where symbolic artificial intelligence methods can give us useful data, and we discuss these next, before moving on to probabilistic models.

SPATIO-TEMPORAL REASONING

Spatial, temporal, and contextual data obey a variety of laws. A drop in temperature from 23 degrees to 22 degrees is not possible without an intermediate temperature of 22.5 degrees, meaning that if the temperature is 23 degrees at time t_1 and 22 degrees at time t_2, and if t_1 is before t_2, then we know that there is a time t_3 between t_1 and t_2 at which the temperature is 22.5 degrees. However, this only holds if the temperatures are observed in the same location. If 23 degrees are measured in the living room at time t_1 and 22 degrees in the kitchen at time t_2, then it is possible that there is no time t_3 at which either the living room or the kitchen has a temperature 22.5 degree.

Researchers have studied universal laws like these for years, particularly in the context of temporal and spatial information. The result is a variety of formalisms that enable us to reason about space and time, which broadly can be put into two categories. One category contains the formalisms used by scientists such as physicists to establish the laws of time and space. We are not primarily interested in this category, but focus on the second category, which contains those formalisms that mimic human reasoning about space and time. The reason for favouring this category is obvious: if a layperson intends to communicate with an ambient intelligence, then it is easier to do so in a language that they are familiar with.

It has been argued in the field of artificial intelligence for many years that the layperson's everyday form of spatio-temporal reasoning is of a qualitative, rather than a quantitative nature; we are not usually interested in precise descriptions of space and time. Coarse and vague qualitative representations frequently suffice to deal with the problems that we want to solve. For example, to know that the meal is prepared in the kitchen (rather than knowing the exact coordinates for this activity) and at lunchtime (rather than 12:03:37) is often enough to decide whether the behaviour is normal or not. For that reason, we have included such vague description in our spatial and temporal layers.

Most qualitative approaches to spatial and temporal reasoning are relational approaches, i.e., approaches that reason about relations between objects (such as regions or time intervals). Of the temporal qualitative approaches, two are dominant and reappear frequently in different scenarios: Allen's temporal logic (Allen, 1983) and Vilain and Kautz's point algebra (1986). The first uses intervals to describe specific events, and a set of 13 basic relations between intervals to describe

interdependencies between events. Allen introduced an algorithm that is based on a composition table to reason about the relations. For example, if we know that the kettle is filled with water before it is switched on and if we further know that the water boils while the kettle is switched on, the algorithm can infer that the kettle is filled with water before the water boils. The algorithm caters for uncertainty in the reasoning process by allowing sets of possible relations in addition to the basic relations. For example, it might be possible for the water to continue boiling after the kettle has been switched off (due to the heating element still being hot), in which case the algorithm would use both the 'during' and 'after' relations in the reasoning process.

Unlike Allen's temporal logic, the point algebra uses time points and three possible relations to describe interdependencies among them: < (precedes), = (same as), and > (follows). These relations can be used in a similar way to interval relations. For example, the time point at which we have finished filling the kettle precedes the time point at which we switch on the kettle. In fact, Vilain and Kautz pointed out that many interval relations can be expressed as point relations by using the starting and finishing endpoints of the intervals.

Both Allen's interval logic and Vilain and Kautz's point algebra can be used for spatial reasoning by interpreting intervals as one-dimensional objects and time points as locations in space, respectively. A multi-dimensional representation can then be achieved by using a coordinate system and describing the spatial relations with respect to the coordinate axes. However, this approach suffers from the fact that reasoning can lead to counterintuitive results when the objects are not aligned with the coordinate axes. Researchers have therefore focused on purely topological approaches that do not depend on such an alignment.

We argue that the topological aspects of space are the most relevant ones for modelling human behaviour in an ambient intelligence. For example, to determine what activity a person might be involved in, it is useful to know which room that person is in (although not necessarily the person's exact position in that room). Or, to determine the possible movements of a person, it is useful to know which rooms are connected to which other rooms. We will therefore focus on the topological aspects of space in our system, but at the same time point out that information about orientation and distance can also play a significant role in the reasoning process.

The topological approach that has gained most attention in the AI community is arguably the Region Connection Calculus (Randell, Cui and Cohn, 1992). The basis of this calculus is a reflexive and symmetric relation, called the connection relation. From this relation, additional relations can be derived, which include the eight jointly exhaustive and pairwise disjoint RCC8 relations. These relations can be used to express, for example, that two regions are disconnected from each other, that they are partially overlapping, or that one is a non-tangential proper part of the other (inside the other region without touching the boundary). Reasoning about space is achieved in the RCC framework in the same way as in Allen's temporal logic: by applying a composition table to pairs of relations.

It should be pointed out that the RCC theory is not the only approach to reasoning about topological relations. Egenhofer and Franzosa (1991) consider the intersections of boundaries and interiors of pairs of regions and derive a formalisation from that, called the 4-intersection calculus (and by adding the complement of the regions to the intersections, the 9-intersection calculus).

BEHAVIOUR RECOGNITION

Until now, we have not discussed the methods by which behaviour recognition and segmentation can be performed. This is because we are considering our temporal and spatial data as separate to the principal behaviour recognition problem, which is based solely on the sensor observations; the categorisation of temporal and spatial data that we have described provides a different way to recognise behaviours, or identify unusual occurrences of behaviours. We do not expect that the exact mechanism used for activity selection will be crucial to this system. Behaviour recognition is one of the more commonly investigated areas of smart home research (Duong *et al.*, 2005; Tapia, Intille and Larson, 2004).

The approach that we favour for behaviour recognition and segmentation is a set of competing HMMs, each of which represents a different behaviour. The HMM (Rabiner, 1989) is the simplest dynamic graphical model, and has been used effectively on many problems. In comparison to many graphical models, it has the benefit that there are computationally tractable algorithms for both training and inference (Marslan, 2009). The challenges of using a set of HMMs for this problem are:

- **training a new HMM on some data** Where data is labelled with the behaviour, this is relatively simple, but without the labels, it is considerably more difficult. The LeZi algorithm (Das *et al.*, 2002) is one approach to this.

- **segmenting the input stream into behaviours** This part of the problem is concerned with recognising behaviours based on the sensory observations, and then parsing the stream of data into individual behaviours. There are methods to do this using competition between HMMs alone (by monitoring the likelihood values of the HMMs on the data, see Chua, Marsland and Guesgen, 2009), but the system that we have described in this chapter can also assist in this segmentation problem by identifying the common spatio-temporal patterns in them.

- **dealing with behaviours that have not been previously seen** It is important that sensory observations that do not match any current behaviour are identified, as it is one aspect of novelty detection to see that the home inhabitant is behaving oddly. However, the ability to then add new HMMs that represent additional behaviours provides a mechanism to allow lifelong learning. It may also enable the system to begin building from a *tabula rasa* state, i.e., the smart home starts with sensors being installed, and without any additional programming and knowledge specific to the particular installation.

Much of the other research on activity segmentation in smart homes has focused on more complicated variants of the HMM, such as the Hierarchical Hidden Markov Model (Nguyen *et al.*, 2005), or Switching Hidden Semi-Markov Model (Duong *et al.*, 2005). In both of these models, a top-level representation of behaviours (e.g., cooking or making coffee) is built up from a set of recognised activities that arise from the individual sensor values. A variant of these methods uses a three level Dynamic Bayesian Network (Liao *et al.*, 2007). These models can be seen as adding complexity to the HMM in order to represent a complete model of behaviours arising from sensor activations. The problem is that more complex models require more data for training, and have higher computational cost.

POSSIBLE IMPLEMENTATIONS OF THE LAYERS

We now discuss a number of possible machine learning algorithms that may be useful in order to realise our idea of different spatial, temporal, and contextual layers for behaviour recognition and novelty detection. We will mention several different types of machine learning algorithm here; the interested reader is directed for more details

to any of the standard texts in the field, such as (Marsland, 2009; Bishop, 2007).

One place where 'maps' like those we have drawn can be seen is in the Self-Organising Map (SOM) of Kohonen (1993), a neural network that has been very widely used in many different application areas. However, in the SOM, the pattern of activations in the network is created by the inputs through the set of weights, which is exactly the opposite of what we are planning here: it is the patterns of activations in the various maps that should assist with behaviour recognition and possibly with the discovery that this behaviour is abnormal or unusual in the time, place, or circumstances in which it is done.

It would be possible to use the entire set of map activations as inputs for a neural network, which would have outputs for the different behaviours that were identified by the system. This could either be done by considering particular patterns as exemplars in a Radial Basis Function Network, and then using a simple linear predictor (such as a Perceptron) to learn the output classes, or just by using the entire set of map activations as inputs to a standard neural network such as the Multi-Layer Perceptron. However, these methods rather remove the benefit of having the patterns in different maps for spatial and temporal data, something that we believe to be important. They also run the risk that massive amounts of data will be required before any useful output can be introduced, and that even when the data is such that no useful output can be generated, they will still produce an output – all of these are well-documented problems with neural networks.

One way of considering the problem that may be useful is to think of the maps as pictures of the spatio-temporal conditions when a particular behaviour occurred. In this case, it is possible to view the problem as one of taking the current set of spatial or temporal data, which may be different in some small way from those previously seen, and identifying a similar set of conditions that are more common. This can be seen as a clustering problem, but it is also the kind of representation that Hopfield networks (Hopfield, 1982) can be used to find very effectively. A probabilistic variant on the Hopfield Network, the Boltzmann Machine (Ackley, Hinton and Sejnowski, 1985), may turn out to be useful for this.

The Boltzmann Machine uses a set of elements that are loosely modelled on neurons (like all artificial neural networks) in that they fire or do not fire based on their inputs. However, unlike most artificial neurons, in the Boltzmann Machine, the input is used to make a stochastic decision about whether or not to fire, with the probability of firing being a function of the strength of the inputs. A set of these neurons are created and connected together using weighted connections, and then the learning problem consists of identifying weights so that the target data is reproduced with high probability. There is a relatively simple, although computationally expensive, algorithm for solving this learning problem, which is described in [26]. While the original Boltzmann Machine used only binary nodes, there are a variety of ways to extend it to use non-binary inputs (Peterson and Anderson, 1987; Sejnowski, 1986).

The neurons of the Boltzmann machine can thus be the various layers of nodes in our system, and the learning problem will be to identify common patterns of activity across the different layers and link them to behaviours. One feature that may make this more useful (although at the cost of increased computational time) is the potential to add latent variables, also known as hidden units. These are nodes that do not have a direct meaning in terms of the layers of the system, but that integrate information from a variety of other nodes, so that non-linear structure in the maps can be learnt.

Another method that could be useful is the Conditional Random Field (CRF) (Lafferty, McCallum and Pereira, 2001; Sutton and McCallum, 2006), a probabilistic graphical model with some similarity to the Hidden Markov Model. Two alternative views of the CRF are as a Boltzmann Machine without hidden nodes, but with higher-order input correlations, and as a Hidden Markov Model with the transition probabilities being defined by probability distributions (conditioned on the states) rather than as constant values.

In this discussion we have presented a large number of different algorithms that can be used for implementation. Our aim is not to suggest which will best match the requirements for the problem, but to provoke research in this area. Certainly, it seems to us that a combination of both symbolic (useful for reasoning about time intervals and spatial relations) and probabilistic (to deal with noise and uncertainty) methods are required. We are currently investigating how these methods can be combined and which algorithms from those discussed above are most suited to the task.

SUMMARY

In this chapter we have considered the problem of interpreting the stream of sensory data coming into a smart home intelligence from its various sensors. We have done this by categorising the various properties of the sensor data. An output by a particular sensor can be interpreted as an assertion that the state of whatever that sensor monitors has changed. There are three types of additional data that we can tag to that observation, concerned with the location of the sensor, the time of the observation, and the environmental variables (themselves the output of other sensors) at that time and in that place. Further, since the sensory data arrives in a temporal stream, we also know activities that are occurring at around the same time, and what behaviours the system has judged those sensory observations to be part of.

Taking these things into account, we have suggested a method by which the various pieces of additional information about the sensory data can be used in conjunction with the data in order to more accurately classify the data into behaviours (or Activities of Daily Living) or, possibly more importantly, to identify abnormal behaviours that could signify illness or otherwise require a carer to be alerted. The system that we propose is based on the representation of the three different forms of additional information in maps, with the pattern of information in the maps being used rather than the individual pieces of data. This enables the system to identify common patterns between different behaviours. We have suggested a variety of artificial intelligence and machine learning techniques that could be used in order to implement such a system.

The maps that represent spatial and temporal information are very different, since the resolution of temporal data can be as fine-grained as you wish (to machine accuracy), while that of spatial data depends on the sensor network. We also separated out relative and absolute time into different maps; the first concerns time from other observations, while the second is based on a system clock; the same separation can be performed with space, bearing in mind the caveats we discussed earlier. We therefore finish by considering some of the properties of data in terms of space, time, and context, and asking questions that need to be considered in order to use our ideas successfully in a smart home.

- **temporal properties**
  - **concurrency:** Do the behaviours correspond to sequences in the sensor stream or are they inter-leaved? If behaviours are interleaved, how can you segment the sensor stream?
  - **multiple scales:** Do the behaviours occur during a short time period or are they extended over longer periods? How does this effect interleaving?
  - **absolute time:** At which time of the day or day of the year does the behaviour occur? Certain activities only occur at certain times or certain days of the year and therefore would not

be reflected in the sensor stream if it does not cover these time periods.
  - **relative time:** Do some behaviours always follow one another? Or never follow each other? Are there some things that can only happen after other events?
- **spatial properties**
  - **absolute space:** Which location or type of location does the sensor reading correspond to? Are there some behaviours that only occur in one room, and some that can occur in every one?
  - **discrete space:** Can the observed space be divided into discrete location or do the sensors provide continous space information (such as distance)? If space is not discrete, associating behaviours with particular locations becomes fuzzy. And does the fact that the sensor locations are specific (and fixed) help or hinder the classification?
- **contextual properties**
  - **behaviour dependencies:** How are the behaviours related to each other? Knowledge about these relationships can be used to prime the recogition process. For example, the tea making behaviour is more likely to be observed if previously the breakfast making behaviour has been recognised. This is linked to the relative time properties discussed above.
  - **environmental influences:** What is the state of the environment and how does it influence behaviours? How can you identify which environmental data is useful, and which is coincidental? For example, breakfast never takes place if the temperature is above 20 degrees Celcius, not because it is too warm, but because in some parts

of the world it never gets that warm before mid-morning.

ACKNOWLEDGMENT

We are very grateful for input from the MUSE group at Massey University.

REFERENCES

Ackley, D., Hinton, G., & Sejnowski, T. (1985). A learning algorithm for Boltzmann machines. *Cognitive Science*, *9*(1), 147–169. doi:10.1207/s15516709cog0901_7

Allen, J. F. (1983). Maintaining knowledge about temporal intervals. *Communications of the ACM*, *26*, 832–843. doi:10.1145/182.358434

Bishop, C. (2007). *Pattern Recognition and Machine Learning*. Berlin: Springer.

Chua, S.-L., Marsland, S., & Guesgen, H. W. Behaviour recognition from sensory streams in smart environments. In *Proc. Australasian Joint Conference on Artificial Intelligence (AI-09)*, Melbourne, Australia, December 2009.

Das, S. K., Cook, D., Bhattacharya, A., Heierman, E. O. III, & Lin, T.-Y. (2002). The role of prediction algorithms in the MavHome smart home architecture. [Special Issue on Smart Homes]. *IEEE Wireless Communications*, *9*(6), 77–84. doi:10.1109/MWC.2002.1160085

Dewey, R. Psychology: An introduction. http://www.intropsych.com/, last visited on 7 July 2009.

Duong, T. V., Bui, H. H., Phung, D. Q., & Venkatesh, S. Activity recognition and abnormality detection with the switching hidden semi-Markov model. In *Proc. CVPR-05*, pages 838–845. 2005.

Egenhofer, M., & Franzosa, R. (1991). Point-set topological spatial relations. *International Journal of Geographical Information Systems*, *5*(2), 161–174. doi:10.1080/02693799108927841

Fine, S., Singer, Y., & Tishby, N. (1998). The hierarchical hidden Markov model: Analysis and applications. *Machine Learning*, *32*, 41–62. doi:10.1023/A:1007469218079

Gahegan, M. (1995). Proximity operators for qualitative spatial reasoning. In Frank, A. U., & Kuhn, W. (Eds.), *Spatial Information Theory: A Theoretical Basis for GIS, Lecture Notes in Computer Science 988*. Berlin, Germany: Springer.

Guesgen, H. W. (2002). Reasoning about distance based on fuzzy sets. [Special Issue on Spatial and Temporal Reasoning]. *Applied Intelligence*, *17*(3), 265–270. doi:10.1023/A:1020087332413

Guesgen, H. W., & Hertzberg, J. (1996). Spatial persistence. [Special Issue on Spatial and Temporal Reasoning]. *Applied Intelligence*, *6*, 11–28. doi:10.1007/BF00117598

Hopfield, J. J. (1982). Neural networks and physical systems with emergent collective computational abilities. *Proceedings of the National Academy of Sciences of the United States of America*, *79*(8), 2554–2558. doi:10.1073/pnas.79.8.2554

V. Jakkula and D. Cook. Anomaly detection using temporal data mining in a smart home environment. *Methods of Information in Medicine*, 2008.

Kohonen, T. (1993). *Self-Organization and Associative Memory* (3rd ed.). Berlin: Springer.

Lafferty, J., McCallum, A., & Pereira, F. Conditional random fields: Probabilistic models for segmenting and labeling sequence data. In *Proc. ICML-01*, pages 282–289, San Francisco, CA, 2001. Morgan Kaufmann.

Levinson, S. C. (2003). *Space in Language and Cognition: Explorations in Cognitive Diversity.* Cambridge, England: Cambridge University Press. doi:10.1017/CBO9780511613609

Liao, L., Patterson, D. J., Fox, D., & Kautz, H. (2007). Learning and inferring transportation routines. *Artificial Intelligence, 171*(5-6), 311–331. doi:10.1016/j.artint.2007.01.006

Marsland, S. (2003). Novelty detection in learning systems. *Neural Computing Surveys, 3,* 157–195.

Marsland, S. (2009). *Machine Learning: An Algorithmic Introduction.* New Jersey, USA: CRC Press.

Nguyen, N. T., Phung, D. Q., Venkatesh, S., & Bui, H. Learning and detecting activities from movement trajectories using the hierarchical hidden Markov model. In *Proc. CVPR-05,* pages 955–960, 2005.

Peterson, C., & Anderson, J. R. (1987). A mean field theory learning algorithm for neural networks. *Complex Systems, 1*(5), 995–1019.

Piaget, J., & Inhelder, B. (1948). *La Représentation de l'Espace chez l'Enfant.* Paris, France: Presses Universitaires de France.

Rabiner, L. R. (1989). A tutorial on hidden Markov models and selected applications in speech recognition. *Proceedings of the IEEE, 77*(2), 257–286. doi:10.1109/5.18626

Randell, D. A., Cui, Z., & Cohn, A. G. A spatial logic based on regions and connection. In *Proc. KR-92,* pages 165–176, Cambridge, Massachusetts, 1992.

F. Rivera-Illingworth, V. Callaghan, and H. Hagras. Detection of normal and novel behaviours in ubiquitous domestic environments. *The Computer Journal,* 2007.

Sejnowski, T. J. (1986). Higher-order Boltzmann machines. *AIP Conference Proceedings, 151*(1), 398–403. doi:10.1063/1.36246

Sutton, C., & McCallum, A. (2006). An introduction to Conditional Random Fields for relational learning. In Getoor, L., & Taskar, B. (Eds.), *Introduction to Statistical Relational Learning* (pp. 93–128). Boston, MA: MIT Press.

Tapia, E. M., Intille, S. S., & Larson, K. Activity recognition in the home using simple and ubiquitous sensors. In *Proc. PERVASIVE-04,* pages 158–175, Linz/Vienna, Austria, 2004.

Vilain, M., & Kautz, H. Constraint propagation algorithms for temporal reasoning. In *Proc. AAAI-86,* pages 377–382, Philadelphia, Pennsylvania, 1986.

KEY TERMS AND DEFINITIONS

Behaviour Recognition: Identifying and classifying activities on the basis of observations such as sensor data streams.

Sensor Observation: Data stream consisting of tokens associated with sensor readings.

Activity of Daily Living: Set of activities performed with a particular aim in mind, often reoccurring on a regular basis.

Novelty Detection: Recognising behaviours that are new or unusual.

Lifelong Learning: The ability of the system to adapt to changes in the environment and in human behaviours.

Hidden Markov Model: Probabilistic graphical model that uses a set of hidden (unknown) states to classify a sequence of observations over time.

Neural Network: Computational model inspired by how neurons work in the brain.

Spatial Calculus: Logic-based approach for reasoning about space.

Temporal Calculus: Logic-based approach for reasoning about time.

Chapter 23
Prototyping Smart Assistance with Bayesian Autonomous Driver Models

Claus Moebus
University of Oldenburg, Germany

Mark Eilers
University of Oldenburg, Germany

ABSTRACT

The Human or Cognitive Centered Design (HCD) of intelligent transport systems requires digital Models of Human Behavior and Cognition (MHBC) enabling Ambient Intelligence e.g. in a smart car. Currently MBHC are developed and used as driver models in traffic scenario simulations, in proving safety asser- tions and in supporting risk-based design. Furthermore, it is tempting to prototype assistance systems (AS) on the basis of a human driver model cloning an expert driver. To that end we propose the Bayesian estimation of MHBCs from human behavior traces generated in new kind of learning experiments: Bayesian model learning under driver control. The models learnt are called Bayesian Autonomous Driver (BAD) models. For the purpose of smart assistance in simulated or real world scenarios the obtained BAD models can be used as Bayesian Assistance Systems (BAS). The critical question is, whether the driving competence of the BAD model is the same as the driving competence of the human driver when generating the training data for the BAD model. We believe that our approach is superior to the proposal to model the strategic and tactical skills of an AS with a Markov Decision Process (MDP). The usage of the BAD model or BAS as a prototype for a smart Partial Autonomous Driving Assistant System (PADAS) is demonstrated within a racing game simulation.

DOI: 10.4018/978-1-61692-857-5.ch023

1 INTRODUCTION

The Human or Cognitive Centered Design (HCD) (Norman, 2007; Sarter et al., 2000) of intelligent transport systems requires digital Models of Human Behavior and Cognition (MHBC) enabling Ambient Intelligence (AMI) e.g. in a smart car. The AMI paradigm is characterized by systems and technologies that are *embedded, context aware, personalized, adaptive,* and *anticipatory* (Zelkha et al., 1998). Models and prototypes we propose here are of that type.

Currently MBHC are developed and used as driver models in traffic scenario simulations (Cacciabue et al., 2007, 2011), in proving safety assertions and in supporting risk-based design. In all cases it is assumed that the conceptualization and development of MHBCs and ambient intelligent assistance systems are *parallel and independent* activities (Flemisch et al., 2008; Löper et al., 2008). In the near future with the need for smarter and more intelligent assistance the *problem of transferring human skills* (Yangsheng et al., 2005) into the envisioned technical systems becomes more and more apparent especially when there is no sound skill theory at hand.

The *conventional approach* to develop smart assistance is to develop control-theoretic or artificial-intelligence-based prototypes (Caccibue et al., 2007, 2011) first and then to evaluate their learnability, usability, and human likeness *ex post.* This makes revision-evaluation cycles necessary which further delay time-to-market and introduce extra costs. An *alternative approach* would be the *handcrafting* of MHBC (Baumann et al., 2009; Gluck et al., 2005; Jürgensohn, 2007; Möbus et al., 2007; Salvucci, 2004, 2007; Weir et al., 2007) on the basis of human behavior traces and their *modification* to prototypes for smart assistance. An ex post evaluation of their human likeness or empirical validity and revision-evaluation cycles remains obligatory, too.

We propose a third *machine-learning alternative.* It is tempting to prototype assistance systems on the basis of a human driver model *cloning* an expert driver. To that end we propose the *Bayesian estimation of* MHBCs *from human behavior traces* generated in new kind of learning experiments: *Bayesian model learning under driver control.* The models learnt are called *Bayesian Autonomous Driver (BAD) models.*

Dynamic probabilistic models are appropriate for this challenge, especially when they are learnt online in *Bayesian model learning under driver control.* For the purpose of smart assistance in simulated or real world scenarios the obtained BAD models can be used as prototypical Bayesian Assistance Systems (BAS). The *critical* question is, whether the driving competence of the BAD model is the same as the driving competence of the human driver when generating the training data for the BAD model.

We believe that our approach is superior to a proposal to model the strategic skills of a PADAS with a *Markov Decision Process (MDP)* (Tango et al., 2011). A MDP needs a reward function. This function has to be derived deductively solving the *inverse reinforcement learning problem* (Abbeel et al., 2004). The deductive derivation of reward function often results in strange nonhuman overall behaviors. The inductive mining of the reward function from car trajectories or behavior traces seems to be a detour and more challenging than our approach.

The two new concepts *Bayesian learning of agent models under human control* and the *usage of a BAD model as a BAS or PADAS* are demonstrated when constructing a prototypical *smart assistance system* for driving stabilization within the racing game simulation TORCS (TORCS, 2011).

BAD models (Eilers et al., 2010a,b, 2011; Möbus et al., 2008, 2009a,b,c, 2010, 2011) are developed in the tradition of Bayesian expert systems (Jensen et al., 2007; Neapolitan, 2004; Pearl, 1988, 2009; Russell et al., 2010), probabilistic robotics (Forbes et al., 1995; Thrun et al., 2005), and Bayesian (robot) programming (BP) (Bessiere, 2003, 2008; Lebeltel et al., 2004; Le Huy et al., 2004). For *Bayesian model learning under driver control* we need concepts from *parameter learning* in Bayesian networks (Neapolitan, 2004). We distinguish *descriptive* and *normative* BAD models. *Descriptive* models can be learnt from the behavior of individuals or groups of drivers. They can be used for simulating human agents in all kinds of traffic scenarios. *Normative* models are learnt from the behavior of *ideal* or *special instructed* human drivers (e.g. driving instructors, racing car drivers). They may be used for the conceptual new *BAS*. Due to their probabilistic nature BAD models or BAS can not only be used for *real-time control* but also for *real-time detection of anomalies* in driver behavior and *real-time generation* of supportive *interventions (countermeasures)*.

2 DISTRIBUTED COGNITION, (PARTIAL) COOPERATION, SMART ASSISTANCE, AND AMBIENT INTELLIGENCE IN DRIVING SCENARIOS

The concept of *distributed cognition* was originated by Edwin Hutchins in the mid 1980s (Hutchins, 1995). He proposed that human knowledge and cognition is not confined to individuals but is also embedded in the objects and tools of the environment. Cognitive processes may be distributed across the members of a social group or the material or environmental structure. With anthropological and non-experimental methods Hutchins studied how crews of ships can function as a *distributed machine*. He preferred studying cognitive systems not as individual agents but which are composed of *multiple agents* and the material world. In later studies he generalized the domains and put an emphasis on airline cockpits crews and human-computer interaction scenarios. In a sense Hutchins anticipated the concepts of ambient intelligence with its *embedded*, *context aware*, *personalized*, *adaptive*, *anticipatory* systems.

Crews on navigation bridges or in aircraft cockpits work in agreement with a *single* principal. Such a scenario is called *cooperative* (Xiang, 2002). A crew forms a cohesive group whose members normally cooperate for longer periods in solving the problems arising from ship or aircraft control. This cooperation includes exchange of complex verbal messages which require a high dimensional state space for the agent models.

Public traffic scenarios are of a fundamentally different kind. Communication, cooperation and the action repertoire of agents is limited in amount and complexity. Agents are their own principals and do not belong to a formal cohesive group. Thus, a scenario is *partial* or *non-cooperative*, when goals are issued by *several different principals* (Xiang, 2002). Traffic agents form ad hoc groups by chance and try to maximize their personal utilities. Internal group norms are substituted by external traffic rules. The solution to a traffic coordination problem is a distributed but synchronized sequence of sets of actions (e.g. collision-free crossing an intersection) emitted by different autonomous agents. Successful problem solutions require (nonverbal) communication and distributed cognition across agents and artifacts.

2.1 Cooperative Scenarios: Crews and In-Vehicle-Dyads

Members of a public traffic scenario with Between-Vehicle Cooperation (BVC) do not form a stable social group but rather an ad hoc group with a limited life time and communication vocabulary. In contrast to that members in a nonpublic traffic scenario (Figure 1) with In-Vehicle Cooperation (IVC) form for a short time period a stable social group similar to a crew.

2.2 Partial Cooperative Scenarios: Ad-Hoc Groups and Shared Space

Shared space describes an approach to the design, management and maintenance of *public spaces*

Figure 1. Driving-school-scenario with in-vehicle-cooperation (graphics from (Möbus et al., 2009b) with kind permission of publisher GZVB and of Springer Science and Business Media)

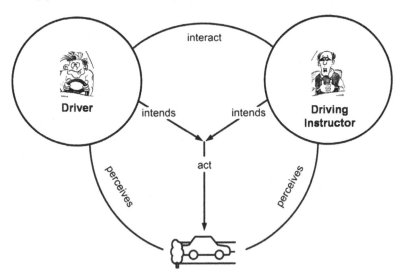

which reduces the adverse effects of conventional traffic engineering by stimulating the *situation awareness* of all traffic agents (Figure 2).

The shared space approach is based on the observation that individuals' behavior in traffic is more positively affected by the built environment of the public space than by conventional traffic control devices (signals, signs, road markings, etc.) or regulations. An explanation for the apparent *paradox* that a reduction in regulation leads to safer roads may be found by studying the *risk compensation effect*: "Shared Space is successful because the *perception of risk* may be a *means* or even a *prerequisite* for increasing public traffic safety. Because when a situation feels unsafe, people are more alert and there are fewer accidents." (SharedSpace, 2011)

2.3 Smart Assistance in Traffic Scenarios

Traffic maneuvers can generate risk anytime. We call risky maneuvers *anomalies* when they have a low probability of occurrence in the behavior stream of experienced drivers and which only

experienced drivers are able to prevent or to anticipate automatically. Other drivers probably cannot and therefore might need support generated by a BAS or PADAS. It is expected from assistance systems that they will enhance *situation awareness*, *cooperation*, and *driving competence* of unskilled or non-cooperative drivers. Thus the design challenge of smart assistance should aim at modeling human traffic agents with their (erroneous) beliefs, expectations, behavior, situation awareness, and their skills to recognize situations, to diagnose and prevent anomalies. These BAD models should then be adapted to BAS or PADAS to solve the *problem of transferring human skills*.

2.4 The Need for Bayesian Assistance in Vehicles with In-Vehicle-Cooperation

As an example for the concept of a BAS we present a scenario based on result of a study of Rizzo (Rizzo et al., 2001) The authors studied the behavior of drivers suffering from Alzheimer disease. At a lane crossing a car incurred from the right (Figure 3). Many maneuvers of the Alzheimer patients

Figure 2. Shared-space with between-agent-, between-vehicle-, and in-vehicle-cooperation (graphics with kind permission of Ben Hamilton-Baillie)

ended in a collision, as they suffered from the *looking without seeing syndrome*. The modeling task should lead to a probabilistic BAS model, which is *diagnosing* and *correcting* the *anomalous* behavior of inexperienced or handicapped drivers. Figure 3 demonstrates the *probabilistic prediction* of hazardous events, anomaly *detection* (1.) and the *anticipatory control* of the driver's behavior by the BAS (2.).

Pink ellipses denote contours of constant density. A driver's behavior is *risky or anomalous* if its behavior is unlikely under the assumption that the driver belongs to a group of normal error-free routine drivers. For *anticipatory planning* the conditional probability of the *NextFutureDrive* under the assumption of the *pastDrive*, the *currentDrive*, and the anticipated *expectedFutureDrive* has to be computed. The BAS gives an advice sampled from this conditional distribution (e.g. the expected value

$$E\left(\begin{array}{l} NextFutureDrive \mid pastDrive,\dots, \\ exptectedFutureDrive \end{array}\right).$$

Figure 4 shows the replacement of the real driving inspector by the corresponding BAS model. Different BAS-types like an experienced Schumacher-racing-style BAS are possible (Figure 5).

How can the BAS be derived by methods of Bayesian driver modeling? We explain this within an obstacle scenario which is known to generate driver *intention conflicts* (Figure 6).

When an obstacle (animal, car) is appearing unexpectedly people *autonomously* react with a maneuver M^- which is *not* recommended by experts. M^- drivers try to avoid collisions even at high velocities by steering to the left or right risking a fatal turnover. The *recommended* maneuver M^+ includes the *hold and brake* sub-maneuvers though most times ending up in a collision.

When drivers are instructed to drive M^+ they generate data which are the training data for the BAS version of the PADAS according to the methods of chapters 4 and 5: *Bayesian learning of agent models under human control.*

Figure 3. Driving behavior of an Alzheimer patient in a simulated intersection incursion (Rizzo et al., 2001), (1.) risk assessment of a the current behavior or trajectory, and (2.) anticipatory planning of a BAS (graphics from (Möbus et al., 2009b) with kind permission of Springer Science and Business Media)

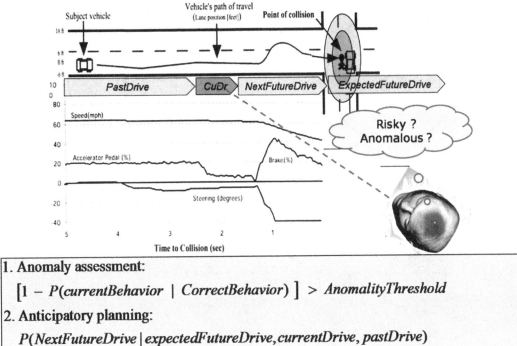

1. **Anomaly assessment:**

$$\left[1 - P(currentBehavior \mid CorrectBehavior) \right] > AnomalityThreshold$$

2. **Anticipatory planning:**

$$P(NextFutureDrive \mid expectedFutureDrive, currentDrive, pastDrive)$$

$$= \sum_{Context} P(NextFutureDrive, Context \mid expectedFutureDrive, currentDrive, pastDrive)$$

With an existing BAS a worst-case scenario can be planned to test the services of the BAS. Drivers are instructed *not* to drive the recommended maneuver $\mathbf{M}^+$. Because of the probabilistic nature of the BAS it is possible to compute the conditional probability $P\left(currentDrive_t \mid \mathbf{M}^+ \right)$. This conditional probability is a measure of the *anomaly* of the driver behavior *under the hypothesis* that the observed actions are generated by a stochastic process which generated the trajectories or behaviors of the correct maneuver $\mathbf{M}^+$.

3 PROBABILISTIC MODELS OF HUMAN BEHAVIOR AND COGNITION IN TRAFFIC SCENARIOS

Computational agent models have to represent perceptions, beliefs, goals, and actions of ego and alter agents. Agent models should

1. predict and generate agent behavior sometimes in interaction with assistance systems
2. identify situations or maneuvers and classify behavior (e.g. anomalous vs. normal) of ego and alter agents
3. provide a robust and valid mapping from human sensory data to human control actions
4. be learnt from time series of raw data or empirical frequency distributions with statistical sound (machine-learning) procedures

Figure 4. Cooperative driving scenario with in-vehicle-cooperation between a non-expert driver and a BAS-prototype Driving Instructor (graphics from (Möbus et al., 2009b) with kind permission of publisher GZVB and of Springer Science and Business Media)

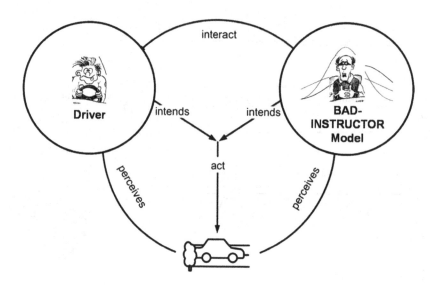

Figure 5. Cooperative driving scenario with in-vehicle-cooperation between a non-expert driver and a BAS-prototype Racing Driver1 (background graphics from (Möbus et al., 2009b) with kind permission of publisher GZVB and of Springer Science and Business Media)

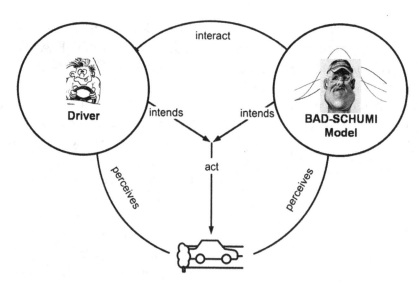

Figure 6. Intention conflict scenario with conflicting behaviors **M⁻** *(incorrect or not recommended maneuver) and* **M⁺** *(correct or recommended maneuver) (graphics from (Möbus et al., 2009b) with kind permission of Springer Science and Business Media)*

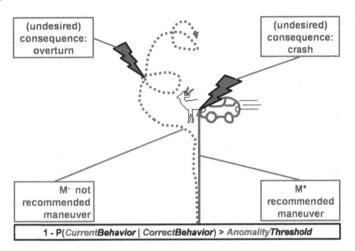

making only a few non-testable ad hoc or axiomatic assumptions

5. be able to learn new patterns of behavior without forgetting already learnt skills (stability-plasticity dilemma (Hamker, 2001)).

A driver is a human agent whose skills and skill acquisition processes can be described by a well-known three-stage model with the *cognitive*, *associative*, and *autonomous* stages or layers (Anderson, 2002; Fitts et al., 1967). Accordingly various modeling approaches are adequate: (1) *production-system* models for the cognitive and associative stage (e.g. models in a cognitive architecture (Anderson et al., 2007, 2008; Baumann et al., 2009; Quin et al., 2007; Salvucci et al., 2004, 2006, 2007)), *control-theoretic* (Bischof, 1995; Jagacinski et al., 2003; Jürgensohn, 2007; Möbus et al., 2007; Salvucci, 2007; Weir et al., 2007), or *probabilistic* models (Chater et al., 2008; Eilers et al., 2010a,b, 2011; Gopnick et al., 2007; Griffiths et al., 2008; Möbus et al., 2008, 2009a,b,c, 2010, 2011; Wickens, 1982) for the autonomous stage.

The great advantage of probabilistic models is that they avoid brittleness and provide *robustness*. This is a great advantage due to the *irre-ducible incompleteness* of knowledge about the environment and the underlying psychological mechanisms (Bessiere et al., 2008). Furthermore probabilistic models of the *Bayesian* type are suited to implement MHBCs which are *embedded, context aware, personalized, adaptive, anticipatory* systems (Figure 3, 6).

3.1 Bayesian Autonomous Driver Models

Due to the variability of human cognition and behavior and the *irreducible* lack of knowledge about latent cognitive mechanisms it seems rational to conceptualizes, estimate and implement *probabilistic* models when modeling human traffic agents. In contrast to other models probabilistic models are not *idiosyncratically handcrafted* but could be *learnt objectively* from human behavior traces. Model validity is either included in the modeling process by *model-driven data-analysis* without any ex-post validation or by our new machine-learning experiments: *Bayesian learning of agent models under human control*.

BAD models describe phenomena on the basis of variables and conditional probability distributions (CPDs). This is in contrast to models in cognitive architectures (e.g. ACT-R) which try to simulate cognitive algorithms and processes on a *granular* and *latent* basis which are difficult to identify even with technical sophisticated methods such as *functional magnetic resonance imaging (FMRI)* methods (Anderson et al., 2008; Quin et al., 2007).

According to the BP (Bessiere, 2003, 2008; Lebeltel et al., 2004; Le Huy et al., 2004) approach BAD models (Eilers et al., 2010a,b, 2011; Möbus et al., 2008, 2009a,b,c, 2010, 2011) are instances of Bayesian belief networks (Jensen et al., 2007; Koller et al., 2009; Pearl, 1988, 2009; Spirtes et al., 2001) using concepts from probabilistic robotics (Forbes et al., 1995, Thrun et al., 2005). BP is a simple and generic framework suitable for the description of human sensory-motor models in the presence of incompleteness and uncertainty. It provides integrated *model-driven data analysis* and *model construction*. In contrast to conventional Bayesian network models BP-models put emphasis on a *recursive structure* and infer concrete motor actions for *real-time control* on the basis of sensory evidence. Actions are sampled from CPDs according various strategies after propagating sensor or task goal evidence. BAD models describe phenomena on the basis of the variables of interest and the decomposition of their joint probability distribution (JPD) into CPD-factors according to the *special chain rule for Bayesian networks*. The underlying CIHs between sets of variables can be tested by standard statistical methods (e.g. the conditional mutual information index (Jensen et al., 2007, p.237)). The parameters of BAD models can be learnt objectively with statistical sound methods by batch learning from multivariate behavior traces or by learning from single cases (Neapolitan, 2004). The latter approach is known as Bayesian estimation (Jensen et al., 2007). We use it for *Bayesian (online) learning of* MHBCs. The learning process runs in a new kind of learning experiments: *Bayesian learning of agent models under human control*.

BAD models could be learnt solely by Bayesian adaption of the model to the real-time behavior of the human driver correcting the BAD model when necessary.

In (Möbus et al., 2008) we described first steps to model lateral and longitudinal control behavior of single and groups of drivers with *reactive Bayesian sensory-motor models*. Then we included the time domain and reported work with *dynamic Bayesian sensory-motor models* (Möbus et al., 2009a,b). Now we together with others work on the idea of behavior hierarchies and mixing behaviors (Bishop et al., 2003; Meila et al., 1995). The goal is a *dynamic* BAD model which is able to decompose complex situations into basic situations and to compose complex behavior from basic motor schemas (*behaviors*, *experts*). This Mixture-of-Behaviors (MoB) model facilitates the management of sensory-motor schemas in a library. Context dependent driver behavior could be generated by mixing pure behavior from different schemas (Eilers et al., 2010, 2011; Möbus et al., 2009c, 2010).

3.2 Basic Concepts of Bayesian Programs

A BP is defined as a mean of specifying a family of probability distributions. By using such a specification it is possible to construct a BAD model, which can effectively control a (virtual) vehicle. The components of a BP are presented in Figure 7 where the analogy to a logic program is helpful.

An *application* consists of a (competence or task model) *description* and a *question*. A *description* is constructed from *preliminary* knowledge and a *data set*. *Preliminary knowledge* is constructed from a set of *pertinent variables*, a *decomposition* of the JPD and a set of *forms*. *Forms* are either *parametric forms* or *questions* to other BPs.

The purpose of a *description* is to specify an effective method to compute a JPD on a set of variables given a set of (experimental) *data* and

Figure 7. Structure of a Bayesian Program (adapted from (Bessiere et al., 2008))

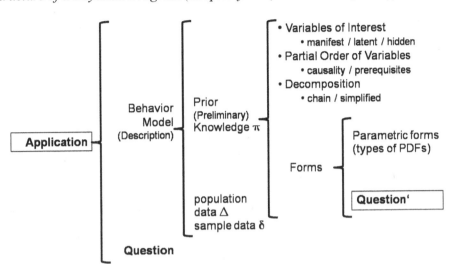

preliminary knowledge. To specify *preliminary knowledge* the modeler must *define the set of relevant variables* on which the JPD is defined, *decompose the JPD* into factors of CPDs according to CIHs, and *define the forms*. Each CPD in the decomposition is a form. Either this is a *parametric form* which parameter are estimated from batch data (behavior traces) or another *question*. Parameter estimation from batch data is the conventional way of estimating the parameters in a BAD model. The *Bayesian estimation* procedure uses only a small fraction of the data (cases) for updating the model parameters. This procedure is described below.

Given a description a *question* is obtained by partitioning the variables into *searched*, *known*, and *unknown* variables. We define a question as the CPD

$$P\left(\begin{array}{l} Searched \mid Known, preliminary \\ knowledge, data \end{array}\right).$$

The selection of an appropriate action can be treated as the inference problem:

$$P\left(\begin{array}{l} Action \mid Percepts, Goals, \\ preliminary\ knowledge, data \end{array}\right).$$

Various *policies* (*Draw, Best, Maximum Aposteriori*, and *Expectation*) are possible whether the

concrete *action* is *drawn* at random, chosen as the *best* action with highest probability, or as the *expected* action. The last strategies are necessary if the BAS should behave deterministically as demanded by industry.

3.3 Classes of Probabilistic Models for Human Behavior and Cognition

Currently we are evaluating the suitability of static and dynamic Probabilistic Graphical Models (Koller et al., 2009).

With the *static* type it is possible to generate reactive (Möbus et al., 2008) and inverse (naïve) (Möbus et al., 2009a) models (Figure 35 – 37 in Appendix A). In practice, naïve Bayesian models can work surprisingly well, even when the independence assumption is not true (Russell et al., 2010, p.499). Our research has shown that static models generate behavior which is too erratic to be similar to human behavior. As a consequence we focus ourselves on the dynamic type of real-time control for simulated cars.

Dynamic models evolve over time. If the model contains discrete time-stamps one can have a model for each unit of time. These local models are called *time-slices* (Jensen et al., 2007). The

Figure 8. Structure of a Dynamic Bayesian Network (DBN) as a Bayesian Program (adapted from (2003); graphics from (Möbus et al., 2009a) with kind permission of Springer Science and Business Media)

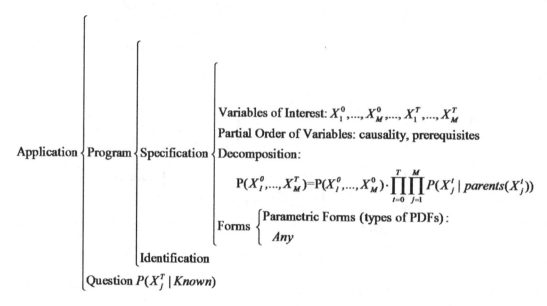

time slices are connected through *temporal links* to give a full model.

A special category of time-stamped model is that of a *Hidden Markov Model* (HMM). They are repetitive temporal models in which the state of the process is described by a *single discrete* random variable. Because of the Markov assumption only temporarily adjacent time slices are linked by a *single* link between the state nodes.

In the case of identical time-slices and *several* identical temporal links we have a *repetitive temporal model* which is called *Dynamic Bayesian Network model* (DBN). The description of the DBN in the BF framework is shown in Figure 8.

In 3.3.1 we present Markov, Hidden Markov Models (HMMs), and their generalization Dynamic Bayesian Networks (DBNs) and then in 3.3.2 we develop and evaluate in a proof of concept a sequence of models culminating in a *psychological motivated sensor-motor model with attention allocation* which could be the basis for a BAS.

3.3.1 Markov, Hidden Markov Models, and Dynamic Bayesian Networks

The *dynamic* type of graphical models (Jensen et al., 2007; Koller et al., 2009) enables the creation of Markov Models (MMs) (Eilers et al., 2010; Möbus et al., 2008, 2009a,b,c, 2010), Hidden Markov Models (HMMs) (Kumugai et al., 2003; Miyazaki, 2001; Oliver et al., 2000) (Figure 38–40 in Appendix A), Input-Output-HMMs (IOHMMs) (Bengio et al., 1996), Reactive IOHMMs (RIOHMMs, Figure 9, Figure 41 in Appendix A), Discrete Bayesian Filters (Koike te al., 2008; Thrun et al., 2005) (Figure 35 in Appendix A), Coupled HMMs (Oliver, 2000)(CHMMs; Figure 43 in Appendix A), and Coupled Reactive HMMS (CRHMMS, Figure 44 in Appendix A). HMMs are sequence classifiers (Jurafsky et al., 2009; Rabiner, 1989) and allow the efficient *recognition* of situations, goals and intentions; e.g. diagnosing driver's intention to stop at a crossroad (Kumugai et al., 2003, 2006). Their suitability for the *generation* of behavior of Belief-Desire-Intention (BDI-) Agents will be evaluated in 3.3.2.

Figure 9. (Reactive) Input-Output HMM (RIOHMM) as Probabilistic Abstraction of Anderson's cognitive ACT-R architecture: 2-time-slices template model (right), 3-time-slices rolled-out model (left) (graphics from (Möbus et al., 2009b) with kind permission of Springer Science and Business Media)

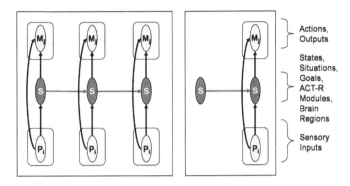

For instance, RIOHMMs (Figure 9, Figure 41 in Appendix A) could in principle implement reactive driver models (e.g. with ACT-R module activations (Anderson et al., 2007, 2008; Qui et al., 2007; Salvucci et al., 2006, 2007)). The two arrows into the random variable nodes M_j denote the combined dependence of actions on sensory inputs and activations of hidden ACT-R modules or brain regions. Even if module activations were known sensory inputs are still necessary to propose specific actions.

CHMMS and CRHMMS permit the modeling of several agents within the HMM formalism. The belief state of each agent depends only on his own history and on the belief state of his partner. Whether this is plausible has to be tested by conditional independence hypotheses (Jensen et al., 2007). Within each agent the model is of the HMM-type.

Whereas (Meila et al., 1995) rely on Hidden Markov Models (HMMs) for learning fine manipulation tasks like grasping and assembly by Markov mixtures of experts we strive for more general dynamic Bayesian Network (DBN) model in learning multi-maneuver driving behavior (Möbus et al., 2009b).

HMMs and DBN are mathematically equivalent. Though, there is a trade-off between estimation efficiency and descriptive expressiveness in

HMMs and DBNs. Estimation in HMMs is more efficient than in DBNs due to algorithms (Viterbi, Baum-Welch (Jurafsky et al., 2009; Russell et al., 2010)) whereas descriptive flexibility is greater in DBNs. At the same time the state-space grows more rapidly in HMMs than in corresponding DBNs. Therefore we focus ourselves on DBNs and try to avoid the *latent* state assumption of HMMs, though it seems to be important to model the *state* of a driver/vehicle with the variables of *position, velocity, lateral and longitudinal (de|ac)-celeration*. This is implemented in the commercial product IPG-Driver (IPG-Driver, 2011). The *state* of the driver/vehicle is important for the definition and description of *undesired events*, the *planning of countermeasures* and *intelligent anticipatory assistance* (Möbus et al., 2010, 2011).

Especially two DBN models influenced our work. The first is the *Switching Linear Dynamic System* (Kumugai et al., 2003)(Figure 45 in Appendix A) and the second is the *Bayesian Filter and Action Model* (Koike et al., 2008, p.180) (Figure 10, Figure 46 in Appendix A). In both models actions are not only dependent on the current process state but also on direct antecedent actions. Thus the generation of erratic behavior is suppressed. Furthermore, the Bayesian Filter and Action Model includes *direct* action effects on the next future process state. This is important when

Figure 10. Bayesian Filter and Action Model (adapted from (Koike te al., 2008, p.180)): 2-time-slices template model (right), 3-time-slices rolled-out model (left)

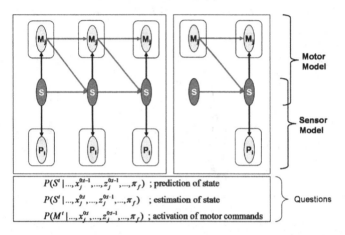

the influence of action effects should be modeled *directly* into the state not making a detour via the environment and the perception of the agent. If the model should make predictions without an embedding (simulation) environment, we have to include a state and an action model. Only when the model can be treated as a self-contained mathematical object the properties of the model need not be evaluated in a simulation within the embedding simulation environment.

Also, including an action effect model is meaningful when some action effects are not perceivable by the agent. E.g. when a night-watchman closes and locks a door, the locking action has a *direct* effect on the state of the door: the door is closed *and* locked. But this *lockedness* is not visible. This could only checked only by further actions.

In our research we strive for the realization of a dynamic Bayesian Autonomous Driver with Mixture-of-Behaviors (BAD-MoB) model. The model is suited to represent the sensor-motor system of individuals or groups of human or artificial agents in the functional *autonomous* layer or stage of Anderson (Anderson, 2002). It is a psychological motivated *mixture-of-experts* (= mixture-of-schema) model with *autonomous and goal-based attention allocation processes.*

The template or class model is distributed across two time slices, and tries to avoid the *latent* state assumptions of HMMs. Learning data are time series or case data of relevant variables: percepts, goals, and actions. Goals are the only latent variables which could be set by commands issued by the higher *associative* layer.

The model propagates information in various directions. When working *top-down*, goals emitted by the associative layer select a corresponding expert (schema), which propagates actions, relevance of areas of interest (AoIs) and perceptions. When working *bottom-up*, percepts trigger AoIs, actions, experts and goals. When the task or goal is defined and the model receives percepts evidence can be propagated *simultaneously* top-down and bottom-up. As a consequence the appropriate *expert* (*schema*) and its *behavior* can be activated.

Thus, the model can be extended (Figure 11) to implement psychological models, e.g. a modified version of the SEEV visual scanning or attention allocation model of (Horrey et al.,2006). In contrast to the SEEV model (Horrey et al., 2006) our model is able to predict the probability of attending a certain AoI on the basis of single, mixed, and even incomplete evidence (goal priorities, percepts, effort to switch between AoIs). In 3.3.2 we show that this architecture is feasible.

Figure 11. Mixture-of-Behaviors (= Mixture-of-Experts) Model with Visual Attention Allocation Extension mapping ideas of (Horrey et al., 2006)into the Dynamic Bayesian Network modeling framework (graphics from (Möbus et al., 2009b) with kind permission of Springer Science and Business Media)

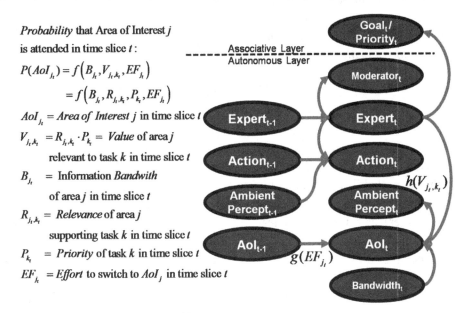

There are various scientific challenges designing and implementing BAD-MoB Models. The *first* main challenge is to describe driver-generated behavior by a *mixture-of-behaviors* architecture. While mixture-of-*experts* approaches are known from pattern classification (Bessiere, 2003; Bishop et al., 2003, 2006) it is the first time that this approach is used in modeling human driver behavior. In a MoB model it is assumed that the behavior can be context-dependent generated as a mixture of ideal schematic *behaviors* (=experts). Thus the stability/plasticity dilemma (Hamker, 2001) of neural network models is avoided. A new behavior will only be learnt in special phases and by adding this new *behavior* to the library of behaviors. Behaviors do not influence each other directly. *Pure* behavior without any additional mixture component is shown only in typical pure situations (e.g. the perception of a hair pin triggers the hair-pin-model *behavior*).

The *second* main challenge is that we want to integrate various perceptional invariants known as tau-measures (Lee, 1976, 2010) from psycho-logical action control theory into a computational human model. In conventional models (Baumann et al., 2009; Cacciabue, 2007; Salvucci et al., 2004, 2007) variables with *different* dimensions (distances, angles, times, changes, etc.) are input to the models. Tau measures transform all non-time measures into the time domain. This seems tob e a very promising approach (Möbus et al., 2011). Some measures are already used in standard engineering: time-to-collision (TTC) or time-to-line-crossing (TTLC).

3.3.2 From Discrete Bayesian Filters to Sensor-Motor Models with Attention Allocation

Now we give a proof of concept. We choose certain model classes and a set of constructed but plausible data and demonstrate that the models show the intended behavior.

In our research we used partial inverted Markov models for modeling the sensory-motor system of the human driver (ch. 4; Figure 24,

25, 31). We discuss what types of DBNs have to be considered when driver state variables (e.g. *lateral and longitudinal (de|ac)-celeration*) are included and when a psychological motivated *mixture-of-behaviors* model with *autonomous and goal-based attention allocation processes* is the ultimate goal (Figure 11).

3.3.2.1 Discrete Bayesian Filter (DBF) and HMMs

We start with the Discrete Bayesian Filter (DBF) (Figure 42 in Appendix A). This is the most fundamental algorithm in probabilistic robotics for estimating state from sensor data. The DBF is a HMM with state, percept and motor variables. The general algorithm consists of two steps in each iteration or recursive call:

1. Prediction step: from the most recent *apriori* belief(state) and the current control (=action) compute a provisional belief(state)
2. Correction step: from the current provisional belief(state) and the current measurements (= percepts) compute *the posteriori* belief(state).

We extended a tutorial example from Thrun (Thrun et al., 2005, ch. 2.4.2) and implemented this DBF in NETICA (Netica, 2010), to show that *state identification* in a DBF works satisfactorily (Figure 47, 48 in Appendix B). Our model represents a night watchman approaching a door in the dark. Before he sees the door his *belief(state)* is uninformed; so the *apriori* belief distribution about the state is flat (Figure 47 in Appendix B). His beliefs are revised when he *pushes* the door (*prediction step*) (Figure 47 in Appendix B). Now he believes with p_open = 0.633 that the door *is open* and with p_closed = (1-p_open) = 0.333 that the door is *closed and locked*. When turning on his flashlight he perceives that the door *is closed* (Figure Figure 48 in Appendix B). This leads in the *correction step* to the posterior *belief(state)* (Figure 48 in Appendix B):

$$P\left(State = is\_open \mid Action = push, ...\right),$$
$$= 0.161$$

$$P\left(\begin{array}{l}State = is\_closed\_and\_locked \mid \\ Action = push, ...\end{array}\right) = 0.763$$

and

$$P\left(\begin{array}{l}State = is\_closed\_and\_unlocked \mid \\ Action = push, ...\end{array}\right) = 0.0763$$

Now we want to show that the DBF is *not* the right model class for the *implementation of a reactive agent*, because the steps in the iteration cycles for the *reactive* agent are different from those of the DBF:

1. Perception step: from the most recent *apriori belief(state)* and the current percept compute a provisional *belief(state)*
2. Action step: from the current provisional *belief(state)* and the current action compute *the posteriori belief(state)*.

In the *perception step* the night watchman sees that the door *is closed* (Figure 49 in Appendix B). He revises his uninformed *apriori* beliefs. He is rather certain that the door *is closed*, but rather uncertain whether the door *is locked*:

$$P\left(\begin{array}{l}State = is\_open \mid Percept = \\ sense\_closed, ...\end{array}\right) = 0.0526,$$

$$P\left(\begin{array}{l}State = is\_closed\_and\_locked \mid \\ Percept = sense\_closed, ...\end{array}\right) = 0.474,$$

and

$$P\left(\begin{array}{l}State = is\_closed\_and\_unlocked \mid \\ Percept = sense\_closed, ...\end{array}\right) = 0.474$$

Now the agent *pushes* the door (Figure 50 in Appendix B). The result is a bit surprising. The door is not opened in the current or next *state*

$$P\left(State = is\_open \mid Action = push, \ldots\right) = 0.161$$

but the belief is that is the door *is closed and locked*

$$P\left(\begin{array}{l} State = is\_closed\_and\_locked \mid \\ Action = push, \ldots \end{array}\right) = 0.763$$

The reason for this puzzling result is, that the belief is consistent with the perception *within* the same time slice, but that the *effect of the action* on the *next* state is *not* modeled by a *direct link* from the action node to the next future state node. Instead time slices are linked only between state nodes. So the action effect on future states is *not directly* included in the model. Action effects enter the model only via the (simulation) environment and the perception of the model. So the effect of actions could not be seen in the model even when the model contains a state variable. This criticism is true for all variants of HMM (Figure 38-44 in Appendix A). It is irrelevant for DBNs *with* action effect models (Figure 10, Figure 46 in Appendix A).

3.3.2.2 DBN-Models with Action Model and Action Effect Prediction

As we discussed in 3.3.1 an *action effect model* is necessary when the properties of the model have to be decoupled from the embedding environment. This is the reason why we discuss these kinds of models here. As an example we implemented the task of the night watchman with a DBN including an action effect model (Figure 51-54 in Appendix B). The *apriori* beliefs are modeled in Figure 51 in Appendix B. The door is perceived as *is closed* (Figure 52 in Appendix B). Then the agent selects the action *push* (Figure 53 in Appendix B). The belief for the next future state is predominantly that the door *is open* then. Despite that belief the

night watchman tries a second glance and sees that the door *is* still *closed* (Figure 54 in Appendix B). Now he revises his belief about the state again. He believes that the door *is closed and locked*. He should check then that belief by a *push action*.

3.3.2.3 Expert-Role, Mixture-of-Behavior, or Schema DBN Model

To the model in 3.3.2.2 we added the possibility that the agent is able to show role-specific or schematic behavior. We call these models Expert-Role, Mixture-of-Behaviors, or Schema Models (Figure 55 in Appendix B). *Top-down* generation of goal-based behavior is possible, when the role node gets evidence by selecting the role or the goal to generate role or schema-specific *behavior*. Furthermore, the model can be used *bottom-up* to infer the role, the *behavior*, or the goal from the percepts and/or the actions. For instance, when the agent *pushes* and *unlocks* the door despite his perception that the door *is closed*, we infer that he is either a *technician* or a *detective* but *not* a *night watchman* (Figure 56 in Appendix B).

3.3.2.4 AoI and Ambient Vision-Role-Model

Next according to Horrey et al. we separated the visual perception into two components: (1) *foveal areas of interest* (AoIs) and (2) *ambient vision* (Figure 57 in Appendix B). If the agent is only interested in the *keyhole* of the door, we infer that he is active in the *detective* role (Figure 58 in Appendix B). The *plausible* actions in the *actual* time slice are *push, explore,* and *unlock,* the *expected* actions in the *next future* time slice are only *push* and *explore.* If we know that the model by its ambient vision component perceives that the door *is open* (Figure 59 in Appendix B), we expect that the agent is still in the *detective* role but with the different action *explore.* If we know for sure that the agent has the same perception as before but is in the role *night watchman* we expect his role-specific behavior is *shut* for the *actual* time slice and *lock* and *go on* in the *next two* future time slices (Figure 60 in Appendix

Figure 12. Left and Right Lane Change Maneuvers

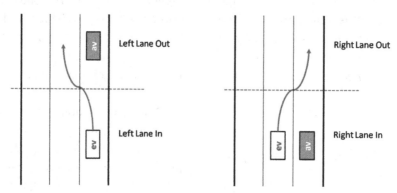

Figure 13. Pass Vehicle or Overtake Maneuver (left) and AoIs viewed from ego vehicle (right)

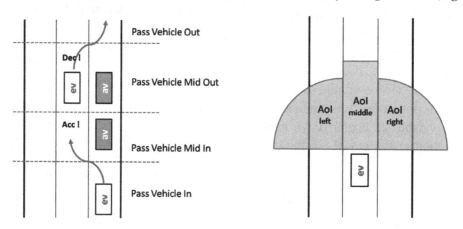

B). When he *shuts* the door but is in the next time slice interested in the *door hinges*, we infer a *role, behavior,* or *intention conflict* because he might be also in the *role* of a *technician* (Figure 61 in Appendix B).

3.3.2.5 Reactive State-Based BAD-MoB-Model with Attention Allocation

Now we return to the driving domain. We developed a NETICA model for a simple scenario with 3 maneuvers and 3 areas of interest (AoIs) (Figure 12-13).

A table describing the model representing levels of competence can be found in Figure 14. The driver and the BAD model are sitting in the *ego* vehicle (ev). Sometimes the driver's perception signals evidence that the AoIs *is_occupied*

depending on the *position* of an *alter* vehicle (av) or the *roadside*.

The 2-time-slices template of the Dynamic *Reactive* BAD-MoB-Model is shown in Figure 15 and a 3-time-slices rolled-out instance of that template model in Figure 16. We call the model *reactive* because the AoIs *directly* influence actions. The model embeds two naïve Bayesian classifiers: One for the *Behaviors* and one for the *States*. This simplifies the structure of the architecture. Time slices are selected so that in each new time slice a new *behavior* is active. A *sequence* of behaviors implements a single *maneuver*. When we replace the reactive submodel for the *Action* variable in Figure 15 by a *third* classifier we can simplify the model and reduce the number of parameters by 78%.

Figure 14. Hierarchy of Driving Skills, Levels of Expertise, and Model Components

Levels of Expertise	Model Component	Hierarchy of Skills, Levels of Expertise
Skills		**Skills** = {..., drivingScenarioSkills, ...}
Scenario Skills		**DrivingScenarioSkills** = { highway, countryRoad, city }
Driving Maneuver Skills	Driving **Maneuver** *Sequence* (horizontally distributed)	**highway.Maneuvers** = { leftLaneChange (lLC), rightLaneChange (rLC), passVehicle (pV), newManeuver }
Driving Behavior Skills	Driving **Behavior** *Layer*	**Behaviors** = { leftLaneIn (lLI), leftLaneOut (lLO), passIn (pI), passMidIn (pMI), passMidOut (pMO), passOut (pO), rightLaneIn (rLI), rightLaneOut (rO), newBehavior } e.g.: **leftLaneChange.Behaviors** = {leftLaneIn, leftLaneOut }
Driving Action Skills	Driving **Actions** *Layer*	**Actions** = { leftCheckLane (lCL), leftSignal (lS), leftTurn (lT), middleStraightAcceleration (mSA), middleStraightDeceleration (mSD), middleLookForward (mLF), rightCheckLane (rCL), rightSignal (rS), rightTurn (rT) } e.g.: **leftLaneIn.Actions** = {lCL, lS, lT, mLF}

Behaviors are placed in the top layer of nodes (Figure 15, 16). We have *behaviors* for each main part of a maneuver (Figure 12-13): *left_lane_in, left_lane_out, pass_in, pass_mid_in, pass_mid_out, pass_out, right_lane_in, right_lane_out*. The next layer of nodes describes the *actions* the model is able to generate: *left_check_lane, left_signal, left_turn, middle_straight_accelerate, middle_straight_decelerate, middle_straight_look, right_check_lane, right_signal, right_turn*.

Below that layer a layer of nodes is describing the *state* (*is_in_left_lane, is_in_middle_lane, is_in_right_lane*) of the vehicle. In the future these state nodes should be augmented by *tau*-, and *tau-dot*-variables describing the driver's state. The three bottom layers contain nodes describing the activation of the three AoIs *AoI_Left, AoI_Middle,* and *AoI_Right*.

When the model is urged to be in the *left_lane_in* behavior by e.g. goal setting from the associative layer (Figure 17, red arrow), we expect in the *same* time-slice that the driver though sometimes *looking forward* his behavior is focused towards the *left* lane. For the AoIs we expect that the middle and right AoI *are occupied* and the left AoI *is empty*. For the *next* time slice we expect the vehicle is *in the left* or *middle lane* and the driver will act

according *left_lane_out* behavior. *Left_lane_out* activated actions in time slice t are a bit different than those before. We expect more forward orientated activities like (de-/ac-)*celebration* and *forward* directed attention.

When the *state* is known (e.g. S = *is_in_middle_*lane) we infer the appropriate expectations (e.g. *left and right lane check, looking forward,* and both *(ac|de)celerations*) (Figure 18).

When the model perceives a combination of AoI evidence, we can infer the *behaviors*. For instance, in Figure 19 the left AoI *is empty* and the middle AoI *is occupied*. We expect that the vehicle *is in the middle* or *right lane* and that the *behaviors* are ambiguous *left_lane_in* or *pass_in*. Their appropriate *actions* are activated as a *mixture of behaviors*. The most probable actions are *mid look forward, left check lane* and *middle deceleration*.

In the case, when all AoIs are occupied (Figure 20) the model *is decelerating* with main attention to the middle AoI (*middle_look_forward*). We call this focusing of attention and narrowing of the attended vision field (sometimes under stress) *Tunnelblick* (tunnel view or tunnel vision2).

Figure 15. a) Dynamic Reactive BAD-MoB-Model with Behavior and State Classifiers, b) Dynamic Submodel, c) Behavior Classifier, d) Reactive Action Submodel, e) State Classifier

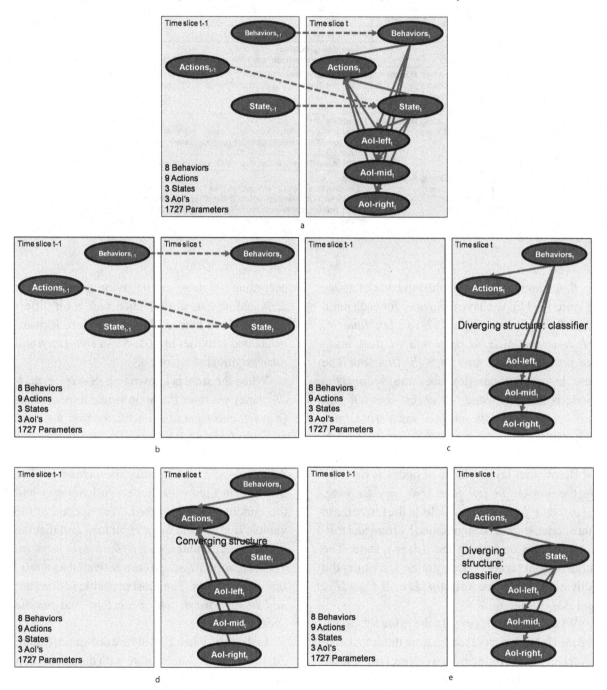

Figure 16. a) Reactive State-Based BAD-MoB-Model with 2 Classifiers and 2 Levels-of-Expertise, b) Blown-up nodes of time-slice t-1 in NETICA-Model of Figure 16.1

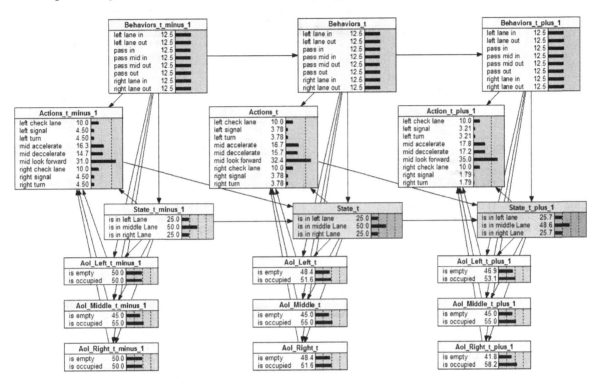

a.

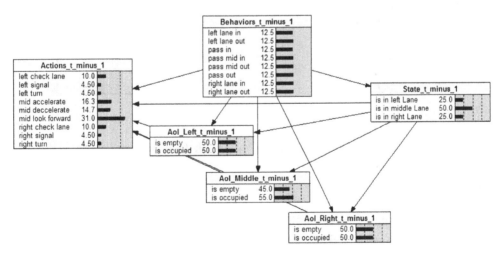

b.

Figure 17. Expectations when BAD-MoB model is in left_lane_in behavior

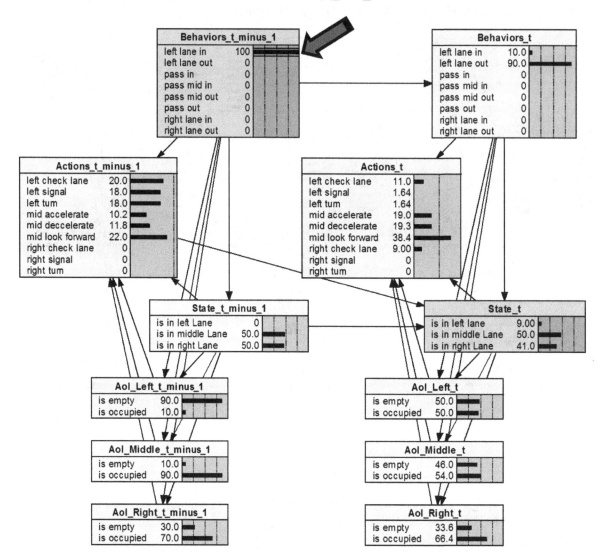

What will happen, if a *goal* is blocked? In Figure 21 this situation is modeled by the appropriate evidence.

The *left_lane_in behavior* is provided with evidence because we assume that a corresponding *goal* in a higher cognitive layer is activated. At the same time the perception all AoIs is set to *is_occupied*. The expected behavior is *looking forward, deceleration, and left_check_lane*, which are indicators for the *Tunnelview* and (helplessly?) looking to the left.

With the rolled-out version of the BAD-MoB-Model it is possible to anticipate hazards (Figure 22). The anticipated hazard is included as percept evidence in time-slice (t+1). *Conditional* to the current state (t-1), the *anticipated* percept evidence (t+1) of the hazard, the *proactively* selected goal-behavior *left_lane_in* (t-1, t), and the *proactively* selected action *left_turn* (t-1, t), we predict that we are able to avoid the hazard in time-slice (t+1) (Figure 23).

Figure 18. Expectations when BAD-MoB model is_in_middle_lane State

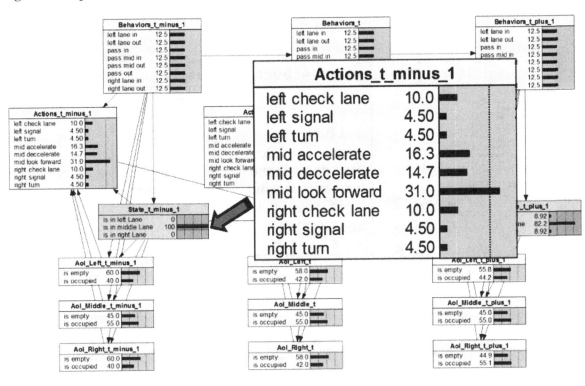

Figure 19. Expectations when BAD-MoB model perceives a combination of AoI evidence

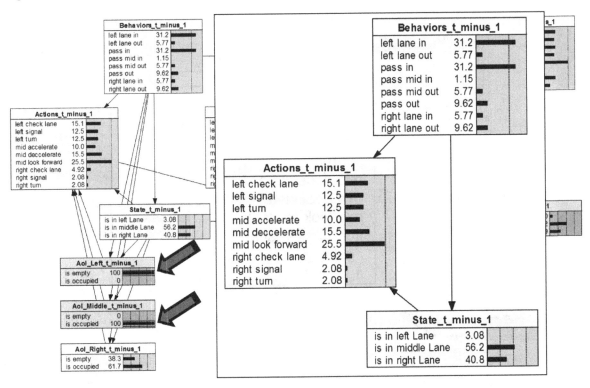

Figure 20. Expectations when BAD-MoB model perceives that all AoIs are occupied. Tunnelview

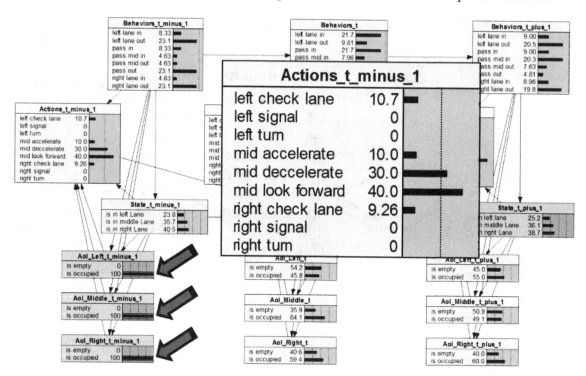

Figure 21. Expectations when BAD-MoB model realizes a blocking of goals or behaviors by a combination of occupied AoIs: Tunnelview

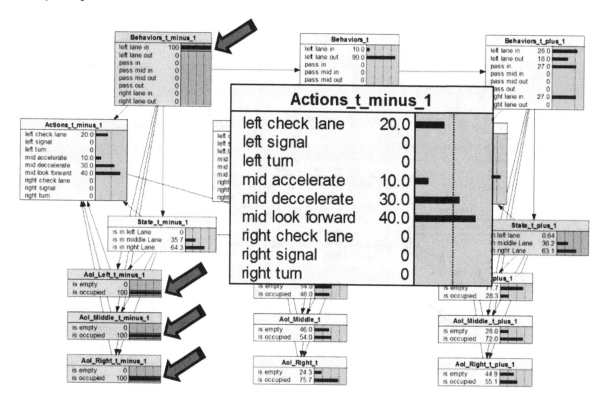

Figure 22. Anticipation of hazards: the BAD-MoB model anticipates in time-slice (t-1) a hazard (relative to the current state) for time-slice (t+1)

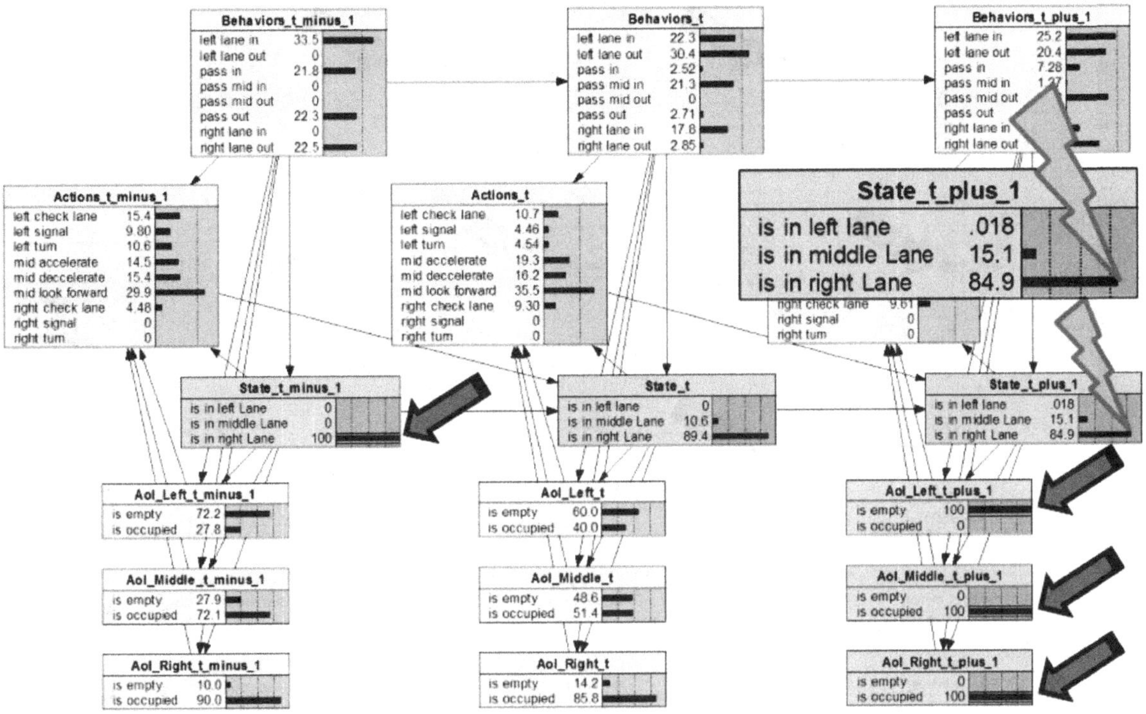

We believe that the proof of concept is convincing: *state-based* BAD-MoB Models are expressive enough to describe and predict a wide range of phenomena including *prediction of hazards*, *anticipatory planning*, and planning of *minimal invasive countermeasures*.

4 EXPERIMENTAL RESULTS

4.1 Use Case for *Autonomous Driving: A Simple BAD Model*

Static reactive or *static inverse* models (Figure 35-37 in Appendix A) have not been satisfactory because they generate behavior which is more erratic and nervous than human behavior is (Möbus et al., 2008). Better results can be obtained by introducing a memory component and using DBNs. In a first step we estimated two DBNs

separately for the lateral and longitudinal control. Our experience is that *partially inverse* models are technically well suited for modelling in the driving domain. (Figure 24, 25).

In an *inverse* model arcs in the directed acyclic graph (DAG) of the graphical model are directed from the *consequence* to the *prerequisites*. The semantics of these arcs are denoted by the conditional probabilities $P\left(Prerequisites \mid Consequence\right)$.

The reasons to use *inverted*

$$P\left(Prerequisites \mid Consequence\right)$$

instead of *reactive* conditional probabilities

$$P\left(Consequence \mid Prerequisites\right)$$

are the possible large number of prerequisites in a reactive model.

Figure 23. Anticipatory Plan: the BAD-MoB model sets as goals the behavior left_lane_in and selects the left_turn action for time-slices (t-1) and t to avoid the hazard in time-slice (t+1)

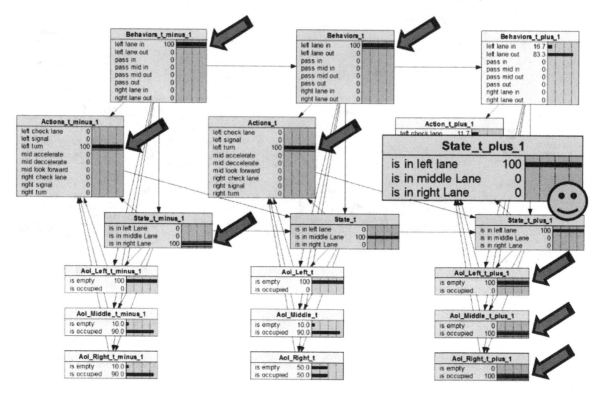

Figure 24. Partially inverse classifier-based DBN of Lateral Control (graphics from (Möbus et al., 2009a) with kind permission of Springer Science and Business Media)

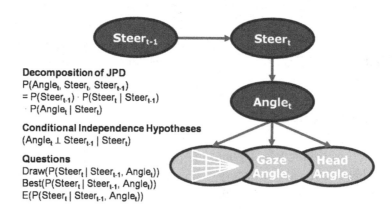

Figure 25. Partially inverse classifier-based DBN of Longitudinal Control (graphics from (Möbus et al., 2009a) with kind permission of Springer Science and Business Media)

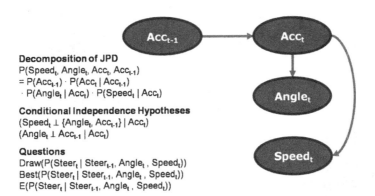

By using *inverted* conditional probability distributions, we significantly reduce the amount of parent nodes for *Consequence*. Furthermore, a conditional probability $P\big(Prerequisites \mid Consequence\big)$ is more robust to possible *unknown* evidence of *Prerequisites*. This occurs almost inevitably due to missing sample data because it is rather unlikely to obtain all possible values of the joint probability distribution $P\big(Consequence, Prerequisites\big)$. Our models are *partially* inverse because most arcs are inverted but the arcs between time slice t-1 and t are in *causal order* from prerequisites to consequences.

The variables of interest are partitioned into *sensor*-variables and *action*-variables. The variables for the partially inverse DBN of *lateral* control are defined as follows: $Steer_t$ and $Steer_{t-1}$ can take 21 different values between -10 (hard right) and +10 (hard left). Variable $Angle_t$ represents the angle between heading vector of the car and the course of the racing track to be reached in 1 second by current speed and can take 21 values between -10 (large positive angle) and +10 (large negative angle). According to Figure 24, the decomposition of their JPD is specified as:

$$P\big(Steer_{t-1}, Steer_t, Angle_t\big) =$$
$$P\big(Steer_{t-1}\big) P\big(Steer_t \mid Steer_{t-1}\big) P\big(Angle_t \mid Steer_t\big).$$

According to the visual attention allocation theory of (Horrey et al., 2006) the perception of the heading angle is influenced by areas in the visual field (*ambient* channel), the head angle and the gaze angle relative to the head. At the present moment light colored nodes in Figure 24 are not included into the driver model. Instead we assumed that drivers are able to compute the aggregate sensory variables *heading angle* and *vehicle speed*. Compared to the lateral control in Salvucci & Gray's model (Möbus et al., 2007; Salvucci et al., 2004) our BAD model is more robust, makes less assumptions about the vision field, and *no* assumptions about gaze-control.

Variables Acc_t and Acc_{t-1} of DBN of *longitudinal* control take 21 different values between -10 (fully depress braking pedal) and +10 (fully depress acceleration pedal). Variable $Angle_t$ represents the angle between heading vector of the car and the course of the racing track to be reached in 2 second by current speed and can take 21 values between -10 (large positive angle) and +10 (large negative angle).

Variable $Speed_t$ represents the perceived longitudinal velocity and takes 10 values between 0 (low speed) and 10 (high speed). The decomposition of their JPD is specified as:

Figure 26. Overview embedding the BAD model in the TORCS driver model

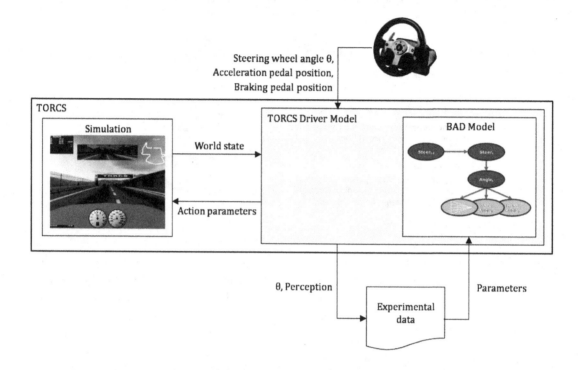

$$P\left(Acc_{t-1}, Acc_t, Angle_t, Speed_t\right) =$$
$$P\left(Acc_{t-1}\right)P\left(Acc_t \mid Acc_{t-1}\right)P\left(Angle_t \mid Acc_t\right)..$$
$$P\left(Speed_t \mid Acc_t\right)$$

All terms of the two decompositions are assumed to have a Gaussian form, whose parameters mean *ì* and standard deviation *ó* need to be obtained from experimental data.

4.1.1 Experimental Settings

To demonstrate the functionality of our BAD models for autonomous lateral and longitudinal control we use the open source racing simulation TORCS. Though considered as a racing game, TORCS accurately simulates car physics and allows the user to implement personal driver models. A driver model controls a vehicle within the TORCS world by action parameters (steering, accelerating, braking etc.) and has access to the current world state of the TORCS simulation. We

developed a driver model, referred as *TORCS driver model*, which is capable to derive action parameters by values read from external controllers, read/write experimental data from/into files and perceive its environment according to the perception component of the BAD model. The BAD model itself is embedded in the TORCS driver model (Figure 26). For implementation and inference of the BAD model we use ProBT©, a Bayesian inference engine and an API for building Bayesian models. ProBT© is published by the ProBAYES company and free available for academic purposes.

As external controller we use the Logitech G25, a controlling device consisting of a force-feedback steering wheel, pedals and a gear box. A human driver can manually control the TORCS vehicle via the steering wheel angle θt and the positions of acceleration- and braking-pedal. To achieve a usable drivability, operative steering wheel angles are limited to thirty percent of the possible steering wheel angles, leading to effective

Figure 27. Bird's eye view of race track with curve radii and rotation angles (graphics from (Möbus et al., 2009a) with kind permission of Springer Science and Business Media)

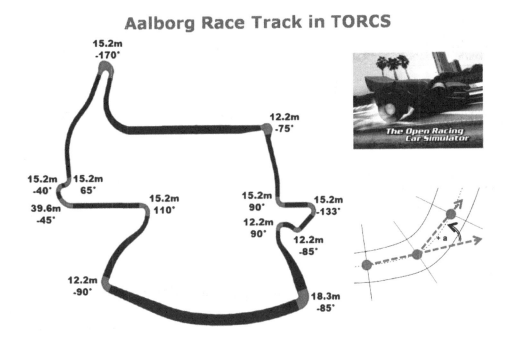

Figure 28. Snapshot of BAD model drive on TORCS race track (graphics from (Möbus et al., 2009a) with kind permission of Springer Science and Business Media)

Figure 29. Comparison of Human and BAD-Model Drives (graphics from (Möbus et al., 2009a) with kind permission of Springer Science and Business Media)

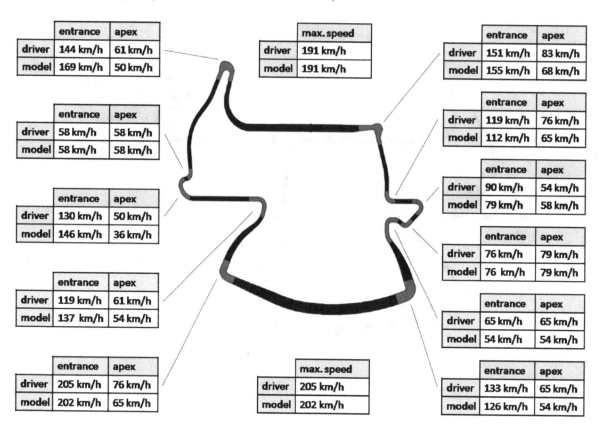

vehicle steering angles between −13.5° and +13.5°. Greater steering wheel movements were possible but would not affect the actual vehicle steering angle.

4.1.2 Recording of Experimental Data

Data were obtained in experimental drives of a single driver on the TORCS racing track "Aalborg". The map of the drive and curve specific measurements are presented in Figure 27. For that purpose, time series of values of *sensory-* and *action*-variables were recorded at an interval of 50ms. The experimental data were used to obtain parameters (means, standard deviations) for the Gaussian parametric forms for each of probability distribution of the BAD model. A single driven

lap was sufficient to obtain the data for estimating parameters stable enough for autonomous control.

4.1.3 Autonomous Driving of BAD Model

Under BAD-Model-control of the vehicle current values for *sensory*-variables of lateral and longitudinal control are sampled every time step t. After inferring the conditional probability distributions $P\left(Steer_t | steer_{t-1}, angle_t\right)$ of the lateral control DBN and $P\left(Acc_t | acc_{t-1}, angle_t, speed_t\right)$ of the longitudinal control DBN, concrete values for $Steer_t$ and Acc_t were randomly drawn from these distributions and used to control the TORCS vehicle. In principle we could choose the conditional expected actions to make the model react deterministically $E\left(Action_t|steer_{t-1},angle_t\right)$.

4.1.4 Results

A snapshot of a BAD model drive is shown in Figure 28 and a comparison of the driver's speed data with model generated speed data in Figure 29. The comparison demonstrates the quality of the simple BAD model but due to collisions with the roadside it is apparent that the capabilities of the BAD model have to be improved.

Improvements are expected by combining the two controllers, by including cognitive constructs like goals and latent states of the driver, and above all segmenting *maneuvers* into context dependent schemas (= *behaviors*).

Using goals (e.g. driving a hairpin or an S-curve) makes it possible to adapt the model to different road segments and situations. We try to use the same model for situation recognition or to situation-adapted control. The modeling idea of a HMM was abandoned because the state variable has to be too fine grained to obtain a high quality in vehicle control. Instead we are guided by the idea of state-based *mixture-of-behaviors* models in 3.3.2.5.

Figure 30. Ambient Vision Field - Position and direction of distance sensors

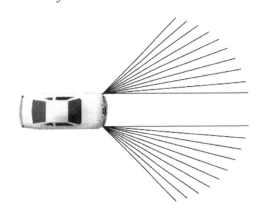

4.2 Use Case for In-Vehicle Cooperation: Driving Under the Lateral Assistance of a BAD Model

As an example for smart assistance, we present the use of a BAD model to assist a human driver's *lateral control*. We decided to change the former perception component of the BAD model from heading angle, represented by variable *Angle_t* in Figure 24 and 25, into twenty distance sensors, represented by variables S_t^0 to S_t^{19}. Positioned at the headlights of a car the distance sensors simulate according to the vision theory of Horrey's (Horrey et al., 2006) an *ambient vision field* with radius 105° (Figure 30).

Figure 31. Partially inverse DBN of Lateral Control for assisted driving

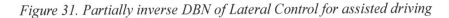

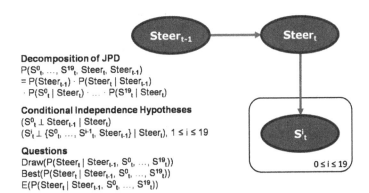

Decomposition of JPD
$P(S_t^0, ..., S_t^{19}, Steer_t, Steer_{t-1})$
$= P(Steer_{t-1}) \cdot P(Steer_t \mid Steer_{t-1})$
$\cdot P(S_t^0 \mid Steer_t) \cdot ... \cdot P(S_t^{19} \mid Steer_t)$

Conditional Independence Hypotheses
$(S_t^0 \perp Steer_{t-1} \mid Steer_t)$
$(S_t^i \perp \{S_t^0, ..., S^{i-1}_t, Steer_{t-1}\} \mid Steer_t), 1 \le i \le 19$

Questions
$Draw(P(Steer_t \mid Steer_{t-1}, S_t^0, ..., S_t^{19}))$
$Best(P(Steer_t \mid Steer_{t-1}, S_t^0, ..., S_t^{19}))$
$E(P(Steer_t \mid Steer_{t-1}, S_t^0, ..., S_t^{19}))$

Figure 32. TORCS driver model with BAD model for lateral assistance

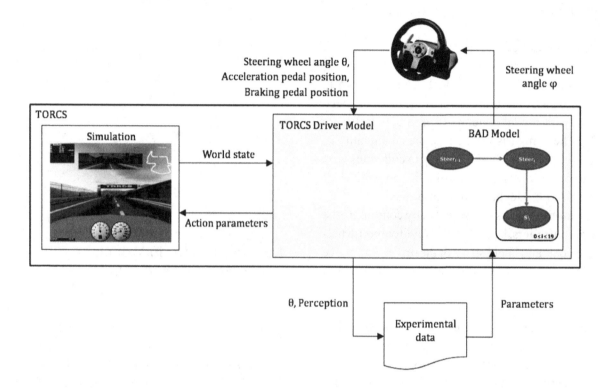

The variables of this BAD model are defined as follows: *Steer*$_t$ and *Steer*$_{t-1}$ can take 21 different values between -10 (hard left) and +10 (hard right). Each of the variables S_t^0 to S_t^{19} can take 20 different values between 0 (short distance) and 19 (long distance). The JPD is decomposed to

$$P\left(Steer_{t-1}, Steer_t, S_t^0, \ldots, S_t^{19}\right) =$$
$$P\left(Steer_{t-1}\right) P\left(Steer_t \mid Steer_{t-1}\right) \prod_{i=0}^{19} P(S_t^i \mid Steer_t).$$

The graphical representation of this decomposition is shown in Figure 31.

4.2.1 Experimental Setting

To demonstrate the functionality of the BAD model for assisted driving the open source rac-

ing simulation TORCS and the Logitech G25 as external controller are used once again.

For lateral assistance an extension of the TORCS driver model was necessary (see Figure 32). To assist the human driver, the BAD model not only must be able to control the simulated TORCS vehicle but also to influence the steering wheel angle $\grave{e}_t$. While the human driver can influence $\grave{e}_t$ simply by turning the steering wheel in the ordinary manner, the BAD model has to control the steering wheel in a different way by applying *force-feedback* commands. These commands are realized by a force-feedback *spring effect* that pushes the steering wheel back toward a certain position $\ddot{o}$ after it has been moved from that position. The strength of the reset force is determined by a function $f\left(\ddot{o}_t - \grave{e}_t\right)$. The variables influencing the effect can be adjusted and therefore can be used to parameterize the strictness and amount of BAD assistance. An overview of the resulting structure is given in Figure 32.

4.2.2 Recording of Experimental Data

To collect the experimental data needed to determine the parameters to define the BAD model, three laps were driven by the second author once again on the racing track "Aalborg" (Figure 27). The experimental data was then used to construct (conditional) probability tables for each of probability distribution of the BAD model. If the probability tables matched the shape of a Gaussian they were discretized with mean $ì$ and standard deviation $ó$.

4.2.3 Driving under Smart Assistance of the BAD Model

While running, at an interval of 50 ms, the TORCS driver model calculates the values for each of the distance sensors to derive current values in the BAD model for $S_{0,t}$ to $S_{19,t}$. Knowing $Steer_{t-1} = steer_{t-1}$ and $S_t^0 = s_t^0, \ldots, S_t^{19} = s_t^{19}$, the conditional probability distribution $P\left(Steer_t | steer_{t-1}, s_t^0, \ldots, s_t^{19}\right)$ will be inferred by the BAD model. By now, the inferred conditional distribution is used for continuous (1.) and temporarily (2.) driving assistance:

1. We created a *highly-automated* approach to assisted driving, letting the BAD model automatically control the steering wheel while a human driver can choose to intervene. To achieve this, a value $steer_t$ was, according to the BP draw strategy, randomly drawn from the distribution $P\left(Steer_t | steer_{t-1}, s_t^0, \ldots, s_t^{19}\right)$ and used to calculate a new center angle $ö$ for the force-feedback spring effect (steering wheel reset force).

2. As a *first* approach to *semi-automated* assisted driving we let a human driver control the vehicle while the BAD model only intervened when the steering movements had a significant low probability in the current situation. This was achieved by inferring $P\left(steer_t = è | steer_{t-1}, s_t^0, \ldots, s_t^{19}\right)$. Once this probability falls below a certain threshold an *anomaly* is detected, therefore a new value $steer_t$ was randomly drawn from the distribution $P\left(Steer_t | steer_{t-1}, s_t^0, \ldots, s_t^{19}\right)$ and used to calculate a new center angle $ö$ for the force-feedback spring effect (steering wheel reset force).

4.2.4. Results

The parameters derived from three driven laps turned out to be sufficient to create a BAD model that was able to assist a human driver. Furthermore, the level of assistance intensity can easily be shifted from rather light to very strict by simply adjusting the parameters of the spring effect and/or the threshold.

5 BAYESIAN LEARNING OF BAYESIAN AGENT MODELS

In order to learn the parameters of the CPDs of the BAD model in an objective manner a set of experimental data is needed. Learning can be done *offline* or *online*.

In *offline* learning as described in chapter 4 the collection of the training data and the testing of the BAD model are temporarily separated activities. Collecting experimental data without real-time reviewing the behavior of the BAD model allows only delayed information about its performance. Furthermore, due to the fact that $P\left(Steer_t | steer_{t-1}, s_t^0, \ldots, s_t^{19}\right)$ has to be inferred *inversely* (Figure 24, 25), an inspection of the

Figure 33. Left: Runtime-visualization showing the driver model and its sensors while approaching a right curve from a bird's eye view. Right: Runtime-visualization of the corresponding apriori uniform conditional probability distribution (red squares) $P\left(Steer_t | steer_{t-1}, s_t^0, \ldots, s_t^{19}\right)$. *Light gray bar shows the human chosen steering wheel angle* $Steer_t = \grave{e}_t$

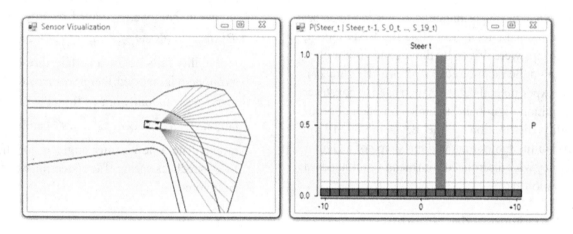

Figure 34. Left: Runtime-visualization showing the driver model and its sensors while approaching a right curve from a bird's eye view. Right: Runtime-visualization of the corresponding Bayesian learned aposteriori conditional probability distribution (red squares) $P\left(Steer_t | steer_{t-1}, s_t^0, \ldots, s_t^{19}\right)$

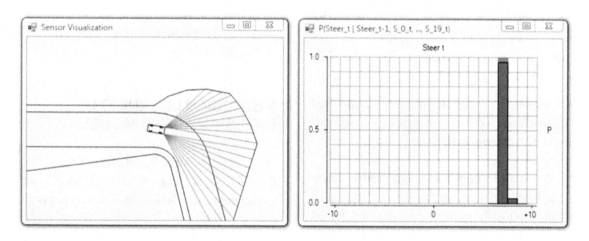

probability distributions of the BAD model is not very informative how to obtain the intended behavior and how to improve the completeness and quality of the model. Offline adapting the BAD model remains a clumsy and subjective procedure similar to the handcrafting of production system models.

A more natural approach would be the *online* learning of the BAD models by Bayesian parameter learning. We propose a *new* methodology: *Bayesian learning of agent models under human control.* The performance of the BAD model is observed by the human driver while the BAD model is driving. New data are learned only when the model behavior is unsatisfying. By *observing and correcting* the actions of the BAD model *only when needed*, problems can be solved, which are nearly impossible to discover by just analyzing its probability distributions. According to Bayesian methodology the old unsatisfactory BAD model is contained in the apriori-hypothesis, which will be revised by new training data to the aposteriori-hypothesis which contains the improved model.

We extended the TORCS driver model to provide the human driver with a learning control in the case of unsatisfactory BAD model behavior. New human experimental data are recorded by pressing a *learning button* attached to the steering wheel. When *learning* at every time step t current percepts provided by TORCS and actions read from the Logitech G25 controller are written into the database, updating the behavioral data and the CPDs. Once the button is pressed again, the data acquisition process is stopped and the conditional probability distributions are modified according to the Bayesian learning methodology.

To test the functionality of this approach, we used an empty database to learn parameters for the BAD model. We started with uniform apriori distributions for each of the conditional probability distributions of the BAD model. At the beginning, the apriori CPD $P\left(Steer_t|steer_{t-1}, s_t^0, \ldots, s_t^{19}\right)$ was uniform and the driving behavior of the BAD model therefore completely random. Figure 33 shows a screenshot of the TORCS driver model approaching a right curve and the corresponding apriori uniform CPD when starting with an empty database.

We then started collecting driving data while correcting the BAD model whenever its actions were not suitable to solve the current situation. The performance of the BAD model improved rapidly and it took only a few standard maneuvers to be able to let the BAD model drive the whole racing track successfully. As an example, Figure

34 shows the driver model approaching the same right curve as showed in Figure 33 after collecting experimental data of one driven right curve, resulting in a very peaked conditional probability distribution $P\left(Steer_t|steer_{t-1}, s_t^0, \ldots, s_t^{19}\right)$.

6 THE CONVERSION OF BAD MODELS TO BAYESIAN ASSISTANCE SYSTEMS

For the purpose of smart assistance in simulated or real world scenarios the obtained BAD models can be used as BAS in principle *as they are*. The only question is, whether the driving competence of the BAD model *is the same* as the driving competence of the human driver controlling the vehicle in the training session.

Our simulation world is so abstract that the sophisticated ambient human perception system can be simulated by a beam of sensors and sensor fusion. In more complicated scenarios we have to refine the model of the vision system (Lebeltel et al., 2004, Koike et al., 2008).

We believe that our approach is superior to a proposal to model the strategic skills of a PA-DAS with a *Markov Decision Process (MDP)* (Tango, 2011). A MDP needs a reward function. This function has to be derived deductively from theoretical concepts or learnt inductively from car trajectories by solving the *inverse reinforcement learning problem* (Abbeel et al., 2004). The deductive derivation of reward function often results in strange nonhuman overall behaviors. The inductive mining of the reward function from car trajectories or behavior traces seems to be a detour and seem more challenging than our approach.

The two new concepts *Bayesian learning of agent models under human control* and the *usage of a BAD model as a BAS or PADAS* are demonstrated here and in (Eilers et al, 2010, 2011; Möbus et al., 2010, 2011).

7 CONCLUSION AND OUTLOOK

We think that dynamic probabilistic models are sufficient expressive to describe and predict a wide range of phenomena. Their subtypes BAD and BAD-MoB models are appropriate for the challenges described in this paper, especially when they are learnt in experiments with *Bayesian learning of agent models under human control*. Next we have to implement further models creating a library of behaviors of various levels of expertise. To that end a careful selected taxonomy of scenarios, maneuvers, behaviors, and control actions without and with alter agents has to be defined and studied. We believe that our approach to use a BAD model as a BAS or PADAS is superior to a proposal to model the strategical and tactical skills of a PADAS with a *Markov Decision Process (MDP)*.

REFERENCES

Abbeel, P., & Ng, A. Y. Apprenticeship learning via inverse reinforcement learning, ACM International Conference Proceeding Series; Vol. 69, Proceedings of the twenty-first international conference on Machine learning, Banff, Alberta, Canada, 2004 Anderson, J.R.: Learning and Memory, John Wiley, 2002

Anderson, J. R. (2007). *How Can the Human Mind Occur in the Physical Universe*. Oxford: Oxford University Press.

Anderson, J. R., Fincham, J. M., Qin, Y., & Stocco, A. (2008). A Central circuit of the mind. *Trends in Cognitive Sciences*, *12*(4), 136–143.

Baumann, M., Colonius, H., Hungar, H., Köster, F., Langner, M., Lüdtke, A., et al. Integrated Modeling for Safe Transportation - Driver modeling and driver experiments, in Th. Jürgensohn, H. Kolrep (Hrsg.), *Fahrermodellierung in Wissenschaft und Wirtschaft, 2. Berliner Fachtagung für Fahrermodellierung*, Fortschrittsbericht des VDI in der Reihe 22 (Mensch-Maschine-Systeme), Nr. 28, 84-99, VDI-Verlag, 2009, ISBN 978-3-18-302822-1

Bengio, Y., & Frasconi, P. (1996). Input/output Hidden Markov Models for Sequence Processing. *IEEE Transactions on Neural Networks*, *7*, 1231–1249.

Bessiere, P. (2003). *Survey: Probabilistic Methodology and Techniques for Artifact Conception and Development, Rapport de Recherche, No. 4730*. INRIA.

Bessiere, P., Laugier, Ch., & Siegwart, R. (Eds.). (2008). *Probabilistic Reasoning and Decision Making in Sensory-Motor Systems*. Berlin: Springer.

Bischof, N. (1995). *Struktur und Bedeutung: Einführung i. d. Systemtheorie*. Bern: Huber.

Bishop, Ch. M. (2006). *Pattern Recognition and Machine Learning*. Heidelberg: Springer.

Bishop, C. M., & Svensén, M. (2003): Bayesian hierarchical mixtures of experts. In: Kjaerulff, U. and C. Meek (Ed.): Proceedings of the Nineteenth Conference on Uncertainty in Artificial Intelligence, pp. 57-64.

Cacciabue, P. C. (Ed.). (2007). *Modelling Driver Behaviour in Automotive Environments*. London: Springer.

Carlo, C. P., Hjälmdahl, M., Luedtke, A., & Riccioli, C. (Eds.). Human Modelling in Assisted Transportation: Models, Tools and Risk Methods, Heidelberg: Springer, 2011, ISBN-13: 978-8847018204

Chater, N., & Oxford, M. (Eds.). (2008). *The Probabilistic Mind: Prospects for Bayesian Cognitive Science*. Oxford, England: Oxford University Press.

Eilers, M., & Möbus, C. (2010). Lernen eines modularen Bayesian Autonomous Driver Mixture-of-Behaviors (BAD MoB) Modells. In Kolrep, H., & Jürgensohn, Th. (Eds.), *Fahrermodellierung - Zwischen kinematischen Menschmodellen und dynamisch-kognitiven Verhaltensmodellen, 3. Berliner Fachtagung für Fahrermodellierung, Fortschrittsbericht des VDI in der Reihe 22 (Mensch-Maschine-Systeme), Nr.32* (pp. 61–74). Düsseldorf: VDI-Verlag.

Eilers, M., & Möbus, C. *Learning of a Bayesian Autonomous Driver Mixture-of-Behaviors (BAD-MoB) Model*, in: Advances in Applied Digital Human Modeling, p. 436-445, Vincent G. Duffy (ed), CRC Press, Taylor & Francis Group, Boca Raton, 2010/2011, ISBN 978-1-4398-3511-1 and in. W. Karwowski and G. Salvendy (eds), 1st International Conference On Applied Digital Human Modeling, 17-20 July, 2010, Intercontinental, Miami Florida, USA, Conference Proceedings, Session Digital Human Modeling in the Bayesian Programming Framework, USA Publishing, ISBN-13: 978-0-9796435-4-5

Eilers, M., & Möbus, C. (2011). Learning the Relevant Percepts of Modular Hierarchical Bayesian Driver Models Using a Bayesian Information Criterion. In Duffy, V. G. (Ed.), *Digital Human Modeling, HCII 2011, LNCS 6777* (pp. 463–472). Springer-Verlag Berlin Heidelberg.

Fitts, P. M., & Posner, M. I. (1967). *Human Performance*. Belmont, CA: Brooks/Cole.

Flemisch, F., Schieben, A., & Kelsch, J. Löper, Chr., and Schomerus, J., Cooperative Control and Active Interfaces for vehicle assistance and automation, FISITA 2008, http://elib.dlr.de/57618/01/FISITA2008_DLR_FlemischEtAl_Cooperative-Control.pdf (visited 27th February, 2010)

Forbes, T., Kanazawa, H. K., & Russell, St. The BATmobile: Towards a Bayesian Automated Taxi, IJCAI'95, Vol 2, 1878 -1885 (1995)

Gluck, K. A., & Pew, R. W. (2005). *Modeling Human Behavior with Integrated Cognitive Architectures*. Mahwah, N.J.: Lawrence Erlbaum Associates.

Gopnik, A. & Tenenbaum, J.B., Bayesian networks, Bayesian learning and cognitive development, Development Science, 10:3, 2007, 281–287

Griffiths, Th. L., Kemp, Ch., & Tenenbaum, J. B. (2008). Bayesian Models of Cognition. In Sun, R. (Ed.), *The Cambridge Handbook of Computational Psychology* (pp. 59–100). Cambridge University Press.

Hamilton-Baillie. http://www.hamilton-baillie.co.uk/ (visited 27th February, 2010)

Hamker, F. H. (2001). RBF learning in a non-stationary environment: the stability-plasticity dilemma. In Howlett, R. J., & Jain, L. C. (Eds.), *Radial Basis Function networks 1: Recent Developments in Theory and Applications; Studies in fuzziness and soft computing (Vol. 66)*. Heidelberg: Physica Verlag.

Horrey, W. J., Wickens, Ch. D., & Consalus, K. P. (2006). Modeling Driver's Visual Attention Allocation While Interacting With In-Vehicle Technologies. *Journal of Experimental Psychology, 12*, 67–78.

Hutchins, E. (1995). *Cognition in the Wild*. Cambridge, Mass.: MIT Press.

IPG-Driver. http://www.ipg.de/32.html (visited 28th February, 2010)

Jagacinski, R. J., & Flach, J. M. (2003). *Control Theory for Humans: Quantitative Approaches to Modeling performance*. Mahwah, N.J.: Lawrence Erlbaum Associates.

Jensen, F. V., & Nielsen, Th. D. (2007). *Bayesian Networks and Decision Graphs* (2nd ed.). Springer.

Jurafsky, D., & Martin, J. H. (2009). *Speech and Language Processing* (2nd ed.). Pearson.

Jürgensohn, Th. Control Theory Models of the Driver, in: Cacciabue (ed), 2007, p. 277 – 292

Koike, C. C., Bessiere, P., & Mazer, E. (2008). Bayesian Approach to Action Selection and Attention Focusing. In Bessiere, P. (Eds.), *Probabilistic Reasoning and Decision Making in Sensory-Motor Systems* (pp. 177–201). Berlin: Springer.

Koller, D., & Friedman, N. (2009). *Probabilistic Graphical Models*. Cambridge, Mass.: MIT Press.

Kumugai, T., & Akamatsu, M. (2006, February). Prediction of Human Driving Behavior Using Dynamic Bayesian Networks, IEICE-Transactions on Info and Systems. *Volume E, 89-D*(2), 857–860.

Kumugai, T., Sakaguchi, Y., Okuwa, M., & Akamatsu, M. Prediction of Driving Behavior through Probabilistic Inference, Proceedings of the 8th International Conference on Engineering Applications of Neural Networks (EANN '03), 117 – 123

Le HyR.ArrigoniA.BessièreP.LebeltelO.; (2004);Teaching Bayesian Behaviours to Video Game Characters; Robotics and Autonom. Systems (Elsevier), Vol. 47: 177-185

Lebeltel, O., Bessiere, P., Diard, J., & Mazer, E. (2004). Bayesian Robot Programming. *Autonomous Robots, 16*, 49–79.

Lee, D. N. (1976). A theory of visual control of braking based on information about time-to-collision. *Perception, 5*, 437–459.

Lee, D. N. How movement is guided (2006) http://www.perception-in-action.ed.ac.uk/publications.htm (visited 28th February, 2010)

Löper, Ch., Kelsch, J., & Flemisch, F. O. Kooperative, Manöverbasierte Automation und Arbitrierung als Bausteine für hochautomatisiertes Fahren, in: Gesamtzentrum für Verkehr Braunschweig (Hrsgb): Automatisierungs-, Assistenzsysteme und eingebettete Systeme für Transportmittel, GZVB, Braunschweig, 2008, S. 215-237Meila, M., Jordan, M.I.: Learning Fine Motion by Markov Mixtures of Experts, MIT, AI Memo No. 1567, 1995

Miyazaki, T., Kodama, T., Furushahi, T., & Ohno, H. Modeling of Human Behaviors in Real Driving Situations, 2001 IEEE Intelligent Transportation Systems Conference Proceedings, 2001, 643 - 645

Möbus, C., & Eilers, M. (2008). First Steps Towards Driver Modeling According to the Bayesian Programming Approach, Symposium Cognitive Modeling, p.59, in: L. Urbas, Th. Goschke & B. Velichkovsky (eds) *KogWis 2008*. Christoph Hille, Dresden, ISBN 978-3-939025-14-6

Möbus, C., & Eilers, M. (2009a). Further Steps Towards Driver Modeling according to the Bayesian Programming Approach, in: *Conference Proceedings, HCII 2009, Digital Human Modeling*, pp. 413-422, LNCS (LNAI), Springer, San Diego, ISBN 978-3-642-02808-3

Möbus, C., & Eilers, M. Garbe, H., and Zilinski, M. (2009b). Probabilistic and Empirical Grounded Modeling of Agents in (Partial) Cooperative Traffic Scenarios, in: *Conference Proceedings, HCII 2009, Digital Human Modeling*, pp. 423-432, LNCS (LNAI), Springer, San Diego, ISBN 978-3-642-02808-3

Möbus, C., & Eilers, M. Zilinski, M. Garbe, H. (2009c). Mixture of Behaviors in a Bayesian Driver Model, in: Lichtenstein, A. et al. (eds), Der Mensch im Mittelpunkt technischer Systeme, p.96 and p.221-226 (CD), Düsseldorf: VDI Verlag, ISBN 978-3-18-302922-8, ISSN 1439-958X

Möbus, C., & Eilers, M. *Mixture of Behaviors and Levels-of-Expertise in a Bayesian Autonomous Driver Model*, in: Advances in Applied Digital Human Modeling, p. 425-435, Vincent G. Duffy (ed), CRC Press, Taylor & Francis Group, Boca Raton, 2010/2011, ISBN 978-1-4398-3511-1 and in. W. Karwowski and G. Salvendy (eds), 1st International Conference On Applied Digital Human Modeling, 17-20 July, 2010, Intercontinental, fMiami Florida, USA, Conference Proceedings, Session Digital Human Modeling in the Bayesian Programming Framework, USA Publishing, ISBN-13: 978-0-9796435-4-5

Möbus, C., Eilers, M., & Garbe, H. (2011). Predicting the Focus of Attention and Deficits in Situation Awareness with a Modular Hierarchical Bayesian Driver Model. In Duffy, V. G. (Ed.), *Digital Human Modeling, HCII 2011, LNCS 6777* (pp. 483–492). Springer-Verlag Berlin Heidelberg.

Möbus, C., Hübner, S., & Garbe, H. (2007). Driver Modelling: Two-Point- or Inverted Gaze-Beam-Steering. In Rötting, M., Wozny, G., Klostermann, A., & Huss, J. (Eds.), *Prospektive Gestaltung von Mensch-Technik-Interaktion, Fortschritt-Berichte VDI-Reihe 22, Nr. 25, 483 – 488.* Düsseldorf: VDI Verlag.

Neapolitan, R. E. (2004). *Learning Bayesian Networks*. Upper Saddle River: Prentice Hall.

NETICA. http://www.norsys.com/ (visited 28th February, 2010)

Norman, D. A. (2007). *The Design of Future Things*. Basic Books.

Oliver, N., & Pentland, A. P. Graphical Models for Driver Behavior Recognition in a SmartCar, IEEE Intl. Conf. Intelligent Vehicles, 7 – 12, 2000

Oliver, N. M. Towards Perceptual Intelligence: Statistical Modeling of Human Individual and Interactive Behaviors, MIT Ph.D. Thesis, 2000

Pearl, J. (1988). *Probabilistic Reasoning in Intelligent Systems*. San Mateo, CA: Morgan Kaufmann.

Pearl, J. (2009). *Causality – Models, Reasoning, and Inference* (2nd ed.). Cambridge University Press.

Quin, Y., Bothell, D., & Anderson, J. R. (2007): ACT-R meets fMRI. In Proceedings of LNAI 4845 (pp. 205-222). Berlin, Germany: Springer.

Rabiner, L. R. (1989). A Tutorial on Hidden Markov Models and Selected Applications in Speech Recognition. *Proceedings of the IEEE, 77*(2), 257–286.

Rizzo, M., McGehee, D. V., Dawson, J. D., & Anderson, S. N. (2001). Simulated Car Crashes at Intersections in Drivers With Alzheimer Disease. *Alzheimer Disease and Associated Disorders, 15*(1), 10–20.

Russell, St., & Norvig, P. (2010). *Artificial Intelligence: A Modern Approach* (3rd ed.). Upper Saddle River, N.J.: Prentice Hall.

Salvucci, D. D. (2006). Modeling Driver Behavior in a Cognitive Architecture. *Human Factors, 48*(2), 362–380.

Salvucci, D. D. (2007). Integrated Models of Driver Behavior. In Gray, W. D. (Ed.), *Integrated Models of Cognitive Systems* (pp. 356–367). Oxford University Press.

Salvucci, D. D., & Gray, R. (2004). A Two-Point Visual Control Model of Steering. *Perception, 33*, 1233–1248.

Sarter, N., Amalberti, R. R., & Amalberti, R. (2000). *Cognitive Engineering in the Aviation Domain*. Lawrence Erlbaum Assoc. Inc.

SharedSpace. http://www.sharedspace.eu/en/home/ (visited 27th February, 2010)Spirtes, P., Glymour, C., and Scheines, R., 2001 (2nd ed.), Causation, Prediction, and Search, Cambridge, Mass: MIT Press

Tango, F., Aras, R., & Pietquin, O. *Learning Optimal Control Strategies from Interactions with a PADAS*, p.119-127, in: Human Modelling in Assisted Transportation: Models, Tools and Risk Methods, Cacciabue Pietro Carlo, Magnus Hjälmdahl, Andreas Luedtke, and Costanza Riccioli (eds), Heidelberg: Springer, 2011, **ISBN-13: 978-8847018204**

Thrun, S., Burgard, W., & Fox, D. (2005). *Probabilistic Robotics*. Cambridge, Mass.: MIT Press.

TORCS. http://torcs.sourceforge.net/ (visited 27th February, 2010)

Weir, D. H., & Chao, K. C. Review of Control Theory Models for Directional and Speed Control, in: Cacciabue, P.C., p. 293 – 311 (2007)

Wickens, Th. D. (1982). *Models for Behavior: Stochastic Processes in Psychology*. San Francisco: Freeman.

Xiang, Y. (2002). *Probabilistic Reasoning in Multiagent Systems - A Graphical Models Approach*. Cambridge: Cambridge University Press.

Xu, Y. (2005). *Ka Keung Caramon Lee, and Ka Keung C. Lee, Human Behavior Learning and Transfer*. CRC Press Inc.

Zelkha, Eli; Epstein, Brian, From Devices to Ambient Intelligence, Digital Living Room Conference, June 1998 (http://www.epstein.org/brian/ambient_intelligence/DLR%20Final%20Internal.ppt; visited 27th February, 2010)

KEY TERMS AND DEFINITIONS

Anomalies: Risky maneuvers are called anomalies when they have a low probability of occurrence in the behavior stream of experienced drivers and which only experienced drivers are able to prevent or to anticipate automatically. A measure of the anomaly of the driver's behavior is the conditional probability of his behavior under the hypothesis that the observed actions are generated by a stochastic process which generated the trajectories or behaviors of the correct maneuver M+.

Anticipatory Planning: For anticipatory planning the conditional probability of the NextFutureDrive under the assumption of the pastDrive, the currentDrive, and the anticipated expectedFutureDrive has to be computed.

Bayesian Assistance Systems (BAS): For the purpose of smart assistance in simulated or real world scenarios the obtained Bayesian Autonomous Driver (BAD) models can be used as prototypical Bayesian Assistance Systems (BAS). Due to their probabilistic nature BAD models or BAS can not only be used for real-time control but also for real-time detection of anomalies in driver behavior and real-time generation of supportive interventions (countermeasures).

Bayesian Autonomous Driver (BAD) model: BAD models describe phenomena on the basis of the variables of interest and the decomposition of their joint probability distribution (JPD) into conditional probability distributions (CPD-factors) according to the special chain rule for Bayesian networks. The underlying conditional independence hypotheses (CIHs) between sets of variables can be tested by standard statistical methods (e.g. the conditional mutual information index. The parameters of BAD models can be learnt objectively with statistical sound methods by batch from multivariate behavior traces or by learning from single cases. Due to their probabilistic nature BAD models or BAS can not only be used for real-time control of vehicles but also for real-time detection of anomalies in driver behavior and real-time generation of supportive interventions (countermeasures).

Bayesian Autonomous Driver with Mixture-of-Behaviors (BAD-MoB) Model: The model is suited to represent the sensor-motor system of individuals or groups of human or artificial agents in the functional autonomous layer or stage of Anderson. In a MoB model it is assumed that the behavior can be context-dependent generated as a mixture of ideal schematic behaviors (= experts). The template or class model is distributed across two time slices, and tries to avoid the latent state assumptions of Hidden Markow Models. Learning data are time series or case data of relevant variables: percepts, goals, and actions. Goals are the only latent variables which could be set by commands issued by the higher associative layer.

Bayesian Filter and Action Model (BFAM): In the Bayesian Filter and Action Model actions are not only dependent on the current process state but also on direct antecedent actions. Thus the generation of erratic behavior is suppressed. Furthermore the BFAM includes direct action effects on the next future process state. This is important when the influence of action effects should be modeled directly into the state not making a detour via the environment and the perception of the agent.

Bayesian Learning of Agent Models under Human Control: The performance of the BAD model is observed by the human driver while the BAD model is driving. New data are learned only when the model behavior is unsatisfying. By observing and correcting the actions of the BAD model only when needed, problems can be solved, which are nearly impossible to discover by just analyzing its probability distributions.

Bayesian (Robot) Programs (BPs): BP is a simple and generic framework suitable for the description of human sensory-motor models in the presence of incompleteness and uncertainty. It provides integrated model-driven data analysis and model construction. In contrast to conventional Bayesian network models BP-models put emphasis on a recursive structure and infer concrete motor actions for real-time control on the basis of sensory evidence. Actions are sampled from CPDs according various strategies after propagating sensor or task goal evidence.

Computational Agent Model: Computational agent models have to represent perceptions, beliefs, goals, and actions of ego and alter agents.

Cooperative Scenario: When goals are issued by one single principal.

Cooperative Driving Scenario: A driving scenario with in-vehicle-cooperation between a human driver and a BAS

Distributed Cognition: Originated by Edwin Hutchins in the mid 1980s. He proposed that human knowledge and cognition is not confined to individuals but is also embedded in the objects and tools of the environment. Cognitive processes may be distributed across the members of a social group or the material or environmental structure.

Dynamic Bayesian Filter (DBF): The DBF is a HMM with state, percept and motor variables. The general algorithm consists of two steps in each iteration or recursive call: 1. Prediction step: from the most recent apriori belief(state) and the current control (= action) compute a provisional belief(state); 2. Correction step: from the current provisional belief(state) and the current measurements (= percepts) compute the posteriori belief(state).

Dynamic Bayesian Network (DBNs): In the case of identical time-slices and several identical temporal links we have a repetitive temporal model which is called Dynamic Bayesian Network model (DBN). DBNs are dynamic probabilistic models. HMMs and DBN are mathematically equivalent. Though, there is a trade-off between estimation efficiency and descriptive expressiveness in HMMs and DBNs. Estimation in HMMs is more efficient than in DBNs due to algorithms (Viterbi, Baum-Welch) whereas descriptive flexibility is greater in DBNs. At the same time the state-space grows more rapidly in HMMs than in corresponding DBNs.

Dynamic Probabilistic Model: Dynamic probabilistic models evolve over time. If the model contains discrete time-stamps one can have a model for each unit of time. These local models are called time-slices. The time slices are connected through temporal links to give a full model.

Hidden Markow Models (HMMs): A special category of time-stamped dynamic probabilistic models is that of a Hidden Markov Model (HMM). They are repetitive temporal models in which the state of the process is described by a single discrete random variable. Because of the Markov assumption only temporarily adjacent time slices are linked by a single link between the state nodes. HMMs are sequence classifiers and allow the efficient recognition of situations, goals and intentions; e.g. diagnosing driver's intention to stop at a crossroad. HMMs and DBN are mathematically equivalent. Though, there is a trade-off between estimation efficiency and descriptive expressiveness in HMMs and DBNs. Estimation in HMMs is more efficient than in DBNs due to algorithms (Viterbi, Baum-Welch) whereas descriptive flexibility is greater in DBNs. At the same time the state-space grows more rapidly in HMMs than in corresponding DBNs.

Partial or Non-Cooperative Scenario: When goals are issued by several different principals.

Shared Space: Based on the observation that individuals' behavior in traffic is more positively affected by the built environment of the public space than by conventional traffic control devices (signals, signs, road markings, etc.) or regulations.

ENDNOTES

[1] http://board.gulli.com/thread/573253-haderer-karikatur-von-michael-schumacher/ (25th, March 2010)

[2] In medical terms, *tunnel vision* is the loss of peripheral vision with retention of central vision, resulting in a constricted circular tunnel-like field of vision (http://en.wikipedia.org/wiki/Tunnel vision, visited 1st March, 2010)

APPENDIX A. DAGS OF STATIC AND DYNAMIC BAYESIAN MODELS

Figure 35. Reactive Bayesian Network (BN); ellipses in plates denote sets of random variables (plate notation(Bishop, 2006))

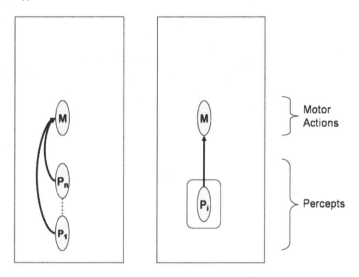

Figure 36. Inverse (naïve) Classifier BN (Le Huy et al., 2004; Russell el al., 2010)

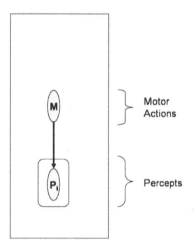

Figure 37. Inverse BN-Model with State Variable (Koike et al, 2008)

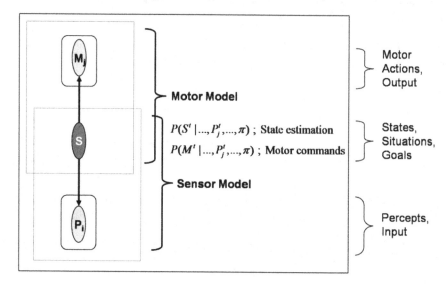

Figure 38. Hidden Markov Model (HMM) (Miyazaki et al., 2001; Kumugai et al., 2003; Rabiner, 1989)

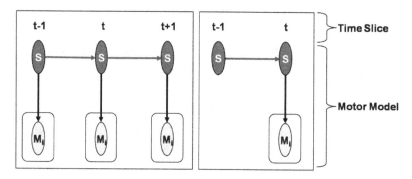

Figure 39. Hidden Markov Model (HMM) with (Inverted) Sensor Model (Le Huy et al., 2004)

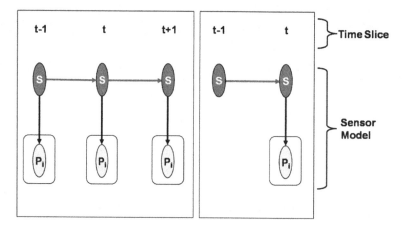

Figure 40. Hidden Markov Model (HMM) with Motor and (Inverted) Sensor Model (Oliver et al., 2000; Miyazaki et al., 2001; Koike et al., 2008)

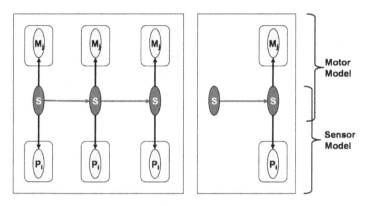

Figure 41. (Reactive) Input-Output HMM (RIOHMM) – slight modification of (Bengio et al., 1996)

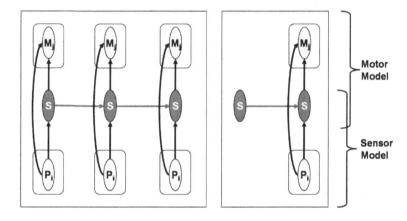

Figure 42. Discrete Bayesian Filter (= HMM with Sensor and Inverted Motor Model) (Thrun et al., 2005; Koike et al., 2008)

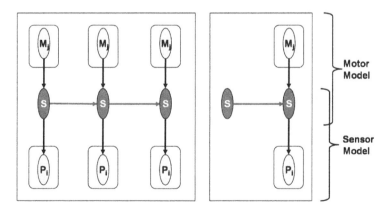

Figure 43. Coupled HMM (CHMM) (Oliver, 2000)

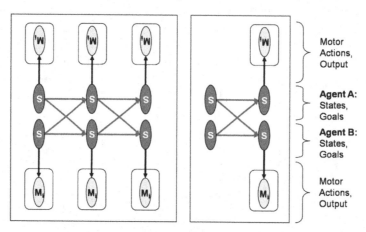

Figure 44. Coupled Reactive HMM (CRHMM)

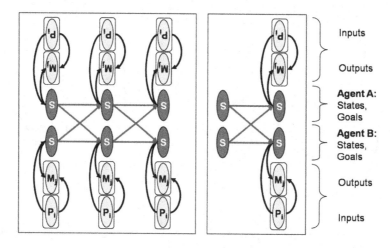

Figure 45. Switching Linear Dynamic System (SLDS) (Kumugai et al., 2003)

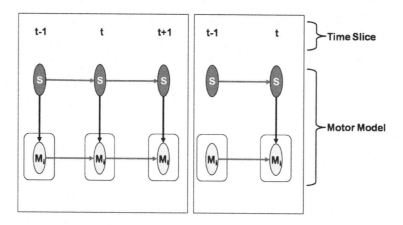

Figure 46. Bayesian Filter and Action Model (Koike et al., 2008)

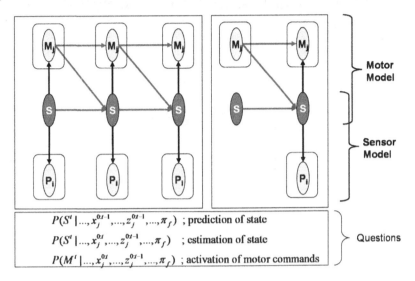

$P(S^t \mid ..., x_j^{0:t-1}, ..., z_j^{0:t-1}, ..., \pi_f)$; prediction of state

$P(S^t \mid ..., x_j^{0:t}, ..., z_j^{0:t-1}, ..., \pi_f)$; estimation of state

$P(M^t \mid ..., x_j^{0:t}, ..., z_j^{0:t-1}, ..., \pi_f)$; activation of motor commands

APPENDIX B. NETICA IMPLEMENTATIONS OF PARADIGMATIC DYNAMIC BAYESIAN AGENT MODELS

Figure 47. Prediction Step in Night Watchman DBF

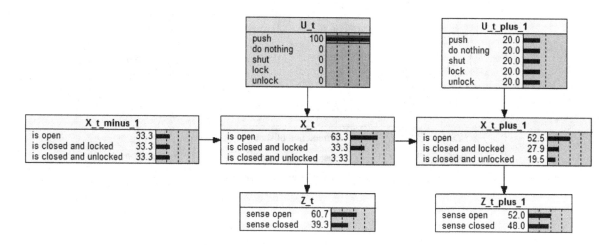

Figure 48. Correction Step in Night Watchman DBF

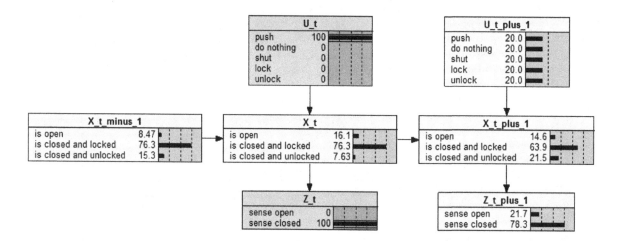

Figure 49. Perception Step in Night Watchman DBF

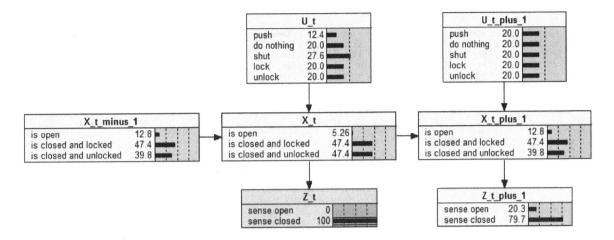

Figure 50. Action Step in Night Watchman DBF

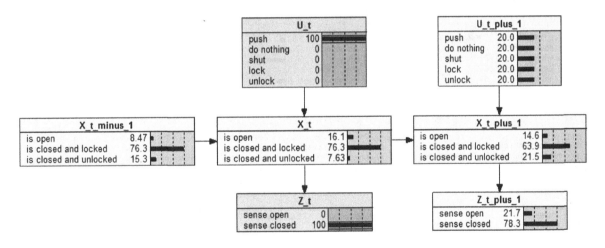

Figure 51. Apriori Beliefs in Expert-Role, Mixed Experts, or Schema DBN Model with Action Effects

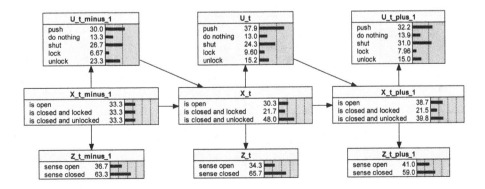

Figure 52. First Perception Step in Night Watchman DBN with Action Effect Model

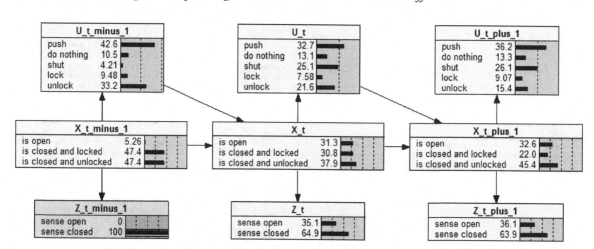

Figure 53. First Action Step in Night Watchman DBN with Action Effect Model

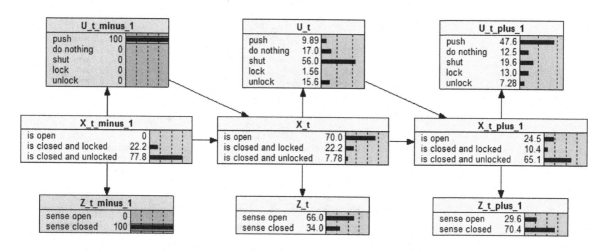

Figure 54. Second Perception Step in Night Watchman DBN with Action Effect Model

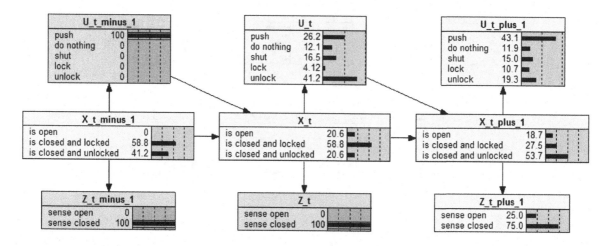

Figure 55. Apriori Beliefs in Expert-Role, Mixed Experts, or Schema DBN Model with Action Effect Model

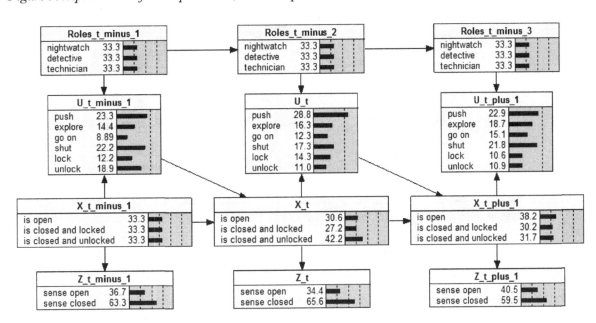

Figure 56. Role, Schema, or Intention Diagnostic in Expert-Role, Mixed Experts, or Schema DBN Model with Action Effect Model

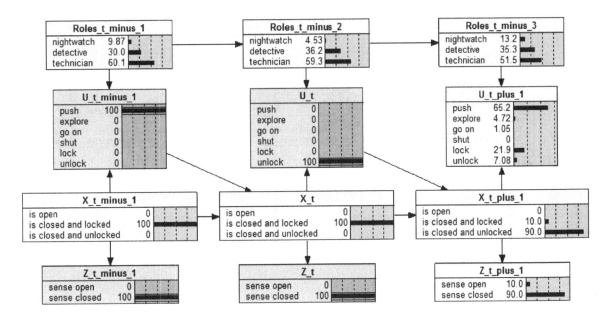

Figure 57. Apriori Beliefs in AoI and Ambient Vision-Role-Model

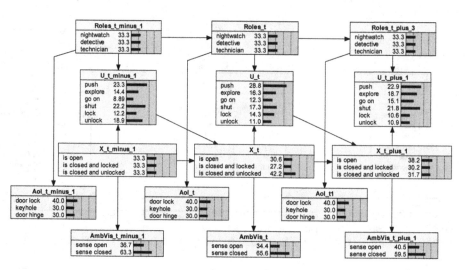

Figure 58. Inference of Intention, Role, and Action in AoI and Ambient Vision-Role-Model

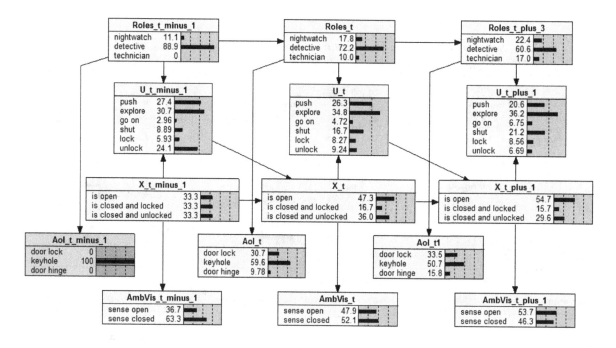

Figure 59. Inference of Intention, Role, and Action in AoI and Ambient Vision-Role-Model

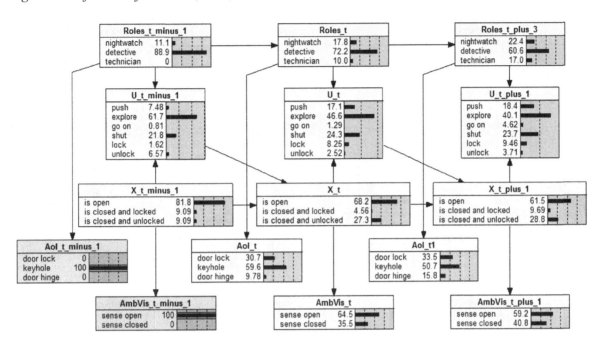

Figure 60. Inference of Role-specific Actions in AoI and Ambient Vision-Role-Model

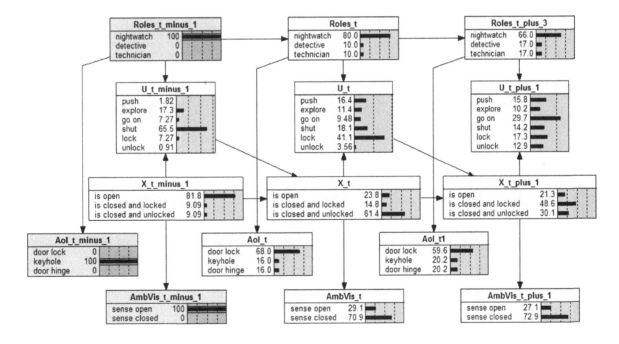

Figure 61. Role or Intention Conflict in AoI and Ambient Vision-Role-Model

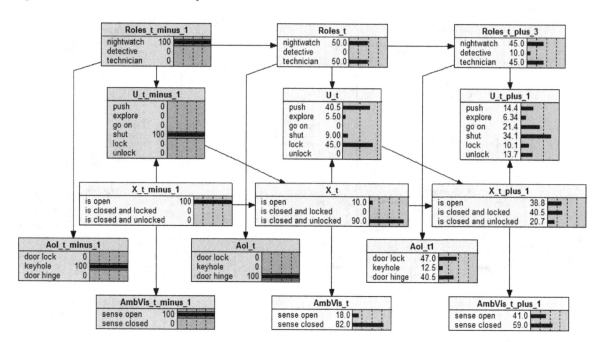

Chapter 24
Context–Sensitive Spatial Interaction and Ambient Control

Bernd Krieg-Brückner
Universität Bremen, Germany

Hui Shi
Universität Bremen, Germany

Bernd Gersdorf
Universität Bremen, Germany

Mathias Döhle
Universität Bremen, Germany

Thomas Röfer
Universität Bremen, Germany

ABSTRACT

In this chapter, we first briefly introduce the setting: mobility assistants (the wheelchair Rolland and iWalker) and smart environment control in the Bremen Ambient Assisted Living Lab. In several example scenarios, we then outline our contributions to the state of the art, focussing on spatial knowledge representation, reasoning and spatial interaction (multi-modal, but with special emphasis on natural language dialogue) between three partners: the user, a mobility assistant, and the smart environment.

1 SENIORS IN SPE IN THE BREMEN AMBIENT ASSISTED LIVING LAB

At the *Bremen Ambient Assisted Living Lab*, BAALL, new technology for Ambient Assisted Living, AAL, will be tested and evaluated for everyday usability. Figure 1 shows the overall layout as it has been realised. The BAALL is

aimed at *seniors in spe* (seniors-to-be), a term referring to people who actively plan their future at a relatively early stage in life (e.g., when choosing an apartment or designing a house for lifelong use) with the expectation to live in this familiar environment for as long as possible. The issue is thus to anticipate the scenarios that may arise from a range of potential age-related physical or cognitive impairments and to plan ways of compensating for these impairments using

DOI: 10.4018/978-1-61692-857-5.ch024

Figure 1. Areas in the BAALL (Simulation)

technological assistance. Such a home should be designed with a basic infrastructure that allows *step-by-step* upgrades to be made subsequently with suitable AAL components as required in order to remove the need for major construction or adaptation work. The appeal of the BAALL is that it looks like an entirely normal apartment, albeit a well equipped one; the technological infrastructure is discreet, if visible at all.

The BAALL contains all standard living areas (home office, bedroom, bathroom and dressing area, living and dining room, kitchenette) within a 60m² apartment suitable for two people to live in on a trial basis, constructed according to the *design for all* principle, modelled as such after the *Casa Agevole* at the Sta. Lucia research hospital in Rome. One challenge in existing buildings is providing mobility assistance in *confined spaces*. Only too often so-called barrier-free developments may be suitable for hand-driven wheelchairs but are not necessarily compatible with power-wheelchairs. With respect to doors, the question is not only whether they are wide enough, but whether they can be remotely controlled since switches may be difficult or impossible to reach. This is just one example where we are looking for ways to evaluate the interaction of mobility assistants described in section 2 (and

thus their users) with the smart environment outlined in section 3, with particular focus on spatial interaction, see sections 4 and 5.

2 MOBILITY ASSISTANTS

Two mobility assistants have been developed: the intelligent wheelchair *Rolland* and the intelligent walker *iWalker*. Both devices provide similar assistance but each for a different target population; see also for a more elaborate description.

2.1 Intelligent Wheelchair *Rolland*

Rolland is based on the commercial power-wheelchair *Xeno*, manufactured by the German company *Otto Bock Mobility Solutions* see Figure 2. We equipped the original wheelchair with two laser range sensors, wheel encoders, and an onboard computer. Various assistants for the wheelchair Rolland are being developed and evaluated to compensate for diminishing physical and cognitive faculties: the *safety assistant* brakes in time; the *driving assistant* avoids obstacles and facilitates passing through a door; the *navigation assistant* guides along a route or drives autonomously; a

Figure 2. Rolland 4 (Xeno, Otto Bock Mobility Solutions)

head joystick and spoken *natural language dialogue* ease interaction, see also Section 5.

2.1.1 Safety Assistant

The *safety assistant* mainly consists of a *safety layer* that ensures that *Rolland* will stop in time before a collision can occur. Either the current driving command is safe, and it can be sent to the wheelchair, or it is not, and the wheelchair has to stop instead. Whether the wheelchair can stop in time depends on the shape of the wheelchair, the actual speeds of the two drive wheels and the current drive command, because it will influence the current speeds in the future, and its current surroundings. The surroundings are measured using the laser range sensors; a model of the environment is maintained in a local obstacle map (see below). Based on the current speeds and the commanded speeds, a *safety area* is searched for obstacles in the map. If the safety area is free of obstacles, the current driving command is safe. Since the shape of such a safety area is rather complex, a large number of safety areas have been pre-computed and stored in a lookup table.

2.1.2 Driving Assistant

The *driving assistant* provides obstacle avoidance for Rolland. The user remains in complete control of the system as long as there are no obstacles on the intended path. Whenever obstacles block the way and would result in the wheelchair being stopped by the *safety layer* (the main part of the safety assistant), the driving assistant takes control and avoids the obstacle with as little deviations from the commands given by the user as possible (cf. Figure 3).

Figure 3. Intervention of the driving assistant. The driver presses the joystick straight ahead (indicated by the thick line); the driving assistant detects an obstacle on the right hand side (cf. safety area in the local obstacle map above), and avoids it to the left (thin line)

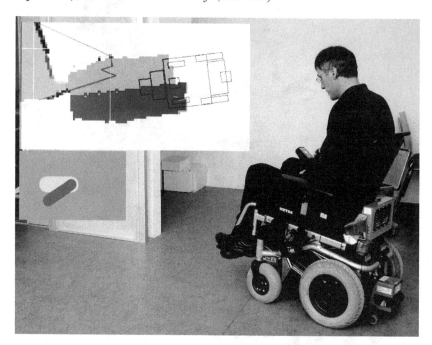

2.1.3 Autonomous Navigation Assistant

The *navigation assistant* is based on a SLAM-based metric representation of the environment with an embedded Route Graph. The aim of the system is to have a robust navigation component for an autonomous wheelchair within a global frame of reference. The user can choose a specific destination – e.g. "Refrigerator", "Entrance", or "Wardrobe" – from a predefined list. Given the destination, the system provides two different use cases: *autonomous navigation* of the robot without any need for interventions by the user, or a *guidance mode* which lets the user drive manually, but provides information about where to navigate next. The key idea behind this assistant is to specify all possible destinations together with all possible routes between these destinations, i.e. the Route Graph. This approach shifts parts of the intelligence needed from the robot to an external human expert who compiles and

edits this representation in advance. Thus it can be assured that necessary routes always exist and lead along paths actually usable by a wheelchair, i.e., without undetectable obstacles; this could not presently be assured for automatic route generation given the current sensorial equipment of the wheelchair *Rolland*.

2.2 Intelligent Walker

The *iWalker* is the prototype of an intelligent walker equipped with electric brakes, wheel encoders, a laser range sensor, and a small control PC, in analogy to Rolland, cf. Figure 4. Thus iWalker provides a similar variety of services to users with walking impairments, also with diminished sight. By controlling the electric brakes, the *walking assistant* is able to help the user keeping clear of obstacles; the *navigation assistant* guides the user along a route, either by displaying an arrow on the screen, or by braking the wheels softly, indicating

Figure 4. Prototype of the iWalker

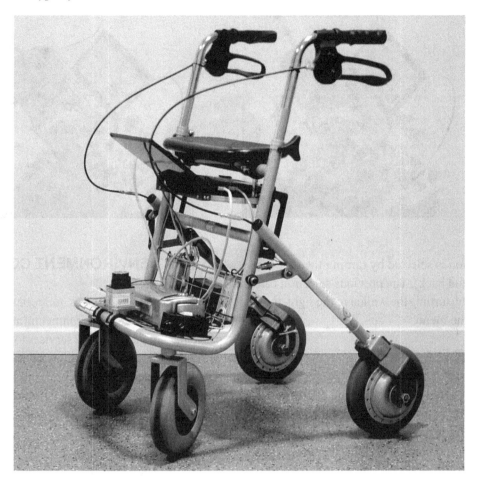

the direction. The functionalities of these assistants are analogous to those of Rolland's driving and navigation assistants. In another variant, iWalker also provides additional safety by regulating the speed when going down a slope, and actively helps going up.

2.2.1 RollScroll Interaction

In contrast to Rolland (and in the absence of additional speech interaction, see section 5.1), the *iWalker* currently has no additional user input device. Moreover, the user's hands are primarily engaged on the handle bars, maintaining a close grip; thus pointing gestures on a touch screen would be cumbersome, or, for some users, even dangerous.

We therefore developed a new interaction device that seems to be quite natural, using odometry as an input channel. When the *iWalker* is standing still, it displays a question mark in the upper left corner of the screen (cf. Figure 5 left). This is a signal that the *iWalker* is ready to accept a "menu" gesture: a quick turn of the *iWalker,* slightly to the left and the right (in any order). Now the selection menu appears on the screen (cf. Figure 5 right). It lists the available options for selecting one of the assistants, and all available target locations in alphabetical order. The user can now scroll through the menu items by moving the *iWalker* forwards and backwards (*RollScroll*, patent pending). A

Figure 5. The guiding arrow and the RollScroll menu

target location is selected by turning the *iWalker* to the left (and back). The previous selection can be retained by turning the walker to the right, thus cancelling the menu.

2.2.2 Guiding Assistant

The *guiding assistant* is a combination of the two assistants described above, i.e. the walking assistant and the navigation assistant. Instead of displaying a guidance arrow on the screen, the user is guided to the target location by braking. The general approach consists of a guidance controller that uses the current driving direction generated by the *navigation assistant* to compute an "intended walking direction" for the *walking assistant*. The latter is then using the brakes of the iWalker to guide the user to the target location. The guidance controller generates motion commands as if the iWalker would drive in free space, i.e., it completely ignores the current obstacle situation. However, since the underlying walking assistant avoids obstacles, the iWalker is able to navigate robustly, even in narrow environments.

3 SMART ENVIRONMENT CONTROL

Another research focus is *interoperability* and safety of systems for environmental assistance, in particular integrated higher services for interaction with smart household appliances and furniture. As we shall see in this section, interaction with the smart environment may be either: direct or indirect, with a mobility assistant being a go-between or mediator.

3.1 Smart Devices

The concept *Smart Device* is frequently used for portable electronic devices with a data interface to the environment (e.g. internet or telephone connections). In an AAL environment, all data exchanging devices are considered smart, e.g. building automation (climate control, light, electric doors, access control), household appliances and kitchen devices (microwave, kitchen scale), multi-media devices, telephone, and furniture with sensors/actors.

In the BAALL, all immobile devices are connected to standard buses for domotic control (building, home control), in particular the KNX bus, via a small server with various plug-ins for these buses that is connected to the Internet. This way, all smart devices, often coming with their individual standard remote control, can now be globally controlled: individually via a centrally located interactive control panel; via a universal mobile control unit (e.g. the iPhone, see section 3.2); or by integration into higher-level services, see section 3.2.2.

3.1.1 Domotic Control

The BAALL is part of a larger office building with preinstalled climate and access control. Heating uses valves with separate temperature sensors, and controls for window openers and venetian blinds, all electrically powered and remote controllable via a LON bus interface. The lights are connected to KNX controlled power rails, sockets, or wires.

3.1.2 Smart Appliances

As an example for household appliances with a CHAIN interface (the KNX protocol via Power

line), the refrigerator in the BAALL is a standard device from Miele with illuminated food compartments. As an add-on in the BAALL, the rack illumination can be activated individually for each food compartment via KNX. This feature is used to direct the user's attention to a specific compartment for storage and retrieval of goods.

3.1.3 Smart Furniture

The kitchen counter (in fact, the whole kitchenette, see Figure 6) can be electrically moved up and down to fit the users preferred height, especially for wheelchair users; in addition, cabinets can be moved downward individually. The original interface uses switches at the front of the kitchen counter for relative motion. For the installation in the BAALL, the kitchen is connected to KNX and now allows remotely controlled absolute positioning (e.g. to reach wheelchair height or the preset height for an individual). Similarly, the wash-basin

Figure 6. Movable kitchenette in the BAALL

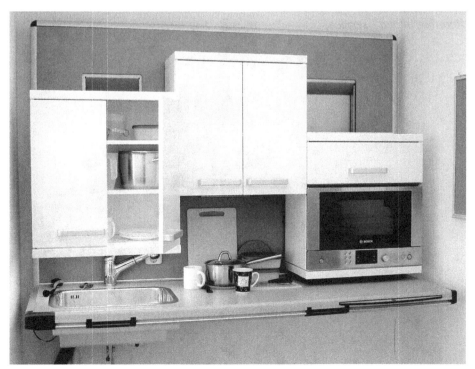

Figure 7. Interfaces for the iPhone

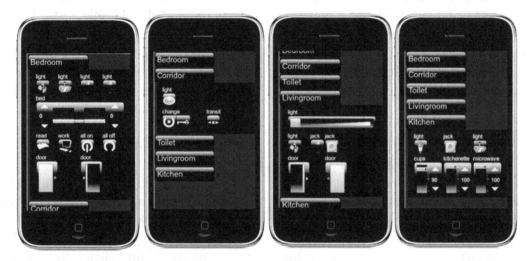

and the toilet can be electrically adjusted for height and have been connected to KNX.

The comfort bed (Figure 8) uses four motors to adjust feet, head, and back. The original IR control has been replaced by a KNX interface for integration into the BAALL.

3.1.4 Doors

Wheelchair and walker users are not able to reach door handles or switches easily, while automatic proximity control is not suitable in complex scenarios or narrow spaces. An elaborate system of sliding doors in the bathroom and dressing area

Figure 8. "Reading in bed" service

of the BAALL apartment (cf. Figure 1) allows for the space available to be used in a variety of ways: the specific requirements of wheelchair users with regards to (right, left, front) toilet access can be met using a minimum of space and the bathroom can be locked (such that open commands from the outside are temporarily disabled), leaving a drive-by corridor through the dressing area; alternatively, a spacious dressing area may be created by leaving the sliding door to bathroom open and locking the access doors to the entire dressing area. Such complex scenarios become possible via the additional KNX control, including sensors for the actual open or closed state of each door.

3.1.5 Security

Security at home is an issue that concerns all, but apparently plays an increasingly dominant role for elderly people. One reason is forgetfulness (have the hotplates been turned off?), see section 3.3.1, another is frailness (fall detection) and declining health. Fortunately, many other groups are doing research in this area, in particular relating to e-health and telemedicine; our sole contribution at present is rather precise localisation and movement information via the mobility assistants and a potentially large body of activity information gathered from the explicit interactions with the environment in the BAALL to aid future research on monitoring Activities of Daily Living.

Fear for physical intrusion is another important concern. We have connected access control/building security systems with domotic control. As an added value, this allows a user-specific profile to be activated when the user in question enters a room or apartment.

3.2 Universal Control

Traditional control devices, like buttons, switches or remote controls, reach their limits when there are too many of them around, see also section 5. In an apartment such as the BAALL, 5 remote controls

would easily come with the original interfaces (bed, wash-basin, paper-less toilet, lights, audio and video equipment). Some vendors of domotic control equipment offer elaborate control panels; these are usually proprietary and restricted to the equipments of one vendor and its partners; addition of another device is quite impossible. Apart from raising the issue of interoperability, such a panel is fixed in one place and thus awkward to use for elderly people who must avoid unnecessary walking, and quite impossible for wheelchair users. Mobile universal control is called for.

3.2.1 Control via the iPhone

One solution that we have temporarily adopted is a universal remote control device connected to the smart environment server via the local WLAN with a user interface supporting icons and touch screen interaction, like the iPhone (or other such mobile devices). In Figure 7, the BAALL interfaces for the bedroom, corridor/dressing area, living-room and kitchen are demonstrated. While such a solution is universal, extensible and "cool" for most of today's users, it may become tedious or impossible to use in advanced age: the icons become too small to read; clicking through a hierarchy of levels becomes too complicated to grasp and clumsy with a shaky touch; etc. Another solution is speech interaction, or a combination of both, see section 5.

3.2.2 Universal Remote Console

An important development trend is the idea of unifying complete control of the environment in one system and employing customer specific software drivers and universal input/output devices in one standard Universal Remote Console, URC; an interface such as the iPhone interface in Figure 7 would then be *generated* rather than custom-made, and tailored to specific impairments. In this context, we will contribute field-bus independent home automation with user specific profiles, e.g. for light scenarios (cf. section 3.3).

3.3 Higher-Level Services

The strict integration of all devices into a single network/server (and possibly the URC standard, see section 3.2.2) allows central control of all individual devices to reach a specific configuration, or to perform a specific task, and additionally to implement combined services. Such "value-added services" and scenarios providing a higher level of assistance (such as the one described in section 3.3.3) require a greater number of basic devices (plus sensors and actors) and services, making the ways in which they interact, and are interdependent on one another, increasingly complex. This may also include the addition of higher-level services (agents) or the creation of a more intelligent environment that reminds users pro-actively to carry out certain activities (such as taking their medicine) or even monitors and supervises certain Activities of Daily Living such as daily liquid consumption.

3.3.1 Combined Services

An example of a combined service is the "reading-in-bed"-service: the main room light dims, the bedside light is switched on, the bed is adjusted into a reading position, and the sliding doors are closed (one could also play the favourite soothing music).

Another typical situation is to leave the home, which requires to close all windows and doors, and to switch off most devices (especially the stove and hotplates). The light at the entrance can be turned off with a small delay.

When using the dressing area, the sliding doors to the neighbouring rooms will be closed and the door to the bathroom will be opened to gain additional space. This service, including the locking of doors to the outside, is available at the touch

of a button on the iPhone interface, see "change" on the Corridor interface in Figure 7.

3.3.2 Context

In the BAALL, applications may depend on the context in various ways: task dependent services (see section 3.3.3 for an example), location and time sensitive services (e.g., turning on the light according to the time-of-day and the current location of the mobile assistant), and intention relevant services (e.g., locking the doors while dressing). The spatial context can always be taken into account, as the position of the mobility assistant is always known. Thus available services can be displayed in a context aware fashion; "switch the light on" always has a context-sensitive meaning. On the iPhone interface in Figure 7, the particular sub-interface for a particular room is automatically opened when the mobility assistant moves to this room. The present state of a device or a complex configuration of devices can be made available to the assistants, enabling deductions to be made: the mobility assistant can, for example, make a detour around the dressing area when it has been locked by another user.

3.3.3 An Activity of Daily Living Scenario

The following scenario demonstrates how a smart environment, the Rolland wheelchair's navigation assistant, and natural language interaction technology can be integrated. The user in question, Mario, is sitting in his wheelchair at the desk of his home office. He says to his Rolland wheelchair "I'd like to eat a pizza". In response to his request, the wheelchair decides on the best way to reach the kitchen and then replies to the user, saying, "OK, I am going to take you to the kitchen". As they make their way there, the sliding doors leading to the kitchen open and close, and lights are switched on as necessary, the kitchen counter is set to the appropriate height for the wheelchair, a kitchen cabinet moves down to the correct level and the light in the kitchen is turned on. Once they have arrived in the kitchen and Mario is at the appropriate place at the kitchen counter, he takes a plate from the cabinet. Mario then asks Rolland, "Where is the pizza?", and Rolland replies by saying, "The pizza is in the fridge, I am taking you to the fridge"; the refrigerator highlights where the pizza is. Mario takes a pizza from the fridge and then asks Rolland to take him to the microwave, which has already moved down to his level. Mario puts the pizza in the microwave and waits until it is done. Finally, Mario asks Rolland to move him to the table. The kitchen light is switched off and the light over the table switched on. Mario is now able to enjoy his pizza.

Due to the known intention, the smart environment can plan appropriate actions pro-actively (such as moving the kitchen counter), and intersperse them with actions enabling navigation (opening doors, planning a detour if they are locked, etc.) on behalf of the mobility assistant.

4 SPATIAL KNOWLEDGE REPRESENTATION, REASONING

An important aspect of AAL is the natural and effective interaction between user, smart devices and other intelligent assistants. Traditional devices, like buttons, switches, or remote controls, are not suitable for controlling ever more complex systems. This is particularly evident when elderly or impaired people are concerned. Apart from barely being able to reach traditional switches or the like, it is almost impossible for a user to negotiate a route with the wheelchair via its joystick, and control all relevant devices at the same time; high-level interaction methods (e.g., natural language, gestures, etc.) are more suitable, see section 5. Moreover, the navigation assistant of an intelligent wheelchair is able to follow a pre-defined route automatically and can thus relieve the user of tedious intermediate

interaction once the general target and intention is known. Besides mobility assistance, there are various activities in an AAL system, which depend on spatio-temporal information representation, such as spatial relations between different objects, or temporal dependence of activities. A cooking activity, for example, might consist of several functions: to negotiate a recipe with the user; to check the availability of various ingredients; to describe the location of a particular ingredient if necessary; and to guide the cooking process.

4.1 Qualitative Knowledge Representation

In contrast to smart devices or intelligent robots, which often use detailed knowledge of the physical world in order to behave in a certain way, humans usually use explicit abstracted knowledge to think or talk about their activities. In Artificial Intelligence, such human conceptualizations are formalized and modeled in the subfield of qualitative representation and reasoning. These kinds of models enable predictions to be made and a person's intention to be checked against existing knowledge representations (for example, "pass the elevator on the left" can be checked against a Route Graph, and a clarification dialogue generated, if the elevator is in fact on the right); on the other hand, they are able to translate system information to the user in an understandable way. In order to enhance natural human interaction with intelligent systems, an additional layer is therefore needed to bridge the gap between human concepts and the physical world of systems. In the context of our application, we distinguish between a *conceptual level* which uses qualitative modeling techniques when interacting with the user, and a *robot level* which uses the metric data of the environment for the mobility assistants.

In this section, we are going to demonstrate our approach through the qualitative representation of the navigation space, a mental construction that is schematised. Some information, such as exact metric data, is often systematically simplified and even distorted. Several empirical studies show that the critical elements of the space for navigation are landmarks and paths, links, and nodes.

In an empirical study is described, which investigates natural language when communicating with the wheelchair Rolland about spatial concepts, such as route navigation and spatial relations between objects. Most route instructions collected in this experiment are *well-structured*, i.e., they contain a temporally ordered sequence of qualitatively described route segments with movements, (re)orientations, and spatial references to landmarks. There are only few cases, in which quantitative spatial information is used, for example "turn ninety degrees right" or "go forward about 20 meters". If a robot is design to assist users in their navigation tasks, a *qualitative* representation of the user's navigation knowledge is needed, in order to understand the user's instructions and to generate clarification dialogues in case a route instruction does not match the robot's internal spatial representation.

In addition, users who require assistance in their daily lives often have difficulties in interacting with the environment in a precise way, e.g. when driving an electronic wheelchair using a joystick. The *qualitative driving assistant* that is currently under development will help such users by interpreting their joystick commands in a qualitative way (e.g. "left" can mean either "turn left immediately" or "turn left at the next intersection" depending on the context).

4.2 Spatial Ontologies, Standardisation

In general, safety, security, and interoperability standards are becoming increasingly important to enable easy (re-)configuration and AAL-readiness in the foreseeable future. There is still a long way to go before semantically controlled interoperability of both basic services as well as higher services (via autonomous virtual "agents")

can be achieved. Interoperability is essential for developing higher services and for re-*configuring* the environment and its control with a minimum of effort when a new device or service is added.

Recently, based on a series of empirical studies, several ontological representations of natural space and spatial actions have been developed. These ontologies define the general concepts of spatial objects and their relations in an application domain, and lay a semantic foundation for human-robot interaction. On the other hand, ontology tools and services can answer queries over the object and their relations or check their consistency.

We believe that a break-through can be achieved with a proper ontological modelling, integrating various ontologies for different aspects into one hyper-ontology emphasising their interrelations; a basis might be the joint ongoing work in cooperation with the OASIS project.

4.3 Spatial Calculi

Specifically, when humans interact with intelligent systems for spatio-temporal tasks (e.g., navigation, localisation, temporally dependent activities), qualitative spatio-temporal calculi, together with ontologies, are useful for interpreting a person's spatial expressions or temporal activities, and reasoning about them. There are two major research directions: approaches based on spatially extended objects (e.g., regions or intervals); and reasoning about the directions and distances of point configurations. Two important examples of the point-based approach are the Cardinal Direction calculus using orientation grids, and the Double-Cross calculus based on relative orientation information. Furthermore, there are some spatial models that cover different layers of representations or combine topological graph structures and spatial calculi. Below we briefly introduce a qualitative spatial model (called *Conceptual Route Graph*), which combines a spatial calculus and a topological graph structure for the interpretation of people's navigation knowledge. In the next subsection, we are going to discuss the reasoning issue using this spatial model.

Motivated by empirical studies, the Conceptual Route Graph is a qualitative conceptualisation of people's navigation knowledge. In the model, the *Double-Cross calculus* is used to represent qualitative orientation information, and distinguishes eight different orientation relations: "left", "right", "front" "back", "right front", "left front", "right back" and "right front", with respect to a directed line. *Route Graphs* have been introduced as a general model of the environment for navigation by various agents in a variety of scenarios. Route Graphs are a special class of graphs. A *node* of a Route Graph, called *place*, has a particular position and has its own "reference system": it may, but need not, be rooted in a (more) global reference system, such as a 2-D geometric system. An *edge* of a Route Graph, called *route segment*, is directed from a source place to a target place, and always has three attributes: an *entry*, a *course* and an *exit*. In the Conceptual Route Graph, each route segment can be used as an orientation frame for interpreting orientation relations. Each place is associated with an original orientation relation with respect to the route segment starting from it, thus the integration of different routes at a common place becomes possible by re-computing the entries and exits of related route segments. The integration of routes into a graph structure allows the identification of *new routes* based on the existing ones.

With the Conceptual Route Graph, we can interpret a set of commonly used route instructions, such as "pass by the lift", "go through the door", "go to the kitchen" or "turn left". In the current definition of the Conceptual Route Graph, landmarks are represented as points; thus to treat route instructions with references to regions (for example, "go along the river", "across the park"), an extension of the Conceptual Route Graph is necessary.

4.4 Qualitative Reasoning

Qualitative knowledge representation bridges the gap between human cognition and the system's physical world, and enables natural interaction. However, the ability to reason qualitatively, for example to derive new knowledge from existing information, to prove the consistency of human expressions, to update knowledge, or to find expected information, should also be supported by an intelligent interaction system. An example could be that the navigation assistant is able to detect any inconsistency in the user's route instructions or to find an optimal route according to the user's route instructions such that suggestive responses or clarification dialogues can be generated by the mobile robot.

Navigation tasks belong to high-level cognitive processes that involve the assessment of environment information, the localisation of spatial objects, and spatial relations between these objects. Describing a route to a robot is a special case of navigation, and may be subject to *knowledge-based mistakes* made by humans. In four categories of spatial knowledge mismatches between users' route instructions and the robot's internal spatial knowledge are defined; we show two examples here. First, a user instructs the wheelchair robot: "Drive past the mailbox room and the copy room", but the mailbox room is in fact located behind the copy room with respect to the robot's current position (cf. Figure 9). In this case the user relates spatial objects incorrectly, a *spatial relation mismatch*. In the second example, the instruction "pass the copy room on the right" is given by the user; consider the following qualitative interpretation of the instruction:

$pos = \mathbf{ab}$

$at(Copyroom, g)$

$\exists \mathbf{p_1 p_2} : Segment \cdot rightFront(g, \mathbf{ab}) \wedge rightBack(g, \mathbf{p_1 p_2}) \wedge front(p_1, \mathbf{ab}) \wedge front(p_2, \mathbf{ab})$

where the current position is represented by the segment ab (i.e. a is the robot's current location, ab the direction), and the copy room is at location g. A segment p_1p_2 in front of the current position (i.e. $front(p_1,ab) \wedge front(p_2,ab)$) should be obtained by formal reasoning (see below), such that the location of the copy room g is on the right front of ab, but right back of p_1p_2. After carrying out the instruction, the new position is p_1p_2.

However, the copy room can only be passed on the left according to the current situation (cf. Figure 9), thus the user locates the lifts falsely, and an *orientation mismatch* occurs. In detecting such knowledge mismatches, spatial reasoning plays a very important role. The composition table of the Double-Cross Calculus provides the Conceptual Route Graph with a theoretical foundation for reasoning about users' navigation instructions, and the reasoning tool SparQ enables automatic reasoning according to the dynamically changed environment state. Considering the example "pass the copy room on the right" again, we get $leftFront(g,ab)$ (i.e. the copy room is at location g, which is in the left front direction of the current position) as the reasoning result. Comparing the interpretation and the result, it is easy to see that there is a spatial relation conflict: $rightFront(g,ab)$ versus $leftFront(g,ab)$, thus no segment can be found in the current situation, which satisfies the condition $rightFront(g,ab)$ in the above interpretation. Detection of such spatial mismatches is important in an interaction system in order to generate a corresponding clarification

dialogue (e.g., "The copyroom can only be passed on the left.").

Spatio-temporal knowledge-based representation and reasoning provides an important theoretical foundation for interaction in AAL applications as well. We take the following situation in the BAALL as an example: Marion is in the dressing room, while Alfred would like to go to the bathroom; however the path to the bathroom is blocked by Marion. A reasonable interaction could be as follows:

Alfred: Take me to the bathroom, please.

iWalker: Please wait a while, since Marion is in the dressing room.

...

iWalker: Now I can take you to the bathroom; should I?

Alfred: Yes, please

In this example, a spatial representation of Alfred's instruction is necessary, including planning the route from his current location to the bathroom. Moreover, to resolve the spatial and temporal conflict: only after the dressing room is free again, the route to the bathroom can be taken; spatio-temporal representation and reasoning are necessary. This requires the modelling and planning of various activities such as "dressing in the

Figure 9. Part of a corridor environment represented as a Conceptual Route Graph

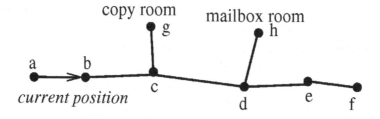

526

dressing room" and "driving to the bathroom", and specification of the temporal and spatial dependency of these activities; for example, changing may happen after the doors to the dressing room are locked, on the other hand the activity "driving to the bathroom" can be started when the doors on the route are unlocked. Some existing spatio-tempral calculi (e.g., the Interval Calculus and its extensions can be applied to represent and to reason about such spatio-temporal relations between activities.

In general, a conceptual model based on formal spatio-temporal calculi is an important foundation for planning and natural interaction with the user. To realise spatio-temporal based natural interaction in an AAL environment, it is necessary to develop and integrate several components to represent the environment using a well-defined qualitative model; to interpret the spatio-temporal relations contained in users' intentions according to the environment state; to justify and plan the effects resulting from the current intention with the help of a qualitative spatio-temporal reasoner; to generate the system's responses; and to carry out necessary actions by mobile assistants or smart devices.

Although qualitative spatio-temporal reasoning is a hard and time-consuming task in a complex environment with a large number of objects and relations, we have successfully used qualitative spatial reasoning in several real-time tasks, such as joint user-wheelchair navigation in a typical office building or supporting some daily living activities in an AAL apartment like BAALL.

5 SPATIAL INTERACTION

Technological progress introduces ever more intelligent systems into people's daily life. Users, in particular the elderly or impaired, are challenged to interact with a variety of smart devices (e.g., sliding doors, lights, kitchen cupboard, etc.) and mobile assistants (e.g., wheelchairs, walkers).

Without well-defined interaction and interaction modalities, such devices and assistants will not provide appropriate support.

5.1 Multi-Modal Interaction

Spoken interaction is usually taken as the first-class modality for the applications we are concerned with; however, rich multimodal interfaces that integrate visualisation and gestures, for example via a touch screen, should be available as well. In an interaction system supporting multiple modalities, the integration (or *fusion*) of user inputs and synthesising output (or *fission*) to different modalities are the two topics to investigate.

5.1.1 Natural Language Dialogue

In the last decade, natural language technology has been used in many different application areas. There are also several research efforts concerning human-robot interaction via natural language, although such work still belongs to fundamental research. For some users with restricted physical capabilities, the interaction via spoken language is particularly promising. We are now evaluating a spoken dialogue system with the navigation assistant of Rolland. Moreover, normal users, who are able to drive an electronic wheelchair with a joystick, may also be better supported when interacting with the wheelchair verbally while they are busy with other tasks (and have their hands full).

5.1.2 Visualisation and Gestures via a Touch Screen

In the spatial context, other modalities may be more efficient when spatial relations or configurations are involved. The ability to choose a navigation route through *touch and draw* gestures or to visualise a complex route on a screen is sometimes more intuitive for users of a navigation assistant. Moreover, existing visualisation functions such as zooming and scrolling provide the user with various views of the environment at different granularity levels.

5.1.3 Head-Joystick

A specific modality for an intelligent wheelchair is the so-called head-joystick we developed: instead of using the standard joystick, users can drive the wheelchair via head movement or nodding gestures; thus it enables a particular group of people to become mobile, even though they are unable to use a normal hand-driven joystick. In addition, the head-joystick provides users with more flexibility when they are engaged in other tasks while driving, and have their hands full.

5.2 Shared Control

An assistance system such as the intelligent wheelchair Rolland is usually shared in it's control by both the user and the highly autonomous system. In addition to knowledge representation mismatches (discussed in section 4.4), mode confusions may occur when the system is in a state different from that expected by the user. The detection of mode-confusion situations is critical in shared-control systems, especially if the shared-control system is safety relevant, since mode confusions may lead to the user's surprise, to the misinterpretation of the system's state, and to unpredictable follow-up actions on behalf of the user.

In a recent user study of our wheelchair with the safety assistant, we observed several mode confusion situations. For example, a user navigated the wheelchair through a narrow place with several obstacles. He kept changing the driving direction in a short temporal sequence. This led to unpredictable orientations of the caster front-wheels and the wheelchair could not follow the user's command in time, though the intended steering commands were appropriate at that moment. As a result, the user behaved unpredictably by giving arbitrary commands. The experiment uncovered another mode confusion phenomenon: almost all subjects acted unusually after the intervention of the safety assistant. They attempted to regain control over the wheelchair, but the wheelchair did not drive in the way they expected; consequently, they tried to give the wheelchair some arbitrary commands via the joystick or pressed the joystick powerfully, which caused even more unexpected movements of the wheelchair. In such situations, clarification dialogues are indispensible, to inform the user about the situation and to help her/him to interact with the system correctly.

5.3 Intention

The Conceptual Route Graph discussed in 4.3 is an example of a conceptual model for representing and reasoning about human knowledge *qualitatively*. In our AAL scenarios, the user can expect the smart environment to plan the control of the smart devices when she/he needs help to carry out Activities of Daily Living (ADLs) such as watching TV, going to bed, or preparing something to eat. To interpret and process such advanced activities, more powerful calculi are needed. This requires more user studies to investigate different activity-orientated *intentions* expressed by natural expressions, and the relations between elementary and complex tasks, such that models for interpreting intentions can be defined and a reasoning process of identifying the user's intentions can be developed. Only in this way can users be served naturally, efficiently and context-sensitively, cf. the „I'd like to eat a pizza" scenario in section 3.3.3; another example would be „I need a plate" – thus the cupboard moves down. Knowledge of the user's intention enables the environment to plan and to relieve the user of otherwise tedious intermittent tasks. The recognition and realisation of peoples' intentions are high challenges in AAL, in particular with regard to Activities of Daily Living, and their eventual supervision. However, we will focus on specific user groups and investigate their expressions and behaviour by pre-defined activities empirically, such that user models can be development for these user groups.

6 FUTURE WORK

The scenario in section 3.3.3, involving Mario being driven around by the Rolland wheelchair and interacting with a smart environment, is perfectly feasible from a technical point of view and already reality at the BAALL. AAL-related services developed in other projects, such as reminders, telemedicine or communication support, should also be integrated into the iWalker and the wheelchair Rolland over time. The universal "little helper" that takes care of everything in the household will perhaps exist one day; we will concentrate on assistance systems concealed in the mobility assistants, common smart devices, and the environment, aiming to make them more intelligent, adaptable to the user's needs, and proactive, fully capable of taking the initiative when necessary. In effect, many of the future AAL assistance systems will be shared between the smart environment and the mobility assistants acting as mediators, such that it becomes immaterial whether an assistance system is realised on one, the other, or shared between them.

Thus the iWalker may well become a rather comprehensive "little helper" eventually, supporting, for example, the whole chain from the smart refrigerator and the shopping list to the department store, and back – carrying the goods as well, of course. At present, the RollScroll menu on the iWalker (see section 2.2.1) is only used to present assistant and navigation selection options; it could easily be extended to also provide commands for interaction with the environment. Note that the menu of currently available options can be made context-dependent, and thus small (see section 3.3.2). While natural language dialogue (see section 5.1.1) is of course more powerful, the particular user may not be able or willing to use speech (or the system not capable of under-

standing, the environment too noisy, etc.), thus RollScroll interaction becomes a natural and rather simple option. We will experiment with a menu of icons on a graphical map, and a button near the handle bar as an alternative to the gestures, in the future. In addition, we will continue the work on the development of the multi-modal interaction framework between users and the intelligent environment of the BAALL, and on the integration of additional qualitative spatio-temporal calculi into the conceptual model, in order to enable users to interact naturally with the environment on broad types of assisted living activities.

Spoken *natural language interaction* appears to be particularly relevant, both via mobility assistants that take on a mediator function, or directly. Users can thus interact directly "with the kitchen" to carry out additional activities even when their hands are not free, such as when they are sticky with dough for example. Interaction can cover controlling kitchen components (e.g. pre-heating the oven), storing and finding items in kitchen cabinets, consulting cook books, as well as steering the wheelchair into the appropriate position and alignment; the head joystick, which was originally developed for the paralyzed, may also be able to help here. A broad field for future research is the idea of "storing and finding" objects, from "managing the refrigerator" to advising users on selecting and combining the clothes in their closet. This would require elaborate dialogue assistance and general AI technology.

We envisage that many of the "side products" arising from assistance systems research will end up playing such an invaluable role in our everyday lives that we will no longer want to do without them; this is clearly the most desirable effect in terms of marketing: a product is appealing since it is *designed for all*. It is important that technical research and development are complemented by findings from other disciplines such as psychology, social sciences, cognitive science, artificial intelligence, computational linguistics, and informatics. It is only in this way that the challenges of appropriate and individually adaptable interfaces for man-machine interaction in multi-modal dialogue can be mastered.

At the BAALL, a major concern is scientifically controlled *evaluation*, whether via initial experiments or long-term evaluation of *everyday usability,* so that this information can be used as a basis for product development. At the BAALL, assistance systems can be tested and evaluated in realistic scenarios. The evaluation results can then, in turn, be used for further modeling and to influence future design.

ACKNOWLEDGMENT

This research has been sponsored by the German Research Council (DFG) in the Transregional Collaborative Research Centre CRC/TR 8 *Spatial Cognition*, and the EU in project *SHARE-it* (FP6-045088), in cooperation with the EU projects *i2home* und *OASIS*. We also thank our sponsors and cooperation partners, in particular OTTO BOCK HEALTHCARE, MIELE, MIDITEC, PRESSALIT CARE, RAUMPLUS, and MÖBELWERKSTÄTTEN SCHRAMM.

REFERENCES

Allen, J. F. Maintaining Knowledge about Temporal Intervals. *CACM*, Vol. 26, Nr. 11, 1983.

AssociationK. N. X.http://www.knx.eu/

Augusto, J. C., & McCullagh, P. Ambient Intelligence: Concepts and Applications. *Computer Science and Information Systems*, Vol. 4 No. 1, 2007. BAALL: http://www.baall.net/

Bateman, J., Hois, J., Ross, R., & Farrar, S. The Generalized Upper Model 3.0: Documentation. *Technical report, Collaborative Research Center for Spatial Cognition, University of Bremen, Germany*, 2006. Bed: http://www.schramm-werkstaetten.de

Bredereke, J., & Lankenau, A. A Rigorous View of Mode Confusion. Proceedings of 21st Int'l Conf. on Computer Safety, Reliability and Security. LNCS 2434, 2002.

Cohn, A. G., Bennett, B., Gooday, J., & Gotts, N. M. (1997). *Qualitative Spatial Representation and Reasoning with the Region Connection Calculus (Vol. 1)*. Geoinformatics.

Cohn, A. G., & Renz, J. (2007). *Qualitative Spatial Representation and Reasoning*. Handbook of Knowledge Representation.

Denis, M: The Description of Routes: A Cognitive Approach to the Production of Spatial Discourse. *Journal of Cahiers de Psychologie Cognitive*, vol. 16, 1997.

Döhle, M. Ein XML Gateway zur Gebäudesteuerung. Diplomarbeit (M.Sc. thesis), Universität Bremen, 2009. EU-Projekt i2home. http://www.i2home.org

EU-Projekt OASIS. http://www.oasis-project.eu

EU-Projekt SHARE-it. http://www.ist-shareit.eu

Fischer, K. (2006). *What Computer Talk Is and Is Not: Human-Computer Conversation as Intercultural Communication (Vol. 17)*. Computational Linguistics.

Frank, A. U. Qualitative Spatial Reasoning with Cardinal Directions. *Proceedings of the Seventh Austrian Conference on Artificial Intelligence*, 1991.

Freksa, C. (1991). *Qualitative Spatial Reasoning*. Cognitive and Linguistic Aspects of Geographic Space.

Gerevini, A., & Nebel, B. Qualitative Spatio-Temporal Reasoning with RCC-8 and Allen's Interval Calculus: Computational Complexity. *Proceedings of ECAI 2002*.

Glover, J., Holstius, D., Manojlovich, M., Montgomery, K., Powers, A., Wu, J., et al. A Robotically-Augmented Walker for Older Adults. Tech. Report CMU-CS-03-170, Carnegie Mellon University, Computer Science Department, Pittsburgh, PA, 2003.

Gollub, J. Umsetzung und Evaluation eines mit Laserscannern gesicherten Rollstuhls als Interaktives Museumsexponat. Diplomarbeit. Fachbereich 3 – Mathematik und Informatik, Universität Bremen (http://www.informatik.uni-bremen.de / agebv/downloads/published /gollub_thesis_07. pdf), 2007.

Grenon, P., Smith, B.: Towards Dynamic Spatial Ontology. *Journal of Spatial Cognition and Computation*, Vol. 4, 2004.

Grisetti, G., Stachniss, C., & Burgard, W. Improved Techniques for Grid Mapping with Rao-Blackwellized Particle Filters. *IEEE Transactions on Robotics*, 2006. Kitchenette: http://www.pressalitcare.de/

Krieg-Brückner, B., Frese, U., Lüttich, K., Mandel, C., Mossakowski, T., Ross, R. J.: Specification of an Ontology for Route Graphs. *Proceedings of Spatial Cognition IV, Germany*. LNAI 3343, 2004.

Krieg-Brückner, B., Gersdorf, B., Döhle, M., & Schill, K. (2009). *Technik für Senioren in spe im Bremen Ambient Assisted Living Lab. 2. Deutscher AAL-Kongress 2009*. VDE-Verlag Berlin-Offenbach.

Krieg-Brückner, B., Krüger, A., Hoffmeister, M., & Lüth, C. (2007).: Kopplung von Zutrittskontrolle und Raumautomation – Eine Basis für die Interaktion mit einer intelligenten Umgebung. *8. Fachtagung Gebäudesicherheit und Gebäudeautomation – Koexistenz oder Integration?* VDI Berichte 2005. VDI Verlag, Düsseldorf, 37-48, 2007.

Krieg-Brückner, B., Röfer, T., Shi, H., & Gersdorf, B. Mobility Assistance in the Bremen Ambient Assisted Living Lab. In Nehmer, J., Lindenberger, U., & Elisabeth Steinhagen-Thiessen, E. (guest editors): *2009 Special Section Technology and Aging: Integrating Psychological, Medical, and Engineering Perspectives*. Gerontology: Regenerative and Technological Gerontology. To appear.

Krieg-Brückner, B., & Shi, H. Orientation Calculi and Route Graphs: Towards Semantic Representations for Route Descriptions. *Proceedings of GIScience 2006*. LNCS 4197, 2006.

Krieg-Brückner, B., Shi, H., Fischer, C., Röfer, T., Cui, J., & Schill, K. (2009). *Welche Sicherheitsassistenz brauchen Rollstuhlfahrer? 2. Deutscher AAL-Kongress 2009*. VDE-Verlag Berlin-Offenbach.

Kuipers, B. (2000). *The Spatial Semantic Hierarchy (Vol. 119)*. Artificial Intelligence.

Kurata, Y., & Shi, H. Interpreting Motion Expressions in Route Instructions Using Two Projection-Based Spatial Models. *Proceedings of 31st Annual German Conference on Artifical Intelligence*, 2008.

Lankenau, A., & Röfer, T. (2001). A Safe and Versatile Mobility Assistant. *IEEE Robotics & Automation Magazine*, 1.

Lauria, S., Kyriacou, T., Bugmann, G., Bos, J., & Klein, E. Converting Natural Language Route Instructions into Robot Executable Procedures. *Proceedings of the IEEE International Workshop on Human and Robot Interactive Communication*, 2002. LON LonMark International. http://www.lonmark.org

Mandel, C., & Frese, U. Comparison of Wheelchair User Interfaces for the Paralysed: Head-Joystick vs. Verbal Path Selection from an offered Route-Set. *Proceedings of the 3rd European Conference on Mobile Robots (ECMR 2007)*, 2007.

Mandel, C., Huebner, K., & Vierhuff, T. Towards an Autonomous Wheelchair: Cognitive Aspects in Service Robotics. *Proceedings of Towards Autonomous Robotic Systems (TAROS 2005)*, 2005.

Mandel, C., Röfer, T., & Frese, U. Applying a 3DOF Orientation Tracker as a Human-Robot Interface for Autonomous Wheelchairs. *Proceedings of the IEEE 10th International Conference on Rehabilitation Robotics (ICORR 2007)*, 2007. Otto Bock Mobility Solutions: http://www.ottobock.de/

Ploennigs, J., Jokisch, O., Kabitzsch, K., Hirschfeld, D.: Generierung angepasster, sprachbasierter Benutzerinterfaces für die Heim- und Gebäudeautomation. *1. Deutscher AAL-Kongress 2008*, VDE-Verlag Berlin-Offenbach. 427-432, 2008.

Ragni, M., & Wölfl, S. Temporalizing Cardinal Directions: From Constraint Satisfaction to Planning. *Proceedings of the Tenth International Conference on Principles of Knowledge Representation and Reasoning*, 2006.

Randell, D., Cui, Z., & Cohn, A. A Spatial Logic Based on Regions and Connection. *Proceedings of 3rd International Conference on Knowledge Representation and Reasoning*, 1992.

Reason, J. *Human Error*. Cambridge University Press, 1990. Refrigerator: http://www.miele.de/

Renz, J., & Nebel, B. (1999). On the Complexity of Qualitative Spatial Reasoning: A Maximal Tractable Fragement of the Region Connection Calculus. *Artificial Intelligence*, *108*(1-2). doi:10.1016/S0004-3702(99)00002-8

Röfer, T., Mandel, C., & Laue, T. Controlling an Automated Wheelchair via Joystick/Head-Joystick Supported by Smart Driving Assistance. In IEEE 11th International Conference on Rehabilitation Robotics (ICORR-2009), to appear.

Rogers, I. (2006). Moving on from Weiser's Vision of Calm Computing: Engaging UbiComp Experiences. In Dourish, P., & Friday, A. (Eds.), *UbiComp 2006. LNCS 4206*. Springer. doi:10.1007/11853565_24

Rushby, J., Crow, J., & Palmer, J. An Automated Method to Detect Potential Mod Confusions. *Proceedings of the 18th IEEE Digital Avionics Systems Conference*, 1999. Security: http://www.miditec.de/ SFB/TR8 Spatial Cognition: http://www.sfbtr8.uni-bremen.de/

Shi, H., & Krieg-Brückner, B. (2008). Modelling Route Instructions for Robust Human-Robot Interaction on Navigation Tasks. *International Journal of Software and Informatics*, *2*(1).

Shi, H., Mandel, C., & Ross, R. J. Interpreting Route Instructions as Qualitative Spatial Actions. *Spatial Cognition V: Reasoning, Action, Interaction: International Conference Spatial Cognition 2006*. LNAI 4387, 2007.

Shi, H., Ross, R., Bateman, J.: Formalising Control in Robust Spoken Dialogue Systems. *Proceedings of Software Engineering and Formal Methods 2005*, IEEE, 2005.

Shi, H., & Tenbrink, T. Telling Rolland Where to Go: HRI Dialogues on Route Navigation (Chapter 13). *Spatial Language and Dialogue*, Oxford University Press, 2009. Sliding doors: http://www.raumplus.de/

Talmy, L. (1983). *How Language Structures Space. Spatial Orientation: Theory*. Research and Application.

Tversky, B. and Lee, P.U.: How Space Structures Language. *Spatial Cognition: An interdisciplinary Approach to Representation and Processing of Spatial Knowledge*. LNAI 1404, 1998.

Vanderheiden, G., & Zimmermann, G. Non-homogenous Network, Control Hub and Smart controller (NCS) Approach to Incremental Smart Homes. In: Proceedings of *HCI International 2007*. LNCS 4555, 2007. Vescovo, F. et al.: http://www.progettarepertutti.org/ progettazione/casa-agevole- fondazione

Wahlster, W. Towards Symmetric Multimodality: Fusion and Fission of Speech, Gesture, and Facial expression. *Proceedings of the 2003 Meeting of the German Conference on Artificial Intelligence*, 2003.

Wallgrün, J. O., Frommberger, L., Wolter, D., Dylla, F., & Freksa, C. A toolbox for qualitative spatial representation and reasoning. *Spatial Cognition V: Reasoning, Action, Interaction: International Conference Spatial Cognition 2006*. LNCS 4387, 2007.

Werner, S., Krieg-Brückner, B., Hermann, T.: Modelling Navigational Knowledge by Route Graphs. *Spatial Cognition II: Integrating Abstract Theories, Empirical Studies, Formal Methods, and Pratical Applications*. LNAI 1849, 2000.

Zimmermann, G. Open User Interface Standards – Towards Coherent, Task-Oriented and Scalable User Interfaces in the Home Environments. Proc. 3rd *IET International Conference on Intelligent Environments (IE07)*, Ulm, Germany. The IET, 2007.

Zimmermann, K., & Freksa, C. (1996). *Qualitative Spatial Reasoning Using Orientation, Distance, and Path Knowledge (Vol. 6)*. Applied Intelligence.

KEY TERMS AND DEFINITIONS

Ambient Assisted Living (AAL): In our context Ambient Assisted Living contains intelligent living environments, mobility assistants, and AI methods und techniques to enable elderly or impaired people to live independently.

Mobility Assistant: A mobility assistant is a mobile robot (e.g., intelligent wheelchair or walker) equipped with assistance components, such as safety assistant, driving assistant or navigation assistant, and provides mobility support for people with physical or cognitive impairments.

Qualitative Representation and Reasoning: The aim of the qualitative representation and reasoning research is to develop methods and techniques to reason about the behaviour of systems or human beings for predefined applications, without precise quantitative information.

Multi-Modal Interaction: Multi-modal interaction enables technical systems and their users to communicate with each other via different channels such as natural language, gesture, or graphic presentation, in addition to conventional buttons, switches, remote control panels, etc.

Spatial Calculus: Spatial calculi are mathematically well-defined models which can be used to represent spatial relations between objects, and to reason about them.

Shared-Control System: A shared-control system is a system independently controlled by a user and an automation component.

Conceptual Route Graph: A qualitative spatial model that combines the Double-Cross Calculus and the Route Graph model for the interpretation of and reasoning about people's navigation knowledge.

Spatial Ontology: A spatial ontology defines the general concepts of spatial objects and their relations for spatial application domains.

Chapter 25
Proactive Assistance in Ecologies of Physically Embedded Intelligent Systems:
A Constraint–Based Approach

Marcello Cirillo
Örebro University, Sweden

Federico Pecora
Örebro University, Sweden

Alessandro Saffiotti
Örebro University, Sweden

ABSTRACT

The main goal of this Chapter is to introduce SAM, an integrated architecture for concurrent activity recognition, planning and execution. SAM provides a general framework to define how an intelligent environment can assess contextual information from sensory data. The architecture builds upon a temporal reasoning framework operating in closed-loop between physical sensing and actuation components in a smart environments. The capabilities of the system as well as possible examples of its use are discussed in the context of the PEIS-Home, a smart environment integrated with robotic components.

1 INTRODUCTION

Intelligent environments are quickly evolving to incorporate sophisticated sensing and actuation capabilities thanks to the progressive inclusion of devices and technologies from the field of autonomous robotics. Several approaches are currently under way that combine insights from the fields of intelligent environments and autonomous robotics, and define different types of smart robotic environments in which several pervasive robotic devices cooperate to perform possibly complex tasks. Instances of this paradigm include network robot systems (Sanfeliu, Hagita, & Saffiotti, 2008), intelligent spaces (Lee & Hashimoto,

DOI: 10.4018/978-1-61692-857-5.ch025

2002), sensor-actuator networks (Dressler, 2006), ubiquitous robotics (Kim, Kim, & Lee, 2004), artificial ecosystems (Sgorbissa & Zaccaria, 2004), and PEIS-Ecology (Saffiotti, et al., 2008). In this chapter, we generically refer to a system of this type as a "smart robotic environment".

In order to provide personalized services to the user, intelligent environments in general and smart robotic environments in particular must employ non-trivial and temporally contextualized knowledge regarding the user's state (Bjorn, Guesgen, & Hubner, 2006). For instance, if a smart home could recognize that the human user is cooking, it could avoid sending a robot to clean the dining room until the subsequent dining activity is over.

The problem we tackle in this chapter is that of realizing a service-providing infrastructure for proactive human assistance in smart robotic environments. Two key capabilities that are often desirable in such intelligent environments are (1) the ability to recognize activities performed by the human user, and (2) the ability to plan and execute the behavior of pervasive devices according to the indications of activity recognition.

This chapter presents SAM[1], a reasoning infrastructure that provides these two key features through the use of temporal constraint reasoning. SAM adopts a knowledge-driven approach to activity recognition, in which requirements expressed in the form of temporal constraints are used to specify how sensor readings correlate to particular states of a set of monitored entities in the environment. These entities can represent, for instance, various levels of abstraction of a human user's daily activities. Also, through the same constraint based formalism, SAM provides a means to specify how the states of monitored entities should lead to the synthesis and execution of plans for service providing actuators. SAM continuously refines its knowledge on the state of affairs in the environment through an on-line inference process. This process concurrently performs abductive inference and proactive plan synthesis, the former providing the capability to recognize

context from current and past sensor readings, the latter providing proactive and contextualized service execution.

SAM is instantiated as part of the PEIS-Home, a testbed environment based on the concept of PEIS-Ecology (Saffiotti, et al., 2008). In a PEIS-Ecology, robotic, sensory and intelligent software components are combined to obtain a context-aware, service-providing environment for the future home. SAM leverages the capabilities of the environment to sense the information required to perform abductive reasoning on user activities as well as its actuation capabilities to enact contextual services.

In the following, we first present the main concepts behind the PEIS-Ecology and its physical instantiation, the PEIS-Home. We then detail SAM, focusing first on how requirements for the abductive reasoning and proactive service planning can be expressed as temporal constraints, and then on how activity recognition and service enactment is performed through SAM's continuous inference process. We also present several case studies in the PEIS-Home, aimed at providing a proof-of-concept validation of the architecture. All case studies are obtained as a result of a human test subject performing daily activities within the PEIS-Home. Finally, we provide a summary of related work, providing examples of other approaches used to obtain similar functionalities as well as architectures similar to SAM employed in different application contexts.

2 THE PEIS-ECOLOGY CONCEPT AND THE PEIS-HOME

The PEIS-Ecology (Saffiotti & Broxvall, 2005) (Saffiotti, et al., 2008) is an approach to realize intelligent environments built around the uniform notion of PEIS (Physically Embedded Intelligent System). In the PEIS-Ecology approach, robots and environment are seen as part of the same system and they work in a symbiotic relationship. The concept of "robot" is taken in its most general interpretation, that is, a robot in the PEIS-Ecology is any computerized system that is able to interact with the environment and with the other components.

A PEIS is generally defined as a set of inter-connected software parts residing in one physical entity. Such physical entities are by definition heterogeneous in nature and each of them can be as simple as temperature sensor and an RFID-tagged object or as complex as a humanoid robot. The communication among these heterogeneous devices is obtained by means of a highly flexible middleware, the PEIS-kernel. The PEIS-kernel allows to exchange information among PEIS and hides their heterogeneity. The PEIS-kernel also maintains a fully decentralized P2P network, performs services like network discovery and dynamic routing of messages, and can cope with a dynamic environment in which PEIS can be added and removed at any time.

A PEIS-Ecology is a collection of inter-connected PEIS, all embedded in the same physical environment. A PEIS ecology aims to obtain a close cooperation among the PEIS pervasively distributed into the environment in order to achieve complex functionalities. This cooperation is also implemented by means of the PEIS-kernel, which provides a distributed tuplespace for exchanging data among PEIS. Each PEIS can use a subscription mechanism to request data from any other PEIS and the underlying architecture will ensure that the tuples containing such data will be routed to the requester. A detailed technical description of the software infrastructure behind the PEIS-Ecology is out of the scope of this chapter. More details about the implementation of the PEIS-kernel can be found in (Broxvall, 2007).

The PEIS-Ecology approach has been instanti-ated in a real domestic setting, the PEIS-Home, deployed at Örebro University Mobile Robotics Lab (Sweden). The real environment reproduces a typical Swedish bachelor apartment (with the only exclusion of the bathroom) consisting of a kitchen, a living room and a bedroom (Figure 1). In the environment, a number of sensors and actuators are deployed (see Figure 2). To describe a real example of how the PEIS vision is reflected

Figure 1. The PEIS-Home

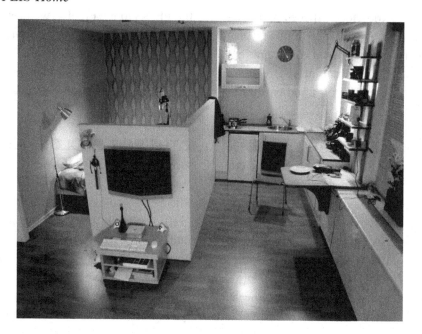

Figure 2. The Actuators and Sensors in the PEIS-Home

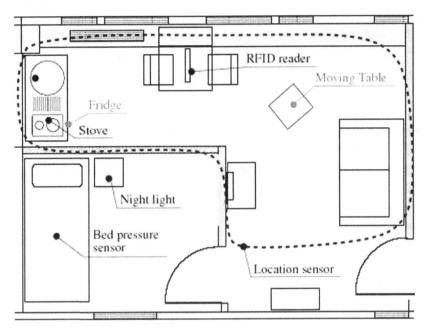

into the real environment, we can think of the following scenario: a user is relaxing on the couch, in the living room, and he requests a drink. As we said, in our approach we do not want to employ a sophisticated robot to perform the task, but we would rather rely on the cooperation of the PEIS present in the apartment. The task, therefore, is achieved by the cooperation of multiple agents: a moving table can navigate to the smart fridge located in the kitchen. A robotic arm, inside the fridge itself, can grasp the drink that is required, while the door of the fridge can open to facilitate the docking of the table. The drink is then placed on the table and can be delivered to the user, while the door closes as soon as the table has completed the un-docking operation[2].

The PEIS-Home has been used as a testbed to address many research problems, such as autonomous configuration of a PEIS-Ecology (Lundh, Karlsson, & Saffiotti, 2008), odor recognition in smart environments (Broxvall, Coradeschi, Loutfi, & Saffiotti, 2006), human-robot interaction (Cirillo, Karlsson, & Saffiotti, 2009), domestic activity monitoring (Ullberg, Loutfi, &

Pecora, 2009) (Pecora & Cirillo, 2009) and cooperative perception (Le Blanc & Saffiotti, 2008).

3 ACTIVITY MONITORING IN A PEIS-ECOLOGY

The purpose of the system described in this paper is to endow a PEIS-Ecology, like the PEIS-Home, with a reasoning infrastructure capable of inferring the current context of a human user given continuous sensor readings. Based on this inferred information, the system should be capable of synthesizing the actuation of service providing devices in the PEIS-Home upon recognizing given occurrences of human behavior. To address this need, we have implemented SAM, a PEIS which can track sensor readings as they are published in the distributed tuple-space of the PEIS-Home and leverages temporal representation and reasoning capabilities to infer context and synthesize plans for service-providing PEIS in the environment.

SAM employs *temporal constraints* to represent sensor activity as it is published in the distributed tuple space of the PEIS-Home. Its inference mechanism is based on temporal constraint reasoning techniques, and the result of inference consists of temporally-constrained symbolic values which can represent both the results of the context deduction and of the contextual services to be enacted in the PEIS-Home. As we explain in this section, the constraint based representation underlying the architecture directly enables the specification of requirements for context recognition and service enactment in a single formalism. SAM builds on the OMPS temporal reasoning framework (which we briefly describe in the following), and leverages its features for realizing a decisional framework that operates in a closed loop with the physical sensing and actuation devices in the PEIS-Home. The fundamental ingredients necessary to implement SAM within the OMPS framework are (1) a *symbolic representation of sensor readings and available actions* which is used to represent the status of sensors and actuators in the PEIS-Home for reasoning by SAM; (2) an *inference module* that carries out the task of inferring the current context and contextual plans; and (3) an *interface between the symbolic representation and the real world* that is used to exchange information between the PEIS-Home's tuplespace and SAM's inference module. In the remainder of this section we describe the symbolic representation used by SAM (1), while SAM's inference and interface features are described in section 4.

3.1 Background: The OMPS Temporal Reasoning Framework

We start by describing the symbolic representation underlying SAM's inference capabilities. SAM is implemented within OMPS[3] (Fratini, Pecora, & Cesta, 2008), a software infrastructure for constraint-based planning and scheduling tool development which has been used to develop a variety of decision support tools, ranging from highly-specialized space mission planning software to classical planning frameworks (Cesta &

Fratini, 2008)[4]. OMPS is grounded on the notion of *component*. A component is an element of a domain theory which represents a logical or physical entity. Components model parts of the real world that are relevant for a specific decisional process, such as complex physical systems or their parts. Components can be used to represent, for example, a robot which can navigate the environment and grasp objects, or an autonomous refrigerator which can open and close its door.

An automated reasoning functionality developed in OMPS consists in a procedure for taking *decisions* on components. Decisions essentially describe an assertion on the possible evolutions in time of a component. For instance, a decision on the fridge component described above could be to open its door no earlier than time instant 30 and no later than time instant 40. More precisely, a decision is an assertion on the value of a component in a given flexible time interval, i.e., a pair $\langle v,[I_s,I_e]\rangle$, where the nature of the value v depends on the specific type of component and I_s, I_e represent, respectively, an interval of admissibility of the start and end times of the decision. In the fridge example, assuming the door takes five seconds to open, the flexible interval is $[I_s=[30,40],I_e=[34,44]]$.

OMPS provides a number of built-in component types, among which *state variables*, which model elements whose state in time is represented by a symbol. OMPS supports disjunctive values for state variables, e.g., a decision on a state variable that models a mobile robot could be $\langle navigate \vee grasp,[I_s,I_e]\rangle$, representing the fact that the robot should be in the process of either navigating or grasping an object during the flexible interval $[I_s,I_e]$. For the purposes of this work, we focus on state variable type components and two custom components that have been developed in SAM to accommodate the needs of the physically instantiated nature of our application domain (described in section Recognizing Activities and Executing Proactive Services in SAM).

The core intuition behind OMPS is the fact that decisions on certain components may entail the need to assert decisions on other components. For instance, the decision to grasp an object in the fridge may require the robot to be docked to the fridge and the fridge door to be open. Such dependencies among component decisions are captured in a domain theory through what are called *synchronizations*. A synchronization is a set of requirements expressed in the form of temporal constraints. Such constraints in OMPS are bounded variants of the relations in Allen's Interval Algebra (Allen, 1984). Specifically, temporal constraints in OMPS enrich Allen's relations with bounds through which it is possible to fine-tune the relative

temporal placement of constrained decisions. For instance, the constraint **A** DURING [3,5][0,∞) **B** states that **A** should be temporally contained in **B**, that the start time of **A** must occur between 3 and 5 units of time after the beginning of **B**, and that the end time of **A** should occur some time before the end of **B**.

3.2 Modeling Constraints Between Actions

Figure 3 (a) shows an example of how temporal constraints can be used to model requirements among PEIS. The three synchronizations involve two PEIS: a robotic table and an intelligent fridge

Figure 3. Three synchronizations in a possible domestic robot planning domain (a), the real actuators available in our intelligent environment (b), and possible timelines for the corresponding components (c)

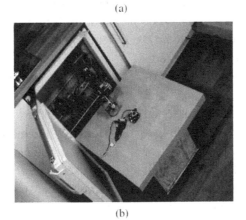

(a)

(b)

(c)

(represented, respectively, by state variables MovingTable and Fridge). The MovingTable can dock and un-dock the Fridge, and navigate to the human user to deliver a drink. The Fridge can open and close its door, as well as grasp a drink inside it and place it on a docked table. The above three synchronizations model three simple requirements of this domain, namely: (1) since the Fridge's door cannot open if it is obstructed by the MovingTable (see Figure 3(b)), and we would like the door to be kept open only when necessary, docking the fridge must occur directly after the fridge door is opened (MET-BY constraint); (2) for the same reasons, the fridge door should close only after the MovingTable has completed the un-docking procedure (BEFORE constraint); and (3) delivering a drink to the human is possible only after the drink has been placed on the table (AFTER constraint).

While temporal constraints express requirements on the temporal intervals of decisions, value constraints express requirements on the value of decisions. OMPS provides the VALUE-EQUALS constraint to model that two decisions should have equal value. For instance, asserting d_1 VALUE-EQUALS d_2 where the two decisions' values are, respectively, $\mathbf{v_1}=\mathbf{A}\vee\mathbf{B}$ and $\mathbf{v_2}=\mathbf{B}\vee\mathbf{C}$, will constrain the value of both decisions to be $\mathbf{B}$ (the intersection of possible values). As for temporal constraints, OMPS provides built-in propagation for value constraints.

Decisions and temporal constraints asserted on components are maintained in a *decision network* (DN), which is at all times kept consistent through *temporal propagation*. This ensures that the temporal intervals underlying the decisions are kept consistent with respect to the temporal constraints while decisions are anchored flexibly in time. In other words, adding a temporal constraint to the DN will either result in the calculation of updated *bounds* for the intervals I_s, I_e for all decisions, or in a propagation failure, indicating that the added constraint or decision is not admissible. Temporal constraint propagation is a polynomial

time operation, as it is based on a Simple Temporal Network (STN) (Dechter, Meiri, & Pearl, 1991).

For each component in the domain, OMPS provides built-in methods to extract the *timeline* of the component. A timeline represents the behavior of a component in time as it is determined by the decisions and constraints imposed on this component in the DN. Figure 3(c) shows a possible timeline for the two components Fridge and MovingTable of the previous example. Notice that, in general, it is possible to extract many timelines for a component, as constraints bound decision start and end times flexibly. In the remainder of this paper we will always employ the earliest start time timeline, i.e., the timeline obtained by choosing the lower bound for all decisions' temporal intervals I_s, I_e.

3.3 Modeling Constraints Between Observations

In the previous example, two state variables are used to model PEIS that have actuation capabilities. Temporal constraints are used to model the requirements that exist between these two "actuator components" in carrying out the task of retrieving a drink from the fridge. These requirements are employed by SAM's continuous inference algorithm to synthesize an operational plan for delivering a drink to the inhabitant of the PEIS-Home. In addition to actuators, however, state variables can be used to represent sensors in the PEIS-Home. While the values of the former represent *commands to be executed*, the values of the latter represent *sensor readings*. Consequently, while temporal constraints among the values of actuator components represent temporal dependencies among commands to be executed that should be upheld in proactive service enactment, temporal constraints among "sensor components" represent temporal dependencies among sensor readings that are the result of specific human activity in the PEIS-Home. For instance, the synchronizations in Figure 4(a) describe possible conditions under which the human activities of **Cooking** and **Eating** can be inferred. The synchronization involves three components, namely a state variable representing the human inhabitant of the PEIS-Home, a state variable representing a stove state sensor, and another state variable representing the location of the human as it is determined by a person localization sensor. The synchronizations model how the relative occurrence of specific values of these components in time can be used as evidence of the human cooking or eating: the former is deduced as a result of the user being located in the **KITCHEN** (DURING constraint) and is temporally equal to the sensed activity of

the Stove sensor; similarly, the requirement for asserting the **Eating** activity consists in the human being having already performed the **Cooking** activity (AFTER constraint) and his being seated at the **KITCHENTABLE**.

A unique feature of SAM is that the same formalism can be employed to express requirements both for enactment and for activity recognition. This is enabled by two specializations of the state variable component type, namely *sensor components* and *actuator components*. As we will see, a single inference algorithm based on temporal constraint reasoning provides a means to

Figure 4. Two synchronizations in a possible domestic activity recognition domain (a), the situations as enacted by a test subject in the PEIS-Home (b), and possible timelines for the corresponding three components (c)

```
Human : Cooking
    EQUALS Stove : ON
    DURING Location [0, ∞)[0, ∞) : KITCHEN
Human : Eating
    AFTER [0, ∞) Human : Cooking
    DURING [0, ∞)[0, ∞) Location : KITCHENTABLE
```

(a)

(b)

Human				
		Cooking		Eating	

Location				
		KITCHEN		KITCHENTABLE	

Stove				
	OFF	ON		OFF	

time

(c)

concurrently deduce context from sensor components and to plan for actuator components.

4 RECOGNIZING ACTIVITIES AND EXECUTING PROACTIVE SERVICES IN SAM

SAM employs three types of components: *state variables*, *sensors* and *actuators*. State variables are employed to model one or more aspects of the user's activities of daily living. For instance, in the examples that follow we will use a state variable whose values are {**Cooking, Eating, InBed, WatchingTV, Out**} to model the human user's domestic activities.

Sensors and actuators are specialized variants of the built-in state variable type which implement

an interface between the real-world sensing and actuation modules and the DN. More specifically, sensors are components that interpret data obtained from the physical sensors deployed in the intelligent environment and represent this information as decisions and constraints in the DN. These decisions and constraints essentially "re-create" the situation as it is assessed by physical sensing. Actuators are components that trigger the execution on a real actuator of a planned decision.

An overall view of the architecture is shown in Figure 5, in which the DN contains decisions and constraints referring to the example given in the previous section. The figure sketches how the key elements of the SAM architecture, namely *a continuous re-planning process* and the *sensor* and *actuator components*, are realized within the OMPS framework.

Figure 5. The SAM architecture and how it is realized within the OMPS framework

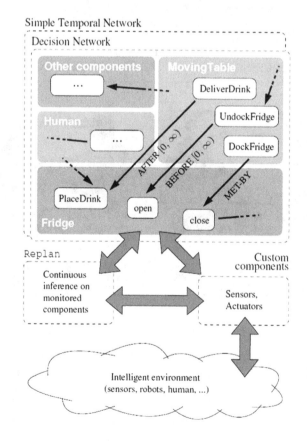

Specifically, a multitude of processes operate concurrently: each sensor is a process that adds decisions and constraints to represent the real-world observations provided by the intelligent environment; in turn, the current DN is manipulated by the iterative re-planning process, which adds decisions and constraints that model the current activity performed by the user and any proactive support operations to be executed by the actuators (such as the plan shown in the previous section); finally, actuators ensure that decisions in the DN that represent operations to be executed are dispatched as commands to the real actuators and that termination of actuation operations are reflected in the DN as they are observed in reality. In a way, the DN acts as a "blackboard" where decisions and constraints re-construct the observed reality as well as the current hypothesis on what the human being is doing. Decisions and constraints are added to the DN in real-time, i.e., as they are observed by the sensors or decided by the activity recognition process. Access to the DN by these processes is scheduled by an overall process scheduler. Each process modifies the DN, thus triggering constraint propagation. Since only one process can access the DN at a time, the rate at which re-planning, sensor and actuator update can occur is determined by the computational load of temporal constraint propagation and inference. Notice that this rate will decrease as decisions are added to the DN, since temporal propagation is $O(n^3)$, where n is the number of time points in the underlying STN.

4.1 Continuous Inference Process

SAM implements a re-planning process which continuously attempts to infer new possible states of the human being and any necessary actuator plans. The process, which relies on the fact that the DN represents at all times the current situation in the real world, possesses two key capabilities: (1) to assess whether the DN contains evidence of sensed values in a given time interval; and (2) to assess whether the DN supports the requirements described in a particular synchronization. Both capabilities can be viewed as ways to *support* decisions. Specifically, supporting a decision means performing one of the two following steps:

Unification. A decision is supported by unification if it is possible to impose a temporal EQUALS constraint and a VALUE-EQUALS constraint between it and another decision which is already supported. For instance, assume that the DN contains a decision d_1 which represents a value **v** sensed in the environment, and that this decision's temporal interval is constrained to occur in time exactly when the sensed value was recorded by a real sensor in the PEIS-Home. Now we introduce an un-supported decision d_2 with value **v** into the DN and attempt to unify it against the already present decision d_1 (i.e., asserting that the two decisions should occur during the same temporal interval and have the same possible values). If the result of imposing these two constraints is successful, then this is an indication that indeed there is an interval of time in which the value **v** has been sensed in the real environment. Notice that if d_2 were also constrained to occur in a certain interval through other constraints, say with the temporal constraint DURATION [5,25], the unification attempt would equate to assessing whether the value **v** has been sensed for at least 5 and at most 25 time units. Overall, unification is the mechanism through which SAM "queries" the DN to assess whether values have been sensed.

Expansion. A decision is supported by expansion of a set of requirements (a synchronization) if it can be bound with the required constraints to the required decisions, and these decisions can be supported. For instance, suppose that we want to support a decision with value v_1 and that the domain contains a synchronization stating the requirements v_1 DURING v_2, v_1 AFTER v_3. SAM will attempt to support the decision with value v_1 by adding two new decisions with values v_2 and v_3, constraining them to the first decision with the prescribed constraints, and recursively attempting to support (either by expansion or by unification) the two new decisions. Notice that these two new decisions could represent sensed values, in which case support by unification would close the chain of support by "confirming" that indeed the values v_2 and v_3 were sensed respectively DURING v_1 and AFTER v_1. Overall, expansion is how SAM assesses whether the current situation of sensor readings in the DN can support a particular hypothesis: it adds a unsupported decision representing the current hypothesis (e.g., that the human being is cooking), and tries to find support for it through the domain theory and existing sensed values in the DN.

The continuous re-planning process implemented in SAM is shown in procedure Replan. The procedure leverages unification and expansion to continuously attempt to support decisions which represent hypotheses on the state of a number of *monitored* components. These components are all those components for which we wish SAM to deduce their current state. In our specific application domain, all these components are state variables which model some aspect of the human user's state. For each monitored component, the procedure adds to the DN a decision whose value is a disjunction of all its possible values (lines 2-4). For instance, if the component in question is the state variable Human described previously, then the new decision to be added will be

$$d_{hyp}^{\text{Human}} =$$
$$\left\langle \left(\textbf{Cooking} \vee \textbf{Eating} \vee \textbf{InBed} \vee \textbf{WatchingTV} \vee \textbf{Out}\right), \left[[0,\infty)[0,\infty)\right]\right\rangle.$$

This decision is marked as un-supported (line 4), i.e., it constitutes a hypothesis on the current activity in which the human user is engaged in. The procedure then constrains this decision to occur after any other decisions on the same component (line 6). This is done in order to avoid that the new decision is trivially supported by unification with a decision supported in a previous call to the Replan procedure (recall that supporting a decision can occur through unification with another decision that is marked as supported). Finally, the procedure triggers a decision supporting algorithm which attempts to support the newly added decisions by recursively expanding synchronizations and unifying the resulting requirements (line 7). In the process of supporting new decisions, their values will be constrained (by VALUE-EQUALS constraints) to take on a specific value. For instance, if the domain theory contains a synchronization stating that the requirements for **Eating** on component Human are a certain set of values on some sensor components, then the un-supported decision is marked as supported, the unary constraint d_{hyp}^{Human} VALUE-EQUALS **Eating** is imposed, and new (un-sup-

Figure 6.

Procedure Replan (DN)
1   **foreach** $c \in MonitoredComponents$ **do**
2       $\mathbf{v} \leftarrow \bigvee_{v_i \in \text{possibleValues}(c)} v_i$
3       $d_{hyp}^c \leftarrow \langle \mathbf{v}, [I_s, I_e] \rangle$
4       $\text{DN} \leftarrow \text{DN} \cup d_{hyp}^c$
5   mark d_{hyp}^c as not supported
6   **foreach** d *on component* c **do** $\text{DN} \leftarrow \text{DN} \cup d_{hyp}^c$ AFTER $[0,\infty)\, d$
7   SupportDecisions (DN)

ported) decisions on the sensor components are added to the DN.

If the decision supporting algorithm terminates successfully, the DN contains the new decisions that have been added by the re-planning procedure, plus all those decisions and constraints that implement support for these decisions. The value of each newly supported decision on monitored components has been constrained to be that required by the synchronization that was used by the SupportDecisions procedure. Since these decisions are linked by temporal constraints to decisions on sensor components, their placement in time will follow the evolution of the DN's decisions on sensor components as time progresses.

Finally (not shown in the procedure), if SupportDecisions fails, the resulting DN is identical to before the re-planning procedure was started, therefore reflecting the fact that no new information was deduced.

4.2 Sensors

The built-in state variable component is not sufficient for realizing an interface with real-world sensors, for the simple reason that it is necessary to build the infrastructure that casts observations in the real world for each sensor into decisions and constraints in the DN. This is more than a simple interface between the observation space of a real sensor and the OMPS framework, as the sensor needs to ensure observed values in the real world are reflected in the DN in the form of decisions and constraints.

In order to realize the interface between OMPS and real-world sensors in the intelligent environment, a new component, the *sensor*, was developed in SAM. A sensor is modeled in the domain for each physical sensor in the intelligent environment. Each sensor component is provided with an interface to the physical sensor, as well as the capability to periodically update the DN with decisions and constraints that model the state of the physical sensor. The process for updating the DN is described in procedure UpdateSensorValues.

Figure 7.

Procedure `UpdateSensorValues`(DN, t_{now})

---

1 $d \leftarrow \langle \mathbf{v}, [[l_s, u_s], [l_e, u_e]] \rangle \in DN \;\; s.t. \;\; u_e = \infty$

2 $\mathbf{v_s} \leftarrow \mathtt{ReadSensor}()$

3 **if** $d = \mathrm{null} \wedge \mathbf{v_s} \neq \mathrm{null}$ **then**

4      $DN \leftarrow DN \cup d' = \langle \mathbf{v_s}, [[0, \infty), [0, \infty)] \rangle$

5      $DN \leftarrow DN \cup d' \; \mathrm{RELEASE} \; [t_{\mathrm{now}}, t_{\mathrm{now}}]$

6      $DN \leftarrow DN \cup d' \; \mathrm{DEADLINE} \; [t_{\mathrm{now}} + 1, \infty]$

7 **else if** $d \neq \mathrm{null} \wedge \mathbf{v_s} = \mathrm{null}$ **then** $DN \leftarrow DN \cup d \; \mathrm{DEADLINE} \; [t_{\mathrm{now}}, t_{\mathrm{now}}]$

8 **else if** $d \neq \mathrm{null} \wedge \mathbf{v_s} \neq \mathrm{null}$ **then**

9   **if** $\mathbf{v_s} = \mathbf{v}$ **then** $DN \leftarrow DN \cup d \; \mathrm{DEADLINE} \; [t_{\mathrm{now}} + 1, \infty]$

10   **else**

11          $DN \leftarrow DN \cup d \; \mathrm{DEADLINE} \; [t_{\mathrm{now}}, t_{\mathrm{now}}]$

12          $DN \leftarrow DN \cup d' = \langle \mathbf{v_s}, [[0, \infty), [0, \infty)] \rangle$

13          $DN \leftarrow DN \cup d' \; \mathrm{RELEASE} \; [t_{\mathrm{now}}, t_{\mathrm{now}}]$

14          $DN \leftarrow DN \cup d' \; \mathrm{DEADLINE} \; [t_{\mathrm{now}} + 1, \infty]$

Specifically, each sensor component's sensing procedure obtains from the DN the decision that represents the value of the sensor at the previous iteration (line 1). This decision, if it exists, is the decision whose end time has an infinite upper bound (u_e). No such decision exists if at the previous iteration the sensor readings were undetermined (d is null, i.e., there is no information on the current sensor value in the DN). The procedure then obtains the current sensor reading from its interface to the physical sensor (line 2). Notice that this could also be undetermined (null in the procedure), as a sensor may not provide a reading at all. At this point, three situations may occur.

New sensor reading. If the DN does not contain an unbounded decision and the physical sensor returns a value, then a decision is added to the DN representing this (new) sensor reading. The start time of this decision is anchored to the current time t_{now} by means of a RELEASE constraint and made to have an unbounded end time (lines 3-6). If the DN contains an unbounded decision that differs from the sensor reading, then the procedure models this fact in the DN as above, and in addition "stops" the previous decision by imposing a DEADLINE constraint, i.e., anchoring the decision's end time to t_{now} (lines 8, 10-14).

Continued sensor reading. If the DN contains an unbounded decision and the physical sensor returns the same value as that of this decision, then the procedure ensures that the increased duration of this decision is reflected in the DN. It does so by updating the lower bound of the decision's end time to beyond the current time by means of a new DEADLINE constraint (lines 8-9). Notice that this ensures that at the next iteration the DN will contain an unbounded decision.

Interrupted sensor reading. If the DN contains an unbounded decision and the physical sensor returns no reading (v_s is null), then the procedure simply interrupts the unbounded decision by bounding its end time to the current time with a DEADLINE constraint (line 7).

In the version of SAM that we present in this chapter and that we used for our case studies, decisions are never removed from the DN. This implies that the performance of the continuous inference process degrades over time, as the number of decisions increases in the DN. In our test runs, described in section 5, this was not a problem, because of the nature and number of sensors employed and because of the relatively limited temporal horizon (below one hour). The study of suitable heuristics and methodologies to remove unnecessary decisions from the DN is part of ongoing work (Ullberg, Loutfi, & Pecora, 2009).

4.3 Actuators

The inference procedure implemented in SAM continuously assesses the applicability of given synchronizations in the current DN by asserting and attempting to support new decisions on monitored components, such as the Human state variable presented earlier. This same mechanism allows to obtain contextualized plan synthesis capabilities through the addition of synchronizations that model how actions carried out by actuators should be temporally related to recognized activities. For instance, in addition to requiring that **Cooking** should be supported by requirements such as "being in the kitchen" and "using the stove", a requirement involving an actuator component can be added, such as "turn on the ventilation over the stove". More in general, for each actuation-capable device in the intelligent environment, an actuator component is modeled in the domain. This component's values represent the possible commands that can be performed by the device. In the domain, these values are added as requirements to the synchronizations of monitored components. As sensor components interface the real world to represent sensor readings in the DN, actuator components interface the real world to trigger commands to real actuators when decisions involving them appear in the DN.

Figure 8.

Procedure `UpdateExecutionState` $(\text{DN}, t_{\text{now}})$

---

1   $D \leftarrow \{\langle \mathbf{v}, [[l_s, u_s], [l_e, u_e]]\rangle \in \text{DN} \;\; s.t. \;\; l_s \leq t_{\text{now}} \wedge u_e = \infty\}$

2   **foreach** $d \in D$ **do**

3   **if** `IsExecuting(v)` **then**

4         $\text{DN} \leftarrow \text{DN} \cup d \; \text{DEADLINE} \, [t_{\text{now}} + 1, \infty]$

5   **else if** $l_s = l_e$ **then**

6   `StartExecuting(v)`

7         $\text{DN} \leftarrow \text{DN} \cup d \; \text{RELEASE} \, [t_{\text{now}}, t_{\text{now}}]$

8   **else** $\text{DN} \leftarrow \text{DN} \cup d \; \text{DEADLINE} \, [t_{\text{now}}, t_{\text{now}}]$

However, in some particular situations we have no means to know with certainty when and for how long a specific command will be executed on a robotic device. For instance, if we issue a command to the moving table to station itself close to the couch, the duration of the movement will depend on the path planning algorithm used and on the number of obstacles that the robot will encounter on its path. For this reason, actuator components also possess a sensory capability that is employed to feed information on the status of command execution back into the DN. As sensor components, actuator components write this information directly into the DN, thus allowing the re-planning process to take into account the current state of execution of the actions.

Actuators execute concurrently with the re-planning and sensing operations described above. The operations performed by actuators are shown in procedure UpdateExecutionState. Each actuator component first identifies all decisions that have an unbounded end time and whose earliest start time falls before or at the current time (line 1). The fact that these decisions are unbounded indicates that they have been planned for execution and their execution has not yet terminated. The fact that their start time lies before or at the current time indicates that they are scheduled to start or have already begun. For each of these decisions, the physical actuator is queried to as-

certain whether the corresponding command is being executed. If so, then the decision is constrained to end at least one time unit beyond the current time (lines 3-4). If the command is not currently in execution, the procedure checks whether the command still needs to be issued to the physical actuator. This is the case if the earliest start and end times of the decision coincide (because the decision's end time was never updated at previous iterations). The procedure dispatches the command to the actuator and anchors the start time of the decision to the current time (lines 5-7). Conversely, if the decision's start and end times do not coincide, then the decision is assumed to be ended, and the procedure imposes the current time as its earliest and latest end time (line 8).

5 CASE STUDIES IN THE PEIS-HOME

Let us illustrate the use of sensor components in SAM with four runs performed in the PEIS-Home. Assume we intend to assess the sleep quality of a person by means of a simple check, that is, how many times and for how long the user turns on his night light when he lies in bed. For this purpose, we employ three physical sensors: a pressure sensor, placed beneath the bed, a luminosity sensor placed close to the night light, and a person tracker based on stereo vision. We then define a domain with three sensor components and the two synchronizations shown in Figure 9 (where omitted, temporal bounds are assumed to be $[0, \infty)$). Note that the human user is modeled by means of two distinct components, Human and HumanAbstract. This allows us to reason at different levels of abstraction on the user: while the decisions taken on component Human are always a direct consequence of sensor readings, synchronizations on values of HumanAbstract describe knowledge that can be inferred from sensor data, what is asserted on the Human component, and from what is asserted on the HumanAbstract component itself. In the example, the first synchronization detects when the user goes to bed. In this case, two decisions should be present in the DN. First, the user should not be observable by the tracking system, since the bedroom is a private area of the apartment and, therefore, outside the field of view of the cameras. Then the pressure sensor beneath the bed should be activated. The resulting **InBed** decision has a duration EQUAL to the one of the positive reading of the bed sensor. The second synchronization grasps the situation in which, although lying in bed, the user is not sleeping. The decision **Awake** on the component HumanAbstract depends therefore on the decision

InBed of the Human and on the sensor readings of NightLight.

This simple domain was employed to test SAM in our intelligent home environment with a human subject. The overall duration of the experiment was 500 seconds, with the concurrent re-planning and sensing processes operating at a rate of about 1 Hz. Figure 12 (a) is a snapshot of the five components' timelines at the end of the run (from top to bottom, the three sensors and the two monitored components).

The outcome of a more complex example is shown in Figure 12(b). In this case the scenario contains four instantiated sensors. Our goal is to determine the afternoon activities of the user living in the apartment, detecting activities like **Cooking**, **Eating** and the more abstract **Lunch**. To realize this example, we define five new synchronizations (Figure 10), three for the Human component and two for the HumanAbstract component. Synchronization (3) identifies the human activity **Cooking**: the user should be in the kitchen and its duration is EQUAL to the activation of the Stove. Synchronization (5) models the **Eating** activity, using both the Location sensor and an RFID reader placed beneath the kitchen table (component KTRfid). A number of objects have been tagged to be recognized by the reader, among which dishes whose presence on the table

Figure 9. Synchronizations defined in our domain for the Human and HumanAbstract components to assess quality of sleep

```
1) Human : InBed
        DURING Location : NOPOS
        EQUALS Bed : ON
2) HumanAbstract : Awake
        DURING Human : InBed
        EQUALS NightLight : ON
```

Figure 10. Synchronizations modeling afternoon activities of the human user

```
1) HumanAbstract : Lunch              2) HumanAbstract : Nap
     STARTED-BY Human : Cooking           AFTER HumanAbstract : Lunch
     FINISHED-BY Human : Eating           EQUALS Human : WatchingTV
     DURING Time : afternoon
3) Human : Cooking                    4) Human : WatchingTV
     DURING Location : KITCHEN            EQUALS Location : COUCH
     EQUALS Stove : ON
5) Human : Eating
     DURING Location : KITCHENTABLE
     EQUALS KTRfid : DISH
```

is required to assert the decision **Eating**. The last synchronization for the Human component (4) correlates the presence of the user on the couch with the activity of **WatchingTV**.

Synchronizations (1) and (2) work at a higher level of abstraction. The decisions asserted on HumanAbstract are inferred from sensor readings (Time), from the Human component and from the HumanAbstract component itself. This way we can identify complex activities such as **Lunch**, which encompasses both **Cooking** and the subsequent **Eating**, and we can capture the fact that after lunch the user, sitting in front of the TV, will most probably fall asleep.

Also this example was executed in the PEIS-Home. The output (Figure 12(b)) highlights how SAM was able to mirror the reality. It is worth mentioning that the decision corresponding to the **Lunch** activity on the HumanAbstract component was identified only when both **Cooking** and **Eating** were asserted on the Human component. Also it can be noted that **Nap** is identified as the current HumanAbstract activity only after the lunch is over and that on the first occurrence of **WatchingTV**, **Nap** was not asserted because it lacked support from the **Lunch** activity.

As an example of how the domain can include actuation as synchronization requirements on monitored components, let us consider the following run of SAM in our intelligent home setting. In this setup, we make use of two devices that are available in our environment, namely the robotic table and autonomous fridge described earlier.

As shown in Figure 11, we use the abductive reasoning to infer when the user is watching TV. In this case, however, we modify the synchronization (4) presented Figure 10 to include the actuators in the loop. The new synchronization (Figure 11, (1)), not only recognizes the **WatchingTV** activity, but also asserts the decision **DeliverDrink** on the MovingTable component. This decision can be supported only if it comes AFTER another decision, namely **PlaceDrink** on component Fridge (synchronization (3)). When SAM's re-planning procedure attempts to support **WatchingTV**, synchronization (5) is called into play, stating that **PlaceDrink** should occur right after (MET-BY) the MovingTable has docked the Fridge and right before the un-docking maneuver (MEETS). The remaining three synchronizations - (2), (4) and (6) - are attempted to complete the chain of support, that is, the Fridge should first grasp the drink with its robotic arm, then open the door before the MovingTable is allowed to dock to it, and finally it should close the door right after the MovingTable has left the docking position. Figure

Figure 11. Synchronizations defining temporal relations between human activities and proactive services

1) Human : **WatchingTV**
   EQUALS Location : **COUCH**
   START MovingTable : **DeliverDrink**
2) MovingTable : **DockFridge**
   MET-BY Fridge : **OpenDoor**
3) MovingTable : **DeliverDrink**
   AFTER Fridge : **PlaceDrink**
4) MovingTable : **UndockFridge**
   BEFORE Fridge : **CloseDoor**
5) Fridge : **PlaceDrink**
   MET-BY MovingTable : **DockFridge**
   MEETS MovingTable : **UndockFridge**
6) OpenDoor : **OpenDoor**
   MET-BY Fridge : **GraspDrink**

13 shows an intuitive diagram of the plan resulting from the abductive reasoning process, along with the connections between real sensors/actuators and the decisions asserted on the components.

This chain of synchronizations leads to the presence in the DN of a plan to retrieve a drink from the fridge and deliver it to the human who is watching TV. Notice that when the planned decisions on the actuator components are first added to the DN, their duration is minimal. Through the actuators' UpdateExecutionState procedure, these durations are updated at every

re-planning period until the devices that are executing the tasks signal that execution has completed. Also, thanks to the continuous propagation of the constraints underlying the plan, decisions are appropriately delayed until their earliest start time coincides with the current time. A complete run of this scenario was performed in our intelligent environment and a snapshot of the final timelines is shown in Figure 12(c).

The previous examples have shown how SAM leverages temporal domain knowledge to monitor the domestic activities of a human from

Figure 12. Timelines resulting from the runs performed in our intelligent home using the sleep monitoring (a), afternoon activities (b) and proactive service (c) domains

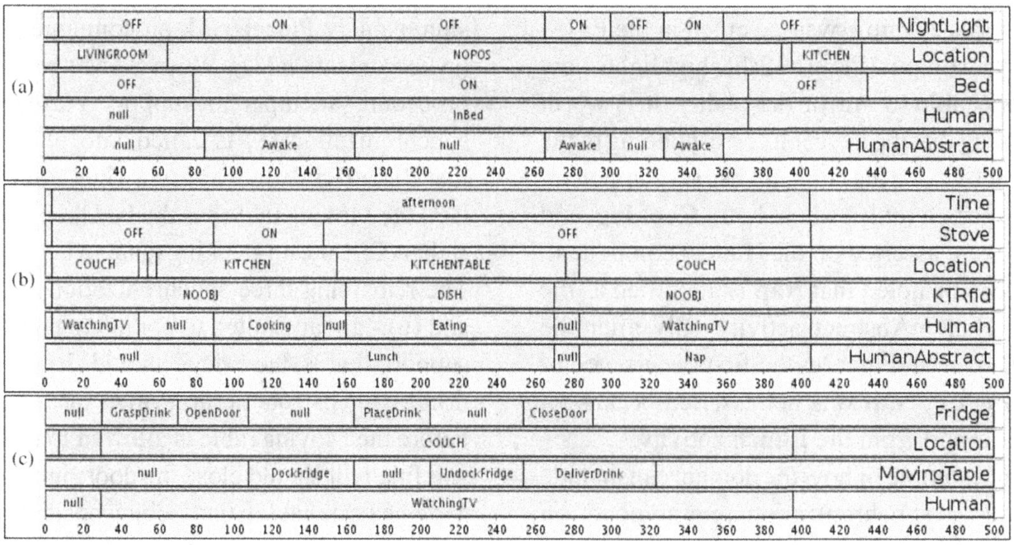

Figure 13. An intuitive sketch of a plan generated by the abductive reasoning process

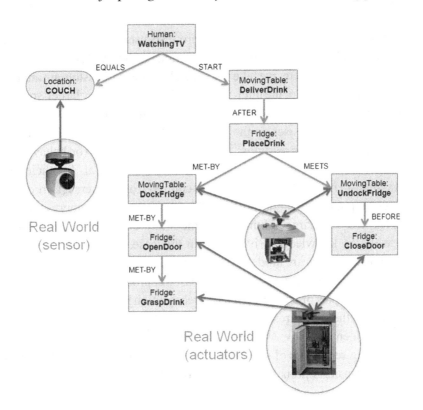

traces of sensor readings. The examples also show its ability to contextually synthesize plans for actuators present in the environment. Both types of requirements are described in a uniform formalism, thus providing a straightforward way to combine the recognition and proactive service providing aspects of the scenario.

The component based nature of SAM directly enables a modular approach to application development. This has been tested by combining the previous test runs into a larger domain which was employed for an extended test case. The previous domains were combined and extended with new sensors and actuators, which required to simply implement the software interface between the new real-world devices and the corresponding sensor/ actuator components in the domain. Modularity is particularly useful for building incremental domains, as it allows to partition the modeling process and to perform validations on separate

sub-domains. The combined domain includes all the synchronizations detailed in the previous three examples, plus a number of variants of the meal eating activities (**Breakfast**, **Lunch** and **Dinner**) and an additional synchronization modeling the absence of the user (**Out**) as a state occurring after the user passes the **Entrance** location and lasting as long as the user is not tracked by the Location component. A twenty minute long run was performed with a user carrying out activities in the PEIS-Home. The run terminated with a DN containing 53 decisions. The brunt of computation is performed by the Simple Temporal Network's propagation procedure, which in this case was therefore dealing with 106 timepoints. All processes ran on a single 2.4 GHz Intel Core2 Quad. The system was capable of maintaining a re-planning period of under eight seconds, therefore guaranteeing a sampling rate well below the minimum duration of meaningful sensor readings.

6 RELATED WORK

Activity recognition has received much attention in the literature and the term has uses employed to indicate a variety of capabilities. Here, we take activity recognition to mean the ability of the intelligent system to deduce temporally contextualized knowledge regarding the state of the user on the basis of a set of heterogeneous sensor readings.

Current approaches to this problem can be roughly categorized as *data-driven* or *knowledge-driven*. In data-driven approaches, models of human behavior are learned from large volumes of data over time, while in knowledge-driven approaches patterns of observations are modeled from first principles rather than learned.

A number of notable examples of data-driven approaches can be found in the literature. Probabilistic methods and Hidden Markov Models have been used for learning sequences of sensor observations with given transition probabilities. Instantiated within the domestic environment setting, these techniques have been leveraged to infer human activities from RFID-tagged object use (Patterson, Fox, Kautz, & Philipose, 2005) as well as vision-based observations (Naeem & Bigham, 2007). An approach based on learning techniques is presented by Lin and Hsu (Lin & Hsu, 2006) who design a framework for activity recognition based on the Lempel-Ziv family of algorithms, using a wearable RFID reader, a WiFi positioning system and the passing of time. Hein and Kirste (Hein & Kirste, 2008) use a combination of an unsupervised learning algorithm (k-Means) and Hidden Markov Models to recognize high level activities from data gathered by wearable sensors. A purely probabilistic approach was explored by Patterson et al. (Patterson, Fox, Kautz, & Philipose, 2005) and by Naeem and Bigham (Naeem & Bigham, 2007). In both approaches, the authors gathered data from a wearable RFID reader in order to recognize Activities of Daily Life (ADL). Patterson et al. processed the data using both Hidden Markov Models and Dynamic Bayesian Networks, while Naeem and Bigham compared two different types of HMMs. HMMs are also used by Kemmotsu et al. (Kemmotsu, Tomonaka, Shiotani, Koketsu, & Iehara, 2005), in combination with histogram analysis, to recognize human behaviors in larger spaces. In particular, HMMs are used to identify short-term motions, while histogram analysis is employed to recognize long-term motions (or behavior). The data used by Kemmotsu et al. for their experiments was acquired from cameras. Large environments and data coming from cameras are also the center of the study of Fusier et al. (Fusier, et al., 2007). In this case, the activities of moving agents are inferred from their trajectories. Lastly, an HMM based approach has also been used to perform plan recognition for human-aware robot task planning in the PEIS-Home, as described in (Cirillo, Karlsson, & Saffiotti, 2009).

In knowledge-driven approaches, patterns of observations are modeled from first principles rather than learned. Abductive processes are employed instead, whereby sensor data is explained by hypothesizing the occurrence of specific human activities.[5] Examples include the work of Goultiaeva and Lespérance (Goultiaeva & Lespérance, 2007), where the Situation Calculus is used to specify very rich plans, as well as approaches based on ontologies (Mastrogiovanni, Scalmato, Sgorbissa, & Zaccaria, 2009) and temporal reasoning approaches in which rich temporal representations are employed to model the conditions under which patterns of human activities occur. An example of the last kind of approach is the study of Jakkula et al. in the framework of the MavHome project (Jakkula, Cook, & Crandall, 2007). A common point between our work and the approach proposed by Jakkula et al. is the use of Allen's temporal relations to model the actions of the user. Finally, scheduling-based solutions for monitoring have been developed in a very similar context to that of the PEIS-Ecology, namely in the RoboCare project (Cesta, Cortellessa, Pecora, & Rasconi, 2007) and in Autominder (Pollack, et al., 2003). An important difference between our work and the two just mentioned is that both Autominder and the RoboCare architecture use pre-compiled (albeit highly flexible) schedules as models for human behavior. In the present work, we employ a planning process to actually instantiate such candidate schedules on-line. It should also be noted that both Autominder and the RoboCare architecture allow proactive behaviors of the system, but in the simple form of reminders to alert the user when violations in the pre-scheduled user plans arise.

Data- and knowledge-driven approaches have complementary strengths and weaknesses. One advantage of the former is the ability to learn patterns of human behavior, rather than having to model them explicitly. This advantage comes at the price of poor scalability and inherent difficulty of providing common-sense knowledge for classifying overwhelming amounts of data (Wu, Osuntogun, Choudhury, Philipose, & Rehg, 2007). In this sense, knowledge-based approaches can provide a powerful means to express such information, which can then be employed by an abductive process that essentially "attaches a meaning" to the observed data. Furthermore, the extent to which models for data-driven approaches can be reused if sensors are removed or added is unclear. In SAM, the specific component based structure of the domain is leveraged to achieve high modularity, thus allowing model composability and reuse.

Indeed, the current literature points to the fact that the data- and knowledge-driven strategies are complementary in scope: the former provide an effective way to recognize elementary activities from large amounts of continuous data; conversely, knowledge-driven approaches are useful when the criteria for recognizing human activities are given by complex but general rules that are clearly identifiable. While the ranges of applicability of the two strategies clearly overlap, knowledge-driven approaches have been less explored in literature.

The activity recognition side of SAM can be categorized as a knowledge-driven approach based on a rich temporal representation and on a planning framework. We have shown how a component-based domain representation grounded on Allen's Interval Algebra can be employed in conjunction with a constraint-based planning and scheduling framework called OMPS (Fratini, Pecora, & Cesta, 2008) to obtain an abductive process for activity recognition. The use of this architecture was demonstrated with several test runs performed in the PEIS-Home.

In addition to specifying the criteria for recognizing human activities from patterns of sensory data, SAM allows to specify how such activities should lead to the synthesis and execution of plans for service-providing actuators. One key feature that allows to achieve this capability is the timeline-based planning infrastructure of OMPS. In this sense, our work is related to the current state-of-the-art in timeline-based planning, which includes CASPER and its predecessors ASPEN and Rax-PS (Knight, Rabideau, Chien, Engelhardt, & Sherwood, 2001) (Jonsson, Morris, Muscettola, Rajan, & Smith, 2000) (Muscettola, Nayak, Pell, & Williams, 1998)[6]. CASPER and its predecessors realize what has been termed as continuous planning, where plans are incrementally expanded rather than planned and executed over limited horizons. These systems share with OMPS a number of features, such as the possibility to plan over state variables and resources as well as temporal and parameter constraints. SAM leverages the state variable metaphor present in all these architectures, as well as the ability to maintain temporally flexible timelines. In addition, SAM leverages the ability of OMPS to employ custom variable types in addition to state variables. This has allowed us to build the sensing and actuation capabilities directly into new variable types which extend the state variable. These variables thus are not only used to represent elements of the domain, but also to implement active processes which operate concurrently with the continuous planning process, providing it with real world data obtained from the intelligent environment.

7 CONCLUSION

In this chapter we have presented SAM, an architecture for concurrent activity recognition, planning and execution. The architecture is built on the OMPS temporal reasoning framework, and leverages its component-based approach to realize a decisional framework that operates in a closed loop with physical sensing and actuation components in an intelligent environment. We have demonstrated the feasibility of the approach with a number of experimental runs in the PEIS-Home, which provides an easy and transparent way to introduce sensors in the environment, managing the low level communication and providing a high level interface to access the data. A rich set of sensors is permanently deployed in the room (stereo cameras, mono cameras, RFID readers, etc.), and new ones can be easily added if needed. The sensor data represents the ground truth for the activity recognition system, and all runs of SAM described in this paper were performed with a human test subject physically triggering sensor readings.

Our main contribution is the novelty of the approach we take to the problem. We use a temporal planning framework to realize an abductive process which explains sensor data by testing the applicability of temporally-constrained requirements on this data. These requirements are given in the form of synchronizations, which are grounded on Allen's Interval Algebra. Also, SAM seamlessly combines human activity recognition and planning for controlling pervasive actuation devices.

SAM provides a general purpose framework to define how an intelligent environment should assess context from sensor readings, as well as how it should react to the assessment. Indeed, our approach is complementary to data-driven methodologies for recognizing activities, and a promising idea for future research is to realize a data-driven activity recognition layer between the physical sensors and SAM's sensing components. This would allow to relax a number of assumptions made on sensing such as the absence of noise, sensor uncertainty and missing information. Indeed, while a temporal constraint based formalism represents an intuitive metaphor for expressing some aspects of household activities, such crisp knowledge about human activities may not always be available. Where it is not, it is reasonable to envisage a data-driven approach for learning how sensor data correlates to human activities.

Finally, we are currently addressing issues related to the response time of SAM's abductive process. Specifically, in (Ullberg, Loutfi, & Pecora, 2009) we provide an optimized inference process which supports activity monitoring over long temporal horizons, i.e. in the order of weeks and months.

ACKNOWLEDGMENT

This work has been partially supported by CUGS (Swedish national computer science graduate school) and by the Swedish KK foundation.

REFERENCES

Allen, J. F. (1984). Towards a General Theory of Action and Time. *Artificial Intelligence, 23*(2), 123–154. doi:10.1016/0004-3702(84)90008-0

Bjorn, G., Guesgen, H. W., & Hubner, S. (2006). Spatiotemporal Reasoning for Smart Homes. In Augusto, J. C., & Nugent, C. D. (Eds.), *Springer*.

Broxvall, M. (2007). A Middleware for Ecologies of Robotic Devices. *Proc. of the First Int. Conf. on Robot Communication and Coordination (ROBOCOMM)*. Athens, Greece.

Broxvall, M., Coradeschi, S., Loutfi, A., & Saffiotti, A. (2006). An Ecological Approach to Odour Recognition in Intelligent Environments. *Proc. of the IEEE Int. Conf. on Robotics and Automation (ICRA)*. Orlando, FL.

Cesta, A., Cortellessa, G., Pecora, F., & Rasconi, R. (2007). Supporting Interaction in the RoboCare Intelligent Assistive Environment. *Proc. of AAAI Spring Symposium on Interaction Challenges for Intelligent Assistants.*

Cesta, A., & Fratini, S. (2008). The Timeline Representation Framework as a Planning and Scheduling Software Development Environment. *Proc. of 27th Workshop of the UK Planning and Scheduling SIG.*

Cirillo, M., Karlsson, L., & Saffiotti, A. (2009). A Human-Aware Robot Task Planner. *Proc. of the Int. Conf. on Automated Planning and Scheduling (ICAPS09).*

Dechter, R., Meiri, I., & Pearl, J. (1991). Temporal Constraint Networks. *Artificial Intelligence, 49*(1-3), 61–95. doi:10.1016/0004-3702(91)90006-6

Dressler, F. (2006). Self-organization in autonomous sensor/actuator networks. *Proc. of the 19th IEEE Int. Conf. on Architecture of Computing Systems.*

Fratini, S., Pecora, F., & Cesta, A. (2008). Unifying Planning and Scheduling as Timelines in a Component-Based Perspective. *Archives of Control Sciences, 18*(2), 231–271.

Fusier, F., Valentin, V., Bremond, F., Thonnat, M., Borg, M., & Thirde, D. (2007). Video Understanding for Complex Activity Recognition. *Machine Vision and Applications, 18*(3), 167–188. doi:10.1007/s00138-006-0054-y

Goultiaeva, A., & Lespérance, Y. (2007). Incremental Plan Recognition in an Agent Programming Framework. *Working Notes of the AAAI Workshop on Plan, Activity, and Intention Recognition (PAIR).*

Hein, A., & Kirste, T. (2008). Towards Recognizing Abstract Activities: An Unsupervised Approach. *BMI '08: Proc. of the 2nd Workshop on Behaviour Monitoring and Interpretation*, (pp. 102-114).

Jakkula, V., Cook, D. J., & Crandall, A. S. (2007). Temporal Pattern Discovery for Anomaly Detection in a Smart Home. *Proc. of the 3rd IET Conf. on Intelligent Environments(IE)*, (pp. 339-345).

Jonsson, A. K., Morris, P. H., Muscettola, N., Rajan, K., & Smith, B. (2000). Planning in Interplanetary Space: Theory and Practice. *Proc. Int. Conf. on AI Planning and Scheduling (AIPS-00).*

Kemmotsu, K., Tomonaka, T., Shiotani, S., Koketsu, Y., & Iehara, M. (2005). Recognizing Human Behaviors With Vision Sensors in Network Robot Systems. *Proc. of The 1st Japan-Korea Joint Symposium on Network Robot Systems (JK-NRS 2005).*

Kim, J. H., Kim, Y. D., & Lee, K. H. (2004). The Third Generations of Robotics: Ubiquitous Robot. *Proc. of the 2nd Int. Conf. on Autonomous Robots and Agents.* Palmerston North, New Zealand.

Knight, R., Rabideau, G., Chien, S., Engelhardt, B., & Sherwood, R. (2001). Casper: Space Exploration through Continuous Planning. *IEEE Intelligent Systems, 16*(5), 70–75.

Le Blanc, K., & Saffiotti, A. (2008). Cooperative Anchoring in Heterogeneous Multi-Robot Systems. *Proc. of the IEEE Int. Conf. on Robotics and Automation (ICRA).* Pasadena, CA.

Lee, J. H., & Hashimoto, H. (2002). Intelligent Space - concept and contents. *Advanced Robotics, 16*(3), 265–280. doi:10.1163/156855302760121936

Lin, C., & Hsu, J. Y. (2006). IPARS: Intelligent Portable Activity Recognition System Via Everyday Objects, Human Movements, and Activity Duration. *AAAI Workshop on Modeling Others from Observations*, (pp. 44-52).

Lundh, R., Karlsson, L., & Saffiotti, A. (2008). Autonomous Functional Configuration of a Network Robot System. *Robotics and Autonomous Systems, 56*(10), 819–830. doi:10.1016/j.robot.2008.06.006

Mastrogiovanni, F., Scalmato, A., Sgorbissa, A., & Zaccaria, R. (2009). Assessing Temporal Relationships Between Events in Smart Environments. *Proc of the Int. Conf. on Intelligent Environments.*

Muscettola, N., Nayak, P. P., Pell, B., & Williams, B. C. (1998). Remote Agent: to Go Boldly Where No AI System Has Gone Before. *Artificial Intelligence, 103*(1–2), 5–48. doi:10.1016/S0004-3702(98)00068-X

Naeem, U., & Bigham, J. (2007). A Comparison of Two Hidden Markov Approaches to Task Identification in the Home Environment. *Proc. of the 2nd Int. Conf. on Pervasive Computing and Applications (ICPCA 2007)*, (pp. 383-388).

Patterson, D. J., Fox, D., Kautz, H., & Philipose, M. (2005). Fine-Grained Activity Recognition by Aggregating Abstract Object Usage. *Proc. of the 9th IEEE International Symposium on Wearable Computers.*

Pecora, F., & Cirillo, M. (2009). A Constraint-Based Approach for Plan Management in Intelligent Environments. *Proc. of the Scheduling and Planning Applications Workshop at ICAPS09.*

Pollack, M. E., Brown, L., Colbry, D., McCarthy, C. E., Orosz, C., & Peintner, B. (2003). Autominder: an Intelligent Cognitive Orthotic System for People with Memory Impairment. *Robotics and Autonomous Systems, 44*(3-4), 273–282. doi:10.1016/S0921-8890(03)00077-0

Rusu, R. B., Bandouch, J., Marton, Z. C., Blodow, N., & Beetz, M. (2005). Action recognition in intelligent environments using point cloud features extracted from silhouette sequences. *Proc. of the Int. Conf. on Smart Objects and Ambient Intelligence (sOc-EUSAI).*

Saffiotti, A., & Broxvall, M. (2005). PEIS ecologies: Ambient intelligence meets autonomous robotics. *Proc. of the Int. Conf. on Smart Objects and Ambient Intelligence (sOc-EUSAI).*

Saffiotti, A., Broxvall, M., Gritti, M., LeBlanc, K., Lundh, R., Rashid, J., et al. (2008). The PEIS-Ecology Project: Vision and Results. *Proc of the IEEE/RSJ Int. Conf. on Intelligent Robots and Systems (IROS)*. Nice, France.

Sanfeliu, A., Hagita, N., & Saffiotti, A. (2008). Special issue on Network robot systems. *Robotics and Autonomous Systems, 56*(10).

Sgorbissa, A., & Zaccaria, R. (2004). The artificial ecosystem: a distributed approach to service robotics. *Proc. of the Int. Conf. on Robotics and Automation (ICRA).*

Shanahan, M. P. (1996). Robotics and the Common Sense Informatic Situation. *Proc. of the European Conf. on Artificial Intelligence (ECAI).*

Ullberg, J., Loutfi, A., & Pecora, F. (2009). Towards Continuous Activity Monitoring with Temporal Constraints. *Proc. of the 4th Workshop on Planning and Plan Execution for Real-World Systems at ICAPS09.*

Wu, J., Osuntogun, A., Choudhury, T., Philipose, M., & Rehg, J. (2007). A Scalable Approach to Activity Recognition Based on Object Use. *Proc. of IEEE Int. Conf. on Computer Vision (ICCV).*

KEY TERMS AND DEFINITIONS

Abductive Inference: A method of logical inference that allows to infer preconditions from consequences.

Activity Recognition: The ability of the intelligent system to infer temporally contextualized knowledge regarding the state of the user on the basis of a set of heterogeneous sensor readings.

Decision Network: A network of constraints between decisions, where decisions represent either sensor readings or human activities.

PEIS: A "Physically Embedded Intelligent System" is a physical device which includes a number of functional components. Each PEIS can use functionalities from other PEIS in the ecology in order to compensate or to complement its own.

PEIS-Ecology: An ecological approach to creating Smart Robotic Environments, which combines concepts from Artificial Intelligence, Robotics and Ubiquitous Computing. The power of a PEIS Ecology does not come from the individual power of its constituent PEIS, but it emerges from their ability to interact and cooperate.

Robot: In the PEIS-Ecology approach, the concept of robot is abstracted by the uniform notion of a PEIS, a physical device which includes a number of functional components.

Smart Robotic Environment: An environment where robots, pervasive sensors and actuators cooperate to provide services for human users.

Temporal Constraint: A temporal relation between variables that represent events in time. In this work, we consider relations in Allen's Interval Algebra.

Temporal Propagation: Propagation of temporal constraints (Temporal Propagation) in a Decision Network is used to assess the temporal consistency of hypotheses inferred on human activities.

ENDNOTES

[1] SAM stands for SAM the Activity Manager.

[2] A video of the scenario can be found at ftp://aass.oru.se/pub/asaffio/movies/peis-ipw08.flv

[3] OMPS = Open Multi-component Planner and Scheduler.

[4] A complete description of OMPS is outside the scope of this chapter. OMPS offers other features than the ones here described, but they are not used by SAM and therefore not relevant in this context.

[5] An approach similar to Shanahan's work (Shanahan, 1996) on inferring information on a robot's environment.

[6] For a more complete comparison between these systems and OMPS, see (Fratini, Pecora, & Cesta, 2008).

Chapter 26
Self–Organizing Mobile Sensor Network:
Distributed Topology Control Framework

Geunho Lee
Japan Advanced Institute of Science and Technology (JAIST), Japan

Nak Young Chong
Japan Advanced Institute of Science and Technology (JAIST), Japan

ABSTRACT

Ambient Intelligence (AmI) is a multidisciplinary approach aimed at enriching physical environments with a network of distributed devices, such as sensors, actuators, and computational resources, in order to support humans in achieving their everyday task. Within the framework of AmI, this chapter presents decentralized coordination for a swarm of autonomous robotic sensors building intelligent environments adopting AmI. The large-scale robotic sensors are regarded as a swarm of wireless sensors mounted on spatially distributed autonomous mobile robots. Therefore, motivated by the experience gained during the development and usage for decentralized coordination of mobile robots in geographically constrained environments, our work introduces the following two detailed functions: self-configuration and flocking. In particular, this chapter addresses the study of a unified framework which governs the adaptively self-organizing processes for a swarm of autonomous robots in the presence of an environmental uncertainty. Based on the hypothesis that the motion planning for robot swarms must be controlled within the framework, the two functions are integrated in a distributed way, and each robot can form an equilateral triangle mesh with its two neighbors in a geometric sense. Extensive simulations are performed in two-dimensional unknown environments to verify that the proposed method yields a computationally efficient, yet robust deployment.

DOI: 10.4018/978-1-61692-857-5.ch026

1. INTRODUCTION

Today the diversities and capabilities of mobile network-enabled systems (e.g., mobile robotic sensors) have been increasing steadily. With the emergence of Ambient Intelligence (AmI) (Remagnino & Foresti, 2005) and the growth of robotic technology, much attention has been paid to increase the availability and use of large-scale robotic sensors with relatively simple capability in the field of robotics. These progresses based on mobile robotic systems have made intelligent environments adopting AmI (i.e., AmI environments) possible, and are gradually becoming more and more pervasive (Dressler, 2007). The robotic sensors allowed to disperse themselves into an unknown area of interest can be used for a wide variety of applications, such as disaster relief searching, environmental or habitat monitoring, surveillance, investigation of water and sewage pipes, etc (Cortes et al., 2004; Choset, 2001; Zecca1 et al., 2009).

Could we provide our societies and many users with smart mobile robotic sensors deploying autonomously that would satisfy their needs, and possibly establish AmI environments composed of those sensors? Technologies to achieve this are already at hand. Advances in building AmI environments have been already achieved for Java programming, for XML (eXtensible Markup Language) documents, and for wireless communications such as WiFi, Bluetooth, RFID (Radio Frequency IDentification), etc. However, such advances are not enough to disperse the mobile robotic sensors into an unknown area of interest, to fill the area, and to establish AmI environmental infrastructures. What we lack is a self-organizable coordination of autonomous robotic sensors that have the capacity to address problems they encounter. This scenario is closely related to the concept of AmI, which hypothesizes a future where people are surrounded by smart mobile sensors. The scenario is not restricted to daily life environment such as homes, schools,

offices, or stores. In order to allow a better support to societies and users, AmI environments can be expanded to include airport facilities, sewage disposal plants, seashores, the Polar Regions, the moon, etc. The ideal scenario we imagine in this chapter considers a nearby future where mobile sensors autonomously establish smart environments, in the sense that they are able to be aware of the needs of societies or users and to provide desirable services satisfying the needs.

The mobile robotic sensors are regarded as a swarm of wireless sensors mounted on spatially distributed autonomous mobile robots (Sahin, 2005). To enable mobile robotic sensors to successfully perform the aforementioned tasks, it is usually required to configure their swarm network and/or maintain their network adapting to geographically constrained environments. To make the scenario of AmI environment, a reality, our goal is to develop the computational basis of unified coordination for supporting various purpose applications of mobile robotic sensors running the same algorithm. For their purpose, this chapter presents the deployment control architecture and algorithms needed to coordinate movements of robotic sensors within their swarm. Specifically, deployment control includes such functions as self-configuration, flocking, and adapting to the environment under the self-organizable framework. Moreover, we try to delineate the requirements that such coordination should satisfy.

The rest of this chapter is organized as follows. In Section 2 the state-of-the-art is summarized and commented in order to provide potential readers with the current trends for the coordination of mobile robot swarms. Section 3 presents the computational robot model and the formal definitions of problems to address. As a solution approach to the unified self-organizable framework, Section 4 describes the local interactions among three neighboring robots and their convergence property. In Sections 5 and 6, based on their framework, we present two kinds of coordination approaches, the adaptive self-configuration and adaptive flocking algorithms, and show the results of simulations. Conclusions are drawn in Section 7.

2. ROBOT SWARM DOMAIN-SPECIFIC COORDINATION CHALLENGES

This section presents an overview of work regarding the coordination of mobile robot swarms in view of building AmI environmental infrastructures. Coordination can be largely divided into the centralized or decentralized coordination strategies. The centralized coordination relies on a specific robot which supervises or organizes the behaviors of the whole swarm through a communication channel. Egerstedt and Hu (Egerstedt & Hu, 2001) employed a virtual reference for the desired trajectory controlled from a remote host with which individual robots maintain their predefined positions. Belta and Kumar (Belta & Kumar, 2002) generated smooth interpolating motion for individual robots so that the total kinetic energy is minimized while certain constraints are satisfied. In general, a heavy computational cost is imposed on the supervising robot which also requires a tight communication with other robots. Heavy computation or failure of a supervising robot leads to vulnerability in the coordination of the overall swarm. Moreover, these strategies lack scalability and become technically unfeasible when a large swarm is considered.

At the opposite end, the decentralized coordination is achieved through the decision of individual robot, and has an advantage in term of robustness and flexibility. The decentralized coordination strategies can be classified as global or local perspective approaches according to a sensing and/or communication boundary. Global perspective approaches (Suzuki & Yamashita, 1999; Lee & Chong, 2009a) may provide a more efficient and cost-effective method exploiting a wide range of sensors, but also exposes robot swarms to lack of scalability. On the other hand, local perspective approaches are based on local rules of behavior observed from social living organisms or physical phenomena.

According to the frameworks of local rules, the local perspective approaches can be further divided into biologically-inspired (Ikemoto et al., 2005; Shimizu et al., 2006), behavior-based (Werger & Mataric, 2001; Balch & Hybinette, 2000), and virtual physics-based frameworks. Many of the behavior-based and virtual physics-based frameworks mainly have used such physical phenomena as van der Waals forces (Zheng & Chen, 2007), gravitational forces (Spears et al., 2004; Spears & Gordon, 1999; Cohen & Peleg, 2006), electric charges (Howard et al., 2002), spring forces (Shucker et al., 2008; McLurkin & Smith, 2004; Fujibayashi et al., 2002), potential fields (Reif & Wang, 1999; Zou & Chakrabarty, 2003; Balch & Hybinette, 2000), line forces (Martison & Payton, 2005), the equilibrium of molecules (Heo & Varshney, 2003), and other virtual forces (Wang et al., 2004). This previous work mostly uses some sort of force balance between inter-individual interactions exerting an attractive or repulsive force on each other. This is mainly because the force-based interaction rules are considered simple but effective, and provide an intuitive understanding of individual behavior. Local rules based on those frameworks mainly focus on the following issues: 1) how to configure robot swarms, 2) how to navigate, or 3) how to adapt to an environment.

First, the self-configuration studies aim at deploying a robot swarm into an area or to achieve pre-determined patterns. Self-configuration by a robot swarm may result in mesh-type networks. These configurations offer high-level coverage and multiple redundant connections, ensuring maximum reliability and flexibility from the standpoint of topology. Depending on whether there are interactions among all robots, the network can be classified into fully or partially-connected topologies (Ghosh et al., 2005; Fowler, 2001). The fully-connected topologies have each robot interact with all of the other robots simultaneously within a certain range. Thus, these topologies impose too tight constraints on robot motion, and increase the computational complexity. Notably, they might lead to deadlocks where some of the robots have become trapped interstitially. These problems arise in most of the previous works (Balch & Hybinette, 2000; Spears et al., 2004; Howard et al., 2002; Fujibayashi et al., 2002; Reif & Wang, 1999; Zou & Chakrabarty, 2003; Heo & Varshney, 2003). Specifically, to cope with deadlocks, a large force needs to be applied to the trapped robots instantly to shuffle the positions (phase transition) (Spears et al., 2004), or all robots have identifiers, and their interrelationship should be pre-specified to selectively activate the force balance (entire attribute specification) (Spears & Gordon, 1999). On the other hand, using the partially-connected topology, robots interact selectively with other robots, but are connected to all robots in the formation. In (Shucker et al., 2008; McLurkin & Smith, 2004), partial graph pairs of robots were proposed that exert virtual forces on each other when their connection is part of the graph. For example, robots may choose to exert forces in a certain direction (McLurkin & Smith, 2004), where this selective interaction helps to prevent them from being too tightly constrained. For a similar reason, robots are enabled to achieve a faster formation without deadlocks (Shucker et al., 2008).

Secondly, flocking enables robot swarms to navigate toward a goal while maintaining their geometric pattern. Reynolds (Reynolds, 1987) originally presented a distributed behavioral flocking model based on fish schools and bird flocks. His work demonstrated that flocking is an example of emergent behavior arising from simple rules. Since his contribution, many flocking studies have been reported in the field of swarm robotics, and can be divided into the leaderless method, where a specific robot is not assigned to conduct its robot swarm, and the leader-follower method. The leaderless methods are based on in-teractions between individual robots. These ideas were mostly borrowed from physical phenomena or living organisms, animals, such as insects, found in nature (Lee & Chong, 2008a; Kerr et al., 2005; Zarzhitsky et al., 2005), or behavior-based approaches (Esposito & Dunbar, 2006). In the leader-follower method, a robot is selected as the moving reference point, and all followers maintain a symmetric geometric pattern with respect to the selected leader while navigating (Tanner et al., 2004; Gervasi & Prencipe, 2003; Balch & Arkin, 1998).

Thirdly, as adaptation works, authors (Ishiguro & Kawakatsu, 2006) introduced emergent behaviors for mobile modular robots reconfiguring their geometric shape according to an environmental condition. Their study was based on coupling between a connectivity control algorithm for connected neighbor robots and nonlinear oscillators generating locomotion, a situation on which each robot locally interacted with their neighbor robots to generate a phase gradient. An adaptive transition technique (Kurabayashi et al., 2006) is presented to enable a swarm of robots to change formation by varying the phase gaps among artificial nonlinear oscillators. Authors (Balch & Hybinette, 2000) proposed a physics-based switching method without a leader, inspired by crystal generation processes. Each robot had several local attachment sites that are attracted to other robots. When the robots encountered an environmental constraint, they could avoid the obstacle depending on behavior-based rules combining the concept of an attractive and repulsive force. Authors (Esposito & Dunbar, 2006) presented a method combining of multiple potential functions that provided a feasible movement direction, under the condition that the robot state vector approached the potential minima. Even though those approaches were shown to be effective to obtain a desired local behavior of individual robots, they required an effort in adjusting specific parameters for successful adaptation.

As mentioned above, many studies have been proposed in recent years to develop such three coordination functions: self-configuration, flocking, and adapting to an environment. In order to make full use of them in unknown areas, it is required to develop a unified framework integrating these functions. The importance of such a framework can be explained as follows: It can 1) consistently control the robot swarm regardless of an environment and/or of task conditions and 2) provide a way for a robot swarm to minimize task overlapping, which can only be achieved by means of coordination. However, many related works cannot provide a generalized solution for integrating these functions. The reason for this is that many studies have focused on developing one coordinated function based on specific robot hardware such as sensors or motors, and did not consider a fully scalable approach under the assumption of a limited visibility range (communication or sensing). In contradiction to those related works, our approach considers the development of a unified framework which governs the adaptively self-organizing processes for a swarm of autonomous robots in the presence of an environmental uncertainty. More specifically, our approach is based on the hypothesis that the motion planning for robot swarms must be controlled under a proposed framework. Therefore, our main motivation is to develop a unified framework for robot swarms, and then to study decentralized control approaches for two functions: adaptive self-configuration and adaptive flocking.

3. DECENTRALIZED COORDINATION PROBLEM

3.1 Computational Robot Model and Notations

In this chapter, we consider a swarm of mobile robots denoted as $r_1,...,r_n$ in a two-dimensional plane (2-D). It is assumed that the initial distribu-

tion of all robots is arbitrary and that their positions are distinct. The robots are 1) autonomous, in the sense that they do not depend on any central control; 2) homogeneous, that is they all have the same computational system, in terms of hardware and software; 3) anonymous, meaning that an identifier is not initially assigned to each robot; 4) oblivious, since they are unable to remember any past actions or observations that would produce an inherently self-stabilizing property (Suzuki & Yamashita, 1999); 5) asynchronous, in that there is no global clock; 6) independent, since they are not connected with one another physically, so that an independent robot is not affected by any other another robot; 7) not allowed to directly communicate with each other (instead of the explicit information exchanging, robots are able to locally interact only by observation), and 8) have their own local coordinate system, but certainly do not reach an initial agreement on a common coordinate system for an overall swarm. Due to a limited observation range, each robot can detect the positions of other robots only within its line-of-sight. At each time, each robot can either be idle or execute its algorithm, repeating recursive activation at each cycle. At each cycle, each robot computes its movement positions (computation), based on the positions of other robots (observation), and moves toward the computed positions (motion).

Next, we introduce notations frequently used in this chapter. Let us consider a robot r_i with local coordinate system $\vec{r}_{x,i}$ and $\vec{r}_{y,i}$ as illustrated in Figure 1-(a). Here, $\vec{r}_{y,i}$ defines the vertical axis of r_i's coordinate system as its heading direction. It is straightforward to determine the horizontal axis $\vec{r}_{x,i}$ by rotating the vertical axis 90 degrees counterclockwise. The position of r_i is denoted as p_i. Note that p_i is (0,0) with respect to its local coordinates. The path is defined as the shortest straight line between p_i and p_j occupied by another robot r_j, and denoted as $\overline{p_i p_j}$. The distance

Figure 1. Illustrating definition and notations with respect to r_i: ((a) r_i's local coordinates, (b) observation set O_i))

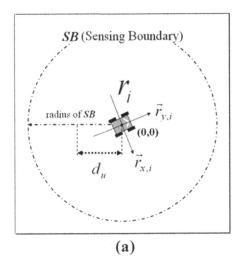

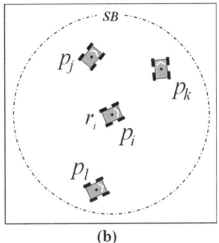

(a)   **(b)**

between p_i and p_j is denoted as $dist(p_i,p_j)$. In particular, we define a uniform distance d_u, the predefined desired distance between r_i and r_j. Moreover, Let $ang(\vec{m},\vec{n})$ denote the angle between two arbitrary vectors $\vec{m}$ and $\vec{n}$. As shown in Figure 1-(b), r_i detects the positions p_j, p_k, and p_l of other robots located within its sensing boundary SB, yielding a set of the positions $O_i(=\{p_j,p_k,p_l\})$

with respect to its local coordinates. When r_i selects two robots r_{s1} and r_{s2} within its SB, we call r_{s1} and r_{s2} the neighbors of r_i, and their position set $\{p_{s1},p_{s2}\}$ is denoted as N_i. Given p_i and N_i, a set of the three distinct positions with respect to r_i is called triangular configuration, denoted as T_i, namely $\{p_i,p_{s1},p_{s2}\}$ (see Figure 6-(b)). In particular, we define an equilateral configuration, denoted by E_i, as a configuration such that all distances be-

Figure 2. Illustration of the local interaction algorithm: ((a) target point computation, (b) moving to the target point)

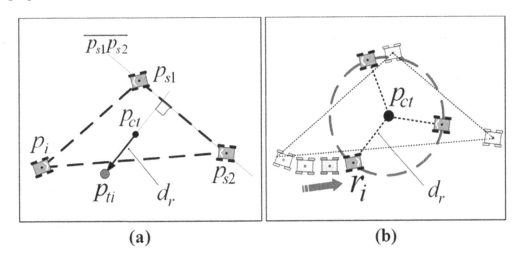

(a)   **(b)**

Figure 3. Illustration of the motion control: ((a) two controls: d_i and αi (b) equilateral triangle E_i)

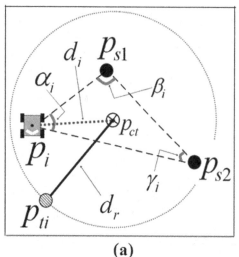
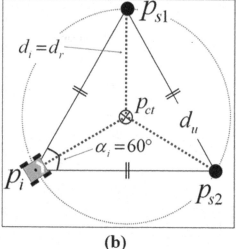

(a)　　　　　　　　　　　　　**(b)**

tween any two of p_i, p_{s1}, and p_{s2} of T_i are equal to d_u (see Figure 3-(b)).

3.2 Problem Definition

Based on the robot model above, we address the decentralized coordination problem for a mobile robot swarm as follows:

How can robot swarms be dispersed in a geographically constrained area in a decentralized coordinated manner?

To solve this problem, our research on dispersing mobile robot swarms considers three directions. First, a unified framework locally interacting with neighboring robots based on a geometrically triangular pattern is proposed to govern the adaptively self-organizing processes for large-scale robots in the presence of an environmental uncertainty. Next, employing the local interaction framework, robots can adaptively self-configure their swarm, and navigate in an unknown environment populated with obstacles. In the remainder of this chapter, therefore, the decentralized coordination problem is defined as the three following problems:

- **Problem-1** (*Local Interaction*): How can one enable r_i to maintain d_u with N_i toward eventually generating E_i?
- **Problem-2** (*Adaptive Self-Configuration*): Given $r_1,...,r_n$ with arbitrarily distinct positions in a 2-D plane, if the plane border of an area and/or of obstacles is detected by any of the robots, how can one enable the robots to self-configure their swarm composed of E_i while conforming to the border?
- **Problem-3** (*Adaptive Flocking*): Given $r_1,...,r_n$ with arbitrarily distinct positions in a 2-D plane, if the plane border of an area and/or of obstacles is detected by any of the robots, how can one enable the robots to navigate toward a goal while adapting their E_i to avoid obstacles?

The solution to the local interaction problem is to enable each robot to eventually converge toward each vertex of E_i (see Section 4). The adaptive self-configuration problem is decomposed into two sub-problems: self-configuration and uniform conformation, each of which is solved based on the local interaction (see Section 5). Moreover, based on the local interaction, we propose that the adaptive flocking can be achieved by solving the following three constituent sub-problems: maintenance, partition, and unification (see Section 6).

From the overall algorithmic standpoint, r_i executes the algorithms with the input (O_i and, if the border is detected, area features of the border) with respect to r_i's local coordinates each time, and outputs r_i's target position p_{ti}. r_i computes p_{ti} (computation), based on O_i (observation) and the area border, and moves toward the computed position (motion). Thus, r_i can either be idle or execute it algorithms, repeating the recursive activation.

3.3 Motivation

The following question that leads to mesh network formation naturally arises: which type of mesh should be used? The reasons for choosing an equilateral triangular mesh rather than other possible meshes are as follows. First, from the geo-metric standpoint of formation, we consider three different regular mesh formations composed of the triangular, rectangular, and hexagonal mesh, with the same side length d_u, respectively. Here we investigate the following features of individual formations: coverage area, coverage density (Iyengar et al., 2005), and the degree of connectivity (Stojmenovic & Wu, 2004). In the triangular mesh formation, the coverage area of each mesh is given by $\frac{3\sqrt{3}}{4}d_u^2$. The position of each robot is a common vertex of six meshes, and each mesh has three vertices. Thus, n robots may form approximately $2n$ meshes whose coverage area is approximately to $\frac{3\sqrt{3}}{2}nd_u^2$. In the rectangular mesh formation, the coverage area of each mesh is d_u^2. The position of each robot is a common vertex of four meshes, and each mesh has four vertices. Thus, n robots may form approximately n meshes whose coverage area is approximately to nd_u^2. Similarly, in the hexagonal mesh formation, the coverage area of each mesh

Table 1. Three network topologies

	triangle-based mesh	square-based mesh	hexagon-based mesh
pattern			
coverage area	$\frac{\sqrt{3}}{2}nd_u^2$	nd_u^2	$\frac{3\sqrt{3}}{4}nd_u^2$
coverage density	$\frac{1.2}{d_u^2}$	$\frac{1}{d_u^2}$	$\frac{0.78}{d_u^2}$
connectivity	6	4	3

is $\dfrac{3\sqrt{3}}{2}d_u^{\ 2}$. n robots may approximately form $\dfrac{n}{2}$ meshes whose coverage area is approximately to $\dfrac{3\sqrt{3}}{4}nd_u^{\ 2}$. Next, the coverage density can be obtained by dividing the number of robots by the coverage area. Thus, the coverage density of the triangular, rectangular, and hexagonal mesh formation is approximately $\dfrac{1.2}{d_u^{\ 2}}$, $\dfrac{1}{d_u^{\ 2}}$, and $\dfrac{0.77}{d_u^{\ 2}}$, respectively. Lastly, the degree of connectivity indicates the number of routing options each robot can have within the uniform distance d_u. Each robot has six, four, and three neighboring robots connected to itself, respectively. It is evident that the hexagonal mesh provides better coverage area than the rectangular and triangular mesh. In contrast, the triangular mesh provides more enhanced coverage density and degree of connectivity while trading off the coverage area. Secondly, among all the possible types of regular meshes, forming the equilateral triangle meshes based on the partially-connected topology can reduce both the computational load and the influence of other robots, due to the limited number of neigh-

bors, and lead to a high scalability. Thirdly, one equilateral triangular mesh network can be built by collecting these triangular patterns. In this chapter, we aim at developing distributed algorithms for dispersing equilateral triangular meshes towards an efficient and robust network covering an assigned area.

4. UNIFIED COORDINATION FRAMEWORK

4.1 Algorithm Description

Algorithm-1 consists of a function $\varphi_{\text{interaction}}$ whose arguments are p_i and N_i. As shown in Figure 2-(a), three robots are configured into T_i whose vertices are p_i, p_{s1}, and p_{s2}, respectively. At each time, r_i finds the centroid p_{ct} in T_i with respect to its local coordinates, and measures the angle ϕ_i between the line $\overline{p_i p_j}$ connecting two neighbors and $\dfrac{d_u}{\sqrt{3}}$. Using p_{ct} and ϕ_i, r_i calculates the target point p_{ti}, which is $\overline{p_{s1} p_{s2}}$, away from p_{ct} along the line d_r

Algorithm 1. Local interaction algorithm

Function: $\phi_{\text{interaction}}\left(p_{s1}, p_{s2}, p_i\right)$

{

    1: centroid p_{ct} calculation

$$\left(p_{ct,x}, p_{ct,y}\right) = \left(\frac{p_{i,x} + p_{s1,x} + p_{s2,x}}{3}, \frac{p_{i,y} + p_{s1,y} + p_{s2,y}}{3}\right)$$

    2: angle φ_i calculation

$$\cos^{-1}\left(\frac{\left(p_{s2,x} - p_{s1,x}\right)}{\sqrt{\left(p_{s2,x} - p_{s1,x}\right)^2 + \left(p_{s2,y} - p_{s1,y}\right)^2}}\right)$$

    3: next target p_{ti} calculation

$$\left(p_{ti,x}, p_{ti,y}\right) = \left(p_{ct,x} + \frac{d_u}{\sqrt{3}}\cos(\varphi_i + \frac{\pi}{2}), p_{ct,y} + \frac{d_u}{\sqrt{3}}\cos(\varphi_i + \frac{\pi}{2})\right)$$

}

perpendicular to $\dfrac{d_u}{\sqrt{3}}$. and then moves to p_{ti} as illustrated in Figure 2-(b). It is important to note that r_i may maintain d_u with its neighbors using Algorithm-1, each time forming an isosceles triangle. By repeatedly executing the algorithm, the three robots will be cooperatively configured into E_i.

4.2 Motion Control

Under Algorithm-1, we describe the motion control enabling the three neighboring robots to generate E_i after starting from an arbitrary T_i. As shown in Figure 3-(a), r_i and its neighbors r_{s1} and r_{s2} are configured into T_i. In T_i, internal angles $\angle p_{s1} p p_{s2}$, $\angle p_{s2} p_{s1} p_i$, and $\angle p_{s1} p_{s2} p_i$ are denoted as αi, βi and γi respectively. In Figure 3-(b), let us consider the circumscribed equilateral triangle $\triangle p_{i_p} s_{1p} s_2$ whose centroid is pc_i and radius is dr To obtain the desired Ei pt_i of ri can be modeled by the distance $di_{(t)}$ from pc_i and $\alpha i(_t)$. First, $di_{(t)}$ is controlled by the following equation:

$$di_{(t)} = -a(di_{(t)} - dr_{)} \tag{1}$$

where a is a positive constant and d_r represents the length $\alpha_i(t) = k'(\dfrac{\pi}{3} - \alpha_i(t))$ Indeed, the solution of (1) is $d_i(t) = |d_i(0)|e^{-at} + d_r$, which converges exponentially to d_r as t leads to infinity. Secondly, $\alpha i_{(t)}$ is controlled by the following equation:

$$\alpha i_{(t)} = k(\beta i_{(t)} + \gamma i_{(t)} - 2\alpha i(_t)) \tag{2}$$

where k is a positive constant. Since the total internal angle of a triangle is 180 degrees, (2) can be rewritten as:

$$\alpha_i(t) = |\alpha_i(0)|e^{-kt} + \dfrac{\pi}{3} \tag{3}$$

where k' is $3k$. Similarly, the solution of (3) is $\dfrac{\pi}{3}$ which converges exponentially to $f_{l,i} = \dfrac{1}{2}(d_i - d_r)^2 - \dfrac{1}{2}(\dfrac{\pi}{3} - \alpha_i)^2$ as t leads to infinity.

In order to show the convergence into any vertex of E_i, we will take advantage of Lyapunov's stability theory (Slotine & Li, 1991). Consider the following scalar function:

$$\alpha_i \neq \dfrac{\pi}{3}. \tag{4}$$

that is always positive definite except $d_i \neq d_r$ and

$$f_{l,i} = -a(d_i - d_r)^2 - k'(\dfrac{\pi}{3} - \alpha_i)^2$$ The derivative of the scalar function is given by:

$$\left[d_i \quad \dfrac{\pi}{3}\right]^{\mathrm{T}} \tag{5}$$

which is obtained using (1) and (3). Eq. (5) is negative definite. The scalar function $f_{l,i}$ is radially unbounded since it tends to infinity as $\|x\| \to \infty$, where x=$[d_i \alpha i_j$T Therefore, the equilibrium state $p_i p_e$, is asymptotically stable, implying that r_i reaches a vertex of E_i.

Finally, compared with the locally interacting frameworks (Shucker et al., 2008; McLurkin & Smith, 2004) mentioned in Section 2, our framework is based on this partially-connected topology for generating an equilateral triangle pattern with side length d_u, which can reduce the number of interacting robots in a given location. More importantly, compared with the computation of virtual spring forces to calculate force balance among neighboring robots, each robot utilizes only the relative distance information of other robots. More details about this convergence proof can be found in our previous work (Lee & Chong, 2008b).

5. ADAPTIVE SELF-CONFIGURATION

Self-configuration is a type of collective behavior of robot swarms, allowing them to disperse in an unknown environment. Self-configurable, low-cost robot swarms can be used for environmental or habitat monitoring by filling an area and establishing *ad hoc* or sensor networks with a group of robots (Cortes et al., 2004; Choset, 2001). It is necessary to develop locally distributed rules to coordinate the positions of robots under geographical constraints of the environment, as illustrated in Figure 4. For example, a liquid water molecule has a partially ordered structure in which hydrogen bonds are constantly formed and broken up, and its shape is an isosceles triangle in which the H-O-H bond angle is approximately 104.5 degrees. The liquid water molecules change their relative positions to conform to the shape of the container they occupy. If the molecules are thought of as mobile robots, each robot can configure its swarm into an area while conforming to the area border. Employing the concept motivated by the liquid water molecules, we attempt to exploit a set of distributed rules forming a mesh network of equilateral triangles over a geographically constrained plane.

5.1 Algorithm Description

Figure 5 shows the flowchart of the adaptive self-configuration algorithm. The input of the algorithm is O_i and area features, if any of robots detect the area border of a geographical constrained environment, with respect to r_i's local coordinates. The output is r_i's next target position p_{ti} using Algorithm-1. Detailly the process, if r_i detects the area border within r_i's SB, as illustrated in Figure 6-(a), r_i defines a point p_e projected from p_i onto the surface with the shortest distance d_e and then computes the tangent $e'(t)$ to the surface at p_e. (For convenience, l_e will be used instead of $e'(t)$). It is obvious that l_e is perpendicular to the vector

$$d_e \leq \frac{\sqrt{3}}{2} d_u.$$ named the surface direction. Let $A(l_e)$ denote the area between the area border and the line passing through p_i and parallel to l_e within SB. In order to determine whether r_i needs to approach the area border, r_i checks if no neighbors exist in $A(l_e)$ or if $\overline{p_i p_{s1}}$ If the condition is satis-

Figure 4. Illustrating the concept of uniformly adaptive self-configuration in an unknown environment

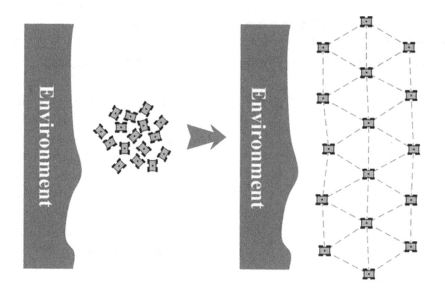

fied, r_i executes the *uniform conformation algorithm*, otherwise, it executes the *self-configuration algorithm*. After forming T_i with the selected N_i, r_i executes Algorithm-1.

First, we introduce the self-configuration algorithm. To form T_i, the first neighbor r_{s1} is simply selected as the one which is located the shortest distance away from r_i in O_i, as shown in Figure 7-(a). As shown in Figure 7-(b), the second neighbor r_{s2} is selected such that the perimeter of T_i is minimized as follows: $\mathbf{min}[dist(p_{s1},p_{s2})+dist(p_{s2},p_i)]$. Following the algorithm, r_i is able to select two neighbors and then form T_i. Secondly, we introduce a uniform conformation algorithm, incorporating a position of the area border as a virtual static robot into T_i. The first neighbor r_{s1} is simply selected as the one which is located the shortest distance away from r_i in O_i, as shown in Figure 6-(a). Next, as illustrated in Figure 6-(b), r_i computes the midpoint p_m of $d_e \leq \dfrac{\sqrt{3}}{2}d_u$, and then defines the virtual point p_v which is projected onto l_e from p_m. Here, p_v is considered as

p_{s2}, and N_i is defined as $\{p_{s1}, p_v\}$. Based on T_i with the selected N_i, r_i executes Algorithm-1 in order to conform to the area border. Note that the currently selected neighbors do not coincide with ones selected the following time due to the assumption of anonymity. Only using the current O_i, T_i is newly formed by r_i.

5.2 Simulation Results and Discussion

To validate the proposed algorithm, we performed two kinds of simulations demonstrating the convergence and robustness of the adaptive self-configuration algorithm. We arrange to SB range that is 2.5 times longer than d_u. Our algorithm terminates when all robots converge to the distance $d_u \pm 1\%$ from their two nearer neighbors.

First, Figure 8 shows that 100 robots have dispersed in an open area with a uniform spatial density from a dense or a sparse configuration. Each of the robots moved according to the adaptive self-configuration algorithm to form E_i. As a

Figure 5. Flowchart of the adaptive self-configuration algorithm

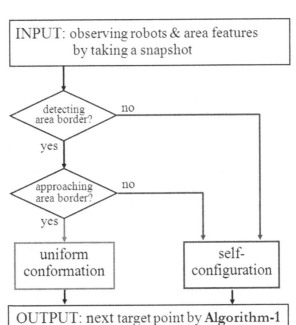

Figure 6. Illustration of the adaptive self-configuration algorithm: ((a) detecting the border, (b) approaching the border)

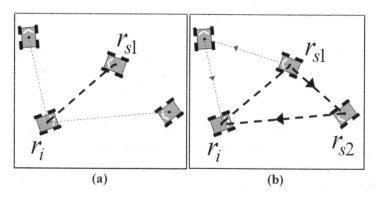

Figure 7. Illustration of the neighbor selection and of the triangular configuration: ((a) r_{s1} selection, (b) r_{s2} selection and triangular configuration T_i)

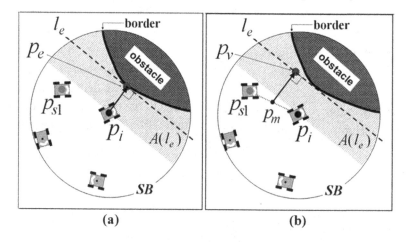

result, each robot in swarms could converge into each vertex of E_i while each time selecting its neighbors. Figure 9 illustrates the simulation on which 100 robots configured themselves into areas with a curved border. We can observe from the simulation that some of the robots detected and approached the border. If the area border is detected by any of the robots, they check whether no neighbors exist in $A(l_e)$ or for $\sum_{i=1}^{n} E_i$. As a result, robots that detect the area border to which they are to conform could incorporate the positions in its neighborhood into T_i. Judging from

the results, the robots could eventually converge into each vertex of E_i, conforming to the border.

Secondly, robustness is verified against sudden disappearances of robot members. Here, we supposed that the robots which disappeared went to another area due to an accidental failure. In task simulations, the total number of robots in a swarm is 100. The simulations begin with the completion state of the self-configuration. In Figure 10-(a), 10 robots disappeared without warning. Due to the loss of several robot members, several holes appeared in the deployment. Each robot checks the existence of the missing robots within *SB*.

Figure 8. Simulation results for the self-configuration of 100 robots over free open environments: ((a) dispersion from a dense configuration, (b) dispersion from a sparse configuration))

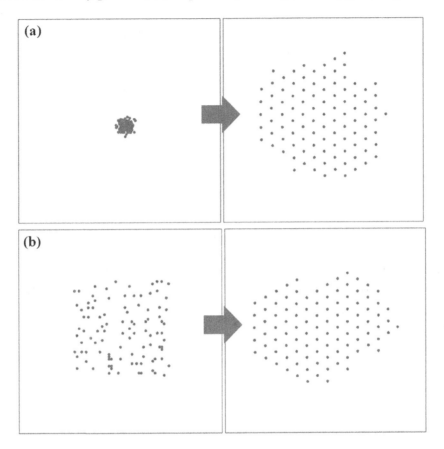

Using the algorithm, the robots started redeploying their swarm as they approached other robots. Figure 10-(b) presents the result of redeployment for 90 robots. In addition, we performed a simulation replacing the number of lost robots as shown in Figure 10-(c) is performed. Figure 10-(d) shows the result of redeployment after replacing the same number of missing robots. Similarly, Figure 11 shows the simulation results for robustness after having replaced 5 robots in an environment with a curved surface. The results indicate that the adaptive self-configuration algorithm is effective in improving the robustness of the self-configuration in spite of environmental uncertainty.

The adaptive self-configuration is regarded as a first step toward real-world implementations of a networking infrastructure under the scenario of AmI environments. Similar visions based on self-configuration are found in many related works as described in Section 2. In this chapter, we intended to adapt the minimal robot model by dropping assumptions such as global knowledge (identification number and common coordinate system), memory, and direct communication as mentioned early. To achieve a successful configuration based on the minimal model, the algorithm relies on the fact that robots can exactly sense the positions of neighbor robots in close proximity, as is the case with sonar reading (Lee & Chong, 2009a) or infrared sensor (IR) reading (Lee et al., 2009c). Based on a partially-connected mesh topology (Ghosh et al., 2005; Fowler, 2001), our self-configuration approach can also take advantage of the redundancy provided by a

Figure 9. Simulation results for the adaptive self-configuration of 100 robots over a curved surface

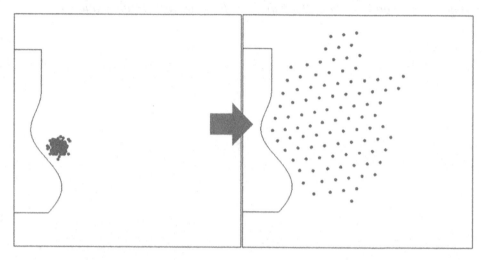

Figure 10. Simulation results for the robustness against the loss of 10 robots: ((a) loss of 10 robots, (b) redeployment with 90 robots, (c) replacement of 10 robots, (d) redeployment with 100 robots)

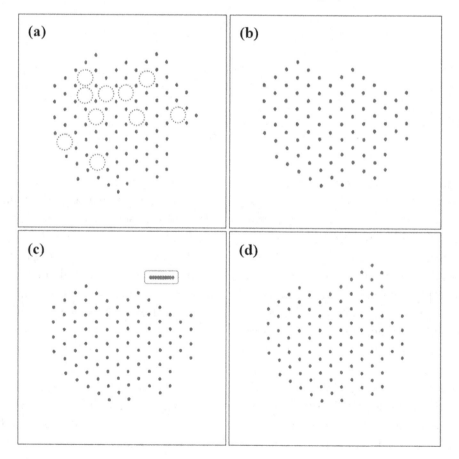

fully connected topology without the expense and complexity of networking processes. In (Shucker et al., 2008; McLurkin & Smith, 2004), partial graph pairs of robots were proposed that exert virtual forces on each other when their connection is part of the graph. Compared with those approaches, the motivation of our research was to construct uniformly spaced equilateral triangles that can more efficiently save the number of neighboring robots in a given location, and to configure a more energy-efficient routing protocol than for randomly deployed robots (Gurumohan & Hui, 2003), while conforming to the border of a constrained area or of obstacles. More details about this adaptive self-configuration algorithm and its convergence proof can be found in our

previous work (Lee & Chong, 2008b), where extensive simulation results were provided to show how a large-scale swarm of robots converge into E_i.

Additionally, based on the self-configuration algorithm, two kinds of research results regarding deployment have been recently presented. First, we proposed the adaptive triangular mesh generation algorithm (Lee et al., 2009d) that enables robot swarms to generate triangular meshes of various sizes, adapting to changing environmental conditions. The main objective was to provide robots with adaptive deployment capabilities to explore an area and to cover the area of interest more efficiently with variable triangular meshes according to sensed area conditions. The properties

Figure 11. Simulation results for the robustness against loss of 5 robots over a curved surface: ((a) loss of 5 robots, (b) redeployment with 95 robots, (c) replacement of 5 robots, (d) redeployment with 100 robots)

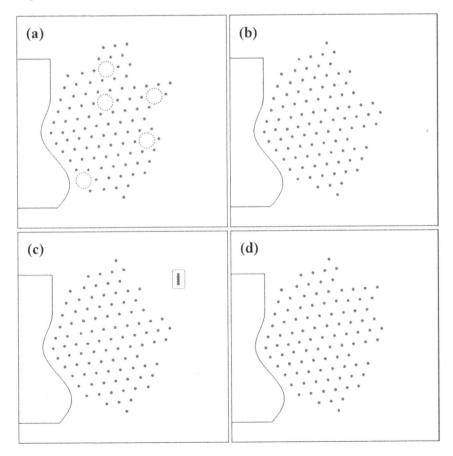

of the proposed algorithm were mathematically explained, and the convergence was proven. We also demonstrated through extensive simulations that a large-scale swarm of robots can establish a triangular mesh network, adapting to varying densities in their self-configured network. Secondly, we addressed the concentric circles generation problem (Lee et al., 2009b) of a mobile robot swarm generating geometric shapes to build wireless *ad hoc* surveillance sensor networks. Robot swarms with limited observation capabilities are required to form different shapes under different task conditions. To solve this problem, our approach had *n* robots generate a circumscribed circle of a regular *n*-polygon based on local interactions with neighboring robots. The approach also enabled a large robot swarm to form concentric circles through consensus. Through this problem, our main objective was to provide a distributed coordination solution in which robots eventually generate circumscribed circles around regular polygons while making the centroids of individual circles coincide. This in turn may shed light on the implementation of other geometric formations with symmetry. We mathematically demonstrated convergence, confirming the feasibility through extensive simulation.

6. ADAPTIVE FLOCKING

This section introduces the adaptive flocking approach that enables robot swarms to split into multiple groups or merge into a single group while navigating through unknown territory. Adaptive flocking is motivated by the observation that schools of fish exhibit a certain group behavior (Wilson, 1976). For instance, when a school of fish is faced with an obstacle, it can avoid collision by being split into a plurality of smaller groups that can be merged after they pass around the obstacle. It is known that local geometric shapes of a school of tuna form a diamond shape (Stocker, 1999), whereby tunas exhibit the following schooling

behaviors: maintenance, partition, and unification. Based on the observation of these habits of schooling fish, we propose several group behavior rules that enable a large swarm of autonomous mobile robots to flock in an unknown environment. The adaptive flocking algorithm provides a way for robot swarms to self-adjust their shape and size according to the environmental condition. From a practical standpoint, the swarm flocking can be considered as a robust *ad hoc* mobile networking model whose connectivity must be maintained in a cluttered environment, as illustrated in Figure 12.

6.1 Algorithm Description

As illustrated in Figure 13, the input for the adaptive flocking algorithm each time is O_i and area features, if any of the robots detect the area border of a geographical constrained environment, with respect to r_i's local coordinates. The output is the target position p_{ti} of each robot. When r_i detects the constraint within its SB, it executes the *partition function* to adapt its position to the constraint. When r_i faces no constraint, but observes other swarms, it executes the *unification function*. Otherwise, it basically executes the *maintenance function* while navigating toward a goal.

The first sub-solution in adaptive flocking is to maintain E_i with neighboring robots while navigating. A swarm is required to maintain a multitude of equilateral triangle meshes, denoted by $\vec{G}$, As illustrated in Figure 14-(a), r_i adjusts $A(\vec{G})$ termed the goal direction, with respect to r_i's local coordinates. Here, let $A(\vec{G})$. denote the area of the goal direction defined within r_i's SB. Next, r_i checks whether there exists a neighbor in $A(\vec{G})$, Within $\vec{f}, r_i$ selects the first neighbor r_{s1} located the shortest distance away from p_i that yields p_{s1}. As shown in Figure 14-(b), the second neighbor r_{s2} is selected such that the total distance from p_{s1} to p_i passing through p_{s2} is minimized. As a result, p_{ti} can be obtained by $\varphi_{interaction}$ in Algorithm 1.

Figure 12. Illustrating concept of adaptive flocking

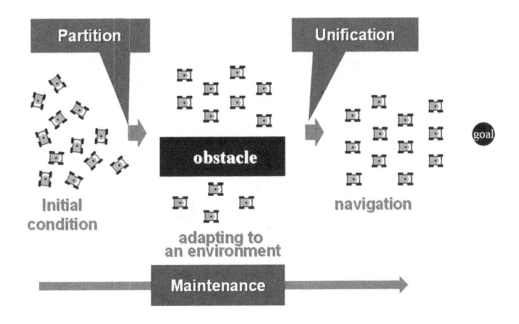

Figure 13. Flowchart of the adaptive flocking algorithm

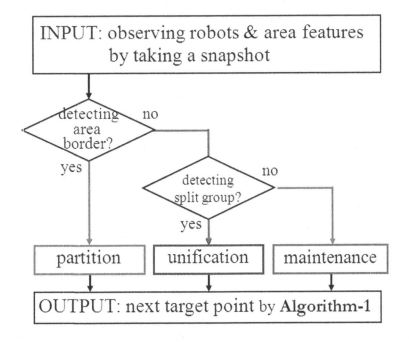

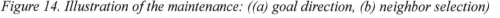

Figure 14. Illustration of the maintenance: ((a) goal direction, (b) neighbor selection)

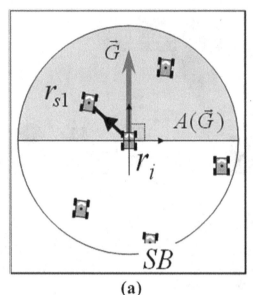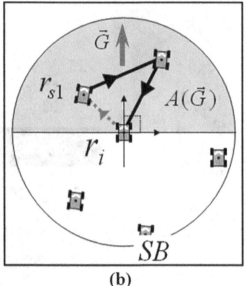

(a)   (b)

Secondly, when a swarm of robots detects the area border of an obstacle in its path, r_i is required to determine its direction toward the goal which avoiding the obstacle. Here, r_i determines its direction by using the relative degree of attraction (Halliday et al., 1997) of the passageway, termed the favorite vector $\left|\vec{f}_j\right| = \left|\dfrac{w_j}{d_j^2}\right|$. whose magnitude is given by $\{\vec{f}_j \mid 1 \leq j \leq n\}$, In Figure 15-(a), s_j denotes the passageway with width w_j, and d_j denotes the distance between the center of w_j and p_i. Note that if r_i cannot exactly measure w_j beyond its SB, w_j may be shortened. Now, the relative degree of attraction of the passageways can be represented by a set of favorite vectors $\vec{f}_j$ and then r_i selects the maximum magnitude of $\left|\vec{f}_j\right|_{\max}$. denoted as $A(\vec{f}_{j\max})$ As shown in Figure 15-(b), r_i defines a maximum favorite area $\left|\vec{f}_j\right|_{\max}$ based on the direction of $A(\vec{f}_{j\max})$, within its SB. Within $ang(\vec{r}_{y,i}, \overline{p_i p_{ref}})$, r_i selects r_{s1} located the

shortest distance away from itself to define p_{s1}. Next, r_{s2} is selected such that the total distance from p_{s1} to p_i passing through p_{s2} is minimized. As a result, p_{ti} can be obtained by $\varphi_{\text{interaction}}$ in Algorithm 1.

The third sub-solution in the adaptive flocking is to enable the multiple swarms in close proximity to merge into a single swarm. Let P_u denote the set of robot positions located within the range of d_u. As shown in Figure 16-(a), r_i selects the reference neighbor p_{ref} in P_u such that the value of $\vec{r}_{y,i}$ where $\overline{p_i p_{ref}}$ indicates r_i's heading direction, is minimized. Then, r_i checks if any neighbors exist in the area obtained by rotating $\overline{p_i p_{ref}}$ 60 degrees clockwise. If there exists one, r_i checks the next neighbor by sweeping another 60 degrees clockwise. r_i continues to check until it does not find any robot, then the last neighbor is defined as p_{rn}. Similarly, r_i attempts to find neighbors by rotating $\overline{p_i p_{rn}}$ counterclockwise, and to locate the last neighbor p_{ln}. The rest area $A(U)$ is defined as the area between $\overline{p_i p_{ln}}$ and $\overline{p_i p_{ln}}$ in SB, where no element of P_u exists. As illustrated in Figure

Figure 15. Illustration of the partition: ((a) favorite vector, (b) neighbor selection)

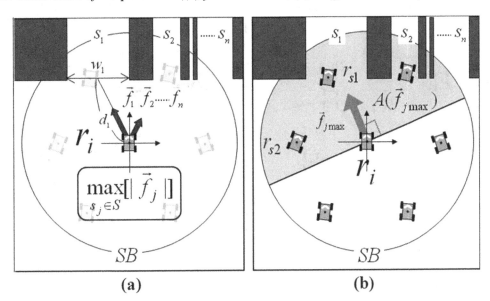

(a)   (b)

16-(b), r_i selects r_{s1} located the shortest distance away from p_i in $A(U)$ as p_{s1}. In $A(U)$, p_{s2} of r_{s2} is defined such that the total distance from p_{s1} to p_i can be minimized through either p_{rn} or p_{ln}. Finally, based on T_i with the selected N_i, p_{ti} can be obtained by using $\varphi_{\text{interaction}}$.

6.2 Simulation Results and Discussion

We present results of simulations that tested the validity of our adaptive flocking algorithm. We consider that a swarm of robots attempts to navigate toward a stationary goal while exploring and adapting to the environment. In such an application scenario, the goal is assumed to be either a light or odor source that can be easily detected by each robot. We arrange to *SB* range that is 3.5 times longer than d_u. The first simulation demonstrates how a swarm of robots adaptively flocks in an environment populated with obstacles. In Figure 17-(a), the swarm navigates toward the goal located at the upper center point. On the way to the goal, some of the robots detect an obstacle that forces the swarm to split into two groups, as

shown in Figure 17-(b). The rest of the robots just follow their neighbors moving ahead. After being split into two swarms, each swarm maintains the same geometric configuration while navigating, as shown in Figure 17-(c). Note that the robots that could not identify the obstacle follow the moving direction of proceeding robots. Figures 17-(d) and (e) show that two swarms merge and/ or split again into smaller swarms due to other obstacles. In Figure 17-(f), the robots have successfully passed through the obstacles. In the second simulation, the proposed algorithm is evaluated in a different environment with narrow passageways, as illustrated in Figure 18 under the same initial conditions. As expected, the observed behavior of the swarm differs from one situation to the other, as can be seen in Figures 17 and 18.

In this section, we presented the adaptive flocking algorithm enabling a swarm of autonomous mobile robots to navigate toward a goal while adapting to a complex environment. Through local interactions by observing the positions of neighboring robots, each robot could maintain a uniform distance with respect to its neighbors, and adapt to the heading direction and the geo-

Figure 16. Illustration of the unification: ((a) unification area, (b) neighbor selection)

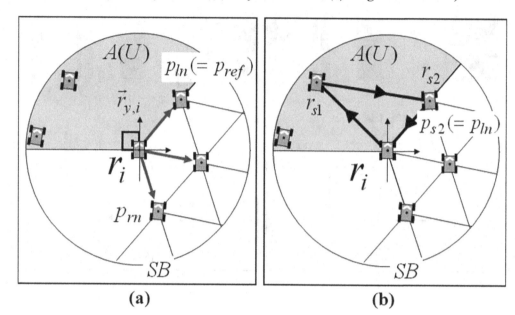

(a)   (b)

Figure 17. Simulation results of adaptive flocking under an unknown environment populated with obstacles

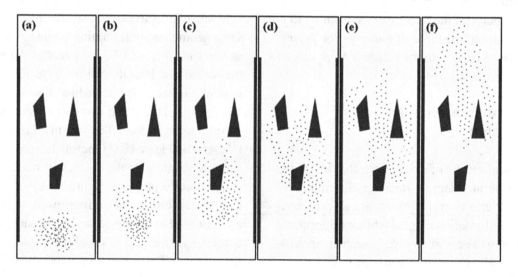

metric shape. Furthermore, we verified the effectiveness of the proposed strategy using our in-house simulator. The simulation results clearly demonstrate that the proposed algorithm is a simple yet robust approach to autonomous navigation of robot swarms in a changing, cluttered environment. In practice, because our robot

model was very simple, this algorithm was easily implementable on a wide variety of resource-constrained mobile robots and platforms. More details about this adaptive flocking algorithm and its convergence proof can be found in our previous work (Lee & Chong, 2008a). We expect that the adaptive flocking algorithm can be effec-

Figure 18. Simulation results of adaptive flocking under an unknown environment with narrow passageways

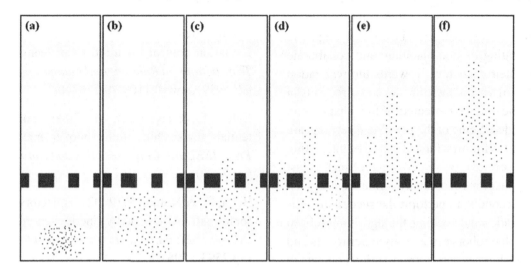

tively applied to mobile robotic sensor networks for surveillance missions or for capturing toxic and hazardous substances.

7. SUMMARY AND FUTURE DIRECTION

In this chapter, in view of constructing AmI environments, we presented the coordination problems for a swarm of mobile robotic sensors in a totally decentralized fashion, and from a mainly algorithmic perspective. The following three problems have been analyzed in detail. First, the unified geometric framework was proposed to enable three neighboring robots to form an equilateral triangle. This framework could provide a unique solution that allows each one to converge into one vertex of an equilateral triangle. Secondly, we addressed the adaptive self-configuration problem of mobile robot swarms in a geographically constrained environment. As the approach to solve the problem, robots were locally allowed to select dynamically and interact with two neighbors within their limited observation range. Under the adaptive self-configuration algorithm, collecting the local behaviors, the swarm could be self-deployed with uniform spatial density while conforming to the area border of a geographical constraint. Moreover, the algorithm also displayed a robust capability of swarm maintenance in spite of the loss of several robots. The simulation results show that the adaptive self-configuration is a simple and efficient approach to the deployment of mobile robotic sensors under condition of environmental uncertainty. Thirdly, we introduced the adaptive flocking algorithm, enabling a swarm of autonomous mobile robots to navigate toward a goal while adapting to an unknown environment. Based on the geometric framework, the robots could maintain a uniform distance between individual robots and adjust their geometric swarm shape while adjusting the heading direction of individual robots. We verified the effectiveness of the algorithm using the in-house simulator. The simulation results clearly demonstrate that the proposed flocking is a simple and efficient approach to autonomous navigation for robot swarms in a changing, cluttered environment. In practice, these proposed approaches are expected to be used in applications such as habitat or environmental monitoring, odor localization, search-and-rescue, and mobile networking.

The field of AmI is featured as sensors-rich environments supported by algorithmic solutions capable of fully leveraging its capacity and of providing distributed services. This requires robotic sensors to coordinate their behaviors, cooperate in generating motion planning, and collaborate during their execution. Towards the realization of the AmI vision, there are several directions in which the studies presented in this chapter can be extended. Some of the possible directions are discussed as follows: First, it would be interesting to study several types of performance such as fast convergence time or optimal distribution in a given space. Secondly, to perform the successful self-configuration and flocking, the algorithms rely on the fact that robots can identify other robots and distinguish other robots from various objects in a given environment within their limited sensing boundary. This engineering issue, which leads to important points that must be addressed before the algorithm can be applicable to practical situations, is left for future work. Thirdly, with respect to solving decentralized coordination problems, we plan to investigate different existences of the robotic model, for instance, if one considers utilizing interaction through communication. As like in the implementation problems of observation, communications in physical robotics will not be free or reliable, and can be constrained by limited bandwidth, range, and interferences. Moreover, it is necessary for robotic sensors to use a previous knowledge such as an individual identification number or global coordinates. Therefore, we will pursue our future work as a starting point for understanding the relation between the capabilities and the difficulties of several communication models. These future directions, as mentioned above, are a nontrivial challenge for the realization of the AmI vision.

REFERENCES

Balch, T., & Arkin, R. C. (1998). Behavior-based formation control for multi-robot teams. *IEEE Transactions on Robotics and Automation, 14*(6), 926–939. .doi:10.1109/70.736776

Balch, T., & Hybinette, M. (2000). Social potentials for scalable multi-robot formations. In *Proc. IEEE Int. Conf. Robotics and Automation,* (pp.73-80).

Belta, C., & Kumar, V. (2002). Trajectory design for formations of robots by kinetic energy shaping. In *Proc. IEEE Int. Conf. Robotics and Automation,* (pp.2593-2598).

Choset, H. (2001). Coverage for robotics - a survey of recent results. *Annals of Mathematics and Artificial Intelligence, 31*(1-4), 113–126. .doi:10.1023/A:1016639210559

Cohen, R., & Peleg, D. (2006). Local algorithms for autonomous robots systems. In P. Flocchini & L. Gasieniec (Eds.), Structural Information and Communication Complexity (LNCS), *4056* (pp. 29–43). New York: Springer. doi:10.1007/11780823_4

Cortes, J., Martinez, S., Karatas, T., & Bullo, F. (2004). Coverage control for mobile sensing networks. *IEEE Transactions on Robotics and Automation, 20*(2), 243–255. .doi:10.1109/TRA.2004.824698

Dressler, F. (2007). *Self-organization in sensor and actuator networks.* John Wiley & Sons. doi:10.1002/9780470724460

Egerstedt, M., & Hu, X. (2001). Formation constrained multi-agent control. *IEEE Transactions on Robotics and Automation, 17*(6), 947–951. .doi:10.1109/70.976029

Esposito, J. M., & Dunbar, T. W. (2006). Maintaining wireless connectivity constraints for swarms in the presence of obstacles. In *Proc. IEEE Int. Conf. Robotics and Automation,* (pp.946-951).

Fowler, T. (2001). Mesh networks for broadband access. *IEE Review, 47*(1), 17–22. .doi:10.1049/ir:20010108

Fujibayashi, K., Murata, S., Sugawara, K., & Yamamura, M. (2002). Self-organizing formation algorithm for active elements. In *Proc. 21st IEEE Sym. Reliable Distributed Systems*, (pp.416-421).

Gervasi, V., & Prencipe, G. (2003). Coordination without communication: the case of the flocking problem. *Discrete Applied Mathematics, 143*(3), 203–223.

Ghosh, S., Basu, K., & Das, S. K. (2005). An architecture for next-generation radio access networks. *IEEE Network, 19*(5), 35–42. .doi:10.1109/MNET.2005.1509950

Gurumohan, P. C., & Hui, J. (2003). Topology design for free space optical networks. In *Proc. 12th Int. Conf. Computer Communications and Networks*, (pp.576-579).

Halliday, D., Resnick, R., & Walker, J. (1997). *Fundamentals of physics* (5th ed.). Wiley.

Heo, N., & Varshney, P. K. (2003). A distributed self spreading algorithm for mobile wireless sensor networks. In *Proc. IEEE Wireless Communication and Networking Conf.*, (pp.1597-1602).

Howard, A., Mataric, M. J., & Sukhatme, G. S. (2002). Mobile sensor network deployment using potential fields: a distributed, scalable solution to the area coverage problem. In *Proc. 6th Int. Sym. Distributed Autonomous Robotic Systems*, (pp.299-308).

Ikemoto, Y., Hasegawa, Y., Fukuda, T., & Matsuda, K. (2005). Graduated spatial pattern formation of robot group. *Information Sciences, 171*(4), 431–445. .doi:10.1016/j.ins.2004.09.013

Ishiguro, A., & Kawakatsu, T. (2006). Self-assembly through the interplay between control and mechanical systems. In *Proc. IEEE/RSJ Int. Conf. Intelligent Robots and Systems*, (pp.631-638).

Iyengar, R., Kar, K., & Banerjee, S. (2005). Low-coordination topologies for redundancy in sensor networks. In *Proc. 6th ACM Int. Sym. Mobile Ad Hoc Networking and Computing*, (pp.332-342).

Kerr, W., Spear, D., Spear, W., & Thayer, D. (2005). Two formal gas models for multi-agent sweeping and obstacle avoidance. In Carbonell, J. G., & Siekmann, J. (Eds.), *Formal Approaches to Agent-Based Systems (LNCS), 3228* (pp. 111–130). Springer.

Kurabayashi, D., Okita, K., & Funato, T. (2006). Obstacle avoidance of a mobile robot group a nonlinear oscillator network. In *Proc. IEEE/RSJ Int. Conf. Intelligent Robots and Systems*, (pp.186-191).

Lee, G., & Chong, N. Y. (2008a). Adaptive flocking of robot swarms: algorithms and properties. *IEICE Transactions on Communications. E (Norwalk, Conn.), 91-B*(9), 2848–2855.

Lee, G., & Chong, N. Y. (2008b). A geometric approach to deploying robot swarms. *Annals of Mathematics and Artificial Intelligence, 52*(2-4), 257–280. .doi:10.1007/s10472-009-9125-x

Lee, G., & Chong, N. Y. (2009a). Decentralized formation control for small-scale robot teams with anonymity. *Mechatronics, 19*(1), 85–105. .doi:10.1016/j.mechatronics.2008.06.005

Lee, G., Chong, N. Y., & Christensen, H. (2009d). Adaptive triangular mesh generation of self-configuring robot swarms. In *Proc. IEEE Int. Conf. Robotics and Automation*, (pp.2737-2742).

Lee, G., Yoon, S., Chong, N. Y., & Christensen, H. (2009b). A mobile sensor network forming concentric circles through local interaction and consensus building. *Journal of Robotics and Mechatronics, 21*(4), 469–477.

Lee, G., Yoon, S., Chong, N. Y., & Christensen, H. (2009c). Self-configuring robot swarms with dual rotating infrared sensors. In *Proc. IEEE/RSJ Int. Conf. Intelligent Robots and Systems*, (pp.4357-4362).

Martison, E., & Payton, D. (2005). Lattice formation in mobile autonomous sensor arrays. In E. Sahin & W. M. Spears (Eds.), Swarm Robotics (LNCS), *3342* (pp. 98–111). New York: Springer. doi:10.1007/978-3-540-30552-1_9

McLurkin, J., & Smith, J. (2004). Distributed algorithms for dispersion in indoor environments using a swarm of autonomous mobile robots. In *Proc. 7th Int. Sym. Distributed Autonomous Robotic Systems*, (pp831-840).

Reif, J., & Wang, H. (1999). Social potential fields: a distributed behavioral control for autonomous robots. *Robotics and Autonomous Systems*, *27*(3), 171–194. .doi:10.1016/S0921-8890(99)00004-4

Remagnino, P. & Foresti, G.L. (2005). Ambient intelligence: a new multidisciplinary paradigm. *IEEE Transactions on Systems, Man, and Cybernetics - Part A*, 35(1), 1-6.

Reynolds, C. W. (1987). Flocks, herds, and schools: a distributed behavioral model. *Computer Graphics*, *21*(4), 25–34. .doi:10.1145/37402.37406

Sahin, E. (2005). Swarm robotics: from sources of inspiration to domains of application. In E. Sahin & W. M. Spears (Eds.), Swarm Robotics (LNCS), *3342* (pp. 10–20). New York: Springer. doi:10.1007/978-3-540-30552-1_2

Shimizu, M., Mori, T., & Ishiguro, A. (2006). A development of a modular robot that enables adaptive reconfiguration. In *Proc. IEEE/RSJ Int. Conf. Intelligent Robots and Systems*, (pp.174-179).

Shucker, B., Murphey, T. D., & Bennett, J. K. (2008). Convergence-preserving switching for topology-dependent decentralized systems. *IEEE Transactions on Robotics*, *24*(6), 1405–1415. .doi:10.1109/TRO.2008.2007940

Slotine, J. E., & Li, W. (1991). *Applied nonlinear control*. Prentice-Hall.

Spears, W., & Gordon, D. (1999). Using artificial physics to control agents. In *Proc. IEEE Int. Conf. Information, Intelligence, and Systems*, (pp.281-288).

Spears, W., Spears, D., Hamann, J., & Heil, R. (2004). Distributed, physics-based control of swarms of vehicles. *Autonomous Robots*, *17*(2–3), 137–162. .doi:10.1023/B:AURO.0000033970.96785.f2

Stocker, S. (1999). Models for tuna school formation. *Mathematical Biosciences*, *156*, 167–190. .doi:10.1016/S0025-5564(98)10065-2

Stojmenovic, I., & Wu, J. (2004). Broadcasting and activity scheduling in ad hoc networks. In S. Basagini, M. Conti, S. Giordano, & I. Stojmenovic (Eds.), *Mobile ad hoc networking* (pp. 205–229). IEEE press. doi:10.1002/0471656895.ch7

Suzuki, I., & Yamashita, M. (1999). Distributed anonymous mobile robots: formation of geometric patterns. *SIAM Journal on Computing*, *28*(4), 1347–1363. .doi:10.1137/S009753979628292X

Tanner, H., Pappas, G., & Kumar, V. (2004). Leader-to-formation stability. *IEEE Transactions on Robotics and Automation*, *20*(3), 443–455. .doi:10.1109/TRA.2004.825275

Wang, G.-L., Cao, G., & Porta, T. L. (2004). Movement-assisted sensor deployment. In *Proc. IEEE Infocom Conf.*, (pp.2469-2479).

Werger, B., & Mataric, M. J. (2001). From insect to internet: situated control for networked robot teams. *Annals of Mathematics and Artificial Intelligence*, *31*(1-4), 173–198. .doi:10.1023/A:1016650101473

Wilson, E. O. (1976). *Sociobiology: the new synthesis*. Harvard University Press.

Zarzhitsky, D., Spears, D., & Spears, W. (2005). Distributed robotics approach to chemical plume tracing. In *Proc. IEEE/RSJ Int. Conf. Intelligent Robots and Systems*, (pp.4034- 4039).

Zecca1, G., Couderc, P., Banatre, M., & Beraldi, R. (2009). Swarm robot synchronization using RFID tags. In *Proc. IEEE Int. Conf. Pervasive Computing and Communications*, (pp.1-4).

Zheng, Y. F., & Chen, W. (2007). Mobile robot team forming for crystallization of protein. *Autonomous Robots*, *23*(1), 69–78. .doi:10.1007/s10514-007-9031-1

Zou, Y., & Chakrabarty, K. (2003). Sensor deployment and target localization based on virtual forces. In *Proc. IEEE Infocom Conf.*, (pp.1293-1303).

KEY TERMS AND DEFINITIONS

Adaptive Flocking: a coordination function enabling mobile robotic sensors to move toward a certain goal as a single unit adapting their shape to environmental constraints.

Adaptive Self-Configuration: a coordination function enabling mobile robotic sensors to build a network adapting to environmental constraints.

Decentralized Coordination: a control strategy for large numbers of autonomous robots, allowing them to control their behavior based only on individual decisions without a supervised specific robot or operator.

Equilateral Triangular Mesh Pattern: a geometric shape composed of equilateral triangular meshes formed by mobile robotic sensors.

Mobile Robotic Sensors: wireless sensors with autonomous mobility capabilities or mounted on autonomous robots.

Self-Organizable Topology Control: a network organization technique by which mobile robotic sensors interact with each other in order to realize a desired network topology.

Chapter 27
Pervasive Computing for Efficient Energy

Mária Bieliková
Slovak University of Technology in Bratislava, Slovakia

Marián Hönsch
Slovak University of Technology in Bratislava, Slovakia

Michal Kompan
Slovak University of Technology in Bratislava, Slovakia

Jakub Šimko
Slovak University of Technology in Bratislava, Slovakia

Dušan Zeleník
Slovak University of Technology in Bratislava, Slovakia

ABSTRACT

Increasing energy consumption requires our attention. Resources are exhaustible, so building new power plants is not the only solution. Since residential expenditure is of major parts of overall consumption, concept of intelligent household has potential to participate on energy usage optimization. In this chapter, we concentrate on software methods, which based on inputs gained from an environment monitor, analyze and consequently reduce non-effective energy consumption. We gave a shape to this concept by description of real prototype system called ECM (Energy Consumption Manager). Besides active energy reduction, the ECM system also has an educative function. User-system interaction is designed to teach the user how to use (electric, in case of our prototype) energy effectively. Methods for the analysis are based on artificial intelligence and information systems fields (neural networks, clustering algorithms, rule-based systems, personalization and adaptation of user interface). The system goes further and gains more effectiveness by exchange of data, related to consumption and appliance behaviour, between households.

DOI: 10.4018/978-1-61692-857-5.ch027

INTRODUCTION

Man and Comfort

Although many inventions in the world were made due to war effort or high tax policies, technical progress is basically powered by mankind's laziness. Through the whole history, there were always smart people, who invested their time into thinking, what would make their work easier or will even substitute them in particular activities. Neolithic agricultural worker used primitive tools made of bones and had indeed a hard time cultivating his field. But inventions such as bronze and iron smelting and forging, draught cattle, crop rotation or watering system meant significant difference in effort/result ratio. With such trends, we can imagine that in not-so-far future, men will no longer be necessary as manual workers. They will act just as administrators of fully automated systems.

Such trends are driven by people's demands for more comfort and less manual work. In the past, it took quite long time for inventions to be developed. But today, we have a powerful weapon: *computers*. We could hardly find a similar thing; computers are versatile and already brought revolution in easing human life. People use computers to fulfil their tasks efficiently. Now it is time for next step: leaving the computers to work independently from human.

Non Effective Energy Consumption: Phenomenon and Problem

Talking about inventions, electricity is also one that pushed world forward significantly and is used widely. Unfortunately, electric energy production is starting to backfire: sources are exhaustible and most of energy is produced with techniques that somehow damage the environment or are potentially dangerous. Thus, we are logically making a senseless hazard, if energy produced is not used efficiently or completely wasted. The more power we use, the more responsible we become. Based

on that, we are pointing towards most developed parts of the world, which are the greatest energy consumers.

We use electric power for different purposes in industry, traffic, offices or households. In this chapter, we focus closely on the consumption of last two groups. A non-efficient consumption can be defined as a usage of energy that does not produce sufficient (if any) return value. In the field of households and offices, energy usage consists of a large and heterogeneous set of cases (heating, lightning, various appliances used for different purposes). Each case has its own set of bad practices that cause non-efficient consumption. Each bad practice can be somehow eliminated, but this requires effort (e.g., a change of habits, introducing a new technology). Large number of cases means a lot of effort. Sometimes the amount of effort seems to be even greater than resulting benefit. Fortunately, besides environmental aspect, people are motivated financially to reduce their consumption.

In major cases like heating (which consumes greatest part of energy in households) or lightning certain solutions have been developed to reduce the energy wasting (e.g., isolations, fluorescent tubes). But there is also a group of "minor" cases of non-efficient energy usage, each often considered "insignificant", but together worth of thinking. This group includes:

- *Old and malfunctioning devices.* Some appliances may do additional consumption when they become older: they start to have various defects (like broken tubing and ice expansion in fridge or water stone in wash machine) or just become outdated (by newer, more effective appliances). Such problems are usually solvable (by repair or buying a new appliance). The problem is their invisibility and ignorance. Major defects can be possibly indicated on energy bill, but minor have a good chance to stay unseen as this kind of problems obviously do not arise suddenly, but aggravate little by little.

- *Standby.* What is the price for device "readiness"? Almost 10% of residential energy consumption is used for silenced-but-prepared state of some appliances, known as standby mode[1]. "Standby modes" use various amounts of electric power (from fractions to tens of watts in extreme cases) and are sometimes used for computations (like watch). But in most cases standby is just to provide more comfort for the user (automatic or remote switching). In fact, most of standby does not provide any other value and are simply wasting energy. Solution is to dutifully turn such appliances off when they are not used. Unfortunately, this requires lot of user's strong will not to forget.
- *"Forgotten appliances".* Empty rooms with turned on lights, TV, computer or iron: non-efficient energy usage, caused by man's business. Like in case of standby, "solution" is simple: turn everything off after action. Easy to say, hard to execute.

Spare Energy, Preserve Comfort

It seems that problems described above need just a little user self denial and everything will be all right. But most users will rather restrict their budget than themselves. But information technologies can help instead; turning off the power supply may be done by simple switches controlled via a computer. What is not so simple is to determine the moment when to intervene. For example, the room with TV turned on may be empty due to toilet break. Appliance with standby must be ready to interact quickly on proper time. Thus, intelligent decisions are needed and artificial intelligence is at hand.

PERVASIVE COMPUTING AND INTELLIGENT HOUSEHOLDS

Intelligent households (as a kind of smart houses where ambient intelligence in the house observes inhabitants in their everyday life and automatically adapts behaviour to their needs) are best known as computer driven enhancements of common activities performed in residential facilities, mostly focused on inhabitants' comfort. Solutions in the field of smart houses include either simple improvements (automatic blinds, interactive fridge or multimedia centre) or complex, integrated systems. We can also say that smart houses are the flagship of pervasive computing. With the methods of artificial intelligence they are constantly trying to optimize processes in housing.

Many ideas for smart houses are still in development. Sometimes they include various features which are not really needed for residential life (they are just PR toys). What is somehow often forgotten is the energetic efficiency. Despite some smart houses offer energy saving features like intelligent light dimming, most of the common non-effective consumption problems remain unsolved (plus there is even more power overhead by smart house system itself). There exist many projects aimed to smart house development carried out by the research groups at universities or/and by industry. In the following, we discuss general issues being solved in the field of intelligent housing together with some examples.

Small, Individual Devices

Sang H. Park et al. introduced their Smart Home Project as a "… set of intelligent home appliances that can provide an awareness of the users' needs, providing them with a better home life experience without overpowering them with complex technologies and intuitive user interfaces. Though some of the ideas are quite quaint (like smart dress table for rapid makeup or artificial flower with built in camera monitoring the entrance), it is a good example of various unilaterally oriented devices that improve various aspects of housing. Purpose of gadgets vary: smart pens for reading and translating, smart mat for footprint-based user identification, pillow monitoring the health of sleeping, door reminder for forgotten things (keys, phone) or smart fridge registering the actual food stock. Most of such devices are based on continuous processing of environment data and generating user interaction events in certain situations.

Healthcare and Life-Easing for Elder People

Smart and automated functions of houses become more important if the inhabitants are older people who are often physically impaired. Such people would be blessed by various devices for home control (such as cooking, light control), communication easing (voice controlled phones, remote control) or health care monitoring. In addition, elders are in most cases not familiar with dealing with new technology: operating complex home-aid system would be a nightmare for them. Therefore, developers of such systems try to delegate as much decisions to artificial intelligence as possible, and to provide the rest of the system controls via advanced interfaces using advanced techniques like voice and gesture recognition. As demand for solutions for elders becomes stronger (US baby boom children are reaching their sixties), the stronger efforts are made for such visions to become a reality.

Rich User Interaction

GENIO is another smart house project, which primary goal is to create almost human like, virtual house servitor, called Majordomos. Authors of the project proposed a prototype system that gathers commands from human user speech, synthesize the answers and sends digital messages further to effectors or for further processing. The conversa-

tion is based on built in scenarios. The aim was not to build smart house itself, but to explore the field of human computer interaction in the field of intelligent households. Problems that lie there consist not only of voice recognition and synthesis quality issues, but also in potential invasive behaviour of smart house systems and "intelligence" of servitor agent acting as a real human (no software has passed the Turing's test yet).

Energy and Effectiveness

Some projects focus primary on energy savings. In an architecture for home network and service logic that is aimed at optimal energy management is proposed. It concentrates on interfaces to appliances, and definition of service architecture of the energy management. The example of real implementation of smart house considering energy saving is the Adaptive House in Colorado. In this project, whole household is covered by various sensors and regulators to maintain stable level of temperature of air and water, while lowering the costs by optimizing the work of devices that influences it (heating, cooling, fans or blinds). At the same time the system tries to ensure the comfort of the users. It adapts to the user needs by expecting his actions (e.g., his presence in room). It requires implemented infrastructure that supports all the monitoring, which constraints immediate applicability of the results achieved.

ENERGY CONSUMPTION MANAGER

We propose Energy Consumption Manager (ECM) as a system that focuses on optimization of electric energy usage in households, offices and other facilities where electric appliances are used. ECM is based on precise measurement of each appliance's power input and other environment parameters. Measured data are statistically processed and offered to a user for reviews. Furthermore, they are analyzed (including the user behaviour analysis) and refined into tips for improving energy consumption effectiveness or warnings about suspicious appliance behaviour. ECM also performs automatic interventions (shutting down of non used appliances).

The ECM system uses special hardware devices designed for easy integration with current appliances. In typical use case, the hardware consists of a set of measurers (one for each appliance) and a small industrial computer, to which the measurers are connected. The whole monitoring is controlled via user interfaces running on PC or mobile device. Besides that there is also an external part of the system (outside household/office) working over the Internet. It provides services for improving the effectiveness of local ECMs by sharing their data with others.

The main contribution of this project is a proposal and design of various analytic methods, algorithms, heuristics and user scenarios with common goal: improvement of energy consumption and related user habits. The system has ambition to become an energy saving complement of existing smart house solutions.

Measuring Power and Collecting Data

Before producing effective advices and actions, ECM needs information about past and current energy consumption situation. Therefore a measuring device is plugged between each appliance and its power source. We also considered building these devices straight in appliances. Measurers are power-supplied directly from that source too. They permanently monitor current state of power and wirelessly report each change to household centre, where they are stored for further processing (see Figure 1). Measurers are also capable of shutting down the energy supply to connected appliance.

Measuring power at home is not a new idea. There are plenty of types of measurers available on market. They vary in functionality they provide and also in price. Our initial effort was focused on including one of existing solutions into our system, at least for prototyping purposes. Most of measurers had an LCD display, but only few provided an electronic interfaces and only one we found, had capability of wireless communication. Unfortunately, it was part of more complex energy measurement system, which was closed to further extensions or other usage, so we could not make the use of its devices[2].

Figure 1. System overview

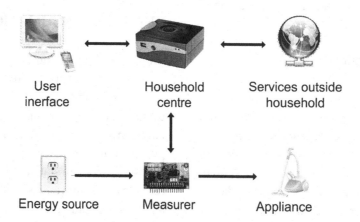

We designed our own wireless measurement device with programmable radio and physical interface. The measures are plugged between the power source and the appliance. The measurers are continuously monitoring the consumption and reporting every change that excites a defined threshold value. The reports consist of messages on how the power consumption changed in time. Tracking the appliance behaviour

Because of huge amount of data coming from measurers it is necessary to process them before next steps of the evaluation. Measured data from measuring apparatus are processed in ECM Analytic module (ECMA). The result of processing is the course of appliance consumption in the time. Later on, the user can view visual representation of these data in various graphs, compare and combine these data in the system main application (Figure 2).

Before actual analysis gathered data are preprocessed. The main point of data preprocessing is to compact them with respect to their information value. It is impossible and undesirable to store all measured data. The system detects periods of consumption with small deviances and replaces these values with arithmetic mean. In this way we are able to significantly reduce amount of stored data (see Figure 3). In the other hand more effective processing method is used to compress data older than one year.

The user is able to easily see, if his appliance (according the appliance's consumption) is used or running in a usual manner. We can see that opened doors at the fridge or during summer's high temperatures, appears in frequently turning on the compressor. Anomalies like not properly closed fridge's door can be easily recognized in short time simply by using the appliance (hours – day). But if something happened with the ap-

Figure 2. ECMA user interface – history of the appliance consumption

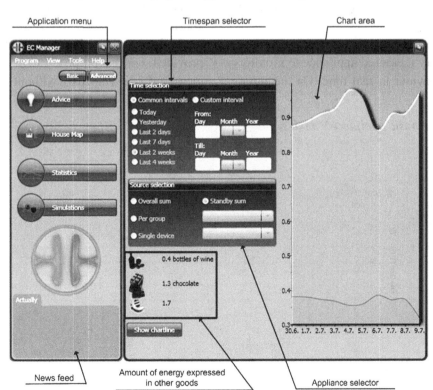

Figure 3. An example of data pre-processing

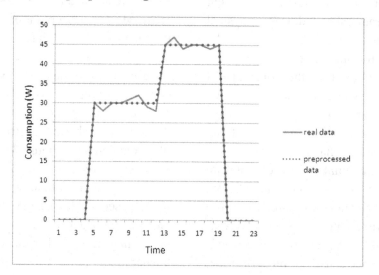

pliance like a damage of the seal or damage of the frigorific mechanism, the user cannot recognize problems immediately or in acceptable time (weeks – months). Situation is usually recognized after high energy bills, which can be on the other hand easily overlooked. The second issue of our system is educating. Users usually don't realize results of putting hot pot into the fridge etc.

It is necessary to be able to recognize each state (mode) of every measured appliance to monitor appliance behaviour in real time. On the input

ECMA gets values measured by measure apparatus in discrete time periods. The system requires the user to provide information about availability of the appliance standby mode. No more information is needed. If the standby mode is available, the system can identify this mode based on measured data without user interaction. This is sophisticated approach of appliance identification, comparing to similar solution where only static division is provided. For this purpose ECMA uses dividing measured values into two (without standby mode)

Figure 4. Three basic appliance's modes

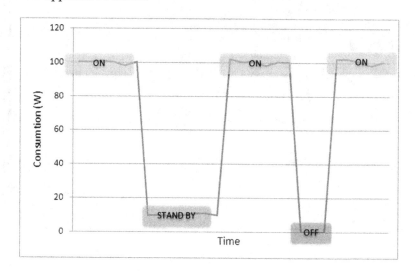

or three groups (with standby mode) based on data mining algorithm k-means, which can provide us required dividing into these groups – on, off and standby. Three appliance's modes are depicted in Figure 4.

In this manner we are able to gain whole history of appliance using based on 24/7 measuring and storing data. History is essential for identifying even smallest deviances in the appliance consumption like behaviour. ECMA compares actual measured data with the historical data stored in the system for every appliance and detects non standard behaviour. Then based on system options it notifies the user about detected problem.

Figure 5 shows an example of non standard fridge behaviour. ECMA is looking for patterns of every connected appliance. In the case of the fridge we are able to discover this pattern based on daily measurement, in case of television it takes a week, but in case of non regularly used appliances like wash machine or hairdryer, the system need not to recognize it at all. These patterns are discovered continuously and ECMA improves them iteratively day by day. As we noticed, finding a pattern of the appliance like a wash machine is difficult, but in this case we can identify length and frequency of weekly usage.

This kind of information can be very useful for identifying appliances anomalies.

The system is able to use these patterns in the other way too. We can easily identify type of appliance without a user interaction. The basic element of the method of appliance identification is measuring and improving the appliance pattern. Appliance behaviour is represented by "time vector". It consists of 48 half-hour periods. Every period contains information about appliance mode. After adding new information the vector is regularly normalized.

It is upon the user to allow ECMA to share anonymous information with other households. If this feature is allowed, system compares gained patterns with other household's appliance's patterns. For this purpose ECMA uses Euclidean distance measurement, based on this, it can find the most similar pattern and in finally to identify type of the appliance.

$$dist = \left(\sum_{i=1}^{n} \left| x_i - y_i \right|^2 \right)^{1/2},$$

where x_i is component of vector x a y_i is component of vector y

Figure 5. An example of non standard behaviour detected (fridge)

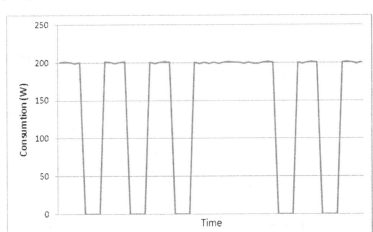

Algorithm 1.

```
For iter = 1 To T
        For e = 1 To N
                x = input for example e
                y = output for example e
                Run x forward thought network, computing all {a_i},{in_i}
                For all weights (j,i) (reverse order)
                        If i is output node
Compute Δ_i=(y_i-a_i)×g'(in_i)
                        Else
                                Compute Δ_i=g'(in_i)∑_k W_{i,k}Δ_k
                W_{j,i}=W_{j,i}+a×a_j×Δ_j
```

ECMA is able to provide appliance consumption predictions too. These are useful to estimate future costs of energy or for estimating appliance replacement return value. This prediction is computed by a neural network with back propagation learning. There are various types of neural networks used to solve classical prediction problem. ECMA is using *Time delay neural network*, where the need of history is solved by enhancing input data with "the history window". Value of this widow was found in the experiment. The learning process is based on back propagation learning, which pseudo code is listed in Algorithm 1.

As training data, ECMA uses measured values, which are aggregated to desired time period. This method is using neural network as universal computing machine, because of continuously changing appliance behaviour and various appliances types.

The whole process of the appliance behaviour tracking can be considered as a process of knowledge discovery. There is large amount of data as an input. Firstly some preprocessing and transformation is needed. Then we use several data mining methods (k-means clustering algorithm, neural network, and Apriori association rules learning algorithm) to discover new patterns. As the output of whole process we have knowledge about the appliance and household's members'

behaviour, which is used to reduce the energy consumption and to educate ECM users towards efficient energy consumption.

Immediate Problem Solving

Important part of every monitoring system, especially in households, is to notify users about actual state of the system. The efficiency of the notifying takes important role in the whole system's effectiveness. In this process monitoring the context (e.g., motion in a room) is important issue. There are plenty of motion sensors in the market. These sensors are designed to turn the light off, where no motion for longer tine is detected. There are also more sophisticated sensors available, but there are much more expensive and they often need some building works.

In the ECMA prototype, we used as a motion detection sensor a common webcam. This webcam can be replaced with various sensor types, not only motion detectors. The advantage over other solutions is that dividing detectors from effectors allows defining various even "crazy" rules without any buildings work needed. For example, when some motion in the kitchen is detected, a television in the living room can be turned on. The system is learning user preferences, and after some time, there is no need to ask users some kind of questions, when ECMA knows an answer from previous notifications (evolving user preferences is important issue which is not considered in current prototype).

When some non-usual behaviour is detected, ECMA can provide various types of messages based on user settings. It can simply do nothing, and store information about actual event, and present it to the user in main application. Other type of information is sending a message to the user's mobile device. This message can consist of a simply text description of detected situation, or it can consists of a question how solving this problem is proposed (e.g., turning lights off). Another way of the notification is using web based interface, which is equivalent to main application. In this case the user can remotely control the system and so his household appliances.

We also see other ways of the user notification. More and more sophisticated or intelligent appliances are used in households day by day. PC is not only tool for work, but takes role of household multimedia and fun centre. This progress gives to software designers more ways how to interact with the users. Another example of this a new opportunity is an intelligent fridge with touch screen. We can easily use this device to notify the user about detected problems or simply stating current state of the appliance.

When we consider also educational aspect, it is essential to recognize household's members, know their habits and discover bad habits in connection with energy and appliance using. Based on this information we personalize the communication, i.e. provide various types of messages to various users. Information about user behaviour can be obtained from sensors in specific rooms. If we know that 15 years old John lives in a specific room, we can easily detect that he often leaves light turned on when he goes to the school. We also personalize showing messages related to particular appliances (e.g. fridge to children).

In the next few years there is a large potential to use new devices like home multimedia centres,

smart appliances. These devices can be easily used as controls of intelligent households systems and can increase comfort of living on the one hand, but on the other hand can save electricity or other energy and costs in the households or other facilities in connection to saving natural resources.

Long Term System Policy

Important part of every electricity bill costs is for standby mode. This mode is useful during the day, when we use appliances in various time intervals. On the other hand, during night or work time there is no need to have appliances in our households in this mode. Currently even governments of many countries see this problem as a big one, and try to reduce it by the law (new appliances cannot exceed 2W limit during standby mode). US department of energy announces: "In the average home, 75% of the electricity used to power home electronics is consumed while the products are turned off." We can find in the history electricity failures caused by high energy consumption of households especially in US.

The easiest solution is to manually unplug appliances from the source. This is very uncomfortable solution, because there are lots of appliances in an average household. Second solution is to buy outlets with switches – this solution can reduce the main problem only in a little. It is essential to remind that users have to turn these appliances on in the mooring. In the ECM system is this problem solved by recognizing user's behaviour patterns, which are used to create stereotypes. The user can easily (calendar and timetable) create his stereotypes for every appliance, room or various groups of appliances. In this manner we manage all appliances in the household in such a way that the user is not overcrowded by the system notifications.

If ECM detects periodically repeated action (television was never turned on between 23:00-04:00), it can after user confirmation of this pattern, turn off the television in this time automatically.

In the case that television is turned on ECM will not cut off the power, but will wait until the user turns the television in the standby mode. This simple rule-based system can be enhanced by adding more sophisticated rules. Because all measurers and effectors are wirelessly connected, the user can set various rules like – when in no light is turned on and I switch off the lamp in my bedroom, cut off every device in standby mode etc. This management is interesting also in aspect of energy producing. Switching off and on could coexist with systems managing energy producing devices such a solar panels, water turbines, etc.

Defined rules can be easily grouped to stereotypes. So the user can create a profile for every common day and situation – weekend, holidays, night etc. Then on the specific day is the chosen stereotype applied, or the user can choose it in the real time via one of the system's user interfaces.

Based on enabled savings explained above we calculated approximately energy consumption decrease that can be achieved by the ECM solution. The model was proofed by measurements in field. An average household in the US consumes about 8,900 kilowatt-hours of electricity per year. By using the ECM system we can save up to 10% of the whole consumption only by reducing the stand-by consumption of the appliances by 70%. We see a lot of additional savings in turning off forgotten appliances, like light or TV, since this statistics are not easy to measure we cannot present current numbers.

ADVANCED FEATURES OF ENERGY CONSUMPTION MANAGER

Model of Data Exchange Among Households

Social networks bring new possibilities to every aspect of our life. People are interested in connecting and sharing information with their friends or even with everyone in the world. While man is essentially a social and competitive creature, obviously he would appreciate to have information where in the rank of saving energy is his household. This opens a space for social networking and distribution of information gained by everyone. We always want to be better than others. Therefore, we proposed a competitive environment by making measurements visible among users with similar habits.

Important issue is anonymity of the data distributed. Actually pervasive computing considers also this aspect as devices around us hold often too much information which could be misused. A GPS locator could help for example thieves who want to stole something while the user is on a vacation and his flat is not guarded. The ECM user does not see information about exact user, but he has the opportunity to find himself in the rank.

ECM is designed to be able to make a comparison between the users based on total energy consumption, consumption of individual appliances or on tend of improvement. The user has to insert basic information such as number of family members. Then ECM is able to compare for example washing machines in households with similar or even the same profiles. It is easy to imagine that in the mentioned case of the washing machine; seemingly irrelevant information has impact on similarity which should be recognized. For example, number of people who are devoted to sport in a household reflects to the frequency of laundering.

To ensure the relevance of comparisons, we propose a method for dynamic clustering of appliances in households. While it does not make sense to compare the usage of the appliance in different households, clustering allows comparing only households with similar habits. The method of comparison in the ECM is based on the tagging of appliances and the dynamic reorganization of the groups of appliances and households. Tagging is made according to the characteristics of the household where the appliance is used and also according to the technical attributes of the appliance. The quality of the groups organizing is also provided by setting the priorities of the various tags. Priority is reduced continuously from the category of appliance to the characteristics of the family itself, in which the appliance is located. Priorities can be changed globally, therefore the organization of appliances could be adjusted either.

Continual insertion of new appliances to the system reorganizes the groups. The system has to maintain the optimum-sized group, and also takes care to combine and merge appliances according to similarities and provided priorities of attributes. The benefit provided for the user is then improved search for an appliance regarding the comparison with similar ones. The output quality increases by time and amount of appliances inserted into system. If the user is very different from the others in the earlier stages, the system nevertheless finds some results. Consequently, dynamic grouping reveals the nearest similar users. Otherwise, if there is already too large number of users in the database who are similar, the system makes use of the tags with the lowest priorities. Accordingly is the division more precise in this case. More appliances in the database mean more relevant results for the user. Ideal situation occurs after identical users and their appliances could be compared.

Picking a New Appliance

Standard purpose of intelligent solutions which help us in everyday life is to solve problems or even better to find problems which we cannot notice and solve them. ECM is a system, which after revealing the problem causing higher consumption of energy, provides also solutions. If the system recommends some solutions and aims to be as much accurate as possible, it has to track user actions and make a model of the user. The advantage of the pervasive computing is that it is where the user is. It means that it can easily store what the user is doing connected to the point of its computing. Basically, the system of devices

may form very clear model of the user and his habits and interests. In the case of the ECM, it is connected to using the appliances. Using the patterns which are then learnt by such a system, notifications or even automatic solutions are made. Solutions may differ according to specification of problems. Sometimes are solution static, but sometimes they have to be fitting for the user and his habits.

As an example, consider the case that the user has just discovered that the refrigerator is not working properly. He is interesting in the replacement. Old refrigerator could consume a disproportionate amount of energy. ECM automatically detects such a failure and proposes a better appliance using comparisons with other users. It search in the database appliance for similar appliance and suggests. The user can browse database manually, and find the appliance that mostly satisfies his preferences.

Manual search is a simple choice of an appliance and additional comparisons with the appliance which is currently used in household or with some different selected appliance out of database. The system knows the habits of the user, and therefore is able to replace appliance and make a simulation to identify potential of the new attributes of this replacement. Such interesting information like return of investment could be provided to help user to decide at the last step of ordering. The system may use information from the first nearest similar user for the estimation of appliance replacement pros and cons. If it is needed it could recalculate some attributes to make more relevant results, for example recalculation for the greater number of persons. However, simulations are performed only for particular and informative results. The task is not to demonstrate simulation for whole period of usage, but it can sum up outputs for few years. Then the users could consider their investment.

To create the best possible estimates for the appliance, exact data should be used. Therefore technical data from distributors and Internet shops are not sufficient. This information is often only approximate, or incomplete, sometimes changed because of marketing targets. An important component of ECM is therefore a database. It is incrementally supplemented by the technical parameters, especially based on data. This data is generated using appliances in real life of customers.

Cooperation with Appliance Producers and Distributors

Mobile devices which are relevant when we want to provide ambient intelligence are often not very cheap. If a user wants to interact with the environment, it is often necessary to buy a sophisticated device. However, starting with mobile phones which could be used to interact with other devices and almost everyone has one. Actually mobile phones are not cheap. One reason why people buy them is because it is almost free if we get it with a contract. It means that providers sell the mobile phone on the cheap to enlarge their profit by providing the services.

Other models like advertisements or selling information discovered in data (by the use of data mining techniques) are interesting because third parties could always make the solution more available. In our case, since the user can afford new device, he would understood that for example advertisement appears in his ECM system. To make this sponsorship more transparent and not so annoying for the user, also different models like selling mined information are available. After tracking behaviour of the users, a sponsor would be interested in such information. In case of ECM, third parties like appliance distributors would pay for information, which appliance is the most famous, or which appliance of their appliance portfolio has most defects. They would be also interested whether their appliance was recommended to the user or not, and if the ECM treat it like environment-friendly appliance.

Of course, this is not only about finances, but more about information exchange. ECM is able to gain information on appliances separately by precise measurement in practice. Otherwise is information not exact and often lofty, made by marketing of distributors. Information on consumption, efficiency, or lifetime and popularity of appliances is obtained from users of the system. It brings overview of the appliances and creates a valuable database, considering the anonymity of each user. This data could be potentially very interesting for vendors and distributors of the appliances.

Similarly, we can assume producers and distributors interest in presenting their advertisements in our system. ECM needs a kind of support in early stages, while the information on appliances (available only in databases of distributors and manufacturers) is too weak. Information such as photos, prices and location where is the appliance available is for the system useful for appliance presentation to the user. It is also interesting in later stages. Usage of this information assists user in selecting a new appliance. From the perspective of distributors it is another option how to present themselves to the customer and enlarge their profits.

We proposed a model (see Figure 6), which creates a harmonious cooperation between ECMs and appliance distributors and vendors. Using web services provided, in general anyone can upload information about the appliance. For inserting and updating information we proposed a concept of virtual money. This concept keeps third parties and our system in the balance.

- Different transactions cost different money.
- For this credit distributors may purchase information about the usage or popularity of the appliances. They can also create ads and offers, which are then submitted and distributed to users.

Figure 6. Exchange model

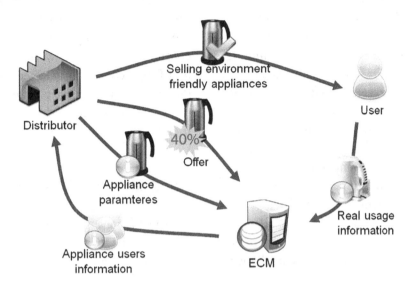

- Advertisement distribution may be limited to specific groups of users. These users are grouped by system according to interest and could be considered interesting by distributors. Groups that distributors want to address are formed on the basis of tags, which are described in the previous section.
- Third parties are able to add tags which are then considered while reorganizing appliances to groups. These tags include specific information which represents their interest in this grouping.
- The user receives a single form of supplementary question. Answering this form is then used for the allocation of tags.

In real usage of ECM it will be necessary to change the values of virtual money for different transactions in for purpose of adjusting the best balance.

Adaptation and ECM Educational Aspects

Methods of ambient intelligence are an effort to reduce our responsibility and increase the feeling of comfort. Actually everyone wants to live in the comfort and when some device enables this we rely on it. Lot of things which we need to solve automatically would be sooner or later solved by some devices. People are lazy in general and we can discuss if this is the best way, and if we really want to calculate everything and solve it automatically. One option is that we would live in a bed. On the other hand, we could predict that people would be so responsible that they would make sports and learn new things. But why to learn math when every nearby device could calculate what would be the price of the items which we want to purchase?

This reveals that we should not only solve problems instead of a user, but also help him how to prevent any of them. Ambient intelligence should help people but there is another point of view – education and keeping them in the form. Educated people would reduce the occurrence of problems and then the systems such as ECM could focus on different fields and increase the comfort of the people in different way. People should solve some tasks themselves. There has to be a balance between the intelligence of the nearby computing devices and human himself. If there is no education, the number of devices and automation would increase exponentially and people would become more addictive on help. Furthermore, there are also fields where any of the nowadays systems can solve the problem. Sometimes it is possible only to notify user, that something is wrong and let him to solve problems by himself.

Users' approach to the ECM may differ in two ways. They can install the system in their household (connect measurers to the appliance, turn on the central unit) and do not care about the rest. The second option is paying attention to the operations of the system. Point is to have an active interest, let's say it is a "game" where the user has to reach maximum effectiveness. Eventually, ECM works in both modes. However, the difference is the speed of convergence to an optimal state of the system (the maximum possible energy saving, using planned and unplanned interventions). In both cases is the system more interactive with the user in early stages (it is necessary to learn home habits first). Reports and questions made by system could be ignored. The system is not invasive and notifies user at ease. During the operation of the system in the household is the frequency of needed interaction decreased. It is limited to special situations when something unexpected happens.

Besides the automatic regulation of the power consumption is another important feature and its educational character. ECM communication with the users is driven by events which are generated when some problem in energy consumption or behaviour occurs. The ECM searches for solutions and provides recommendations to save energy. Although these are often available from other sources, through the ECM is user notified and educated exactly on the time, in appropriate situations, when he deals with specific problems. In case of

ECM, common recommendations are annotated (by the concepts to which they relate). When the user does an activity (e.g., views statistics of the consumption of the lamp in the bathroom) the best fitting concept is selected from the base of the recommendations (e.g., that frequent turning on and off of the light may cause of shortening its lifetime and could be counterproductive).

IMPLEMENTATION ASPECTS

To proof the concept of our project we implemented parts of the ECM system. The advanced ideas are all based on results from continuous energy observations of appliance's energy consumption. As first we needed a representative dataset of measurements to validate all other ideas. On the market were no acceptable measurement devices with open physical interface that could meet our requirements. Therefore, we decided to create our own measurer.

In ECM prototype it consists of multimeter Voltcraft VC-230 for power measurement purposes and EnOcean TCM 120, a wireless communication module. Both components are connected with RS-232 interface. EnOcean is a relatively new transmission standard. We picked it because of low energy consumption of its communication modules (power input of TCM module is below 0.1 W). Besides wireless connectivity, we also used TCM 120 for implementation of measurer logic, which includes detecting the power input change (multimeter constantly sends the actual power level value through RS-232) and storing data about changes until the next opportunity to send them to central unit.

While designing the communication between measurers and central unit, we had to deal with low robustness of EnOcean transmission standard and limited computing capabilities of TCM devices, which are an exchange for modest energy costs. Communication in EnOcean is based on sending short telegrams. Telegram consists of:

- 4-byte field for source device identifier,
- 4-byte field for transferred data,
- Control checksum, enabling to detect broken telegrams on receiver side.

However, the standard does not support any kind of unicast communication (since it has no destination identifier field). It also lacks the mechanism of reliable telegram delivery (such as TCP protocol in RM-OSI) and a support for multiplexed communication (only one device can broadcast at one time, otherwise messages could be damaged as a result of interference). Further complications were caused by TCM computing limits (2kB EEPROM, 32kB flash memory, 265B RAM), which blocked any attempts to put more communication control functions into measurers. Moreover, TCMs appeared to have problems with CPU timers that caused synchronization problems after longer periods of running. However, a household system can consist of dozens of communicating devices, and their synchronization is crucial (to avoid interference).

For reasons described above, we designed highly centralized, time planning based communication control mechanism running at central unit. Its primary goal is to compute a time plan that determines when a certain measurer may transfer its data. The plan covers the span about 20 – 120 seconds in length and is repeated periodically. Each measurer is given a time frame for its transmission. The plan period has not to be too long since power changes have to be delivered in shortest possible time (certain system functions depend on actual information). On the other hand, all devices must have opportunity to transmit their data during one period.

However, time needs for transmitting all changes can vary through time (some appliances can suddenly create lot of power flow changes). Also, new measurers may be plugged in; others can malfunction or be plugged out (without notifying the central unit). The plan must also count with significant overheads for plan announcements,

frames reserved for registering new devices and inter-transmission gaps.

Communication control should also supervise synchronization behaviour of measurers within the plan. This can be done only by confronting real transmission times of measurers with planned ones. Devices that start slipping out of their time frames are forced to reset their timers by special signal. However, there is no mechanism of recovery of data, lost due to signal interference during wrong transmission times.

Matching the real transmission times with the plan is in fact the only source of knowledge about how the plan is being fulfilled (there is no kind of "control" or "diagnostic" telegrams sent by measurers). According to that the percentage of time frame usage of each device can be determined and used as hint for expanding or contracting the frame. For example, if a usage of frame drops under 30%, the assigned frame is not likely used effectively and should be shortened. On the other hand, usage over 90% might indicate the device need more time space.

Our communication planning considers all this factors. We designed it as a rule-based system, transforming the plan (divided into atomic length chunks) into optimal form using the hints described above and some special commands (from higher level of the system). We chose an approach of continuous plan modifications (preserving as much of the plan as possible) rather than recomposing the whole plan as new (even if it could be better). The reason is that the more changes are made, the longer the announcement overhead has to be. Several strategies are implemented to maintain the plan flexibility. For example, rules ensure placing time frames to preserve space economy, starting to put frames with regular density.

The results we reached with this framework were good enough to create a representative dataset for further testing and validation of our concepts. However the development in embedded wireless communication is now emerging, further measurers will be able to operate with greater bandwidth and advanced signal modulations. So the concept of measurement has the potential to reach simplicity, better data transfer rates, greater communication distances and simultaneous collision free transfers.

CONCLUSION

In this chapter we have shown our opinions on the future role of pervasive computing in intelligent households, particularly in the field of saving energy. We presented our visions which were transformed into prototype system Energy Consumption Manager aimed at automatic reduction of non-efficient electric energy consumption in households (or offices) including advisory functions for the user, so the user can improve his energy saving practices. ECM measures power input of appliances by special hardware devices. Measured data are afterwards processed. Results are statistics about energy consumption (graphs, trends, expected costs, consumption predictions), notifications about suspicious appliance behaviour (indications of malfunction or wrong use), acute warnings and interventions (forgotten light) and long-term consumption plans (standby elimination).

ECM is an adaptive system based on monitoring user behaviour. This enables the system to work with minimum user interaction. Further enhancements of analytical abilities of ECM as well as other services (remote control, appliance market) for user are provided within external system layer, working over the Internet.

REFERENCES

Bohn, J., Coroamă, V., Langheinrich, M., Mattern, F., & Rohs, M. (2005). Social, Economic, and Ethical Implications of Ambient Intelligence and Ubiquitous Computing. In Weber, W., Rabaey, J. M., & Aarts, E. (Eds.), *Ambient Intelligence* (pp. 5–29). Berlin, Heidelberg: Springer. doi:10.1007/3-540-27139-2_2

Dunham, M. H. (2003). *Data mining – Introductory and Advanced topics*. Upper Saddle River, NJ: Pearson Education Inc.

Garate, A., Herrasti, N., & Lopez, A. (2005). GENIO: An ambient intelligence application in home automation and entertainment environment. *Proceedings of the Joint soc-eusai conference* (pp. 241 - 245).

Goschnick, J., & Körber, R. (2002). Condition Monitoring For Intelligent Household Appliances. In Lahrmann, A., & Tschulena, G. (Eds.), *Sensors in Household Appliances* (*Vol. 5*, pp. 52–68). Weinheim: Wiley-VCH.

Helal, S., Mann, W., El-Zabadani, H., King, J., Kaddoura, Y., & Jansen, E. (2005, March). The Gator tech smart house: a programmable pervasive space. *Computer*, *38*, 50–60. doi:10.1109/MC.2005.107

Hönsch, M., Kompan, M., Šimko, J., & Zeleník, D. (2008). Intelligent Household: Energy consumption manager, In B. Mannová, P. Šaloun & M. Bieliková (Eds.), *Proc. of CZ ACM Student Research Competition* (pp. 33-41). Prague: CZ ACM.

Kadar, P. (2008). Making the power system intelligent. *Proceedings of the Icrepq'08 international conference on renewable energy and power quality* (pp. 12-14).

Maes, P., & Mistry, P. (2009). The Sixth sense. *Proceedings of the Ted talk in reframe session* Long Beach, CA: TED 2009.

Mozer, M. C. (2005). Lessons from an Adaptive House. In Cook, D. J., & Das, S. K. (Eds.), *Smart environments: technologies, protocols, and applications* (pp. 273–294). Hoboken, NJ: John Wiley & Sons, Inc.

Park, S., Won, S., Lee, J., & Kim, S. (2003). Smart home – digitally engineered domestic life. *Personal and Ubiquitous Computing, 7*(3-4), 189–196. doi:10.1007/s00779-003-0228-9

Šimún, M., Andrejko, A., & Bieliková, M. (2008) Maintenance of Learner's Characteristics by Spreading a Change. In M. Kendall & B. Samways (Eds.), IFIP International Federation for Information Processing, World Computer Congress, Volume 281: *Learning to Live in the Knowledge Society* (pp. 223-226). Boston: Springer.

Stefanov, D. H., Bien, Z., & Bang, W. (2004). The Smart house for older persons and persons with physical disabilities: structure, technology arrangements, and perspectives. *IEEE Transactions on Neural Systems and Rehabilitation Engineering, 12*, 228–250. doi:10.1109/TNSRE.2004.828423

Tompros, S., Mouratidis, N., Caragiozidis, M., Hrasnica, H., & Gavras, A. (2008). A pervasive network architecture featuring intelligent energy management of households. *Proceedings of the 1st int. Conf. on Pervasive Technologies Related To Assistive Environments* (pp. 1-6). New York: ACM.

Witten, I., & Frank, E. (2005). *Data mining: practical machine learning tools and techniques* (2nd ed.). San Francisco: Morgan Kaufmann.

ENDNOTES

1   Mémento sur l'energie /Energie Data Book, CEA, Edition 2002, p. 38.

2   Elektronik-Versandhaus. http://www.elv.de/output/controller.aspx?cid=74&detail=10&detail2=15910

Chapter 28
Ambient Intelligence and Immersive Geospatial Visual Analytics

Raffaele De Amicis
Fondazione Graphitech, Italy

Giuseppe Conti
Fondazione Graphitech, Italy

ABSTRACT

The unprecedented success of 3D geobrowsers, mostly due to the user-friendliness typical of their interfaces and to the extremely wide set of information available, has undoubtedly marked a turning point within the geospatial domain, clearly departing from previous IT solutions in the field of Geographical Information Systems (GIS). This technological great leap forward has paved the way to a new generation of GeoVisual Analytics (GVA) applications capable to ensure access, filtering and processing of large repositories of geospatial information. Within this context we refer to GeoVisual Analytics as the set of tools, technologies and methodologies which can be deployed to increase situational awareness by helping operators identify specific data patterns within a vast information flow made of multidimensional geographical data coming from static databases as well as from sensor networks.

This shift, which essentially follows the vision of a "Digital Earth" envisaged by Al Gore a decade ago, is bringing concepts typical of Ambient Intelligence to a wider, environmental scale. The chapter first presents the reasons that have brought to such a radical shift and it highlights the profound implications at the technological and societal level. We further demonstrate how the domain of ambient intelligence is not any more spatially limited and how the combined use of AMI and VA can be essential for the development of new tools to be used at environmental scale level.

The convergence of technologies typical of GVA with those of AmI can be crucial within several mission-critical tasks, typical for instance of environmental management and control. These can be significantly enhanced through the adoption of paradigms and technologies typical of Ambient Intelligence including

DOI: 10.4018/978-1-61692-857-5.ch028

sensor networks, pervasive computing, advanced visualization technologies, human-computer interaction, data processing, indexing, intelligent extraction and filtering of data.

The following sections then showcase the results of an extensive research effort by the authors (De Amicis et al., 2009; De Amicis et al., 2009a; Conti et al. 2009) whereby functionalities typical of AmI have been integrated within a GeoVisual Analytics (GVA) framework. The research work foster a paradigm shift. The typical AmI paradigm sees users interact in a seamless manner objects and systems that surround them. Object and systems used in our daily lives exchange information to react or proactively support users in the most transparent way. However this requires both a controlled space and devices or objects being fitted specific technologies or sensors. The authors instead propose a paradigm shift whereby the user can related to a boundary-free space interacting with its virtual instance that represents it.

The solution developed has been designed to support the decision making processes in the context of environmental management where it is essential to provide the means for tasks such as early warning system, risk assessment and mapping, risk management and last but not least, simulation. The system has been designed to allow operators to interactively access and create geographical information as well as to visually build, encapsulate, and parameterize complex geo-processing models within a 3D interactive environment.

The chapter finally concludes by highlighting a number of open challenges brought by the convergence between GVA and AmI which need to be addressed by the research community in the next future.

BOUNDARY-FREE SPATIAL DOMAIN FOR AMBIENT INTELLIGENCE

In recent years the focus of many research efforts in the field of Ambient Intelligence (AmI) has brought to several anticipatory, context-aware and custom-tailored IT solutions capable to respond to a number of case studies often limited in terms of spatial domain as home or office. Extensive use of hardware technologies such as RFID, sensor networks, wireless communications as well as software techniques including software agents, affective computing paradigms or human-centered interfaces have been mostly confined to indoor scenarios or to well-controlled outdoor contexts.

However the ubiquitous adoption of geospatial technologies has made it possible to expand the domain of Ambient Intelligence from the boundaries of enclosed spaces and from scenarios such as home automation, health care and elderly assistance, to a much wider perspective which is embracing the environment that surrounds us.

In a perhaps provocative and certainly challenging claim, the world itself has become the domain of Ambient Intelligence (AmI). In fact the convergence between Geospatial Technologies and AmI opens up radically new scenarios of globalised intelligent applications capable to benefit from heterogeneous distributed information on the spatial environment surrounding us. These include data from spatial databases, web-based repositories, a large variety of real-time information continuously delivered by ground sensors and by satellite-born technologies and, last but not least, an increasingly large amount of geo-referenced information generated by user communities which are available on the web.

This information-rich scenario is at the base of a number of intelligent services essential to numerous daily activities. In fact, in our everyday lives, we already make extensive use of Geospatial Technologies and of Geographic Information (GI) in a number of applications which today are regarded as a commodity, including:

- Personal navigation systems;
- Media-rich web 2.0 "mash-ups", using a variety of mapping APIs (Application Programming Interfaces) as components of larger modular web 2.0 applications;
- 3D web-based applications, known as 3D Geobrowsers or "spinning globes", the most famous of which being perhaps Google Earth, Microsoft Virtual Earth or NASA WorldWind.

In our daily practice we use a number of web-based applications delivering real-time geo-referenced information for instance on road traffic and congestion, on localized weather forecast, on pollution distribution. We constantly use geographical information and personal satellite navigation technologies for pedestrian and vehicle navigation and routing systems, to find the closest shop or to track the path of a user playing outdoor sports[1].

More specifically the use of geospatial technologies is extremely important in the context of environmental management and planning. Operators need to be able to access data on the ecosystem, to perform spatial analysis of environmental data, to extract environmental indicators of interest. For this reason it is essential to provide them with the most effective tools, based on geospatial technologies, to ensure the best possible management of the environment as well as to ensure the wellbeing of those living within it.

Within a wider context it is therefore important to provide operators with technologies to access, manage, process GI from basic cartographic data, to information on wildlife, population and pollutant distribution, to classification of industrial sites as well as localization of other anthropogenic activities. The type of data being managed may include among other, high-resolution areal or satellite imagery, digital raster or vector maps, alphanumerical information on economic, social, demographic indicators, real-time sensor data on pollution to name but a few.

Within this context the adoption of an Ambient Intelligence-oriented methodology can significantly improve analysis and monitoring of ecosystem services. This is extremely important as planning or management of environmental resources can have severe long-lasting consequences both in terms of public health and more generally in terms of people's quality of live and welfare.

Today operators in charge of environmental management and control rely on concurrent use of different specifically designed applications to access real-time data, to process information, to simulate events as well as to coordinate response to unexpected events. Such an diverse set of information needs to be accessed and processed by a very heterogeneous team, made of administrators and operators, characterized by distinct professional expertise and by different levels of IT literacy. Traditional Computer Supported Cooperative Work (CSCW) and groupware solutions are unfit to cope with such an articulated scenario. A number of legacy IT systems, including GIS-based solutions, are used within an often fragmented workflow, which requires the operators to switch between different software solutions, with limited integration and interoperability. As a result, great attention has been paid by both the scientific and industrial community to the development of IT infrastructural solutions capable of delivering Enterprise Application Integration (EAI), typically through the deployment of complex distributed Service Oriented Architectures (SOA), based on open transport protocol developed by standardization bodies such as OGC (Open Geospatial Consortium) and ISO (International Organization for Standardization).

Nonetheless commercial enterprise-level software lack of the most essential AmI capabilities since they offer limited support to personalization and adaptation to user requirements, they are not capable of providing context awareness neither of behaving in an anticipatory manner to support the operator in a proactive way. Therefore, as a matter of fact, there is a consistent lack of integrated solutions able to support the entire operator's workflow and capable to offer, within a single software solution, the functionalities offered by different IT systems today in use. In particular there is clear need for IT solutions capable to help operators extract knowledge from the data they manage, to acquire awareness and finally to extract intelligence essential for their decision making process.

An Historical Perspective

Until a few years ago it was not possible, as it is today, to access large repositories of Geographic Information in a dynamic manner from any location, provided that a connection to the Internet is available. In the recent past a widespread information campaign has captured the public attention, focusing towards our planet and towards the environment that surrounds us, most of the hype being on how the wide set of information related to the environment could be made accessible in the most transparent way. This awareness rising process was fuelled by both the scientific community as well as by some of the largest IT enterprises including Google and Microsoft.

When we analyze the historical reasons for such an interest shift, we observe two major milestones which have occurred in the last ten years. The first milestone was certainly set on 31 January 1998 by Al Gore, at the time vice president of the USA, during a very well known talk he gave at the California Science Center in Los Angeles titled: "The Digital Earth: Understanding our planet in the 21st Century". Al Gore envisioned that the development of IT technologies would allow scientists and people alike to use a virtual representation of the Earth as the interface to access digital information of various nature.

This certainly visionary talk attired global attention from both the political, scientific community as well as from the general public. Without further illustrating the scientific achievements available prior and after Al Gore's visionary talk,

to the general public this vision came true only few years later, when on June 28 2005 Google released their "3D Mapping and Search Product" based on former Keyhole technology. The product, branded as Google Earth, represent the second major milestone which brought to the aforementioned technological shift, and it has had an extremely significant impact at the societal level.

In fact at that time neither common people nor most experts operating within the traditional geospatial domain could have foreseen such a radical leap forward. Today, only a few years after the release of Google Earth, we are now all aquatinted with consumer 3D Geobrowsers such as Google Earth, Microsoft Virtual Earth or NASA WorldWind to name but a few. These technologies are used daily for the most diverse personal and professional use and are regarded as a commodity by the majority of internet users.

Within a very few years 3D Geobrowsers, mapping APIs and, more generally, geospatial technologies have therefore evolved from a limited community of experts to the subject of public interest. As a result geospatial technologies are now ubiquitous, as everybody can access different forms of geospatial data and services from their PC as well as from their mobile devices such as Smartphones or PDAs.

If we carefully analyze the societal implications brought by this change we observe that a profound paradigm shift has occurred. In fact, unlike any other previous form of digital representation, for the first time in IT history the information on the environment directly describes the environment itself without any form of mediation. Users can now directly navigate a "virtual Earth", they do not access images, maps or other data which may be available from a webpage or from a wiki. Users do not need to access a media container that, through a number of different communication channels, reproduces some information on the Earth, be this satellite imagery or the result of a spatial analysis plotted on a graph.

Rather, following the footsteps of Gore's far looking vision the "Virtual Earth" itself becomes the means to access information on the planet. Other traditional information channels including text, images, videos etc can be also made available within the virtual Earth and related to a location in space. Users associate pictures, videos, documents to a geographical locations. Analogously they can see the output of an environmental sensor directly rendered on top of the 3D representation of their surrounding environment.

Social Networking and Ambient Intelligence

Nonetheless, paradoxically, following this major step forward, the technological vision behind the concept of today's 3D Geobrowsers has not seen any significantly evolution.

In recent times the emerging trend of social networking has shown that knowledge can be built from the capability of several people to share an interest or to discuss a topic. The internet is rich in examples of social network communities creating geospatial knowledge of interest for our surrounding environment, for our security, for our daily life. During exceptional environmental events social networking, backed by with the availability of web-based geospatial technologies, has proved essential to catalyze geographical data of relevance. Two examples are the use of web-based geospatial technologies in relation to the disasters caused by hurricane Katrina[2] and by the earthquake that struck Italy[3] on summer 2009. In both cases technologies such as Google Map or Google Earth, were used as social networking platform. People were providing vital localized information to support disaster relief agencies as well as to plan rescue operations. Following the well known metaphor of "people as sensors" (Laituri et al., 2008) these examples show how the entire community can become a sensor network providing qualified information of public interest.

Without focusing on issues typical of social networking on the quality and reliability of geographical information, whose analysis is beyond the scope of this chapter and which are subject of a number of scientific activities[4], it is worth noting how the two examples clearly denote the importance of techno-cultural attitude at social level. This is clearly emerging when comparing the staggering amount of information which was produced by the user community in the aftermath the hurricane, to identify damages as well as to support coordination of rescue activities.

The social network trend clearly shows how it is possible to transform the digital representation of the territory into an virtual "environment" through a constant user centric production of information thanks to the involvement of the wider user community. However, rather paradoxically, today such a social networking attitude heavily relies on the willingness of a service providers, as for instance Google, to ensure public access to the information coming from the community.

This gap must be filled both from the technological and from the scientific point of view. While the following sections will thoroughly illustrate how this issue has been tackled from the technological point of view, let us focus for a moment on its cultural shortcomings.

In fact the biggest challenge perhaps is to make people aware of the enormous potential and implications of such a social-driven approach to geospatial information. Although the user is not within an intelligent environment as referred to in the literature, where they can interact with sensors and actuators indeed, using geospatial technologies people can now interact with a wide range of information on their surrounding environment[5].

This user-centered vision has been at the basis of the research effort illustrated in the reminder of the chapter. The work has focused on the use of AmI and geospatial technologies in the specific context of environmental management.

Information on the environment ubiquitously available from the network becomes accessible to a large community of users, who can further contribute to it by extending the nodes the network in made of. In the research carried on by the authors this issue was tackled by adopting strategies typical of GeoVisual Analytics (GVA). The field of GeoVisual Analytics (De Amicis et al., 2009) has recently emerged as result of extended use of geospatial technologies within a Visual Analytics context (VA) (Thomas et al., 2005). VA has emerged to deliver methods and technologies to access to and interact with the widest and most diverse sets of digital information as well as to extract, either through automatic or semi-automatic processes, specific data patterns and information of interest for the analyst.

The challenge faced by the authors was to integrate approaches typical of VA within a geospatial context, wherever the information being manipulated has a clear reference to a real position on earth. The ultimate goal of was therefore to create tools and technologies that could help operators gain both a global comprehension of large-scale environmental phenomena as well as to be able to focus on specific data patterns, together with their relevant attributes, at local scale.

The research activity has brought to the development of a Service Oriented Architecture that exposing a number of functionalities to access, analyze and process geospatial data through interoperable standards. The web-based client has been implemented following the idea of a node that can be used by people to add new information within a wider network made of interoperable web services. The result has been the development of a framework that could be used by everybody to create other nodes of the network. The user can use a web-service to access and deliver information in an interactive manner. Furthermore the system can be used to deliver processing functionalities over the web, made available as web services, to create an intelligent environments capable to ensure ubiquitous, context aware, personalized interaction with geospatial data. The user can perform *ex-ante* and *ex-post* analysis through a

user-friendly 3D client developed on top of the WorldWind[6] technology. The latter is the result of a far looking vision brought forward by NASA which admirably has released its source code to the entire community, a gift to the community whose implications perhaps have not fully realized.

RELATED WORK

The use of the Ambient Intelligence paradigm in outdoor scenarios has been, perhaps indirectly, addressed by a number of previous research works. When compared to the traditional concept of Ambient Intelligence extending its boundaries to the surrounding environment rises a number of issues mainly related to collection and management of information. In fact the variety of data available which include satellite or airborne data (orthophoto, LIDAR, etc.), data available within databases requires new approach to information management and to intelligent filtering and visualization. With regards to this a significant contribution has been achieved by Visual Analytics (Thomas et al., 2005) a discipline that delivers techniques to intelligently extract information pattern within complex, multi-dimensional data flows. The concept of Visual Analytics in turn has been tailored to the specific requirements of geographical information yielding the discipline of GeoVisual Analytics (De Amicis et al., 2009). In both VA and GVA an important role is played by the interactivity of the system whereby users need to perceive as being immersed within the environment itself (Conti et al. 2005; Santos et al., 2007; Conti et al., 2008).

With specific regard to collection of real time information several works have extensively explored the use of sensor networks. Most interestingly however (Laituri et al., 2008) brings forward the idea of sensor networks, highlighting how users can be considered as living probes, behaving as sensors within the real world. In their work the authors introduce the metaphor of "people as sensors" showing how the broad community can be considered and dealt with as a sensor network providing extremely qualified information of great relevance for the community itself.

Collecting information from users rises the issue of trust management (Beth et al., 1994; Blaze et al., 1998; Blaze et al., 1996), that is to which degree user- generated information can be considered reliable and how to provide means for automatic cross-validation of information by the user's community through so-called recommendation systems (Abdul-Rahman et al., 1997; Herlocker et al., 2004; Weeks et al., 2001).

However tracking the position of users and their preferences rises serious concerns in terms of user privacy as location, user preferences are potentially known to the system and therefore need to be adequately protected. A number of international initiatives are specifically focusing on this very issue with particular regards to the Communications, Air-interface, Long and Medium range (CALM) project (http://www.calm.hu/) endorsed by ISO TC204 Working Group 16.

INTELLIGENT GEOSPATIAL VISUAL ANALYTICS

The framework developed by the authors was designed to respond to a number of activities related to the governance and management of a territory. Public administrators, civil servants, risk managers and security operators all use Geographical Information Systems (GIS) for their daily activities. In fact since the majority of the activities carried out by public administration is related to the territory, or it requires referring to a location in space, most public administration units and departments make already use of GIS software for their activities. Geospatial technologies are used for very different purposes ranging from infrastructure development and planning, to land use management, to cadastre, to emergency management, to e-government.

In particular the framework responds to the emerging trend to provide interoperable software solution capable to manage geospatial information of environmental relevance. In particular it is worth mentioning the Directive 2007/2/EC of the European Parliament and of the Council which promotes the development of Infrastructure for Spatial Information in the European Community often referred to as INSPIRE[7]. The Directive, entered into force in May 2007, represent a major step forward in terms of harmonization and interoperability in the geospatial domain.

In line with the approach proposed by INSPIRE, the GVA framework developed relies on use of Service Oriented Architecture, traditionally intertwined with the use of intelligent service-oriented technologies. The SOA approach has brought to a flexible infrastructure whereby different atomic functionalities, i.e. services, can be accessed over the network through a standard interface, to deliver web-based interoperable access to geospatial information.

Specifically the framework heavily relies on very heterogeneous IT ecosystem where several Web Services (WS) are used as the building blocks of federated infrastructures, also known as Spatial Data Infrastructures or SDIs, capable to expose a very heterogeneous set of services including:

- registry and discovery services, capable to ensure unique identification of geospatial data/services available within the infrastructure;
- services for visualisation and download, capable to provide standardized approach to different types of geographical information;
- transformation services, ensuring transparent access to data available within different formats, projections and reference systems and, last but not least,
- processing services, essential to analysis, simulation and more generally to decision making.

From a software engineering perspective, providing intelligent and transparent access to all the aforementioned geospatial services has been perhaps the most important challenge faced during the development of the system. The main goal in fact was to create an infrastructure capable of ensuring user-friendly discovery and search for data and services on top of existing registry and catalogue services yet ensuring compliancy with the Open Geospatial Consortium (OGC) standard CSW (Catalog Service Web).

More specifically the main challenge was to deliver a mechanism ensuring context awareness that could filter, out of massively distributed datasets and services, those of interest for the user. In fact as geospatial data and services grow in number, at an estimated pace of 60% per year, more connected users need to access this data and services in the easiest way. Although several of international initiatives are currently promoting interoperability and harmonization in the field of geospatial information, little effort is being paid

to classical Ambient intelligence Issues including service auto configuration or intelligent service matching.

Figure 1 illustrates, from the architectural point of view, the GVA framework implemented. A lightweight Java-based client can accesses different types of services available within the service ecosystem. The communication among the different components of the SOA relies on use of shared data models as defined by the International Standard Organization (ISO) within the ISO19100 standard series. Furthermore communication is based upon the adoption of common open transport protocols at HTTP level, defined by the Open Geospatial Consortium (OGC), including the so-called Web Mapping Service (WMS), Web Feature Service (WFS) and Web Processing Service (WPS) to name but a few.

To allow for maximum scalability most of the business logic functionalities have been transferred to the server level. This logical shift requires that the business-logic functionalities needed by

Figure 1. A diagram showing the overall interaction between the client application and the service ecosystem

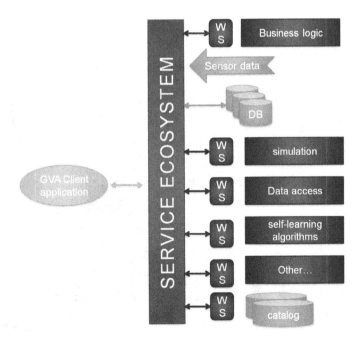

the operator are made available within the SOA in different ways:

- through an Enterprise Service Bus (ESB),
- via simple HTTP post/get,
- through a standard Web Service interface as Simple Object Access Protocol (SOAP) qualified by a WSDL (Web Service Description Language) descriptor,
- through Remote Method Invocation (RMI) or
- through asynchronous messaging-based strategies deployed as Java Messaging Service (JMS).

The concurrent availability of several communication channels ensures that the most suitable choice can be made according to the level of performance vs. interoperability required by the application.

As it will be highlighted in following sections, besides ensuring significant scalability, this approach has brought a further key benefit in that deployment of ubiquitous AmI/GVA solutions can be achieved regardless of the type of application the operator is using. In the framework developed

the client can in fact be deployed as standard standalone Java application, web-based applet, Java WebStart or Java mobile client[8]. Since a number of components responsible for the business logic have been developed at server side, these can be accessed as a service by a variety a different clients ensuring ubiquitous access to data and functions regardless of the device adopted.

Ambient Intelligence and Immersive Geospatial Visual Analytics

The GVA client developed, and illustrated in Figure 2, features a human-centric interface where interactivity and user friendliness, typical of interactive 3D applications, clearly departs from previous IT solutions in the field of Geographical Information Systems (GIS). Here we should highlight that to ensure the best possible immersiveness the authors have not merely adopted a 3D interface but they have dealt with a number of cognitive factors (Conti et al. 2005; Santos et al., 2007; Conti et al., 2008) that could influence the operators' level of consciousness in feeling "immersed" within the three-dimensional environment. The experience from previous work in fact had clearly shown that

Figure 2. A screenshot of the GVA client application

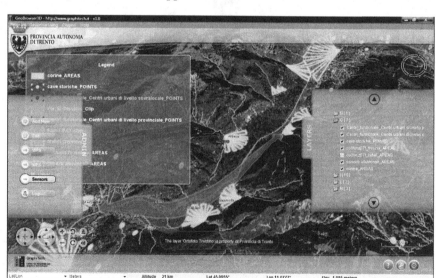

the use of interactive visualization technologies, as well as the use of effective interaction metaphors capable to support in a very transparent manner the workflows typical of the operators during their daily activities, can significantly contribute to the sense of immersiveness provided to the operator without necessarily making use of advanced 3D stereoscopic visualization technologies. To this extent, in the context of geospatial applications, immersiveness becomes an important amplifier of the user's level of context awareness in that it allows effective control over a number of functionalities essential to the environmental managing process.

Following such a user-centered approach the system features a streamlined interfaces capable to build on top of the operator' workflows to significantly reduce their cognitive burden required to interact with a large geospatial datasets. This choice proved essential to let them acquire "context awareness" over the information they are dealing with. Such an improved "conceptual bandwidth" was based upon the use of complementary conceptual channels that allow conveying a greater amount of information regarding the spatial and semantic nature of the actions and functionalities available to the user.

Specifically the GVA client features a 3D interface designed to support the different stages of the operators' workflow. The interface, illustrated in Figure 2, has been designed to automatically adapt to the different services available through the network, according to the specific privileges granted to the user. A list of different information layers is automatically built according to the different services available. Layers are used as logically-stack hierarchical data containers for raster, 2D/3D vector data as well as for any other alphanumerical information that may be required to the operators. A layer in fact can contain for instance maps, vector data describing the street network or 3D placeholders used to describe a specific features.

An *ad hoc* service provides re-projection, if necessary, in a completely transparent manner. This way whenever the users upload new geographical features, they do not need to deal with complex re-projection issues, typical of the geospatial domain, as the system automatically caters for it automatically.

Sensor Networks

As mentioned earlier in the chapter, delivering a user centric approach was considered essential. In a typical GVA scenario, the operator needs to monitor in real time the territory to identify any arising information. The challenge was therefore to develop non-invasive messages and graphical representations that could allow the operator to identify in the shortest possible time portions of the territory potentially subject to risks or environmental crisis. The operators needed to be able to identify the relevant portions of the territory and to visualize, at much higher detail, a specific feature or data pattern within these areas.

Within the system developed Geographic Information coming from sensor networks is accessed in a user friendly way. As shown in Figure 3 in information coming from sensor networks, both available as live data stream or as http/xml formatted data, is collected by a server component which is then responsible to expose the relevant information as web service to the GVA applications.

As shown in Figure 4 the information available from the sensor network is rendered according to the different compound data collected by the physical sensor unit. In the case illustrated a number of compound meteorological data such as air temperature, pressure, humidity are visualized as different segments of a 3D pie menu. In order to support operators in their monitoring activities the operators can set automatic triggering conditions defining specific data patterns of relevance. This information is used to render an arrow next to the relevant pie chart sector indicating a safe (green) pattern or, else, a red arrow

Figure 3. The mechanism used to deliver information coming from environmental sensors

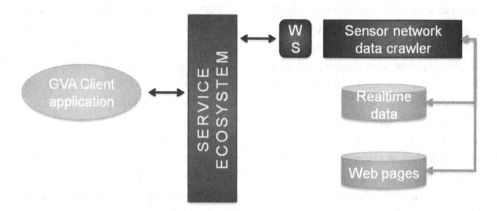

Figure 4. A screenshot of the GVA client showing data from a weather sensor

marking a potentially critical trend. Whenever the operator selects the pie chart segment the system sends automatically a request to a service in charge of plotting data and, as a result, the corresponding time-series is plotted in as overlay chart.

Additionally real-time multimedia data can be shown within the 3D scene. Following a message-driven approach, each user can register to a specific message topic of interest (e.g. emergencies,

traffic, weather, earthquakes). As illustrated in Figure 5, a non-invasive placeholder is created within the 3D scene, within a given geographical position, every time a new information item becomes available. The underlying messaging system is based on the use of GeoRSS feeds, a geographical-aware extensions of standards Really Simple Syndication, or RSS, feeds.

Figure 5. A screenshot of the GVA client showing real-time multimedia data distributed as GeoRSS feeds

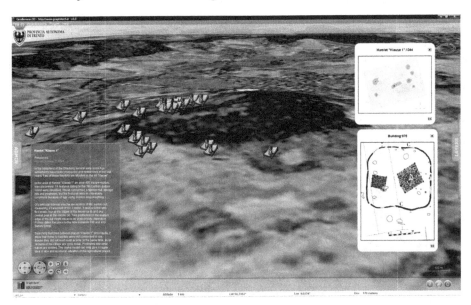

Geo-referenced feeds, containing multimedia information, are used to provide a real time message-based visualization of geo-located pieces of intelligence or information coming from sensors as well as from other users. Data coming from GeoRSS feeds can be made available as yet another data layer. As illustrated in Figure 5 this way it becomes possible to visualize multimedia localized information related to a given point in space.

User Centered Interface

Besides providing user friendly access to geographical information, the GVA environment needed to provide access to processing of geographical information through a well-defined and easy to use 3D graphical interface based on a well known interaction dialogue. In this case it was essential that the operator could be able to interact with the system to simulate and extract intelligence out of existing information. The approach followed by the authors is based upon the visual programming paradigm. When the operator logs into the system he/she is granted

access to a number of geo-processes according to the operator's privileges. The interface is then automatically configured according to the different processing functionalities that may be available to the operator from different servers, as OGC compliant Web Processing Service (WPS). As a result the graph-based interface illustrated in Figure 6 is automatically created with the services available to the user. The metadata associated to each service is used to hierarchically arrange the services within the graph.

To deploy a processing service the operator drags one of the WPS components, represented as a node of the graph, directly onto the 3D scene. Each component is rendered within the environment and it represents an independent processing unit available at server level and which can be executed asynchronously. The metadata associated with each node is used to automatically create a graphical representation of each component.

As illustrated in Figure 7 each processing service is represented by a main component and a number of connectors attached to it, which are rendered within the 3D scene. Within the core

Figure 6. A screenshot of the GVA client showing the graph-based representation of different processes

Figure 7. A screenshot of the GVA client showing a compound process chain

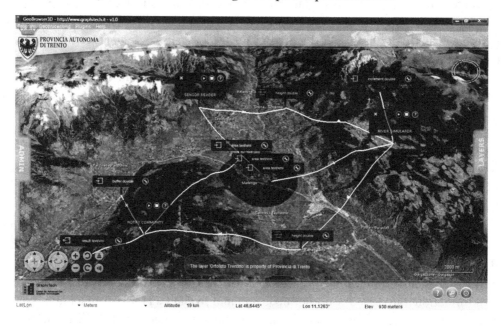

component, the system renders some ancillary information such as the process name, as well as all the necessary graphical user interface elements including buttons to start, pause or stop the process. Attached to the core block, the system renders a number of connectors which are automatically created according to the different input and output data respectively being required by the process and being produced as result of the processing. The meta-information associated to each processing web-service is also used to render a symbol, rendered within each connector, that represents the corresponding data type. This could be for instance an image, text, or same numerical value of different type (a double, a float etc.).

Following this approach if the users wished for instance to create a buffering zone around a given feature, they should select the appropriate processing component (e.g. performing a buffering operation based on the Douglas-Peucker algorithm) and drag the input slot over a given vector feature. As soon as the user presses the play button the information on the vector feature is transmitted to the server which hosts the processing service, which in turn calculates the buffer and sends it back to the client which eventually renders it within the 3D scene. A message based system among client and server components ensures that once the process is invoked the result is notified back to the client whenever the result is available.

The entire process is absolutely transparent to the final user who only needs to drag the icon from the graph onto the scene, connect the relevant information and invoke the service.

This web-service oriented approach has been pushed even further in that the system allows operators to create complex compound processes through chaining of elementary processing units. The different components in fact are considered by the GVA application as the building block of a workflow. A complex workflow can be easily created by the operator by dragging an output connector over an input connector (or vice versa). If the two connectors have compatible data types the system automatically snaps the two creating a direct link within the two processing units. This process can be repeated indefinitely with an arbitrary number of processes. Most importantly it should be noted that the different processes can represent web processing services (WPS) physically residing on different servers. As this is completely transparent to the user he/she is not aware of the underlying service infrastructure.

As soon as the operator invokes the process chain the system automatically caters for the orchestration of the different process, passing the requested data from one process to the next. In an AmI perspective, interoperability between services, based on the use of the WPS transport protocol and data type matching, is automatically enforced through the meta-information available for each of the process. This ensures that the user can focus on the specific workflow without being forced to deal with the technicalities of each different process.

Future Challenges

As illustrated in the previous sections the convergence of the paradigms at the base of both AmI and GVA is essential for the development of new techniques and tools capable to provide better insight to operators in need to access, manage and process vast repositories of Geographic Information(GI), who need to identify key data pattern of environmental interest. This convergence will affect on a number of domains such as eEnvironment, eEnergy, eTraffic, eDisaster Management, homeland security and environmental security.

On the other hand the nature of environmental management requires that GVA interfaces and tools need to be capable to provide a seamless operational functionality regardless of the device adopted, thus extending the concept of "adaptability" to different input/output configurations. This is essential for instance to ensure that the operator can move from a desktop based GVA application to a high-resolution tessellated screen, typical of control room, or to a mobile application required for *in situ* operations, in a fully transparent way. Following a context aware paradigm, the convergence of a number of key enabling technologies, including interoperable web services, broadband and wireless communication infrastructures, localization technologies (including GPS) and easy-to-use mobile IT solutions, is paving the way to a new generation of intelligent mobile GVA/

AmI solutions. This aspect represents one of the major challenges to be faced in the near future.

Building on top of Location Based Services (LBS) future developments will have to focus on how to deliver ubiquitous GVA solutions which transcend the boundaries of the physical device being adopted. Context-aware access to GVA solution, is essential to support on site operators in the process of synthesis and analysis of information. Clearly this requires specific new approaches to deliver the best (and safest) experience to a user in the most different conditions (both environmental and hardware), in order to make geospatial data interactively available to a onsite users looking for decision support. For this reason future research will particularly focus on intelligent complexity reduction. This is essential to minimize interaction with the device (e.g. to better analyze a give situation) and therefore to maximize its efficacy.

A first prototype of mobile client, visible in Figure 8, has been already developed by the authors. The challenge to be faced now is make use of Geographic Information to deliver personalized and context-aware GVA solutions available from mobile devices.

The challenge open to GVA is therefore to embrace AmI paradigms is to ensure full adaptability of the system (in terms of interface, feedback, dialogue) to the various operating conditions (driving, walking etc.) as automatically identified by the system. To be able draw conclusions from a combination of evidences and assumptions in a GVA context within a nomadic scenario will in fact require addressing a number of issues such as screen size, interaction devices (mouse, touchpad, touch screen, voice interfaces), feature restrictions imposed by the use of portable hardware, as well as variable environmental conditions, most notably presence of glare or string sunlight. This includes support of appropriate interaction techniques according various operating contexts (e.g. a walking onsite operator may want to receive only voice alerts from the system through the headset without being forced to watch the screen).

Figure 8. A screenshot of the first prototype of mobile system

For this reason the main challenge to be addressed in the near future is clearly the development of GVA interactions metaphors specifically targeted to mobile users, based on small screens and limited interaction capabilities, which will allow interactive support for interpretation of crucial information patterns. Data as well as user preferences need then to be saved in a centralized form and made accessible to mobile users in a fully transparent way while operating on site. It will be also necessary to investigate how to support analytic reasoning, note taking, presentation and dissemination of Geographic Information on mobile devices. New visual representations will have to be studied to ensure appropriate analysis of complex and heterogeneous information stemming from information system, e.g. based on data sources like on-site sensors, databases and geographic information systems. The integrated views of the data will have to support and improve perception and evaluation of complex situations by not forcing the user to perceptually and cognitively integrate multiple separate elements.

For this reason a number of alternative interactions and representations will have to be investigated to allow creating of different representations to support the user identifying, in the shortest possible time, the best option. The challenge is to allow users to analyse the data available, to identify sections of the environment subject to alerts, to select the relevant sections and to visualise further information at higher detail. For this reason further work will explore interactive construction and processing of spatial queries through mobile interfaces as well as on the creation of the relevant visualizations. Not only the representations will need to be refined, but so will the interactions with respect to the precision and the type of incremental, intermediate results of the reasoning process.

Given its focus, a dialog-based interaction needs to be developed ensuring provision of non-invasive educational interventions to help the operator comprehend the most critical issues and to promote safe choices. Dialog will be essential for instance when a sequence of critical situations is detected, to inform the operator and to create a higher awareness on security-related issues.

In a different context, if we compare the proposed framework with existing smart home architectures, information coming from the surrounding environment could be used by the intelligent home. For instance in case of emergency, when a sudden reduction in the production of electricity would require power cuts, a GVA system could be used to rapidly assess the areas potentially affected by the shortage and to send instructions to smart homes to have them automatically cut unnecessary power drainage, powering only essential home services.

Taking an opposite perspective we could even consider that information originating from single smart homes (e.g. on energy consumption) could be used within the general infrastructure,

in fact thus transforming each smart home into an additional node or of the sensor network and thus introducing a further level of distributed intelligence within the system (e.g. to optimize energy production).

Last but not least it should be noted how the system presented in fact could be used in a range of very different applications, for instance as the basis for a social networking tool for elderly and persons living alone at their homes, by providing them easy and interactive access to the information related to their community or their interests, thus introducing yet another social-aware application to the system presented.

ACKNOWLEDGMENT

Part of the achievements discussed in this paper, as well as some of the data used for the illustrations, are the result of the project "Sviluppo di una Infrastruttura Dati Territoriali secondo una Architettura Orientata ai Servizi" commissioned by Autonomous Provincia of Trento, Italy. Additionally part of the work presented in this chapter has been developed in the context of the project i-Tour "intelligent Transport system for Optimized URban trips". Specifically the research leading to these results has received funding from the European Community's Seventh Framework Programme (FP7/2007-2013) under the Grant Agreement n. 234239. Further some of the material shown within the images of this chapter have been produced in the context of the EU project Net-Connect "Connecting European Culture Through New Technology" financed by the Cultura 2000 programme. Finally the authors wish to thank Prof. Licia Capra at UCL London for providing an extensive reference list on trust management within social networks and Mr Stefano Piffer, Mr Bruno Simões, Mr Daniele Magliocchetti from Graphitech for their support on the development of the PC and mobile client.

REFERENCES

Abdul-Rahman, A., & Hailes, S. (1997) Using Recommendations for Managing Trust in Distributed Systems. In Proc. of IEEE Malaysia International Conference on Communication (MICC'97), Kuala Lumpur, Malaysia.

Beth, T., Borcherding, M., & Klein, B. (1994) Valuation of Trust in Open Networks. In Proc. of the 3rd European Symposium on Research in Computer Security, pp.3–18, Brighton, UK.

Blaze, M., Feigenbaum, J., & Keromytis, A. D. (1998) KeyNote: Trust Management for Public-Key Infrastructures. In Proc. of 6th International Workshop on Security Protocols, vol. 1550 of LNCS, pag. 59–63, Cambridge, UK, Springer-Verlag.

Blaze, M., Feigenbaum, J., & Lacy, J. (1996) Decentralized Trust Management. In Proc. of IEEE Symposium on Security and Privacy, pag. 164–173, Oakland, Ca.

Capra, L. (2004) Engineering Human Trust in Mobile System Collaborations. In Proc. of the 12th International Symposium on the Foundations of Software Engineering (SIGSOFT 2004/FSE-12), pag. 107–116, Newport Beach, CA, USA, ACM Press.

Carbone, M., Nielsen, M., & Sassone, V. (2003) A Formal Model for Trust in Dynamic Networks. In Proc. of First International Conference on Software Engineering and Formal Methods (SEFM'03), pag. 54–63, Brisbane, Australia.

Conti, G., Disperati, S. P., & De Amicis, R. (2009). The use of Geobrowser, Virtual Worlds and Visual Analytics in the context of developing countries' Environmental Security. In Proceedings of 6th Digital Earth Conference, Beijing, 2009.

Conti, G., Ucelli, G., & De Amicis, R. (2005). "Verba Volant Scripta Manent" a false Axiom within Virtual Environments. A semi-Automatic Tool for Retrieval of Semantics Understanding for Speech Enabled VR Applications". In Computers & Graphics vol. 30, no. 4, Volume 30, Issue 4, August 2006, pp. 619-628. http://dx.doi.org/10.1016/j.cag.2006.03.004. ISSN: 0097-8493.

Conti, G., Witzel, M., & De Amicis, R. (2008). A user-driven experience in the design of a multi-modal interface for industrial design review. In Khong C. W., Wong C. Y., von Niaman B. (Eds), Proceedings Human Factors in Telecommunication, 17-20 March 2008, Kuala Lumpur Malaysia, Prentice Hall, New York, pp. 351-358.

De Amicis, R., Conti, G., Piffer, S., & Simões, B. (2009a). Geospatial Visual Analytics, Open Challenges And Protection Of The Environment. In De Amicis R., Stojanovic R., Conti G. (Eds) (2009). GeoVisual Analytics: Geographical Information Processing and Visual Analytics for Environmental Security. Springer.

De Amicis, R., Conti, G., Simões, B., Lattuca, R., Tosi, N., Piffer, S., & Pellitteri, G. (2009). *Geo-Visual Analytics in the Future Internet. In the International Journal on Interactive Design and Manufacturing (IJIDeM), March 2009.* Springer.

Grandison, T., & Sloman, M. (2003) Trust Management Tools for Internet Applications. In Proc. of the 1st International Conference on Trust Management (iTrust), Crete.

Herlocker, J., Konstan, J., Terveen, L., & Riedl, J. (2004) Evaluating collaborative filtering recommender systems. In ACM Transactions on Information Systems, 22, pag. 5–53.

Laituri, M., & Kodrich, K. (2008). On Line Disaster Response Community: People as Sensors of High Magnitude Disasters Using Internet GIS. *Sensors (Basel, Switzerland), 8(5),* 3037–3055. doi:10.3390/s8053037

Liu, J., & Issarny, V. (2004) Enhanced Reputation Mechanism for Mobile Ad Hoc Networks. In Proc. of the 2nd International Conference on Trust Management (iTrust), vol. 2995, pag. 48–62, Oxford, UK, LNCS.

Quercia, D., Hailes, S., & Capra, L. (2006) B-trust: Bayesian Trust Framework for Pervasive Computing. In Proc. of the 4th International Conference on Trust Management, pag. 298–312, Pisa, Italy, LNCS.

R. De Amicis R. R. Stojanovic R., G. Conti G. (2009). GeoVisual Analytics: Geographical Information Processing and Visual Analytics for Environmental Security. Springer.

Santos, P., Stork, A., Gierlinger, T., Pagani, A., Paloc, C., Barandarian, I., et al. (2007). IMPROVE: An innovative application for collaborative mobile mixed reality design review. In International Journal on Interactive Design and Manufacturing, Springer Paris. ISSN: 1955-2513 (Print) 1955-2505 (Online).

Thomas, J. J., & Cook, K. A. (Eds.). (2005). *Illuminating the Path: The Research and Development Agenda for Visual Analytics*. IEEE CS Press.

Weeks, S. (2001) Understanding Trust Management Systems. In Proc. IEEE Symposium on Security and Privacy, pag. 94–105, Oakland, CA.

KEY TERMS AND DEFINITIONS

API: Application Programming Interface
ESB: Enterprise Service Bus
GI: Geographic Information
GIS: Geographical Information Systems
GVA: GeoVisual Analytics
OGC: Open Geospatial Consortium, Inc.
SDI: Spatial Data Infrastructure
SOA: Service Oriented Architecture

ENDNOTES

[1] A notable example is the Nokia Sports Tracker available at: http://sportstracker.nokia.com/nts/main/index.do

[2] For online access refer to http://earth.google.com/katrina.html

[3] For online access refer to http://www.google.it/landing/terremoto_abruzzo.html (in Italian)

[4] For a more comprehensive study on the issues of trust within social network refer to Grandison et al. (2003), Blaze et al. (1996), Blaze et al. (1998), Abdul-Rahman et al. (1997), Liu et al. (2004), Capra (2004), Quercia et al. (2006), Carbone et al. (2003), Beth et al. (1994), Weeks (1997), Herlocker et al. (2004).

[5] A notable example of this is the ShopSavvy application (available at: http://www.biggu.com/). When user takes a picture of a product's bar code with his/her mobile phone the software locates the shop offering the same product in the vicinities. The user can then select the shop (e.g. offering the lowest price). Using the GPS receiver fitted within the Smartphone the system provides then the necessary directions to reach the shop.

[6] NASA WorldWind, available at: http://worldwind.arc.nasa.gov/

[7] More information on the INSPIRE Directive are available from: http://inspire.jrc.ec.europa.eu/

[8] A first implementation of the mobile client has been developed over the Android™ SDK for Smartphones developed by the Open Handset Alliance. More information on Adroid can be found at: http://www.android.com/

Chapter 29
Possibilities of Ambient Intelligence and Smart Environments in Educational Institutions

Peter Mikulecký
University of Hradec Králové, Czech Republic

Kamila Olševičová
University of Hradec Králové, Czech Republic

Vladimír Bureš
University of Hradec Králové, Czech Republic

Karel Mls
University of Hradec Králové, Czech Republic

ABSTRACT

The objective of the chapter is to identify and analyze key aspects and possibilities of Ambient Intelligence (AmI) applications in educational processes and institutions (universities), as well as to present a couple of possible visions for these applications. A number of related problems are discussed as well, namely agent-based AmI application architectures. Results of a brief survey among optional users of these applications are presented as well.

INTRODUCTION

The vision of Ambient Intelligence (*AmI*) offers the conception of environment that will be sensitive and responsive to the presence of humans. The AmI vision builds on advanced results of interdisciplinary research. The development of AmI applications is a complex task and all their features and functioning can be hardly predefined or forecasted because of different emergent or synergic effects.

We believe that educational institutions in general, and especially universities are one of promising application domains where experiment-

DOI: 10.4018/978-1-61692-857-5.ch029

ing with AmI solutions could be quite fruitful. This is a knowledge-rich environment, where an intelligent support from the side of the environment can be very beneficial for all types of expected users, and users (students, lecturers, and other staff) are expected to be open to new technologies and approaches.

In this chapter, written by a group of scholars from the University of Hradec Kralove, Czech Republic, the authors wish to present some ideas supporting the vision of smart environments for university education. More precisely, this vision aims at envisaging a truly intelligent environment for education where the access to relevant knowledge or information will be as easy as possible, relevant to the level of students' skills and suitably supporting the teachers' lectures or practices. In such an environment, any kinds of information- and knowledge-related problems would be solved immediately using an appropriate support from the environment, and where studying will be a real pleasure.

The organization of this chapter is as follows. Ambient Intelligence, Net Generation and Homo Zappiens concepts are explained firstly. Then the AmI vision for universities is described through three application visions. Possible AmI application agent-based architectures for these visions are suggested. The results of a brief questionnaire are presented. The objective of the questionnaire survey was to obtain feedback from the expected users of the proposed applications, namely from students, teachers and administrative staff of the university.

AMBIENT INTELLIGENCE

As it is well known, the concept of Ambient Intelligence (*AmI*) was introduced in the ISTAG report (ISTAG, 2001) and interpreted e.g. by (Alcaniz and Rey, 2005; Remagnino et al, 2005; Bohn et al, 2005; Snijders, 2005) and others. This concept provides a vision of society of the future, where

the people will find themselves in an environment of intelligent and intuitively usable interfaces, ergonomic space in a broad sense, encompassing better, secure and active living environment around them, capable of aiding them with daily chores and professional duties by recognizing the presence of individuals, reacting to it in a non-disturbing, invisible way, fully integrated into the particular situation. Nearly synonymous concepts of disappearing computing or calm computing express the technology diffused into everyday objects and settings (Russell et al, 2005). From the technological point of view, *AmI* bears ship to the conception of ubiquitous computing (*Ubi-Comp*), the term firstly used by Mark Weiser in 1998 (Alcaniz and Rey, 2005; Bohn et al, 2005). The *UbiComp* is defined as the use of computers everywhere and is determined by interactions that are not channelled through a single workstation.

The *AmI* environment is characterized by merging of physical and digital space. It means that tangible objects and physical surroundings are acquiring a digital representation (Kameas et al, 2005). The *AmI* environment is considered to host several *UbiComp* applications.

The *AmI* artefact (also smart object, smart device) is an element of the *AmI* environment that has got following properties and abilities (Kameas et al, 2005):

- information processing,
- interaction with environment,
- autonomy,
- collaboration,
- composeability,
- changeability.

The building of an *AmI* artefact from any common object consists of two phases: embedding hardware modules into the object and installing software. Hardware components are especially batteries, sensors, processors, wireless modules and screens. Software components are those of hardware drivers, networking subsystems, operating system, and middleware for integration of artefact in distributed systems. The *AmI* hardware builds on four components: distributed processing, hierarchical storage, tangible interfaces and ubiquitous communication. Snijders (2005) presents a three-level hierarchy of devices with different functionalities, designed to solve the energy constraint that is a determinative factor of any *AmI* application.

Technical features of UbiComp systems and the main tasks to be solved by AmI technological background are summarized e.g. by (Alcaniz and Rey, 2005). The crucial research domains related to the AmI vision are suggested in (ISTAG, 2005). The main areas to be evolved are those of:

- development of necessary hardware, especially
  - smart materials that enable mass storage, emit light, process data, active and passive tagging or access to networks,
  - specific devices for particular applications, with limited processor and hard disk requirements and therefore of low cost,
  - sensor technology bridging the physical world and the cyberworld,
  - interfaces with a good display quality and responsiveness to user input, supporting natural interaction that combines speech, vision, gesture and facial expression,
- defining of new software architectures and appropriate software, mainly
  - 'invisible' file systems, that allow user to access data under the principle of "produce one, present anywhere", but without knowing specific file names, locations and formats,
  - automatic installation mechanisms and migration of programs from one computer to another with ability of self-managing and self-adjusting,

but without requiring fundamental changes in configurations,
- working on contextual awareness and personalization of information that is tailored to user's requirements and location observed by networks of sensors and cameras, status tracking, interactions, user modelling etc.,
- working on privacy issues, both practically in sense of encryption techniques to ensure security and theoretically (see (Lessig, 1999) for the concept of personal privacy in cyberspace).

The concept of AmI is strongly motivated by economic aspects – probably economic motivation is the most significant incentive in this area. A discussion about real time or now-economy has been presented in (Bohn et al, 2005), where more and more entities in the economic processes, such as goods, factories, and vehicles, are being enhanced with comprehensive methods of monitoring and information extraction. The authors point out how two technologies, the ability to track real-world entities and the introspection capabilities of smart objects, will change both business models and consumers' behaviour.

The societal acceptance of AmI vision depends on such features of AmI applications as:

- ability to facilitate human contact,
- orientation towards community and cultural enhancement,
- ability to inspire trust and confidence,
- supporting citizenship and consumer choice,
- consistence with long term sustainability both at personal, societal and environmental levels,
- controllability by ordinary people (ISTAG 2002).

The psychological theories of different types of intelligence can help to understand human reasoning, and human interaction with machines. Each individual possesses diverse intelligences (see e.g. logical, linguistic, musical, spatial, interpersonal and other intelligences provided by (Gardner, 1985), or analytic, creative and practical intelligences offered by (Stemberg, 1985)) in different percentages. As (Bettiol and Campi, 2005) notices, this mixture of intelligences determines the learning style and motivations of each individual, therefore the AmI application must adopt itself dynamically to peculiarities of its users. Other psychological factor, that has to be taken into account when designing AmI environments, is that people tend to continue their habits; therefore the applications should respect the natural behaviour patterns of humans.

The political impacts of the vision of AmI have their starting point in the resolution adopted at the Lisbon congress of the EU in 2000, on the basis of which the European Commission resolved to secure Europe's leading role in the field of generic and applied technologies for creation of the knowledge society. The new technologies must not be the cause for excluding some groups of citizens from society, but it must ensure universal and equal access to its digital knowledge sources. The most controversial, breath-taking implications of the AmI vision, especially those that seem to attack the freedom of choice of humans, or to increase our dependence on the correct functioning of numerous artificial systems are logically related to its psychological dimension and represent the main barrier that can at least slow-down the acceptance of AmI approach.

NET GENERATION AND HOMO ZAPPIENS

Net Generation is the generation of youth which is growing up with modern information and communication technologies shaping strongly their mental models, i.e. views on the world around them. While using several technologies they are learning to develop new skills and exhibiting new behaviour patterns. This generation is also called Homo Zappiens (Veen, van Staalduinen, 2009) or Millennials (Howe, Strauss, 2000). According to (Veen, van Staalduinen, 2009) and (Oblinger and Oblinger, 2005) characteristics of this generation are:

- preference for images and symbols as an enrichment of plain text,
- seemingly effortless adoption of technology,
- cooperation and sharing in networks,
- usage of technology in a functional manner, i.e. not touching what they can't use, and increasingly, this generation seems to take exploration and learning, discovering the world, into their own hands,
- gravity toward group activity,
- identification with parents' values and feel close to their parents,
- fascination by new technologies.

Characteristics of Homo Zappiens are summarized in the table 1.

Although Net Generation is usually described in terms of age, it is not exclusively related to this attribute. Exposure to technology is the other one which has the same relevance. For example, as stated by Oblinger and Oblinger (2005) individuals who are heavy users of IT tend to have characteristics similar to the Net Generation. In fact, the pervasiveness of technology in our professions and in our personal lives ensures that most individuals gradually assume some Net Generation characteristics.

TECHNOLOGY IN EDUCATIONAL PROCESS

In the research presented by Kvavik (2005) students said that they used a computer first for doing classroom activities and studying (mean of

Table 1. Characteristics of Homo Zappiens (adapted from (Veen, van Staalduinen, 2009))

Characteristic	Description
Iconic preferences	Homo Zappiens' preference for icons is a very necessary attitude to survive in an era where older generations are confronted with 'information overload', yet Homo Zappiens seems capable of handling this phenomenon (Veen and Vrakking, 2006). In its communications with peers Homo Zappiens uses icons and abbreviations as well. Lindström and Seybold (2003) have labelled this language of shortcuts 'TweenSpeak'.
Technology is air	Homo Zappiens is merely interested in technology if it works and will just as easily pick up something else if that suits their needs better. They often have little understanding of the fundamentals of the technology they are using, yet they can explain the functions that make a tool useful. Tapscott (1998) formulates this perception of technology as: "It doesn't exist. It's like the air".
Inversed education	Up to about the age of five, children seem to ask their parents how to use a personal computer. From the age of six most children have learned how to use the personal computer and will often first resort to asking friends, before them asking their parents. From the age of eight upwards, this generation is educating their parents on how to use PC's (Veen and Jacobs, 2005). This 'inverse education' is typical for this generation.
Networking is their lifestyle	To the Net Generation, living in networks is as normal as breathing. Homo Zappiens' networks include both virtual and physical networks. They are almost constantly connected to electronic networks, through which they stay in contact with their friends and a wide source of information available.
Cooperation	Homo Zappiens has made the use of networks a lifestyle. They use their network of contacts to provide them with the information they need and if this network does not suffice, they ask an online community consisting of many individuals they do not know, but who are willing to help. For the Homo Zappiens, knowledge sharing is common even with those who you do not know at all.
Virtual is real	Youth today does not make the same distinction between the 'real' world and the 'virtual' world that so much of society still does. To them, when they communicate with a friend through chat or in a game, this communication is not less real than a physical meeting. Communities and social networks appear to be physical, virtual and hybrid at the same time. (Oblinger and Oblinger, 2005).
Multiple identities	Homo Zappiens has online and face-to-face identities as is illustrated by a boy describing a friend: "Online he is okay, but at school he is a nerd". (Veen and Jacobs, 2005). Young people are accustomed to playing with different characters or roles and feel the consequences of these different roles as other gamers react on them.
Multitasking	These children seem to be online, watch TV, talk on the phone, listen to the radio and write a document, apparently all at the same time. (Oblinger and Oblinger, 2005). Children seem to divide their attention across the different information flows, focusing only on one, but keeping a lower level of attention on the others. By using their attention flexibly, Homo Zappiens seems capable of handling much more information than previous generations (Veen and Vrakking, 2006).
Critical evaluation	As a consequence of multitasking, they instantly and almost subconsciously value different streams of information to decide where to place their attention. Homo Zappiens is confronted with a lot of information, not all of it to be taken at face value. Critical evaluation is what children do when selecting and filtering information flows.
Zapping	Homo Zappiens seems to show a zapping behaviour that is specifically aiming at filtering information from different programs at a time. The purpose is to get the message in order to understand (Veen and Vrakking, 2006). It allows them to select only those bits of information from each channel that are critical for understanding what the program is all about.
Instant pay-off	The Net Generation has little patience and short attention spans. Their skills are aimed at processing various flows of different information quickly, but they have also come to expect this kind of high-density information streaming; anything less and they will become bored. Lindström and Seybold (2003) label them the 'Instant Generation'.
Self-confidence through self-direction	ICT offers youth control over, not just devices, but communication, networks and situations as well; situations which they will often have to master as adults (Tapscott, 1998). Through the use of technology, this generation has added options for exploring their own individualism. Games are a prime example of this, as they allow any gamer an infinite number of tries to attempt to reach certain goals.

4.01). Students used the computer approximately 2–5 hours a week for writing documents, surfing the Internet for pleasure, e-mailing, using instant messaging, using an electronic device at work or downloading/listening to music or videos. Other activities such as completing a learning activity, playing games, creating spreadsheets, and creating

Table 2. The future of higher education (adapted from (Veen, van Staalduinen, 2009))

Element	Consequences
Primary process	Teaching: Teachers will take up the role of expert and coach. They still have a thorough understanding of subject matter, and matching skills, but their expertise will mostly be used to coach and coax students into finding out for themselves. Learning: Will become more individual. Student will go 'shopping through universities', using blended learning, and keeping a personal portfolio about their education and experience. Initial education will still be 100% class-taught, but as students progress, more of their learning will occur through (professional) practice.
Content	Content will increasingly make use of standardized formats for learning modules that are internationally accepted. Universities and organizations will offer distributed and discontinuous learning modules; so-called 'learning bricks'.
Inter-organizational	Increasing multi-disciplinarity in subject matter will force faculties and other educational institutions to restructure and reorganize. Individual students will be assessed in terms of competences and experience, as presented through their (online) portfolios. As all students have individual, and thus unique, learning histories (Diepstraten, 2006), this will impact the value of diplomas granted by individual universities, obsolescing them eventually.
Intra-organizational	Relationships between universities, corporations, and governmental organizations will become more and more important. Valorisation of research, through partaking in organizational networks, will become the key focus of higher education. In order to make this happen, higher education will more boldly have to take up the role of contributing to the understanding of societal processes and problems (such as obesity and cancer), and its solutions. Higher education will do this by providing scientifically valid analyses that can be trusted by society.

presentations (including Web sites) occupied an average student's time less than 2 hours per week.

Kvavik found that the most frequent computer use was in support of academic activities and that presentation software was driven primarily by the requirements of the students' study field and the curriculum. Factors that explain hours of use fall into the following categories: academic requirements, class status, gender, and age. Academic usage is strongly related to the student's academic study field and class status (senior/freshman). Communications and entertainment are significantly related to gender and age (Kvavik, 2005).

For the above mentioned reasons it is inevitable for universities, which perform as basic elements in the higher educational system, to bear in mind these characteristics and be prepared for related changes. These have to be effectively incorporated in learning, research and administrative processes. Educating people with these characteristics (regardless of the study type, study field or students´ age) leads to necessity to fulfil requirements in technology deployment and technology learning. Related issues shaping future of educational institutions are outlined in the table 2.

As stated by McNeely (2005) the main feature that Net Generation expects from technology is interactivity, regardless if it is with a computer, a professor, or a classmate. Traditional lectures are not fulfilling their learning potential. Contemporary usage of technology in form of learning management systems or online courses doesn't work well with them. The development of Web 2.0 causes that the social component of learning with the help of technology is required (Husband, Bair, 2007). Kvavik (2005) found that most students (over 70%) prefers moderate or extensive usage of IT in their classes (see table 3). Very important fact is that required changes are "field-independent", i.e. the utilization of technology is expected in different disciplines. It is obvious from the table 4 that engineering students have the highest preference for technology in the classroom (67.8%), followed by business students (64.3%). Moreover, the preference of no technology is low in every study fields (Kvavik, 2005).

Surely, the usage of technology is possible in most courses, but it depends on teachers' technical skills and their willingness to look for new ways of utilizing IT in the education. In (Olševičová, Poulová, 2009) the statistics of

Table 3. Student Preference for Use of IT in Classes (adapted from (Kvavik, 2005))

Student preference of...	Number of students	%
Classes with no IT	128	2.9
Classes with limited IT	991	22.7
Classes with moderate IT	1802	41.2
Classes with extensive IT	1346	30.8
Entirely online classes	97	2.2

Table 4. Preferences for technology by study fields (adapted from (Kvavik, 2005))

Discipline	Prefer no technology	Prefer limited technology	Prefer extensive technology
Engineering	4.8%	24.4%	67.8%
Business	1.3%	28.2%	64.3%
Life sciences	4.8%	35.3%	56.3%
Physical sciences	5.7%	40.9%	51.8%
Social sciences	7.9%	44.4%	44.2%
Education	3.5%	47.9%	42.9%
Humanities	7.7%	47.9%	40.2%
Fine arts	9.0%	46.9%	39.3%

growing numbers of e-learning courses provided by the faculty of authors of this chapter was presented. (Approx. 70% of teachers participate on e-courses' design. There were 187 e-courses in use in academic year 2008-2009, with nearly 21 000 seats in total. Each student accessed around 4 e-courses per term.)

To sum up, Clayton-Pedersen (2005) claims that there are several implications for educational institutions related to education of Net Generation. These are closely related to development and existence of smart environments in the universities:

- Faculty's understanding of the teaching and learning power of technology needs to be increased.
- Increasing the use of technology will increase demands for technological tools to be effectively integrated into the curriculum to enhance student learning.

- Tools need to be developed to help faculty integrate technology into the curriculum.
- Much of the learning technology innovation in higher education has been focused on K–12 teacher preparation and development. More focus needs to be placed on preparing existing faculty for the future Net Generation students who will populate the 21st-century classroom.
- Institutions need to establish greater expectations for maximizing their investment in technology by exploring and assessing the best use of technology for learning.
- Greater investments may be needed in faculty professional development in the effective use of technology for learning.
- Faculty's effort to infuse technology into the curriculum requires support in developing strategies and in resolving technical difficulties. This means more than the technical help desk. What is needed is assis-

tance for using technology to achieve the teaching and learning outcomes we desire.

- There is a need for integrating technology that is in the service of learning throughout the curriculum.
- More intentional use of technology to capture what students know and are able to integrate in their learning is needed.
- The extent to which technology is a tool for learning and a tool for assessment of learning will facilitate faculty's increasing comfort in integrating technology into the curriculum.

AMI VISIONS FOR UNIVERSITY

The university is a good place where experiments with innovative applications can be performed, because there is a lot of people who are interested in and open to new technologies. The intelligent applications can be built upon the existing ICT infrastructure, such as university intranet applications, learning management systems, information systems and databases of libraries etc. Also there are numerous types of hardware devices that can be involved, e.g. ID cards, equipment of computer labs, private notebooks and mobile phones of students, teachers and administrative staff etc.

We focused first on defining particular visions of AmI applications for supporting activities of people at the university. As an example we describe the optional realization of visions at the University of Hradec Kralove, which is a typical example of a small Czech university, and we discuss pros and cons of these applications introduction.

To study the impact of AmI applications to human privacy, and to learn how time-consuming and disturbing would be the usage of the AmI extension of existing systems, we suggested following visions for identification of key aspects worth to be tackled in an educational environment.

Vision No 1: Application for Locating People in the University Campus

University campus consists of several buildings where people migrate from place to place during the day. From time to time, student looks for his/her teacher, teachers look for their colleagues, people need to know who is or is not occupied, who went for lunch and will be back soon, who left for the conference abroad etc. New students in the first semester still do not know names of all teachers, or locations of their offices etc. Part-time students often take long journey to get to the campus and they could be really disappointed in case the consultation with the teacher was not possible.

There are different information systems that can provide partial, more or less fuzzy and more or less formal information about locating people. Online scheduling system informs about schedules of all teachers, this information is formal and available to everyone, and the schedule is valid for the whole semester. Actual changes, like lessons cancelled due to illness of teacher, are highlighted on a special website that is available after logging to the university intranet. Study agenda information system contains information about bindings of teachers and students to courses, and about dates of exams (if classroom is booked for the exam, information is available in the scheduling system, too). Learning management system (e.g. WebCT) involves information about online status of students and teaching assistants of e-courses. Some teachers (and students) use communication channels like ICQ or Skype to be in touch, so there are groups of people who can deduce whether their colleagues are or are not in their offices from the on-line status, but this information is private and not available to people who are not in contact lists. Also private mobile phones numbers are not available to everyone. If more teachers share the same office, they can provide information about locating their roommates to optional visitors. Some information can be available to the management of the faculty, like information from the security cameras in classrooms and labs, or information from ID cards of students and employees who entered the building. If people use MS Outlook or Google calendar to organize meetings, these application contain location information, too.

We can simply imagine a university smart environment as a system that would integrate data from all previously mentioned systems to provide information about current location of all the people in the campus. The application would be accessed mainly using a mobile phone and/or ID card with a smart chip and would inform whether the position of a given person is known or is not, and if the person is available to the searcher. E.g. the teacher is available to students of his courses anytime from Monday to Wednesday, but to other students only on Friday, while Thursday is dedicated to project meetings with colleagues (so the teacher does not provide consultations to students). The profile of availability of each person would be described by the set of rules, defining activity, location, list of searchers who can see this status and list of searchers who are always welcomed. Part of rules can be generated automatically from existing systems and databases (like scheduling system or WebCT), part of them would be managed by the owner of the profile.

Certainly we can believe that people at the university campus are open to meet other students or teachers. The question is whether these people would accept our smart environment solution or would perceive it like something what endangers them. Also updating profiles would be time-consuming in comparison with simpler but not so powerful methods (e.g., yellow paper sticks on doors).

Vision No 2: Intelligent Assistant for Graduation Thesis Preparation

Final thesis is central for students' graduation at universities. It is obvious that the selection of topic and supervisor are of a crucial importance.

However, at this stage the student is usually lacking knowledge of the domain, knowledge of the academic personnel, and is not experienced enough in searching and using relevant resources. Sometimes the reality is far away from students' applying systematic approaches and working effectively. In the phase of the thesis objective formulation, students may be heavily dependent on supervisors who might not be aware of students' personal profiles and interests. In later phases, students tend to postpone their work due to other duties, due to being overloaded with resources or – contrary – due to having very few resources, being confused with irrelevant resources, being unable to differentiate between serious electronic resources and Wikipedia-like articles etc.

There are various factors influencing the final decision about the objective of the thesis. These are sometimes handled in a random and messy manner. This is caused mainly by the lack of information available at the right time.

The topic selection is tightly interlinked with selecting the supervisor. At Czech universities, there are several possibilities:

- Student gets interested in some topic from the list of available topics published by the faculty.
- Student can address the assumed supervisor with his/her own topic or idea, based on private interests, preferences, working experiences etc.
- Student can address the assumed supervisor and ask him for research-related topics, typically future PhD students do this.
- Supervisor can suggest the topic to particular student in relation to his/her achievements during the course of the study.

Only a minority of students are picked directly by the supervisor. The rest of the students are dependent on the availability of the topic and the supervisor. Typically, the lists of available topics are presented on whiteboards or electronically in the study agenda system, on teachers' websites etc. Some topics are defined by external authorities (companies, public administration, various organizations) in cooperation with internal supervisors (members of the academic staff). Other topics are related to the research projects of the faculty. Certain themes require passing particular courses and students would need to be informed about it in advance to be able to pass these courses before they decide about their final topic.

It is assumed that the supervisor will be open to accept most students' own proposals for topics and thus there is not in every case the direct link from the topic to the supervisor. Instead the student might have a topic and would be trying to find an appropriate supervisor to held auspices of that topic. With regard to this and other alternatives, there is a presumption that the student is able to judge about the suitability of the topic. This however appears to be flawed. Correspondingly, assuming that the supervisor is able to judge the students' capability and interests to gauge and tailor the topic appropriately is flawed as well. While the student is usually able to decide the broader domain be it e.g. marketing, finance, programming, etc. they are not always able to judge the exact topic since the student is not usually aware about interestingness, topicality or difficulty of particular topics. The experience also indicates that it is not appropriate to assign a random topic suggested by the supervisor to a hesitating student. In such a case the student is starting from the scratch and/or lacks the motivation to permeate into the topic to a necessary depth so that the thesis is of a good quality. Further, the decision about the topic might be influenced by the awareness of opportunities such as participation on a project, continuation of someone else's research or job opportunity. Considering these pieces of information is rather difficult at the time of topic selection since the student is pressed to make decision as soon as possible so that the topic and the supervisor are not booked by someone else.

Except the process of assigning topics to students, also other problems related to theses deserves attention. The limited extent of bachelor or master thesis means that students usually cannot finish the whole project; typically bachelor theses in computer science and informatics are reduced to the analysis of intended information system, but the application is not implemented, or some application is implemented, but is not tested well etc. Teams of two or three students could achieve better results. But this requires providing better information about the progress and results of particular theses. At the moment it is not easy for students to find out (a) what topics were covered by their colleagues in previous years, (b) whether it is possible to continue in someone's work, (c) if any of their current classmates look for cooperation etc.

Moreover, some students are not experienced in collecting, organizing and processing resources. They are not well informed about scientific databases and digital libraries subscribed by the faculty and they are not used to browse current issues of journals to be in touch with the newest trends. If they were more informed about the scientific papers, possibly they could better evaluate the choice of final themes and could find out something really interesting for them. During the thesis preparation, the personalized recommendation system could assist student and help him/her to find and manage electronic resources.

In particular the intended application should provide following main functionalities:

- Creating user profiles – the user is student and his/her profile has to contain information that will be taken into account during the topic recommendation. Part of profile data can be obtained from the study agenda system where students' scores from the course of their study are stored. Part of the profile should be edited by user, who could answer questions about his/her preferences, interests, work experiences etc.

- Automated topic recommendation – the application provides the list of suitable topics, generated according the user's profile.
- Recommendation on demand – the user can insert queries to browse database of topics, supervisors and resources registered by the application.
- Navigating student through the process of thesis preparation – student is informed about official deadlines, about newly published relevant documents such as dean's instruction about theses submission etc.. This application works with the optimized scenario of thesis' workflow (e.g. it plans the meeting with supervisor once a month and arranges it, after communication with supervisor's calendar etc.).

The application has to be integrated with existing information systems and databases, namely study agenda system, internet and intranet of the faculty, learning management system, personal websites and calendars of teachers, university library gateway to subscribed databases, web sites of partners of the university (organizations participating in regional industrial cluster that suggest theses' topics and look for promising graduates) etc.

Vision No 3: Lecture Hall

This vision has been adapted from (Brown 2005) as an example of a modern Lecture Hall utilization. Such a model example of a lecture hall could be taken from many modern universities, however this vision has been so well written that we decided to adapt it in order to show a superb illustration of a vision which we are sharing as well.

Described vision is situated in a lecture hall where 150 students take their psychology class. During this class the professor uses two projection screens for showing course material and the displaying the familiar "voting" screen. This lecture hall is of relatively recent vintage; its seats

and paired tables make it much easier to deploy and use "tools," which include printouts of the day's reading, as well as a small laptop computer. Every student is using some device to access the course's Web site - some with laptops, others with tablet computers, still others with handheld computers. Using wireless connections, they all access the course's Web site and navigate to the site's "voting" page.

The professor commences her lecture. In one of the older lecture halls, she might have been tied to the lectern so that she could click through her PowerPoint slides. Or she might have abandoned her slides in order to write on the blackboard while her students scribbled notes in their notebooks. But in this newly renovated lecture hall, she and her students have many more options. She has what is called a "magic wand," a radio-frequency controller that enables her to operate her computer - as well as many of the classroom's functions - wirelessly, from any point in the room. She can capture anything she writes on the blackboard and make it available to her students on the course Web site. Freed from needing to take extensive notes, the students are able to participate more fully in the class discussion. Finally, the professor is carrying a small recorder that captures her lecture, digitizes the audio, and uploads it to the course Web site for the students to review when they prepare for finals.

This lecture hall enables professor to circulate through the room, using aisles that create paths through the students' seats. As she roams, she calls on students to share reactions to the readings. She encourages other students to offer additional comments. Soon there is some debate about the reading, which is facilitated by the room's rows of paired tables and swivel chairs, making it possible to maintain eye contact with nearly everyone in the room. This type of discussion can be supported by students' diagrams sketched on their laptops that can help to explain the concepts being discussed. It is also possible to show it to the class by launching the classroom's screen sharing application. Within a few seconds, students' computer's screen can be projected on the room's main screen. Consequently, the class discussion can focus on this diagram, and the professor, using a virtual pencil, is capable to make notes on the diagram. The diagram and notes are captured and placed on the class Web site for review.

The lecture hall also enables students to access the Web page from laptops, handhelds, or wireless IP-based phones. In this way they can complete a poll or submit their responses in few minutes. Apparently, results can be quickly tabulated and displayed.

Video conferencing is an integral functionality of the lecture hall. Thus, the class can have a conversation with the expert, who appears on the right-hand screen and is at large research institution more than 400 miles away. Students listen to the expert's comments and are able to pose questions using one of the three cordless microphones available to the class. On the left-hand screen, the visiting professor shows some images and charts that help explain the concepts under discussion.

POSSIBLE AMBIENT INTELLIGENCE APPLICATION ARCHITECTURES

As Cook (2009) pointed out, only recently the related research has advanced to the point where the dream about smart environments can become a reality. Because the software architecture for smart environments is very complex, and because there may be multiple entities (e.g., residents) that the environment is serving, such a small environment can be naturally viewed as a multi-agent system.

In what follows we outline briefly some possible agent-based architectures for the visions presented in the previous chapter. These are just a sketch; however such architecture, when evolved further on, could be a real basis for functioning smart educational environments.

Agent-Based Architecture for Vision No. 1

People tracking and localizing problem presented in the first vision is very common in real world situations. While up to date systems, proposed to the problem solution are generally based on utilization of some specific technology (e.g. video cameras and processing the video signal (Fiore et al, 2008), (Petrushin et al, 2006), sound and infrared signals analysis (Sekmen et al., 2002), single purpose detectors (Sogo et al., 2000), RFIDs, GPS and others), our approach appears mainly from the available sources of data – existing information systems within the university.

According to (Petrushin et al, 2006), many approaches to people localizing problem may be proposed. In our context, various aspects such as number of information sources used, their time/space sensitivity and freshness of provided data plays the vital role in the output results accuracy. Another possible problem may arise due to two or more systems output overlapping and indicating controversial data on particular user's presence in a campus. On the other hand, there is a nonzero probability of no relevant information available when required. In such situations, taking into consideration the domain knowledge about both environment conditions and user's preferences and behavioural patterns can essentially improve the accuracy of the system.

To fulfil the expected functionality, following components and features will be further studied:

- Simple and efficient agents that collaboratively analyze and aggregate data from multiple sources.
- Knowledge about the environment and user's behaviour to bridge the temporal lack of data.
- Data redundancy to improve the accuracy of the system responses.

Data from particular information source will be processed by a specific agent, while the final decision about user's availability will result from negotiation between information agents and decision making agent. Personal models, based on individual profiles and recent priorities, and common domain knowledge as well will help in finding the most credible results of the system.

Agent-Based Architecture for Vision No. 2

With respect to the objective of the 2nd vision, which is to assist students who complete their theses, the dispersed nature of the information and resources suitable and necessary for topic selection as well as the differences in their processing directs to the adoption of cooperative information filtering (Oates et al., 1994). In particular the multi-agent approach is being applied in which agents are orchestrated to make personalized and context aware recommendation for the user. The use of the multi-agent systems for the purposes of recommendation has already been covered in a number of articles studied for example in (Carbo and Molina, 2004; Lorenzi, 2007; Miao et al, 2007; or Wang and Benbasat, 2007). The major challenges in multi-agent approach are the management of interdependences between the activities so that the shared goal can be achieved. In this aspect there is a distinction between distributed processing and distributed problem solving. Oates (1994) suggests that distributed processing involves independent subtasks during which the agents do not have to communicate to achieve shared goal (Oates et al., 1994). The shared goal is a composition of results returned by individual agents. In distributed problem solving the agents need to interact extensively since the existence of interdependences between subtasks. The agents are solving the task locally and exchanging information while arriving to partial or tentative results and thus reducing the uncertainty about the shared goal (Oates et al., 1994).

The shared goal in our case is to recommend the most suitable topics and the supervisors while considering the student experience, past projects and assignments, explicit personal preferences, available resources in digital libraries. The nature of the problem points to the distributed processing approach in which there is only a dependency between the topic and the supervisor who has to be familiar with the topic and available at the same time.

The central agent, from student's point of view, is his/her thesis assistant agent. Instances of these agents are assigned to individual students. Agents generate user profiles and reuse it for recommendations. Other types of agents exist only in one instance and each agent manages its resource (study agenda system, internet and intranet of the faculty, learning management system, personal websites and calendars, subscribed databases, web sites of partners of the university).

In our approach each agent operates with the three types of knowledge models:

- model of the information resource,
- domain knowledge model
- local knowledge base.

The model of the information resource serves for performing the particular tasks. In most cases this means executing queries to obtain some data or documents from a database or a library. The domain knowledge model will consists of some heuristics in the form of rules and constraints that apply when performing the given tasks i.e. constructing the queries and filtering the results. The local knowledge base will contain the previously performed tasks together with other recognized aspects that influenced the task execution and also the value of the payoff function. The domain knowledge model together with the local knowledge base would prevent the need for an exhaustive search considering rather the time and resources (Lorenzi, 2007).

Agent-Based Architecture for Vision No. 3

The objective of the vision No. 3 is, among other its features, to provide students during the teaching process by tailored knowledge and information, supporting intelligently the teacher's lecturing. Here multi-agent e-learning systems could be utilized as appropriate supporting solutions. As an example we can mention a rather recent system ISABEL (Garruzzo et al., 2007).

The ISABEL is a new sophisticated multi-agent e-learning system, where the basic idea is in partitioning the students in clusters of students that have similar profiles, where each cluster is managed by a tutor agent. When a student visits an e-learning site using a given device (say, a notebook, or a smart phone), a teacher agent associated with the site collaborates with some tutor agents associated with the student, in order to provide him with useful recommendations. Generally, these systems use a profile of the student to represents his interests and preferences, and often exploit software agents in order to construct such a profile. More in particular, each student is associated to a software agent which monitors his Web activities, and when the student accesses an e-learning site, his agent exploits the student's profile interacting with the site. In this interaction, the site can use both content-based and collaborative filtering techniques to provide recommendations to the student's agent by adapting the site presentation.

Another possibility here is to use the idea of so-called ad hoc agent environment (Misker, Veenman, and Rothkrantz, 2004). An ad hoc agent environment is an alternate way for users to interact with an ambient intelligent environment. Agents are associated with every device, service or content. The user interacts with his environment as a whole, instead of interacting with individual applications on individual devices. Devices and services in the environment have to be more or less independent, but in the case when they have to react to some users' action or other

initiative, they start to organize themselves in a sense to create a group or a collection of those agents which are relevant to the solution of the given situation or problem. A serious problem here could lie in the tension between the user being in control and the autonomy of agents. The notion of cooperating groups is introduced (Misker, Veenman, and Rothkrantz, 2004) as a way for users to gain control over which agents collaborate. Users can then establish connections between devices and content that are meaningful to them, in the context of their task.

RESULTS OF BRIEF SURVEY OF VISIONS ACCEPTANCE

In order to find out opinions related to the acceptability of created visions by the Net Generation members, a semi-structured interview with students at the Faculty of Informatics and Management at the University of Hradec Králové (FIM UHK), Czech Republic, was conducted. 79 respondents expressed their opinions on particular visions. They belonged to both major groups of students at FIM UHK – future managers (61% of respondents) and IT specialists (39% of respondents), and was enrolled in bachelor, master and doctoral study programs. Number of respondents enables to generalise and use results as a good indicator which shows that ideas condensed in particular visions are reasonable and worthwhile to further development and implementation.

The interview started with two simple questions. Students were asked if they knew the concepts of Ambient Intelligence and Smart Environment before. Figure 1 shows results - the majority of students did not hear about these concepts yet, only approximately one fifth of respondents have clear or rough idea what are these concepts about. It is not surprising that these respondents are currently either at a master or doctoral degree – respondents studying at a bachelor degree mostly didn't know about these concepts, only

Figure 1. Introductory questions

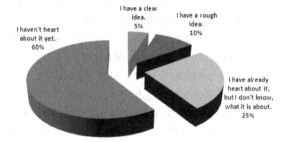

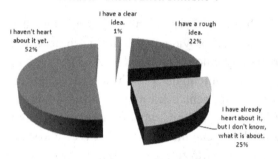

11 of them had a rough idea. It is obvious from figure 1 that respondents were more familiar with the concept of Smart Environment than with the Ambient Intelligence one. Moreover, there was only one student who had clear idea about both AmI and Smart Environment concept (see fig. 1).

Afterwards, questions related to particular visions were asked. Students reacted to three basic statements, which were focused on feasibility of a particular vision, interest in provided services included in the described environment, and accessibility of utilisable technologies. Results are shown in figures 2, 3, and 4 respectively. It is very important from the acceptance point of view that "strongly disagree" as one alternative for respondents' reaction appeared only seven times in connection with all visions. Moreover, the Intelligent Assistant didn't receive any answer of this type. In case of the Lecture Hall vision all "strongly disagree" answers were given by students from the first year at the bachelor degree. Furthermore, two strongly negative answers in the Campus Location vision were closely related to concerns connected with "big brother phenomenon". Negative reaction expressed by a "disagree" answer could be found only at insignificant level in case of the Campus Location vision. Lecture Hall and Intelligent Assistant visions are connected with relatively higher number of "disagree" answers with respect to their feasibility and tech-

nology accessibility. Nevertheless, respondents would use services provided by these visions.

There are some general conclusions and comments that can be derived from obtained results. For instance, although all visions are closely connected with students' life at the university, the Intelligent Assistant received the most positive reaction from their side – 57% strongly agreed with corresponding statement (the highest rank reached in comparison to all "strongly agree" answers). On the other hand respondents thought that this vision together with the Lecture Hall vision is technologically too complicated to be realised at this time. The Campus Location vision got the highest number of positive answers ("agree" and "strongly agree"). This result probably reflects the fact that problems related to searching for a particular person are quite common and very actual for students. The Lecture Hall vision was the most successful vision with respect to "strongly agree" answers related to feasibility and technological preparedness. The reason lies in the fact that this vision is the closest to the existing reality at FIM UHK, where new multimedia supported lecture hall with some elements introduced in the vision was open year ago.

Figure 2. Questions about Campus Locations Vision

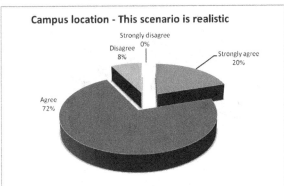

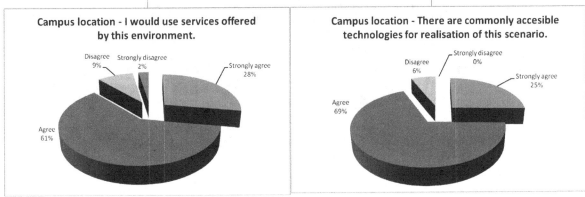

Figure 3. Questions about Intelligent Assistant Vision

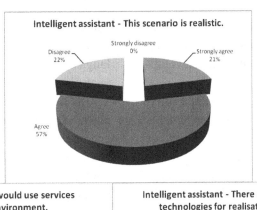

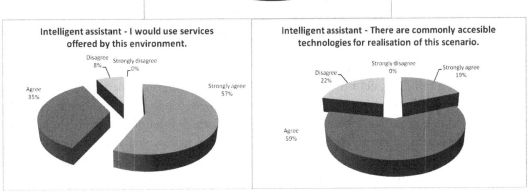

Figure 4. Questions about Lecture Hall Vision

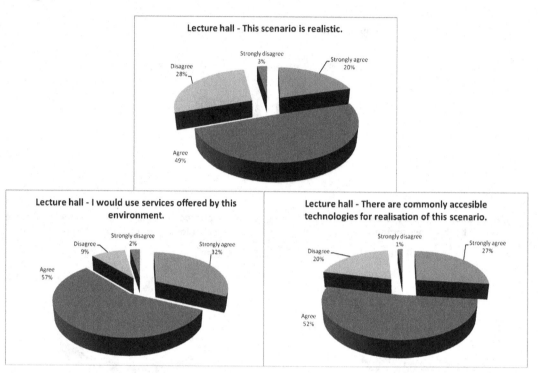

CONCLUSION

From the visions described in this chapter it is obvious that introduction of Ambient Intelligence in educational institutions is possible and can bring us new experiences utilizable in further development of AmI applications.

If an environment should demonstrate intelligent behaviour, it has to contain some elements which can ensure this type of demonstration. For instance, in the case of the first and third visions the development of an intelligent environment can be supported by data mining from databases. The huge amounts of data that are generated in institutions' applications and stored in data warehouses represent a potential for finding interesting hidden information. Knowledge discovery in databases in general and data mining in particular constitute a set of principles, approaches, tools and techniques that can be used to distil valuable information from the vast amounts of stored data (e.g. about persons moving behaviour or patterns, or about different educational content interrelations). In case of the second vision the information extraction from text or knowledge ontology can be utilized. The lecturing hall can be also supported by natural language processing, which can ensure a lecture analysis and transformation to a course web site.

Taking advantage of knowledge technologies educational institutions can become real "knowledge centers", which are formed by intelligent applications, devices, or technologies. This technologically intensive system, whose significant attribute is an intelligent interaction with its users, can support following activities:

- plan classroom instructions in a way that increases individual attention in critical areas,
- identifying training needs for teachers,
- aim to develop competence (opposed to awarding certificates, which cannot be an end in itself),

- enable students to learn in a way that promotes interest in learning & continue learning in environments other than the institute which imparts formal education etc.

The idea of enhancing university educational environment by suitably chosen AmI solutions has been already sketched e.g. in (Mikulecky, Olševičová, 2005) and for the case of e-learning education more elaborated in (Olševičová, Mikulecký, 2008). In this chapter, we presented three visions, which certainly can by typical for any modern university which uses nowadays information and communication technology. The basic principle of any further development lies, apart of its technological basis, also in understanding and using new approaches, new ways how these technologies can be utilized. We are deeply convinced, that AmI solutions are, among other recent approaches, the right way of achieving a new quality of educational process, which will be both motivating and effective for anyone.

ACKNOWLEDGMENT

The research has been partially supported by the Czech Grant Foundation, grants No. 402/09/0662 and No. 406/09/0346.

REFERENCES

Alcaniz, M., & Rey, B. (2005) New Technologies For Ambient Intelligence. In: *Ambient Intelligence*. IOS Press [online] http://www.ambientintelligence.com

Bettiol, C., & Campi, C. (2005) Challenges for Ambient Intelligence: Empowering the Users. In: *Ambient Intelligence*. IOS Press [online] http://www.ambientintelligence.com

Bohn, J. (2005). Social, Economic and Ethical Implications of Ambient Intelligence and Ubiquitous Computing. In *Ambient Intelligence* (pp. 5–29). Springer-Verlag. doi:10.1007/3-540-27139-2_2

Brown, M. (2005). Learning Spaces. In *Educating the Net Generation*. Washington, DC: Educause.

Carbo, J., & Molina, J. M. (2004). Agent-based collaborative filtering based on fuzzy recommendations, In. *International Journal of Web Engineering and Technology*, *1*, 414–426. doi:10.1504/IJWET.2004.006267

Čech, P., & Olševičová, K. (2009) The Proposal of Intelligent Assistant for Graduation Thesis Preparation. In: *Workshops Proceedings of the 5th International Conference on Intelligent Environments*. IOS Press, pp. 321-327

Clayton-Pedersen, A., & O'Neil, N. (2005) *Curricula Designed to Meet 21ˢᵗ-Century Expectaions*. In: *Educating the Net Generation*. Washington, DC, Educause

Cook, D. J. (2009). Multi-agent Smart Environments, In. *J. of Ambient Intelligence and Smart Environments*, *1*, 51–56.

Diepstraten, I. (2006). *De Nieuwe Leerder: Trendsettende leerbiografieën in een kennissamenleving*. Tilburg: F&N Boekservice.

Fiore, L., Fehr, D., Bodor, R., Drenner, A., & Somasundaram, G. (2008). Multi-Camera Human Activity Monitoring. *Journal of Intelligent & Robotic Systems*, *52*(Issue 1), 5–43. doi:10.1007/s10846-007-9201-6

Gardner, H. (1985). *Frames of mind: the theory of multiple intelligences*. New York: Basic Books Inc.

Garruzzo, S., Rosaci, D., & Sarné, G. M. L. (2007) ISABEL: A Multi Agent e-Learning System That Supports Multiple Devices. In: *IEEE/WIC/ACM International Conference on Intelligent Agent Technology*, IEEE, New York, pp. 485-488.

Howe, N., & Strauss, W. (2000). *Millennials Rising: The Next Greatest Generation*. New York: Vintage Books.

Husband, J., & Bair, J. (2007). *Making Knowledge Work – The Arrival of Web 2.0*. London: Ark Group.

ISTAG. (2001) *Scenarios for Ambient Intelligence in 2010*. Luxembourg

ISTAG. (2002). *Strategic Orientations and Priorities for IST in FP6*. Luxembourg.

ISTAG. (2005) Ambient Intelligence: from Vision to Reality. In: *Ambient Intelligence*. IOS Press [online] http://www.ambientintelligence.com

Kameas, A., et al. (2005) Computing in Tangible: Using Artifacts as Components of Ambient Intelligence Environments. In *Ambient Intelligence*. In: *Ambient Intelligence*. IOS Press [online] http://www.ambientintelligence.com

Kvavik, R. (2005). *Convenience, Communications, and Control: How Students Use Technology* (Oblinger, D., & Oblinger, J. L., Eds.). Educause, Washington: Educating the Net Generation.

Lessig, L. (1999). *Code and Others Laws of Cyperspace*. New York: Basic Books.

Lindström, M., & Seybold, P. (2003). *Brandchild: remarkable insights into the minds of today's global kids and their relationships with brands*. London: Kogan Page.

Lorenzi, F. (2007) A multiagent knowledge-based recommender approach with truth maintenance. In: *Proceedings of the 2007 ACM conference on Recommender systems*, Minneapolis

McNeely, B. (2005). Using Technology as a Learning Tool, Not Just the Cool New Thing. In Oblinger, D., & Oblinger, J. L. (Eds.), *Educating the Net Generation*. Educause, Washington.

Miao, C., Yang, Q., Fang, H., & Goh, A. (2007). A cognitive approach for agent-based personalized recommendation. *Knowledge-Based Systems*, *20*, 397–405. doi:10.1016/j.knosys.2006.06.006

Mikulecky, P., & Olševičová, K. (2005). University Education as an Ambient Intelligence Scenario. *In Proceedings of 4ᵗʰ European Conference on E-learning 2005*, Amsterdam, The Netherlands, pp. 333-341.

Misker, J. M. V., Veenman, C. J., & Rothkrantz, L. J. M. (2004). Groups of Collaborating Users and Agents in Ambient Intelligent Environments. In: *Proc. ACM AAMAS 2004*, pp. 1318-1319

Mls, K., Salmeron, J., Olševičová, K., & Husáková, M. (2009). Cognitive Personal Software Assistants within AmI environment. In *Ambient Intelligence Perspectives II* (pp. 167–174). IOS Press Amsterdam.

Oates, T., Prasad, M. V. N., & Lesser, V. R. (1994). *Cooperative Information Gathering: A Distributed Problem Solving Approach*. University of Massachusetts.

Oblinger, D., & Oblinger, J. L. (Eds.). (2005). *Educating the Net Generation*. Washington: Educause.

Olševičová, K., & Mikulecký, P. (2008) Learning Management System as Ambient Intelligence Playground. In: *Int. Journal of Web Based Communities*, Vol. 4, No. 3, pp. 348-358

Olševičová, K., & Poulová, P. (2009) Analysis of E-learning Courses Used at FIM UHK With Respect to Constraint Satisfaction Problems. In: *Proceedings of ECEL, 8th European Conference on E-learning*. ACL, Reading, pp. 425-430

Petrushin, V. A., Wei, G., & Gershman, A. V. (2006). Multiple-camera people localization in an indoor environment. *Knowledge and Information Systems*, *10*(Issue 2), 229–241. doi:10.1007/s10115-006-0025-7

Remagnino, P. (2005). *Ambient Intelligence: A Gentle Introduction. Ambient Intelligence: A Novel Paradigm* (pp. 1–14). Springer-Verlag.

Russell, D. M., Streitz, N. A., & Winograd, T. (2005). Building Disappearing Computers. *Communications of the ACM, 48*(3), 42–48. doi:10.1145/1047671.1047702

Sekmen, A. S., Wilkes, M., & Kawamura, K. (2002). An application of passive human-robot interaction: human tracking based on attention distraction. *IEEE Transactions on Systems, Man, and Cybernetics. Part A, Systems and Humans, 32*(2), 248–259. doi:10.1109/TSMCA.2002.1021112

Snijders, F. (2005). *Ambient Intelligence Technology: An Overview. Ambient Intelligence* (pp. 255–270). Springer.

Sogo, T., Ishiguro, H., & Trivedi, M. M. (2000) *Real-Time Target Localization and Tracking by N-Ocular Stereo*. IEEE Workshop on Omnidirectional Vision (OMNIVIS'00), p.153

Stemberg, R. J. (1985). *Beyond IQ: A Triarchic Theory of Human Intelligence*. New York: Cambridge Unviersity Press.

Tapscott, D. (1998). *Growing up Digital: The Rise of the Net-Generation*. New York: McGraw-Hill.

Veen, W., and van, Staalduinen, J.P. (2009) Homo Zappiens and its Impact on Learning in Higher Education. *IADIS International Conference e-Learning 2009*, Algarve, Portugal

Veen, W., & Jacobs, F. (2005). *Leren van Jongeren: Een literatuuronderzoek naar nieuwe geletterdheid*. Utrecht: Stichting SURF.

Veen, W., & Vrakking, B. (2006). *Homo Zappiens: Growing Up in a Digital Age*. London: Network Continuum Education.

Wang, W., & Benbasat, I. (2007). Recommendation Agents for Electronic Commerce: Effects of Explanation Facilities on Trusting Beliefs. *Information Systems, 23*, 217–246.

KEY TERMS AND DEFINITIONS

Agent: As defined in artificial intelligence literature is anything that can be viewed as perceiving its environment through sensors and acting upon the environment through actuators. Some authors extend the definition of agent by enumerating required features of agent such as mobility, correctness, reactivity and pro-activity, goal-orientation, adaptability, robustness, benevolence, sociability, ability to communicate with other agents etc.

Agent-Based Model: A model that involves numerous interacting autonomous agents, homogeneous or heterogeneous. The objective of agent-based modeling is to help us to understand effects and impacts of interactions of individuals. The AmI environment can be seen as composed of particular agents.

Ambient Intelligence: A vision of society of the future, where the people will find themselves in an environment of intelligent and intuitively usable interfaces, ergonomic space in a broad sense, encompassing better, secure and active living environment around them, capable of aiding them with daily chores and professional duties by recognizing the presence of individuals, reacting to it in a non-disturbing, invisible way, fully integrated into the particular situation.

Net Generation: The generation of youth which is growing up with modern information and communication technologies shaping strongly their mental models, i.e. views on the world around them. While using several technologies they are learning to develop new skills and exhibiting new behavior patterns.

The AmI Artefact: (also smart object, smart device) is an element of the AmI environment that has got properties and abilities such as information processing, interaction with environment, autonomy, collaboration, composeability, or changeability.

The AmI Environment: Characterized by merging of physical and digital space. It means that tangible objects and physical surroundings are acquiring a digital representation.

Chapter 30

Opinion Mining and Information Retrieval:
Techniques for E-Commerce

Shishir K. Shandilya
Devi Ahilya University, India

Suresh Jain
KCB Technical Academy, India

ABSTRACT

E-commerce is an increasingly pervasive element of ambient intelligence. Ambient intelligence promotes the user-centered system where as per the feedback of user, the system changes itself to facilitate the transmission and marketing of goods and information to the appropriate e-commerce market. Ambient Intelligence ensures that the e-commerce activities generate good confidence level among the customers. The confidence occurs when the customers feel that the product can be relied upon to act in their best interest and knowledge. It affects the decision that whether a customer decides to buy the product or not.

With the rapid expansion in the field of E-Commerce, most of the people are buying products on the web and also writing reviews about their experiences with the products. Popular products are receiving large number of reviews every day. New customers who want to buy the product are now firstly looking to have an unbiased summary of product reputation in market based on opinions of existing customers. Opinion Mining is an exciting research area that is currently under rapid development. It uses techniques from well-established technologies like Natural Language Processing, Data Mining, Machine Learning and Information Retrieval. This quick growth of online information on web has attracted increasing interest in technologies for automatically mining personal opinions from Web documents. Such technologies would benefit e-commerce community for advertising and promotion to target their potential buyers.

The explosive increase in Internet usage has attracted technologies for automatically mining the user-generated contents (UGC) from Web documents. These UGC-rich resources have raised new opportunities and challenges to carry out the opinion extraction and mining tasks for opinion summaries. The

DOI: 10.4018/978-1-61692-857-5.ch030

technology of opinion extraction allows users to retrieve and analyze people's opinions scattered over Web documents. Opinion mining is a process which is concerned with the opinions generated by the consumers about the product. Opinion Mining aims at understanding, extraction and classification of opinions scattered in unstructured text of online resources. The search engines performs well when one wants to know about any product before purchase, but the filtering and analysis of search results often complex and time-consuming. This generated the need of intelligent technologies which could process these unstructured online text documents through automatic classification, concept recognition, text summarization, etc. These tools are based on traditional natural language techniques, statistical analysis, and machine learning techniques. Automatic knowledge extraction over large text collections like Internet has been a challenging task due to many constraints such as needs of large annotated training data, requirement of extensive manual processing of data, and huge amount of domain-specific terms. Ambient Intelligence (AmI) in wed-enabled technologies supports and promotes the intelligent e-commerce services to enable the provision of personalized, self-configurable, and intuitive applications for facilitating UGC knowledge for buying confidence. In this chapter, we will discuss various approaches of Opinion Mining which combines Ambient Intelligence, Natural Language Processing and Machine Learning methods based on textual and grammatical clues.

INTRODUCTION

Based on the convergence of ubiquitous computing, ubiquitous communication and intelligent user-friendly interfaces, Ambient Intelligence is future of E-commerce. Ubiquitous Computing deals with the integration of microchips into everyday objects while Ubiquitous Communication facilitates these objects to communicate with each other and intelligent user interfaces provides the control to the users to control and interact with the AmI environment effectively. The present global market scenario facilitates and promotes the e-commerce and e-business activities, making them a strong catalyst for economic development. The rapid and remarkable development in the field of information and communication technologies has increased the consumer participation and customization in almost every business. Users now actively write their experiences, choices, and recommendations on blogs, discussion boards or websites etc. Product reviews exist in various formats on the Internet. Like the websites dedicated to a specific type of product (such as TV or

Fridge), sites for newspapers and magazines that may feature reviews (like Consumer Reports), and sites that collects professional or user reviews in specific domains (like codeguru.com in computers), or the broader domains (wikipedia.com and yahoo.com). The business community has welcomed this change in market and dealing with it perfectly by strategically positioning their products or services. The company determines the upcoming opportunities and work for them with appropriate strategies and policies while exploiting the latest technologies, which is encouraging the use e-commerce technologies in advertising, marketing and promotion. The Internet is used as a medium for enhanced customer satisfaction. The companies are providing more detailed information about the product or service to increase the trust and determination of potential buyers.

The recent developments in web technologies have increased the acceptability and reliability ratio of e-commerce. Customers are now relying more trustfully on the information of web which plays remarkable role in their decisions of purchase. Customers now tend to search the web thoroughly before any purchase/decision. They prefer to know the opinions and suggestions of existing customers to portray the image of product or service accordingly. Fast technological advancements in digitization of user-generated content, natural language processing and web mining have paved a way for mining directly the opinions scattered over web about a particular product/service. This requires fast and efficient procedure which can also represent the results in simple and lucid manner. Using such technologies makes the e-commerce competitive enough to provide greater satisfaction to the potential customers.

The business strategies and future action plans relies greatly on the consumer's opinions and experiences who have used the product at least once. As the businesses are rapidly transferring over web, so the opinions and suggestions are. The user-generated text is growing enormously, making it prospective area of mining. This needs to automatically extract and analyze personal opinions from web documents. Such technologies can be an alternative to conventional questionnaire-based surveys and would also benefit Web users who seek reviews on certain consumer products of their interest.

FEATURE-BASED OPINION MINING

The classification based on evaluative text is practically quite not feasible because if the author or the reviewer is writing some positive opinions about the product, it does not implies that he is completely satisfied and positive about the product and if he is writing some negative comments, it is not necessary that he do not like the product at all. Usually, any reviewer writes some positive and some negative opinions about the product. In the mixture of positive/negative sentences, it may have best comments on some property or part of the product while having strong criticizing opinion about the other parts. Although, the

Figure 1. Representation of sample sentence

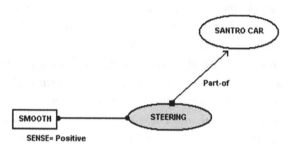

general view of the reviewer need to be process and understood carefully before classifying the document. To achieve this, individual sentences of the documents are to be treated as separate entities to process them for their polarity as positive, negative or neutral in some cases. It needs extraction and identification of product features from individual sentences. Such sentences may be about any service, person, or product etc., so to make it generalized, it is termed as object which may have set of components (parts) and set of attributes (properties). Now, for every sentence in the document, the object could be defined in the terms of a hierarchical set of components and attributes. It may be represented as tree, in which object is at the root and each non-root node is either a component or sub-component liked with "parts of" relationships. Each node is also associated with a set of attributes.

E.g. The smooth steering of Santro car.

The representation of the above sentence in shown in Figure 1.

Feature-based opinion mining gives promising results (over 90%) for the documents related to specific domains of product, services, and person etc. Feature Identification and Extraction are the two main tasks for feature-based opinion mining. The opinions are generally found in user reviews on web. To analyze and process these reviews, one must understand the basic formats of reviews which are generally found on web.

Figure 2. Types of formats of reviews

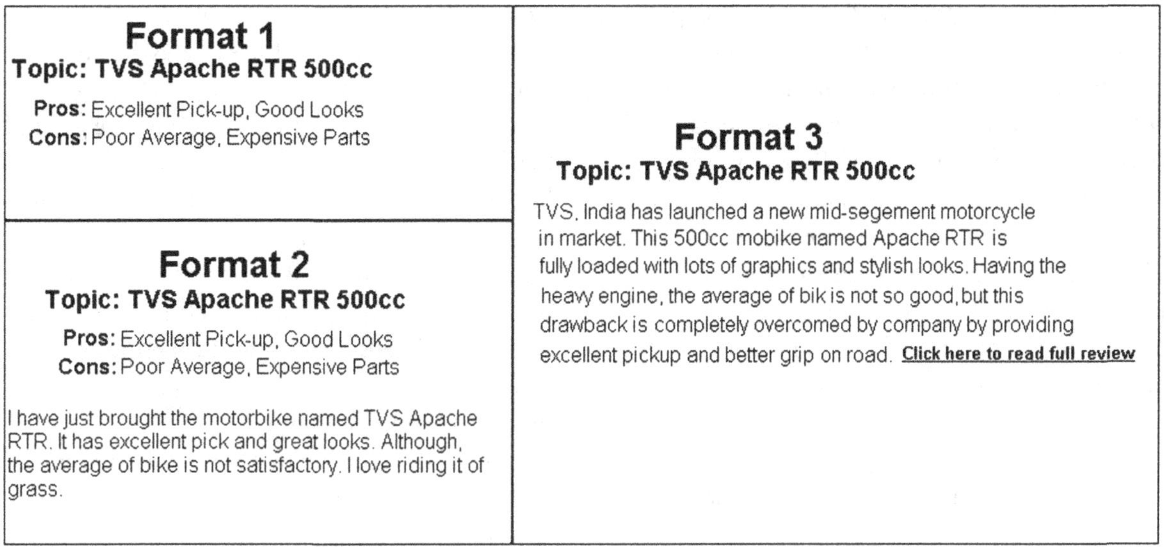

Format 1 **Topic: TVS Apache RTR 500cc** **Pros:** Excellent Pick-up, Good Looks **Cons:** Poor Average, Expensive Parts	
Format 2 **Topic: TVS Apache RTR 500cc** **Pros:** Excellent Pick-up, Good Looks **Cons:** Poor Average, Expensive Parts I have just brought the motorbike named TVS Apache RTR. It has excellent pick and great looks. Although, the average of bike is not satisfactory. I love riding it of grass.	**Format 3** **Topic: TVS Apache RTR 500cc** TVS, India has launched a new mid-segement motorcycle in market. This 500cc mobike named Apache RTR is fully loaded with lots of graphics and stylish looks. Having the heavy engine, the average of bik is not so good, but this drawback is completely overcomed by company by providing excellent pickup and better grip on road. **Click here to read full review**

1. **Format 1:** Pros & Cons: In this type of format, the reviewer only describes the keywords under the category of pros and cons. There is no other detail of product except these heads. The keywords generally describe the main positive and negative properties.
2. **Format 2:** Pros & Cons with detailed review: In this format, along with the separate pros and cons, the reviewer is asked to write the brief text about the product.
3. **Format 3:** Free Format: In this type of format, the reviewer can freely writes the text about the product and there is no category of pros and cons.

The example of these formats are shown in Figure 2.

In format 1 & 2, as there are separate pros and cons heads, the sentiment classification is easier as only features are to be identified. While, in case of format 3, the complete review is to be processed for features recognition and corresponding evaluations.

USER GENERATED CONTENT

User-generated Content (UGC) refers to the content produced by the consumers on various types of media. It is also known as Consumer Generated Media or User-created Content. With the growing number of reviews, it is becoming cumbersome for potential customers to analyze all the reviews of web to have an overall view and for the company to keep track of their customer's choice, experiences, difficulties and suggestions about their products over the span of time. Search for the user-generated content is becoming increasingly popular as they results the user-opinions about products, person, service, organization, commonly termed as object of interest for market intelligence, advertisement placement and customization etc. Initially, the search engines were crawling the web for user-generated content exactly as they does in case of simple keyword search, then displaying only selected data. In this, the results returned by search engines were again processed for classification which makes it time-consuming and complex. Now, due to the advancements in web technologies, search engines specifically for user-generated content are evolving. The data on web is usually unstructured and search engines crawls data from various sources which maintains the data in different formats and structures. One bottleneck is to maintain the results in one standard format for further processing. To resolve this, the researchers have used opinion-frames, to save the retrieved opinions from heterogeneous web. The resultant data of web search query is stored in these opinion frames which are used as an input for further process of opinion mining. These opinion frames may have following constituents,

1. **Opinion holder:** A person who is making an evaluation (an author or an unspecified person).
2. **Subject:** A named entity (product or company) of a given particular class of interest (e.g. a car model name in the automobile domain).
3. **Part:** A part, member or related object of the subject with respect to which evaluation is made (engine, interior, etc. in the automobile domain)
4. **Attribute:** An attribute (of a part) of the subject with respect to which evaluation is made (size, color, design, etc.)
5. **Evaluation:** An evaluative or subjective phrase used to express an evaluation or the opinion holder's mental/emotional attitude (good, poor, powerful, stylish etc.)
6. **Condition:** A condition under which the evaluation applies (on accident, in rain, etc.)
7. **Support:** An objective fact or experience described as a supporting factor of the evaluation (500cc, 35kms etc.)

As per this assumption, the opinion extraction can be defined as the task of filling the slots of opinion frame as per the understood meaning/recognition of given text.

As described in Figure 3, the free text are processed and transformed into opinion frames for more detailed and structured processing.

These opinion frames provides a structure to the data for better and more controlled analysis. The constituents of opinion frames may be added or removed according to the need and choice of analyzer to have better results. However, the general concept will be same in every case. As there could be enormous amount of resultant data the opinion frames could be of huge number, making it difficult for the analyzer to read and understand all of them. Therefore, the opinion frames are represented graphically. This graphical representation provides quick comparative information about the object as shown in Figure 4.

The opinions on web are temporal as the choice and the taste of consumers varies time to time, therefore the time axis is generally desired by the analyzer to know the pattern of consumer-experiences over time.

The web contains huge amount of such unstructured text which may turn into vital information if processed properly. The extraction and analysis of text on web may be used to have public opinions. This task needs to incorporate the technologies of Artificial Intelligence, Natural Language Processing, and Web Mining. It is commonly known as opinion mining, which allows Web users to retrieve and summarize people's opinions scattered over Web documents. Opinion Mining Systems can identify the words and their meaning in the sentence with their relationships in texts for further processing. Opinion Extraction aims to produce richer information pool, useful for in-depth analysis of opinions and to train the algorithms. The technology of opinion extraction allows users to retrieve and analyze people's opinions scattered over Web documents. Quick analysis and understanding of unstructured text of web is becoming increasingly important with the huge increase in the number of digital documents available. This unstructured text may turn into vital information after tactical processing that in turn could be useful for various analysis and study purposes like analysis of product creditability and brand identity. The classification of text documents or

Figure 3. Opinion Frames

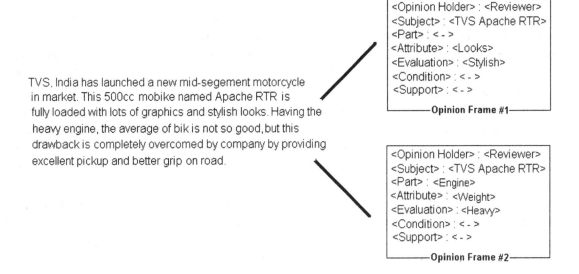

Figure 4. Graphical Representation of Opinion Summary

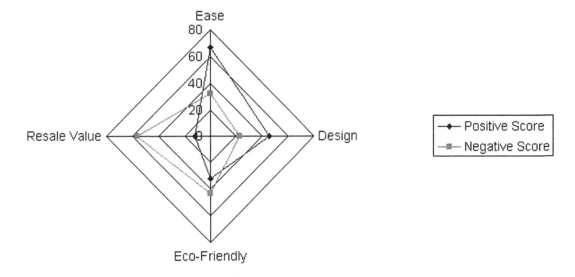

passages into positive vs. negative classes could be done according to their sentiment classification. Thereafter, the NLP techniques may be used to extract the opinions by using the information about the aspects of interest.

SENTIMENT CLASSIFICATION

Sentiment Classification gives a quick determination of the prevailing opinion on an object. It determines whether a sentence or document has either positive or negative orientation. Sentiment Classification deals with the task of classifying the text in negative or positive sets. It is done by reading and understanding the evaluative text and grouping them into positive or negative set with respect to its meaning of text. For example, given a product review, the sentiment classification procedure will determine whether the review expresses a positive or a negative sentiment about the product. The document-level classification provides no details about what the reviewer liked or disliked. To know this, the classification is required at the sentence-level. In sentence-level classification, the features of object (product or service) are evaluated separately using the sentiment words like great, excellent, amazing, horrible, bad, worst, etc. The approaches for sentiment classification can be classified as: the unsupervised approach (Turney, 2002) and the supervised approach (Pang *et al.*, 2002).

Church and Hanks in 1989 have introduced the point-wise mutual information (PMI) between *ph* and *words* is defined as follows,

$$PMI\left(ph, words\right) = \log_2\left(\frac{\mathrm{p(ph~\&~words)}}{\mathrm{p(ph).p(words)}}\right)$$

where *p(ph,words)* is the probability that *ph* and *words* co-occur. If the words are statistically independent, the probability that they co-occur is given by the product *p(ph).p(words)*. The ratio between *p(ph, words)* and *p(ph, words)* is a measure of the degree of statistical dependence between *phrase* and *words*. PMIs are calculated based on the number of web pages returned by search engines (hits), when the phrase and the word are queried together.

The model of Turney in 2002, the Sentiment Orientation of the phrase ph is calculated according to the adjectives and adverbs present in the document. It is defined as follows,

SO(ph) = PMI(ph, pos words) – PMI(ph, neg words)

where *pos words* represents pre-defined positive words such as *excellent, good*", and *neg words* represents pre-defined negative words such as \ *poor, bad*". Given the semantic orientation of each phrase, the document is classified as positive if the average of the *SO(phrases)* > 0 and negative otherwise.

Sentiment Classification can also be done by using the supervised machine learning methods. Various text categorization methods could be used to classify text into one of predefined categories using information from training texts. This process requires some pre-defined knowledge about the categories and some training runs on sample training data. The Naive Bayes Classification, Maximum Entropy Classification and Support Vector Machines are proven useful techniques for sentiment classification. These methods use the Part-of-speech (POS) information for analysis of the phrases and for the determination of adjectives and adverbs. As if adjectives are used for the prediction of the features of words, it gives high correlation between the presence of adjectives and sentence subjectivity. This finding has been taken as evidence that adjectives are good indicators of sentiments and can be used as guide for feature selection for sentiment classification. Likewise, the other indicators like noun, verbs and adverbs etc also plays significant role in sentences and are very useful to understand the sentence.

The algorithm proposed by Turney in 2002 extracts the phrases containing adjectives or adverbs, considering that they are good indicators of opinions. It extract two consecutive words, where first could be an adjective/adverb and the other is a context word if their POS tags conform to any pattern. Then, it estimate the semantic orientation of the extracted phrases using the point-wise mutual information as discussed above. The semantic orientation (SO) of a phrase is computed based on its association with the positive reference word and its association with the negative reference word. The probabilities are calculated by issuing queries to a search engine and collecting the number of

Figure 5. Opinion Mining System

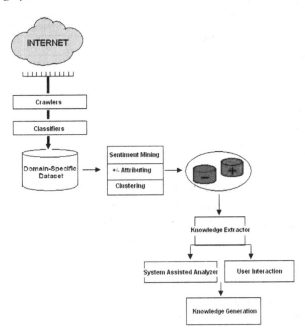

hits. By searching the two terms together and separately, we can estimate the probabilities. To avoid division by zero, 0.01 is added to the hits. Given a review, the algorithm computes the average *SO* of all phrases in the review, and classifies the review as recommended if the average *SO* is positive, not recommended otherwise.

For example, in the review document of any car, for the pattern is [- JJ (Adjective) - NN (Noun) -], it may return "good pickup", "bad sound", "smooth steering", "nice looks" etc which are further processed for PMI and SO as described above to finally termed the review as positive (if calculated Semantic Orientation is positive) or negative (if calculated Semantic Orientation is negative). The machine learning methods are also commonly used method for sentiment classification.

Review could be also in the form of comparative evaluations between the similar kind of products. The comparisons are obvious in user's review and provide better evaluative information. The comparative statements also provide the relative polarity (positive or negative) of particular object as compared to another product. For example, "*Mobile A is 50 grams lighter than Mobile B*" or "*Mobile A has 3.0 megapixel camera while Mobile B has 4.1 megapixel*". The Comparative sentences could formally be defined as the sentences that imply relation(s) based on the similarities and differences of tow or more object. These comparative statements may be classified into two classes, gradable and non-gradable. Gradable comparisons are those which have the relationships like greater or lesser, equal to, or bigger or smaller than etc. While the non-gradable comparisons compares the features of two or more products without grading them.

A dictionary of customer reviews could be created and expanded it manually with external resources including publicly available ordinal thesauri. As a result, the dictionary will be having good number of entries for further processing. The process stars by the execution of the filtration submodule to standardize the data collected. Every website, forum or blog have a different structure for the opinions. The crawlers extract text by separating and identifying the specific domain. The classifiers are then trained to categories the hierarchical classification of web e.g. the text submitted gets base class (Figure 5).

Specialized classifiers are trained for base categories such as Automobile, Music and specific product etc. Finally the dataset containing the domain specific data is formed for further processing. Now the lingual knowledge subroutine makes the system able to work on language rules and to perform sentiment analysis. Sentiment analysis deals with predicting the sense of text.

For example, if user has written a book review then sentiment analyzer tries to predict whether user is criticizing the book or he is praising the book. The text from database are extracted and clustered to increase accuracy of prediction of opinions. The text is also processed based on the fact that opinion could be positive or negative. Therefore, the resultant datasets were clustered and identified as positive and negative sets, each holding the opinions in standard format. Now the knowledge extractor smartly extracts these opinions as per the fired query for analysis and analyzes each opinion in the light of user (optional) presence and intelligent modules designed for same task.

Finally the knowledge is extraction from the system about the opinions extracted from web. The results may be evaluated by recall R and precision P which are defined as follows,

R = correctly extracted relations/total number of relations.

P = correctly extracted relations/total number of relations found by the system.

For example, see Table 1.

It is clearly shown in the table that both the precision and recall are found in higher number by manual study which is obvious, while the results found by the system needs to be refined by introducing various factors and parameters. Also, as

Table 1. Sample Results

Found by System		Found by Study	
P	R	P	R
410	338	466	456

per the current research, the results are refining but for specific domains only. For an example, the opinion mining system for mining the reviews of automobile industry may give best results, but the similar system is behaving unacceptably for another domain like restaurants. It is mainly due to the use of specific domain dictionaries. Therefore, having a standard system which works well on cross domain-level is still an open area for research.

APPLICATION OF OPINION MINING IN E-COMMERCE

With the advent of Ambient Intelligence, customers can easily access the user-generated information. With this new information, customers may analyze the reviews of other customers. The concept of ambient Intelligence (AmI) provides a vision as to how information technology, in which users are empowered through e-commerce, will get higher level of satisfaction and confidence in purchasing. In the context of e-commerce, purchase may dramatically influenced by confidence between an E-commerce consumer view and vendor's brand identity. All the related factors of confidence are revolved around the opinions and their comparative analysis. An effective E-Commerce in Ambient Intelligent Environment cannot be completed without an unbiased opinion mining analysis.

E-Commerce is using the Internet technologies to transform the way business processes are performed. The field of Opinion Mining is well-suited to various types of E-Commerce technologies. The intelligent e-commerce technologies are becoming the utmost need of corporate world to remain fully-equipped and prepared for latest technology changes and digital divide. In the present scenario, the customers are offered with lots of personalized and customized options for every product. Also the features and costs are marginally varies. The present cyber and technological revolution has changed the market flow and so the customer's behaviors. Everyday, more companies and users gain access to the web and everyday, more purchases are transacted electronically. Hence, there is an ultimate need to thoroughly examining and analyzing the customer's reviews. Sentiment-analysis technologies for extracting opinions from unstructured web documents would be excellent tools for handling many e-business intelligence tasks. With the use of such technologies, along with the extracting of information about the company's brand image, its reputation and after sales service, one may also have the information for the current trends, customer's expectations and sales predictions and act accordingly.

The area of Opinion Mining and Sentiment Analysis is getting increasing interest in the field of E-Commerce due to following factors,

- High degree of competition and higher level of customization on products makes ii necessary for company to be well-alert about the current and future trends of the market which is only driven by the users by their choices, experiences and sometimes by influences and recommendations.
- The incremental growth of machine learning methods in natural language processing and information retrieval;
- Due the advent of review specific websites, the training datasets for machine learning algorithms are easily available.

Companies on Internet are discovering assisting technologies to succeed in E-Commerce. They are looking forward for tools and technologies for mining and analyzing the user's opinions and to represent them in such a format which could be easier to understand.

SUMMARY

Products or services are often discussed by customers on the Web in their reviews. In the recent scenario of business, the e-commerce technologies are using the extracted knowledge from reviews available on web to have an idea about recent trend, customer's satisfaction and customer's behavior in future to prepare the strategic plan(s) accordingly. Current search engines are unable to satisfy any complex queries such as analysis, prediction, integration, etc. An example of such integration-based tasks is opinion mining regarding products or services. The search engines still work on keyword search technologies. Recent research on opinion mining has focused on sentiment classification and analysis on various formats of reviews using feature-based opinion search. There are some successes in opinion mining which concentrated more on sentiment classification and analysis. Natural Language Processing (NLP) and machine learning techniques have been utilized in supervised or unsupervised modes allowing the extraction and classification of sentiment and opinions. It now in demand by the users and companies to have an unbiased opinion summary about the object of interest to know what specific aspects of any object (product, person, service or organization etc) are being described on web in positive or negative terms and to have these results in collective and organized manner. The inter-relationship between generating user's confidence through opinion analysis and e-commerce interfaces requires the strategic logic formulation in Ambient Intelligent environment.

REFERENCES

Agarwal, A., & Bhattacharyya, P. (2005). Sentiment analysis: A new approach for effective use of linguistic knowledge and exploiting similarities in a set of documents to be classified, in *Proceedings of the International Conference on Natural Language Processing (ICON)*.

Agrawal, R., Rajagopalan, S., Srikant, R., & Xu, Y. "Mining newsgroups using networks arising from social behavior," in *Proceedings of WWW*, pp. 529–535, 2003.

Airoldi, E. M., Bai, X., & Padman, R. (2006). *"Markov blankets and meta-heuristic search: Sentiment extraction from unstructured text," Lecture Notes in Computer Science (Vol. 3932,* pp. 167–187). Advances in Web Mining and Web Usage Analysis.

Akerlof, G. A. (1970). "The market for "Lemons": Quality uncertainty and the market mechanism. *The Quarterly Journal of Economics, 84,* 488–500. doi:10.2307/1879431

Al Masum, S. M., Prendinger, H., & Ishizuka, M. "SenseNet: A linguistic tool to visualize numerical-valence based sentiment of textual data," in *Proceedings of the International Conference on Natural Language Processing (ICON)*, pp. 147–152, 2007. (Poster paper).

Allan, J. (2002). Introduction to topic detection and tracking. In Allan, J. (Ed.), *Topic Detection and Tracking: Event-based Information Organization* (pp. 1–16). Norwell, MA, USA: Kluwer Academic Publishers.

Alm, C. O., Roth, D., & Sproat, R. "Emotions from text: Machine learning for text-based emotion prediction," in *Proceedings of the Human Language Technology Conference and the Conference on Empirical Methods in Natural Language Processing (HLT/EMNLP)*, 2005.

Anagnostopoulos, A., Broder, A. Z., & Carmel, D. (2006). Sampling search-engine results. *World Wide Web (Bussum), 9,* 397–429. doi:10.1007/s11280-006-0222-z

Andreevskaia, A., & Bergler, S. "Mining WordNet for a fuzzy sentiment: Sentiment tag extraction from WordNet glosses," in *Proceedings of the European*

Argamon, S., Karlgren, J., & Shanahan, J. G. (Eds.). *Proceedings of the SIGIR Workshop on Stylistic Analysis of Text For Information Access.* ACM, 2005.

Aue, A., & Gamon, M. "Automatic identification of sentiment vocabulary: Exploiting low association with known sentiment terms," in *Proceedings of the ACL Workshop on Feature Engineering for Machine Learning in Natural Language Processing,* 2005.

Aue, A., & Gamon, M. (2005). Customizing sentiment classifiers to new domains: A case study. In *Proceedings of Recent Advances in Natural Language Processing.* RANLP.

Bautin, M., Vijayarenu, L., & Skiena, S. "International sentiment analysis for news and blogs," in *Proceedings of the International Conference on Weblogs and Social Media (ICWSM)*, 2008.

Benamara, F., Cesarano, C., Picariello, A., Reforgiato, D., & Subrahmanian, V. S. "Sentiment analysis: Adjectives and adverbs are better than adjectives alone," in *Proceedings of the International Conference on Weblogs and Social Media (ICWSM)*, 2007. (Short paper).

Branavan, S. R. K., Chen, H., Eisenstein, J., & Barzilay, R. "Learning document-level semantic properties from free-text annotations," in *Proceedings of the Association for Computational Linguistics (ACL)*, 2008.

Breck, E., Choi, Y., & Cardie, C. "Identifying expressions of opinion in context," in *Proceedings of the International Joint Conference on Artificial Intelligence(IJCAI)*, Hyderabad, India, 2007.

comScore/the Kelsey group, "Online consumer-generated reviews have significant impact on offline purchase behavior," Press Release, http://www.comscore.com/ press/release. asp?press=1928, November 2007.

Conrad, J. G., & Schilder, F. "Opinion mining in legal blogs," in *Proceedingsof the International Conference on Artificial Intelligence and Law (ICAIL)*,pp. 231–236, New York, NY, USA: ACM, 2007.

Das, S. R., & Chen, M. Y. (2007). Yahoo! for Amazon: Sentiment extraction from small talk on the Web. *Management Science*, *53*, 1375–1388. doi:10.1287/mnsc.1070.0704

Dave, K., Lawrence, S., & Pennock, D. M. "Mining the peanut gallery: Opinion extraction and semantic classification of product reviews," in *Proceedings of WWW*, pp. 519–528, 2003.

Devitt, A., & Ahmad, K. "Sentiment analysis in financial news: A cohesionbased approach," in *Proceedings of the Association for Computational Linguistics(ACL)*, pp. 984–991, 2007.

Ding, X., Liu, B., & Yu, P. S. "A holistic lexicon-based approach to opinion mining," in *Proceedings of the Conference on Web Search and Web Data Mining (WSDM)*, 2008.

Fukuhara, T., Nakagawa, H., & Nishida, T. "Understanding sentiment of people from news articles: Temporal sentiment analysis of social events," in *Proceedings of the International Conference on Weblogs and Social Media (ICWSM)*, 2007.

Goldberg, A. B., Zhu, X., & Wright, S. (2007). Dissimilarity in graph-based semisupervised classification. In *Artificial Intelligence and Statistics*. AISTATS.

Higashinaka, R., Walker, M., & Prasad, R. (2007). *"Learning to generate naturalistic utterances using reviews in spoken dialogue systems,"* ACM *Transactions on Speech and Language Processing*. TSLP.

Jin, X., Li, Y., Mah, T., & Tong, J. "Sensitive webpage classification for content advertising," in *Proceedings of the International Workshop on Data Mining and Audience Intelligence for Advertising*, 2007.

Jindal, N., & Liu, B. "Mining comparative sentences and relations," in *Proceedingsof AAAI*, 2006.

Kobayashi, N. (2007). *Opinion Mining from Web documents: Extraction and Structurization*, (Doctoral Dissertation), Nara Institute of Science and Technology.

Liu, J., Cao, Y., Lin, C.-Y., Huang, Y., & Zhou, M. "Low-quality product review detection in opinion summarization," in *Proceedings of theJoint Conference on Empirical Methods in Natural Language Processing andComputational Natural Language Learning (EMNLP-CoNLL)*, pp. 334–342, 2007.

KEY TERMS AND DEFINITIONS

Opinion Mining: Opinion mining is a process of information retrieval, which is concerned not with the topic, a document is about, but with the opinion it expresses.

Point-wise Mutual Information (PMI): It is a measure of association used in information theory and statistics. The PMI of a pair of outcomes of random variables quantifies the discrepancy between the probabilities of their coincidences.

Semantic Orientation: The semantic orientation of an opinion on a feature f states whether the opinion is positive, negative or neutral.

Sentiment Analysis: It aims to determine the attitude of a writer on the topic. The attitude may be their judgment or evaluation, or the intended emotional communication on the topic.

Sentiment Classification: Given a set of evaluative documents D, it determines whether each document $d \in D$ expresses a positive or negative opinion (or sentiment) on an object.

Targeted Opinion Detection: In the targeted opinion detection stage, opinion detection ventured outside the relatively structured review data of the Web to consider associations between opinions and their corresponding topics.

User Generated Content: It refers to the content produced by the consumers on various types of media like forum, blogs, discussion boards, and social networking sites.

Compilation of References

Aarts, E., Harwig, R., & Schuurmans, M. (2001). Ambient Intelligence. In Denning, P. (Ed.), *The Invisible Future* (pp. 235–250). New York: McGraw Hill.

Aarts, E., & de Ruyter, B. New research perspectives on Ambient Intelligence. In Journal of Ambient Intelligence and Smart Environments, 1, 5-14, IOS Press, 2009.

Aarts, E., & Eggen, B. (Eds.). *Ambient Intelligence in Home-Lab*. Neroc, Eindhoven, The Netherlands, 2002.

Aarts, E., Collier, R., van Loenen, E., & de Ruyter, B. (Eds.). (2003). *Ambient Intelligence. Proc. of the First European Symposium, EUSAI 2003*. Lecture Notes in Computer Science, vol. 2875. Springer Verlag, 2003, pp. 432.

Abbeel, P., & Ng, A. Y. Apprenticeship learning via inverse reinforcement learning, ACM International Conference Proceeding Series; Vol. 69, Proceedings of the twenty-first international conference on Machine learning, Banff, Alberta, Canada, 2004 Anderson, J.R.: Learning and Memory, John Wiley, 2002

Abbott, R., White, L., Ross, W., Masaki, K., Curb, D., & Petrovitch, H. (2004). Walking and dementia in physically capable elderly men. *Journal of the American Medical Association, 292*(12), 1447–1453.

Abdul-Rahman, A., & Hailes, S. (1997) Using Recommendations for Managing Trust in Distributed Systems. In Proc. of IEEE Malaysia International Conference on Communication (MICC'97), Kuala Lumpur, Malaysia.

Abe, A. (2009). Cognitive chance discovery. In S. C. (Ed.), *Universal Access in HCI, Part I, HCII2009 (LNCS5614)* (pp. 315–323). Berlin/New York: Springer.

Abellard, A., Khelifa, M. M. B., et al. (2003) A Wheelchair Neural Control. In: Mokhtari M (ed) Independent Living for Persons with Disabilities and Elderly People, ICOST'2003 1st International Conference on Smart Homes and Health Telematics, Assistive Technology Research Series 12, IOS Press, pp. 128–133.

Acampora, G., & Vincenzo, L. (2008). A Proposal of Ubiquitous Fuzzy Computing for Ambient Intelligence. *Information Sciences, 178*(Issue 3), 631–646. doi:10.1016/j.ins.2007.08.023

Ackley, D., Hinton, G., & Sejnowski, T. (1985). A learning algorithm for Boltzmann machines. *Cognitive Science, 9*(1), 147–169. doi:10.1207/s15516709cog0901_7

Ackoff, R. L. (1989). From Data to Wisdom. *Journal of Applied Systems Analysis, 16*, 3–9.

Adam, C. (2007). *Emotions: from psychological theories to logical formalization and implementation in a BDI agent*. PhD thesis, INP Toulouse, France.

Adam, C., Gaudou, B., Herzig, A., & Longin, D. (2006) A logical framework for an emotionally aware intelligent environment. In *AITAmI'06*. IOS Press.

Adam, C., Herzig, A., & Longin, D. *A logical formalization of the OCC theory of emotions.* In: *Synthese*, Springer, Vol. 168:2, p. 201-248, 2009.

Adam, C., & Longin, D. Endowing emotional agents with coping strategies: from emotions to emotional behaviour. In *IVA'2007*, (LNCS 4722, pp 348–349). Springer, 2007.

Adamatti, D & Bazzan, A. (2003). AFRODITE – *A Framework for Simulation of Agents with Emotions*. ABS 2003 – Agent Based Simulation 2003. Montpelier, France, 28-30 April.

Addlesee, M., R., C., Hodges, S., Newman, J., Steggles, P., Ward, A., et al. (2001). Implementing a Sentient Computing System. *IEEE Computer, 34* (8), 50-56.

Adlam, T., Carey-Smith, B., Evans, N., Orpwood, R., Boger, J., & Mihailidis, A. (2009). Implementing Monitoring and Technological Interventions in Smart Homes for People with Dementia, Case Studies. In Gottfried, B., & Aghajan, H. (Eds.), *Behaviour Monitoring and Interpretation, Smart Environments*. IOS Press.

Aghajan, H., & Wu, C. (2007). From Distributed Vision Networks to Human Behaviour Interpretation. In Gottfried, B. (Ed.), *BMI'07 (Vol. 296*, pp. 129–143). CEUR.

Agarwal, A., & Bhattacharyya, P. (2005). Sentiment analysis: A new approach for effective use of linguistic knowledge and exploiting similarities in a set of documents to be classified, in *Proceedings of the International Conference on Natural Language Processing* (ICON).

Agrawal, R., Rajagopalan, S., Srikant, R., & Xu, Y. "Mining newsgroups using networks arising from social behavior," in *Proceedings of WWW*, pp. 529–535, 2003.

Ahlgren, B., et al. "Ambient Networks: Bridging Heterogeneous Network Domains." Personal, Indoor and Mobile Radio Communications, 2005. PIMRC 2005. IEEE 16th International Symposium on (2005). Page(s): 937-941

Aiello, M., Marchese, M., Busetta, P., & Calabrese, G. (2004). Opening the home: a web service approach to domotics. Technical Report DIT-04-109, University of Trento.

Airoldi, E. M., Bai, X., & Padman, R. (2006). *"Markov blankets and meta-heuristic search: Sentiment extraction from unstructured text," Lecture Notes in Computer Science (Vol. 3932*, pp. 167–187). Advances in Web Mining and Web Usage Analysis.

Akerlof, G. A. (1970). "The market for "Lemons": Quality uncertainty and the market mechanism. *The Quarterly Journal of Economics, 84*, 488–500. doi:10.2307/1879431

Al Masum, S. M., Prendinger, H., & Ishizuka, M. "SenseNet: A linguistic tool to visualize numerical-valence based sentiment of textual data," in *Proceedings of the International Conference on Natural Language Processing (ICON)*, pp. 147–152, 2007. (Poster paper).

Albrecht, D. W., Zukerman, D. W., & Nicholson, A. E. (1998). Bayesian Models for Keyhole Plan Recognition in an Adventure Game. *User Modeling and User-Adapted Interaction, 8*(1-2), 5–47.

Alcaniz, M., & Rey, B. (2005) New Technologies For Ambient Intelligence. In: *Ambient Intelligence*. IOS Press [online] http://www.ambientintelligence.com

Alder, G. S. (1998). Ethical Issues in Electronic Performance Monitoring: A Consideration of Deontological and Teleological Perspectives. *Journal of Business Ethics, 17*, 729–743. doi:10.1023/A:1005776615072

Alia, M., Eide, V., Paspallis, N., Eliassen, F., Hallsteinsen, S., & Papadopoulos, G. (2007). A Utility-based Adaptivity Model for Mobile Applications. In Proceedings of the 21st International Conference on Advanced Information Networking and Applications Workshops-Volume 02, IEEE Computer Society Washington, DC, USA, 556–563.

Allan, J. (2002). Introduction to topic detection and tracking. In Allan, J. (Ed.), *Topic Detection and Tracking: Event-based Information Organization* (pp. 1–16). Norwell, MA, USA: Kluwer Academic Publishers.

Allen, J. F. (1983). Maintaining knowledge about temporal intervals. *Communications of the ACM, 26*, 832–843. doi:10.1145/182.358434

Allen, J. F. (1984). Towards a General Theory of Action and Time. *Artificial Intelligence, 23*(2), 123–154. doi:10.1016/0004-3702(84)90008-0

Allen, J. F. Maintaining Knowledge about Temporal Intervals. *CACM*, Vol. 26, Nr. 11, 1983.

Alm, C. O., Roth, D., & Sproat, R. "Emotions from text: Machine learning for text-based emotion prediction," in *Proceedings of the Human Language Technology Conference and the Conference on Empirical Methods in Natural Language Processing (HLT/EMNLP)*, 2005.

Alur, R., & Henzinger, T. (2002). Alternating-time temporal logic. *Journal of the ACM, 49*, 672–713. doi:10.1145/585265.585270

Amal El Fallah-Seghrouchni and Alexandru Suna. (2005). CLAIM and SyMPA: A Programming Environment for Intelligent and Mobile Agents. In *Multi-Agent Programming* (pp. 95–122). Languages, Platforms and Applications.

Ambient devices: Information at a Glance. (2004). Retrieved 08 16, 2008, from http://www.ambientdevices.com.

Amft, O., Kusserow, M., & Troster, G. (2007). Probabilistic parsing of dietary activity events. In *Proc. of BSN07, 13*, 242–7.

Amigo. (2009). Project web site. Retrieved on November 16, 2009, from http://www.hitech-projects.com/ euprojects/ amigo

Anagnostopoulos, A., Broder, A. Z., & Carmel, D. (2006). Sampling search-engine results. *World Wide Web (Bussum), 9*, 397–429. doi:10.1007/s11280-006-0222-z

Anand, S. Rao and Michael P. Georgeff. Modeling rational agents within a BDI-architecture. In *Proceedings of KR'91*, pages 473–484. Morgan Kaufmann Publishers, 1991.

Anastasopoulos, M., et al. "Towards a Reference Middleware Architecture for Ambient Intelligence." 20th ACM Conference on Object-Oriented Programming Systems, Languages and Applications. San Diego, 2005. Pages: 12-17.

Anderson, J. R. (2007). *How Can the Human Mind Occur in the Physical Universe*. Oxford: Oxford University Press.

Anderson, J. R., Fincham, J. M., Qin, Y., & Stocco, A. (2008). A Central circuit of the mind. *Trends in Cognitive Sciences, 12*(4), 136–143.

Anderson, C., Burt, P., & Van Der Wal, G. (1985). Change detection and tracking using pyramid transformation techniques. Proceedings from SPIE CIRC '85: *The Conference on Intelligent Robots and Computer Vision, 579*, 72-78.

Andreevskaia, A., & Bergler, S. "Mining WordNet for a fuzzy sentiment: Sentiment tag extraction from WordNet glosses," in *Proceedings of the European*

Andrés del Valle, A. C., & Opalach, A. (2005). The Persuasive Mirror: Computerized Persuasion for Healthy Living. In *Proceedings of the Eleventh International Conference on Human-Computer Interaction.*

Andrew Ortony, G. L. (1988). *Clore, and A. Collins. The cognitive structure of emotions.* Cambridge, MA: Cambridge University Press.

Andrienko, N., & Andrienko, G. Extracting patterns of individual movement behaviour from a massive collection of tracked positions. In: BMI 2007, vol. 296, 1–16. CEUR (2007)

Antonio R. Damasio. *Descartes' error: emotion, reason, and the human brain*. Putnam pub. group, 1994.

Antoniou, G., & Wagner, G. (2003). Rules and defeasible reasoning on the semantic web. In Proc. of RuleML Workshop 2003 (pp. 111-120).

Aoidh, E. M., & Bertolotto, M. (2007). Improving Spatial Data Usability by Capturing User Interactions. In Fabrikant, S. I., & Wachowicz, M. (Eds.), *The European Information Society, Leading the Way with Geo-information, Lecture Notes in Geoinformation and Cartography*. Springer.

Appelt, D. E., & Pollack, M. E. (1992, March). Weighted Abduction for Plan Ascription. *Journal of User Modeling and User-Adapted Interaction, 2*(1-2), 1–25.

Argamon, S., Karlgren, J., & Shanahan, J. G. (Eds.). *Proceedings of the SIGIR Workshop on Stylistic Analysis of Text For Information Access*. ACM, 2005.

Armstrong, N., Nugent, C., Moore, G., & Finlay, D. (2009). Mapping user needs to smartphone services for persons with chronic disease. In *Proc. of ICOST09.*

Ashley, P. *Authorization for a Large Heterogeneous Multi-Domain System*. In Proceedings of the AUUG

AssociationK. N. X.http://www.knx.eu/

Aue, A., & Gamon, M. (2005). Customizing sentiment classifiers to new domains: A case study. In *Proceedings of Recent Advances in Natural Language Processing*. RANLP.

Aue, A., & Gamon, M. "Automatic identification of sentiment vocabulary: Exploiting low association with known sentiment terms," in *Proceedings of the ACL Workshop on Feature Engineering for Machine Learning in NaturalLanguage Processing*, 2005.

Augusto, J. C., & Shapiro, D. (2007). *Advances in Ambient Intelligence. In the Frontiers of Artificial Intelligence and Application (FAIA) Series* (*Vol. 164*). IOS Press.

Augusto, J. C., & McCullagh, P. Ambient Intelligence: Concepts and Applications. *Computer Science and Information Systems*, Vol. 4 No. 1, 2007. BAALL: http://www.baall.net/

Augusto, J. C., & Nugent, C. D. (2006). Designing Smart Homes: The Role of Artificial Intelligence. Vol. 4008 of LNAI, Springer-Verlag, Berlin, Germany.

Augusto, J.-C., McCullagh, P., McClelland, V., & Walkden, J.-A. Enhanced healthcare provision through assisted decision making in a smart home environment. In *AITAmI07*, 2007.

Avancha, S., Patel, C., & Joshi, A. (2004). Ontology-driven Adaptive Sensor Networks. In Proc of MobiQuitous'04 (pp. 194-202).

Avrahami-Zilberbrand, D., & Kaminka, G. A. (2005). Fast and complete symbolic plan recognition. Proc. of the 19th Int. Joint Conference on Artificial Intelligence (IJCAI'05), International Joint Conference on Artificial Intelligence, pp.653–658.

Avrahami-Zilberbrand, D., & Kaminka, G. A. (2007). Incorporating Observer Biases in Keyhole Plan Recognition (Efficiently!), in: Proc. of the Twenty-Second AAAI Conference on Artificial Intelligence, AAAI Press, Menlo Park, CA, pp.944–949.

Avrahami-Zilberbrand, D., & Kaminka, G.A. (2005). Fast and complete symbolic plan recognition. In Proceedings of the International Joint Conference on Artificial Intelligence (IJCAI) 2005, 653-658.

Avrahami-Zilberbrand, D., & Kaminka, G. A. (2007). Incorporating observer biases in keyhole plan recognition (efficiently!). In Proceedings of the Twenty-Second National Conference on Artificial Intelligence (AAAI-07), 2007.

Avtanski, A., Stefanov, D., & Bien, Z. Z. (2003) Control of Indoor-operated Autonomous Wheelchair. In: Mokhtari M (ed) Independent Living for Persons with Disabilities and Elderly People, ICOST'2003 1st International Conference on Smart Homes and Health Telematics, Assistive Technology Research Series 12, IOS Press, pp. 120–127.

Ayers, (2001). Monitoring Human Behavior from Video Taken in an Office Environment. *Image and Vision Computing*, *19*(23), 833–846. doi:10.1016/S0262-8856(01)00047-6

Aylett, R., Louchart, S., Dias, J., Paiva, A., & Vala, M. (2005). FearNot! - An Experiment in Emergent Narrative. [Springer.]. *IVA*, *2005*, 305–316.

Aziz, A. A., Klein, M. C. A., & Treur, J. (2010). An Integrative Ambient Agent Model for Unipolar Depression Relapse Prevention. *Journal of Ambient Intelligence and Smart Environments*, *2*, 5–20.

Aztiria, A., Izaguirre, A., Basagoiti, R., & Augusto, J. C. (2008). Autonomous Learning of User's Preferences improved through User Feedback. In Gottfried, B., & Aghajan, H. (Eds.), *BMI'08* (*Vol. 396*, pp. 87–101). CEUR.

Baader, F., Calvanese, D., McGuinness, D. L., Nardi, D., & Patel-Schneider, P. F. (2007). *The Description Logic Handbook: Theory, Implementation, and Applications* (2nd ed.). Cambridge University Press.

Badii, A., Adetoye, A., Thiemert, D., & Hoffman, M. (2008). A Framework Architecture for Semantic Resolution of Security for AMI. Proceedings of the Hydra Consortium Internet of Things & Services (IOTS) Workshop at the TrustAMI08 Conference 18-19 September 2008, Sophia Antipolis, http://www.strategiestm.com/ conferences/ internet-of-things/ 08/ proceedings.htm

Badler, N. I., Phillips, C. B., & Webber, B. L. (1993). *Simulating Humans*. New York: Oxford University Press.

Balch, T., & Arkin, R. C. (1998). Behavior-based formation control for multi-robot teams. *IEEE Transactions on Robotics and Automation*, *14*(6), 926–939. .doi:10.1109/70.736776

Balch, T., & Hybinette, M. (2000). Social potentials for scalable multi-robot formations. In *Proc. IEEE Int. Conf. Robotics and Automation*, (pp.73-80).

Ballard, D. H., & Brown, C. M. (1982). *Computer Vision*. Prentice-Hall Inc New Jersey.

Ballesteros, F. J., Soriano, E., Algara, K. L., & Guardiola, G. (2007, September). Plan B: Using Files instead of Middleware Abstractions for Pervasive Computing Environments. *IEEE Pervasive Computing / IEEE Computer Society [and] IEEE Communications Society, 6*(Issue 3), 58–65. doi:10.1109/MPRV.2007.65

Ballesteros, F. J., Guardiola, G., Soriano, E., & Leal, K. *Traditional Systems can Work Well for Pervasive Applications. A Case Study: Plan 9 from Bell Labs Becomes Ubiquitous*. IEEE Intl. Conf. on Pervasive Computing and Communications 2005.

Ballesteros, F. J., Soriano, E., Leal, K., & Guardiola, G. *Plan B: An Operating System for Ubiquitous Computing Environments*. IEEE Intl. Conf. on Pervasive Computing and Communications 2006b.

Balzert, H. Lehrbuch der Software-Technik - Software Entwicklung, 2. Auflage ed. Heidelberg; Berlin: Spektrum, Akademischer Verlag, 2000.

Barger, T., Alwan, M., Kell, S., Turner, B., Wood, S., & Naidu, A. (2002) Objective remote assessment of activities of daily living: analysis of meal preparation patterns. Poster presentation, http://marc.med.virginia.edu /pdfs/library/ADL.pdf, 2002.

Baron-Cohen, S. (1995). *Mindblindness*. MIT Press.

Barron, J. L., Fleet, D. J., & Beauchemin, S. S. (1994). Performance of Optical Flow techniques. Proceedings from IJCV '94. *International Journal of Computer Vision, 12*(1), 43–77. doi:10.1007/BF01420984

Bartelt, C., Fischer, T., Niebuhr, D., Rausch, A., Seidl, F., & Trapp, M. "Dynamic Integration of Heterogeneous Mobile Devices." the Workshop in Design and Evolution of Autonomic Application Software (DEAS 2005), ICSE 2005, May 21, 2005, St. Louis, Missouri, USA. Pages: 1 - 7

Bartneck, C. (2002). emuu - an embodied emotional character for the ambient intelligent home.

Barwise, J., & Perry, J. (1983). *Situations and Attitudes. Bradford Books*. MIT Press.

Barwise, J., & Perry, J. (1980). The situation underground. Working Papers in Semantics vol. 1, Stanford University Press.

Bateman, J., Hois, J., Ross, R., & Farrar, S. The Generalized Upper Model 3.0: Documentation. *Technical report, Collaborative Research Center for Spatial Cognition, University of Bremen*, Germany, 2006. Bed: http://www.schramm-werkstaetten.de

Bates, J. (1994). *The Role of Emotion in Believable Agents*. Communications of the ACM, Special Issue on Agents, July.

Bauchet, J., & Mayers, A. (2005). Modelisation of ADLs in its Environment for Cognitive Assistance, In: Proc. of the 3rd International Conference on Smart homes and health Telematic, ICOST'05, Sherbrooke, Canada, pp. 3-10.

Bauchet, J., Giroux, S., & Pigot, H. LussierDesrochers, D., & Lachapelle, Y. (2008). Pervasive assistance in smart homes for people with intellectual disabilities: A case study on meal preparation. In *International Journal of ARM, 9*.

Baum, C., & Edwards, D. F. (1993). Cognitive Performance in Senile Dementia of the Alzheimer's type: The Kitchen Task Assessment. *The American Journal of Occupational Therapy., 47*, 431–436.

Baumann, M., Colonius, H., Hungar, H., Köster, F., Langner, M., Lüdtke, A., et al. Integrated Modeling for Safe Transportation - Driver modeling and driver experiments, in Th. Jürgensohn, H. Kolrep (Hrsg.), Fahrermodellierung in Wissenschaft und Wirtschaft, 2. Berliner Fachtagung für Fahrermodellierung, Fortschrittsbericht des VDI in der Reihe 22 (Mensch-Maschine-Systeme), Nr. 28, 84-99, VDI-Verlag, 2009, ISBN 978-3-18-302822-1

Baumberg, A. M. (1995). *Learning Deformable Models for Tracking Human Motion*. Doctoral dissertation, University of Leeds, Leeds, UK.

Baumberg, A., & Hogg, D. (1995). An adaptative eigenshape model. Proceedings from BMVC '95: *British Machine Vision Conference*, 87-96, Birmingham, UK.

Baurley, S., Brock, P., Geelhoed, E., & Moore, A. (2007). Communication-Wear: User Feedback as Part of a Co-Design Process. In I. Oakley & S. Brewster (Eds.), *the Second International Workshop on Haptic and Audio Interaction Design* (pp. 55-68). Springer-Verlag Berlin Heidelberg.

Bautin, M., Vijayarenu, L., & Skiena, S. "International sentiment analysis for news and blogs," in *Proceedings of the International Conference on Weblogsand Social Media (ICWSM)*, 2008.

Beauche, S., & Poizat, P. (2008). Automated Service Composition with Adaptive Planning. In Bouguettaya, A., Krueger, I. & Margaria, T. (Eds.), *Proceedings of the 6th International Conference on Service-Oriented Computing* (pp. 530-537). Berlin, Heidelberg: Springer-Verlag.

Becchio, C., Pierno, A., Mari, M., Lusher, D., & Castello, U. (2007). Motor contagion from eye gaze. the case of autism. *Brain, 130,* 2401–2411. doi:10.1093/brain/awm171

Becker C. and Nicklas D., "Where do spatial context-models end and where do ontologies start? A proposal of a combined approach," 2004.

Beigl, M., Gellersen, H. W., & Schmidt, A. (2001). MediaCups: Experience with Design and Use of Computer-Augmented Everyday Objects. *Computer Networks. Special Issue on Pervasive Computing, 35*(4), 401–409.

Beigl, M., Gellersen, H. W., & Schmidt, A. (2001). Mediacups: Experience with design and use of computer-augmented everyday objects. *Computer Networks, 35*(4), 401–409. doi:10.1016/S1389-1286(00)00180-8

Beigl, M., & Gellersen, H. W. (2003). Smart-Its: An embedded platform for smart objects. In Proc. of the Smart Objects Conference, Grenoble, France.

Bellavista, P., Küpper, A., & Helal, S. (2008). Location-Based Services: Back to the Future. *IEEE Pervasive Computing / IEEE Computer Society [and] IEEE Communications Society, 7*(2), 85–89. doi:10.1109/MPRV.2008.34

Belle, S., Burgio, L., Burns, R., Coon, D., Czaja, S., & Gallagher-Thompson, D. (2006). Enhancing the quality of life of dementia caregivers from different ethnic or racial groups: A randomized, controlled trial. *Annals of Internal Medicine, 145*(10), 727–738.

Bellinger, G., Castro, D., & Mills, A. (2004) Data, Information, Knowledge, & Wisdom. Retrieved November 10, 2009, from http://www.systems-thinking.org/ dikw/ dikw.htm.

Belnap, N., Perloff, M., & Xu, M. (2001). *Facing the future: agents and choices in our indeterminist world.* New York: Oxford University Press.

Belta, C., & Kumar, V. (2002). Trajectory design for formations of robots by kinetic energy shaping. In *Proc. IEEE Int. Conf. Robotics and Automation,* (pp.2593-2598).

Ben Mokhtar, S., Georgantas, N., & Issarny, V. (2007). COCOA: COnversation-based Service Composition in PervAsive Computing Environments with QoS Support. *Journal of Systems and Software, 80*(12), 1941–1955. doi:10.1016/j.jss.2007.03.002

Ben Mokhtar, S., Raverdy, P.-G., Urbieta, A., & Speicys Cardoso, R. (2008). Interoperable Semantic and Syntactic Service Matching for Ambient Computing Environments. *Proceedings of the 1st International Workshop on Ad-hoc Ambient Computing,* France.

Benamara, F., Cesarano, C., Picariello, A., Reforgiato, D., & Subrahmanian, V. S. "Sentiment analysis: Adjectives and adverbs are better than adjectives alone," in *Proceedings of the International Conference on Weblogs and Social Media (ICWSM)*, 2007. (Short paper).

Bengio, Y., & Frasconi, P. (1996). Input/output Hidden Markov Models for Sequence Processing. *IEEE Transactions on Neural Networks, 7,* 1231–1249.

Bengtsson, J., & Savenstedt, S. (2007). *Evaluation of field trial 1. Technical report.* COGKNOW Consortium.

Benta, K., Cremene, M., & Todica, V. (2009). Towards an affective aware home. In *Proc. of ICOST09.*

Berekoven, L. (1974). *Dienstleistungsbetrieb.* Wiesbaden: Wesen - Struktur - Bedeutung.

Berkman, L. F., Glass, T., Brissette, I., & Seeman, T. E. (2000). From social integration to health: Durkheim in the new millennium. *Social Science & Medicine, 51,* 843–857. doi:10.1016/S0277-9536(00)00065-4

Berners-Lee, T., Hendler, J., & Lassila, O. (2001). The Semantic Web. *Scientific American, 284*(5), 34–43. doi:10.1038/scientificamerican0501-34

Bernier, O. (2006). Real-Time 3D Articulated Pose Tracking using Particle Filters Interacting through Belief Propagation. Proceedings from ICPR ' 06: *The 18th International Conference on Pattern Recognition, 1,* 90-93.

Bernier, O., & Cheung-Mon-Chang, P. (2006). Real-time 3D articulated pose tracking using particle filtering and belief propagation on factor graphs. Proceedings from BMVC '06: *British Machine Vision Conference*, 01, 5-8.

Bertolino, A., De Angelis, G., Frantzen, L., & Polini, A. (2008). The PLASTIC Framework and Tools for Testing Service-Oriented Applications. *In Software Engineering. International Summer Schools, ISSSE, 2006-2008*, 106–139.

Bertolino, A., De Angelis, G., & Polini, A. (2009). Online Validation of Service Oriented Systems in the European Project TAS3. *Proceedings of the ICSE Workshop on Principles of Engineering Service Oriented Systems, PESOS* (pp. 107-110). Washington DC: IEEE Computer Society.

Bessiere, P. (2003). *Survey: Probabilistic Methodology and Techniques for Artifact Conception and Development, Rapport de Recherche, No. 4730*. INRIA.

Bessiere, P., Laugier, Ch., & Siegwart, R. (Eds.). (2008). *Probabilistic Reasoning and Decision Making in Sensory-Motor Systems*. Berlin: Springer.

Beth, T., Borcherding, M., & Klein, B. (1994) Valuation of Trust in Open Networks. In Proc. of the 3rd European Symposium on Research in Computer Security, pp.3–18, Brighton, UK.

Bettiol, C., & Campi, C. (2005) Challenges for Ambient Intelligence: Empowering the Users. In: *Ambient Intelligence*. IOS Press [online] http://www.ambientintelligence.com

Bickmore, T. W., & Picard, R. W. (2005). Establishing and Maintaining Long-Term Human-Computer Relationships. *ACM Transactions on Computer-Human Interaction*, *12*(2), 293–327. doi:10.1145/1067860.1067867

Bihler, P., Brunie, L., & Scuturici, V. M. (2005) Modeling User Intention in Pervasive Service Environments. IFIP International Conference on Embedded and Ubiquitous Computing, 977-986.

Bilmes, J. (1998). A gentle tutorial on the EM algorithm and its application to parameter estimation for Gaussian mixture and hidden Markov models. Technical Report ICSI-TR-97-021. Technical report, University of Berkeley.

Bischof, N. (1995). *Struktur und Bedeutung: Einführung i. d. Systemtheorie*. Bern: Huber.

Bishop, Ch. M. (2006). *Pattern Recognition and Machine Learning*. Heidelberg: Springer.

Bishop, C. M., & Svensén, M. (2003): Bayesian hierarchical mixtures of experts. In: Kjaerulff, U. and C. Meek (Ed.): Proceedings of the Nineteenth Conference on Uncertainty in Artificial Intelligence, pp. 57-64.

Biswas, J. F. Naumann F., & Qiu Q. (2006). Assessing the completeness of sensor data, *Proceedings of the Second International Conference on Database Systems for Advanced Applications*, (pp. 717-732)

Bjorn, G., Guesgen, H. W., & Hubner, S. (2006). Spatio-temporal Reasoning for Smart Homes. In Augusto, J. C., & Nugent, C. D. (Eds.), *Springer*.

Blackburn, S., Brownsell, S., & Hawley, M. (2006). Assistive technology for independence (at4i) executive summary.

Blake, A., & Isard, M. (1998). Active Contours. Springer-Verlag. Bogaert, M., Chleq, N., Cornez, P., Regazzoni, C., Teschioni, A., & Thonnat, M. The passwords project. Proceedings from ICIP '96: *The International Conference on Image Processing*, Vol 3, 675-678.

Blaze, M., Feigenbaum, J., & Keromytis, A. D. (1998) KeyNote: Trust Management for Public-Key Infrastructures. In Proc. of 6th International Workshop on Security Protocols, vol. 1550 of LNCS, pag. 59–63, Cambridge, UK, Springer-Verlag.

Blaze, M., Feigenbaum, J., & Lacy, J. (1996) Decentralized Trust Management. In Proc. of IEEE Symposium on Security and Privacy, pag. 164–173, Oakland, Ca.

Boger, J., Poupart, P., Hoey, J., Boutilier, C., Fernie, G., & Mihailidis, A. (2006). A planning system based on markov decision processes to guide people with dementia through activities of daily living. *IEEE Transactions on Information Technology in Biomedicine*, *10*(2), 323–333.

Boger, J., Poupart, P., Hoey, J., Boutilier, C., Fernie, G., & Mihailidis, A. (2005). A decision theoretic approach to task assistance for persons with dementia. In Proceedings of the 19th International Joint Conference on Artificial Intelligence (IJCAI05), pp. 1293–1299.

Bohlen, M., & Frei, H. (2009). Ambient Intelligence in the city. overview and new perspectives. In Nakashima, H., Aghajan, H., & Augusto, J. (Eds.), *Handbook of Ambient Intelligence and Smart Environments* (pp. 911–938). Heidelberg: Springer.

Böhmann, T. Modularisierung von IT-Dienstleistungen, 1 ed. Wiesbaden: 2004.

Bohn, J. (2005). Social, Economic and Ethical Implications of Ambient Intelligence and Ubiquitous Computing. In *Ambient Intelligence* (pp. 5–29). Springer-Verlag. doi:10.1007/3-540-27139-2_2

Bohn, J., Coroamă, V., Langheinrich, M., Mattern, F., & Rohs, M. (2005). Social, Economic, and Ethical Implications of Ambient Intelligence and Ubiquitous Computing. In Weber, W., Rabaey, J. M., & Aarts, E. (Eds.), *Ambient Intelligence* (pp. 5–29). Berlin, Heidelberg: Springer. doi:10.1007/3-540-27139-2_2

Bookman, A., Harrington, M., Pass, L., & Reisner, E. (2007). *Family Caregiver Handbook*. Cambridge, MA: Massachusetts Institute of Technology.

Bordini, R., Hübner, J., & Vieira, R. (2005) Jason and the Golden Fleece of agent-oriented programming.

Bosse, T., Jonker, C. M., van der Meij, L., & Treur, J. (2007). A Language and Environment for Analysis of Dynamics by SimulaTiOn. *International Journal of AI Tools*, *16*(issue 3), 435–464. doi:10.1142/S0218213007003357

Bosse, T., van Lambalgen, R., van Maanen, P. P., & Treur, J. (2011). (in press). A System to Support Attention Allocation: Development and Application. *Web Intelligence and Agent Systems Journal*.

Bosse, T., van Maanen, P.-P., & Treur, J. (2009). Simulation and Formal Analysis of Visual Attention. *Web Intelligence and Agent Systems Journal*, *7*, 89–105.

Bosse, T., Hoogendoorn, M., Klein, M. C. A., & Treur, J. (2008). A Component-Based Agent Model for Assessment of Driving Behaviour. In: Sandnes, F.E., Burgess, M., and Rong, C. (eds.), *Proceedings of the Fifth International Conference on Ubiquitous Intelligence and Computing, UIC'08*. Lecture Notes in Computer Science, vol. 5061, Springer Verlag, 2008, pp. 229-243.

Bosse, T., van Maanen, P. P., & Treur, J. (2006). A Cognitive Model for Visual Attention and its Application. In: Nishida, T., Klusch, M., Sycara, K., Yokoo, M., Liu, J., Wah, B., Cheung, W., and Cheung, Y.-M. (eds.), *Proceedings of the Sixth International Conference on Intelligent Agent Technology, IAT '06*. IEEE Computer Society Press, pp. 255-262.

Botelho, L. M., & Coelho, H. (2001). Machinery for artificial emotions. *Cybernetics and Systems*, *32*(5), 465–506.

Both, F., Hoogendoorn, M., Klein, M. C. A., & Treur, J. (2009). Design and Analysis of an Ambient Intelligent System Supporting Depression Therapy. In: Luis Azevedo and Ana Rita Londral (Eds), *Proceedings of the Second International Conference on Health Informatics, HEALTHINF '09*. INSTICC Press, pp. 142-148.

Bottaro, A., & Anne, G. "Extended Service Binder: Dynamic Service Availability Management in Ambient Intelligence." First Workshop on Future Research Challenges for Software and Service. Vienna, Austria, April 2006. Pages: 383-388

Bottaro, A., Bourcier, J., Escofier, C., & Lalanda, P. (2007a). Context-Aware Service Composition in a Home Control Gateway. *Proceedings of the IEEE International Conference on Pervasive Services*. Washington DC: IEEE Computer Society.

Bottaro, A., Gerodolle, A., & Lalanda, P. (2007b). Pervasive Service Composition in the Home Network. *Proceedings of the 21th International Conference on Advanced Networking and Applications AINA'07* (pp. 596-603), Washington, DC: IEEE Computer Society.

Bouchard, B., Giroux, S., & Bouzouane, A. (2006). A smart home agent for plan recognition of cognitively-impaired patients. *Journal of Computers*, *1*(5), 53–62.

Bouchard, B., Bouzouane, A., & Giroux, S. (2005). Plan recognition in Smart Homes: An approach based on the Lattice Theory [HWRS]. *Journal of Human-friendly Welfare Robotic Systems*, *6*(4), 29–45.

Bouchard, B. (2006). Un modèle de reconnaissance de plan pour les personnes atteintes de la maladie d'Alzheimer basé sur la théorie des treillis et sur un modèle d'action en logique de description, Ph.D. Thesis, University of Sherbrooke, Canada, 268 pages.

Bouchard, B., Bouzouane, A., & Giroux, S. (2007). A Keyhole Plan Recognition Model for Alzheimer's Patients: First Results, Journal of Applied Artificial Intelligence (AAI), Taylor & Francis publisher, Vol. 22 (7), July 2007, pp. 623–658.

Boudy, J., Baldinger, J.-L., et al. (2006) Telemedicine for Elderly Patient at Home: the TelePat Project. In: Nugent C and Augusto JC (eds) Smart Homes and Beyond, ICOST'2006 4th International Conference on Smart Homes and Health Telematics, Assistive Technology Research Series 19, IOS Press, pp. 74–81.

Bouzouane, A., Bouchard, B., & Giroux, S. (2005). Action Description Logic for Smart Home Agent Recognition, in Proc. of the International conference on Human-Computer Interaction (HCI) of the The International Association of Science and Technology for Development (IASTED), Phoenix, USA, pp 185-190.

Branavan, S. R. K., Chen, H., Eisenstein, J., & Barzilay, R. "Learning document-level semantic properties from free-text annotations," in *Proceedings of the Association for Computational Linguistics (ACL)*, 2008.

Bratman, M. E. (1987). *Intentions, Plans, and Practical Reason*. Cambridge, MA: Harvard University Press.

Breck, E., Choi, Y., & Cardie, C. "Identifying expressions of opinion in context," in *Proceedings of the International Joint Conference on Artificial Intelligence(IJCAI)*, Hyderabad, India, 2007.

Bredereke, J., & Lankenau, A. A Rigorous View of Mode Confusion. Proceedings of 21st Int'l Conf. on Computer Safety, Reliability and Security. LNCS 2434, 2002.

Bregler, C., & Malik, J. (1998). Tracking people with twists and exponential maps. Proceedings from ICCVPR '98: *The International Conference on Computer Vision and Pattern Recognition*, 8-15.

Bremond, F. (1997). *Environnement de résolution de problèmes pour l'interprétation de séquences d'images*. Doctoral dissertation, University of Nice Sophia-Antipolis, France.

Brewster, S., & Goodman, J. (2008). Hci and the older population.

Brickley, D., & Guha, R. (2000). Resource Description Framework (RDF) schema specification, from http://www.w3.org/ TR/ RDF-schema.

Brisson, A., Dias, J., & Paiva, A. (2007). *From Chinese Shadows to Interactive Shadows: Building a storytelling application with autonomous shadows*. Workshop on Agent Based Systems for Human Learning and Entertainment (ABSHLE) AAMAS ACM 2007, 11-02-2007

Broekstra, J., Kampman, A., & Harmelen, F. (2002). Sesame: A Generic Architecture for Storing and Querying RDF and RDF Schema. In Proc. of the 1st International Semantic Web Conference (pp. 54-68), Sardinia, Italy.

Broersen, J., Herzig, A., & Troquard, N. (2006). Embedding alternating-time temporal logic in strategic STIT logic of agency. *Journal of Logic and Computation, 16*(5), 559–578. doi:10.1093/logcom/exl025

Bromberg, Y.-D., & Issarny, V. (2005). INDISS: Interoperable Discovery System for Networked Services. In Alonso, G. (Ed.) *Proceedings of the of ACM/IFIP/USENIX 5th International Conference on Middleware Middleware '05* (pp. 164-183), New York, NY: Springer-Verlag New York.

Brown, M. (2005). Learning Spaces. In *Educating the Net Generation*. Washington, DC: Educause.

Broxvall, M. (2007). A Middleware for Ecologies of Robotic Devices. *Proc. of the First Int. Conf. on Robot Communication and Coordination (ROBOCOMM)*. Athens, Greece.

Broxvall, M., Coradeschi, S., Loutfi, A., & Saffiotti, A. (2006). An Ecological Approach to Odour Recognition in Intelligent Environments. *Proc. of the IEEE Int. Conf. on Robotics and Automation (ICRA)*. Orlando, FL.

Brumitt, B., & Meyers, B. (2000). *EasyLiving: Technologies for Intelligent Environments. Hand-held and Ubiquitous Computing 2000, LNCS 1927*. Springer-Verlag.

Brumitt, B., Meyers, B., Krumm, J., Kern, A., & Shafer, S. (2000). EasyLiving: Technologies for Intelligent Environments. In P. Thomas & H.-W. Gellersen (Eds.), *the Second International Symposium on Handheld and Ubiquitous Computing* (pp. 12-29). Springer-Verlag Berlin Heidelberg.

Buford, J., Kumar, R., & Perkins, G. (2006). Composition Trust Bindings in Pervasive Computing Service Composition. *Proceedings of the 4th IEEE International Conference on Pervasive Computing and Communications Workshops,* Washington DC: IEEE Computer Society.

Bui, H. H. (2003). A general model for online probabilistic plan recognition. In Proceedings of the International Joint Conference on Artificial Intelligence (IJCAI) 2003.

Bullinger, H.-J., & Meiren, T. (2001). In Bruhn, M., & Meffert, H. (Eds.), *"Service Engineering. Entwicklung und Gestaltung von Dienstleistungen,"* in *Handbuch Dienstleistungsmanagement* (pp. 149–177). Wiesbaden: Gabler.

Burton, D. (1988). Do anxious swimmers swim slower? Reexamining the elusive anxiety-performance relationship. *Journal of Sport & Exercise Psychology, 10*, 45–61.

Cacciabue, P. C. (Ed.). (2007). *Modelling Driver Behaviour in Automotive Environments*. London: Springer.

Caffier, P. P., Erdmann, U., & Ullsperger, P. (2003). Experimental evaluation of eye-blink parameters as a drowsiness measure. *European Journal of Applied Physiology, 89*(3-4), 319–325. doi:10.1007/s00421-003-0807-5

Cai, Q., Mitiche, A., & Aggarwal, J. K. (1995). Tracking human motion in an indoor environment. Proceedings from ICIP '95: *The Second International Conference on Image Processing*, 215-218.

Cai, Y. How Many Pixels Do We Need to See Things? Lecture Notes in Computer Science (LNCS), ICCS Proceedings, 2003. Arnheim, R. Visual Thinking. University of California Press, 1969 Allport, A. Visual Attention. MIT Press, 1993.

camunda service GmbH, "BPM-Pakete haben Nachholbedarf," Berlin: camunda service GmbH, 2008.

Canelli, N., Pisetta, A., et al. (2003) Development of a "Domotic Cell" for Health Care Delivery in Prison Environments. In: Mokhtari M (ed) Independent Living for Persons with Disabilities and Elderly People, ICOST'2003 1st International Conference on Smart Homes and Health Telematics, Assistive Technology Research Series 12, IOS Press, pp. 96–103.

Capezio, F., Giuni, A., Mastrogiovanni, F., Sgorbissa, A., Vernazza, P., Vernazza, T., & Zaccaria, R. (2007). Perspectives of Ambient Intelligence. In *Journal of the Italian AEIT Association* (pp. 42–49). Sweet Home.

Caporuscio, M., Raverdy, P.-G., & Issarny, V. (2009). (To appear). *ubi*SOAP: A Service Oriented Middleware for Ubiquitous Networking. *Journal of Transactions on Service Computing*.

Capra, L., Emmerich, W., & Mascolo, C. (2003). CARISMA: Context-Aware Reflective mIddleware System for Mobile Applications. *IEEE Transactions on Software Engineering, 29*(10), 921–945. doi:10.1109/TSE.2003.1237173

Capra, L. (2004) Engineering Human Trust in Mobile System Collaborations. In Proc. of the 12th International Symposium on the Foundations of Software Engineering (SIGSOFT 2004/FSE-12), pag. 107–116, Newport Beach, CA, USA, ACM Press.

Carberry, S. (2001). Techniques for plan recognition. *User Modeling and User-Adapted Interaction, 11*(1–2), 31–48.

Carberry, S., & Elzer, S. (2007). In Paliouras, G. (Ed.), *Exploiting evidence analysis in plan recognition. UM 2007, LNAI 4511 C. Conati, K. McCoy* (pp. 7–16). Springer-Verlag Berlin Heidelberg.

Carbo, J., & Molina, J. M. (2004). Agent-based collaborative filtering based on fuzzy recommendations, In. *International Journal of Web Engineering and Technology, 1*, 414–426. doi:10.1504/IJWET.2004.006267

Carbone, M., Nielsen, M., & Sassone, V. (2003) A Formal Model for Trust in Dynamic Networks. In Proc. of First International Conference on Software Engineering and Formal Methods (SEFM'03), pag. 54–63, Brisbane, Australia.

Carlo, C. P., Hjälmdahl, M., Luedtke, A., & Riccioli, C. (Eds.). Human Modelling in Assisted Transportation: Models, Tools and Risk Methods, Heidelberg: Springer, 2011, **ISBN-13:** 978-8847018204

Carriero, N., & Gelernter, D. (1989, April). Linda in context. *Communications of the ACM, 32*(issue 4), 444–458. doi:10.1145/63334.63337

Castaneda, H. N. (1975). *Thinking and Doing*. Dordrecht: D. Reidel.

Castelfranchi, C., & Lorini, E. Cognitive anatomy and functions of expectations. In *IJCAI03 Workshop on Cognitive Modeling of Agents and Multi-Agent Interactions*. Morgan Kaufmann Publisher, 2003.

Castellot, R. (2007a). *Report detailing state-of-the-art in cognitive reminder devices. Technical report.* COGKNOW Consortium.

Castellot, R. (2007b). *Report detailing state-of-the-art in services supporting cognitive disabilities. Technical report.* COGKNOW Consortium.

Čech, P., & Olševičová, K. (2009) The Proposal of Intelligent Assistant for Graduation Thesis Preparation. In: *Workshops Proceedings of the 5th International Conference on Intelligent Environments*. IOS Press, pp. 321-327

Cesta, A., & Fratini, S. (2008). The Timeline Representation Framework as a Planning and Scheduling Software Development Environment. *Proc. of 27th Workshop of the UK Planning and Scheduling SIG.*

Cesta, A., Cortellessa, G., Pecora, F., & Rasconi, R. (2007). Supporting Interaction in the RoboCare Intelligent Assistive Environment. *Proc. of AAAI Spring Symposium on Interaction Challenges for Intelligent Assistants.*

Chai D. & Ngan K. N. (1999). Face Segmentation Using Skin Color Map in Videophone Applications, *IEEE Trans. Circuits and Systems for Video Technology,* (9) (pp. 551-564)

Chan, E., Bresler, J., Al-Muhtadi, J., & Campbell, R. *Gaia Microserver: An Extendable Mobile Middleware Platform.* Proceedings of 3rd IEEE Intl. Conf. on Pervasive Computing and Communications, 2005, p. 309-313.

Chan, M., Campo, E., et al. (2004) Monitoring Elderly People Using a Multisensor System. Proc. 2nd International Conference on Smart Homes and Health Telematics (ICOST 2004), Vol. 14, pp. 162–169.

Chang, C. K., Jiang, H., Ming, H., & Oyama, K. (2009). Situ: A Situation-Theoretic Approach to Context-Aware Service Evolution. *IEEE Transactions on Services Computing, 2*(3), 261–275. doi:10.1109/TSC.2009.21

Chang, C. K., Oyama, K., Jaygarl, H., & Ming, H. (2008). On distributed run-time software evolution driven by stakeholders of smart home development. Second International Symposium on Universal Communication (ISUC'08), 59-66.

Chang, T. H., & Gong, S. (2001). Tracking multiple people with a multicamera system. Proceedings from IEEE ICCV '01: *The International Conference on Computer Vision Workshop on Multi-Object Tracking,* 19-26, Vancouver.

Chantzara, M., Anagnostou, M., & Sykas, E. (2006). Designing a Quality-Aware Discovery Mechanism for Acquiring Context Information. *Proceedings of the 20th International Conference on Advanced Information Networking and Applications, 1(6), AINA'06.* Washington DC: IEEE Computer Society.

Charniak, E., & Goldman, R. (1993). A Bayesian Model of Plan Recognition. *Artificial Intelligence Journal, 64,* 53–79.

Charniak, E., & McDermott, D. (1985). *Introduction to artificial intelligence.* Reading, MA: Addison Wesley.

Chater, N., & Oxford, M. (Eds.). (2008). *The Probabilistic Mind: Prospects for Bayesian Cognitive Science.* Oxford, England: Oxford University Press.

Chellas, B. F. (1980). *Modal logic: an introduction.* Cambridge: Cambridge University Press.

Chen, H., Finin, T., & Joshi, A. (2003). An ontology for context-aware pervasive computing environments. Special Issue on Ontologies for Distributed Systems. *The Knowledge Engineering Review, 18*(3), 197–207. doi:10.1017/S0269888904000025

Chen, H., Finin, T., Joshi, A., Perich, F., Chakraborty, D., & Kagal, L. (2004a). Intelligent agents meet the semantic web in smart spaces. *IEEE Internet Computing, 19*(5), 69–79. doi:10.1109/MIC.2004.66

Chen, H., Perich, F., Finin, T., & Joshi, A. (2004b). Standard ontology for ubiquitous and pervasive applications. In *Proc. of the MobiQuitous'04.* SOUPA.

Chen Wu and Hamid Aghajan. Context-aware gesture analysis for speaker hci. In *AITAmI'08,* 2008.

Chen, L., Nugent, C., Mulvenna, M., Finlay, D., Hong, X., & Poland, M. (2008a). Using event calculus for behaviour reasoning and assistance in a smart home. In *Proc. of ICOST08*, 81–9.

Chen, L., Nugent, C., Mulvenna, M., Finlay, D., Hong, X., & Poland, M. (2008b). A logical framework for behaviour reasoning and assistance in a smart home. In *International Journal of ARM, 9.*

Chen, Y., & Rui, Y. (2004). Real-time Speaker Tracking Using Particle Filter Sensor Fusion. *Proceeding of the IEEE '04*, 92(3), 485-494.

Chen, Y., Rui, Y., & Huang, T. S. (2001). JPDAF based HMM for real-time contour tracking. Proceedings from ICCVPR '01: *The International Conference on Computer Vision and Pattern Recognition*, Vol. 1, 543-550.

Cheng, D. C., & Thawonmas, R. (2004). Case-based plan recognition for real-time strategy games. In Proceedings of the 5th Game-On International Conference (CGAIDE) 2004, Reading, UK, Nov. 2004, 36-40.

Chleq, N., & Thonnat, M. (1996). Realtime image sequence interpretation for videosurveillance. Proceedings from ICIP '96: *The International Conference on Image Processing*, 801-804.

Choi, S., Seo, Y., Kim, H., & Hong, K. (1997). Where are the ball and players? soccer game analysis with color-based tracking and image mosaik. Proceedings from ICIAP '97: *The Ninth International Conference on Image Analysis and Processing*, 196-203.

Choset, H. (2001). Coverage for robotics - a survey of recent results. *Annals of Mathematics and Artificial Intelligence, 31*(1-4), 113–126. .doi:10.1023/A:1016639210559

Choudhury, T., Philipose, M., Wyatt, D., & Lester, J. (2006). Towards activity databases: Using sensors and statistical models to summarize people's lives, in IEEE Data Engineering Bulletin, pp. 49-58.

Christensen, H. B. (2002), "Using Logic Programming to Detect Activities in Pervasive Healthcare", Proc. Int'l Conf. Logic Programming, Springer, pp. 185–199.

Chua, S.-L., Marsland, S., & Guesgen, H. W. Behaviour recognition from sensory streams in smart environments. In *Proc. Australasian Joint Conference on Artificial Intelligence (AI-09)*, Melbourne, Australia, December 2009.

Cichoki, A., Washizawa, Y., Rutkowski, T., Bakardjian, H., Phan, A., & Choi, S. (2008). Noninvasive BCIs: Multiway Signal-Processing Array Decompositions. *Computer, 41*(10), 34–42. doi:10.1109/MC.2008.431

Cirillo, M., Karlsson, L., & Saffiotti, A. (2009). A Human-Aware Robot Task Planner. *Proc. of the Int. Conf. on Automated Planning and Scheduling (ICAPS09).*

Clark, A. (2008). *Supersizing the Mind: Embodiment, Action and Cognitive Extension*. Oxford: Oxford University Press.

Clark, A., & Chalmers, D. J. (1998). The extended mind. *Analysis, 58*, 10–23. doi:10.1111/1467-8284.00096

Clayton-Pedersen, A., & O'Neil, N. (2005) *Curricula Designed to Meet 21st-Century Expectaions*. In: *Educating the Net Generation*. Washington, DC, Educause

Clifford Neuman, B., & Ts'o, T. (1994, September). Kerberos: An Authentication Service for Computer Networks. *IEEE Communications, 329*, 33–38. doi:10.1109/35.312841

Cohen, P. R., & Levesque, H. J. (1990). Intention is choice with commitment. *Artificial Intelligence, 42*(2-3), 213–261. doi:10.1016/0004-3702(90)90055-5

Cohen, P. R., Perrault, C. R., & Allen, J. F. (1981). Beyond question answering. In Lehnert, W., & Ringle, M. (Eds.), *Strategies for Natural Language Processing* (pp. 245–274). Hillsdale, NJ: Lawrence Erlbaum Associates.

Cohen, R., & Peleg, D. (2006). Local algorithms for autonomous robots systems. In P. Flocchini & L. Gasieniec (Eds.), Structural Information and Communication Complexity (LNCS), *4056* (pp. 29–43). New York: Springer. doi:10.1007/11780823_4

Cohn, A. G., Bennett, B., Gooday, J., & Gotts, N. M. (1997). *Qualitative Spatial Representation and Reasoning with the Region Connection Calculus (Vol. 1)*. Geoinformatics.

Cohn, A. G., & Renz, J. (2007). *Qualitative Spatial Representation and Reasoning*. Handbook of Knowledge Representation.

comScore/the Kelsey group, "Online consumer-generated reviews have significant impact on offline purchase behavior," Press Release, http://www.comscore.com/ press/release.asp?press=1928, November 2007.

Conlin, J. (2009). Getting around: making fast and frugal navigation decisions. In Markus Raab, M., Johnson, J., & Heekeren, H. (Eds.), *Mind and Motion: The Bidirectional Link between Thought and Action (Vol. 174*, pp. 109–117). Elsevier. doi:10.1016/S0079-6123(09)01310-7

Conrad, J. G., & Schilder, F. "Opinion mining in legal blogs," in *Proceedings of the International Conference on Artificial Intelligence and Law (ICAIL)*, pp. 231–236, New York, NY, USA: ACM, 2007.

Conte, R., & Castelfranchi, C. (1995). *Cognitive and social action*. London: London University College of London Press.

Conti, G., Disperati, S. P., & De Amicis, R. (2009). The use of Geobrowser, Virtual Worlds and Visual Analytics in the context of developing countries' Environmental Security. In Proceedings of 6th Digital Earth Conference, Beijing, 2009.

Conti, G., Ucelli, G., & De Amicis, R. (2005). "Verba Volant Scripta Manent" a false Axiom within Virtual Environments. A semi-Automatic Tool for Retrieval of Semantics Understanding for Speech Enabled VR Applications". In Computers & Graphics vol. 30, no. 4, Volume 30, Issue 4, August 2006, pp. 619-628. http://dx.doi.org/10.1016/j.cag.2006.03.004. ISSN: 0097-8493.

Conti, G., Witzel, M., & De Amicis, R. (2008). A user-driven experience in the design of a multi-modal interface for industrial design review. In Khong C. W., Wong C. Y., von Niaman B. (Eds), Proceedings Human Factors in Telecommunication, 17-20 March 2008, Kuala Lumpur Malaysia, Prentice Hall, New York, pp. 351-358.

Cook, D., Augusto, J., & Vikramaditya, R. (2009). Ambient Intelligence: Technologies, applications, and opportunities. *Pervasive and Mobile Computing, 5*, 277–298. doi:10.1016/j.pmcj.2009.04.001

Cook, J., & Das, S. (2007). How smart are our environments? an updated look at the state of the art. *Pervasive and Mobile Computing, 3*(2), 53–73. doi:10.1016/j.pmcj.2006.12.001

Cook, D. J. (2009). Multi-agent Smart Environments, In. *J. of Ambient Intelligence and Smart Environments, 1*, 51–56.

Cook, D. J., Youngblood, M., & Das, S. K. (2006). A Multi-Agent Approach to Controlling a Smart Environment. In Augusto, J. C., & Nugent, C. (Eds.), *Designing Smart Homes The Role of Artificial Intelligence, Lecture Notes in Artificial Intelligence 4008* (pp. 165–182). Springer-Verlag Berlin Heidelberg.

Cook, D., Schmitter-Edgecombe, M., Crandall, A., Sanders, C., & Thomas, B. (2009). Collecting and disseminating smart home sensor data in the casas project. In *Proc. of CHI09 Workshop on Developing Shared Home Behavior Datasets to Advance HCI and Ubiquitous Computing Research*.

Cootes, T. F., Taylor, C. J., & Graham, J. Training models of shape from sets of examples. In D.C. Hogg and R.D. Boyle, editors, Processings of the Bristish Machine Vision Conference, Leeds, UK, pages 9-18, Springer Verlag, London, 1992

Cortes, J., Martinez, S., Karatas, T., & Bullo, F. (2004). Coverage control for mobile sensing networks. *IEEE Transactions on Robotics and Automation, 20*(2), 243–255. .doi:10.1109/TRA.2004.824698

Costa, P. T. Jr, & McCrae, P. R. (1992). *NEO PI-R Professional Manual*. Odessa, Fla: Psychological Assessment Resources.

Costa, P. T. Jr, & McCrae, R. R. (1992). *NEO PI-R professional manual*. Odessa, FL: Psychological Assessment Resources, Inc.

Coutaz, J., Dearle, A., Dupuy-Chessa, S., Kirby, G., Lachenal, C., Morrison, R., et al. (2003). Working document on GloSS ontology. Technical Report IST-2000-26070 Report D9.2, Global Smart Spaces Project.

Cox, I. J., & Hingorani, S. L. (1996). An Efficient Implementation of Reid's Multiple Hypothesis Traking Algorithm and Its Evaluation for the Propose of Visual Traking. *IEEE Transactions on Pattern Analysis and Machine Intelligence, 18*(2), 138–150. doi:10.1109/34.481539

Cox, M.T., Kerkez, B. (2006).Case-Based Plan Recognition with Novel Input. *International Journal of Control and Intelligent Systems 34*(2), 2006, 96-104.

Cox, R., Grosse, E., Pike, R., Presotto, D., & Quinlan, S. *Security in Plan 9*. In Proceedings of the USENIX 2002 Security Symposium, p. 3-16, 2002.

Crowley, J. L., & Demazeau, Y. (1993). Principles and Techniques for Sensor Data Fusion. *Journal of Signal Processing Systems, 32*, 5–27.

Cupillard, F., Avanzi, A., Bremond, F., & Thonnat, M. (2004). Video understanding for metro surveillance. *The IEEE International Conference on Networking, Sensing and Control, 1*, 186-191.

Cynthia Breazeal and B. Scassellati. (1999). *How to build robots that make friends and influence people*. IROS.

Dahlke, B., & Kergaßner, R. (1996). In Kleinaltenkamp, M., Fließ, S., & Jacob, F. (Eds.), *"Customer Integration und die Gestaltung von Geschäftsbeziehungen," in Customer Integration. Von der Kundenorientierung zur Kundenintegration* (pp. 177–192). Wiesbaden.

Damásio, A. (1994). O *erro de Descartes: emoção, razão e cérebro humano*. Publicações Europa-América.

Darling, J. (February 2006). Remote supervision system and method.

Das, S. K., Cook, D., Bhattacharya, A., Heierman, E. O. III, & Lin, T.-Y. (2002). The role of prediction algorithms in the MavHome smart home architecture. [Special Issue on Smart Homes]. *IEEE Wireless Communications, 9*(6), 77–84. doi:10.1109/MWC.2002.1160085

Das, S. R., & Chen, M. Y. (2007). Yahoo! for Amazon: Sentiment extraction from small talk on the Web. *Management Science, 53*, 1375–1388. doi:10.1287/mnsc.1070.0704

Dave, K., Lawrence, S., & Pennock, D. M. "Mining the peanut gallery: Opinion extraction and semantic classification of product reviews," in *Proceedings of WWW*, pp. 519–528, 2003.

Davidyuk, O., Selek, I., Duran, J. I., & Riekki, J. (2008b). Algorithms for Composing Pervasive Applications. *International Journal of Software Engineering and Its Applications, 2*(2), 71–94.

Davidyuk, O., Sánchez, I., Duran, J. I., & Riekki, J. (2008a). Autonomic Composition of Ubiquitous Multimedia Applications in REACHES. *Proceedings of the 7th International ACM Conference on Mobile and Ubiquitous Multimedia, MUM'08* (pp. 105-108). New York, NY: ACM.

Davidyuk, O., Sánchez, I., Duran, J. I., & Riekki, J. (2009). CADEAU: Collecting and Delivering Multimedia Information in Ubiquitous Environments. In Kamei K. (Ed.) *Adjunct Proceedings of the 7th International Conference on Pervasive Computing PERVASIVE'09* (pp. 283-287).

Davis, G., Wiratunga, N., et al. (2003) Matching Smart-House technology to needs of the elderly and disabled. Proc. Workshops of the 5th International Conference on Case-Based Reasoning (ICCBR 03), pp. 29–36.

Day, R. L., Laland, K., & Odling-Smee, J. (2003). Rethinking adaptation. The niche-construction perspective. *Perspectives in Biology and Medicine, 46*(1), 80–95. doi:10.1353/pbm.2003.0003

De Amicis, R., Conti, G., Simões, B., Lattuca, R., Tosi, N., Piffer, S., & Pellitteri, G. (2009). *Geo-Visual Analytics in the Future Internet. In the International Journal on Interactive Design and Manufacturing (IJIDeM), March 2009*. Springer.

De Amicis, R., Conti, G., Piffer, S., & Simões, B. (2009a). Geospatial Visual Analytics, Open Challenges And Protection Of The Environment. In De Amicis R., Stojanovic R., Conti G. (Eds) (2009). GeoVisual Analytics: Geographical Information Processing and Visual Analytics for Environmental Security. Springer.

De Guzman, E. S. (2004). *Presence Displays: Tangible Peripheral Displays for Promoting Awareness and Connectedness*. Unpublished master thesis, University of California at Berkeley.

de Leon, D. (2002). Cognitive task transformations. *Cognitive Systems Research, 3*, 349–359. doi:10.1016/S1389-0417(02)00047-5

de Rosis, F., Pelachaud, C., Poggi, I., Carofiglio, V., & Carolis, B. D. (2003). From Greta's mind to her face: modelling the dynamics of affective states in a conversational embodied agent. *International Journal of Human-Computer Studies, 59*(1-2), 81–118. doi:10.1016/S1071-5819(03)00020-X

Dechter, R., Meiri, I., & Pearl, J. (1991). Temporal Constraint Networks. *Artificial Intelligence, 49*(1-3), 61–95. doi:10.1016/0004-3702(91)90006-6

Demolombe, R., & Fernandez, A. M. O. (2006). Intention recognition in the situation calculus and probability theory frameworks. In *Proceedings of Computational Logic in Multi-agent Systems*. CLIMA.

Denis, M: The Description of Routes: A Cognitive Approach to the Production of Spatial Discourse. *Journal of Cahiers de Psychologie Cognitive*, vol. 16, 1997.

Dennett, D. C. (1987). *The Intentional Stance*. Cambridge, Mass.: MIT Press.

Deriche, R. (1987). Using Canny's criteria to derive a recursively implemented optimal edge detector. *International Journal of Computer Vision*, 2, 167–187. doi:10.1007/BF00123164

Descheneaux, C., & Pigot, H. (2009). Interactive calendar to help maintain social interactions for elderly people and people with mild cognitive impairments. In *Proc. of ICOST2009*.

Deutscher, J., Blake, A., & Reid, I. (2000). Articulated body motion capture by annealed particle filtering. Proceedings from ICCVPR '00: *The International Conference on Computer Vision and Pattern Recognition*, 2, 126-133.

Devitt, A., & Ahmad, K. "Sentiment analysis in financial news: A cohesionbased approach," in *Proceedings of the Association for Computational Linguistics(ACL)*, pp. 984–991, 2007.

Dewey, R. Psychology: An introduction. http://www.intropsych.com/, last visited on 7 July 2009.

Dey, A. K. (2001). Understanding and Using Context. *Personal and Ubiquitous Computing*, 5(1), 4–7. doi:10.1007/s007790170019

Dey, A. K. (2000, December). Providing Architectural Support for Building Context-Aware Applications. PhD thesis, College of Computing, Georgia Institute of Technology.

Dey, A. K., & Abowd, G. D. (2000). Towards a better understanding of context and context-awareness. In Proc. of CHI'2000.

Dias, J., & Paiva, A. (2005). Feeling and Reasoning: a Computational Model for Emotional Agents, in EPIA Affective Computing Workshop, Covilhã, 2005 Springer

Diepstraten, I. (2006). *De Nieuwe Leerder: Trendsettende leerbiografieën in een kennissamenleving*. Tilburg: F&N Boekservice.

Dimitrijevic, M., Lepetit, V., & Fua, P. (2005). Human body pose recognition using spatio-temporal templates. Proceedings from ICCV: *The International Conference on Computer Vision workshop on Modeling People and Human Interaction*, Beijing, China.

DIN. "Service Engineering - Entwicklungsbegleitende Normund und Dienstleistungen,", DIN-Fachbericht 75 ed Berlin, Wien, Zürich: DIN - Deutsches Institut für Normung, 1998.

Ding, X., Liu, B., & Yu, P. S. "A holistic lexicon-based approach to opinion mining," in *Proceedings of the Conference on Web Search and Web Data Mining (WSDM)*, 2008.

Do, J.-H., Jung, S. H., et al. (2006) Gesture-Based Interface for Home Appliance Control in Smart Home. In: Nugent C and Augusto JC (eds) Smart Homes and Beyond, ICOST'2006 4th International Conference on Smart Homes and Health Telematics, Assistive Technology Research Series 19, IOS Press, pp. 23–30.

Doctorow, E. L. (1980). *Loon Lake*. New York: Random House.

Döhle, M. Ein XML Gateway zur Gebäudesteuerung. Diplomarbeit (M.Sc. thesis), Universität Bremen, 2009. EU-Projekt i2home. http://www.i2home.org

Dolog, P. (2004). *Model-driven navigation design for semantic web applications with the UML-guide*. Engineering Advanced Web Applications.

Donald, M. (1998). Hominid enculturation and cognitive evolution. In Renfrew, C., Mellars, P., & Scarre, C. (Eds.), *Cognition and Material Culture: the Archeology of External Symbolic Storage* (pp. 7–17). Cambridge: The McDonald Institute for Archaeological Research.

Doucet, A., De Freitas, J. F. G., & Gordon, N. J. (2001). *Sequential Monte Carlo methods in practice*. Springer.

Dragoni, A. F., Giorgini, P., & Serafini, L. (2002). Mental States Recognition from Communication. *Journal of Logic Computation*, 12(1), 119–136.

Dressler, F. (2006). Self-organization in autonomous sensor/actuator networks. *Proc. of the 19th IEEE Int. Conf. on Architecture of Computing Systems.*

Dressler, F. (2007). *Self-organization in sensor and actuator networks.* John Wiley & Sons. doi:10.1002/9780470724460

Du, L., Sullivan, G., & Baker, K. (1993). Quantitative analysis of the view point consistency constraint in mode-based vision. Proceedings from ICCV '93: *The International Conference on Computer Vision*, 632-639.

Dubois, D. Uncertainty in Knowledge Representation and Reasoning: A partial bibliography, Institut de Recherche en Informatique de Toulouse (IRIT) www.irit.fr/~Didier.Dubois

Duchowski, A. T. (2004). Gaze-Contingent Displays: A Review. *Cyberpsychology & Behavior, 7*(6). doi:10.1089/cpb.2004.7.621

Dunham, M. H. (2003). *Data mining – Introductory and Advanced topics.* Upper Saddle River, NJ: Pearson Education Inc.

Duong, T. V., Bui, H. H., Phung, D. Q., & Venkatesh, S. Activity recognition and abnormality detection with the switching hidden semi-Markov model. In *Proc. CVPR-05*, pages 838–845. 2005.

Duval, T. S. (1972). *A theory of objective self-awareness.* Academic Press.

Edvardsson, B., & Olsson, J. (1996). Key Concepts for New Service Development. *The Service Industries Journal, 16*(2), 140–164. doi:10.1080/02642069600000019

Egenhofer, M., & Franzosa, R. (1991). Point-set topological spatial relations. *International Journal of Geographical Information Systems, 5*(2), 161–174. doi:10.1080/02693799108927841

Egerstedt, M., & Hu, X. (2001). Formation constrained multi-agent control. *IEEE Transactions on Robotics and Automation, 17*(6), 947–951. .doi:10.1109/70.976029

Ehrlenspiel, K. (1995). *Integrierte Produktentwicklung - Methoden für Prozessorganisation, Produktherstellung und Konstruktion.* München: Carl Hanser Verlag.

Eilers, M., & Möbus, C. (2011). Learning the Relevant Percepts of Modular Hierarchical Bayesian Driver Models Using a Bayesian Information Criterion. In Duffy, V. G. (Ed.), *Digital Human Modeling, HCII 2011, LNCS 6777* (pp. 463–472). Springer-Verlag Berlin Heidelberg.

Eilers, M., & Möbus, C. (2010). Lernen eines modularen Bayesian Autonomous Driver Mixture-of-Behaviors (BAD MoB) Modells. In Kolrep, H., & Jürgensohn, Th. (Eds.), *Fahrermodellierung - Zwischen kinematischen Menschmodellen und dynamisch-kognitiven Verhaltensmodellen, 3. Berliner Fachtagung für Fahrermodellierung, Fortschrittsbericht des VDI in der Reihe 22 (Mensch-Maschine-Systeme), Nr.32* (pp. 61–74). Düsseldorf: VDI-Verlag.

Eilers, M., & Möbus, C. *Learning of a Bayesian Autonomous Driver Mixture-of-Behaviors (BAD-MoB) Model*, in: Advances in Applied Digital Human Modeling, p. 436-445, Vincent G. Duffy (ed), CRC Press, Taylor & Francis Group, Boca Raton, 2010/2011, ISBN 978-1-4398-3511-1 and in. W. Karwowski and G. Salvendy (eds), 1st International Conference On Applied Digital Human Modeling, 17-20 July, 2010, Intercontinental, Miami Florida, USA, Conference Proceedings, Session Digital Human Modeling in the Bayesian Programming Framework, USA Publishing, ISBN-13: 978-0-9796435-4-5

Ekman, P. (1992). An argument for basic emotions. *Cognition and Emotion, 6,* 169–200. doi:10.1080/02699939208411068

Ekman, P. (1999). Basic emotions. In Dalgleish, T., & Power, M. (Eds.), *Handbook of cognition and emotion.* John Wiley & Sons.

el al Sonka, M. (1998). *Image Processing, Analysis, and Machine Vision* (2nd ed.). PWS Publishing.

El Jed, M., Pallamin, N., Dugdale, J., & Pavard, B. Modelling character emotion in an interactive virtual environment. In *AISB 2004 Symposium: Motion, Emotion and Cognition*, 2004.

Elliot, C. (1992). *The Affective Reasoner A process model of emotions in a multi-agent systems.* PhD dissertation. Northwestern University, USA, (1992)

Elliott, C., Rickel, J., & Lester, J. (1999). Lifelike pedagogical agents and affective computing: An exploratory synthesis. *LNCS, 1600,* 195–211.

El-Nasr, M., Yen, J., & Ioerger, T. R. (2000). FLAME -Fuzzy Logic Adaptive Model of Emotions. *Autonomous Agents and Multi-Agent Systems*, 3, 217–257. doi:10.1023/A:1010030809960

El-Nasr, M. S., Ioerger, T. R., & Yen, J. (1999). *PETEEI: a PET with evolving emotional intelligence*. In Proceedings of the Third Annual Conference on Autonomous Agents (Seattle, Washington, United States). O. Etzioni, J. P. Müller, and J. M. Bradshaw, Eds. AGENTS '99. ACM, New York, NY, 9-15.

Eng, K., Douglas, R., & Verschure, P. (2005). An interactive space that learns to influence human behavior. *Systems, Man and Cybernetics, Part A, IEEE Transactions on, 35*(1), 66–77.

Engelhardt, W., Kleinaltenkamp, M., & Reckenfelderbäumer, M. "Leistungsbündel als Absatzobjekte. Ein Ansatz zur Überwindung der Dichotomie von Sach- und Dienstleistungen,", 45 ed 1993, pp. 395-426.

Ernst, T. (2004) E-Wheelchair: A Communication System Based on IPv6 and NEMO. Proc. 2nd International Conference on Smart Homes and Health Telematics (ICOST 2004), Vol. 14, pp. 213–220.

Erol, K., Hendler, J., & Nau, D. (1994). Semantics for hierarchical task-network planning. CS-TR-3239, University of Maryland, 1994.

Esposito, J. M., & Dunbar, T. W. (2006). Maintaining wireless connectivity constraints for swarms in the presence of obstacles. In *Proc. IEEE Int. Conf. Robotics and Automation*, (pp.946-951).

Eugster, P., Felber, P., Guerraoui, R., & Kermarrec, A. (2003). The Many Faces of Publish/Subscribe. *ACM Computing Surveys, 35*(2), 114–131. doi:10.1145/857076.857078

EU-Projekt OASIS. http://www.oasis-project.eu

EU-Projekt SHARE-it. http://www.ist-shareit.eu

Eysenck, H. J. (1991). Dimensions of personality: 16, 5, or 3?--Criteria for a taxonomic paradigm. *Personality and Individual Differences, 12*, 773–790. doi:10.1016/0191-8869(91)90144-Z

F. J. Ballesteros, E. Soriano, G. Guardiola, K. Leal. *The Plan B OS for Ubiquitous Computing. Voice, Security, and Terminals as Case Studies*. Elsevier Pervasive and Mobile Computing Journal. 2(2006), pp. 472-488. 2006a.

F. Rivera-Illingworth, V. Callaghan, and H. Hagras. Detection of normal and novel behaviours in ubiquitous domestic environments. *The Computer Journal*, 2007.

Facial Identification CatalogFBI, Nov. 1988

Fahy, P., & Clarke, S. (2004). CASS: a middleware for mobile context-aware applications. In Proceedings of the Workshop on Context Awareness, MobiSys 2004.

Farrington, D. P. (1993). Understanding and preventing bullying. In Tonny, M., & Morris, N. (Eds.), *Crime and Justice (Vol. 17)*. Chicago: University of Chicago Press.

Feki, M., Biswas, J., & Tolstikov, A. (2009). Model and algorithmic framework for detection and correction of cognitive errors. *Technology and Health Care Journal, 17*(3), 203–219.

Felzenszwalb, P. F., & Huttenlocher, D. P. (2000). Efficient Matching of Pictorial Structures. Proceedings from IC-CVPR '00: *The International Conference on Computer Vision and Pattern Recognition*, 66-75.

Ferri, C., Prince, M., Brayne, C., Brodaty, H., Fratiglioni, L., & Ganguli, M. (2005). Global prevalence of dementia: a delphi consensus study. *Lancet, 366*, 2112–2117.

Ferscha, A., Emsenhuber, B., & Schmitzberger, H. (2006). Aesthetic Awareness Displays. In T. Pfifer, A. Schmidt, W. Woo, G. Doherty, F. Vernier, K. Delaney, B. Yerazunis, M. Chalmers, & J. Kiniry (Eds.), *Adjunct Proceedings of the 4th International Conference on Pervasive Computing* (pp. 161-165). Osterreichische Computer Gesellschaft.

Fieguth, P., & Terzopoulos, D. (1997). Color-based tracking of heads and other mobile objects at video frame rates. Proceedings from ICCVPR '97: *The International Conference on Computer Vision and Pattern Recognition*, 21.

Fine, S., Singer, Y., & Tishby, N. (1998). The hierarchical hidden Markov model: Analysis and applications. *Machine Learning, 32*, 41–62. doi:10.1023/A:1007469218079

Fiore, L., Fehr, D., Bodor, R., Drenner, A., & Somasundaram, G. (2008). Multi-Camera Human Activity Monitoring. *Journal of Intelligent & Robotic Systems, 52*(Issue 1), 5–43. doi:10.1007/s10846-007-9201-6

Fischer, K. (2006). *What Computer Talk Is and Is Not: Human-Computer Conversation as Intercultural Communication* (*Vol. 17*). Computational Linguistics.

Fitts, P. M., & Posner, M. I. (1967). *Human Performance*. Belmont, CA: Brooks/Cole.

Flemisch, F., Schieben, A., & Kelsch, J. Löper, Chr., and Schomerus, J., Cooperative Control and Active Interfaces for vehicle assistance and automation, FISITA 2008, http://elib.dlr.de/57618/01/FISITA2008_DLR_FlemischEtAl_CooperativeControl.pdf (visited 27th February, 2010)

Fließ, S., & Kleinaltenkamp, M. (2004). Blueprinting the service company: Managing service processes efficiently. *Journal of Business Research, 57*(4), 392–404. doi:10.1016/S0148-2963(02)00273-4

Fließ, S., Lasshof, B., & Mekel, M. (2004). *Möglichkeiten der Integration eines Zeitmanagements in das Blueprinting von Dienstleistungsprozessen*. Douglas-Stiftungslehrstuhl für Dienstleistungsmanagement.

Fließ, S. Prozessorganisation in Dienstleistungsunternehmen. Stuttgart: 2006.

Flinn, M., Geary, D., & Ward, C. (2002). Ecological dominance, social competition, and coalitionary arms races why humans evolved extraordinary intelligence. *Evolution and Human Behavior, 26*(1), 10–46. doi:10.1016/j.evolhumbehav.2004.08.005

Flury, T., Privat, G., & Ramparany, F. (2004). OWL-based location ontology for context-aware services. Proceedings of the Artificial Intelligence in Mobile Systems (AIMS 2004), 52-57.

Fook, V. F. S., Hao, S., et al. (2006) Fiber Bragg Grating Sensor System for Monitoring and Handling Bedridden Patients. In: Nugent C and Augusto JC (eds) Smart Homes and Beyond, ICOST'2006 4th International Conference on Smart Homes and Health Telematics, Assistive Technology Research Series 19, IOS Press, pp. 239–246.

Fook, V. F. S., Tee, J. H., et al. (2007) Smart Mote-Based Medical System for Monitoring and Handling Medication among Persons with Dementia. 5th International Conference on Smart Homes and Health Telematics (ICOST 2007), LNCS 4541, Springer, pp. 54–62.

Forbes, T., Kanazawa, H. K., & Russell, St. The BATmobile: Towards a Bayesian Automated Taxi, IJCAI'95, Vol 2, 1878 -1885 (1995)

Forest, F. (2006). *Astrid Oehme, Karim Yaici, and Celine Verchere-Morice. Psycho-social aspects of context awareness in ambient intelligent mobile systems. In 15th*. IST Summit.

Forney, G. D. (1973). The Viterbi algorithm. *Proceedings of the IEEE, 61*, 268–278.

Forsyth, D. A., & Fleck, M. M. (1997). Body plans. Proceedings from ICCVPR '97: *The International Conference on Computer Vision and Pattern Recognition*, 678-683.

Fouquet, Y., Vuillerme, N., & Demongeot, J. (2009). Pervasive informatics and persistent actimetric information in health smart homes. In *Proc. of ICOST09*.

Fowler, T. (2001). Mesh networks for broadband access. *IEE Review, 47*(1), 17–22. .doi:10.1049/ir:20010108

Fox, A., Johanson, B., Hanrahan, P., & Winograd, T. (2000, May-June). Integrating information appliances into an interactive workspace. *IEEE Computer Graphics and Applications, 20*(Issue 3), 54–65. doi:10.1109/38.844373

François, A. R. J., & Kang, E.-Y. E. (2003). A handheld mirror simulation. In *Proceedings of 2003 International Conference on Multimedia and Expo* Vol.1. (pp. 745-748), IEEE Computer Society, Washington DC, USA.

Frank, A. U. Qualitative Spatial Reasoning with Cardinal Directions. *Proceedings of the Seventh Austrian Conference on Artificial Intelligence*, 1991.

Fratini, S., Pecora, F., & Cesta, A. (2008). Unifying Planning and Scheduling as Timelines in a Component-Based Perspective. *Archives of Control Sciences, 18*(2), 231–271.

Freksa, C. (1991). *Qualitative Spatial Reasoning*. Cognitive and Linguistic Aspects of Geographic Space.

Friedman-Hill, E. (2007). Jess, the Rule Engine for the Java Platform, from http://www.jessrules.com/ jess index.shtml.

Frijda, N. H. (1986). *The Emotions*. Cambridge University Press.

Frischen, A., Bayliss, A., & Tipper, S. (2007). Gaze-cueing of attention: Visual attention, social cognition and individual differences. *Psychological Bulletin, 133*(4), 694–724. doi:10.1037/0033-2909.133.4.694

Frischen, A., Loach, D., & Tipper, S. P. (2009). Seeing the world through another person's eyes: Simulating selective attention via action observation. *Cognition, 111*(2), 212–218. doi:10.1016/j.cognition.2009.02.003

Fujibayashi, K., Murata, S., Sugawara, K., & Yamamura, M. (2002). Self-organizing formation algorithm for active elements. In *Proc. 21st IEEE Sym. Reliable Distributed Systems*, (pp.416-421).

Fujinami, K., & Inagawa, N. (2009). Page-Flipping Detection and Information Presentation for Implicit Interaction with a Book. *International Journal of Ubiquitous Multimedia Engineering, 4*(4), 93–112.

Fujinami, K., & Nakajima, T. (2006). Bazaar: A Middleware for Physical World Abstraction. *Journal of Mobile Multimedia, 2*(2), 124–145.

Fujinami, K., & Nakajima, T. (2005). Sentient Artefact: Acquiring User's Context Through Daily Objects. In T. Enokido, L. Yan, B. Xiao, D. Kim, Y. Dai, & L. T. Yang (Eds.), *the Second International Symposium on Ubiquitous Intelligence and Smart Worlds* (pp. 335-344). International Federation for Information Processing.

Fujinami, K., & Riekki, J. (2008). A Case Study on an Ambient Display as a Persuasive Medium for Exercise Awareness. In H.O.-Kukkonen, P. Hasle, M. Harjumaa, K. Segerståhl, & P. Øhrstrøm (Eds.), *the Third International Conference on Persuasive Technology* (pp. 266-269). Springer-Verlag Berlin Heidelberg.

Fukuhara, T., Nakagawa, H., & Nishida, T. "Understanding sentiment of people from news articles: Temporal sentiment analysis of social events," in *Proceedings of the International Conference on Weblogs and Social Media (ICWSM)*, 2007.

Fung, T. H., & Kowalski, R. (1997). The IFF Proof Procedure for Abductive Logic Programming. *The Journal of Logic Programming*, 1997.

Funk, C., Kuhmünch, C., & Niedermeier, C. "A Model of Pervasive Services for Service Composition", CAMS05: OTM 2005 Workshop on Context-Aware Mobile Systems, Agia Napa, Cyprus, October 2005. Pages 215-224.

Fusier, F., Valentin, V., Bremond, F., Thonnat, M., Borg, M., & Thirde, D. (2007). Video Understanding for Complex Activity Recognition. *Machine Vision and Applications, 18*(3), 167–188. doi:10.1007/s00138-006-0054-y

Gahegan, M. (1995). Proximity operators for qualitative spatial reasoning. In Frank, A. U., & Kuhn, W. (Eds.), *Spatial Information Theory: A Theoretical Basis for GIS, Lecture Notes in Computer Science 988*. Berlin, Germany: Springer.

Garate, A., Herrasti, N., & Lopez, A. (2005). GENIO: An ambient intelligence application in home automation and entertainment environment. *Proceedings of the Joint soc-eusai conference* (pp. 241 - 245).

Garcia, C., & Tziritas, G. (1999). Face Detection Using Quantized Skin Color Regions Merging and Wavelet Packet Analysis. *IEEE Transactions on Multimedia*, (1): 264–277. doi:10.1109/6046.784465

Gärdenfors, P. (2003), *How Homo Became Sapiens: On The Evolution Of Thinking*. Oxford University Press, 2003.

Gardner, H. (1985). *Frames of mind: the theory of multiple intelligences*. New York: Basic Books Inc.

Garruzzo, S., Rosaci, D., & Sarné, G. M. L. (2007) ISA-BEL: A Multi Agent e-Learning System That Supports Multiple Devices. In: *IEEE/WIC/ACM International Conference on Intelligent Agent Technology*, IEEE, New York, pp. 485-488.

Gavrila, D. M., & Davis, L. S. (1996). 3-D Model-Based Tracking of Humans in Actions: A Multi-View Approach. Proceedings from ICCVPR '96: *The International Conference on Computer Vision and Pattern Recognition*, 73-80.

Geib, C. (2007). Plan recognition. In Kott, A., & McEneaney, W. M. (Eds.), *Adversarial Reasoning: Computational Approaches to Reading the Opponent's Mind* (pp. 77–100). Chapman & Hall/CRC.

Geib, C. W., & Steedman, M. (2003). On natural language processing and plan recognition. In Proceedings of the International Joint Conference on Artificial Intelligence (IJCAI) 2003, 1612- 1617.

Geib, C., & Goldman, R. (2001). Plan Recognition in Intrusion Detection Systems. In DARPA Information Survivability Conference and Exposition (DISCEX).

Geib, C., & Goldman, R. (2005) Partial Observability and Probabilistic Plan/Goal Recognition, in: Proc. of the IJCAI-05 workshop on Modeling Others from Observations (MOO-05), Int. Joint Conference on Artificial Intelligence, pp. 1–6.

Geisler, W. S., & Perry, J. S. (1998). Real-time foveated multiresolution system for low-bandwidth video communication. In *Proceedings of Human Vision and Electronic Imaging*. Bellingham, WA: SPIE.

Gelfond, M., & Lifschitz, V. (1992). Representing actions in extended logic programs. In Proceedings of the International Symposium on Logic Programming, 1992.

Georgantas, N., Issarny, V., Ben Mokhtar, S., Bromberg, Y.-D., Bianco, S., & Thomson, G. (2009To appear). Middleware Architecture for Ambient Intelligence in the Networked Home. In Nakashima, H., Augusto, J. C., & Aghajan, H. (Eds.), *Handbook of Ambient Intelligence and Smart Environments*. Springer.

Georgeff, M. P. (1987), Planning, in Annual Reviews, Palo Alto, California.

Gerevini, A., & Nebel, B. Qualitative Spatio-Temporal Reasoning with RCC-8 and Allen's Interval Calculus: Computational Complexity. *Proceedings of ECAI 2002*.

Gervasi, V., & Prencipe, G. (2003). Coordination without communication: the case of the flocking problem. *Discrete Applied Mathematics, 143*(3), 203–223.

Ghasem-Aghaee, N., & Oren, T. I. (2004). Effects of cognitive complexity in agent simulation: Basics. *SCS, 04*, 15–19.

Ghosh, S., Basu, K., & Das, S. K. (2005). An architecture for next-generation radio access networks. *IEEE Network, 19*(5), 35–42. .doi:10.1109/MNET.2005.1509950

Gibson, E., & Pick, A. (2000). *An ecological approach to perceptual learning and development*. Oxford: Oxford University Press.

Gibson, J. J. (1979). *The Ecological Approach to Visual Perception*. Boston, MA: Houghton Mifflin.

Giovannetti, T., Libon, D. J., Buxbaum, L. J., & Schwartz, M. F. (2002). Naturalistic action impairments in dementia. *Neuropsychologia, 40*(8), 1220–1232.

Giroux, S., Bauchet, J., Pigot, H., Lusser-Desrochers, D., & Lachappelle, Y. (2008). Pervasive behavior tracking for cognitive assistance. The 3rd International Conference on Pervasive Technologies Related to Assistive Environments, Petra'08, Greece, July 15-19, 2008.

Giroux, S., Leblanc, T., Bouzouane, A., Bouchard, B., Pigot, H., & Bauchet, J. (2009) The Praxis of Cognitive Assistance in Smart Homes, IOS Press book chapter, pp. 1-30, (to appear).

Glover, J., Holstius, D., Manojlovich, M., Montgomery, K., Powers, A., Wu, J., et al. A Robotically-Augmented Walker for Older Adults. Tech. Report CMU-CS-03-170, Carnegie Mellon University, Computer Science Department, Pittsburgh, PA, 2003.

Gluck, K. A., & Pew, R. W. (2005). *Modeling Human Behavior with Integrated Cognitive Architectures*. Mahwah, N.J.: Lawrence Erlbaum Associates.

Godfrey-Smith, P. (1998). *Complexity and the Function of Mind in Nature*. Cambridge: Cambridge University Press.

Goecke, R., & Stein, S. "Marktführerschaft durch Leistungsbündelung und kundenorientiertes Service Engineering,", 13 ed 1998, pp. 11-13.

Goldberg, A. B., Zhu, X., & Wright, S. (2007). Dissimilarity in graph-based semisupervised classification. In *Artificial Intelligence and Statistics*. AISTATS.

Goldman, A. I. (2006). *Simulating Minds: The Philosophy, Psychology and Neuroscience of Mind Reading*. Oxford University Press.

Goldman, R. P., Geib, C. W., & Miller, C. A. (1999). A new model of plan recognition. In Proceedings of the Conference on Uncertainty in Artificial Intelligence, San Francisco, CA (USA), 245–254.

Gollub, J. Umsetzung und Evaluation eines mit Laserscannern gesicherten Rollstuhls als Interaktives Museumsexponat. Diplomarbeit. Fachbereich 3 – Mathematik und Informatik, Universität Bremen (http://www.informatik. uni-bremen.de/agebv/downloads/published/gollub_thesis_07.pdf), 2007.

Gonzalez, R. C., & Woods, R. E. (1992). *Digital Image Processing, 2nd Ed.* (pp 626-629).

Goodman, B. A., Litman, D. J. (1992). On the interaction between plan recognition and intelligent interfaces. User Modeling and User-Adapted Interactions 2(1-2), 1992, 83-115.

Gopnik, A. & Tenenbaum, J.B., Bayesian networks, Bayesian learning and cognitive development, Development Science, 10:3, 2007, 281 – 287

Goschnick, J., & Körber, R. (2002). Condition Monitoring For Intelligent Household Appliances. In Lahrmann, A., & Tschulena, G. (Eds.), *Sensors in Household Appliances* (*Vol. 5*, pp. 52–68). Weinheim: Wiley-VCH.

Goshorn, R., Goshorn, D., & Kölsch, M. The Enhancement of Low-Level Classifications for Ambient Assisted Living. In B. Gottfried and H. Aghajan, editors, *2nd Workshop on Behaviour Monitoring and Interpretation (BMI'08)*, volume 396, pages 87–101. CEUR Proceedings, 2008.

Gottfried, B., & Aghajan, H. (Eds.). (2009). *Behaviour Monitoring and Interpretation – Smart Environments.* IOS Press.

Gottfried, B. (2009). (b). Locomotion Activities in Smart Environments. In Nakashima, H., Augusto, J. C., & Aghajan, H. (Eds.), *Handbook of Ambient Intelligence and Smart Environments. To appear at Springer.*

Gottfried, B. (a). Behaviour Monitoring and Interpretation: A Computational Approach to Ethology. In B. Mertsching et al. (eds.). *The German Conference on Artificial Intelligence.* LNAI 5803, 572-580, Springer, 2009.

Gottfried, B. (Ed.). *Behaviour Monitoring and Interpretation*, volume 296. CEUR Proceedings, 2007.

Gottfried, B., & Aghajan, H. (Eds.). *Behaviour Monitoring and Interpretation*, volume 396. CEUR Proceedings, 2008.

Gouin-Vallerand, C., & Giroux, C. (2007). Managing and Deployment of Applications with OSGi in the Context of Smart Homes. Proc. of the Third IEEE International Conference on Wireless and Mobile Computing, Networking, and Communications (WiMob 2007).

Goultiaeva, A., & Lespérance, Y. (2007). Incremental Plan Recognition in an Agent Programming Framework. *Working Notes of the AAAI Workshop on Plan, Activity, and Intention Recognition (PAIR).*

Grace, P., Blair, G. S., & Samuel, S. (2005). A reflective framework for discovery and interaction in heterogeneous mobile environments. *ACM SIGMOBILE Mobile Computing and Communications Review*, *9*(1), 2–14. doi:10.1145/1055959.1055962

Grace, P., Blair, G. S., & Samuel, S. "ReMMoC, A Reflective Middleware to Support Mobile Client Interoperability", International Symposium on Distributed Objects and Applications (DOA), Catania, Sicily, Italy, November 2003. pp. 1170-1187.

Grandison, T., & Sloman, M. (2003) Trust Management Tools for Internet Applications. In Proc. of the 1st International Conference on Trust Management (iTrust), Crete.

Gratch, J., & Marsella, S. (2006). Evaluating a computational model of emotion. *Journal of Autonomous Agents and Multiagent Systems*, *11*(1), 23–43. doi:10.1007/s10458-005-1081-1

Gratch, J., Okhmatovskaia, A., Lamothe, F., Marsella, S., Morales, M., van der Werf, R. J., & Morency, L. (2006). Virtual Rapport. 6th International Conference on Intelligent Virtual Agents, Marina del Rey, CA.

Gratch, J., Wang, N., Gerten, J., Fast, E., & Duffy, R. (2007). Creating Rapport with Virtual Agents. 7th International Conference on Intelligent Virtual Agents, Paris, France.

Green, D. J. (2005). Realtime Compliance Management Using a Wireless Realtime Pillbottle – A Report on the Pilot Study of SIMPILL. In: *Proc. of the International Conference for eHealth, Telemedicine and Health, Med-e-Tel'05*, 2005, Luxemburg.

Greenfield, A. (2006). *Everyware: the dawning age of ubiquitous computing.* New Riders. 11–12.

Gregory, R. (1997). *Mirrors in Mind*. W. H. Freeman and Company.

Grenon, P., Smith, B.: Towards Dynamic Spatial Ontology. *Journal of Spatial Cognition and Computation*, Vol. 4, 2004.

Griffiths, Th. L., Kemp, Ch., & Tenenbaum, J. B. (2008). Bayesian Models of Cognition. In Sun, R. (Ed.), *The Cambridge Handbook of Computational Psychology* (pp. 59–100). Cambridge University Press.

Grisetti, G., Stachniss, C., & Burgard, W. Improved Techniques for Grid Mapping with Rao-Blackwellized Particle Filters. *IEEE Transactions on Robotics*, 2006. Kitchenette: http://www.pressalitcare.de/

Grob H.L and Seufert S., "Vorgehensmodelle bei der Entwicklung von CAL-Software,", Arbeitsbericht 5 ed Münster: Universität Münster, 1996.

Gruber, T. R. (1993). A translation approach to portable ontologies. *Knowledge Acquisition*, 5(2), 199–220. doi:10.1006/knac.1993.1008

Gu, T., Pung, H. K., & Zhang, D. Q. (2005, January). A service-oriented middleware for building context-aware services. *Journal of Network and Computer Applications*, 28(1), 1–18. doi:10.1016/j.jnca.2004.06.002

Guesgen, H. W. (2002). Reasoning about distance based on fuzzy sets. [Special Issue on Spatial and Temporal Reasoning]. *Applied Intelligence*, 17(3), 265–270. doi:10.1023/A:1020087332413

Guesgen, H. W., & Hertzberg, J. (1996). Spatial persistence. [Special Issue on Spatial and Temporal Reasoning]. *Applied Intelligence*, 6, 11–28. doi:10.1007/BF00117598

Guo, B., Sun, L., & Zhang, D. (2010). *The Architecture Design of a Cross-Domain Context Management System* (pp. 499–504). Germany: In Proc. of Percom Workshops.

Guo, B., Zhang, D., & Imai, M. (2011). Towards a Co-operative Programming Framework for Context-Aware Applications. *Personal and Ubiquitous Computing*, 15(3), 221–233. doi:10.1007/s00779-010-0329-1

Guo, B., Satake, S., & Imai, M. (2006). Sixth-Sense: Context Reasoning for Potential Objects Detection in Smart Sensor Rich Environment. In Proc. of the IAT-06 (pp. 191-194), Hong Kong, China.

Gurumohan, P. C., & Hui, J. (2003). Topology design for free space optical networks. In *Proc. 12th Int. Conf. Computer Communications and Networks*, (pp.576-579).

Gutwirth, S. (2009). Beyond identity. *IDIS*, 1, 123–133. doi:10.1007/s12394-009-0009-3

Haarslev, V., & Moller, R. (2001). Racer system description. In Proc. of the International Joint Conference on Automated Reasoning (pp. 701-705).

Haigh, K., Kiff, L., & Ho, G. (2006). The Independent LifeStyle Assistant: Lessons Learned. *Assistive Technology*, 18(1), 87–106.

Halevy, A. (2005). Why Your Data Wont Mix. *ACM Queue; Tomorrow's Computing Today*, 3(8), 50–58. doi:10.1145/1103822.1103836

Hall, R. S., & Cervantes, H. (2004). Challenges in Building Service-Oriented Applications for OSGi. *IEEE Communications Magazine*, (May): 144–149. doi:10.1109/MCOM.2004.1299359

Halliday, D., Resnick, R., & Walker, J. (1997). *Fundamentals of physics* (5th ed.). Wiley.

Hamilton-Baillie. http://www.hamilton-baillie.co.uk/ (visited 27th February, 2010)

Hamker, F. H. (2001). RBF learning in a non-stationary environment: the stability-plasticity dilemma. In Howlett, R. J., & Jain, L. C. (Eds.), *Radial Basis Function networks 1: Recent Developments in Theory and Applications; Studies in fuzziness and soft computing* (Vol. 66). Heidelberg: Physica Verlag.

Hampson, S. (1999). State of the Art: Personality. *The Psychologist*, 12(6), 284–290.

Han, M., Xu, W., & Gong, Y. (2007). Multi-object trajectory tracking. *Journal of Machine Vision and Applications*, 18, 221–232. doi:10.1007/s00138-007-0071-5

Handte, M., Herrmann, K., Schiele, G., & Becker, C. (2007). Supporting Pluggable Configuration Algorithms in PCOM. *Proceedings of International Workshop on Pervasive Computing and Communications* (pp. 472-476).

Hansen, K. M., Zhang, W., & Fernandes, J. (2008). Flexible Generation of Pervasive Web Services Using OSGi Declarative Services and OWL Ontologies. 15th Asia-Pacific Software Engineering Conference (APSEC 2008), 3-5 December 2008, Beijing, China. IEEE, 135-142.

Hanser, F., Gruenerbl, A., Rodegast, C., & Lukowicz, P. (2008). *Design and real life deployment of a pervasive monitoring system for dementia patients* (pp. 279–280). Proc. of Pervasive Computing Technologies for Healthcare.

Hardian, B., Indulska, J., & Henricksen, K. (2008). Exposing Contextual Information for Balancing Software Autonomy and User Control in Context-Aware Systems, *Proceedings of the Workshop on Context-Aware Pervasive Communities: Infrastructures, Services and Applications CAPC '08.* (Sydney, May, 2008).

Haritaoglu, I., Harwood, D., & Davis, L. S. (2000). W^4: Real-Time Surveillance of People and Their Activities. *IEEE Transactions on Pattern Analysis and Machine Intelligence*, 22(8), 809–830. doi:10.1109/34.868683

Haritaoglu, I., Harwood, D., & Davis, L. S. (1998). Ghost: A human body part labeling system using silhouettes. Proceedings from ICPR '98: *The Fourteenth International Conference on Pattern Recognition*, 1, 77-82.

Haritaoglu, I., Harwood, D., & Davis, L. (2000). W4: Real-Time Surveillance of People and Their Activities. *IEEE Trans. on Pattern Recognition and Machine Intelligence, 22*(8).

Harnad, S. (1990). The Symbol Grounding Problem. *Physica D. Nonlinear Phenomena, 42*, 335–346. doi:10.1016/0167-2789(90)90087-6

Harper, R., Randall, D., Smyth, N., & Evans, C. (2007). Thanks for the memory. In *Proc. of HCI07.* Heledd. L., & Moore. R.

Harter, A., Hopper, A., Steggles, P., Ward, A., & Webster, P. (1999). The Anatomy of a Context-Aware Application. In *Proceedings of the Fifth Annual ACM/IEEE International Conference on Mobile Computing and Networking* (pp. 59-68). ACM, New York, USA.

Hein, A., & Kirste, T. (2008). Towards Recognizing Abstract Activities: An Unsupervised Approach. *BMI '08: Proc. of the 2nd Workshop on Behaviour Monitoring and Interpretation*, (pp. 102-114).

Heine, C., & Kirn, S. Adapt at agent.hospital - agent based support of clinical processes. In Proceedings of the 13th European Conference on Information Systems, The European IS profession in the Global Networking Environment, ECIS, Turku, Finland, June 14–16, p 14, 2004.

Heinze, C., Goss, S., Lloyd, I., & Pearce, A. (1999). Plan Recognition in Military Simulation: Incorporating Machine Learning with Intelligent Agents. In Proc of IJCAI-99 Workshop on Team Behaviour and Plan Recognition, pp. 53–64.

Helal, A., & Mann, W. (2005). *Gator Tech Smart House: A Programmable Pervasive Space* (pp. 64–74). IEEE Computer Magazine.

Helal, S., Mann, W., El-Zabadani, H., King, J., Kaddoura, Y., & Jansen, E. (2005, March). The Gator tech smart house: a programmable pervasive space. *Computer, 38*, 50–60. doi:10.1109/MC.2005.107

Helmut Prendinger and Mitsuru Ishizuka. (2001). Social role awareness in animated agents. In. *Proceedings of AGENTS, 2001*, 270–277.

Helmut Prendinger and Mitsuru Ishizuka. (2005). Human physiology as a basis for designing and evaluating affective communication with life-like characters. *IEICE Transactions on Information and Systems. E (Norwalk, Conn.), 88-D*(11), 2453–2460.

Henricksen, K., & Indulska, J. (2006, Feb.). Developing context-aware pervasive computing applications: Models and approach. *Journal of Pervasive and Mobile Computing, 2*(1), 37–64. doi:10.1016/j.pmcj.2005.07.003

Henricksen, K. (2002). *02.* Modeling Context Information in Pervasive Computing Systems. In Proc. of Pervasive.

Heo, N., & Varshney, P. K. (2003). A distributed self spreading algorithm for mobile wireless sensor networks. In *Proc. IEEE Wireless Communication and Networking Conf.,* (pp.1597-1602).

Herlocker, J., Konstan, J., Terveen, L., & Riedl, J. (2004) Evaluating collaborative filtering recommender systems. In ACM Transactions on Information Systems, 22, pag. 5–53.

Heschong, L. (2002). Daylighting and human performance. *ASHRAE Journal, 44*(6), 65–67.

Hesselman, C., Tokmakoff, A., Pawar, P., & Iacob, S. (2006). Discovery and Composition of Services for Context-Aware Systems. *Proceedings of the 1st IEEE European Conference on Smart Sensing and Context* (pp. 67-81). Berlin: Springer-Verlag.

Heusinger, W. Das Intelligente Haus, 1 ed. Frankfurt am Main, Berlin, Bern, Bruxlles, New York, Oxford, Wien: Peter Lang GmbH - Europäischer Verlag der Wissenschaft, 2004.

Heuvelink, A., & Both, F. (2007). BoA: A cognitive tactical picture compilation agent. In: *Proceedings of the 2007 IEEE/WIC/ACM International Conference on Intelligent Agent Technology, IAT 2007*. IEEE Computer Society Press, 2007, pp. 175-181.

Higashinaka, R., Walker, M., & Prasad, R. (2007). *"Learning to generate naturalistic utterances using reviews in spoken dialogue systems," ACM Transactions on Speech and Language Processing*. TSLP.

Hildebrandt, M. (2008a). Ambient intelligence, criminal liability and democracy. *Criminal Law and Philosophy, 2*(2), 163–180. doi:10.1007/s11572-007-9042-1

Hildebrandt, M. (2008b). A vision of ambient law. In Brownsword, R., & Yeung, K. (Eds.), *Regulating Technologies: Legal Futures, Regulatory Frames and Technological Fixes* (pp. 175–191). Oxford: Hart Publishing.

Hilke, W. (1989). In Hilke, W. (Ed.), *"Grundprobleme und Entwicklungstendenzen des Dienstleistungs-Marketing,"* in *Dienstleistungs-Marketing* (pp. 5–44). Wiesbaden.

Hinckley, K., Pierce, J., Horvitz, E., & Sinclair, M. (2005). Foreground and Background Interaction with Sensor-Enhanced Mobile Devices. *ACM Transactions on Computer-Human Interaction, 12*(1), 1–22. doi:10.1145/1057237.1057240

Hintikka, J. (1962). *Knowledge and Belief*. New York: Cornell University Press.

Hitachi Human-Interaction Laboratory. (2006). *Miragraphy*. Retrieved 08 17, 2008, from http://hhil.hitachi.co.jp/products/miragraphy.html (in Japanese)

Hoey, J., von Bertoldi, A., Poupart, P., & Mihailidis, A. (2007). Assisting persons with dementia during handwashing using a partially observable markov decision process. In *Proc. of ICVS07*.

Hoey, J., Von Bertoldi, A., Craig, T., Poupart, P. & Mihailidis, A. (in press, 2009). Automated Handwashing Assistance For Persons With Dementia Using Video and A Partially Observable Markov Decision Process. Computer Vision and Image Understanding (Special Issue on Computer Vision Systems).

Hofer, T., Schwinger, W., Pichler, M., Leonhartsberger, G., & Altmann, J. (2003). Context-awareness on mobile devices – the hydrogen approach, in: Proceedings of the 36th Annual Hawaii International Conference on System Sciences (HICSS'2003), Big Island, Hawaii, USA.

Hogg, D. (1983). Model-based vision: A program to see a walking person. *Image and Vision Computing, 1*, 5–20. doi:10.1016/0262-8856(83)90003-3

Hohm D., Jonuschat H., Scharp M., Scheer D., and Scholl G., "Innovative Dienstleistungen "rund um das Wohnen" professionell entwickeln - Service Engineering in der Wohnungswirtschaft," GdW Bundesverband deutscher Wohnungsunternehmen e.V., 2004.

Holmquist, L., & Skog, T. (2003). Informative art: Information visualization in everyday environments. In *Proceedings of the First International conference on Computer graphics and interactive techniques in Australasia and South East Asia* (pp. 229-235). ACM.

Hong, X., Nugent, C., Mulvenna, M., Mcclean, S., Scotney, B., & Devlin, S. (2008). Assessment of the impact of sensor failure in the recognition of activities of daily living. In *Proc. of ICOST08*, 136–144.

Hongseng, S., Nevatia, R., & Bremond, F. (2004). Video-based event recognition: activity representation and probabilistic recognition methods. *Computer Vision and Image Understanding*, (96): 129–162. doi:10.1016/j.cviu.2004.02.005

Hönsch, M., Kompan, M., Šimko, J., & Zeleník, D. (2008). Intelligent Household: Energy consumption manager, In B. Mannová, P. Šaloun & M. Bieliková (Eds.), *Proc. of CZ ACM Student Research Competition* (pp. 33-41). Prague: CZ ACM.

Hopfield, J. J. (1982). Neural networks and physical systems with emergent collective computational abilities. *Proceedings of the National Academy of Sciences of the United States of America, 79*(8), 2554–2558. doi:10.1073/pnas.79.8.2554

Horrey, W. J., Wickens, Ch. D., & Consalus, K. P. (2006). Modeling Driver's Visual Attention Allocation While Interacting With In-Vehicle Technologies. *Journal of Experimental Psychology, 12*, 67–78.

Horrocks, I., Patel-Schneider, P. F., Boley, H., Tabet, S., Grosof, B., & Dean, M. (2004). SWRL: A Semantic Web Rule Language Combining OWL and RuleML, from http://www.daml.org/ 2004 /04/ swrl.

Horty, J. F., & Belnap, N. (1995). The deliberative STIT: A study of action, omission, and obligation. *Journal of Philosophical Logic, 24*(6), 583–644. doi:10.1007/BF01306968

Howard, P. J., & Howard, J. M. (2004). *The BIG FIVE Quickstart: An introduction to the five-factor model of personality for human resource professionals*. Charlotte, NC: Center for Applied Cognitive Studies.

Howard, A., Mataric, M. J., & Sukhatme, G. S. (2002). Mobile sensor network deployment using potential fields: a distributed, scalable solution to the area coverage problem. In *Proc. 6th Int. Sym. Distributed Autonomous Robotic Systems*, (pp.299-308).

Howe, N., & Strauss, W. (2000). *Millennials Rising: The Next Greatest Generation*. New York: Vintage Books.

http://en.wikipedia.org/wiki/Mental_rotation

http://en.wikipedia.org/wiki/Visual_search

http://en.wikipedia.org/ wiki/ Ambient_intelligence retrieved on May 27, 2009.

Hu, W., Tan, T., Wang, L., & Maybank, S. (2005). A survey on visual surveillance of object motion and behaviors. *IEEE Transactions on Systems, Man, and Cybernetics. Part C, 34*(3), 334–352.

Hu, P. H., Son, T. C., & Baral, C. (2007, July). Reasoning and planning with sensing actions, incomplete information, and static laws using Answer Set Programming. *Theory and Practice of Logic Programming, 7*(4), 377–450.

Hu, N., Huang, W., & Surendra, R. (2006). Robust Attentive Behavior Detection by Non-Linear Head Pose Embedding And Estimation, *ECCV'06* (LNCS 3953). (pp.356-367)

Hudlicka, E. (2006). Depth of Feelings: Alternatives for Modeling Affect in User Models. *TSD, 2006*, 13–18.

Hume, D. *A Treatise of Human Nature*. Clarendon Press, Oxford, 1978. Selby-Bigge, L. A. & Nidditch, P. H. (Eds.).

Hurley A. C., et. al. (1999). Measurement of Observed Agitation in Patients with Dementia of the Alzheimer Type, *Journal of Mental Health and Aging*

Husband, J., & Bair, J. (2007). *Making Knowledge Work – The Arrival of Web 2.0*. London: Ark Group.

Hutchins, E. (1995). *Cognition in the Wild*. Cambridge, Mass.: MIT Press.

Hye, P. (2007). *A design study of pedestrian space as an interactive space*. (presented at IASDR 2007)

Ikemoto, Y., Hasegawa, Y., Fukuda, T., & Matsuda, K. (2005). Graduated spatial pattern formation of robot group. *Information Sciences, 171*(4), 431–445. .doi:10.1016/j.ins.2004.09.013

Ingstrup, M., & Hansen, K. M. (2009). Modeling Architectural Change: Architectural scripting and its applications to reconfiguration. In Proceedings of WICSA/ECSA 2009, 337-340.

Intille, S. S. (2002). *Designing a home of the future* (pp. 80–86). IEEE Pervasive Computing, Vol. April-June.

Intille, S. S. (2006) The Goal: Smart People, Not Smart Homes. In: Nugent C and Augusto JC (eds) Smart Homes and Beyond, ICOST'2006 4th International Conference on Smart Homes and Health Telematics, Assistive Technology Research Series 19, IOS Press, pp. 3–6.

Intille, S., Larson, K., Tapia, E., Beaudin, J., Kaushik, P., Nawyn, J., & Rockinson, R. (2006). Using a live-in laboratory for ubiquitous computing research. In *Proc. of PERVASIVE06*, 349–365.

Ioffe, S., & Forsyth, D. A. (2001). Human tracking with mixtures of trees. Proceedings from ICCV '01: *The International Conference on Computer Vision*, 690-695.

IPG-Driver. http://www.ipg.de/32.html (visited 28th February, 2010)

Isard, M., & Blake, A. (1998). Condensation-Conditional Density Propagation for Visual Tracking. *International Journal of Computer Vision, 29*(1), 5–28. doi:10.1023/A:1008078328650

Isard, M., & Blake, A. (1996). Contour tracking for stochastic propagation of conditional density. Proceedings from ECCV '96: *The European Conference on Computer Vision*, 343-356.

Isherwood, C. (1952). *Goodbye to Berlin*. Signet.

Ishiguro, A., & Kawakatsu, T. (2006). Self-assembly through the interplay between control and mechanical systems. In *Proc. IEEE/RSJ Int. Conf. Intelligent Robots and Systems*, (pp.631-638).

ISTAG. (2002). *Strategic Orientations and Priorities for IST in FP6*. Luxembourg.

ISTAG. (2001) *Scenarios for Ambient Intelligence in 2010*. Luxembourg

ISTAG. (2005) Ambient Intelligence: from Vision to Reality. In: *Ambient Intelligence*. IOS Press [online] http://www.ambientintelligence.com

Itti, L., & Koch, C. (2001). Computational Modeling of Visual Attention. *Nature Reviews. Neuroscience, 2*(3), 194–203. doi:10.1038/35058500

ITU-T Recommendation J.190, Architecture of Media-HomeNet that supports cable-based services, 2002.

Iwabuchi, E., Nakagawa, M., & Siio, I. (2009). Smart Makeup Mirror: Computer-Augmented Mirror to Aid Makeup Application. In J. A. Jacko (Ed.), *the Thirteenth International Conference on Human-Computer Interaction. Part IV: Interacting in Various Application Domains.* (pp.495-503). Springer-Verlag Berlin Heidelberg. Jordan, P.W. (2002). *Designing Pleasurable Products: An Introduction to the New Human Factors.* Taylor & Francis.

Iyengar, R., Kar, K., & Banerjee, S. (2005). Low-coordination topologies for redundancy in sensor networks. In *Proc. 6th ACM Int. Sym. Mobile Ad Hoc Networking and Computing*, (pp.332-342).

Jabri, S., Duric, Z., Wechsler, H., & Rosenfeld, A. (2000). Detection and location of people in video images using adaptive fusion of color and edge information. Proceedings from ICPR '00: *The International Conference on Pattern Recognition, 4*, 627-630.

Jagacinski, R. J., & Flach, J. M. (2003). *Control Theory for Humans: Quantitative Approaches to Modeling performance*. Mahwah, N.J.: Lawrence Erlbaum Associates.

Jain, R., Martin, W., & Aggarwal, J. (1979). Segmentation throught the detection of changes due to motion. *Computer Graphics and Image Processing, 2*, 13–34. doi:10.1016/0146-664X(79)90074-1

Jakkula, V. R., & Cook, D. J. (2007). Using temporal relations in Smart Environment Data for Activity Prediction, in: Proc. of the 24th International Conference on Machine Learning, Corvallis, Oregon, pp. 1–4.

Jakkula, V., Cook, D. J., & Crandall, A. S. (2007). Temporal Pattern Discovery for Anomaly Detection in a Smart Home. *Proc. of the 3rd IET Conf. on Intelligent Environments(IE)*, (pp. 339-345).

Jammes, F., & Smit, H. Service-oriented paradigms in industrial automation. IEEE Transactions on Industrial Informatics. v1. 2005. pp. 62-70.

Jaques, P. A., & Vicari, R. M. (2007). A BDI approach to infer student's emotions in an intelligent learning environment. *Computers & Education, 49*(2), 360–384. doi:10.1016/j.compedu.2005.09.002

Jarvis, P. A., Lunt, T. F., & Myers, K. L. (2005). Identifying terrorist activity with AI plan-recognition technology. *AI Magazine, 26*(3), 73–81.

Jarvis, P., Lunt, T., & Myers, K. (2004). Identifying terrorist activity with AI plan recognition technology. In the Sixteenth Innovative Applications of Artificial Intelligence Conference (IAAI 04), AAAI Press, 2004.

Javed, O., Rasheed, Z., Shafique, K., & Shah, M. (2003). Tracking across multiple cameras with disjoint views. Proceedings from ICCV '03: *The International Conference on Computer Vision*, 1-6.

Jensen, F. V., & Nielsen, Th. D. (2007). *Bayesian Networks and Decision Graphs* (2nd ed.). Springer.

Jiang, L., Liu, D.-Y., & Yang, B. (2004) Smart Home Research. Proc. of the Third International Conference on Machine Learning and Cybernetics, pp. 659–663.

Jianqi, Y., & Lalanda, P. (2008). Integrating UPnP in a Development Environment for Service-Oriented Applications. *Proceedings of the IEEE International Conference on Industrial Technology ICIT, 08*, 1–5.

Jiehan Zhou, J. Riekki, Y. Changrong, and E. Kärkkäinen. AmE framework: a model for emotion-aware ambient intelligence. In *ACII2007: Doctoral Consortium*, 2007.

Jimison, H., Pavel, M., McKanna, J., & Pavel, J. (2004). Unobtrusive monitoring of computer interactions to detect cognitive status in elders. *IEEE Transactions on Information Technology in Biomedicine, 8*, 248–252.

Jin, X., Li, Y., Mah, T., & Tong, J. "Sensitive webpage classification for content advertising," in *Proceedings of the International Workshop on Data Mining and Audience Intelligence for Advertising*, 2007.

Jindal, N., & Liu, B. "Mining comparative sentences and relations," in *Proceedings of AAAI*, 2006.

Johanson, B., Fox, A., & Winograd, T. (2002, April). The Interactive Workspaces Project: Experiences with Ubiquitous Computing Rooms. *IEEE Pervasive Computing / IEEE Computer Society [and] IEEE Communications Society, 1*(Issue 2), 67–74. doi:10.1109/MPRV.2002.1012339

Johanson, B., Winograd, T., & Fox, A. (2003, April). Interactive Workspaces. *IEEE Computer, 36*(Issue 4), 99–101.

Johanson, B., & Fox, A. *The Event Heap: a coordination infrastructure for interactive workspaces*. Mobile Computing Systems and Applications, 2002. Proceedings Fourth IEEE Workshop, p. 83 – 93, June 2002.

John, F. (2001). *Horty. Agency and Deontic Logic*. Oxford: Oxford University Press.

John, R. (1983). *Searle. Intentionality: An essay in the philosophy of mind*. Cambridge University Press.

John, O. P., & Srivastava, S. (1999). The Big Five Trait Taxonomy: History, Measurement, and Theoretical Perspectives. In Pervin, L. A., & John, O. P. (Eds.), *Handbook of personality: Theory and research* (Vol. 2, pp. 102–138). New York: Guilford Press.

Johnson, N. (1998). *Learning Object Behaviour Models*. Doctoral dissertation, University of Leeds, Leeds, UK. Retrieved from http://www.scs.leeds.ac.uk/neilj/ps/thesis.ps.gz

Johnson-Laird, P. N., & Oatley, K. (1992). Basic emotions: a cognitive science approach to function, folk theory and empirical study. *Cognition and Emotion, 6*, 201–223. doi:10.1080/02699939208411069

Jonathan Gratch and Stacy Marsella. (2004). A Domain-independent Framework for modelling Emotions. *Journal of Cognitive Systems Research, 5*(4), 269–306. doi:10.1016/j.cogsys.2004.02.002

Jones M. J. & Rehg J. M. (2002). Statistical Color Models with Application to Skin Detection, *International Journal of Computer Vision*, (46.) (pp. 81-96)

Jonsson, A. K., Morris, P. H., Muscettola, N., Rajan, K., & Smith, B. (2000). Planning in Interplanetary Space: Theory and Practice. *Proc. Int. Conf. on AI Planning and Scheduling (AIPS-00)*.

Jordan, M. I., Sejnowski, T. J., & Poggio, T. (2001). *Graphical Models: Foundations of Neural Computation*. MIT Press.

Ju, S., Black, M., & Yacoob, Y. (1996). Cardboard people: A parameterized model of articulated image motion. Proceedings from ICAFGR '96: *The International Conference on Automatic Face and Gesture Recognition*, 38-44.

Junestrand, S., Keijer, U., & Tollmar, K. (2001) Private and Public Digital Domestic Spaces. International Journal of Human-Computer Studies, Vol. 54, No. 5, Academic Press, London, England, pp. 753–778.

Jurafsky, D., & Martin, J. H. (2009). *Speech and Language Processing* (2nd ed.). Pearson.

Jürgensohn, Th. Control Theory Models of the Driver, in: Cacciabue (ed), 2007, p. 277 – 292

K. Ducatel, M. Bogdanowicz, F. Scapolo, J. Leijten, and J.-C. Burgelman. Scenarios for ambient intelligence in 2010. ISTAG report at IPTS Seville, 2001.

Kadar, P. (2008). Making the power system intelligent. *Proceedings of the Icrepq'08 international conference on renewable energy and power quality* (pp. 12-14).

Kaddoura, Y., King, J., and Helal, (Sumi) A. (2005). Cost-Precision Tradeoffs in Unencumbered Floor-based Indoor Location Tracking. In: Giroux S and Pigot H (eds) From Smart Homes to Smart Care, ICOST'2005 3rd International Conference on Smart Homes and Health Telematics, Assistive Technology Research Series 15, IOS Press, pp.75–82.

Kahn, J., Katz, R. and Pister K., Emerging Challenges: Mobile Networking for 'Smart Dust', J. Comm. Networks, Sept. 2000, pp. 188-196.

Kahneman, D. (1995). Varieties of counterfactual thinking. In Roese, N. J., & Olson, J. M. (Eds.), *What might have been: the social psychology of counterfactual thinking* (pp. 375–396). Mahwah, NJ: Erlbaum.

Kakas, A., Kowalski, R., & Toni, F. (1998). The role of logic programming in abduction. In: Gabbay, D., Hogger, C.J., Robinson, J.A. (Eds.): Handbook of Logic in Artificial Intelligence and Programming 5, Oxford University Press, 1998, 235-324.

Kalasapur, S., Kumar, M., & Shirazi, B. A. (2005). Personalized Service Composition for Ubiquitous Multimedia Delivery. *Proceedings of the 6th IEEE International Symposium on a World of Wireless Mobile and Multimedia Networks WoWMoM'05* (pp. 258-263), Washington, DC: IEEE Computer Society.

Kameas, A., et al. (2005) Computing in Tangible: Using Artifacts as Components of Ambient Intelligence Environments. In *Ambient Intelligence*. In: *Ambient Intelligence*. IOS Press [online] http://www.ambientintelligence.com

Karaulova, I. A., Hall, P. M., & Marshall, A. D. (2000). A hierarchical model of dynamics for tracking people with a single video camera. Proceedings from BMVC '00: *The British Machine Vision Conference*, 352-361.

Karlsson, B., Ciarlini, A. E. M., Feijo, B., & Furtado, A. L. (2006). Applying a plan-recognition/plan-generation paradigm to interactive storytelling, The LOGTELL Case Study. Monografias em Ciência da Computação Series (MCC 24/07), ISSN 0103-9741, Informatics Department/ PUC-Rio, Rio de Janeiro, Brazil, September 2007. Also in the Proceedings of the ICAPS06 Workshop on AI Planning for Computer Games and Synthetic Characters, Lake District, UK, June, 2006, 31-40.

Katz, S., Ford, A. B., Moskowitz, R. D., Jackson, B. A., & Jaffe, M. W. (1963). Studies of illness in the aged. The index of ADL: A standardized measure of biological and psychosocial function. *Journal of the American Medical Association, 185*(12), 914–919.

Kautz, H. (1987). A Formal Theory of Plan Recognition. PhD thesis, Dept. of Computer Science, University of Rochester.

Kautz, H. (1991) A Formal Theory of Plan Recognition and its Implementation, Reasoning About Plans, Allen J., Pelavin R. and Tenenberg J. eds., Morgan Kaufmann, San Mateo, C.A., 69-125.

Kautz, H., & Allen, J. F. (1986). Generalized plan recognition. In Proceedings of the Fifth National Conference on Artificial Intelligence (AAAI-86), 1986, 32-38.

Kautz, H., Arnstein, L., Borriello, G., Etzioni, O., & Fox, D. (2002). An overview of the assisted cognition project. In *Proc. of AAAI02 Workshop on Automation as Caregiver: The Role of Intelligent Technology in Elder Care*.

Kautz, H., Etzioni, O., Fox, D., & Weld, D. (March 2003). Foundations of assisted cognition systems. Technical report, University of Washington. Intel Research Lab. (2009). Technology for independent living.

Kawar, F., Fujinami, K., & Nakajima, T. (2005). Augmenting Everyday Life with Sentient Artefacts. In *Proceedings of the 2005 Joint Conference on Smart Objects and Ambient Intelligence: Innovative Context-Aware Services: Usages and Technologies* (pp. 141-146). ACM, New York, USA.

Kawsar, F., Nakajima, T., & Fujinami, K. (2008). Deploy Spontaneously: Supporting End-Users in Building and Enhancing a Smart Home. *Proceedings of the 6th International Conference on Ubiquitous Computing, UbiComp'08* (pp. 282-291). Vol. 344. New York, NY: ACM.

Kazuyoshi Wada and Takanori Shibata. Robot therapy in a care house -its sociopsychological and physiological effects on the residents. In *ICRA06*, 2006.

Kelley, J. F. (1984). An iterative design methodology for user-friendly natural language office information applications. *ACM Transactions on Office Information Systems, 2*(1), 26–41. doi:10.1145/357417.357420

Kelley, R., Tavakkoli, A., King, C., Nicolescu, M., Nicolescu, M., & Bebis, G. (2008), Understanding human intentions via hidden markov models in autonomous mobile robots. Third ACM/IEEE international conference on Human robot interaction (HRI'08), 367-374.

Kemmotsu, K., Tomonaka, T., Shiotani, S., Koketsu, Y., & Iehara, M. (2005). Recognizing Human Behaviors With Vision Sensors in Network Robot Systems. *Proc. of The 1st Japan-Korea Joint Symposium on Network Robot Systems (JK-NRS 2005).*

Kephart, J., & Chess, D. (1996). *The Vision of Autonomic Computing. 2003. Mitchell, M.: An Introduction to Genetic Algorithms.* Bradford Books.

Kernighan, B. W., & Pike, R. (1984). *The UNIX Programming Environment.* Prentice-Hall.

Kerr, W., Spear, D., Spear, W., & Thayer, D. (2005). Two formal gas models for multi-agent sweeping and obstacle avoidance. In Carbonell, J. G., & Siekmann, J. (Eds.), *Formal Approaches to Agent-Based Systems (LNCS), 3228* (pp. 111–130). Springer.

Khan, S., Javed, O., Rasheed, Z., & Shah, M. (2001). Human tracking in multiple cameras. Proceedings from ICCV '01: *The International Conference on Computer Vision*, 331-336.

Kidd, C. D., Orr, R. J., et al. (1999) The Aware Home: A Living Laboratory for Ubiquitous Computing Research. Proc. of the Second International Workshop on Cooperative Buildings - CoBuild'99.

Kiefer, P., & Schlieder, C. (2007). Exploring Context Sensitivity in Spatial Intention Recognition. In Gottfried, B. (Ed.), *BMI'07 (Vol. 296*, pp. 102–116). CEUR.

Kim, D., & Kim, D. (2006) An Intelligent Smart Home Control Using Body Gestures. Proc. of 2006 International Conference on Hybrid Information Technology (ICHIT 2006), vol.II, pp. 439–446.

Kim, J. H., Kim, Y. D., & Lee, K. H. (2004). The Third Generations of Robotics: Ubiquitous Robot. *Proc. of the 2nd Int. Conf. on Autonomous Robots and Agents.* Palmerston North, New Zealand.

Kingman-Brundage, J. "The ABC's of Service System Blueprinting," Chicago: Bitner M.J.; Crosby L.A., 1989, pp. 30-33.

Kirk, R., & Newmarch, J. "A Location-aware, Service-based Audio System", IEEE Consumer Communications and Networking Conference, 2005. pp. 343-347.

Kirsh, D., & Maglio, P. (1994). On distinguishing epistemic from pragmatic action. *Cognitive Science, 18*, 513–549. doi:10.1207/s15516709cog1804_1

Knight, R., Rabideau, G., Chien, S., Engelhardt, B., & Sherwood, R. (2001). Casper: Space Exploration through Continuous Planning. *IEEE Intelligent Systems, 16*(5), 70–75.

Ko, E. J., Lee, H. J., & Lee, J. W. (2006). Ontology-Based Context Modeling and Reasoning for U-HealthCare. The Institute of Electronics [IEICE]. *Information and Communication Engineers, 90-D*(8), 1262–1270.

Kobayashi, N. (2007). *Opinion Mining from Web documents: Extraction and Structurization,* (Doctoral Dissertation), Nara Institute of Science and Technology.

Koehler, J., & Nebel, B. Ho_mann, J., & Dimopoulos, Y. (1997): Extending planning graphs to an adl subset. In: Recent Advances in AI Planning. Volume 1348 of Lecture Notes in Computer Science. Springer, 273-285

Kohonen, T. (1993). *Self-Organization and Associative Memory* (3rd ed.). Berlin: Springer.

Kohtake, N., Ohsawa, R., Yonezawa, T., Matsukura, Y., Masayuki, I., Thakashio, K., et al. (2005). u-Texture: Self-Organizable Universal Panels for Creating Smart Surroundings. In M. Beigl, S. Intille, J. Rekimoto, & H. Tokuda (Eds.), *the Seventh International Conference on Ubiquitous Computing* (pp. 19-36). Springer-Verlag Berlin Heidelberg.

Koike, C. C., Bessiere, P., & Mazer, E. (2008). Bayesian Approach to Action Selection and Attention Focusing. In Bessiere, P. (Eds.), *Probabilistic Reasoning and Decision Making in Sensory-Motor Systems* (pp. 177–201). Berlin: Springer.

Koller, D., Daniilidis, K., & Nagel, H.-H. (1993). Model-based object tracking in monocular image sequence of road trafic scenes. *International Journal of Computer Vision, 3*(10), 257–281.

Koller, D., & Friedman, N. (2009). *Probabilistic Graphical Models*. Cambridge, Mass.: MIT Press.

Konolige, K., & Pollack, M. E. (1989). Ascribing plans to agents: preliminary report. In 11th International Joint Conference on Artificial Intelligence, Detroit, MI, 1989, 924-930.

Korpipaa, P., Mantyjarvi, J., Kela, J., Keranen, H., & Malm, E. (2003). Managing context information in mobile devices. *Pervasive Computing, IEEE, 2*(3), 42–51. doi:10.1109/MPRV.2003.1228526

Kortum, P. and Wilson Geisler, "Implementation of a foveated image coding system for image bandwidth reduction", in SPIE Proceedings, 2657, Page 350-360, 1996.

Kowalski, R. A., & Sadri, F. (2009). Integrating logic programming and production systems in abductive logic programming agents. Invited papers, the 3rd International Conference on Web Reasoning and Rule Systems, October 25-26, Chantilly, Virginia, USA, 2009.

Kowalski, R., & Sergot, M. (1986). A logic-based calculus of events. In *New Generation Computing*, Vol. 4, No.1, February 1986, 67-95.

Kramer, J., & Magee, J. (2007). Self-Managed Systems: an Architectural Challenge. International Conference on Software Engineering, 259-268

Kranz M., Schmidt A., Bogdan Rusu R., Maldoado A., Beetz M., Hörnler B., and Rigoll G., "Sensing Technologies and the Player-Middleware for Context-Awareness in Kitchen Environments," 2007.

Krieg-Brückner, B., Gersdorf, B., Döhle, M., & Schill, K. (2009). *Technik für Senioren in spe im Bremen Ambient Assisted Living Lab. 2. Deutscher AAL-Kongress 2009*. VDE-Verlag Berlin-Offenbach.

Krieg-Brückner, B., Shi, H., Fischer, C., Röfer, T., Cui, J., & Schill, K. (2009). *Welche Sicherheitsassistenz brauchen Rollstuhlfahrer? 2. Deutscher AAL-Kongress 2009*. VDE-Verlag Berlin-Offenbach.

Krieg-Brückner, B., & Shi, H. Orientation Calculi and Route Graphs: Towards Semantic Representations for Route Descriptions. *Proceedings of GIScience 2006*. LNCS 4197, 2006.

Krieg-Brückner, B., Frese, U., Lüttich, K., Mandel, C., Mossakowski, T., Ross, R. J.: Specification of an Ontology for Route Graphs. *Proceedings of Spatial Cognition IV, Germany*. LNAI 3343, 2004.

Krieg-Brückner, B., Krüger, A., Hoffmeister, M., & Lüth, C. (2007).: Kopplung von Zutrittskontrolle und Raumautomation – Eine Basis für die Interaktion mit einer intelligenten Umgebung. *8. Fachtagung Gebäudesicherheit und Gebäudeautomation – Koexistenz oder Integration?* VDI Berichte 2005. VDI Verlag, Düsseldorf, 37-48, 2007.

Krieg-Brückner, B., Röfer, T., Shi, H., & Gersdorf, B. Mobility Assistance in the Bremen Ambient Assisted Living Lab. In Nehmer, J., Lindenberger, U., & Elisabeth Steinhagen-Thiessen, E. (guest editors): *2009 Special Section Technology and Aging: Integrating Psychological, Medical, and Engineering Perspectives*. Gerontology: Regenerative and Technological Gerontology. To appear.

Kritzler, M., Lewejohann, L., & Krüger, A. (2007). Analysing Movement and Behavioural Patterns of Laboratory Mice in a Semi Natural Environment Based on Data collected via RFID-Technology. In Gottfried, B. (Ed.), *BMI'07 (Vol. 296*, pp. 17–28). CEUR.

Krummenacher, R., & Strang, T. (2007) Ontology-Based Context Modeling. Third Workshop on Context Awareness for Proactive Systems.

Kschischang, F. R., Frey, B. J., & Loeliger, H.-A. (2001). Factor graphs and the sum-product algorithm. *IEEE Transactions on Information Theory, 47*(2), 498–519. doi:10.1109/18.910572

Kuipers, B. (2000). *The Spatial Semantic Hierarchy (Vol. 119)*. Artificial Intelligence.

Kukhun, D. A., & Sedes, F. (2008) Adaptive Solutions for Access Control within Pervasive Healthcare Systems. Proc. 6th International Conference on Smart Homes and Health Telematics (ICOST 2008), LNCS 5120, Springer, pp. 42–53.

Kumugai, T., & Akamatsu, M. (2006, February). Prediction of Human Driving Behavior Using Dynamic Bayesian Networks, IEICE-Transactions on Info and Systems. *Volume E, 89-D*(2), 857–860.

Kumugai, T., Sakaguchi, Y., Okuwa, M., & Akamatsu, M. Prediction of Driving Behavior through Probabilistic Inference, Proceedings of the 8th International Conference on Engineering Applications of Neural Networks (EANN '03), 117 – 123

Kurabayashi, D., Okita, K., & Funato, T. (2006). Obstacle avoidance of a mobile robot group a nonlinear oscillator network. In *Proc. IEEE/RSJ Int. Conf. Intelligent Robots and Systems*, (pp.186-191).

Kurata, Y., & Shi, H. Interpreting Motion Expressions in Route Instructions Using Two Projection-Based Spatial Models. *Proceedings of 31st Annual German Conference on Artifical Intelligence*, 2008.

Kvasnica, M. (2004) Six-DoF Force-Torque Wheelchair Control by Means of the Body Motion. Proc. 2nd International Conference on Smart Homes and Health Telematics (ICOST 2004), Vol. 14, pp. 247–252.

Kvavik, R. (2005). *Convenience, Communications, and Control: How Students Use Technology* (Oblinger, D., & Oblinger, J. L., Eds.). Educause, Washington: Educating the Net Generation.

Lafferty, J., McCallum, A., & Pereira, F. Conditional random fields: Probabilistic models for segmenting and labeling sequence data. In *Proc. ICML-01*, pages 282–289, San Francisco, CA, 2001. Morgan Kaufmann.

Laituri, M., & Kodrich, K. (2008). On Line Disaster Response Community: People as Sensors of High Magnitude Disasters Using Internet GIS. *Sensors (Basel, Switzerland)*, 8(5), 3037–3055. doi:10.3390/s8053037

Laland, K., & Brown, G. (2006). Niche construction, human behavior, and the adaptive-lag hypothesis. *Evolutionary Anthropology*, 15, 95–104. doi:10.1002/evan.20093

Laland, K., Odling-Smee, J., & Feldman, M. (2000). Niche construction, biological evolution and cultural change. *The Behavioral and Brain Sciences*, 23(1), 131–175. doi:10.1017/S0140525X00002417

Lampe, M., & Strassner, M. (2003). *The potential of RFID for moveable asset management*. Seattle: In Proc. of UbiComp.

Lamsweerde, A. v. (2001). Goal-oriented Requirements Engineering: A Guided Tour. Fifth IEEE International Symposium on Requirements Engineering (RE'01), 249-262.

Lamsweerde, A. v., & Letier, E., From Object Orientation to Goal Orientation: A Paradigm Shift for Requirements Engineering. Radical Innovations of Software and Systems Engineering, 325-340.

Lankenau, A., & Röfer, T. (2001). A Safe and Versatile Mobility Assistant. *IEEE Robotics & Automation Magazine*, 1.

Larkin, J. H., & Simon, H. A. (1987). Why a diagram is (sometimes) worth 10,000 words. *Cognitive Science*, 11, 65–100. doi:10.1111/j.1551-6708.1987.tb00863.x

Lashina, T. I. (2004). Intelligent Bathroom. In *European Symposium on Ambient Intelligence, Workshop "Ambient Intelligence Technologies for Wellbeing at Home"*. Eindhoven University of Technology, The Netherlands.

Lauria, S., Kyriacou, T., Bugmann, G., Bos, J., & Klein, E. Converting Natural Language Route Instructions into Robot Executable Procedures. *Proceedings of the IEEE International Workshop on Human and Robot Interactive Communication*, 2002. LON LonMark International. http://www.lonmark.org

Lauterbach, C., Glaser, R., Savio, D., Schnell, M., Weber, W., Kornely, S., & Stöhr, A. (2005). *A Self-Organizing and Fault-Tolerant Wired Peer-to-Peer Sensor Network for Textile Applications, Engineering self-organizing applications* (pp. 256–266). Berlin Heidelberg, Germany: Springer-Verlag.

Lay, D. C. (2005). *Linear Algebra and its Applications, 3rd updated Edition*. New York: Addison-Wesley.

Le Blanc, K., & Saffiotti, A. (2008). Cooperative Anchoring in Heterogeneous Multi-Robot Systems. *Proc. of the IEEE Int. Conf. on Robotics and Automation (ICRA)*. Pasadena, CA.

Le Hy R. Arrigoni A. Bessière P. Lebeltel O.; (2004); Teaching Bayesian Behaviours to Video Game Characters; Robotics and Autonom. Systems (Elsevier), Vol. 47: 177-185

Lebeltel, O., Bessiere, P., Diard, J., & Mazer, E. (2004). Bayesian Robot Programming. *Autonomous Robots, 16*, 49–79.

Lee, Y., Chun, S., & Geller, J. (2004). Web-Based Semantic Pervasive Computing Services. *Proceedings of the IEEE Intelligent Informatics Bulletin, 4*(2).

Lee, D. N. (1976). A theory of visual control of braking based on information about time-to-collision. *Perception, 5*, 437–459.

Lee, J. H., & Hashimoto, H. (2002). Intelligent Space - concept and contents. *Advanced Robotics, 16*(3), 265–280. doi:10.1163/156855302760121936

Lee, G., & Chong, N. Y. (2008a). Adaptive flocking of robot swarms: algorithms and properties. *IEICE Transactions on Communications. E (Norwalk, Conn.), 91-B*(9), 2848–2855.

Lee, G., & Chong, N. Y. (2008b). A geometric approach to deploying robot swarms. *Annals of Mathematics and Artificial Intelligence, 52*(2-4), 257–280. .doi:10.1007/s10472-009-9125-x

Lee, G., & Chong, N. Y. (2009a). Decentralized formation control for small-scale robot teams with anonymity. *Mechatronics, 19*(1), 85–105. .doi:10.1016/j.mechatronics.2008.06.005

Lee, G., Yoon, S., Chong, N. Y., & Christensen, H. (2009b). A mobile sensor network forming concentric circles through local interaction and consensus building. *Journal of Robotics and Mechatronics, 21*(4), 469–477.

Lee, D. N. How movement is guided (2006) http://www.perception-in-action.ed.ac.uk/publications.htm (visited 28th February, 2010)

Lee, G., Chong, N. Y., & Christensen, H. (2009d). Adaptive triangular mesh generation of self-configuring robot swarms. In *Proc. IEEE Int. Conf. Robotics and Automation*, (pp.2737-2742).

Lee, G., Yoon, S., Chong, N. Y., & Christensen, H. (2009c). Self-configuring robot swarms with dual rotating infrared sensors. In *Proc. IEEE/RSJ Int. Conf. Intelligent Robots and Systems*, (pp.4357-4362).

Lee, H., Bien, Z., & Kim, Y. (2005). 24-hour Continuous Health Monitoring System for the Disabled and the Elderly in Sensor Network Environment. In: Giroux S and Pigot H (eds) From Smart Homes to Smart Care, ICOST'2005 3rd International Conference on Smart Homes and Health Telematics, Assistive Technology Research Series 15, IOS Press, pp. 116–123.

Lee, J.-S., Park, K.-S., & Hahn, M.-S. (2005) WindowActive: An Interactive House Window On Demand. 1st Korea-Japan Joint Workshop on Ubiquitous Computing and Networking Systems (UbiCNS 2005), pp. 481–484.

Lee, M. S. (2000). Detecting People in cluttered indoor scenes, *Proceedings of the. IEEE International Conference on Computer Vision and Pattern Recognition.* (1) (pp. 804-809)

Lee, M. W., & Cohen, I. (2004). Proposal maps driven MCMC for estimating human body pose in static images. Proceedings from ICCVPR '04: *The International Conference on Computer Vision and Pattern Recognition, 2*, 334-341.

Leignel, C. (2006). *Modèle 2D du corps pour l'analyse des gestes par l'image via une architecture de type tableau noir: Application aux interfaces homme-machine évoluées.* Doctoral dissertation, University of Rennes 1, Rennes, France.

Leignel, C., & Viallet, J.E. (2004). A blackboard architecture for the detection and tracking of a person. Proceedings from RFIA '04: *Reconnaissance de Formes et Intelligence Artificielle*, 334-341.

Lesh, N., Rich, C., & Sidner, C. L. (1998). *Using plan recognition in Human-Computer Collaboration, TR-98-23, Mitsubishi Electric Research Lab.* MERL.

Lesh, N., Rich, C., & Sidner, C. L. (1999). Using plan recognition in human-computer collaboration. In Proceedings of the Seventh International Conference on User Modelling, Canada, July 1999, 23-32.

Lesser, V., Atighetchi, M., et al. (1999) A Multi-Agent System for Intelligent Environment Control. UMass Computer Science Technical Report 1998-40.

Lessig, L. (1999). *Code and Others Laws of Cyperspace.* New York: Basic Books.

Levesque, H., Reiter, R., Lesperance, Y., Lin, F., & Scherl, R. (1997). GOLOG: A Logic Programming Language for Dynamic Domains. *The Journal of Logic Programming, 31*, 59–84.

Levesque, H., Pirri, F., & Reiter, R. (1998). Foundations for the situation calculus. Electronic Transactions on Artificial Intelligence, 159–178.

Levinson, S. C. (2003). *Space in Language and Cognition: Explorations in Cognitive Diversity*. Cambridge, England: Cambridge University Press. doi:10.1017/CBO9780511613609

Lewis, M., Cameron, D., Xie, S., & Arpinar, B. (2006). ES3N: A Semantic Approach to Data Management in Sensor Networks. In Proc. of 5th International Semantic Web Conference (ISWC'06).

Li, Q. (2003). *Rosa, Michael De, Rus, Daniela: Distributed Algorithms for Guiding Navigation across a Sensor Network*. Dartmouth Department of Computer Science.

Li, S., & Hayashi, A. (1998, October). *Robot navigation in outdoor environments by using GPS information and panoramic views,* International Conference on Intelligent Robots and Systems, (Volume 1, Issue, 13-17 Oct 1998 Page(s): 570 - 575 vol.1)

Liao, L., Patterson, D. J., Fox, D., & Kautz, H. (2007). Learning and inferring transportation routines. *Artificial Intelligence, 171*(5-6), 311–331. doi:10.1016/j.artint.2007.01.006

Lin, C., & Hsu, J. Y. (2006). IPARS: Intelligent Portable Activity Recognition System Via Everyday Objects, Human Movements, and Activity Duration. *AAAI Workshop on Modeling Others from Observations*, (pp. 44-52).

Lin, F., Yang, C., & Yang, Y. (2007). Supporting Learning Context-aware and Auto-notification Mechanism on an Examination System. In T. Bastiaens & S. Carliner (Eds.), Proceedings of World Conference on E-Learning in Corporate, Government, Healthcare, and Higher Education 2007, pp. 352-359.

Lindström, M., & Seybold, P. (2003). *Brandchild: remarkable insights into the minds of today's global kids and their relationships with brands*. London: Kogan Page.

Lipton, A. J., Fujiyoshi, H., & Patil, R. S. (1998). Moving target classification and tracking from real-time video. Proceedings from WACV '98: *The Fourth IEEE Workshop on Applications of Computer Vision*, 129-136.

Liu, A., Hile, H., Borriello, G., Kautz, H., Brown, P., Harniss, M., & Johnson, K. (2009). *Informing the design of an automated wayfinding system for individuals with cognitive impairments*. Proc. of Pervasive Computing Technologies for Healthcare.

Liu, C. D., Chung, P. C., Chung, Y. N., & Thonnat, M. (2007). Understanding of human behaviors from videos in nursing care monitoring systems. *Journal of. High Speed Networks, 16*(1), 91–103.

Liu, J., & Issarny, V. (2004) Enhanced Reputation Mechanism for Mobile Ad Hoc Networks. In Proc. of the 2nd International Conference on Trust Management (iTrust), vol. 2995, pag. 48–62, Oxford, UK, LNCS.

Liu, J., Cao, Y., Lin, C.-Y., Huang, Y., & Zhou, M. "Low-quality product review detection in opinion summarization," in *Proceedings of the Joint Conference on Empirical Methods in Natural Language Processing and Computational Natural Language Learning (EMNLP-CoNLL)*, pp. 334–342, 2007.

Liyuan, L., Huang, W., Gu, I. Y. H., & Tian, Q. (2004). Statistical Modeling of Complex Backgrounds for Foreground Object Detection. *IEEE Transactions on Image Processing, 13*(11), 1459–1472. doi:10.1109/TIP.2004.836169

Liyuan, L., Huang, W., Tian, Q., Gu, I. Y. H., & Hleung, M. (2004). Human Detection and Behavior Classification for Intelligent Video Surveillance, *International Symposium on Computer Vision, Object Tracking and Recognition*

Logan, B., Healey, J., Philipose, M., Tapia, E., & Intille, S. (2007). A long-term evaluation of sensing modalities for activity recognition. In *Proc. of Ubicomp07*, 483–500.

Logsdon, R., Gibbons, L., McCurry, S., & Teri, L. (2002). Assessing quality of life in older adults with cognitive impairment. *Psychosomatic Medicine, 64*, 510–519.

Loke, S. W. (2004). Logic programming for context-aware pervasive computing: Language support, characterizing situations, and integration with the Web. In Proc. of the IEEE/WIC/ACM Conference on Web Intelligence (WI'04).

Löper, Ch., Kelsch, J., & Flemisch, F. O. Kooperative, Manöverbasierte Automation und Arbitrierung als Bausteine für hochautomatisiertes Fahren, in: Gesamtzentrum für Verkehr Braunschweig (Hrsgb): Automatisierungs-, Assistenzsysteme und eingebettete Systeme für Transportmittel, GZVB, Braunschweig, 2008, S. 215-237Meila, M., Jordan, M.I.: Learning Fine Motion by Markov Mixtures of Experts, MIT, AI Memo No. 1567, 1995

Lorenzi, F. (2007) A multiagent knowledge-based recommender approach with truth maintenance. In: *Proceedings of the 2007 ACM conference on Recommender systems*, Minneapolis

Lorini, E., & Castelfranchi, C. (2007). The cognitive structure of surprise: looking for basic principles. *Topoi: an International Review of Philosophy*, 26(1), 133–149.

Lorini, E., & Herzig, A. (2008). A logic of intention and attempt. *Synthese*, *163*(1), 45–77. doi:10.1007/s11229-008-9309-7

Lorini, E., & Schwarzentruber, F. A Logic for Reasoning about Counterfactual Emotions. In C. Boutilier (Eds.), Proceedings of the Twenty-first International Joint Conference on Artificial Intelligence (IJCAI'09), AAAI Press, pages 867-872, 2009.

Lu"hr, S., West, G., & Venkatesh, S. (2007). Recognition of emergent human behaviour in a smart home: A data mining approach. *Pervasive and Mobile Computing*, *3*(2), 95–116.

Lugmayr, A. (2006). "The Future is 'Ambient'", Proceedings- Spie The International Society. *Optical Engineering (Redondo Beach, Calif.)*, *6074*, 60740J.

Lundh, R., Karlsson, L., & Saffiotti, A. (2008). Autonomous Functional Configuration of a Network Robot System. *Robotics and Autonomous Systems*, *56*(10), 819–830. doi:10.1016/j.robot.2008.06.006

M. Boronowsky, O. Herzog, M. Lawo. Wearable Computing: Information and Commun-ication Technology Supporting Mobile Workers (Wearable Computing: Informations- und Kommunikationstechnologie zur Unterstützung mobiler Arbeiter). it - Information Technology 50(1), 30-39, 2008.

MacCormick, J., & Isard, M. (2000). Partitioned sampling, articulated objects, and interface-quality hand tracking. Proceedings from ECCV '00: *The European Conference on Computer Vision*, 2, 3-19.

Maes, P., & Mistry, P. (2009). The Sixth sense. *Proceedings of the Ted talk in reframe session* Long Beach, CA: TED 2009.

Magnani, L. (2007). *Morality in a Technological World. Knowledge as Duty*. Cambridge: Cambridge University Press. doi:10.1017/CBO9780511498657

Magnani, L. (2009). *Abductive Cognition. The Epistemological and Eco-Cognitive Dimensions of Hypothetical Reasoning*. Berlin, Heidelberg: Springer-Verlag.

Magnani, L., & Bardone, E. (2008). Sharing representations and creating chances through cognitive niche construction. The role of affordances and abduction. In Iwata, S., Oshawa, Y., Tsumoto, S., Zhong, N., Shi, Y., & Magnani, L. (Eds.), *Communications and Discoveries from Multidisciplinary Data* (pp. 3–40). Berlin: Springer. doi:10.1007/978-3-540-78733-4_1

Mahoney, D., Purtilo, R., Webbe, F., Alwan, M., Bharucha, A., & Adlam, T. (2007). In-home monitoring of persons with dementia: Ethical guidelines for technology research and development. *Alzheimer's & Dementia*, *3*, 217–226.

Maier, E., & Kempter, G. (2009). Aladin - a magic lamp for the elderly? In Hideyuki Nakashima, H., Aghajan, H., & Augusto, J. (Eds.), *Handbook of Ambient Intelligence and Smart Environments* (pp. 1201–1227). Heidelberg: Springer.

Majaranta, P., & Raiha, K. J. (2002). Twenty years of eye typing: systems and design issues. In: Eye Tracking Research and Applications (ETRA) Symposium. New Orleans: ACM.

Mancarella, P., Terreni, G., Sadri, F., Toni, F., & Endriss, U. (2009). The CIFF proof procedure for abductive logic programming with constraints: theory, implementation and experiments. *Theory and Practice of Logic Programming, 9*, 2009.

Mandel, C., & Frese, U. Comparison of Wheelchair User Interfaces for the Paralysed: Head-Joystick vs. Verbal Path Selection from an offered Route-Set. *Proceedings of the 3rd European Conference on Mobile Robots (ECMR 2007)*, 2007.

Mandel, C., Huebner, K., & Vierhuff, T. Towards an Autonomous Wheelchair: Cognitive Aspects in Service Robotics. *Proceedings of Towards Autonomous Robotic Systems (TAROS 2005)*, 2005.

Mandel, C., Röfer, T., & Frese, U. Applying a 3DOF Orientation Tracker as a Human-Robot Interface for Autonomous Wheelchairs. *Proceedings of the IEEE 10th International Conference on Rehabilitation Robotics (ICORR 2007)*, 2007. Otto Bock Mobility Solutions: http://www.ottobock.de/

Mankoff, J., Dey, A. K., Hsieh, G., Kientz, J., Lederer, S., & Ames, M. (2003). Heuristic evaluation of ambient displays. In *Proceedings of the SIGCHI Conference on Human Factors in Computing Systems (pp. 169—176)*. ACM, New York, USA.

Mao, W., & Gratch, J. (2004). Decision-Theoric Approach to Plan Recognition. Technical Report ICT-TR-01-2004, Institute for Creative Technologies (ICT).

Mao, W., & Gratch, J. (2004). A utility-based approach to intention recognition. AAMAS 2004 Workshop on Agent Tracking: Modelling Other Agents from Observations.

Marcus, G. (2004). *The Birth of the Mind: How a Tiny Number of Genes Creates the Complexities of Human Thought*. New York: Basic Books.

Marr, D. (1982). *Vision*. San Francisco, CA: Freeman.

Marreiros, G., Machado, J., Ramos, C., Neves, J.(2009) *Argumentation-based Decision Making in Ambient Intelligence Environments*. International Journal of Reasoning-based Intelligent Systems, 2009.

Marreiros, G., Santos, R., Freitas, C., Ramos, C., Neves, J., Bulas-Cruz, J. (2008). *LAID - a Smart Decision Room with Ambient Intelligence for Group Decision Making and Argumentation Support considering Emotional Aspects.* International Journal of Smart Home Vol. 2, No. 2, Published by Science &Engineering Research Support Center, 2008.

Marreiros, G., Santos, R., Ramos, C., Neves, J., Novais, P., Machado, J., & Bulas-Cruz, J. (2007). *Ambient Intelligence in Emotion Based Ubiquitous Decision Making*, Proc. Artificial Intelligence Techniques for Ambient Intelligence, IJCAI'07 – Twentieth International Joint Conference on Artificial Intelligence. Hyderabad, India.

Marreiros, G., Santos, R., Ramos, C., Neves, J., Novais, P., Machado, J., & Bulas-Cruz, J. Ambient intelligence in emotion based ubiquitous decision making. In *AITAmI'07*, 2007.

Marsella, S., Gratch, J., & Rickel, J. (2003). *Expressive Behaviors for Virtual Worlds. Life-like Characters Tools, Affective Functions and Applications*. Springer Cognitive Technologies Series.

Marsella, S. C., & Gratch, J. (2009). EMA: A process model of appraisal dynamics. *Cognitive Systems Research, 10*(1), 70–90. doi:10.1016/j.cogsys.2008.03.005

Marsella, S., & Gratch, J. (2006). EMA: A Computational Model of Appraisal Dynamics. In Trappl, R. (Ed.), *Cybernetics and Systems 2006* (pp. 601–606). Vienna: Austrian Society for Cybernetic Studies.

Marsland, S. (2003). Novelty detection in learning systems. *Neural Computing Surveys, 3*, 157–195.

Marsland, S. (2009). *Machine Learning: An Algorithmic Introduction*. New Jersey, USA: CRC Press.

Marti, P., & Lund, H. H. (2004). Emotional Constructions in Ambient Intelligence Environments. *Intelligenza Artificiale, 1*(1), 22–27.

Marti, P., Giusti, L., Pollini, A., & Rullo, A. (2005). Experiencing the flow: design issues in human-robot interaction. In *Proceedings of the 2005 Joint Conference on Smart Objects and Ambient intelligence: innovative Context-Aware Services: Usages and Technologies* (Grenoble, France, October 12 - 14, 2005). sOc-EUSAI '05, vol. 121. ACM, New York, NY, 69-74.

Martison, E., & Payton, D. (2005). Lattice formation in mobile autonomous sensor arrays. In E. Sahin & W. M. Spears (Eds.), Swarm Robotics (LNCS), *3342* (pp. 98–111). New York: Springer. doi:10.1007/978-3-540-30552-1_9

Mastrogiovanni, F., Sgorbissa, A., & Zaccaria, R. (2009). On the Problem of Describing Activities in Context-Aware Environments. In The International Journal of Assistive Robotics and Mechatronics, vol. 9, no. 4.

Mastrogiovanni, F., Sgorbissa, A., & Zaccaria, R. (2008). Understanding events relationally and temporally related: Context assessment strategies for a smart home. Second International Symposium on Universal Communication, 217-224.

Mastrogiovanni, F., Scalmato, A., Sgorbissa, A., & Zaccaria, R. (2009). Assessing Temporal Relationships Between Events in Smart Environments. *Proc of the Int. Conf. on Intellingent Environments.*

Masuoka, R., Parsia, B., & Labrou, Y. (2003). Task Computing - the Semantic Web meets Pervasive Computing. In Fensel D. et al (Eds.) *Proceedings of the 2nd International Semantic Web Conference ISWC'03* (pp. 866-881), Lecture Notes In Computer Science, Vol. 2870. Berlin, Heidelberg: Springer-Verlag.

Mateas, M. (2002). *Interactive Drama, Art and Artificial Intelligence*. Carnegie Mellon University.

Matsuoka, K. (2004). Aware Home Understanding Life Activities, *Proceedings of 2nd International Conference On Smart Home and Health Telematics*, IOS press (pp. 186-193)

Matsuyama, T. (1998). Cooperative distributed vision. Proceedings from DIU '98: *The Darpa Image Understanding Workshop, 1*, 365-384.

Matthews, T., Dey, A. K., Mankoff, J., Carter, S., & Rattenbury, T. (2004). A Toolkit for Managing User Attention in Peripheral Displays. In *Proceedings of the Seventeenth annual ACM symposium on User Interface Software and Technology* (pp. 247-256). ACM, New York, USA.

Mavrommati, I., & Darzentas, J. (2007). End User Tools for Ambient Intelligence Environments: An Overview. In Jacko, J. (Ed.), *Human-Computer Interaction Part II* (pp. 864–872). Springer-Verlag Berlin Heidelberg.

Mavrommati, I., & Kameas, A. (2003). End-User Programming Tools in Ubiquitous Computing Applications. In Stephanidis C. (Ed.), *Proceedings of International Conference on Human-Computer Interaction* (pp. 864-872). London, UK: Lawrence Erlbaum Associates.

Mayfiled, J. (2000). Evaluating plan recognition systems: three properties of a good explanation. *Artificial Intelligence Review, 14*, 351–376.

McCann, H. (1991). Settled objectives and rational constraints. *American Philosophical Quarterly, 28*, 25–36.

McCarthy, J. (1980). Circumscription – A Form of Non-Monotonic Reasoning. *Artificial Intelligence, 13*, 27–39.

McCarthy, J., Costa, T., Huang, E., & Tullio, J. (2001). Defragmenting the organization: Disseminating community knowledge through peripheral displays. *Workshop on Community Knowledge at the 7th European Conference on Computer Supported Cooperative Work* (pp. 117-126).

McDowell, I., & Newell, C. (1996). *Measuring Health: A Guide to Rating Scales and Questionnaires* (2nd ed.). New York: Oxford University Press.

McHugh, M., & Smeaton, A. F. (2006) Event Detection Using Audio in a Smart Home Context. In: Nugent C and Augusto JC (eds) Smart Homes and Beyond, ICOST'2006 4th International Conference on Smart Homes and Health Telematics, Assistive Technology Research Series 19, IOS Press, pp. 23–30.

McLurkin, J., & Smith, J. (2004). Distributed algorithms for dispersion in indoor environments using a swarm of autonomous mobile robots. In *Proc. 7th Int. Sym. Distributed Autonomous Robotic Systems*, (pp831-840).

McNeely, B. (2005). Using Technology as a Learning Tool, Not Just the Cool New Thing. In Oblinger, D., & Oblinger, J. L. (Eds.), *Educating the Net Generation*. Educause, Washington.

McRae, R. R., & Costa, P. T. (1987). Validation of the five-factor model of personality across instruments and observers. *Journal of Personality and Social Psychology, 52*, 81–90. doi:10.1037/0022-3514.52.1.81

Mea, V. D., Pittaro, M., & Roberto, V. Knowledge management and modelling in health care organizations: The standard operating procedures. In: M. Wimmer (ed), Knowledge Management in Electronic Government, 5th IFIP International Working Conference, KMGov Krems, Austria, May 17–19, 2004, Proceedings, Springer, LNCS, vol 3035, 136–146, 2004.

Meffert, H., & Bruhn, M. (2000). *Dienstleistungsmarketing*. Wiesbaden: Grundlagen - Konzepte - Methoden.

Megret, R. (2003). *Structuration spatio-temporelle de séquences vidéo*. Doctoral dissertation, INSA Lyon, Lyon, France.

Mehdi Dastani, M. (2005). Birna van Riemsdijk, and John-Jules Ch. Meyer. Programming Multi-Agent Systems in 3APL. In *Multi-Agent Programming* (pp. 39–67). Languages, Platforms and Applications.

Mehrabian, A. (1995). Framework for a comprehensive description and measurement of emotional states. *Genetic, Social, and General Psychology Monographs, 121*, 339–361.

Mehrabian, A. (1996). Analysis of the Big-Five Personality Factors in Terms of the PAD Temperament Model. *Australian Journal of Psychology, 48*(2), 86–92. doi:10.1080/00049539608259510

Meiren, T., Hofmann, H. R., & Klein, L. "Vorgehensmodelle für das Service Engineering,", 13 ed 1998, pp. 20-25.

Meis, J., & Schöpe, L. "ServiceDesign with the ServiceBlueprint,". Knowledge Systems Institute Graduate School, Ed. San Francisco, Calif.: Knowledge Systems Institute Graduate School, 2006, pp. 708-713.

Mendez, G., Rickel, J., & Antonio, A. (2003). *Steve meets Jack: the integration of an intelligent tutor and a virtual environment with planning capabilities. Computer science school*. Technical University Madrid.

Meneguzzi, F. R., & Luck, M. Composing high-level plans for declarative programming. Workshop on Declarative Agent Languages and Technologies, 69-85.

Menser B. & Wien M. (2000). Segmentation and Tracking of Facial Regions in Color Image Sequences, *SPIE Visual Communication and Image Processing*, (4067), (pp. 731-740).

Messer, A., Kunjithapatham, A., Sheshagiri, M., Song, H., Kumar, P., Nguyen, P., & Yi, K. H. (2006). InterPlay: A Middleware for Seamless Device Integration and Task Orchestration in a Networked Home. *Proceedings of the Annual IEEE International Conference on Pervasive Computing PerCom '06* (pp. 296-307), Washington DC: IEEE Computer Society. Ben Mokhtar, S. (2007). *Semantic Middleware for Service-Oriented Pervasive Computing*. Doctoral dissertation, University of Paris 6, Paris, France.

Meyer, A. Dienstleistungs-Marketing: Erkenntnisse und praktische Ergebnisse. München, Augsburg: 1992.

Meyer, S., & Rakotonirainy, A. (2003). A survey of research on context-aware homes. In Proc. of the ACSW frontiers 2003 (pp. 159-168), Adelaide, Australia.

Miao, C., Yang, Q., Fang, H., & Goh, A. (2007). A cognitive approach for agent-based personalized recommendation. *Knowledge-Based Systems, 20*, 397–405. doi:10.1016/j.knosys.2006.06.006

Mihailidis, A., Carmichael, B., & Boger, J. (2004). The use of computer vision in an intelligent environment to support aging-in-place, safety, and independence in the home. *IEEE Transactions on Information Technology in Biomedicine, 8*, 238–247.

Mihailidis, A., Boger, J., Craig, T., & Hoey, J. (2008). The COACH prompting system to assist older adults with dementia through handwashing: An efficacy study. *BMC Geriatrics, 8*(28), 1–18.

Mihailidis, A., Fernie, G. R., & Barbenel, J. (2001). The use of artificial intelligence in the design of an intelligent cognitive orthosis for people with dementia. *Assistive Technology, 13*, 23–39.

Mihailidis, A., Boger, J., Canido, M., & Hoey, J. (2007). The use of an intelligent prompting system for people with dementia: A case study. *ACM Interactions (Special issue on Designing for seniors: innovations for graying times), 14*(4):34–7.

Mikulecky, P., & Olševičová, K. (2005). University Education as an Ambient Intelligence Scenario. *In Proceedings of 4th European Conference on E-learning 2005*, Amsterdam, The Netherlands, pp. 333-341.

Millonig, A., Brändle, N., Ray, M., Bauer, D., & van der Spek, S. (2009). Pedestrian Behaviour Monitoring: Methods and Experiences. In Gottfried, B., & Aghajan, H. (Eds.), *Behaviour Monitoring and Interpretation, Smart Environments*. IOS Press.

Millonig, A., & Gartner, G. (2007). Shadowing, Tracking, Interviewing: How to Explore Human ST-Behaviour Patterns. In Gottfried, B. (Ed.), *BMI'07* (*Vol. 296*, pp. 29–42). CEUR.

Ming, H., Oyama, K., & Chang, C. K. (2008). Human-Intention Driven Self Adaptive Software Evolvability in Distributed Service Environments. Twelfth IEEE International Workshop on Future Trends of Distributed Computing Systems (FTDCS'08), 51-57.

Misker, J. M. V., Veenman, C. J., & Rothkrantz, L. J. M. (2004). Groups of Collaborating Users and Agents in Ambient Intelligent Environments. In: *Proc. ACM AAMAS 2004*, pp. 1318-1319

Mistry, P., Maes, P., & Chang, L. (2009). Wuw wear ur world: a wearable gestural interface. In *Proc. of Human factors in Computing Systems*, 4111–6.

Mithen, S. (1999). Handaxes and ice age carvings: hard evidence for the evolution of consciousness. In Hameroff, A. R., Kaszniak, A. W., & Chalmers, D. J. (Eds.), *Toward a Science of Consciousness III. The Third Tucson Discussions and Debates* (pp. 281–296). Cambridge: MIT Press.

Miyazaki, T., Kodama, T., Furushahi, T., & Ohno, H. Modeling of Human Behaviors in Real Driving Situations, 2001 IEEE Intelligent Transportation Systems Conference Proceedings, 2001, 643 - 645

Mizoguchi, R., Welkenhuysen, J. v., & Ikeda, M. (1995). Task ontology for reuse of problem-solving knowledge. Second International Conference on Knowledge Building and Knowledge Sharing (KB & KS'95), 46-57.

Mls, K., Salmeron, J., Olševičová, K., & Husáková, M. (2009). Cognitive Personal Software Assistants within AmI environment. In *Ambient Intelligence Perspectives II* (pp. 167–174). IOS Press Amsterdam.

Möbus, C., Eilers, M., & Garbe, H. (2011). Predicting the Focus of Attention and Deficits in Situation Awareness with a Modular Hierarchical Bayesian Driver Model. In Duffy, V. G. (Ed.), *Digital Human Modeling, HCII 2011, LNCS 6777* (pp. 483–492). Springer-Verlag Berlin Heidelberg.

Möbus, C., Hübner, S., & Garbe, H. (2007). Driver Modelling: Two-Point- or Inverted Gaze-Beam-Steering. In Rötting, M., Wozny, G., Klostermann, A., & Huss, J. (Eds.), *Prospektive Gestaltung von Mensch-Technik-Interaktion, Fortschritt-Berichte VDI-Reihe 22, Nr. 25, 483 – 488*. Düsseldorf: VDI Verlag.

Möbus, C., & Eilers, M. (2008). First Steps Towards Driver Modeling According to the Bayesian Programming Approach, Symposium Cognitive Modeling, p.59, in: L. Urbas, Th. Goschke & B. Velichkovsky (eds) *KogWis 2008*. Christoph Hille, Dresden, ISBN 978-3-939025-14-6

Möbus, C., & Eilers, M. (2009a). Further Steps Towards Driver Modeling according to the Bayesian Programming Approach, in: *Conference Proceedings, HCII 2009, Digital Human Modeling*, pp. 413-422, LNCS (LNAI), Springer, San Diego, ISBN 978-3-642-02808-3

Möbus, C., & Eilers, M. Garbe, H., and Zilinski, M. (2009b). Probabilistic and Empirical Grounded Modeling of Agents in (Partial) Cooperative Traffic Scenarios, in: *Conference Proceedings, HCII 2009, Digital Human Modeling*, pp. 423-432, LNCS (LNAI), Springer, San Diego, ISBN 978-3-642-02808-3

Möbus, C., & Eilers, M. Zilinski, M. Garbe, H. (2009c). Mixture of Behaviors in a Bayesian Driver Model, in: Lichtenstein, A. et al. (eds), Der Mensch im Mittelpunkt technischer Systeme, p.96 and p.221-226 (CD), Düsseldorf: VDI Verlag, ISBN 978-3-18-302922-8, ISSN 1439-958X

Möbus, C., & Eilers, M. *Mixture of Behaviors and Levels-of-Expertise in a Bayesian Autonomous Driver Model*, in: Advances in Applied Digital Human Modeling, p. 425-435, Vincent G. Duffy (ed), CRC Press, Taylor & Francis Group, Boca Raton, 2010/2011, ISBN 978-1-4398-3511-1 and in. W. Karwowski and G. Salvendy (eds), 1st International Conference On Applied Digital Human Modeling, 17-20 July, 2010, Intercontinental, fMiami Florida, USA, Conference Proceedings, Session Digital Human Modeling in the Bayesian Programming Framework, USA Publishing, ISBN-13: 978-0-9796435-4-5

Modayil, J., Levinson, R., Harman, C., Halper, D., & Kautz, H. (2008). Integrating sensing and cueing for more effective activity reminders. In *AI in Eldercare. New Solutions to Old Problems*.

Modayil, J., Bai, T., & Kautz, H. (2008). Improving the recognition of interleaved activities. In the Proceedings of the 10th International Conference on Ubiquitous Computing (UbiComp), 2008, 40-43.

Mokhtar, S. B., Bromberg, Y.-D., & Georgantas, N. (2006, March). Amigo middleware core: Prototype implementation and documentation. Deliverable 3.2, European Amigo Project.

Móra, M., Lopes, J., Vicari, R., & Coelho, H. (1999). BDI models and systems: Bridging the gap, Workshop on Intelligent Agents V, Agent Theories, Architectures, and Languages (ATAL'98), London, UK, 11-27.

Mori, G., & Malik, J. (2002). Estimating human body configurations using shape context matching. Proceedings from ECCV '02: *The European Conference on Computer Vision*, 666-680.

Mori, G., Ren, X., Efros, A. A., & Malik, J. (2004). Recovering human body configurations: Combining segmentation and recognition. Proceedings from ICCVPR '04: *The International Conference on Computer Vision and Pattern Recognition*, 2, 326-333.

Mori, T., et al. (2001) Accumulation and Summarization of Human Daily Action Data in One-Room-Type Sensing System. Proc. IROS (IEEE/RSJ International Conference on Intelligent Robots and Systems) 2001, pp. 2349–2354.

Mori, T., Noguchi, H., et al. (2004) Sensing Room: Distributed Sensor Environment for Measurement of Human Daily Behavior. Proc. of First International Workshop on Networked Sensing Systems (INSS2004), pp. 40–43.

Morris, M., Lundell, J., & Dishman, E. (2004). Catalyzing social interaction with ubiquitous computing: a needs assessment of elders coping with cognitive decline. In *Proc. of CHI04*, 1151–4.

Morris, M., Lundell, J., Dishman, E., & Needham, B. (2003). New perspectives on ubiquitous computing from ethnographic study of elders with cognitive decline. In *Proc. of UbiComp03*, 227–242.

Mozer, M. C. (2005). Lessons from an Adaptive House. In Cook, D., & Das, R. (Eds.), *Smart Environments: Technologies, Protocols, and Applications* (pp. 273–294). Hoboken, NJ: J. Wiley & Sons.

Mulder, F., & Voorbraak, F. (2003). A formal description of tactical plan recognition. *Information Fusion*, *4*, 47–61.

Murphy, K. P. (2002). Dynamic Bayesian Networks: Representation, Inference and learninng, *Ph.D. Thesis, University of California*, Berkeley Jensen F. V. (2001). Bayesian Networks and Decision Graphs, S*pringer Verlag, New York, Inc., New York*, NY, USA

Muscettola, N., Nayak, P. P., Pell, B., & Williams, B. C. (1998). Remote Agent: to Go Boldly Where No AI System Has Gone Before. *Artificial Intelligence*, *103*(1–2), 5–48. doi:10.1016/S0004-3702(98)00068-X

Myers, K. L. (1997). Abductive completion of plan sketches. In Proceedings of the 14th National Conference on Artificial Intelligence, American Association of Artificial Intelligence (AAAI-97), Menlo Park, CA: AAAI Press, 1997.

Myers, K. L., Jarvis, P. A., Tyson, W. M., & Wolverton, M. J. (2003). A mixed-initiative framework for robust plan sketching. In Proceedings of the 13thInternational Conferences on AI Planning and Scheduling, Trento, Italy, June, 2003.

Mynatt, E., Rowan, J., Jacobs, A., & Craighill, S. (2001). Digital Family Portraits: Supporting Peace of Mind for Extended Family Members. In *Proceedings of the SIGCHI conference on Human factors in computing systems* (pp. 333-340). ACM, New York, USA.

Naeem, U., & Bigham, J. (2007). A Comparison of Two Hidden Markov Approaches to Task Identification in the Home Environment. *Proc. of the 2nd Int. Conf. on Pervasive Computing and Applications (ICPCA 2007)*, (pp. 383-388).

Nagel, H.-H. (1998). The representation of situations and their recognition from image sequences. Proceedings from RFIA '98: *Reconnaissance de Formes et Intelligence Artificielle*, 1221-1229.

Nakano, Y., Takemoto, M., Yamato, Y., & Sunaga, H. (2006). Effective Web-Service Creation Mechanism for Ubiquitous Service Oriented Architecture. *Proceedings of the 8th IEEE International Conference on E-Commerce Technology and the 3rd IEEE International Conference on Enterprise Computing, E-Commerce, and E-Services CEC/EEE'06* (p. 85). Washington DC: IEEE Computer Society.

Nakazawa, J., Yura, J., & Tokuda, H. (2004). Galaxy: a Service Shaping Approach for Addressing the Hidden Service Problem. *Proceedings of the 2nd IEEE Workshop on Software Technologies for Future Embedded and Ubiquitous Systems* (pp. 35-39).

Narayanaswami, C., & Raghunath, M. T. (2000). Application design for a smart watch with a high resolution display. In *Proceedings of the Fourth IEEE International Symposium on Wearable Computers* (pp. 7 - 14), IEEE Computer Society, Washington DC, USA.

National Conference, Brisbane, Qld., September 1997.

Natsoulas, T. (2004). To see is to perceive what they afford: James J. Gibson's concept of affordance. *Mind and Behaviour*, *2*(4), 323–348.

Neapolitan, R. E. (2004). *Learning Bayesian Networks*. Upper Saddle River: Prentice Hall.

Nerzic, P. (1996). Two Methods for Recognizing Erroneous Plans in Human-Machine Dialogue. In AAAI'96 Workshop: Detecting, Repairing and Preventing Human-Machine Miscommunication.

NETICA. http://www.norsys.com/ (visited 28th February, 2010)

Nguyen, N. T., Phung, D. Q., Venkatesh, S., & Bui, H. Learning and detecting activities from movement trajectories using the hierarchical hidden Markov model. In *Proc. CVPR-05*, pages 955–960, 2005.

Niebert, N. (2004). Ambient Networks - An Architecture for Communication Networks Beyond 3G. *IEEE Wireless Communications*, *11*(Issue 2), 14–22. doi:10.1109/MWC.2004.1295733

Noguchi, H. Mori, T. and Sato, T. Network middleware for utilization of sensors in room, in Proceedings of IEEE/RSJ International Conference on Intelligent Robots and Systems, vol. 2, 2003, pp. 1832-1838.

Nokia. (2009). *Point&Find API for mobile phones*. Retrieved on November 16, 2009, from http://pointandfind.nokia.com

Noriega, P. (2007a). *Modèle du corps pour le suivi du haut du corps en monoculaire*. Doctoral dissertation, University of Nancy 1, Nancy, France.

Noriega, P. (2007b). Multiclues 3D Monocular Upper Body Tracking Using Constrained Belief Propagation. Proceedings from BMVC '07: *The British Machine Vision Conference*, 10-13.

Norman, D. A. (2003). *Emotional Design: Why We Love (or Hate) Everyday Things*. Basic Books.

Norman, D. A. (2007). *The Design of Future Things*. Basic Books.

Nyan, M. N., Tay, F. E. H., & Seah, K. H. W. (2004) Segment Extraction Using Wavelet Multiresolution Analysis for Human Activities and Fall Incidents Monitoring System. Proc. 2nd International Conference on Smart Homes and Health Telematics (ICOST 2004), Vol. 14, pp. 177–185.

O'Donoghue, J., Herbert, J., & Stack, P. (2006) Remote Non-Intrusive Patient Monitoring. In: Nugent C and Augusto JC (eds) Smart Homes and Beyond, ICOST'2006 4th International Conference on Smart Homes and Health Telematics, Assistive Technology Research Series 19, IOS Press, pp. 23–30.

O'Rourke, J., & Badler, N. (1980). Model-based image analysis of human motion using constraint propagation. *IEEE Transactions on Pattern Analysis and Machine Intelligence*, *2*(6), 522–536.

Oates, T., Prasad, M. V. N., & Lesser, V. R. (1994). *Cooperative Information Gathering: A Distributed Problem Solving Approach*. University of Massachusetts.

Object Management Group. (2004). *Business Process Modeling Notation (BPMN)*. Information.

Object Management Group, "Business Process Modeling Notation Specification - Version 1.2," Object Management Group, 9 A.D..

Oblinger, D., & Oblinger, J. L. (Eds.). (2005). *Educating the Net Generation*. Washington: Educause.

O'Connor, M. J., & Das, A. K. (2009). OWL: Experiences and Directions (OWLED), Fifth International Workshop, Chantilly, VA. In Press.

O'Connor, M. J., Knublauch, H., & Tu, S. W. (2005). Supporting rule system interoperability on the semantic web with SWRL. In Proc. of the 4th International Semantic Web Conference.

Odling-Smee, F., Laland, K., & Feldman, M. (2003). *Niche Construction. A Neglected Process in Evolution*. New York, NJ: Princeton University Press.

Oehme, A., Herbon, A., Kupschick, S., & Zentsch, E. Physiological correlates of emotions. In *AISBO07*, 2007.

Oh, Y., & Woo, W. (2004) A Unified Application Service Model for ubiHome by Exploiting Intelligent Context-Awareness. Proc. of Second Intern. Symp. on Ubiquitous Computing Systems (UCS2004), pp. 117–122.

Ohno, T., & Mukawa, N. (2004). A free-head, simple calibration, gaze tracking system that enables gaze-based interaction. In *Proceedings of the 2004 Symposium on Eye Tracking Research & Applications* (pp. 115-122). ACM, New York, USA.

Oliver, N. M. Towards Perceptual Intelligence: Statistical Modeling of Human Individual and Interactive Behaviors, MIT Ph.D. Thesis, 2000

Oliver, N., & Pentland, A. P. Graphical Models for Driver Behavior Recognition in a SmartCar, IEEE Intl. Conf. Intelligent Vehicles, 7 – 12, 2000

Olševičová, K., & Mikulecký, P. (2008) Learning Management System as Ambient Intelligence Playground. In: *Int. Journal of Web Based Communities*, Vol. 4, No. 3, pp. 348-358

Olševičová, K., & Poulová, P. (2009) Analysis of E-learning Courses Used at FIM UHK With Respect to Constraint Satisfaction Problems. In: *Proceedings of ECEL, 8th European Conference on E-learning*. ACL, Reading, pp. 425-430

Ortony, A., Clore, C., & Collins, A. (1998). *The Cognitive Structure of Emotions*. Cambridge University Press.

Ortony, A. (2003). On making believable emotional agents believable. In Trappl, R., Petta, P., & Payr, S. (Eds.), *Emotions in Humans and Artifacts* (pp. 189–212). MIT Press.

Ortony, A. (2003). *On making believable emotional agents believable*. In R. P. Trapple, P. (Ed.), Emotions in humans and artefacts. Cambridge: MIT Press.

OSGi Alliance. (2007). OSGi service platform - service compendium. Technical Report Release 4, Version 4.1, OSGi.

Oyama, K., Jaygarl, H., Xia, J., Chang, C. K., Takeuchi, A., & Fujimoto, H. (2008a). A Human-Machine Dimensional Inference Ontology that Weaves Human Intentions and Requirements of Context Awareness Systems. Computer Software and Applications Conference (COMPSAC'08), 287-294.

Oyama, K., Jaygarl, H., Xia, J., Chang, C. K., Takeuchi, A., & Fujimoto, H. (2008b). Requirements Analysis Using Feedback from Context Awareness Systems. Second IEEE International Workshop on Requirements Engineering For Services (REFS '08), 625-630.

Oyama, K., Takeuchi, A., & Fujimoto, H. (2006). CA-PIS Model Based Software Design Method for Sharing Experts' Thought Processes. Computer Software and Applications Conference (COMPSAC'06), 307-316.

Paiva et. al. (2001). *SAFIRA- Supporting Affective Interactions in Real-time Applications*. CAST - Living in mixed realities, Special Issue of netzpannung.org/journal.

Paluska, J. M., Pham, H., Saif, U., Chau, G., Terman, C., & Ward, S. (2008). Structured decomposition of adaptive applications. *International Journal of Pervasive and Mobile Computing*, *4*(6), 791–806. doi:10.1016/j.pmcj.2008.04.006

Panina, D., & Aiello, J. R. (2005). Acceptance of electronic monitoring and its consequences in different cultural contexts: A conceptual model. *Journal of International Management*, *11*, 269–292. doi:10.1016/j.intman.2005.03.009

Paradiso, J., Hsiao, K. Y., & Benbasat, A. (2000). Interfacing to the foot: Apparatus and applications. In *Extended Abstracts on Human Factors in Computing Systems* (pp. 175–176). New York, USA: ACM. doi:10.1145/633292.633389

Park, S., Won, S., Lee, J., & Kim, S. (2003). Smart home – digitally engineered domestic life. *Personal and Ubiquitous Computing*, *7*(3-4), 189–196. doi:10.1007/s00779-003-0228-9

Park, K.-H., & Bien, Z. Z. (2003) Intelligent Sweet Home for Assisting the Elderly and the Handicapped. In: Mokhtari M (ed) Independent Living for Persons with Disabilities and Elderly People, ICOST'2003 1st International Conference on Smart Homes and Health Telematics, Assistive Technology Research Series 12, IOS Press, pp. 151–158.

Park, S., & Kautz, H. (2008). Hierarchical recognition of activities of daily living using multi-scale, multi-perspective vision and RFID. The 4th IET International Conference on Intelligent Environments, Seattle, WA, July 2008a.

Park, S., & Kautz, H. (2008). Privacy-preserving recognition of activities in daily living from multi-view silhouettes and RFID-based training, AAAI Fall 2008 Symposium on AI in Eldercare: New Solutions to Old Problems, Washington, DC, November 7 - 9, 2008b.

Patterson, D. J., Kautz, H. A., Fox, D., & Liao, L. (2007). Pervasive computing in the home and community. In Bardram, J. E., Mihailidis, A., & Wan, D. (Eds.), *Pervasive Computing in Healthcare* (pp. 79–103). CRC Press.

Patterson, D. J., Fox, D., Kautz, H., & Philipose, M. (2005). Fine-Grained Activity Recognition by Aggregating Abstract Object Usage. *Proc. of the 9th IEEE International Symposium on Wearable Computers.*

Patterson, D. J., Liao, L., Fox, D., & Kautz, H. (2003). Inferring High-Level Behavior from Low-Level Sensors. In Proceeding of UBICOMP 2003: The 5th International Conference on Ubiquitous Computing, Anind Dey, Albrecht Schmidt, et Joseph F. McCarthy, editors, volume LNCS 2864, 73–89. Springer-Verlag.

Pauly, M. (2002). A modal logic for coalitional power in games. *Journal of Logic and Computation*, *12*(1), 149–166. doi:10.1093/logcom/12.1.149

Pauwels, E. J., Salah, A. A., & Tavenard, R. (2007). Sensor Networks for Ambient Intelligence, in: Proc. of the 9th IEEE workshop on MultiMedia Signal Processing, Chania, Crete, Greece pp. 13–16.

Pearl, J. (1988). *Probabilistic Reasoning in Intelligent Systems*. San Mateo, CA: Morgan Kaufmann.

Pearl, J. (2009). *Causality – Models, Reasoning, and Inference* (2nd ed.). Cambridge University Press.

Pecora, F., & Cirillo, M. (2009). A Constraint-Based Approach for Plan Management in Intelligent Environments. *Proc. of the Scheduling and Planning Applications Workshop at ICAPS09.*

Peirce, C. S. (1958). *The collected papers of Charles Sanders Peirce* (Burks, A. W., Ed.). *Vol. VII-VIII*). Cambridge, MA: Harvard.

Peirce, C. S. (1967). *The Charles S. Peirce Papers: Manuscript Collection in the Houghton Library*. Worcester, MA: The University of Massachusetts Press. (Annotated Catalogue of the Papers of Charles S. Peirce. Numbered according to Richard S. Robin. Available in the Peirce Microfilm edition. Pagination: CSP = Peirce / ISP = Institute for Studies in Pragmaticism.)

Pentland, A. P. (1995). Machine understanding of human action. Proceedings from IFFTT '95: *Seventh International Forum on Frontier of Telecommunication Technology*, 757-764, Tokyo: ARPA Press.

Pentney, W., Philipose, M., Bilmes, J., & Kautz, H. (2007). Learning large scale common sense models of everyday life. In *Proc. of AAAI07*, 465–470. Christopher Presley. (2009). Health triangle. Information Society Technologies (IST) Program. (2009). Helping people with mild dementia to navigate their day.

Pepels W., Marketing, 4 ed Oldenbourg Wissenschafts-verlag, 2004.

Pereira, L. M., & Anh, H. T. (2009a). Intention Recognition via Causal Bayes Networks plus Plan Generation, in: L. Rocha et al. (eds.), Progress in Artificial Intelligence, Procs. 14th Portuguese International Conference on Artificial Intelligence (EPIA'09), Springer LNAI, October 2009.

Pereira, L. M., & Anh, H. T. (2009b). Elder care via intention recognition and evolution prospection, in: S. Abreu, D. Seipel (eds.), Procs. 18th International Conference on Applications of Declarative Programming and Knowledge Management (INAP'09), Évora, Portugal, November 2009.

Pereira, L. M., Alferes, J. J., & Aparicio, J. N. (1992). Contradiction removal semantics with explicit negation. In Proceedings of the Applied Logic Conference, 1992.

Pereira, L.M., Saptawijaya, A. (2009). Modelling Morality with Prospective Logic, in International Journal of Reasoning-based Intelligent Systems (IJRIS), 1(3/4): 209-221.

Perez, P., Hue, C., Vermaak, J., & Gangnet, M. (2002). Color-based probabilistic tracking. Proceedings from ECCV '02: *The European Conference on Computer Vision*, 661-675.

Peterson, C., & Anderson, J. R. (1987). A mean field theory learning algorithm for neural networks. *Complex Systems*, *1*(5), 995–1019.

Petrushin, V. A., Wei, G., & Gershman, A. V. (2006). Multiple-camera people localization in an indoor environment. *Knowledge and Information Systems*, *10*(Issue 2), 229–241. doi:10.1007/s10115-006-0025-7

Petta, P. (1999) Principled Generation of Expressive Behavior in an Interactive Exhibit. In Juan D Velásquez (ed.): *Workshop: Emotion-Based Agent Architectures (EBAA'99) at the 3rd International* Conference *on Autonomous Agents, Seattle WA USA, May 1 1999*, pp. 94–98, 1999. 34, 36, 110, 14

Petta, P., Staller, A., Trappl, R., Mantler, S., Szalavári, S., Psik, T., & Gervautz, M. (1999) Towards Engaging Full-Body Interaction. In H.-J. Bullinger and P.H. Vossen (eds.): *Adjunct Conference Proceedings, HCI International '99, 8th International Conference on Human-*Computer *Interaction, jointly with 15th Symposium on Human Interface, Munich Germany, August 22-27 1999*, pp. 280–281. Fraunhofer IRB Verlag, 1999. 36, Picard, R. (1995). *Affective Computing*. M.I.T Media Laboratory Perceptual Computing Section Technical Report Nº 321, Nov. 26, (1995). [on-line] Available: http://vismod.media.mit.edu/tech-reports/TR-321.pdf

Philip R. Cohen and Hector J. Levesque. Intention is choice with commitment. *Artificial Intelligence Journal*, 42(2– 3):213–261, 1990.

Philipose, M., Fishkin, K. P., Perkowitz, M., Patterson, D. J., Hahnel, D., Fox, D., & Kautz, H. (2005). Inferring ADLs from interactions with objects. *IEEE Pervasive Computing / IEEE Computer Society [and] IEEE Communications Society*, 2005.

Philipose, M. (2009). Technology for long-term care: Scaling eldercare to the next billion. In *Proc. of ICOST09*.

Philipose, M., Fishkin, K. P., Perkowitz, M., Patterson, D. J., Fox, D., Kautz, H., & Hähnel, D. (2004). Inferring Activities from Interactions with Objects, IEEE Pervasive Computing: mobile and Ubiquitous Systems, 3(4), pp. 50–57.

Philips Electronics, N. V. (2007). Mirror TV Multi-media Display with latest high definition LCD display. Retrieved November 19, 2009, from http://www.p4c.philips.com/files/3/30hm9202_12/30hm9202_12_pss_eng.pdf

Phua, C., Foo, V., Biswas, J., Tolstikov, A., Aung, A., Maniyeri, J., et al. (2009). 2-layer erroneous-plan recognition for dementia patients in smart homes. In *Proc. of HealthCom09*.

Phung, S. L., Chai, D., & Bouzerdoum, A. (2001). A Universal and Robust Human Skin Color Model Using Neural Networks, *Proceedings of the INNS-IEEE International Joint Conference on Neural Networks*, (4). (pp. 2844-2849)

Piaget, J. (1974). *Adaption and Intelligence*. Chicago: University of Chicago Press.

Piaget, J., & Inhelder, B. (1948). *La Représentation de l'Espace chez l'Enfant*. Paris, France: Presses Universitaires de France.

Pigot, H., Lefebvre, B., et al. (2003) The Role of Intelligent Habitats in Upholding Elders in Residence. 5th International Conference on Simulations in Biomedicine.

Pigot, H., Mayers, A., & Giroux, S. (2003). The intelligent habitat and everyday life activity support, in: Proc. Int. Conf. in Simulations in Biomedicine, Slovenia, pp.507–516.

Pike, R., Presotto, D., Thompson, K., & Trickey, H. (1990). Plan 9 from Bell Labs. *EUUG Newsletter, 10*, 3.

Pingali, G., Pinhanez, C., Levas, A., Kjeldsen, R., Podlaseck, M., Chan, H., & Sukaviriya, N. (2003). Steerable Interfaces for Pervasive Computing Spaces. In *Proceedings of the First IEEE International Conference on Pervasive Computing and Communications*, (pp. 315-322). IEEE Computer Society, Washington DC, USA.

Pinhanez, C., & Podlaseck, M. (2005). To frame or not to frame: The role and design of frameless display in ubiquitous applications. In M. Beigl, S. Intille, J. Rekimoto, & H. Tokuda (Eds.), *the Seventh International Conference on Ubiquitous Computing* (pp. 340-357). Springer-Verlag Berlin Heidelberg.

Plänkers, R., & Fua, P. (2003). Articulated soft objects for multiview shape and motion capture. *IEEE Transactions on Pattern Analysis and Machine Intelligence, 25*(9), 1182–1187. doi:10.1109/TPAMI.2003.1227995

PLASTIC. (2009). *Project's web site*. Retrieved on November 16, 2009, from http://www-c.inria.fr/plastic/the-plastic-middleware

Ploennigs, J., Jokisch, O., Kabitzsch, K., Hirschfeld, D.: Generierung angepasster, sprachbasierter Benutzerinterfaces für die Heim- und Gebäudeautomation. *1. Deutscher AAL-Kongress 2008*, VDE-Verlag Berlin-Offenbach. 427-432, 2008.

Pollack, M. E. (2005). Intelligent Technology for an Aging Population: The Use of AI to Assist Elders with Cognitive Impairment. *AI Magazine, 26*(2), 9–24.

Pollack, M. E., Brown, L., Colbry, D., McCarthy, C. E., Orosz, C., & Peintner, B. (2003). Autominder: an Intelligent Cognitive Orthotic System for People with Memory Impairment. *Robotics and Autonomous Systems, 44*(3-4), 273–282. doi:10.1016/S0921-8890(03)00077-0

Pollack, M., Brown, L., Colbry, D., McCarthy, C., Orosz, C., Peintner, B., Ramakrishnan, S., & Tsamardinos, I. (2003). Autominder: An intelligent cognitive orthotic system for people with memory impairment.

Poole, D. (1993). Probabilistic Horn abduction and Bayesian networks. *Artificial Intelligence, 64*(1), 81–129.

Poupart, P. (2005). Exploiting Structure To Efficiently Solve Large Scale Partially Observable Markov Decision Processes. PhD thesis, University of Toronto, Toronto, Canada.

Pousman, Z., Romero, M., Smith, A., & Mateas, M. (2008). Living with Tableau Machine: a longitudinal investigation of a curious domestic intelligence. In *Proceedings of the Tenth International Conference on Ubiquitos Computing* (pp. 370-379). ACM, New York, USA.

Power, M. (June 2004). System for monitoring patients with alzheimer's disease or related dementia.

Preuveneers, D., & Berbers, Y. (2005a). Automated Context-Driven Composition of Pervasive Services to Alleviate Non-Functional Concerns. *International Journal of Computing and Information Sciences, 3*(2), 19–28.

Preuveneers, D., & Berbers, Y. (2005b). Semantic and Syntactic Modeling of Component-Based Services for Context-Aware Pervasive Systems Using OWL-S. *Proceedings of the 1st International Workshop on Managing Context Information in Mobile and Pervasive Environments* (pp. 30-39).

Prud'hommeaux, E., & Seaborne, A. (2006). SPARQL Query Language for RDF, from http://www.w3.org/TR/rdf-sparql-query/.

Psik, T., Matkovic, K., Sainitzer, R., Petta, P., & Szalavári, Z. (2003). The Invisible Person Advanced Interaction Using an Embedded Interface. In Deisinger J. and Kunz A. (eds.): Proceedings of the 7. International Immersive Projection Technologies Workshop and the 9.Eurographics Workshop on Virtual Environments (IPT/EGVE 2003), May 22-23 2003, Zurich Switzerland, pp. 29–37. ACM Press New York USA, 2003. 36, 144

Py, D. (1992). Reconnaissance de plan pour l'aide à la démonstration dans un tuteur intelligent de la géométrie. Thèse de Doctorat, Université de Rennes 1.

Qi, G., Pan, J. Z., & Ji, Q. (2007). Extending description logics with uncertainty reasoning in possibilistic logic. In Proc of the 9th ECSQARU, Hammamet, Tunisia, pp. 828-839.

Quaresma, P., & Lopes, J. G. (1995). Unified Logic Programming Approach to the Abduction of Plans and Intentions in Information-Seeking Dialogues. *Journal of Logic Programming, Elsevier Science Inc.*, *1995*, 103–119.

Quercia, D., Hailes, S., & Capra, L. (2006) B-trust: Bayesian Trust Framework for Pervasive Computing. In Proc. of the 4th International Conference on Trust Management, pag. 298–312, Pisa, Italy, LNCS.

Quin, Y., Bothell, D., & Anderson, J. R. (2007): ACT-R meets fMRI. In Proceedings of LNAI 4845 (pp. 205-222). Berlin, Germany: Springer.

Quinlan, J. R. (1993). *C4.5: Programs For Machine Learning*. San Mateo, CA: Morgan Kaufmann.

R. De Amicis R. R. Stojanovic R., G. Conti G. (2009). Geo-Visual Analytics: Geographical Information Processing and Visual Analytics for Environmental Security. Springer.

Rabiner, L. R. (1989). A Tutorial on Hidden Markov Models and Selected Applications in Speech Recognition. *Proceedings of the IEEE*, *77*(2), 257–286.

Ragni, M., & Wölfl, S. Temporalizing Cardinal Directions: From Constraint Satisfaction to Planning. *Proceedings of the Tenth International Conference on Principles of Knowledge Representation and Reasoning*, 2006.

Ramaswamy R., "Design and Management of Service Process," 1996.

Rammal, A., & Trouilhet, S. (2008). Keeping elderly people at home: A multi-agent classification of monitoring data. In *Proc. of ICOST08*, 145–152.

Ramos, C., Marreiros, G., Santos, R., & Freitas, C. (2009). *Smart Offices and Intelligent Decision Rooms. Handbook of Ambient Intelligence and Smart Environments* (Nakashima, H., Augusto, J., & Aghajan, H., Eds.). Springer.

Ramos, C, Augusto, J.C., Shapiro, D. (2008). *Ambient Intelligence: the next step for AI*, IEEE Intelligent Systems magazine, vol 23, n. 2, pp.15-18.

Randell, D., Cui, Z., & Cohn, A. A Spatial Logic Based on Regions and Connection. *Proceedings of 3rd International Conference on Knowledge Representation and Reasoning*, 1992.

Ranganathan, A., & Campbell, R. H. (2004). Pervasive Autonomic Computing Based on Planning. [Washington, DC: IEEE Computer Society.]. *Proceedings of the IEEE International Conference on Autonomic Computing ICAC*, *04*, 80–87. doi:10.1109/ICAC.2004.1301350

Ranganathan, A., McGrath, R. E., Campbell, R. H., & Mickunas, M. D. (2003). Use of ontologies in a pervasive computing environment. *The Knowledge Engineering Review*, *18*(3), 209–220. doi:10.1017/S0269888904000037

Ranganathan, A., & Campbell, R. H. (2004). Autonomic pervasive computing based on planning. Autonomic Computing, 2004. Proceedings. International Conference on In Proceedings of International Conference on Autonomic Computing. 80-87.

Ranganathan, A., & Campbell, R. H. (2003). A Middleware for Context-Aware Agents in Ubiquitous Computing Environments. In Proc. of ACM/IFIP/USENIX International Middleware Conference, Brazil.

Rank, S. (2004). Affective Acting: An Appraisal-Based Architecture for Agents As Actors. Master's thesis, Institute for Medical Cybernetics and Artificial Intelligence, Medical University of Vienna & Vienna University of Technology, Diplomarbeit, 2004. URL http://www.ofai. at/~stefan.rank/StefanRank-AAAThesis.pdf. 13, 36,145

Rank, S. (2007). Building a computational model of emotion based on parallel processes and resource management. In Proc. of the Doctoral Consortium at ACII2007, Lisbon Portugal EU, Sept. 12-14, 2007.

Rank, S., & Petta, P. (2007). *Basing artificial emotion on process and resource management*. In Paiva A. et al. (eds.): 2nd Int'l. Conf. on Affective Computing and Intelligent Interaction (ACII2007), Lisbon Portugal, Sept.12–14, 2007, LNCS 4738, Springer, 350 361, 2007.

Rao, A. S. (1993) Reactive plan recognition. Technical Report 46, Australian Artificial Intelligence Institute, Carlton, Australia.

Rao, A. S., & Georgeff, M. P. (1991). Modeling rational agents within a BDI-architecture. Second International Conference on Principles of Knowledge Representation and Reasoning (KR'91), 473-484.

Rao, M., & Georgeff, P. (1995). BDI-agents: from theory to practice. In the Proceedings of the First International Conference on Multi-agent Systems (ICMAS'95), 1995.

Raverdy, P., & Issarny, V. "Context-aware Service Discovery in Heterogeneous Networks", IEEE International Symposium on a World of Wireless, Mobile and Multimedia Networks (WoWMoM 2005), June 2005. Pages: 478 – 480.

Raverdy, P.-G., Issarny, V., Chibout, R., & de La Chapelle, A. (2006). A Multi-Protocol Approach to Service Discovery and Access in Pervasive Environments. *Proceedings the 3rd Annual International Conference on Mobile and Ubiquitous Systems: Networks and Services MOBIQUITOUS'06* (pp. 1-9), Washington DC: IEEE Computer Society.

Reason, J. *Human Error*. Cambridge University Press, 1990. Refrigerator: http://www.miele.de/

Reid, D. B. (1979). An algorithm for tracking multiple targets. *IEEE Transactions on Automatic Control, 24*(6), 843–854. doi:10.1109/TAC.1979.1102177

Reif, J., & Wang, H. (1999). Social potential fields: a distributed behavioral control for autonomous robots. *Robotics and Autonomous Systems, 27*(3), 171–194. .doi:10.1016/S0921-8890(99)00004-4

Reisberg, B., Doody, R., Stffler, A., Schmitt, F., Ferris, S., & Mbius, H. (2003). Memantine in moderate-to-severe alzheimer's disease. *The New England Journal of Medicine, 348*(14), 1333–1341.

Reisberg, B., Ferris, S., Anand, R., Leon, M., Schneck, M., Buttinger, C., & Borenstein, J. (1984). Functional staging of dementia of the alzheimer type. *Annals of the New York Academy of Sciences, 435*, 481–483.

Reiter, R. (2001). *Knowledge in action: logical foundations for specifying and implementing dynamical systems.* MIT Press, 2001.

Remagnino, P., Foresti, G., & Ellis, T. (2005). *Ambient Intelligence*. New York: Springer. doi:10.1007/b100343

Remagnino, P., Shihab, A., & Jones, G. (2004). Distributed intelligence for multi-camera visual surveillance. *Pattern Recognition: Special Issue on Agent-Based Computer Vision, 37*(4), 675–689.

Remagnino, P. (2005). *Ambient Intelligence: A Gentle Introduction. Ambient Intelligence: A Novel Paradigm* (pp. 1–14). Springer-Verlag.

Remagnino, P. & Foresti, G.L. (2005). Ambient intelligence: a new multidisciplinary paradigm. *IEEE Transactions on Systems, Man, and Cybernetics - Part A, 35*(1), 1-6.

Ren, X., Berg, A. C., & Malik, J. (2005). Recovering human body configurations using pairwise constraints between parts. Proceedings from ICCV '05: *The International Conference on Computer Vision, 1*, 824-831.

Renz, J., & Nebel, B. (1999). On the Complexity of Qualitative Spatial Reasoning: A Maximal Tractable Fragement of the Region Connection Calculus. *Artificial Intelligence, 108*(1-2). doi:10.1016/S0004-3702(99)00002-8

Reynolds, C. W. (1987). Flocks, herds, and schools: a distributed behavioral model. *Computer Graphics, 21*(4), 25–34. .doi:10.1145/37402.37406

Ribeiro, P. C., & Jose Santos-Victor, J. (2005). Human Activity Recognition from Video: modeling, feature selection and classification architecture, International Workshop on Human Activity Recognition and modeling, (pp.61-70).

Richard, S. (1991). *Lazarus. Emotion and Adaptation.* Oxford University Press.

Richardson, R., Paradiso, J., Leydon, K., & Fernstrom, M. (2004). Z-tiles: Building blocks for modular pressure-sensing. In *Proc. of Conf. on Human Factors in Computing Systems Chi04*, (pp. 1529-1532).

Richerson, P. J., & Boyd, R. (2005). *Not by Genes Alone. How Culture Transformed Human Evolution*. Chicago, London: The University of Chicago Press.

Rickel, J., & Lewis Johnson, W. Steve: an animated pedagogical agent for procedural training in virtual environments. *Animated Interface Agents: making them intelligent*, pages 71–76, 1997.

Riekki, J. (2007). RFID and Smart Spaces. *International Journal of Internet Protocol Technology*, 2(3-4), 143–152. doi:10.1504/IJIPT.2007.016216

Rigole, P., Vandervelpen, C., Luyten, K., Berbers, Y., Vandewoude, Y., & Coninx, K. (2005). A Component-Based Infrastructure for Pervasive User Interaction. *Proceedings of the International Conference on Software Techniques for Embedded and Pervasive Systems* (pp. 1-16).

Rigoll, G., Eickeler, S., & Muller. (2000). Person Tracking in Real-World Scenarios Using Statistical Methods. Proceedings from ICAFGR '00: *The Fourth International Conference on Automatic Face and Gesture Recognition*, 342-347.

Riva, G., & Vatalaro, F. Davide, F., and Alcañiz, M. (eds). (2005). *Ambient Intelligence*. IOS Press, 2001.

Rizzo, M., McGehee, D. V., Dawson, J. D., & Anderson, S. N. (2001). Simulated Car Crashes at Intersections in Drivers With Alzheimer Disease. *Alzheimer Disease and Associated Disorders*, 15(1), 10–20.

Rodenstein, R. (1999). Employing the Periphery: The Window as Interface. In *Extended abstracts on SIGCHI Conference on Human Factors in Computing Systems* (pp. 204-205). ACM, New York, USA.

Röfer, T., Mandel, C., & Laue, T. Controlling an Automated Wheelchair via Joystick/Head-Joystick Supported by Smart Driving Assistance. In IEEE 11th International Conference on Rehabilitation Robotics (ICORR-2009), to appear.

Roger Jianxin Jiao. (2007). *Qianli Xu, and Jun Du. Affective human factors design with ambient intelligence*. HAI.

Rogers, W. A., Meyer, B., Walker, N., & Fisk, A. D. (1998). Functional limitations to daily living tasks in the aged: focus groups analysis. *Human Factors*, 40, 111–125.

Rogers, I. (2006). Moving on from Weiser's Vision of Calm Computing: Engaging UbiComp Experiences. In Dourish, P., & Friday, A. (Eds.), *UbiComp 2006. LNCS 4206*. Springer. doi:10.1007/11853565_24

Röhrig, C., & Kunemund, F. (2007). Mobile Robot Localization using WLAN Signal Strengths. In *Proceedings of the 4th IEEE Workshop on Intelligent Data Acquisition and Advanced Computing Systems: Technology and Applications*, IDAACS 2007, (pp 704-709).

Rolland, C., Kaabi, R. S., & Kraïem, N. (2007). On ISOA: Intentional Services Oriented Architecture. International Conference on Advanced information Systems Engineering (CAISE), 158-172.

Ronfard, R., Schmid, C., & Triggs, B. (2002). Learning to Parse Pictures of People. Proceedings from ECCV '02: *The European Conference on Computer Vision*, 700-714.

Rosalind, W. (1997). *Picard. Affective Computing*. MIT Press.

Rosalind, W. (2001). Picard, E. Vyzas, and J. Healey. Toward machine emotional intelligence: analysis of affective physiological state. *IEEE Transactions on Pattern Analysis and Machine Intelligence*, 23(10).

Roseman, I., Spindel, M., & Jose, P. (1990). Appraisals of emotion-eliciting events: Testing a theory of discrete emotions. *Journal of Personality and Social Psychology*, 59, 899–915. doi:10.1037/0022-3514.59.5.899

Rosemann, M., & Recker, J. (2006). *Context-aware process design: Exploring the extrinsic drivers for process flexibility. T. Latour & M. Petit, 18th international conference on advanced information systems engineering. Proceedings of workshops and doctoral consortium* (pp. 149–158). Luxembourg: Namur University Press.

Rota, N. A. (1998). *Système adaptatif pour le traitement de séquences d'images pour le suivi de personnes*. Master's thesis, University of Paris IV, DEA IARFA Laboratoire d'informatique, INRIA Sophia-Antipolis.

Rouvoy, R., Eliassen, F., Floch, J., Hallsteinsen, S., & Stav, E. (2008). Composing Components and Services using a Planning-based Adaptation Middleware. In Pautasso, C., & Tanter, É. (Eds.), *SC 2008. LNCS* (*Vol. 4954*, pp. 52–67). Heidelberg: Springer.

Rouvoy, R., Barone, P., Ding, Y., Eliassen, F., Hallsteinsen, S., Lorenzo, J., et al. (2009). MUSIC: Middleware Support for Self-Adaptation in Ubiquitous and Service-Oriented Environments. In Cheng, B. H. et al (Eds.) *Software Engineering For Self-Adaptive Systems* (pp. 164-182), Lecture Notes In Computer Science, Vol. 5525. Berlin, Heidelberg: Springer-Verlag.

Rouvoy, R., Eliassen, F., Floch, J., Hallsteinsen, S., & Stav, E. (2008). Composing Components and Services Using a Planning-Based Adaptation Middleware. *Proceedings of the 7th Symposium on Software Composition SC'08* (pp. 52-67).

Roy, D. (1999). Learning from Sights and Sounds: A Computational Model. Ph.D. In *Media Arts and Sciences*. MIT.

Roy, P. C., Bouchard, B., Bouzouane, A., & Giroux, S. (2009b): Ambient Activity Recognition: A Possibilistic Approach, Proc. of the IADIS International Conference on Intelligent Systems and Agents 2009 (ISA'09), Algarve, Portugal, 21 - 23 June 2009, pp. 1-5.

Roy, P., Bouchard, B., Bouzouane, A., & Giroux, S. (2007). A hybrid plan recognition model for Alzheimer's patients: interleaved-erroneous dilemma. IEEE/WIC/ACM International Conference on Intelligent Agent Technology, 2007, 131- 137.

Rushby, J., Crow, J., & Palmer, J. An Automated Method to Detect Potential Mod Confusions. *Proceedings of the 18th IEEE Digital Avionics Systems Conference*, 1999. Security: http://www.miditec.de/ SFB/TR8 Spatial Cognition: http://www.sfbtr8.uni-bremen.de/

Russell, S., & Norvig, P. (2003). *Artificial Intelligence: A Modern Approach* (2nd ed.). Englewood Cliffs, NJ: Prentice-Hall.

Russell, St., & Norvig, P. (2010). *Artificial Intelligence: A Modern Approach* (3rd ed.). Upper Saddle River, N.J.: Prentice Hall.

Russell, D. M., Streitz, N. A., & Winograd, T. (2005). Building Disappearing Computers. *Communications of the ACM, 48*(3), 42–48. doi:10.1145/1047671.1047702

Russo, J., Sukojo, A., et al. (2004) SmartWave – Intelligent Meal Preparation System to Help Older People Live Independently. Proc. 2nd International Conference on Smart Homes and Health Telematics (ICOST 2004), Vol. 14, pp. 122–135.

Rusu, R. B., Bandouch, J., Marton, Z. C., Blodow, N., & Beetz, M. (2005). Action recognition in intelligent environments using point cloud features extracted from silhouette sequences. *Proc. of the Int. Conf. on Smart Objects and Ambient Intelligence (sOc-EUSAI)*.

Saffiotti, A., & Broxvall, M. (2005). PEIS ecologies: Ambient intelligence meets autonomous robotics. *Proc. of the Int. Conf. on Smart Objects and Ambient Intelligence (sOc-EUSAI)*.

Saffiotti, A., Broxvall, M., Gritti, M., LeBlanc, K., Lundh, R., Rashid, J., et al. (2008). The PEIS-Ecology Project: Vision and Results. *Proc of the IEEE/RSJ Int. Conf. on Intelligent Robots and Systems (IROS)*. Nice, France.

Sahin, E. (2005). Swarm robotics: from sources of inspiration to domains of application. In E. Sahin & W. M. Spears (Eds.), Swarm Robotics (LNCS), *3342* (pp. 10–20). New York: Springer. doi:10.1007/978-3-540-30552-1_2

Saif, U., Pham, H., Paluska, J. M., Waterman, J., Terman, C., & Ward, S. (2003). A Case for Goal-oriented Programming Semantics. *Proceedings of the System Support for Ubiquitous Computing Workshop at Ubicomp'03*.

Salber, D., Dey, A. K., & Abowd, G. D. (1999). The Context Toolkit: Aiding the development of context-enabled applications. In Proc. of CHI'99 (pp. 434–441).

Salvucci, D. D. (2006). Modeling Driver Behavior in a Cognitive Architecture. *Human Factors, 48*(2), 362–380.

Salvucci, D. D., & Gray, R. (2004). A Two-Point Visual Control Model of Steering. *Perception, 33*, 1233–1248.

Salvucci, D. D. (2007). Integrated Models of Driver Behavior. In Gray, W. D. (Ed.), *Integrated Models of Cognitive Systems* (pp. 356–367). Oxford University Press.

Sanchez, D., Tentori, M., & Favela, J. (2008). Activity recognition for the smart hospital. *IEEE Intelligent Systems*, (March/April): 50–57.

Sánchez, I., Cortés, M., & Riekki, J. (2007). Controlling Multimedia Players using NFC Enabled mobile phones. *Proceedings of the 6th International Conference on Mobile and Ubiquitous Multimedia MUM'07*, Vol. 284. (pp.118-124). New York, NY: ACM.

Sanfeliu, A., Hagita, N., & Saffiotti, A. (2008). Special issue on Network robot systems. *Robotics and Autonomous Systems*, *56*(10).

Santos, P., Stork, A., Gierlinger, T., Pagani, A., Paloc, C., Barandarian, I., et al. (2007). IMPROVE: An innovative application for collaborative mobile mixed reality design review. In International Journal on Interactive Design and Manufacturing, Springer Paris. ISSN: 1955-2513 (Print) 1955-2505 (Online).

Sarter, N., Amalberti, R. R., & Amalberti, R. (2000). *Cognitive Engineering in the Aviation Domain*. Lawrence Erlbaum Assoc. Inc.

Satyanarayanan, M. (2001). Pervasive computing: vision and challenges. [see also IEEE Wireless Communications]. *Personal Communications, IEEE*, *8*(4), 10–17. doi:10.1109/98.943998

Scarantino, A. (2003). Affordances explained. *Philosophy of Science*, *70*, 949–961. doi:10.1086/377380

Scharp, M., Halme, M., & Jonuschat, H. Nachhaltige Dienstleistungen der Wohnungswirtschaft, Arbeitsbericht 9 ed. Berlin: IZT - Institut für Zukunftsstudien und Technologiebewertung, 2004.

Scheer, R. (2004). The 'Mental State' Theory of Intentions. *Philosophy (London, England)*, *79*(1), 121–131. doi:10.1017/S0031819104000087

Scheer, A.-W., & Thomas, O. "Geschäftsprozessmodellierung mit der ereignisgesteuerten Prozesskette,", 8 ed 2005, pp. 1069-1079.

Scherer, K. R. (2001). *Appraisal Processes in Emotion: Theory, Methods, Research, chapter Appraisal Considered as a Process of Multilevel Sequential Checking* (pp. 92–120). New York: Oxford University Press.

Scheuing, E. E., & Johnson, E. M. (1989). A proposed model for new service development. *Journal of Services Marketing*, *3*(2), 25–34. doi:10.1108/EUM0000000002484

Schiller, C. A., Gan, Y. M., et al. (2003) Exploring New Ways of Enabling e-Government Services for the Smart Home with Speech Interaction. In: Mokhtari M (ed) Independent Living for Persons with Disabilities and Elderly People, ICOST'2003 1st International Conference on Smart Homes and Health Telematics, Assistive Technology Research Series 12, IOS Press, pp.151–158.

Schmidt, A. (2000). Implicit human computer interaction though context. *Personal Technologies*, *4*(2-3), 191–199. doi:10.1007/BF01324126

Schmidt, C., Sridharan, N., & Goodson, J. (1978). The plan recognition problem: an intersection of psychology and artificial intelligence. *Artificial Intelligence*, 45–83.

Schmidt, B. (2002), *How to give Agents a Personality*, In: Proc. 3rd Workshop on Agent- Based Simulation, Passau, Germany, April 7-9, 2002. SCS-Europe, Ghent, Belgium, pp. 13-17.

Schneider, K., Wagner, D., & Behrens, H. "Vorgehensmodelle zum Service Engineering," in Service Engineering - Entwicklung und Gestaltung innovativer Dienstleistungen. Bullinger H.-J. and Scheer A.-W., Eds. Berlin, Heidelberg, New York, Barcelona, Hongkong, London, Mailand, Paris, Tokio: Springer, 2003, pp. 117-141.

Schüller, A. Dienstleistungsmärkte in der Bundesrepublik Deutschland. Köln: 1967.

Scmidt, C., Sridharan, N., & Goodson, J. (1978). The plan recognition problem: an intersection of psychology and artificial intelligence. *Artificial Intelligence*, *11*, 45–83.

Sebe, N., Cohen, I., Gevers, T., & Huang, T. S. Multimodal approaches for emotion recognition: a survey. In *SPIE: Internet Imaging*, pages 56–67, 2004.

Sejnowski, T. J. (1986). Higher-order Boltzmann machines. *AIP Conference Proceedings*, *151*(1), 398–403. doi:10.1063/1.36246

Sekmen, A. S., Wilkes, M., & Kawamura, K. (2002). An application of passive human-robot interaction: human tracking based on attention distraction. *IEEE Transactions on Systems, Man, and Cybernetics. Part A, Systems and Humans*, *32*(2), 248–259. doi:10.1109/TSMCA.2002.1021112

Sellen, A., Fogg, A., Hodges, S., & Wood, K. (2007). Do life-logging technologies support memory for the past? an experimental study using sensecam. In *Proc. of CHI07*.

Sgorbissa, A., & Zaccaria, R. (2004). The artificial ecosystem: a distributed approach to service robotics. *Proc. of the Int. Conf. on Robotics and Automation (ICRA)*.

Shanahan, M. P. (1996). Robotics and the Common Sense Informatic Situation. *Proc. of the European Conf. on Artificial Intelligence (ECAI)*.

SharedSpace. http://www.sharedspace.eu/en/home/ (visited 27th February, 2010)Spirtes, P., Glymour, C., and Scheines, R., 2001 (2nd ed.), Causation, Prediction, and Search, Cambridge, Mass: MIT Press

Shaw, M., & Garlan, D. (1996). *Software Architecture: Perspectives on an Emerging Discipline*. Prentice Hall.

Shell, J. S., Selker, T., & Vertegaal, R. (2003). Interacting with groups of computers. *Communications of the ACM*, *46*, 40–46. doi:10.1145/636772.636796

Shen, J., & Castan, S. (1992). An optimal linear operator for step edge detection. Proceedings from CVGIP '92. *Computer Vision Graphics and Image Processing, 54*(2), 13–17.

Shi, H., & Krieg-Brückner, B. (2008). Modelling Route Instructions for Robust Human-Robot Interaction on Navigation Tasks. *International Journal of Software and Informatics, 2*(1).

Shi, H., & Tenbrink, T. Telling Rolland Where to Go: HRI Dialogues on Route Navigation (Chapter 13). *Spatial Language and Dialogue*, Oxford University Press, 2009. Sliding doors: http://www.raumplus.de/

Shi, H., Mandel, C., & Ross, R. J. Interpreting Route Instructions as Qualitative Spatial Actions. *Spatial Cognition V: Reasoning, Action, Interaction: International Conference Spatial Cognition 2006*. LNAI 4387, 2007.

Shi, H., Ross, R., Bateman, J.: Formalising Control in Robust Spoken Dialogue Systems. *Proceedings of Software Engineering and Formal Methods 2005*, IEEE, 2005.

Shimizu, M., Mori, T., & Ishiguro, A. (2006). A development of a modular robot that enables adaptive reconfiguration. In *Proc. IEEE/RSJ Int. Conf. Intelligent Robots and Systems*, (pp.174-179).

Shostack, G. L. (1982). How to Design a Service. *European Journal of Marketing, 16*(January/February), 49–63. doi:10.1108/EUM0000000004799

Shostack, G. L. "Designing Services that deliver,", 62 ed 1984, pp. 133-139.

Shucker, B., Murphey, T. D., & Bennett, J. K. (2008). Convergence-preserving switching for topology-dependent decentralized systems. *IEEE Transactions on Robotics, 24*(6), 1405–1415. .doi:10.1109/TRO.2008.2007940

Sidenbladh, H., & Black, M. (2003). Learning the statistics of people in images and video. *International Journal of Computer Vision, 54*(13), 183–209.

Sidenbladh, H., Black, M. J., & Fleet, D. J. (2000). Stochastic tracking of 3D human figures using 2D image motion. Proceedings from ECCV '00: *The European Conference on Computer Vision*, 702-718.

Siebel, N. T. (2003). *Design and Implementation of People Tracking Algorithms for Visual Surveillance Applications*. Doctoral dissertation, University of Reading, Reading, UK.

Siebert, J., Cao, J., Zhou, Y., Wang, M., & Raychoudhury, V. (2007). Universal Adaptor: A Novel Approach to Supporting Multi-Protocol Service Discovery in Pervasive Computing. In Kuo T.-W. et al (Eds.), *Proceedings of the International Conference on Embedded and Ubiquitous Computing EUC'07* (pp. 683-693), Lecture Notes In Computer Science, Vol. 4808. Berlin, Heidelberg: Springer-Verlag.

Siegel J., "BPMN und BPDM - Die OMG-Spezifikationen zur Modellierung von Geschäftsprozessen," 2007.

Siegel J., "BPMN unter der Lupe," 2008.

Sigal, L., Isard, M., Sigelman, B. H., & Black, M. (2004). Attractive people: Assembling loose-limbed models using non-parametric belief propagation. *Journal of Advances in Neural Information Processing Systems, 16*, 1539–1546.

Sigal, L., Bhatia, S., Roth, S., Black, M. J., & Isard, M. (2004). Tracking loose-limbed people. Proceedings from ICCVPR '04: *The International Conference on Computer Vision and Pattern Recognition, 1*, 421-428.

Siio, I., Rowan, J., Mima, N., & Mynatt, E. (2003). Augmented everyday things. In *Proc. of the Canadian annual conference Graphics Interface*. Digital Decor.

Šimún, M., Andrejko, A., & Bieliková, M. (2008) Maintenance of Learner's Characteristics by Spreading a Change. In M. Kendall & B. Samways (Eds.), IFIP International Federation for Information Processing, World Computer Congress, Volume 281: *Learning to Live in the Knowledge Society* (pp. 223-226). Boston: Springer.

Sindlar, M. P., Dastani, M. M., Dignum, F., & Meyer, J.-J. Ch. (2008). *Mental state abduction of BDI-based agents. Declarative Agent Languages and Technologies*. Estoril, Portugal: DALT.

Sixsmith, A., Gibson, G., Orpwood, R., & Torrington, J. (2007). Developing a technology wish-list to enhance the quality of life of people with dementia. *Gerontechnology (Valkenswaard)*, *6*, 2–19.

Slotine, J. E., & Li, W. (1991). *Applied nonlinear control*. Prentice-Hall.

Sminchisescu, C., & Telea, A. (2002). Human pose estimation from silhouettes - A consistent approach using distance levels sets. Proceedings from WSCG '02: *International Conference on Computer Graphics, Visualization and Computer Vision*, 413-420.

Sminchisescu, C., & Triggs, B. (2001). Covariance scaled sampling for monocular 3D body tracking. Proceedings from IEEE CCVPR '01: *The IEEE Conference on Computer Vision and Pattern Recognition*, *1*, 447-454.

Smith, C. A., & Lazarus, R. S. (1990). Emotion and adaptation. In Pervin, L. A. (Ed.), *Handbook of personality: Theory and research* (pp. 609–637). New York: Guilford.

Smith, M., Welty, C., & McGuinness, D. (2003). Web ontology language (OWL) guide Version 1, from http://www.w3.org/ TR/ owl-guide/.

Snijders, F. (2005). *Ambient Intelligence Technology: An Overview. Ambient Intelligence* (pp. 255–270). Springer.

Sobottaka, K., & Pitas, I. (1998). Novel method for Automatic Face Segmentation, Facial Feature Extraction and Tracking. *Signal Processing Image Communication*, *12*(3), 263–281. doi:10.1016/S0923-5965(97)00042-8

Soede, M. (2003) The Home and Care Technology for Chronically Ill and Disabled Persons. In: Mokhtari M (ed) Independent Living for Persons with Disabilities and Elderly People, ICOST'2003 1st International Conference on Smart Homes and Health Telematics, Assistive Technology Research Series 12, IOS Press, pp. 3–9.

Sogo, T., Ishiguro, H., & Trivedi, M. M. (2000) *Real-Time Target Localization and Tracking by N-Ocular Stereo*. IEEE Workshop on Omnidirectional Vision (OMNIVIS'00), p.153

Solso, R. L. (1993). *Cognition and the Visual Arts*. The MIT Press.

Son, Y., Ku, T., Park, J., & Moon, K. (June 2008). Inference-based home network error handling system and method.

Song, S.-K., Jang, J., & Park, S. (2008) An Efficient Method for Activity Recognition of the Elderly Using Tilt Signals of Tri-axial Acceleration Sensor. Proc. 6th International Conference on Smart Homes and Health Telematics (ICOST 2008), LNCS 5120, Springer, pp. 99–104.

Song, Y. J., Tobagus, W., Leong, D. Y., Johanson, B., & Fox, A. *iSecurity: A Security Framework for Interactive Workspaces*. Tech Report, Stanford University, 2004.

Soriano, E., Ballesteros, F. J., & Guardiola, G. (2007a). Human-to-Human Authorization for Resource Sharing in SHAD: Roles and Protocols. *Elsevier Pervasive and Mobile Computing.*, *3*(6), 607–738.

Soriano, E., Ballesteros, F. J., & Guardiola, G. *SHAD: A Human Centered Security Architecture for the Plan B Operating System*. IEEE Intl. Conf. on Pervasive Computing and Communications 2007b.

Sousa, J. P., Poladian, V., Garlan, D., Schmerl, B., & Shaw, M. (2006). *Task-Based Adaptation for Ubiquitous Computing, (Tech. Rep.)*. Pittsburgh, PA: Carnegie Mellon University, School of Computer Science.

Sousa, J. P., Schmerl, B., Steenkiste, P., & Garlan, D. (2008a). Activity-Oriented Computing. In Mostefaoui, S., Maamar, Z., & Giaglis, G. M. (Eds.), *Advances in Ubiquitous Computing: Future Paradigms and Directions* (pp. 280–315). Hershey, PA: IGI Publishing. doi:10.4018/978-1-59904-840-6.ch011

Sousa, J. P., Schmerl, B., Steenkiste, P., & Garlan, D. (2008b). uDesign: End-User Design Applied to Monitoring and Control Applications for Smart Spaces. *Proceedings of the 7ᵗʰ IEEE/IFIP Conference on Software Architecture* (pp. 72-80). Washington, DC: IEEE Computer Society.

Spears, W., Spears, D., Hamann, J., & Heil, R. (2004). Distributed, physics-based control of swarms of vehicles. *Autonomous Robots*, *17*(2–3), 137–162. .doi:10.1023/B:AURO.0000033970.96785.f2

Spears, W., & Gordon, D. (1999). Using artificial physics to control agents. In *Proc. IEEE Int. Conf. Information, Intelligence, and Systems*, (pp.281-288).

Sperandio, P., Bublitz, S., & Fernandes, J. (2008). Wireless Device Discovery and Testing Environment. Technical Report D5.9, Hydra Consortium. IST 2005-034891.

Stacy Marsella and Jonathan Gratch. Modeling coping behavior in virtual humans: don't worry, be happy. In *AAMAS-2003,* pages 313–320. ACM, 2003.

Staller, A., & Petta, P. (2001). Introducing *Emotions into the Computational Study of Social Norms: A First Evaluation. Journal of Artificial Societies and Social Simulation,* 4*(1).

Staller, A., & Petta, P. (1998). *Towards a Tractable Appraisal Based Architecture for Situated Cognizers.* In Lola Cañamero, Chisato Numaoka, and Paolo Petta (eds.): Workshop Notes, 5th International Conference of the Society for Adaptive Behaviour (SAB98), Zürich Switzerland, August 21, pp. 56–61, 1998. 36, 149

Stasko, J., Miller, T., Pousman, Z., Plaue, C., & Ullah, O. (2004). Personalized Peripheral Information Awareness through Informative Art. In N. Davis, E. Mynatt, & I. Siio (Eds.), *the Sixth International Conference on Ubiquitous Computing* (pp. 18-35). Springer-Verlag Berlin Heidelberg.

Stefanov, D. H., Bien, Z., & Bang, W. (2004). The Smart house for older persons and persons with physical disabilities: structure, technology arrangements, and perspectives. *IEEE Transactions on Neural Systems and Rehabilitation Engineering*, *12*, 228–250. doi:10.1109/TNSRE.2004.828423

Steinhage, A., & Lauterbach, C. (2008). *(n.d.). Monitoring Movement Behaviour by Means of a Large-Area Proximity Sensor Array in the Floor* (pp. 15–27). BMI.

Steinhage, A., & Schöner, G. (1997). Self-calibration based on invariant view recognition: Dynamic approach to navigation. *Robotics and Autonomous Systems*, *20*, 133–156. doi:10.1016/S0921-8890(96)00072-3

Stemberg, R. J. (1985). *Beyond IQ: A Triarchic Theory of Human Intelligence.* New York: Cambridge Unviersity Press.

Steunebrink, B., Dastani, M., & Meyer, J.-J. A logic of emotions for intelligent agents. In R.C. Holte and A.E. Howe, editors, *Proc. AAAI-07*, pages 142–147, 2007. AAAI Press.

Stocker, S. (1999). Models for tuna school formation. *Mathematical Biosciences*, *156*, 167–190. .doi:10.1016/S0025-5564(98)10065-2

Stojmenovic, I., & Wu, J. (2004). Broadcasting and activity scheduling in ad hoc networks. In S. Basagini, M. Conti, S. Giordano, & I. Stojmenovic (Eds.), *Mobile ad hoc networking* (pp. 205–229). IEEE press. doi:10.1002/0471656895.ch7

Stormont, D. P., Abdallah, C. T., Byrne, R. H., & Heileman, G. L. (1998). *A Survey of Mobile Robot Navigation Methods Using Ultrasonic Sensors.* In *Proceedings of the Third ASCE Specialty Conference on Robotics for Challenging Environments.* Albuquerque, New Mexico, April 26-30, 1998, (pp. 244-250).

Strang, T., & Linnhoff-Popien, L. (2004). A Context Modeling Survey. In Proc. of the First International Workshop on Advanced Context Modeling, Reasoning and Management (pp. 33-40).

Streitz, N. A., Rocker, C., Prante, T., Alphen, D. V., Stenzel, R., & Magerkurth, C. (2005). Designing Smart Artifacts for Smart Environments. *IEEE Computer*, *38*(3), 41–49.

Sudderth, E. B., Ihler, A. T., Freeman, W. T., & Willsky, A. S. Nonparametric belief propagation. Proceedings from ICCVPR '03: *The International Conference on Computer Vision and Pattern Recognition*, 1, 605-612.

Sukthankar, G., & Sycara, K. (2008). Robust and efficient plan recognition for dynamic multi-agent teams (Short Paper). In Proceedings of 7th International Conference on Autonomous Agents and Multi-agent Systems (AAMAS 2008), International Foundation for Autonomous Agents and Multi-agent Systems, May 2008.

Sukthankar, G., & Sycara, K. (2008). Hypothesis Pruning and Ranking for Large Plan Recognition Problems. In Proceedings of the Twenty-Third AAAI Conference on Artificial Intelligence (AAAI-08), June 2008.

Susskinda, J. M., Littlewortb, G., Bartlettb, M. S., Movellanb, J., & Anderson, A. K. (2007). Human and computer recognition of facial expressions of emotion. *Neuropsychologia, 45*, 152–162. doi:10.1016/j.neuropsychologia.2006.05.001

Sutton, C., & McCallum, A. (2006). An introduction to Conditional Random Fields for relational learning. In Getoor, L., & Taskar, B. (Eds.), *Introduction to Statistical Relational Learning* (pp. 93–128). Boston, MA: MIT Press.

Suzić, R., & Svenson, P. (2006). Capabilities-based plan recognition. In Proceedings of the 9th International Conference on Information Fusion, Italy, July 2006.

Suzuki, I., & Yamashita, M. (1999). Distributed anonymous mobile robots: formation of geometric patterns. *SIAM Journal on Computing, 28*(4), 1347–1363. .doi:10.1137/S009753979628292X

Swartout, W., Gratch, J., Hill, R. W., Hovy, E., Marsella, S., Rickel, J., & Traum, D. R. (2004). Toward virtual humans. In *Working notes of the AAAI Fall symposium on Achieving Human-Level Intelligence through Integrated Systems and Research*, Crystal City, VA, USA.

Tabachneck-Schijf, H. J. M., Leonardo, A. M., & Simon, H. A. (1997). CaMeRa: A computational model of multiple representations. *Cognitive Science, 21*, 305–350. doi:10.1207/s15516709cog2103_3

Takemoto, M., Oh-ishi, T., Iwata, T., Yamato, Y., Tanaka, Y., & Shinno, K. (2004). A Service-Composition and Service-Emergence Framework for Ubiquitous-Computing Environments. [Washington DC: IEEE Computer Society.]. *Proceedings of International Symposium on Applications and the Internet, SAINT, 04-W*, 313–318.

Talmy, L. (1983). *How Language Structures Space. Spatial Orientation: Theory.* Research and Application.

Tamura, T., & Togawa, T. (1998). Fully automated health monitoring system in the home. *Medical Engineering & Physics, 20*, 573–579. doi:10.1016/S1350-4533(98)00064-2

Tamura, T., Masuda, Y., et al. (2004) Application of Mobile Phone Technology in the Elderly – A Simple Telecare System for Home Rehabilitation. Proc. 2nd International Conference on Smart Homes and Health Telematics (ICOST 2004), Vol. 14, pp. 278–282.

Tangney, J. P. (1999). The self-conscious emotions: shame, guilt, embarrassment and pride. In Dalgleish, T., & Power, M. (Eds.), *Handbook of cognition and emotion*. John Wiley & Sons.

Tango, F., Aras, R., & Pietquin, O. *Learning Optimal Control Strategies from Interactions with a PADAS*, p.119-127, in: Human Modelling in Assisted Transportation: Models, Tools and Risk Methods, Cacciabue Pietro Carlo, Magnus Hjälmdahl, Andreas Luedtke, and Costanza Riccioli (eds), Heidelberg: Springer, 2011, **ISBN-13:** 978-8847018204

Tanner, H., Pappas, G., & Kumar, V. (2004). Leader-to-formation stability. *IEEE Transactions on Robotics and Automation, 20*(3), 443–455. .doi:10.1109/TRA.2004.825275

Tapia, E. M., Intille, S. S., & Larson, K. Activity recognition in the home using simple and ubiquitous sensors. In *Proc. PERVASIVE-04*, pages 158–175, Linz/Vienna, Austria, 2004.

Tapscott, D. (1998). *Growing up Digital: The Rise of the Net-Generation*. New York: McGraw-Hill.

Teng, E., Hasegawa, K., Homma, A., Imai, Y., Larson, E., & Graves, A. (1994). The cognitive abilities screening instrument (casi): a practical test for cross-cultural epidemiological studies of dementia. *International Psychogeriatrics, 6*(1), 45–58.

Terzic, K., Hotz, L., & Neumann, B. (2007). Division of Work During Behaviour Recognition. In Gottfried, B. (Ed.), *BMI'07 (Vol. 296*, pp. 144–159). CEUR.

Thang, V., Qiu, Q., Aung, A. P. W., & Biswas, J. (2006). *Applications of Ultrasound Sensors in a Smart Environment*. Pervasive and Mobile Computing Journal.

Theeuwes, J. (1992). Perceptual selectivity for color and form. [Fulltext]. *Perception & Psychophysics, 51,* 599–606. doi:10.3758/BF03211656

Thirde, D., Borg, M., Ferryman, J., Fusier, F., Valentin, V., Brémond, F., & Thonnat, M. A Real-Time Scene Understanding System for Airport Apron Monitoring. *IEEE International Conference on Computer Vision Systems (ICVS 2006)* in New York City, USA, 2006.

Thomas, N. J. T. (1999). Are theories of imagery theories of imagination? An active perception approach to conscious mental content. *Cognitive Science, 23,* 207–245. doi:10.1207/s15516709cog2302_3

Thomas, J. J., & Cook, K. A. (Eds.). (2005). *Illuminating the Path: The Research and Development Agenda for Visual Analytics.* IEEE CS Press.

Thrun, S., Burgard, W., & Fox, D. (2005). *Probabilistic Robotics.* Cambridge, Mass.: MIT Press.

Tolstikov, A., Biswas, J., Tham, C., & Yap, P. (2008). Eating activity primitives detection a step towards adl recognition. In *Proc. of HealthCom08,* 35–41.

Tolstikov, A., Tham, C. K., Wendong, X., & Biswas, J. (2007). Information quality mapping in resource-constrained multi-modal data fusion system over wireless sensor network with losses, *Proceedings of the International Conference on Information, Communications and Signal Processing,* December 2007.

Tomasello, M., Carpenter, M., Call, J., Behne, T., & Moll, H. (2005). Understanding and Sharing Intentions: The Origins of Cultural Cognition. *The Behavioral and Brain Sciences, 28*(5), 675–691. doi:10.1017/S0140525X05000129

Tompros, S., Mouratidis, N., Caragiozidis, M., Hrasnica, H., & Gavras, A. (2008). A pervasive network architecture featuring intelligent energy management of households. *Proceedings of the 1st int. Conf. on Pervasive Technologies Related To Assistive Environments* (pp. 1-6). New York: ACM.

TORCS. http://torcs.sourceforge.net/ (visited 27th February, 2010)

Tränkler, H.-R. (2001). In Tränkler, H.-R., & Schneider, F. (Eds.), *"Zukunftsmarkt intelligentes Haus," in Das intelligente Haus - Wohnen und Arbeiten mit zukunftsweisender Technik* (pp. 17–34). Pflaum Verlag.

Treisman, A., & Gelade, G. (1980). A feature integration theory of attention. *Cognitive Psychology, 12,* 97–136. doi:10.1016/0010-0285(80)90005-5

Treuil, J.-P., Drogoul, A., & Zucker, J.-D. (2008). *Modélisation et simulation à base d'agents: Approches particulaires, modèles à base d'agents, de la mise en pratique aux questions théoriques.* Dunod.

Treur, J. (2008). On Human Aspects in Ambient Intelligence. In: *Proceedings of the First International Workshop on Human Aspects in Ambient Intelligence.* Published in: M. Mühlhäuser, A. Ferscha, and E. Aitenbichler (eds.), *Constructing Ambient Intelligence: AmI-07 Workshops Proceedings.* Communications in Computer and Information Science (CCIS), vol. 11, Springer Verlag, pp. 262-267.

Treur, J. On human aspects in ambient intelligence. In *Proceedings of First International Workshop on Human Aspects in Ambient Intelligence (HAI),* pages 5–10, 2007.

Turner, K., Oberdorf, V., Raja, G., Maestas, G., & Epstein, H. (July 2003). System and method for facilitating the care of an individual and dissemination of information.

Tversky, B. and Lee, P.U.: How Space Structures Language. *Spatial Cognition: An interdisciplinary Approach to Representation and Processing of Spatial Knowledge.* LNAI 1404, 1998.

UKARI. (n.d.). *Website of the UKARI project.* Retrieved from http://open-ukari.nict.go.jp/Ukari-Project-e.html

Ullberg, J., Loutfi, A., & Pecora, F. (2009). Towards Continuous Activity Monitoring with Temporal Constraints. *Proc. of the 4th Workshop on Planning and Plan Execution for Real-World Systems at ICAPS09.*

Updike, J. (1996). *The Rabbit is Rich.* Ballantine Books.

Urban, C. (2000). PECS – A Reference Model for the Simulation of Multi-Agent Systems. In Suleiman, R., Troitzsch, K. G., & Gilbert, G. N. (Eds.), *Tools and Techniques for Social Science Simulation.* Heidelberg: Physica Verlag.

Uschold, M., & Gruninger, M. (1996). Ontologies: Principles, Methods and Applications. *The Knowledge Engineering Review, 11*(2), 93–136. doi:10.1017/S0269888900007797

V. Jakkula and D. Cook. Anomaly detection using temporal data mining in a smart home environment. *Methods of Information in Medicine*, 2008.

Valcour, V., Masaki, K., Curb, J., & Blanchette, P. (2000). The detection of dementia in the primary care setting. *Archives of Internal Medicine, 160*, 2964–2968.

Valera, M., & Velastin, S. (2005). Intelligent distributed surveillance systems: a review. Proceedings from IEE-VISP '05: *Vision. Image and Signal Processing, 152*(2), 192–204. doi:10.1049/ip-vis:20041147

van Breemen, A. iCat: Experimenting with animabotics. In *AISB*, pages 27–32, University of Hertfordshire, Hatfield, UK, April 2005.

Vanderheiden, G., & Zimmermann, G. Non-homogenous Network, Control Hub and Smart controller (NCS) Approach to Incremental Smart Homes. In: Proceedings of *HCI International 2007*. LNCS 4555, 2007. Vescovo, F. et al.: http://www.progettarepertutti.org/ progettazione/ casa-agevole- fondazione

Vasilakos, A., & Pedrycz, W. (2006). *Ambient Intelligence, Wireless Networking, and Ubiquitous Computing*, (pp. 12–45). Norwood, MA: Artech House, Inc.

Vastenburg, M., Keyson, D., & de Ridder, H. (2007). Measuring User Experiences of Prototypical Autonomous Products in a Simulated Home Environment. *International Journal of Human-Computer Interaction*, (2): 998–1007.

Veen, W., & Jacobs, F. (2005). *Leren van Jongeren: Een literatuuronderzoek naar nieuwe geletterdheid*. Utrecht: Stichting SURF.

Veen, W., & Vrakking, B. (2006). *Homo Zappiens: Growing Up in a Digital Age*. London: Network Continuum Education.

Veen, W., and van, Staalduinen, J.P. (2009) Homo Zappiens and its Impact on Learning in Higher Education. *IADIS International Conference e-Learning 2009*, Algarve, Portugal

Vehvilainen, L., Zielstorff, R., Gertman, P., Tzeng, M., & Estey, G. (2002). Alzheimer's caregiver internet support system (aciss): Evaluating the feasibility and effectiveness of supporting family caregivers virtually. In *American Medical Informatics Association 2002 Symposium*.

Velasquez, J., & Maes, P. (1997) *Cathexis: A Computational Model of Emotions*. Proceedings of the First International Conference on Autonomous Agents, pp. 518-519.

Verbeek, P.-P. (2008). Cyborg intentionality: Rethinking the phenomenology of human–technology relations. *Phenomenology and the Cognitive Sciences, 7*, 387–395. doi:10.1007/s11097-008-9099-x

Verbeek, P.-P. (2009). The moral relevance of technological artifacts. In Sollie, P., & Duwell, M. (Eds.), *Evaluating New Technologies* (pp. 63–77). Berlin: Springer. doi:10.1007/978-90-481-2229-5_6

Verghese, J., Lipton, R., Katz, M., Hall, C., Derby, C., & Kuslansky, G. (2003). Leisure activities and the risk of dementia in the elderly. *The New England Journal of Medicine, 348*, 2508–2616.

Verghese, P. (2001). Visual search and attention: A signal detection theory approach. *Neuron, 31*, 523-535(13).

Vicente, K. J. (2003). Beyond the lens model and direct perception: toward a broader ecological psychology. *Ecological Psychology, 15*(3), 241–267. doi:10.1207/S15326969ECO1503_4

Victorino, A. C., & Rives, P. (2004). *Bayesian segmentation of laser range scan for indoor navigation*, Intelligent Robots and Systems, 2004. (IROS 2004). In *Proceedings. 2004 IEEE/RSJ* (pp. 2731- 2736 vol.3).

Vilain, M., & Kautz, H. Constraint propagation algorithms for temporal reasoning. In *Proc. AAAI-86*, pages 377–382, Philadelphia, Pennsylvania, 1986.

Vurgun, S., Philipose, M., & Pavel, M. (2007). A statistical reasoning system for medication prompting. In *Proc. of Ubicomp07*, 1–18.

Waern, A., & Stenborg, O. (1995). Recognizing the plans of a replanning user. In Proc. of the IJCAI-95 Workshop on The Next Generation of Plan Recognition Systems, Montréal, Canada, pp. 113–118.

Wahlster, W. Towards Symmetric Multimodality: Fusion and Fission of Speech, Gesture, and Facial expression. *Proceedings of the 2003 Meeting of the German Conference on Artificial Intelligence*, 2003.

Wallgrün, J. O., Frommberger, L., Wolter, D., Dylla, F., & Freksa, C. A toolbox for qualitative spatial representation and reasoning. *Spatial Cognition V: Reasoning, Action, Interaction: International Conference Spatial Cognition 2006.* LNCS 4387, 2007.

Wang, X. H., Zhang, D. Q., Dong, J. S., Chin, C. Y., & Hettiarachchi, S. (2004). Semantic Space: An infrastructure for smart spaces. *IEEE Pervasive Computing / IEEE Computer Society [and] IEEE Communications Society, 3*(3), 32–39. doi:10.1109/MPRV.2004.1321026

Wang, H., & Chang, S. F. (1997). A Highly Efficient System for Automatic Face Detection in Mpeg Video. *IEEE Transactions on Circuits and Systems for Video Technology,* (7): 615–628. doi:10.1109/76.611173

Wang, W., & Benbasat, I. (2007). Recommendation Agents for Electronic Commerce: Effects of Explanation Facilities on Trusting Beliefs. *Information Systems, 23,* 217–246.

Wang, G.-L., Cao, G., & Porta, T. L. (2004). Movement-assisted sensor deployment. In *Proc. IEEE Infocom Conf.,* (pp.2469-2479).

Wang, X., Zhang, D. Q., Gu, T., & Pung, H. K. (2007). Ontology Based Context Modeling and Reasoning using OWL. Workshop on Context Modeling and Reasoning at IEEE Fifth International Conference on Pervasive Computing and Communications Workshops (PerComW'07), 14-19.

Want, R., Hopper, A., Falcao, V., & Gibbons, J. (1992). The Active Badge Location System. *ACM Transactions on Information Systems, 10*(1), 91–102. doi:10.1145/128756.128759

Web Ontology Language (OWL) homepage (2009-11-3). http://www.w3.org/ 2004/ OWL/.

Web site. 2007: http://en.wikipedia.org/wiki/Spline_ (mathematics)

Weeks, S. (2001) Understanding Trust Management Systems. In Proc. IEEE Symposium on Security and Privacy, pag. 94–105, Oakland, CA.

Weinzierl, T., & Schneider, F. (2001). In Tränkler, H.-R., & Schneider, F. (Eds.), *"Gebäudesystemtechnik," in Das intelligente Haus - Wohnen und Arbeiten mit zukunfts-weisender Technik* (pp. 349–361). Pflaum Verlag.

Weir, D. H., & Chao, K. C. Review of Control Theory Models for Directional and Speed Control, in: Cacciabue, P.C., p. 293 – 311 (2007)

Weiser, M. (1991). The Computer for the 21st Century. *Scientific American, 265*(3), 94–104. doi:10.1038/scientificamerican0991-94

Weiser, M. (1991). The Computer for the 21st Century, *Scientific American.*

Weitzel, M., Smith, A., Lee, D., Deugd, S., & Helal, S. (2009). Participatory medicine: Leveraging social networks in telehealth solutions. In *Proc. of ICOST2009.*

Wells, A. J. (2002). Gibson's affordances and Turing's theory of computation. *Ecological Psychology, 14*(3), 141–180. doi:10.1207/S15326969ECO1403_3

Werger, B., & Mataric, M. J. (2001). From insect to internet: situated control for networked robot teams. *Annals of Mathematics and Artificial Intelligence, 31*(1-4), 173–198. .doi:10.1023/A:1016650101473

Werner, S., Krieg-Brückner, B., Hermann, T.: Modelling Navigational Knowledge by Route Graphs. *Spatial Cognition II: Integrating Abstract Theories, Empirical Studies, Formal Methods, and Pratical Applications.* LNAI 1849, 2000.

West, G., Newman, K., & Greenhill, S. (2005). Using a Camera to Implement Virtual Sensors in a Smart House. In: Giroux S and Pigot H (eds) From Smart Homes to Smart Care, ICOST'2005 3rd International Conference on Smart Homes and Health Telematics, Assistive Technology Research Series 15, IOS Press, pp. 75–82.

White, S. A. (2007). *Introduction to BPMN.* IBM Co-operation.

Wickens, C. D. (2002). Situation awareness and workload in aviation. *Current Directions in Psychological Science, 11*(4), 128–133. doi:10.1111/1467-8721.00184

Wickens, Th. D. (1982). *Models for Behavior: Stochastic Processes in Psychology.* San Francisco: Freeman.

Wilensky, R. (1983). *Planning and understanding.* Reading, MA: Addison Wesley.

Wilson, E. O. (1976). *Sociobiology: the new synthesis.* Harvard University Press.

Wilson, D., & Philipose, M. (2005). Maximum A Posteriori Path Estimation with Input Trace Perturbation: Algorithms and Application to Credible Rating of Human Routines. In IJCAI-05, Proceedings of the Nineteenth International Joint Conference on Artificial Intelligence, Edinburgh, Scotland, UK, pp. 895–901.

Wisner, P., & Kalofonos, D. N. (2007). A Framework for End-User Programming of Smart Homes Using Mobile Devices. *Proceedings of the 4th IEEE Consumer Communications and Networking Conference CCNC'07* (pp. 716-721), Washington DC: IEEE Computer Society.

Wisneski, C., Ishii, H., Dahley, A., & Gorbet, M. B., S., U., et al. (1998). Ambient Displays: Turning Architectural Space into an Interface between People and Digital Information. In N.A.Streitz, S.Konomi, & H.-J. Burkhardt (Eds.), *the First International Workshop on Cooperative Buildings* (pp. 22-32). Springer-Verlag Berlin Heidelberg.

Witten, I., & Frank, E. (2005). *Data mining: practical machine learning tools and techniques* (2nd ed.). San Francisco: Morgan Kaufmann.

Wittgenstein, L. Philosophische Untersuchungen, Oxford 1953.

Wobcke, W. (2002). Two Logical Theories of Plan Recognition. *Journal of Logic Computation, 12*(3), 371–412.

Wolfe, J. M. (1998). Visual Search. In Pashler, H. (Ed.), *Attention*. East Sussex, UK: Psychology Press. Fulltext.

Wren, C. R., Azarbayejani, A., Darrell, T., & Pentland, A. P. (1997). PFINDER: Real-time tracking of the human body. *IEEE Transactions on Pattern Analysis and Machine Intelligence, 19*(7), 780–785. doi:10.1109/34.598236

Wu, J., Osuntogun, A., Choudhury, T., Philipose, M., & Rehg, J. (2007). A Scalable Approach to Activity Recognition Based on Object Use. *Proc. of IEEE Int. Conf. on Computer Vision (ICCV)*.

Wu, T., & Matsuyama, T. (2003). Real-time active 3D shape reconstruction for 3D video. Proceedings from ISPA '03: *The third International Symposium on Image and Signal Processing and Analysis, 1*, 186-191.

Xiang, Y. (2002). *Probabilistic Reasoning in Multiagent Systems - A Graphical Models Approach*. Cambridge: Cambridge University Press.

Xu, Y. (2005). *Ka Keung Caramon Lee, and Ka Keung C. Lee, Human Behavior Learning and Transfer*. CRC Press Inc.

Yamada, T., Iijima, T., & Yamaguchi, T. (2005). Architecture for Cooperating Information Appliances Using Ontology. In Proc. of 19th Annual Conference of the Japanese Society for Artificial Intelligence.

Yamazaki, T. (2007). The Ubiquitous Home. *International Journal on Smart Homes, 1*(1), 17–22.

Yamazaki, T. (2005) Ubiquitous Home: Real-Life Testbed for Home Context-Aware Service. Proc. Tridentcom 2005 (First International Conference on Testbeds and Research Infrastructures for the DEvelopment of NeTworks and COMmunities), pp. 54–59.

Yamazaki, T. (2006) Human Action Detection and Context-Aware Service Implementation in a Real-life Living Space Test Bed. Proc. TridentCom 2006 (Second International Conference on Testbeds and Research Infrastructures for the DEvelopment of NeTworks and COMmunities).

Yamazaki, T., & Toyomura, T. (2008). Real-life experimental data acquisition in smart home and data analysis tool development. *In International Journal of ARM, 9*.

Yamazaki, T., Ueda, H., et al. (2005). Networked Appliances Collaboration on the Ubiquitous Home. In: Giroux S and Pigot H (eds) From Smart Homes to Smart Care, ICOST'2005 3rd International Conference on Smart Homes and Health Telematics, Assistive Technology Research Series 15, IOS Press, pp. 135–142.

Yang, J., Schilit, B., & McDonald, D. (2008, April). Activity recognition for the digital home. *Computer, 41*(4), 102–104.

Yang M. H. & Ahuja N., (1999). Gaussian Mixture Model for Human Skin Color and Its Applications in Image and Video Databases, *SPIE Storage and Retrieval for Image and Video Databases*, (3656). (pp. 45-466).

Yang, Y., & Mahon, F. Williams. M.H. & Pfeifer, T. (2006). Context-Aware Dynamic Personalized Service Re-composition in a Pervasive Service Environment. In Ma J. et al (Eds.) *Proceedings of the 3rd International Conference on Ubiquitous Intelligence and Computing UIC'06* (pp. 724-735). Berlin, Heidelberg: Springer-Verlag.

Yap, K. K., Srinivasan, V., & Motani, M. (2005). Max: Human-centric search of the physical world. In Proc. of SenSys'05 (pp. 166–179).

Yarbus, A. L. (1967). *Yarbus. Eye Movements during Perception of Complex Objects*. New York, New York: Plenum Press.

Yarbus, (1967) A. L. Yarbus. Eye Movements during Perception of Complex Objects. Plenum Press, New York, New York, 1967

Yau, S. S., Gong, H., Huang, D., Gao, W., & Zhu, L. (2008). Specification, Decomposition and Agent Synthesis for Situation-Aware Service-Based Systems. *Journal of Systems and Software, 81*(10), 1663–1680. doi:10.1016/j.jss.2008.02.035

Yau, S. S., & Liu, J. (2006). Hierarchical Situation Modeling and Reasoning for Pervasive Computing. Third Workshop on Software Technologies for Future Embedded & Ubiquitous Systems (SEUS), 5-10.

Yedidia, J. S., Freeman, W. T., & Weiss, Y. (2001). *Understanding belief propagation and its generalizations*. Technical Report TR2001-22, Mitsubishi Electric Research Laboratory.

Zarzhitsky, D., Spears, D., & Spears, W. (2005). Distributed robotics approach to chemical plume tracing. In *Proc. IEEE/RSJ Int. Conf. Intelligent Robots and Systems*, (pp.4034- 4039).

Zecca1, G., Couderc, P., Banatre, M., & Beraldi, R. (2009). Swarm robot synchronization using RFID tags. In *Proc. IEEE Int. Conf. Pervasive Computing and Communications*, (pp.1-4).

Zeelenberg, M., & van Dijk, E. (2005). On the psychology of "if only": regret and the comparison between factual and counterfactual outcomes. *Organizational Behavior and Human Decision Processes, 97*(2), 152–160. doi:10.1016/j.obhdp.2005.04.001

Zelkha, Eli; Epstein, Brian, From Devices to Ambient Intelligence, Digital Living Room Conference, June 1998 (http://www.epstein.org/brian/ambient_intelligence/DLR%20Final%20Internal.ppt; visited 27th February, 2010)

Zentralverband Elektrotechnik- und Elektroindustrie. Handbuch: Gebäudesystemtechnik. Grundlagen, 4 überarbeitete Auflage ed. Frankfurt am Main: Zentralverband Elektrotechnik- und Elektronikindustrie/Zentralverband der Deutschen Elektrohandwerke, 1997.

Zhang, J. (1997). The nature of external representations in problem solving. *Cognitive Science, 21*(2), 179–217. doi:10.1207/s15516709cog2102_3

Zhang, J., & Patel, V. L. (2006). Distributed cognition, representation, and affordance. *Cognition & Pragmatics, 2*, 333–341. doi:10.1075/pc.14.2.12zha

Zhang, D., Wang, X., et al. (2003) OSGi Based Service Infrastructure for Context Aware Connected Homes. In: Mokhtari M (ed) Independent Living for Persons with Disabilities and Elderly People, ICOST'2003 1st International Conference on Smart Homes and Health Telematics, Assistive Technology Research Series 12, IOS Press, pp. 81–88.

Zhang, D., Yu, Z., & Chin, C. Y. (2004). Context-aware infrastructure for personalized HealthCare. International Workshop on Personalized Health, IOS Press, 154-163.

Zhang, W., & Hansen, K. M. (2008a) Semantic web based Self-management for a Pervasive Service Middleware. Second IEEE International Conference on Self-Adaptive and Self-Organizing Systems. IEEE Computer Society. Venice, Italy. 245-254.

Zhang, W., & Hansen, K. M. (2008b). An OWL/SWRL based Diagnosis Approach in a Web Service-based Middleware for Embedded and Networked Systems. Proceedings of The 20th International Conference on Software Engineering and Knowledge Engineering (SEKE 2008). Redwood City, San Francisco Bay, USA, 893-898.

Zhang, W., & Hansen, K. M. (2008c). Towards Self-managed Pervasive Middleware using OWL/SWRL ontologies. Fifth International Workshop on Modeling and Reasoning in Context (MRC 2008), Held together with HCP 08, Delft, The Netherlands, 9-12 June 2008, 1-12.

Zhang, W., & Hansen, K. M. (2009). Evaluation of NSGA-II and MOCell Genetic Algorithms for Self-management Planning in a Pervasive Service Middleware. 14th IEEE International Conference on Engineering of Complex Computer Systems (ICECCS 2009), Potsdam. Germany. June 2-4 *2009*, 192-201.

Zhang, W., Hansen, K. M., & Kunz, T. (2009). Enhancing Intelligence of a Product Line Enabled Pervasive Middleware. Pervasive and Mobile Computing Journal. Elsevier. June, 2009. http://dx.doi.org/ 10.1016/ j.pmcj.2009.07.002.

Zhang, W., Schütte, J., Ingstrup, M., & Hansen, K. M. (2009). A Genetic Algorithms-based Approach for Optimized Self-protection in a Pervasive Service Middleware. The Seventh International Conference on Service Oriented Computing (ICSoC 2009), Stockholm. Sweden Nov 24-27 *2009*. Springer LNCS 5900, 404-419.

Zhao, T., Nevatia, R., & Lv, F. (2001). Segmentation and Tracking of Multiple Humans in Complex Situations. Proceedings from ICCVPR '01: *The International Conference on Computer Vision and Pattern Recognition*, 2, 194-201.

Zheng, Y. F., & Chen, W. (2007). Mobile robot team forming for crystallization of protein. *Autonomous Robots*, *23*(1), 69–78. .doi:10.1007/s10514-007-9031-1

Zhou, H., & Hou, K. M. (2005). Real-time Cardiac Arrhythmia Tele-Assistance and Monitoring Platform: RECATA. In: Giroux S and Pigot H (eds) From Smart Homes to Smart Care, ICOST'2005 3rd International Conference on Smart Homes and Health Telematics, Assistive Technology Research Series 15, IOS Press, pp. 99–106.

Zhuang, X. (2004). Vehicle detection and segmentation in dynamic traffic image sequences with scale-rate, *Proceedings of the IEEE International Conference on Intelligent Transportation Systems*, (pp. 570-574).

Zimmermann, K., & Freksa, C. (1996). *Qualitative Spatial Reasoning Using Orientation, Distance, and Path Knowledge* (*Vol. 6*). Applied Intelligence.

Zimmermann, G. Open User Interface Standards – Towards Coherent, Task-Oriented and Scalable User Interfaces in the Home Environments. Proc. 3rd *IET International Conference on Intelligent Environments (IE07)*, Ulm, Germany. The IET, 2007.

Zollner, G. (1995). *Kundennähe in Dienstleistungsunternehmen*. Wiesbaden: Empirische Analyse von Banken.

Zou, Y., & Chakrabarty, K. (2003). Sensor deployment and target localization based on virtual forces. In *Proc. IEEE Infocom Conf.*, (pp.1293-1303).

Zukow-Goldring, P., & Arbib, M. (2007). Affordances, effectivities, and assisted imitation: Caregivers and the directing of attention. *Neurocomputing*, *70*(13-15), 2181–2193. doi:10.1016/j.neucom.2006.02.029

About the Contributors

Nak-Young Chong received his B.S., M.S., and Ph.D. from Hanyang University, Seoul, Korea in 1987, 1989, and 1994, respectively. From 1994-98, he was a senior researcher at Daewoo Heavy Industries. After Daewoo, he spent 1 year at KIST. From 1998-2007, he was on the research staff of AIST. In 2003, he joined the JAIST faculty as Associate Professor of Information Science. Dr. Chong serves as Associate Editor of the IEEE Transactions on Robotics, and International Journal of Assistive Robotics and Systems. He will serve as Program Chair/Co-Chair for the ICAM 2010, IEEE-ROMAN 2011, and IEEE-CASE 2012. He served as Co-Chair of the IEEE RAS Technical Committee on Networked Robots in 2004-06, and Fujitsu Scientific Systems WGs in 2004-08. He was a visiting scholar at AIST in 1995-96, Northwestern University in 2001 and Georgia Tech in 2008-09. He is a director of KROS, and a member of IEEE, RSJ, and SICE.

Fulvio Mastrogiovanni's research activities expand in various branches of Robotics, specifically Humanoid Robotics, Ambient Intelligence, and Distributed Robotics. Fulvio received his Master degree in Computer Engineering from the University of Genova in 2003, with a Thesis about task planning in Distributed service Robotics. In 2008, he received the Ph.D. degree in Robotics from the University of Genoa, with the Dissertation "Context Awareness: Another Step Towards Ubiquitous Robotics". Currently, Fulvio is interested in large scale tactile sensing for Humanoids, cognitive representation mechanisms, and neural-based planning and control. Fulvio received the Best Paper Awards at DARS 2008 and at IEEE RO-MAN 2010.

* * *

Carole Adam defended her PhD in Artificial Intelligence in Toulouse in 2007 under the supervision of Andreas Herzig and Dominique Longin; her topic was the logical formalisation of emotions and their implementation in an agent. In 2008 she has been working as a postdoc in Orange Labs (Lannion, France) on an extension for JADE accounting for electronic institutions for B2B applications. She is now working in RMIT University in Melbourne, Australia on a project to develop intelligent interactive toys that can engage children in long-term relationships. Her research interests include BDI logics, in particular for the formalisation of emotions for Embodied Conversational Agents, but also logics of social norms or group mental attitudes, and formalisations of dialogue.

Hamdi Aloulou is a Masters student at the National School of Engineers of Sfax, Tunisia. He is doing his Masters training at the Institute for Infocomm Research (I2R), Singapore in the topic of healthcare and aging assistance. He is graduated as a computer engineer from the National School of Engineers of Sfax, Tunisia.

Francisco J. Ballesteros (http://lsub.org/who/nemo) got his MS in CS on 1993 and his PhD on CS on 1998, at Technical University of Madrid. He worked for several telecommunications companies, doing systems software. He is the (co)author of LiS, a STREAMS framework for Linux. Since 1995 he has been a professor at several Spanish Universities where he has been teaching and developing Operating Systems. He developed the Off++ kernel, for the 2K Operating System jointly with the SRG at University of Illinois at Urbana Champaign. He has been working in R&D on topics related to Plan 9 from Bell Labs, including the Plan B and Octopus Operating Systems. He is also the author of the Omero window system. Currently he is the head of the Systems Lab at URJC. (http://lsub.org)

Emanuele Bardone received his PhD in Philosophy from the University of Pavia. He teaches Philosophy of Cognition at the University of Pavia and is currently working at the Computational Philosophy Laboratory of the same university. He is author of Seeking Chances: From Biased Rationality to Distributed Cognition (forthcoming). His main research interests include distributed cognition, ecological psychology, and abductive reasoning.

Mária Bieliková received her Master's degree (with summa cum laude) in 1989 and her PhD degree in 1995, both from the Slovak University of Technology. Since 2005, she has been a full-time professor at the same university. She has (co-) authored five books, several teaching materials and more than 130 scientific papers. She is a member of the Editorial Board of the Int. Journal of Intelligent Information and Database Systems and the editor of 31 proceedings of scientific conferences, five of them published by Springer. Her research interests are in the areas of ambient intelligence and web-based adaptive systems, user modeling, and especially the social dimension of personalized web-based systems (a list of selected publications presenting research results is available at http://www.fiit.stuba.sk/~bielik/). She is senior member of ACM, senior member of IEEE and its Computer Society and a member of the executive committee of the Slovak Society for Computer Science.

Jit Biswas is a Senior Scientist in the Networking Protocols Department at the A*Star Institute for Infocomm Research (I^2R), Singapore, where he is the involved in Healthcare projects under the "Assistive ICM for Health Monitoring and Rehabilitation" Program. His recently concluded projects include agitation monitoring for dementia patients using ambient sensors (with Alexandra Hospital, Singapore) and data fusion and feature extraction for sleeping posture determination and bedsore prevention (with ITRI, Taiwan), studying circadian patterns in sleep activity patterns among elderly patients and a core wireless sensor networking project in the area of UWB sentient computing – architecture and middleware. Dr Biswas has a Bachelors degree in Electrical Engineering from BITS, India and a Ph.D. degree in Computer Science from the University of Texas at Austin, USA.

Tibor Bosse is an Assistant Professor in the Agent Systems Research Group in the Artificial Intelligence Department of the Vrije Universiteit of Amsterdam. His research focuses on computational (agent-based) modelling of human-related (mostly cognitive) processes, such as attention, emotion, and workload. These models can be used both for theoretical purposes (to gain a deeper understanding of the processes themselves) and for practical applications, e.g., in the domains of ambient intelligence, human factors design, and virtual environments.

Bruno Bouchard works as researcher and professor since 2007 at the Department of Computer Science and Mathematic of the Université du Québec à Chicoutimi (UQAC), Canada. He received a Ph.D. in computer science from the University of Sherbrooke, Canada, in 2006. He has co-founded at UQAC, in 2008, the LIAPA laboratory together with the professor Abdenour Bouzouane. This laboratory specifically explores ambient intelligence and recognition techniques for assistance to elders in loss of autonomy. His research actually has the financial support of the Natural Sciences and Engineering Research Council of Canada (NSERC), the Quebec Research Funds on Nature and Technology (FQRNT), and the Foundation UQAC.

Abdenour Bouzouane is professor since 1997 at the Department of Computer Science and Mathematic of the Université du Québec à Chicoutimi (UQAC), Canada. He received a Ph.D. in computer science from Ecole Centrale de Lyon, France, in 1993. His main research interests are multi-agent systems, temporal reasoning, description logic and ambient computing. His research actually has the financial support of the Natural Sciences and Engineering Research Council of Canada (NSERC), the Quebec Research Funds on Nature and Technology (FQRNT), and the Foundation UQAC.

Vladimír Bureš works as an associate professor at the Department of Information Technologies at the Faculty of Informatics and Management of the University of Hradec Kralove, Czech Republic. He also cooperates with other educational institutions (the Vysoká škola manažmentu/City University Bratislava, Slovakia, or Faculty of Economics and Administration at the University of Pardubice, Czech Republic). His scientific interests include files such as knowledge management, ambient intelligence, or systems theory. He has participated and holds experience on a couple of European research projects (e.g. 6th and 7th Framework Program of the EU – Enhanced Learning Unlimited, and VitalMind respectively) and the Czech Science Foundation projects (406/04/2140 KNOMEDIAS or 402/03/1325 AMIMADES). He is a member of some Czech and international conference program committees (e.g. International Congress on Pervasive Computing and Management, IADIS Web-Based Communities, or Ambient Intelligence Forum).

Yang Cai is Director of Instinctive Computing Lab and Senior Scientist of Cylab, College of Engineering at Carnegie Mellon University, Pittsburgh, PA, USA. He is also an adjunct Professor at School of Computer Engineering, Zhejiang University, Hangzhou, China. He received post-doctorial education from Professor Herbert Simon at Carnegie Mellon, Ph.D. in Robotic Transportation from West Virginia University, MS degree in Management Science and BS degree in Control Engineering from Zhejiang University. He was a NASA Faculty Fellow and participated projects of the Space Shuttle Rewaterproofing Robot, satellite onboard processing and scientific visualization studio. Cai created new theories of Instinctive Computing, Empathic Machines and Ambient Diagnostics, published by Springer in LNAI volumes. His current research projects include video analytics and visualization of sensor web. He won the Best Paper award at ICCS in May, 2009.

Carl K. Chang is Professor and Chair of the Department of Computer Science at Iowa State University. He received a PhD in computer science from Northwestern University in 1982, and worked for GTE Automatic Electric and Bell Laboratories before joining the University of Illinois at Chicago in 1984. He joined Iowa State University in 2002. His research interests include requirements engineering, software architecture, net-centric computing and services computing. Chang is 2004 President of IEEE Computer Society. Previously he served as the Editor-in-Chief for IEEE Software (1991-94). He received the Computer Society's Meritorious Service Award, Outstanding Contribution Award, the Golden Core recognition, and the IEEE Third Millennium Medal. In 2006 he received the prestigious Marin Drinov Medal from the Bulgarian Academy of Sciences, and was recognized by IBM with the IBM Faculty Award in 2006, 2007 and 2009. In January 2007 he became the Editor-in-Chief of IEEE Computer, the flagship publication of IEEE Computer Society. Chang is a Fellow of IEEE and of AAAS.

Marcello Cirillo is currently a Ph.D. Student at the Mobile Robotics Lab of the AASS center, at the School of Science and Technology of Örebro University. In 2004 he has been awarded a grant for Master's studies at IRIDIA, Université Libre de Bruxelles (B). He has received his M.Sc. in Computer Science Engineering from Politecnico di Milano (I) in 2005. His research interests include planning under uncertainty, sensor networks and ambient intelligence. His current focus is on human-aware robot task planning and human activity recognition in intelligent environments. He is expected to defend his Ph.D. thesis at the end of 2010.

Alvin Kok-Weng Chu is currently studying for the Diploma of Computer Engineering at School of Engineering, Temasek Polytechnic, Singapore.

Giuseppe Conti (1974) (Italy) is senior researcher at GT since 2002, he received a MEng in Civil Engineering from University of Palermo (Italy) and a PhD degree at the ABACUS (Architecture and Building Aids Computer. Unit Strathclyde) University of Strathclyde, Glasgow (UK). He has worked for international engineering consultants UK. He has been involved in several EU and industrial projects dealing with issues related to the use of Virtual Reality to planning, large terrain 3D visualization and in the context of geobrowsers. He is the board of AM/FM GIS Italy association. He has been invited speaker at a number of international event including a number of NATO initiatives on environmental protection and GIS. He has more than 80 publications in the field of computer graphics and applications.

Oleg Davidyuk received his MSc degree in Information Technology from Lappeenranta University of Technology, Finland, in 2004. He is currently working towards his PhD degree in MediaTeam Oulu Research Group in University of Oulu, Finland. His research interests include application and service composition, user interaction design, middleware and ubiquitous computing. Oleg's publications can be found at www.mediateam.oulu.fi/publications/?search=davidyuk.

Raffaele De Amicis (1970) (Italy) is Director of GT, he holds a MEng in Mechanical Engineering from University of Calabria (Italy), a Ph.D. on Surface Modelling in Virtual Environments from University of Bologna (Italy). He has been research fellow at the Industrial Applications Department of Fraunhofer Institute, Darmstadt (Germany) and senior researcher at the Interactive Graphics Systems Group, University of Darmstadt. He has been involved in several EU and Industrial projects. He was co-director of a NATO Advanced Research Workshop on "Geographical Information Processing and Visual Analytics for Environmental Security" and coordinator of a ICT-PSP project. His research interests are in CAD, virtual reality, computer supported cooperative work in engineering, virtual reconstructions.

Mathias Döhle received a Diploma in Computer Science from Universität Bremen in 2009 and is presently a Software Engineer at the German Research Center for Artificial Intelligence (DFKI). He was instrumental in realising the technical infrastructure of the Bremen Ambient Assisted Living Lab, BAALL.

Mark Eilers, born in Westerstede, Germany in 1982, studies computer science at the Department of Computing Science of the Carl von Ossietzky University of Oldenburg and will receive his master degree in June 2010. Since 2007 he works as a student assistant at the Institute for Information Technology OFFIS in the field of research of digital human modelling. In 2008, he became a member of the special research group Learning and Cognitive Systems of Prof. Dr. Claus Möbus at the University of Oldenburg and focused his research on Bayesian programming and Bayesian (driver) models.

Mohamed Ali Feki is a senior researcher at the Ambient Media Department of Bell-Labs Belgium. Dr Feki research focus is on Ambient Intelligence and Internet of Things, as technology enablers to realize the vision of Ambient Assistive Living. Dr Feki received his Ph.D. from Telecom Sud-Paris France and a Masters in advanced Networking and Telecommunication from University of Troyes, France.

Victor Siang-Fook Foo is a Senior Research Officer in Networking Protocols Department at the A*Star Institute for Infocomm Research (I2R), Singapore. His research interests include context awareness, mobile/wireless/optical/sensor networks and machine learning. He received the Bachelor of Engineering degree and a Master of Engineering degree from National University of Singapore in 1997 and 1999 respectively.

Kaori Fujinami is an associate professor in the Department of Computer and Information Sciences, Tokyo University of Agriculture and Technology. He worked for Nippon Telegraph and Telephone Corp. (NTT) and NTT COMWARE Corp. as a software engineer/researcher from April 1995 to April 2003. He received his MS in Electrical Engineering and Ph.D. in Computer Science from Waseda University in March 1995 and March 2005, respectively. His research interests include smart object systems, human-computer interaction and activity recognition. Dr. Fujinami can be contacted at kaori.fujinami@gmail.com, http://www.tuat.ac.jp/~fujinami/lab/English.html

Benoit Gaudou has obtained his PhD from the University of Toulouse in 2008 under the supervision of Andreas Herzig and Dominique Longin, about the use of modal logic to characterize group mental attitudes, and their application to the definition of an Agents Communication Language semantics. During this period, he has also worked on a logical formalization of emotions. He is now a post-doctoral fellow in the UMI UMMISCO international research team and working in the IFI school in Hanoi. In addition to modal logics for knowledge represensation (such as mental attitudes, speech acts or emotions), his research interests also include the use of the notion of trust and trust nets in the context of distributed information systems with perturbed agents (that are agents communicating false information) and the study of the collaborative (agent-based) modelling and simulation of complex systems.

Nikolaos Georgantas received his Ph.D. in 2001 in Electrical and Computer Engineering from the National Technical University of Athens. He is currently a researcher at INRIA with the ARLES research project-team. His research interests relate to distributed systems, middleware, ubiquitous computing systems and service and network architectures for telecommunication systems. He is or has been involved in a number of European projects and several industrial collaborations.

Bernd Gersdorf received a Diploma and Dr. Ing. in computer science from Universität Bremen, Germany, in 1986 and 1992, resp. In 1995, he started to work in industry in software development for process automation, and later for flight safety systems. He joined the Research Department on Safe and Secure Cognitive Systems at the German Research Center for Artificial Intelligence (DFKI), Bremen, in early 2007 to develop intelligent mobility platforms for elderly people.

Sylvain Giroux is a well established scientist currently working as professor at the Department of Computer Science of the University of Sherbrooke, Canada. He received a Ph.D. in computer science from the University of Montreal, Canada, in 1993. Several years ago, he worked in close collaboration with the Center for Advanced Studies, Research and Development in Sardinia (CRS4), Italy. His main research interests are mobile computing, pervasive computing, distributed artificial intelligence, multi-agent systems, user modelling, intelligent tutoring systems. Sylvain Giroux has co-founded the DOMUS laboratory of the University of Sherbrooke with the professor Hélène Pigot, in 2005. He contributed to application domains as varied as cognitive assistance, advisor systems, task-support systems, geophysics, e-commerce, bio-medical applications, public administration and tourism. He receives financial support from many public and private organisms and societies, such as NSERC, FQRNT, France Telecom, Ericsson, SAP, etc.

Björn Gottfried works at the Centre for Computing Technologies at the University of Bremen, Germany, where he received his doctoral degree in the context of spatial reasoning. He is a research scientist and lecturer in the context of Artificial Intelligence, in particular image processing, and spatial and diagrammatic reasoning. Over the last ten years he published mainly in the context of image processing, spatial and temporal reasoning, and ambient intelligence, numerous journal and conference papers, has been invited as course lecturer at international conferences, and holds 1 European patent and 1 US patent. He is member in the programme committees of several conferences and workshops about ambient intelligence, spatial reasoning and related fields and organises the annual BMI workshop on behaviour monitoring and interpretation.

Gorka Guardiola obtained his MS in Telecommunication Engineering from the Universidad Carlos III de Madrid (2003). He obtained his Ph. D. in October 2007 at the Universidad Rey Juan Carlos de Madrid. His research areas are is distributed operating systems and ambient intelligence. He has done several internships and collaborations with Bell Labs, Alcatel-Lucent, and Austin Research Lab, IBM. As a member of the Systems Lab, he has worked on the Plan B, Octopus and Plan 9 operating systems.

Hans W. Guesgen is a Professor of Computer Science in the School of Engineering and Advanced Technology at Massey University in Palmerston North, New Zealand. He holds a diploma in computer science and mathematics of the University of Bonn, a doctorate in computer science of the University of Kaiserslautern, and a higher doctorate (Habilitation) in computer science of the University of Hamburg, Germany. He worked as a research scientist at the German National Research Center of Computer Science (GMD) at Sankt Augustin from 1983 to 1992. During this period he held a one-year post-doctoral fellowship at the International Computer Science Institute in Berkeley, California. In 1992 he joined the Computer Science Department of the University of Auckland, where he worked until moving to Massey University in 2007. His research interests include ambient intelligence and spatio-temporal reasoning.

Bin Guo is currently a post-doctorate researcher at the Institute Telecom & Management SudParis, France. His research interests include wireless sensor networks (WSNs), context-aware middleware, pervasive gaming, semantic web and end-user programming in smart environments. He received his Ph.D. degree in computer science from Keio University, Tokyo, Japan, in 2009. He has worked in several projects in the field of ambient intelligence, including EU's Feel@Home CELTIC Project in 2009 and Japan's 21st COE Program "System Design: Paradigm Shift from Intelligence to Life" during 2006-2007. He is a member of IEEE.

Klaus Marius Hansen is a (full) Professor of Software Engineering at University of Iceland, Iceland. He was most recently scientific manager of the infastructure group of the Danish national network for "pervasive communication" (KomIAlt), manager of the "software" research area of the Alexandra Institute, and member of the steering committee of the MiNEMA ESF Scientific Program. Since 1997, Klaus Marius Hansen has been involved in numerous research and development projects with partners from industry and in EU projects. His current research interests include software architecture and experimental software engineering in particular in relation to pervasive computing, middleware, and dependable systems. Klaus Marius Hansen has published more than 50 papers in international journals, conferences, and workshops.

Marián Hönsch is a master student in Software Engineering study program at the Slovak University of Technology in Bratislava. He received Bachelor's degree in Informatics from the same university in 2008. He stayed for one year work and research practicum in EnOcean GmbH Munich, Germany. His actual research interests are in the areas of identifying and working with virtual communities, enabling collaborative filtering based on different domains of virtual communities. In 2008, he participated on Energy Consumption Manager project, which placed 2nd at Microsoft Imagine Cup 2008 worldwide finals in Software Design category.

Mark Hoogendoorn is an assistant professor at the VU University Amsterdam, Department of Artificial Intelligence. Before starting as an assistant professor he has been a visiting researcher at the University of Minnesota, Department of Computer Science and Engineering. He obtained his PhD degree from the VU University in 2007. In his PhD research he focused on organizational change within multi-agent systems, applying his research in projects in various domains, including incident management, logistics, and the naval domain. His current research interests include multi-agent systems, cognitive modeling, and ambient intelligence. cinnamon

Weimin Huang is a research scientist in the Institute for Infocomm Research (I2R), Singapore. His current research focuses on computer vision technology for image processing, object detection, tracking, event/human activity recognition. He won a TEC Innovator Award for the project "Intelligent CCTV", which is designed for the public security monitoring for abnormal event and behavior detection, and also worked as work package leader for project ASTRALS on home monitoring aiming for homecare and healthcare. His research interests include pattern recognition, medical image processing, computer vision, and statistical learning.

Michita Imai is Associate Professor of Faculty of science and technology at Keio university, a researcher at ATR Intelligent Robot Laboratories, and a visiting researcher at the University of Chicago. He received his Ph.D. degree in Computer Science from Keio Univ. in 2002. In 1994, he joined NTT Human Interface Laboratories. He joined the ATR Media Integration & Communications Research Laboratories in 1997. His research interests include autonomous robots, human-robot interaction, speech dialogue systems, humanoid, and spontaneous behaviors. He is a member of Information and Communication Engineers Japan (IEICE-J), the Information Processing Society of Japan, the Japanese Cognitive Science Society, the Japanese Society for Artificial Intelligence, Human Interface Society, IEEE, and ACM.

Valérie Issarny got her PhD and "Habilitation à diriger des recherches" in computer science from the University of Rennes I, France, in 1991 and 1997 respectively. She currently holds a "Directeur de recherche" position at INRIA. Since 2002, she is the head of the INRIA ARLES research project-team at INRIA-Rocquencourt. Her research interests relate to distributed systems, software engineering, mobile wireless systems, middleware and ubiquitous computing. Further information about Valérie's research interests and her publications can be obtained from http://www-rocq.inria.fr/arles/members/issarny.html.

Suresh Jain is currently associated with KCB Technical Academy, Indore as the Director of the Institute. He is Professor (on leave) of Computer Engineering Department at Institute of Engineering & Technology (IET) of Devi Ahilya University (DAVV). He received his Masters and Doctorate degree in Computer Engineering from DAVV and B.E. degree from MANIT, Bhopal. He is a Ph.D. supervisor in the area of Web Mining & Information Retrieval. Also, he has organized various national level technical events. He is the life member of Computer Society of India, Indian Society of Technical Education and various international/national societies. He has authored several quality research papers in the international & national journals & conferences.

Jean-Michel Jolion received the Diplôme d'ingénieur in 1984, and the PhD in 1987, both in Computer Science from the Institut National des Sciences Appliquées (INSA) of Lyon (France). From 1987 to 1988, he was a staff member of the Computer Vision Laboratory, University of Maryland. From 1988 to 1994 he was appointed as an Associate Professor at the University Lyon 1. From 1994 to 2007, he has been with INSA as Professor of computer science. From 2007, he is in charge of the Université de Lyon as its executive director. His research interests belong to pattern recognition. He has studied robust statistics applied to clustering, multiresolution and pyramid techniques, graphs based representations... mainly applied to the image retrieval domain. Professor Jolion is a member of IAPR and IEEE.

David Kaufer is Professor of Rhetoric in the Department of English at Carnegie Mellon. His research focuses on written language as representational composition, ways in which written words on the page systematically structure worlds of experience for readers. He studies how words structure such interactive experience both in stand-alone texts and in rich multi-media contexts at the interface of words and images. Among his books are (with Brian Butler) Rhetoric and Arts of Design (1996), Designing Interactive Worlds with Words: Principles of Writing as Representational Composition (2000), and (with Suguru

Ishizaki, Brian Butler, and Jeff Collins) The Power of Words: Unveiling the Speaker and Writer's Hidden Craft. (2004). He co-founded a joint Master of Design program between the Department of English and the School of Design that investigates the interface of words and images.

Fahim Kawsar is a researcher in the Computing Department of Lancaster University. He received his Ph.D. in Computer Science from Waseda University in March, 2009. His research evolves around ubiquitous computing with specific focus on smart object systems, human-centric system infrastructures and tangible interfaces. Dr. Kawsar has published in the areas of distributed middleware, smart objects, personalization, and physical interfaces. He is a recipient of 2006-08 Microsoft Research (Asia) fellowship and a member of ACM and IEEE. Dr. Kawsar can be contacted at fahim.kawsar@gmail.com, http://www.fahim-kawsar.net.

Michel Klein is an assistant professor in the Agent Systems Research Group at the VU University Amsterdam since 2007. In 1996 he received his master Business Informatics at the same university. In 2004, he obtained a PhD degree with a thesis on Change Management for Distributed Ontologies. He has large experience in the field of Artificial Intelligence, especially in the areas of knowledge representation and modelling, knowledge-based systems and dynamic modelling. His current research theme is the application of artificial intelligence techniques to supporting human functioning. Specifically, he focuses on technology that support people in healthcare applications, such as intelligent self-management systems for chronic patients or support systems for people with mental disorders such as depression.

Michal Kompan is a master student in Software Engineering study program at the Slovak University of Technology in Bratislava. He received Bachelor's degree in Informatics from the same university in 2008. His research interests are in the areas of personalized web-based systems, analyzing similarity of web documents and user modeling. In 2008, he participated on Energy Consumption Manager project, which placed 2nd at Microsoft Imagine Cup 2008 worldwide finals in Software Design category.

Bernd Krieg-Brückner (M.S. in Computer Science, Cornell University, 1971; Dr.rer.nat, TU München, 1978) was a principal contributor to the design of the programming language ADA. After a stay at UC Berkeley and Stanford University, he became a professor at Universität Bremen in 1982. Since 2005, he has been director of the Research Department on Safe and Secure Cognitive Systems of the German Research Center for Artificial Intelligence (DFKI) at Bremen. His research has concentrated, on the one hand, on the development of formal languages, methods, and tools for the development of correct software as a basis for abstraction and reuse of formal developments, and safety and security. Since 1990, formal specification, formal ontology, and formal safety have been applied to cognitive science, robotics, and Ambient Assisted Living, now in the DFG-funded Transregional Collaborative Research Center SFB/TR 8 Spatial Cognition.

Christl Lauterbach founded Future-Shape GmbH, Höhenkirchen-Siegertsbrunn, Germany in 2005 and has been Managing Director since then. From 2003 to 2005 she was Project Manager for large-area smart textiles at Infineon's Corporate Research, Munich, Germany. From 1999 to 2003 she worked as a Senior Staff Engineer at Infineon Technologies AG, Emerging Technologies Group, Munich, Germany, looking for new applications for microelectronics and developing wearable electronics. From 1994 to 1999 she was with Corporate Technology, Siemens AG, Munich, Germany, with the Microelectronics Group designing circuits for chip cards, flash memories and adiabatic circuits. She was working at Corporate Research and Technology, Photonics Group, Siemens AG, Munich, Germany, on research and development of semiconductor technology and photonic devices from 1977 to 1994. Christl Lauterbach holds an assistant degree for Communication Engineering and has more than 150 patents and patent pendings.

Geunho Lee received his B.S. in Electronics Engineering from Seoul National University of Technology (SNUT), Seoul, Korea, M.S. in Electrical and Electronic Eng. from Yonsei University, Seoul, Korea, and Ph.D. in Information Science from Japan Advanced Institute of Science and Technology (JAIST), Ishikawa, Japan, in 1999, 2002, and 2008, respectively. Dr. Lee is currently a Research Assistant Professor with the School of Information Science, JAIST. His research interests include decentralized controls for robot swarms and welfare robots.

Christine Leignel received the Master's degree in Physics in 1993, and the PhD degree in Signal Processing and Telecommunications from Rennes I University in October 2006. Her subject's thesis, « a 2D body model for the analysis of gestures by the image via an architecture of type blackboard », was conducted at France telecom Lannion (France). She joined in 2007 the LIRIS Laboratory (INSA Lyon France) as a postodoctoral researcher in the Project Canada « behavior: Analysis, Detection, Alert ». Since 2008, she is currently a postdoctoral researcher at the IGEAT laboratory at the University Libre of Brussels. Her research interests lie in teledetection within the Armurs project, « Automatic Recognition for Map Update by Remote Sensing », whose aim is to detect changes between an old database and a recent satellite image.

Dominique Longin studied computer science in Toulon and Toulouse. In 1999 he obtained a Ph.D. in Computer Science at Paul Sabatier University in Toulouse. In his thesis he proposed a formalization of the evolution of agents' beliefs in cooperative dialogues. In 2001-2002 he was a post-doctoral fellow at the LTC ('Laboratoire Travail et Cognition' – Work and Cognition Laboratory) at the University Toulouse-Le-Mirail, where he worked on cognitive architectures in the framework of a `Cognitique' project. He has been a CNRS researcher (Chargé de Recherche CNRS) since 2002, and is currently member of the LILaC group (Logic, Interaction, Language, and Computation) of IRIT (Institut de Recherche en Informatique de Toulouse, UMR 5505 - Computer Science Research Institute of Toulouse, France). His main research topic is the formal characterization of intention, speech acts, individual beliefs and group beliefs. More recently he has worked on theories of emotions, norms, acceptance, and speech acts. He is currently the project manager of the French ANR project CECIL (www.irit.fr/CECIL).

Emiliano Lorini has been a CNRS researcher (Chargé de Recherche CNRS) at IRIT since 2009. He defended a PhD in Cognitive Science in 2007 at the University of Siena (Italy). His current research interests are in emotion theory, trust and reputation, logics for modelling multi-agent systems, deontic logic, speech act theory and game theory. He authored more than 60 articles in journals and international conferences in the AI field and in the cognitive science field. He was a PC member of major conferences in the AI field such as ECAI 2008, AAMAS 2009, and AAMAS 2010. He is or was involved in several research projects such as the European project MindRaces (2004-2007), the French ANR project ForTrust (2007-2010), the French ANR project CECIL (2009-2011).

Lorenzo Magnani is Professor of Philosophy of Science at the University of Pavia and the director of its Computational Philosophy Laboratory. He taught at the Georgia Institute of Technology (USA) and The City University of New York (USA). He is currently Visiting Professor at the Sun Yat-sen University, Guangzhou (Canton), P.R.China. He is the author of Abduction, Reason, Science (2001), Morality in a Technological World. Knowledge as Duty (2007), Abductive Cognition. The Eco-Cognitive Dimension of Hypothetical Reasoning (2009), and Philosophy of Violence (forthcoming). In 1998 he started the series of International Conferences on Model-Based Reasoning.

Jayachandran Maniyeri is a Senior Research Officer at the Networking Protocols Department of Institute for Infocomm Research (I²R), Singapore. His current research interests are in healthcare and service oriented systems. He has a Master of Technology degree from Indian Institute Science, Bangalore, India.

Goreti Marreiros is professor of Computer Engineering at the Institute of Engineering – Polytechnic of Porto (ISEP/IPP) and researcher at the Knowledge Engineering and Decision Support Research Group (GECAD). She received her PhD in informatics from the University of Minho. Her research interests include Multi-Agent Systems, Emotional Agents, Persuasive Argumentation, Group Decision Support Systems.

Stephen Marsland is an associate professor in computer science and the the postgraduate director in the School of Engineering and Advanced Technology at Massey University in Palmerston North, New Zealand. He has a degree in mathematics from Oxford University (1998) and a PhD from Manchester University (2002). Before moving to Massey University in 2004, he held postdoc positions in the UK, the USA, and Germany. His research interests are in mathematical computing, principally shape spaces, Euler equations, machine learning, and algorithms.

Jochen Meis studied business information systems at the University of Münster and graduated in 2003. After his study he started at the Fraunhofer Institute for Software and System Engineering. His research topics are service engineering, home automation and business process perspectives of IT-based services. The work focuses to model services and to support services with an IT-infrastructure. For this sensors and actuators are directly modeled as an active part of the service process.

Peter Mikulecký is a professor of Managerial Informatics at the Faculty of Informatics and Management at the University of Hradec Kralove, Czech Republic, and the head of the Department of Information Technologies. Recently he is also a member of the Accreditation Commission of the Government of Slovak Republic (since 2004) responsible for accreditations of Slovak higher educational institutions. Research of Professor Mikulecký covers ambient intelligence, artificial intelligence, knowledge-based systems and technologies, knowledge management, as well as human – computer interaction. He has published more than 150 papers in various journals and conference proceedings in these areas, as well as a number of books and book chapters. He was one of the founders of a regular series of events called Ambient Intelligence Forum; he is also a member of programme committees for a number of international conferences.

Simanta Mitra received his master's and Ph.D. in Computer Science from Iowa State University. His research interests include software maintenance, software architecture, programming languages, and web services. He worked at Newmonics Inc. before joining Iowa State University in 1999 where he is currently employed as a Senior Lecturer at the department of Computer Science.

Karel Mls works as an assistant professor at the Department of Information Technologies at the Faculty of Informatics and Management, University of Hradec Kralove, Czech Republic. He professes Decision Support Systems and Computational Intelligence, his research interests cover fuzzy logic, concept and cognitive mapping and modeling of decision making in human and artificial systems, among others. Recently he is involved in the research projects on optimization, autonomous systems and intelligent software assistants.

Claus Möbus studied psychology at the Technical University Braunschweig and at the University of Heidelberg. In 1970 he received a Diploma and in 1974 a PhD degree in Psychology. He then became a postdoc researcher (C1) at the Psychological Institute in Heidelberg. His research focussed on knowledge acquisition, knowledge and skill assessment and formal quantitative methods. In 1978 he finished his Habilitation and received a venia legendi for psychology. From 1977 to 1978 he worked as a Professor for Psychology at the Institute for Psychology in the Department of Educational Sciences at the Free University of Berlin (FU Berlin). In 1978 he became a Professor for Applied Informatics in the Faculty of Natural Sciences and Mathematics in the University of Oldenburg. In 1984 he was one of the founding members of the new Faculty of Informatics (Fachbereich Informatik). In 1986 he established a special research group Learning and Knowledge-Based Systems (LKS). In 2008 the group refocused the research activities towards modelling learning and cognitive systems. Now he is leader of the group Learning and Cognitive Systems in the Department of Computing Sciences at the University of Oldenburg (http://www.lks.uni-oldenburg.de/). Depending on research grants the group is working in four research areas: (1) Applied Artificial Intelligence (Knowledge Acquisition, Data Mining, Decisions under Uncertainty and Risk), (2) Content, Knowledge, and Cognitive Engineering, (3) Learner, User and Operator Modelling, (4) Engineering of Innovative Learning and Problem-Solving Environments.

Atulya K. Nagar is the Foundation Professor of Computer and Mathematical Sciences at Liverpool Hope University and is Head of Department of Computer Science. He received a prestigious Commonwealth Fellowship for pursuing his Doctorate in applied non-linear mathematics, which he earned from the University of York, UK, in 1996. He holds BSc (Hons.), MSc and MPhil (with distinction) degrees, in Mathematical Sciences, from the MDS University of Ajmer, India. Prior to joining Liverpool Hope University, Prof. Nagar has worked for several years as a Senior Research Scientist, on various EPSRC sponsored research projects, in the department of Mathematical Sciences, and later in the department of Systems Engineering, at Brunel University. He holds a Visiting Professorship at the University of Madras. His multidisciplinary research interests include: Mathematical Physics, Nonlinear Differential Equations, Systems Engineering, Computational Optimisation, Bio-informatics, and Intelligent Systems. He is a member of the IEEE.

Kamila Olševičová works as an assistant professor at the Department of Information Technologies at the Faculty of Informatics and Management, University of Hradec Kralove, Czech Republic. Her research interests include multi-agent modeling and simulations, knowledge-based technologies and knowledge representation. She participated on several European research projects and the Czech Grant Agency projects focused on knowledge management and e-learning.

Katsunori Oyama received his BS, ME, and PhD degrees in Computer Science from Nihon University, Japan in 2001, 2003, and 2007, respectively. He is currently a postdoctoral researcher in the Department of Computer Science, Iowa State University, USA. He received the Special Model Award in 2005 and Silver Model Award in 2007 for having attended the Embedded Technology Software Design Robot Contest (http://www.etrobo.jp/). His primary research interests concentrate on Design Thought Process, Ontology Engineering, Object Oriented Analysis and Design, Situation-Awareness and Software Evolution. He is a member of IPSJ.

Federico Pecora is a post-doctoral research fellow at Örebro University. He has a Ph.D. (defended in 2007, with honourable mention) in Computer Science Engineering from the University of Rome "La Sapienza", Italy. He has previously been a researcher with the Italian National Research Council and a visiting scholar at the Robotics Institute of Carnegie Mellon University. He has been involved in high-profile national and European research projects such as RoboCare, in which he developed robotic solutions for elderly care, and APSI, in which he has contributed to the development of the OMPS framework for continuous support to mission planning activities at the European Space Agency. Federico is an active developer of the PEIS-Home at AASS, and currently focuses on multi-agent and multi-robot systems, constraint-based reasoning, planning and scheduling.

Clifton Phua is a Research Fellow at the Data Mining Department of Institute for Infocomm Research (I²R), Singapore. His current research interests are in healthcare and security data mining. He has a Ph.D. and a Bachelor of Business Systems (Honours) from Monash University, Australia.

Thomas Röfer received a Diploma in computer science and Dr. Ing. from Universität Bremen, Germany, in 1993 and 1998, respectively. He is a member of the Transregional Collaborative Research Center SFB/TR 8 "Spatial Cognition" at Bremen, funded by the Deutsche Forschungsgemeinschaft (DFG), and of the Executive Committee of the RoboCup Federation. He is currently Senior Researcher at the German Research Center for Artificial Intelligence (DFKI) in the Research Department on Safe and Secure Cognitive Systems at Bremen, Germany. His research interests include rehabilitation robotics, robot soccer, real-time computer vision, world modelling, and humanoid robots.

Carlos Ramos is the director of Gecad (the Knowledge Engineering and Decision Support Research Centre) and coordinator professor at the Polytechnic of Porto's Institute of Engineering. His main areas of interest are ambient intelligence, knowledge-based systems, decision support systems, multiagent systems, and planning. He received his PhD in electrical and computer engineering from the University of Porto

Jukka Riekki is professor at the University of Oulu, in the Department of Electrical and Information Engineering. He leads together with his colleague the Intelligent Systems Group. His main research interests are in context-aware systems serving people in their everyday environment. Currently he studies in several projects physical user interfaces, context recognition, and service composition. In these projects he cooperates with research groups from China, Japan, and Sweden. He is a member of IEEE.

Patrice C. Roy is currently a Ph.D. candidate at University of Sherbrooke, Sherbrooke City, Canada. He received his M.Sc. in computer science from the Université du Québec à Montréal, Montréal city, Canada, in 2007, and his B.Sc.A. in computer science from the Université du Québec à Chicoutimi, Saguenay city, Canada, in 2006. He also received a Graduate Diploma in Earth Sciences and a B.Sc. in Geology from the Université du Québec à Chicoutimi, Saguenay city, Canada. He is currently a member of the DOMUS laboratory of the Université de Sherbrooke. His research interests include cognitive assistance in smart homes, activity recognition, multi-agent systems, and artificial intelligence methods in geology. He received postgraduate scholarships from the Natural Sciences and Engineering Research Council of Canada (NSERC) for his M.Sc. and Ph.D research.

Fariba Sadri studied BSc in mathematics and computer science at the University of London, and later PhD in logic programming in the Department of Computing at Imperial College London, where she has remained since. Her earlier work was on integrity checking in deductive databases and temporal reasoning. In more recent times her work concerns agent technologies and multi-agent systems. She has worked on logic-based agent models, reasoning, dynamic belief revision, and inter-agent communication and negotiation. She is on the Steering Committee of several regular events in multi-agent systems and ambient intelligence.

Alessandro Saffiotti (Ph.D. in Applied Science from Université Libre de Bruxelles, Belgium) is full professor of computer science at Örebro University. He has previously been a researcher with the University of Pisa (I), SRI International (USA), and the Université Libre de Bruxelles (B). In 1998 he founded the AASS Mobile Robotics Lab, which he heads since then. Alessandro Saffiotti's research interests encompass autonomous robotics, soft computing, and non standard logics for common-sense reasoning. He edited a book and authored more than 120 research papers in international peer-reviewed journals and conferences. He has chaired many international events, and served as program co-chair for IJCAI (the premier international conference on Artificial Intelligence) in 2005. He is a senior member of IEEE, and a member of AAAI and ECAI.

Ricardo Santos is professor of Computer Engineering at School of Technology and Management of Felgueiras – Polytechnic of Porto (ESTGF/IPP) and researcher at the Knowledge Engineering and Decision Support Research Group (GECAD). He is a PhD student in informatics at the University of Trás-os-Montes e Alto Douro. His research interests include Multi-Agent Systems, Emotion, Personality, Ubiquitous computing, Persuasive Argumentation and Group Decision Support Systems.

Shishir K. Shandilya, Head, Post Graduate Department Computer Science & Engineering, NIIST; PhD (Computer Engineering), M.Tech (CSE), MBA (HR); He got the Young Scientist Award for consecutive two years (2005 & 2006) by Indian Science Congress & MP Council of Science & Technology for Computer Engineering; He also carries various awards like Computer Wizard-2002 and Excellent Mentor-2009; He has written five international-fame books and published over 40 quality research papers in international & national journals & conferences; He is actively steering the international conferences as Conference Chair and international journals as Reviewer & Coordinator; He is an active member of over 20 international professional bodies; He is working as Principal Investigator on various research projects; He is actively involved in assisting various IT companies as Sr. Consultant.

Hui Shi received a Master's Degree in Computer Science at Changsha Institute of Technology in China and a Dr.-Ing. from Universität Bremen in 1994. She is a lecturer in the Department of Mathematics and Informatics at Universität Bremen, member of the Transregional Collaborative Research Center SFB/TR 8 "Spatial Cognition" at Bremen (funded by the Deutsche Forschungsgemeinschaft, DFG), and a Senior Researcher in the Research Department on Safe and Secure Cognitive Systems at the German Research Center for Artificial Intelligence (DFKI), Bremen. Her research areas include Formal Methods, Shared Control Systems, User Modelling, Dialogue Modelling, and Human-Robot Interaction.

Jakub Šimko is a master student in Software Engineering study program at the Slovak University of Technology in Bratislava. He holds a Bachelor's degree in Informatics (2008) from the same university. His research interests are in the areas of exploratory web search, improving user experience, user modeling and personalized recommendations. He participates on the project Factic involving development of semantic web browser, including graph visualization library. In 2008, he participated on Energy Consumption Manager project, which placed 2nd at Microsoft Imagine Cup 2008 worldwide finals in Software Design category.

Enrique Soriano works as a lecturer at the Rey Juan Carlos University of Madrid (Spain). He earned a degree on computer engineering in 2002. He received a four year full-time research grant from the Spanish Ministry of Science and obtained his PhD in October, 2006. He has been involved in the design and development of the Plan B Operating System and the Octopus. He designed and implemented SHAD, a security architecture for Plan B. He is currently working on the development of new drivers for Plan 9 and the design of new file system protocols. His research work is centered in operating systems, security, and pervasive computing.

Axel Steinhage is Director R&D at Future-Shape GmbH, Höhenkirchen-Siegertsbrunn, Germany since 2006. From 2001 to 2005 he lead the Man-Machine-Interaction group as a Senior Staff Engineer at the Corporate Research Department of Infineon Technologies AG, Munich. Between 1997 and 2000 he worked as a Postdoc and leader of the Anthropomorphic Robotics Group and the Behavioural Dynamics Group at the Institute for Neuroinformatics in Bochum, Germany and at the CNRS in Marseille, France. In 1997 he received a doctor's degree in theoretical physics. Dr. Steinhage has written over 60 international publications in the research fields of robotics, neurobiology, navigation, dynamical systems, speech recognition, image processing, remote sensing, mobile communication and ambient assisted living.

Andrei Tolstikov is currently a Research Fellow at the Networking Protocols Department of Institute of Infocomm Research (I2R), Singapore. His current research interests are in the area of Information Quality for Ambient Intelligence and sensor networks applications. He has a Masterd degree from the Moscow Institute of Physics and Technology and currently finishing his studies for Ph.D. degree in the National University of Singapore.

Jan Treur received his Ph.D. in Mathematics and Logic in 1976 from Utrecht University. Since 1986 he works in Artificial Intelligence, from 1990 as a full professor and head of the Department of Artificial Intelligence at the Vrije Universiteit Amsterdam, currently consisting of about 50 researchers. More in particular he is leading the Agent Systems Research Group within the Department of Artificial

Intelligence at the VUA. He is an internationally well-recognized expert in agent technology, cognitive modeling and knowledge engineering. His research concerns the analysis of dynamics in biological, psychological, social and artificial systems. Psychological applications address, for example, dynamics of beliefs, desires, intentions, emotions, mood, trust, and practical reasoning.

Rianne van Lambalgen is a PhD student at the Agent Systems Research group of the department Artificial Intelligence at the Vrije Universiteit. In 2005 she obtained her Master degree Cognitive Psychology and in 2007 she obtained her Master degree in Artificial Intelligence (Cognitive Science). In her current PhD she focuses on cognitive processes in dynamic circumstances such as cognitive workload, attention and task engagement. Her research is aimed at the question of how (computational) cognitive models can be used to predict human functioning when task demands are high and how automation support can be given to achieve an optimal human performance.

Peter-Paul van Maanen obtained his Bachelor and Master of Science degree (with distinction) in Computer Science with specialization in Artificial Intelligence at the Utrecht University. He wrote his thesis at the Institute of Cognitive Sciences and Technologies in Rome. Since September 2004 Peter-Paul is researcher and project manager at TNO Human Factors, The Netherlands, and is currently working at the Department of Cognitive Systems Engineering. In addition to his work at TNO, Peter-Paul is researcher-lecturer at the Department of Artificial Intelligence, Vrije Universiteit Amsterdam, since September 2007. Peter-Paul was or is closely involved in more than 15 research projects at both TNO and the Vrije Universiteit (in cases as project manager) and co-authored (award winning) scientific publications in a wide range of research areas such as cognitive modeling, adaptive autonomy and support, trust and reliance decision making, emotion, serious games, attention, ethical decision making, unmanned vehicles, and stress.

Aung Aung Phyo Wai is a Research Engineer in Networking Protocols Department of Institute for Infocomm Research (I²R), Singapore. His current research interests include Wireless Sensor Network, Ambient Intelligence and Multimodal Sensors Fusion. He received his Bachelor of Engineering and Master of Science degree from Yangon Technological University, Myanmar and Nanyang Technological University, Singapore respectively.

Baoying Wang is currently an assistant professor in Waynesburg University, PA, USA. She received her PhD degree in computer science from North Dakota State University. Her Master's degree is from Minnesota State University, St. Cloud. Her research interests include data mining, bioinformatics, and high performance computing. She serves as a reviewer/committee member of several international conferences including, the ACM Symposium of Applied Computing (SAC 2008) Bioinformatics Track, IADIS (International Association for Development of the Information Society) Applied Computing 2008, the 2009 International Conference on Data Mining (DMIN'09), etc. She is a member of ACM, ISCA and SIGMOD.

Rebecca L. Willard graduated summa cum laude from Waynesburg University in 2009 with a Bachelor of Science in Computer Forensics and a Bachelor of Arts in Criminal Justice Administration.

Manfred Wojciechowski received his diploma in Computer Science from the University of Dortmund, Germany, in 1996. Since then he is working as a researcher at the Fraunhofer Institute for Software and System Engineering. His research interest is currently focused on context modeling and its practical application.

Duangui Xu is currently studying for the Diploma of Computer Engineering at School of Engineering, Temasek Polytechnic, Singapore.

Tatsuya Yamazaki received the B.E., M.E., and Ph.D degrees in information engineering from Niigata University in 1987, 1989, and 2002, respectively. He joined the Communications Research Laboratory (currently the National Institute of Information and Communications Technology) as a researcher in 1989. Since 2009, he has been an executive researcher at NICT. From 1992 to 1993 and 1995 to 1996, he was a visiting researcher at the National Optics Institute, Canada. From 1997 to 2001, he was a senior researcher at ATR Adaptive Communications Research Laboratories. His areas of interest include adaptive QoS management, QoS/QoE/QoL mapping, statistical image processing, pattern recognition, ubiquitous computing, and networks. He is a member of the IEEE.

Dušan Zeleník is a master student in Software Engineering study program at the Slovak University of Technology in Bratislava. He holds a Bachelor's degree in Informatics (2008) from the same university. His research interests are in the areas of personalized systems, recommendation systems, user modeling and neural networks. He is part of the project focused on generating recommendations using content similarities of web news. In 2008, he participated on Energy Consumption Manager project, which placed 2nd at Microsoft Imagine Cup 2008 worldwide finals in Software Design category.

Daqing Zhang is a professor on Ambient Intelligence and Pervasive System Design at Institute TELECOM SudParis, France. He initiated and led the research in Smart Home, Healthcare/Elderly care and context-aware computing from 2000 to 2007 at the Institute for Infocomm Research (I2R), Singapore. Dr. Zhang was the Program Chair of the First International Conference of Smart Home and Health Telematics (ICOST2003) in Paris, France and the Sixth International Conf. on Ubiquitous Intelligence and Computing (UIC 2009) in Brisbane, Australia. He is the associate editor of ACM Transactions on Intelligent Systems and Technology, Journal of Ambient Intelligence and Humanized Computing (Springer), etc. He also served in the technical committee for conferences such as UbiComp, Pervasive, PerCom, etc. Dr. Zhang's research interests include pervasive elderly care, service-oriented computing, sensor based activity recognition, context-aware systems etc. He has published more than 100 papers in referred journals, conferences and books. Daqing Zhang obtained his Ph.D. from University of Rome "La Sapienza" and University of L'Aquila, Italy in 1996.

Weishan Zhang is a Research Associate Professor since June, 2007 at Computer Science Department, University of Aarhus. His research is part of the EU Hydra project. He was a visiting scholar of Department of Systems and Computer Engineering, Carleton University, Canada (Jan. 2006 - Jan.2007). He was an Associate Professor at School of Software Engineering, Tongji University, Shanghai, China (Aug. 2003- June 2007). He was a NSTB post-doctoral research fellow at Department of Computer Science, National University of Singapore (Sept. 2001 to Aug. 2003). His research interests include Semantic web enabled software engineering including semantic software architecture, Intelligent pervasive/mobile middleware including context-awareness and context modeling; Semantic web and context-awareness based self-management; Software reuse including Frame technology, software architecture, and software product line; Software evolution and software system re-engineering; Applied computational intelligence for software engineering and pervasive computing, including software agents and genetic algorithm. He has published 2 book chapters, 3 journal papers, 28 conferences papers and 15 technical reports.

Index

Symbols

3D/2D Articulated Model 39
3D geobrowsers 602, 604-606
3D (three dimensional) 18-20, 22, 24, 26-30, 32-33, 35, 38-39, 81, 99, 111, 273, 280, 602-607, 610-613, 615
2D (two dimensional) 18-19, 22, 24-27, 37, 39, 611

A

Abductive Inference 535, 557
Access Control List (ACL) 191, 196, 638, 650-651
Acoustic map 49
Actions Tendencies 127
Action terms 8
Activities of Daily Living (ADLS) 301-304, 306-309, 315-318, 320-323, 330, 337, 340-342, 371, 373-374, 392, 421, 430, 444, 456, 521-522, 528, 542
Activity Monitoring 48, 54-55, 537, 554, 556, 637
Activity-oriented computing (AOC) 197-198, 200, 204, 214
Activity Recognition 57, 179, 182, 307-308, 317, 319-339, 341, 343-345, 349-350, 369, 374, 392, 394, 396-397, 409, 414, 420-422, 440, 458-459, 534-535, 541, 543, 552-557
Actuator components 540-542, 546-547, 550-551
Adaptability 2, 4, 7, 165, 196, 615-616, 639
Adaptive flocking 559, 562, 564-565, 574-579, 581, 583
Adaptive flocking algorithm 574-575, 577-579
Adaptive self-configuration 559, 562, 564, 568-573, 579, 583
Advanced Industrial Science and Technology (AIST) 173
Advanced Telecommunication Research Institute International (ATR) 99
Affordance 1-2, 4, 7-10, 12-13, 15-16

Agent 4, 9-11, 52, 72, 90-91, 95-99, 101-105, 110-127, 131-133, 139-140, 142-146, 148, 154-155, 157-159, 161-164, 171-172, 285, 288, 309, 316, 321, 323-325, 327-329, 331-334, 337-339, 342, 344-351, 353-355, 360-366, 368-369, 373-375, 379-382, 384-386, 388, 390-391, 442, 461-462, 464-465, 467-468, 471-472, 474-475, 491-494, 499, 506, 555-556, 587, 632-633, 637, 639
Agent-Based Model 639
Agent Based Simulation for Group Decision (AB-S4GD) 103
Agitation behavior recognition 403, 405, 409
Agreeableness (A) 2-137, 139-140, 142-146, 148-149, 151, 153-228, 230-232, 234-310, 312-345, 347-382, 384-422, 424-436, 438-501, 513-574, 576-625, 627-634, 636-642, 644-652
AI methods 425, 439, 533
ALADIN 11, 15
Allocentric monitoring 434-436
Alzheimer's Carer Internet Support System (ACISS) 304, 319
Alzheimer's disease 301, 318, 428
Ambient Activity Recognition 320, 329, 344
Ambient Assisted Living (AAL) 41-42, 55, 433, 437, 440-441, 513-514, 518, 523, 526-529, 531, 533
Ambient Intelligence (AmI) 1-2, 4-7, 9-15, 41, 75, 78, 88-89, 99, 101-103, 105-113, 119, 122-126, 128-132, 148, 162-164, 182, 198, 212, 216-217, 219, 239-250, 269, 300, 302, 319-322, 324-325, 328-329, 331-332, 334-335, 338-339, 341-342, 344, 346, 348, 351, 369, 371, 387, 392-394, 420, 422, 424-425, 438-441, 443-445, 449, 451, 453, 460-462, 498, 530, 556, 558-560, 571, 579-580, 582, 586, 596-597, 600, 602-604, 606-607, 609-610, 615-616, 620-623, 627, 631, 633-634, 636-641, 648
AmI Artefact 621, 639